Insect Pheromone Biochemistry and Molecular Biology

The biosynthesis and detection of pheromones and plant volatiles

Insect Pheromone Biochemistry and Molecular Biology

The biosynthesis and detection of pheromones and plant volatiles

Edited by
Gary Blomquist
University of Nevada,
Biochemistry Department, Reno, USA

and

Richard Vogt
Department of Biological Sciences,
University of South Carolina, Columbia, USA

ELSEVIER
ACADEMIC
PRESS

Amsterdam • Boston • Heidelberg • London • New York • Oxford • Paris
• San Diego • San Francisco • Singapore • Sydney • Tokyo

This book is printed on acid-free paper

Elsevier Academic Press
84 Theobald's Road, London WC1X 8RR, UK
http://www.academicpress.com

Elsevier Academic Press
525 B Street, Suite 1900 San Diego, California 92101-4495, USA
http://www.academicpress.com

ISBN 0-12-107151-0

Library of Congress Catalog Number:

A catalogue record for this book is available from the British Library

Typeset by Replika Press Pvt Ltd, India
Printed and bound in Great Britain by Biddles Ltd, Guildford and Kings Lynn

03 04 05 06 07 08 B 9 8 7 6 5 4 1

Contents

Part 2: Pheromone Detection

Contributors

Blomquist, Gary J., Department of Biochemistry, University of Nevada, Reno, Reno, NV 89557, USA

Breer, Heinz, University of Hohenheim, Institute of Physiology, Garbenstrasse 30, 70593 Stuttgart, Germany

Bruyne, Marien de, Neurobiologie, Freie Universität Berlin, Königin-Luise-Strasse 28-30, D 14195 Berlin, Germany

Carlsson, Mikael A., Department of Crop Science, Swedish University of Agricultural Sciences, SE-230 53 ALNARP, Sweden

Eisner, Thomas, Department of Neurobiology and Behavior and, Department of Chemistry and Chemical Biology, Cornell University, Ithaca, NY 14853, USA

Fan, Yongliang, Department of Entomology, Campus Box 7613, North Carolina State University, Raleigh, North Carolina 27695-7613, USA

Hansson, Bill S., Department of Crop Science, Swedish University of Agricultural Sciences, SE-230 53 ALNARP, Sweden

Howard, Ralph W., USDA-ARS, 1515 College Avenue, Manhattan, KS 66502, USA

Jacquin-Joly, Emmanuelle, INRA Unité de Phytopharmacie et des Médiateurs, Chimiques-Route de Saint-Cyr, 78026, Versailles, France

Jallon, Jean-Marc, CNRS URA 1491 Neurobiologie, University of Paris XI South, 91405 Orsay, Cedix, France

Jurenka, Russell, Department of Entomology, Iowa State University, Ames, IA 50011, USA

Knipple, Douglas C., Cornell University, Department of Entomology, New York State Agricultural Experiment Station, Geneva, NY 14456, USA

Krieger, Jürgen, University of Hohenheim, Institute of Physiology, Garbenstrasse 30, 70593 Stuttgart, Germany

Leal, Walter S., Honorary Maeda-Duffey Laboratory, Department of Entomology, University of California, Davis, CA 95616, USA

Loudon, Catherine, Department of Ecology and Evolutionary Biology, University of Kansas, Lawrence KS 66045, USA

Ma, Peter W.K., Department of Entomology and Plant Pathology, Mississippi State University, Mississippi State, MS 39762, USA

Meinwald, Jerrold, Department of Neurobiology and Behavior and Department of Chemistry and Chemical Biology, Cornell University, Ithaca, NY 14853, USA

Nagnan-Le Meillour, Patricia, INRA-Unité de Phytopharmacie et des Médiateurs Chimiques Bâtiment A-Route de Saint-Cyr, F-78026, Versailles, France

Picimbon, Jean-François, Department of Ecology, University of Lund, Sölvegatan 37, SE 223 62 Lund, Sweden

Plettner, Erika, Simon Fraser University, Dept. of Chemistry, Burnaby, B.C. V5A 1S6, Canada

Rafaeli, Ada, Department of Stored Products, ARO Volcani Center, Bet Dagen, Israel

Raguso, Robert A., Department of Biological Sciences, 700 Sumter Street, University of South Carolina, Columbia, SC 29208, USA

Ramaswamy, Sonny, Department of Entomology, Kansas State University, Manhattan, KS 66506, USA

Roelofs, Wendell L., Cornell University, Department of Entomology, New York State Agricultural Experiment Station, Geneva, NY 14456, USA

Schal, Coby, Department of Entomology, Campus Box 7613, North Carolina State University, Raleigh, North Carolina 27695, USA

Seybold, Steven J., Chemical Ecology of Forest Insects, USDA Forest Service, Pacific Southwest Research Station, 720 Olive Drive, Suite D, Davis, CA 95616, USA

Tittiger, Claus, Chemical Ecology of Forest Insects, USDA Forest Service, Pacific Southwest Research Station, 720 Olive Drive, Suite D, Davis, CA 95616, USA

Vanderwel, Désirée, Department of Chemistry, University of Winnipeg, 515 Portage Avenue, Winnipeg, Manitoba R3B 2E9, Canada

Vogt, Richard G., Department of Biological Sciences, 700 Sumter Street, University of South Carolina, Columbia, SC 29208, USA

Vosshall, Leslie B., Laboratory of Neurogenetics and Behavior, The Rockefeller University, New York, NY 10021, USA

Wicker-Thomas, Claude, CNRS URA 1491 Neurobiologie, University of Paris XI South, 91405 Orsay, Cedix, France

Zwiebel, Lawrence J., Department of Biological Sciences, 6270 Medical Research Building III, Vanderbilt University, VU Station B 3582, Nashville, TN 37235, USA

Preface

The 1987 book *Pheromone Biochemistry* (Prestwich and Blomquist) summarized what was then known about the production and reception of insect pheromones. Remarkable advances in our understanding of pheromone production have occurred in the last one and a half decades, which is mirrored by similar advances in our understanding of pheromone reception. This progress is detailed herein by selected authors who are the leaders in the field. We have assembled contributed chapters from experts who are at the frontiers of pheromone chemistry, neurobiology, chemical ecology, molecular biology and biochemistry.

The increase in awareness of work on pheromone production and reception has advanced from an occasional symposium in these areas to where symposia are now routine at many insect science meetings. Despite this increased interest and understanding, the field still seems in its infancy, as studies on pheromone production and reception have concentrated on only a few species and selected systems. The powerful tools of genomics and proteomics have only very recently and in limited scope been brought into play in furthering our understanding of pheromone production and reception, and they will undoubtedly result in even faster advances over the next decade. The *Drosophila* and other insect genome projects have proved a gold mine for increasing our understanding of pheromone production and reception, and this information is rapidly transferring to other insects.

Our knowledge of pheromone production is one area in which scientists who study insects are ahead of their counterparts who study vertebrate systems. This is in part due to the clear-cut and readily identifiable response that insects have to semiochemicals, which has allowed the identification of pheromones from over three thousand species. Insect odor reception also stands out from vertebrate systems because insect olfactory neurons are contained in sensilla. This compartmentalization has allowed the establishment of biochemical networks serving specific classes of sensory neurons. These networks include odor receptors (ORs), odorant binding proteins (OBPs) and odor degrading enzymes (ODEs) which interact, possess unique binding sites for the same odors, and belong to multigene families, all providing a molecular dimensionality that far surpasses that found in olfactory systems of other organisms.

This book is designed to provide the reader with an up-to-date compilation of our understanding of pheromone production and reception and hopefully to serve as a source book for the next decade. It is divided into two parts. The first deals with the biosynthesis and endocrine regulation of pheromone production. It emphasizes studies done on moths, and is balanced by chapters on beetles, flies, cockroaches and social insects. The second part covers odor reception, focusing on those proteins known to be uniquely expressed in the antennae and with likely roles in processing pheromone and other odorant signals. These processes are placed in a broader context in chapters addressing the biomechanics of how antennae are designed to capture odors and the physiological responses in the antenna and brain that result from odor reception. Much is known from a variety of moth species, from *Drosophila melanogaster* and considerable information is also available from species distributed throughout the Neoptera, including cockroaches, bugs, honey bees, ants, beetles, and assorted Diptera. Expression of olfactory reception genes have been characterized in juvenile and adult stages, and in response to developmental hormones.

It appears that insect pheromones arose from co-opting molecules that had other purposes for signal functions. Components were then shaped by selective forces acting on pre-existing structures and evolved so that the signal might be more easily discriminated. Thus, cuticular hydrocarbons evolved to be used extensively as pheromones in Diptera and other species, monoterpenoids evolved to be used as pheromone components in some Coleoptera, and fatty acids were modified to become pheromones in Lepidoptera. The biosynthetic pathways used by insects to produce pheromone usually involve only slight modifications of "normal" metabolism to produce highly specific chemical signals. For example, the judicious use of a unique δ-11 desaturase and highly specific chain-shortening reactions account for the carbon skeleton of many lepidopteran pheromone components. Similarly, insects have co-opted hormones that are used for other purposes to regulate pheromone production. In many insect groups, including the beetles and cockroaches, the developmental hormone, juvenile hormone, functions to induce pheromone production. In some Diptera, the other major developmental hormone, 20-hydroxyecdysone, has evolved to regulate pheromone production in adult females. Juvenile hormone and ecdysteroids appear to regulate pheromone production by up-regulating the message for specific genes, whereas the third major pheromone regulatory hormone, pheromone biosynthesis activating neuropeptide (PBAN), acts by affecting enzyme activity through second messenger systems. The details of these processes are described in appropriate chapters herein. The studies on pheromone biosynthesis and endocrine regulation are still embryonic, and it is hoped that this book will serve to stimulate further work.

In insects, odor molecules are thought to be transported to ORs by OBPs and subsequently degraded by ODEs, though there is considerable uncertainty about the nature of the interactions between these three protein classes. OBPs and ORs comprise multigene families and there are multiple ODEs. The distinct odor sensitive phenotypes of different sensilla are presumed to be based on the differential expression of specific combinations of ORs, OBPs and ODEs which have ligand binding sites tuned to the relevant odor molecule(s). It appears that pheromone sensitive sensilla and the underlying mechanisms of pheromone detection is a specialized version of odor detection in general, allowing us to extrapolate recent advances in the identification of non-pheromonal ORs from sequenced genomes to pheromone systems. As of this writing no bona fide insect pheromone receptor has been identified. Pheromone specific OBPs have been identified, and these pheromone binding proteins (PBPs) are clearly homologous with OBPs expressed in sensilla which respond to plant volatiles or so called "general odorants". ODEs have been characterized which are pheromone specific, but an aldehyde oxidase (AOX) has been characterized which associates with both pheromone and plant volatile sensitive sensilla. Thus, until we learn otherwise, we assume a commonality in the underlying mechanisms of pheromone and plant volatile detection. The complexity and diversity of this system raise many interesting questions concerning genomic organization and how gene family members are singled out for expression, how genes evolve and diverge, the determination of sensilla phenotypes and the combinatorial coordination of multiple genes expressed by different sensillar cells to express those phenotypes, the structural basis of odor binding, the dynamics of a biochemical network, and the organization of a network that serves as a molecular interface between organism and environment.

The mechanisms of pheromone biosynthesis and reception are among the molecular pathways underlying insect reproductive behavior. Odor reception systems have evolved in the context of behaviorally relevant olfactory signals such as those mediating responses to oviposition cues. Coevolution has presumably occurred for scent based plant-insect interactions which satisfy essential pollination needs of the plant and nutritional needs of the insect. Coevolution must also be ongoing between pheromone biosynthesis and the biochemistry of pheromone detection; these two systems clearly must complement each other, but the mechanisms maintaining linkage between the two systems are unknown. As genomes of insects and plants are being laid bare, this book documents where we stand today in our understanding of the molecular genetics of pheromones and odor detection.

Many of the target insects for pheromone biosynthesis and reception studies are pests of agricultural or medical importance. The intellectual exercise of unraveling the mechanisms involved in the production and reception of pheromone molecules is designed ultimately to lead to the development of the behavior-modifying agrochemicals of the future.

Gary J. Blomquist
Richard G. Vogt
April 2003

Acknowledgments

We thank our wives Cheri and Gay not only for their support and understanding, but also for their help in editing a number of chapters. We appreciate the editing help of Anna Gilg-Young. We are very grateful to the authors for their preparation of the up-to-date, exciting and visually stimulating chapters in a timely manner. We thank our mentors, colleagues, post-doctoral fellows and students for making this work fun and rewarding. We thank Charles Crumly at Academic Press for his help in initiating this project and Andy Richford and Jackie Holding for their excellent help in seeing it through to completion.

The editors acknowledge financial support to their laboratories by the National Science Foundation (GJB, RGV), USDA-NRI (GJB, RGV), the National Institutes of Health (RGV) and the Nevada Agricultural Experiment Station (GJB) that made the research we report possible.

Part 1

Pheromone Production

1

Biosynthesis and detection of pheromones and plant volatiles – introduction and overview

Gary J. Blomquist and Richard G. Vogt

1.1 Introduction and overview

The first half of this book deals with the production of pheromones, primarily in female insects, and the second half deals with reception of pheromones and other odorants, the former primarily in male insects and the latter in males, females and juveniles. Most of the work on pheromone production and reception is recent, all occurring in the past three decades. The emphasis in this book is on work done since 1987, when *Pheromone Biochemistry* (Prestwich and Blomquist, 1987) was published. From the work presented in this edition, it can readily be seen that the field has undergone tremendous advances in the last one and a half decades.

Our understanding of pheromone production has evolved from identifying biochemical pathways towards unraveling the molecular biology of key regulatory enzymes, and in one system, a genomics approach has been initiated. Our understanding of the regulation of pheromone production has similarly advanced from simply knowing that a particular hormone increased pheromone production in a few species to developing an understanding of which enzymes are regulated at the molecular level.

Our understanding of pheromone reception had undergone dramatic change just prior to 1987 with the proposal that Pheromone Binding Proteins (PBPs) and pheromone degrading enzymes transported and inactivated pheromonal signals within the sensilla (see Figure 1.2B). The general framework of this process is now known to be widespread for most insects and for the reception of pheromones, plant volatiles and other odorants. Biochemical transduction pathways have been identified, including the all important olfactory receptor proteins. A marriage of molecular genetics and genomics with biomechanics, behavior, anatomy and physiology is giving new understanding of how pheromones are detected, and raising many new questions in the process.

1.2 Pheromone production: biosynthesis of pheromones

The first insect sex pheromone identified was bombykol, (*E*, *Z*)-10,12-hexadecadien-1-ol (Butenandt *et al.*, 1959) from the silkworm moth, *Bombyx mori* (L.). The elucidation of the structure spanned 20 years and required a half million female abdomens. A few years later, (*Z*)-7-dodecenyl acetate was identified as the sex pheromone of the cabbage looper, *Trichoplusia ni* (Berger, 1966). By 1970, following the pioneering work by Silverstein on bark beetles, in which three terpenes were identified as a synergistic pheromone blend for *Ips paraconfusus* (Silverstein *et al.*, 1966), it became recognized that most insect pheromones consisted of multicomponent blends. This has since been shown to be true for most insects, and single-component pheromones are rare. Over the past four decades, extensive research on insect pheromones has resulted in the chemical and/or behavioral elucidation of pheromone components from several thousand species of insects, with much of the work concentrating on sex pheromones from economically important pests.

One of the early issues addressed in pheromone production was the origin of pheromone components. Ultimately, all precursors for pheromone biosynthesis can be traced through dietary intake. A question asked in several systems was whether pheromone components were derived from dietary components that were altered only minimally or whether they were synthesized *de novo*. This simple question proved surprisingly difficult to answer, and different answers were obtained for different groups of insects. By the mid-1980s, isotope studies had demonstrated that in the Lepidoptera, most of the sex pheromone components were synthesized *de novo* (Bjostad *et al.*, 1987), with some exceptions (see Eisner and Meinwald, Chapter 12). Early studies in the boll weevil, *Anthonomus grandis*, gave evidence that the monoterpenoid pheromone components could arise both from modification of dietary precursors (Thompson and Mitlin, 1979) and from *de novo* biosynthesis (Mitlin and Hedin, 1974). The relative contribution of each has still not been fully resolved (Tillman *et al.*, 1999). Early studies in bark beetles yielded convincing evidence that aggregation pheromone components such as 2-methyl-6-methylene-7-octen-4-ol (ipsenol), 2-methyl-6-methylene-2,7-octadien-4-ol (ipsdienol) and cis- and trans-verbenol were synthesized by the slight modification, usually hydroxylation, of host tree- derived monoterpenoid precursors (Hughes, 1974 Byers *et al.*, 1979; Byers, 1981). Indeed, Hendry *et al.* (1980) convincingly demonstrated that deuterium labeled myrcene was directly converted to ipsenol and ipsdienol. This led to the widely accepted dogma that bark beetles obtain their pheromones differently than most other insects, where *de novo* biosynthesis is the norm. Not until the last decade did it become apparent that the situation in bark beetles was more complicated, and radiotracer studies demonstrated that some bark beetle pheromone components are synthesized *de novo* (Seybold *et al.*, 1995). This led to renewed interest in both the biochemistry (Seybold and Vanderwel, Chapter 6) and molecular biology (Tittiger, Chapter 7) of bark beetle pheromone production.

By 1987, when *Pheromone Biochemistry* (Prestwich and Blomquist, 1987) was published, the biosynthetic pathways of pheromones for a number of species had been determined, and work was progressing toward the characterization of some of the unique enzymes involved. It became apparent that the products of normal metabolism, particularly those of the fatty acid and isoprenoid pathways, were modified by a few pheromone gland-specific enzymes to produce the myriad of pheromone molecules. The elegant work of the Roelofs laboratory (Bjostad *et al.*, 1987) demonstrated that many of the lepidopteran pheromones could be formed by the appropriate interplay of highly selective chain shortening and unique desaturases followed by modification of the carboxyl carbon. This work has been extended, and a clear understanding of the biosynthetic pathways for many of the lepidopteran pheromones is now known (Jurenka, Chapter 3). Also, the delta-11 and other pheromone-specific desaturases in Lepidoptera have been characterized at the molecular level (Knipple and Roelofs Chapter 4). Chain shortening of fatty acids is also involved in producing the queen pheromone in honeybees (Plettner *et al.*, 1996, 1998) and this is reviewed in Chapter 11 (Blomquist and Howard). In some insects, fatty acid elongation followed by decarboxylation produce the hydrocarbon pheromones, and these include examples of lepidopterans (Jurenka, Chapter 3), dipterans (Blomquist, Chapter 8; Jallon and Wicker-Thomas, Chapter 9), the German cockroach (Schal *et al.*, Chapter 10) and the social insects (Blomquist and Howard, Chapter 11). More recent work in bark beetles has shown that *Ips* and *Dendroctonus* spp. produce their monterpenoid-derived pheromones ipsenol, ipsdienol and frontalin by modifications of isoprenoid pathway products (Seybold and Vanderwel, Chapter 6). Until that work, it was considered very rare for animals to produce monoterpenoids (C10 isoprenoids).

Sex pheromone gland cells can be individual cells or clusters of cells forming glandular tissue, and these structures can be located almost everywhere externally on insects, including antennae, the head, thorax, legs and abdomen (Ma and Ramaswamy Chapter 2). Recent work at the molecular level has shown that midgut tissue (Hall *et al.*, 2002a, 2002b; Nardi *et al.*, 2002) in *I. pini* and

Dendroctonus jeffreyi upregulate key mevalonate pathway enzymes during the induction of pheromone biosynthesis.

The powerful tools of molecular biology have recently been applied to studies on pheromone production. Knipple and Roelofs discuss their work on these unique desaturases involved in lepidopteran pheromone production in Chapter 4, and Tittiger reviews the work on the characterization of the mevalonate pathway enzymes in bark beetle pheromone production in Chapter 7. The sequencing of the *Drosophila* genome has resulted in advances in our understanding of the genetics in *Drosophila* and, combined with biochemical and molecular approaches, has led to new insights in a host of areas, including pheromone production (Jallon and Wicker-Thomas, Chapter 9).

1.3 Endocrine regulation of pheromone production

The production and/or release of sex pheromones is influenced by a variety of environmental factors (Shorey, 1974). In general, insects do not release pheromones until they are reproductively competent, although exceptions occur. Pheromone production is usually age related and coincides with the maturation of ovaries or testes, and in some cases with feeding. The observation that females of certain species have repeated reproductive cycles and that mating occurs only during defined periods of each cycle led to the proposal that pheromone production might be under hormonal control (Barth, 1965). Early work on cockroaches established that females require the presence of functional corpora allata in order to produce sex pheromones. Allatectomized females produce no pheromone and this prevents successful mating (Barth, 1961, 1962). Within a few years, the role of juvenile hormone (JH) in regulating pheromone production was established, and JH was shown to regulate pheromone production in a number of cockroach and beetle species (Hughes and Renwick, 1977a, b; Vanderwel and Oehlschlager, 1987; Vanderwel, 1994; Blomquist and Dillwith, 1983; Schal, Chapter 10; Seybold and Vanderwel, Chapter 6).

A unifying theme of this work was that the same hormone that regulated ovarian maturation (JH) also regulated pheromone production, coordinating sexual maturity with mating. Thus, in retrospect, it was not surprising that ovarian-produced 20-hydroxyecdysone, which plays an important role in reproduction in female Diptera, was shown to be the key hormone inducing sex pheromone production in the female housefly, *Musca domestica* (Adams *et al.*, 1984; Dillwith *et al.*, 1983; Blomquist *et al.*, 1987). This work has been extended to show that 20-hydroxyecdysone regulates the fatty acyl-CoA elongase enzymes to induce muscalure production (Blomquist, Chapter 8).

It was recognized by the mid-1980s that Lepidoptera regulated pheromone production through a different mechanism than flies, cockroaches and beetles

(Raina and Klun, 1984), but it wasn't until 1989 that the structure of the pheromone biosynthesis activating neuropeptide (PBAN) was elucidated (Raina *et al.*, 1989). The rapid advances in this area in the last decade are chronicled by Rafaeli and Jurenka (Chapter 5).

The three hormones that regulate pheromone production in insects are shown in Figure 1.1. PBAN alters enzyme activity through second messengers at one or more steps during or subsequent to fatty acid synthesis during pheromone production (Rafaeli and Jurenka, Chapter 5). In contrast, 20-hydroxyecdysone and JH induce or repress the synthesis of specific enzymes at the transcription level (Tittiger, Chapter 7; Blomquist, Chapter 9).

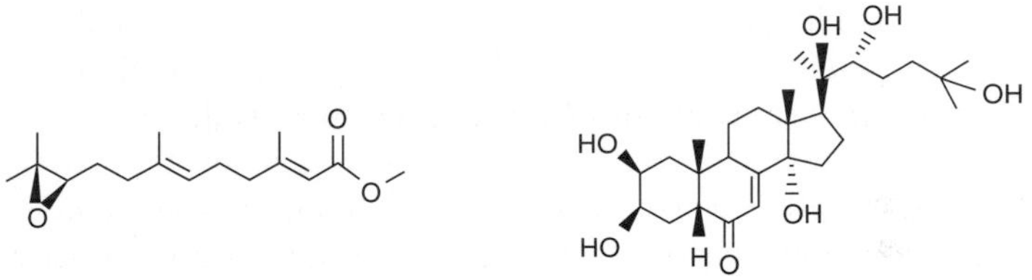

Juvenile hormone III
A

20-Hydroxyecdysone
B

1 5 10 15
Hez-PBAN Leu-Ser- Asp-Asp-Met-Pro-Ala-Thr-Pro-Ala-Asp-Gln-Glu-Met-Tyr-Arg-Gln-
Boam-PBAN Leu- Ser-Glu-Asp-Met-Pro-Ala-Thr-Pro-Ala-Asp-Gln-Glue-Met-Tyr-Gln-Pro-
Lyd-PBAN Leu-Ala-Asp-Asp-Met-Pro-Ala-Thr-Met-Ala-Asp-Gln-Glu-Val-Try-Arg-Prog-

20 25 30
PBAN
Asp-Pro-Glu-Gln-Ile- Asp-Ser-Arg-Thr-Lys-Tyr-Phe-Ser-Pro-Arg-Leu-NH_2
Asp-Pro-Glu-Glu-Met-Glu-Ser-Arg-Thr-Arg-Tyr-Phe-Ser-Pro-Arg-Leu-NH_2
Glu-Pro-Glu-Gln- Ile- Asp-Ser-Arg-Asn-Lys-Tyr-Phe-Ser-Pro-Arg-Leu-NH_2

Amidated C-terminal pentapeptide

Figure 1.1 The three major types of hormones that regulate pheromone production in insects. A Juvenile Hormone III (C16 JH), B 20-Hydroxyecdysone and C PBANs from the corn earworm, *Helicoverpa zea* (Raina *et al.*, 1989), the silkworm moth *Bombyx mori* (Kitamura *et al.*, 1989) and the gypsy moth, *Lymantira dispar* (Masler *et al.*, 1994). The minimum sequence (pentapeptide) required for activity is indicated.

In no model pheromone biosynthetic system is the molecular mechanism of hormonal regulation completely understood. The mechanism of action of JH and the nature of its receptor remain one of the mysteries of insect science, and the clear-cut action of JH by itself in inducing specific genes in pheromone production in bark beetles offers an excellent model for study. A better understanding of the PBAN receptor and the second messenger system it triggers as well as the steps regulated in pheromone biosynthesis is also needed. The next several years should see some of the key questions answered in model insects.

1.4 Detection of pheromones and plant volatiles

One reads that olfaction is the oldest sense, although this is generally stated from a human/vertebrate perspective and with the implication that olfaction was invented in fish. Chemoreception is certainly one of the oldest senses, present in bacteria and presumably one of those life essentials that was required of the earliest of chemotrophic single celled organisms. Life would be quite irrelevant in the absence of sensory input: no food, no escaping those who want to eat you, and no sex. There would be no learning since there would be no input through which one could learn anything, and no memory since there would be no means of experiencing. Chemoreception presumably became established in prokaryotes near the onset of life on earth, and became increasingly diversified and specialized through the evolution of eukaryotes. Plant, fungus and animal lineages diverged when organisms were still single celled; multicellularity developed independently within each of these lineages. Animals developed unique means to communicate between cells (cell–cell interactions and endocrine systems) and to coordinate their body movements and behaviors (nervous systems and endocrine systems). The detection of chemicals external to the animal body presumably became transformed into what we now think of as taste and smell. Chemoreception is not necessarily neuronal, but what we consider as smell and taste are clearly neuronal processes. The mechanisms underlying smell and taste are those common to the nervous system: neurons respond to external chemical stimuli (neurotransmitters and neuropeptides) via receptor proteins in their membranes which activate ion channels either directly (many receptors are themselves ion channels) or via second messenger transductory systems (e.g. G-protein coupled receptors). These processes are common to all those organisms that we consider animals, except for sponges, which are somewhat transitional between single and multicellularity and lack nervous systems. But even jellyfish have nervous systems that perform the same cellular functions as our own.

So, at the onset of this new millennium, we quite readily accept that olfaction and taste work via sensory neurons and that odor and taste molecules stimulate these neurons by binding to receptor proteins. But this is a remarkably recent view, certainly not preceding the experiments which elucidated the mechanisms underlying acetylcholine stimulation of nerve-muscle synapses (Eccles *et al.*, 1941; Fatt and Katz, 1951, 1952; del Castillo and Katz, 1954). It is perhaps, then, not surprising that Vincent Dethier (e.g. Dethier, 1962) and Dietrich Schneider (e.g. Schneider, 1969) should have been in the position to establish insect taste and olfaction as neurobiological systems. There is no doubt that Schneider and his colleagues are responsible for establishing the basics of how pheromones and other odors are detected by insects (e.g. Schneider, 1957; Boeckh, 1962; Schneider *et al.*, 1964; Boeckh *et al.*, 1965; Kaissling and Priesner, 1970; Steinbrecht and Kasang, 1972; Kaissling 1974a).

In 1980, there were no identified gene products that functioned in the detection of odors, pheromone or otherwise. This is not to say there had been no efforts to identify olfactory proteins. The degradation of pheromone molecules had been characterized in. *B. mori* (e.g. Kasang, 1971, 1972, 1974; Kasang and Kaissling, 1972) and *T. ni* (Mayer, 1975; Ferkovich *et al*., 1973a, b, 1980). Structure–activity studies suggested pheromones were detected by receptor proteins (e.g. Kikuchi, 1975; Kafka and Neuwirth, 1975), and chemicals that disrupt protein structure had been used to uncouple odor response pathways (e.g. Villet, 1974; Frazier and Heitz, 1975). Other studies suggested roles of second messengers in olfactory transduction (e.g. Villet, 1978; see Wieczorek and Schweikl, 1985). But these studies yielded no actual proteins or genes. In 1974, Karl-Ernst Kaissling (Kaissling, 1974b) proposed a model for pheromone detection in silk moths in which pheromone molecules were transported to neuronal receptor proteins via pore-tubules (after Steinbrecht and Müller, 1971; also see Steinbrecht, 1997) and were subsequently inactivated by some rapid but non-enzymatic process (after Kasang, 1973) (see Figure 1.2a). In 1980, however, matters were about to change.

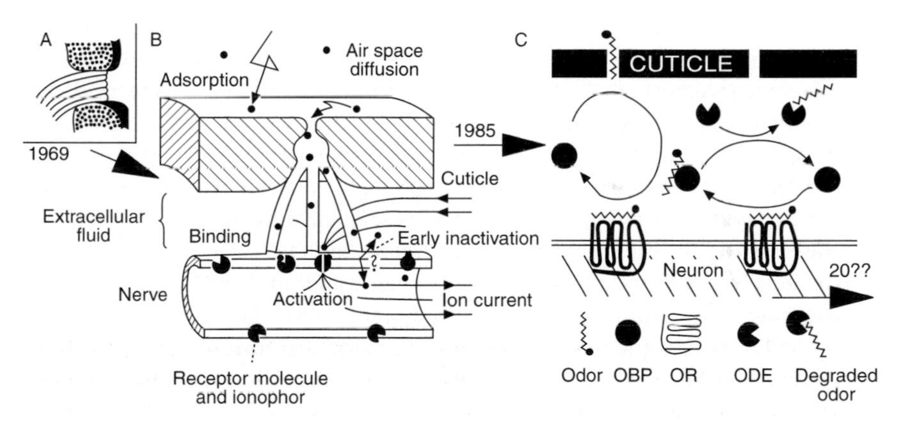

Figure 1.2 Historical models of insect odor detection. 'A' is taken from a figure in Slifer *et al.* (1959) and depicts neuronal cilia making contact with air through a pore penetrating the cuticle hair wall. In 1969, K. D. Ernst published images showing that these 'cilia' were not neuronal, but rather were tubular extensions from the cuticular pores (pore-tubules) (Ernst, 1969; Steinbrecht, 1997). 'B' is taken from a figure of Kaissling (1974) and depicts pore-tubules serving as conduits to transport pheromone molecules from the air to the neuronal membrane. 'C' is after Vogt *et al.* (1985) and depicts OBPs serving to transport odor molecules from pores to ORs, and ODEs as odor inactivators (see Chapter 14 for more information). '20??' is to imply the coming emergence of new understanding of the mechanisms underlying odor detection.

In 1981, the pheromone binding protein (PBP) and sensilla esterase (SE) of *Antheraea polyphemus* were identified (Vogt and Riddiford, 1981); and in 1985 a new model for pheromone detection was proposed. In this model, PBPs transported pheromone to receptor proteins (replacing pore-tubules in this role) and SE rapidly inactivated pheromone by enzymatic degradation (Vogt *et al.*, 1985) (Figure 1.2b). In 1987, when the previous edition of this book was published, not a great deal more was known. PBPs had been characterized in the gypsy moth, *Lymantria dispar* (Vogt, 1987; see Vogt *et al.*, 1989), and Glenn Prestwich was designing pheromone analogs to chemically dissect the biochemical pathways of pheromone detection (Prestwich, 1987a, b). In retrospect, this new model for pheromone reception (Vogt *et al.*, 1985) was as much a watershed for biochemical studies of insect olfaction as the previous edition of this book was a description of the calm before a coming storm that has yet to show any sign of letting up.

Much has occurred since 1987. PBPs have become established as only a subclass of a much larger family of insect OBPs that are represented at least throughout the neopterous insects, from cockroach to honeybee. The SE of *A. polyphemus* is known to be only one type of odor degrading enzyme (ODE) (and has recently been cloned: Ishida and Leal, 2002); a variety of ODEs are now known from diverse species. Odor receptors (ORs) have been characterized from *D. melanogaster, Anopheles gambiae* and *Heliothis virescens*, and the transductory processes that follow OR activation have been characterized. The differential expression of many olfactory genes has been described, providing explanations for the diverse functional phenotypes of olfactory sensilla. And *Drosophila* has become the "new kid on the block," contributing its genome and genetic manipulations as important new tools for elucidating olfactory mechanisms and interpreting the olfactory genomes of other insects.

The chapters of Part 2 reflect well the expansion of molecular genetic studies in insect odor detection that has occurred since 1987. The part is introduced (Chapter 13) with a thought-provoking review of the current state and future opportunities of the field. This part then presents discussions about several pre-receptor proteins, including odor degrading enzymes and odorant binding proteins as well as proteins we refer to as if in some secret code as CSPs and SNMPs (Chapters 14–18), followed by discussions of odor receptors and olfactory transduction mechanisms (Chapters 19, 20). The part closes with chapters providing critical context in which to view olfactory biochemistry, including discussions of antennal form and function (Chapter 21), the biosynthesis and ecology of plant volatiles that serve as pollination cues (Chapter 22), and the physiology and genetics underlying olfactory behaviors (Chapters 23, 24).

The contributing authors represent considerable expertise for the subject. Larry Zwiebel (Chapter 13) has opened the malaria mosquito *A. gambiae* to studies of olfactory molecular genetics, with profound implications for human health (e.g. Fox *et al.*, 2002; Hill *et al.*, 2002). Richard Vogt (Chapter 14) identified the first OBPs and ODEs and continues to provide the odd tidbit now and then (e.g. Vogt and Riddiford, 1981; Rogers *et al.*, 1999, 2001; Vogt *et al.*, 2002). Walter Leal (Chapter 15) and Erika Plettner (Chapter 16) continually make ground-breaking advances in our understanding of the mechanisms underlying OBP function (e.g. Wojtasek and Leal, 1999; Sandler *et al.*, 2000; Plettner *et al.*, 2000; Kowcun *et al.*, 2001). Patricia Nagnan-Le Meillour and Emmanuelle Jacquin-Joly (Chapter 17) have focused on the olfactory biochemistry of *Mamestra brassicae*, providing important insights into a lepidopteran family (Noctuidae) of profound agricultural importance (Maïbèche-Coisné *et al.*, 1997; Bohbot *et al.*, 1998; Campanacci *et al.*, 2001). Jean-François Picimbon (Chapter 18) has expanded our understanding and appreciation of the CSP/SAP family of proteins, some of which may have OBP-like functions, and is as well exploring the evolution of PBPs in noctuiid moths (Picimbon and Leal, 1999; Picimbon *et al.*, 2000; Picimbon and Gadenne, 2002).

Leslie Vosshall (Chapter 19) was among the first to identify and characterize OR genes in insects, using *Drosophila*, and has provided important insights into the mechanisms underlying the conveyance of olfactory information from ORs to the brain (Vosshall *et al.*, 1999, 2000; Vosshall, 2000, 2001). Vosshall has additionally made significant contributions to the field of circadian behavior (Vosshall *et al.*, 1994; Vosshall, and Young, 1995), a subject of historical importance to insect olfactory biology (e.g. Rau and Rau, 1929). Jürgen Krieger and Heinz Breer (Chapter 20) were among the first to clone insect OBPs, and have made major contributions to our understanding of pre-receptor, receptor and transductory mechanisms in both insects and vertebrates (Raming *et al.*, 1989; Freitag *et al.*, 1995; Krieger and Breer, 1999; Bette *et al.*, 2002). Recently, they were the first to characterize OR genes in Lepidoptera (Krieger *et al.*, 2002).

Catherine Loudon (Chapter 21) is renewing studies of how the design of an insect's antennae relates to its ability to capture odor molecules (Loudon and Koehl, 2000). This subject has been kicked around several times in our history (e.g. Adam and Delbrück, 1968; Vogel, 1983), but until Loudon's work, was not fully addressed. Robert Raguso (Chapter 22) has been a student of floral scent chemistry and insect pollination behavior. His studies are informing us that plant volatiles contain olfactory signals as specific as pheromones in eliciting behavioral responses from insects (Raguso and Pichersky, 1995; Raguso *et al.*, 1996; Raguso and Roy, 1998; Levin *et al.*, 2001). Marien de Bruyne (Chapter 23) has made important contributions to olfactory coding in *Drosophila* at the periphery, characterizing odor responsiveness and sensitivities of specific classes of olfactory sensilla, with a strong orientation towards genetics and genomics (de Bruyne *et al.*, 1999, 2001; Clyne *et al.*, 1999; Warr *et al.*, 2001). Mikael Carlsson and Bill Hansson (Chapter 24) have an extensive history with olfactory coding in moths and orthopteroids, and are in a unique position to understand olfactory coding in the brain in a manner that relates across the neopterous insects (Anton and Hansson, 1995; Carlsson *et al.*, 2002). Hansson in particular was the first to characterize the topographic organization of the interface between primary and secondary olfactory neurons in any animal (Hansson *et al.*, 1991, 1992).

This book is designed as a sourcebook for the next decade of research, and we hope it fills this expectation. Chapters have been assembled from experts who are at the frontiers of pheromone physiology, biochemistry, morphology, neurobiology and molecular biology. Ultimately, just as behavioral chemicals themselves have been extended to pest management, research on pheromone biosynthesis, hormonal regulation and reception may be directed toward application and ultimately used in insect control.

References

Adam G. and Delbrück M. (1968) Reduction of dimensionality in biological diffusion processes. In *Structural Chemistry and Molecular Biology* eds. A. Rich and N. Davidson, pp. 198–215. W. H. Freeman, San Francisco.

Adams T. S., Dillwith J. W. and Blomquist G. J. (1984) The role of 20-hydroxyecdysone in housefly sex pheromone biosynthesis. *J. Insect Physiol.* 30, 287–294.

Anton S. and Hansson B.S. (1995) Sex pheromone and plant-associated odour processing in antennal lobe interneurons of male *Spodoptera littoralis* (Lepidoptera: Noctuidae). *J. Comp. Physiol. A.* **176**, 773–789.

Barth R. H., Jr (1961) Hormonal control of sex attractant production in the Cuban cockroach. *Science* **133**, 1598–1599.

Barth R. H., Jr (1962) The endocrine control of mating behavior in the cockroach *Byrsotria fumigata* (Guerin). *Gen. Comp. Endocrinol.* **2**, 53–69.

Barth R. H., Jr (1965) Insect mating behavior: endocrine control of a chemical communication system. *Science* **149**, 882–883.

Berger R. S. (1966) Isolation, identification and synthesis of sex attractant of the cabbage looper, *Trichoplusia ni. Ann. Entomol. Soc. Amer.* **59**, 767–771.

Bette S. Breer H. and Krieger J. (2002) Probing a pheromone binding protein of the silkmoth Antheraea polyphemus by endogenous tryptophan fluorescence. *Insect Biochem. Mol. Biol.* **32**, 241–246.

Bjostad L. B., Wolf W. A. and Roelofs W. L. (1987) Pheromone biosynthesis in lepidopterans: desaturation and chain shortening. In *Pheromon Biochemistry*, eds G. J. Blomquist and G. D. Prestwich, pp. 77–120. Academic Press, Orlando, FL.

Blomquist G. J. and Dillwith J. W. (1983) Pheromones: biochemistry and physiology. In *Endocrinology of Insects*, eds R. G. H. Downer and H. Laufer, pp. 527–542. Alan R. Liss, Inc., New York, NY.

Blomquist G. J., Dillwith J. W. and Adams T. S. (1987) Biosynthesis and endocrine regulation of sex pheromone production in Diptera. In *Pheromone Biochemistry*, eds G. J. Blomquist and G. D. Prestwich, pp. 217–250. Academic Press, Orlando, FL.

Boeckh J. (1962) Elektrophysiologische Untersuchungen an einzelnen Geruchsrezeptoren auf den Antennnen des Totengräbers (*Necrophorus*, Coleoptera). *Z. Vergl. Physiol.* **46**, 212–248.

Boeckh J., Kaissling K. E. and Schneider D. (1965) Insect olfactory receptors. *Cold Spring Harbor Symposium on Quantitative Biology* **30**, 263–280.

Bohbot J., Sobrio F., Lucas P. and Nagnan-Le Meillour P. (1998) Functional characterization of a new class of odorant-binding proteins in the moth *Mamestra brassicae*. *Biochem Biophys. Res. Commun.* **253**, 489–494.

Butenandt A., Beckmann R., Stamm D. and Hecker E. (1959) Uber den Sexual- Lockstoff des Seidenspinners *Bombyx mori*. Reindarstellung und Konstitution. *Z. Naturforsch.* **14**, 283–284.

Byers J. A. (1981) Pheromone biosynthesis in the bark beetle, *Ips paraconfusus*, during feeding or exposure to vapours of host plant precursors. *Insect Biochem.* **11**, 563–569.

Byers J. A. (1983) Bark beetle conversion of a plant compound to a sex-specific inhibitor of pheromone attraction. *Science* **220**, 624–626.

Byers J. A., Wood D. L., Browne L. E., Fish R. H., Piatek B. and Hendry L. B. (1979) Relationship between a host plant compound, myrcene, and pheromone production in the bark beetle, *Ips paraconfusus*. *J. Insect Physiol.* **25**, 477–482.

Campanacci V., Mosbah A., Bornet O., Wechselberger R., Jacquin-Joly E., Cambillau C., Darbon H. and Tegoni M.. (2001) Chemosensory protein from the moth *Mamestra brassicae*. Expression and secondary structure from ^{1}H and ^{15}N NMR. *Eur. J. Biochem.* **268**, 4731–4739.

Carlsson M. A., Galizia C. G. and Hansson B. S. (2002) Spatial representation of odours in the antennal lobe of the moth *Spodoptera littoralis* (Lepidoptera: Noctuidae). *Chem. Senses.* **27**(3), 231–244.

Clyne P. J., Certel S., de Bruyne M., Zaslavsky L., Johnson W., Carlson J. R. (1999) The odor-specificities of a subset of olfactory receptor neurons are governed by *acj6*, a POU domain transcription factor. *Neuron* **22**, 339–347.

de Bruyne M., Clyne P. J., and Carlson J. R. (1999) Odor coding in a model olfactory organ: the *Drosophila* maxillary palp. *J. Neurosci.* **19**, 4520–4532.

de Bruyne M., Foster K. and Carlson J. R. (2001) Odor coding in the *Drosophila* antenna. *Neuron* **30**, 537–552.

del Castillo J. and Katz B. (1954) Quantal components of the end-plate potential. *J. Physiol.* **124**, 560–573.

Dethier, V. 1962. *To Know a Fly*. Holden-Day, San Francisco.

Dillwith J. W., Adams T. S. and Blomquist G. J. (1983) Correlation of housefly sex pheromone production with ovarian development. *J. Insect Physiol.* **29**, 377–386.

Eccles J. C., Katz B. and Kuffler S. W. (1941) Nature of the "endplate potential" in curarized muscle. *J. Neurophysiol.* **4**, 362–387.

Ernst, K. D. (1969) Die Feinstruktur von Reichsensilen auf der Antenne des Aaskäfers *Necrophorus* (Coleoptera). *Z. Zellforsch. Mikrosk. Anat.* **94**, 72–102.

Fatt P. and Katz B. (1951) An analysis of the end-plate potential recorded with an intracellular electrode. *J. Physiol.* **115**, 320–370.

Fatt P. and Katz B. (1952) Spontaneous subthreshold activity at motor nerve endings. *J. Physiol.* **117**, 109–128.

Ferkovich S. M., Mayer M. S. and Rutter R. R. (1973a) Conversion of the sex pheromone of the cabbage looper. *Nature* **242**, 53–55.

Ferkovich S. M., Mayer M. S. and Rutter R. R. (1973b) Sex pheromone of the cabbage looper: reactions with antennal proteins *in vitro*. *J. Insect Physiol.* **19**, 2231–2243.

Ferkovich S. M., Van Essen F. and Taylor T. R. (1980) Hydrolysis of sex pheromone by antennal esterases of the cabbage looper, Trichoplusia ni. Chem. Senses Flavour **5**, 33–45.

Fox A. N., Pitts R. J. and Zwiebel L. J. (2002) A cluster of candidate odorant receptors from the malaria vector mosquito, *Anopheles gambiae*. *Chem. Senses* **27**, 453–459.

Frazier J. L. and Heitz J. R. (1975) Electrophysiological studies of the interactions of sulfhydryl reagents with insect olfactory receptors. *J. Miss. Acad. Sci* **19**, 188.
Freitag J., Krieger J., Strotmann J. and Breer H. (1995) Two classes of olfactory receptors in *Xenopus laevis. Neuron* **15**, 1383–1392.
Hall G. M., Tittiger C., Andrews G., Mastick G., Kuenzli M., Luo X., Seybold S. J. and Blomquist G. J. (2002a) Male pine engraver Beetles, *Ips pini*, synthesize the Naturwissenschaften **89**, 79–83.
Hall G. M., Tittiger C., Blomquist G. J., Andrews G., Mastick G., Barkawi L. A., Bengoa C. S. and Seybold S. J. (2002b) Male Jeffrey Pine Beetles, *Dendroctonus jeffreyi*, synthesize the pheromone component frontalin in anterior midgut tissue. *Insect Biochem. Mol. Biol.* **32**, 1525–1532.
Hansson B. S., Christensen T. A. and Hildebrand J. G. (1991). Functionally distinct subdivisions of the macroglomerular complex in the antennal lobe of the male sphinx moth *Manduca sexta. J. Comp. Neurol.* **312**, 264–278.
Hansson B. S., Ljungberg H., Hallberg E. and Löfstedt, C. (1992). Functional specialization of olfactory glomeruli in a moth. *Science* **256**, 1313–1315.
Hendry L. B., Piatek B., Browne L. E., Wood D. L., Byers J. A., Fish R. H. and Hicks R. A. (1980) *In vivo* conversion of a labelled host plant chemical to pheromones of the bark beetle, *Ips paraconfusus. Nature* **284**, 485.
Hill C. A., Fox A. N., Pitts R. J., Kent L. B., Tan P. L., Chrystal M. A., Cravchik A., Collins FH., Robertson H. M. and Zwiebel L. J. (2002) G protein-coupled receptors in *Anopheles gambiae. Science* **298**, 176–178.
Hughes P. R. (1974) Myrcene: a precursor of pheromones in *Ips beetles. J. Insect Physiol.* **20**, 1271–1275.
Hughes P. R. and Renwick J. A. A. (1977a) Hormonal and host factors stimulating pheromone synthesis in female western pine beetles, *Dendroctonus brevicomis. Physiol. Entomol.* **2**, 289–292.
Hughes P. R. and Renwick J. A. A. (1977b) Neural and hormonal control of pheromone biosynthesis in the bark beetle, *Ips paraconfusus. Physiol. Entomol.* **2**, 117–123.
Ishida Y. and Leal S. (2002) Cloning of putative odorant-degrading enzyme and integumental esterase cDNAs from the wild silkmoth, *Antheraea polyphemus. Insect Biochem. Mol. Biol.* **32**, 1775–1780.
Kafka W. A. and Neuwirth J. (1975) A model of pheromone molecule-acceptor interaction. *Z. Naturforsch.* **30**, 278–282.
Kaissling K. E. (1974) Sensory transduction in insect olfactory receptors. In *Biochemistry of Sensory Functions*. eds L. Jaenicke. Springer, Berlin, pp. 243–273.
Kaissling K.-E. and Priesner E. (1970) Die Riechschwelle des Seidenspinners. *Naturwissenschaften* **57**, 23–28.
Kasang G. (1971) Bombykol reception and metabolism on the antennae of the silkmoth *Bombyx mori. In Gustation and Olfaction,* eds Ohloff G. and Thomas A. F., pp. 245–250. Academic Press, New York.
Kasang G. (1973) Physikomichemische Vorange beim Riechen des Seidenspinners. *Naturwissenschaften* **60**, 95–101.
Kasang G. (1974) Uptake of the sex pheromone 3H-bombykol and related compounds by male and female *Bombyx* antennae. *J. Insect Physiol.* **20**, 2407–2422.
Kasang G. and Kaissling K. E. (1972) Specificity of primary and secondary olfactory processes in *Bombyx* antennae. In *Int. Symp. Olfaction and Taste IV*, ed. D. Schneider 200–206. Verlagsgesellschaft, Stuttgart.
Kikuchi T. (1975) Correlation of moth sex pheromone activities with molecular characteristics involved in conformers of bombykol and its derivatives. *Proc. Natl. Acad. Sci. USA* **72**, 3337–3341.

Kitamura A., Nagasawa H., Kataoka H., Inoue T., Matsumoto S., Ando T. and Suzuki A. (1989) Amino acid sequence of pheromone biosynthesis activating neuropeptide (PBAN) of the silkworm, *Bombyx mori. Biochem. Biophys. Res. Commun.* **163**, 520–526.

Kowcun A., Honson N. and Plettner E. (2001) Olfaction in the gypsy moth, *Lymantria dispar*: effect of pH, ionic strength, and reductants on pheromone transport by pheromone-binding proteins. *J. Biol. Chem.* **276**, 44770–44776.

Krieger J. and Breer H. (1999) Olfactory reception in invertebrates. *Science* **286**, 720–723.

Krieger J., Raming K., Dewer Y. M., Bette S., Conzelmann S. and Breer H. (2002) A divergent gene family encoding candidate olfactory receptors of the moth *Heliothis virescens. Eur. J. Neurosci.* **16**(4), 619–628.

Levin R. A., Raguso R. A. and McDade L. A. (2001) Fragrance chemistry and pollinator affinities in Nyctaginaceae. *Phytochemistry* **58**, 429–440.

Loudon, C. and Koehl, M. A. R. (2000) Sniffing by a silkworm moth: wing fanning enhances air penetration through and pheromone interception by antennae. *Journal of Experimental Biology* **203**, 2977–2990.

Maïbèche-Coisné M., Sobrio F., Delaunay T., Lettere M., Dubroca J., Jacquin-Joly E. and Nagnan-LeMeillour P. (1997) Pheromone Binding Proteins of the moth *Mamestra brassicae*: specificity of ligand binding. *Insect Biochem. Mol. Biol.* **27**, 213–221.

Masler E. P., Raina A. K., Wagner R. M. and Kochansky J. P. (1994) Isolation and identification of a pheromonotropic neuropeptide from the brain–subesophageal ganglion complex of *Lymantria dispar*: a new member of the PBAN family. *Insect Biochem. Mol. Biol.* **24**, 829–836.

Mayer M. S. (1975) Hydrolysis of sex pheromone by the antennae of *Trichoplusia ni.* Experentia **31**, 452–454.

Mitlin N. and Hedin P. A. (1974) Biosynthesis of grandlure, the pheromone of the boll weevil, *Anthonomus grandis*, from acetate, mevalonate, and glucose. *J. Insect Physiol.* **20**, 1825–1831.

Nardi J. B., Gilg Young A., Ujhelyi E., Tittiger C., Lehane M. J. and Blomquist G. J. (2002) Specialization of midgut cells for synthesis of male isoprenoid pheromone in two scolytid beetles, *Dendroctonus jeffreyi* and *Ips pini. Tissue and Cell.* **226**, 221–231.

Picimbon J. F. and Gadenne C. (2002) Evolution and noctuid pheromone binding proteins: identification of PBP in the black cutworm moth, *Agrotis ipsilon. Insect Biochem. Molec. Biol.* **32**, 839–846.

Picimbon J. F. and Leal W. S. (1999) Olfactory soluble proteins of cockroaches. *Insect Biochem. Mol. Biol.*, **29**, 973–978.

Picimbon J. F., Dietrich K., Breer H. and Krieger J. (2000). Chemosensory proteins of Locusta migratoria (Orthoptera: Acrididae). *Insect Biochem. Mol. Biol.* **30**, 233–241.

Plettner E., Slessor K. N., Winston M. L. and Oliver J. E. (1996) Caste-selective pheromone biosynthesis in honeybees. *Science* **271**, 1851–1853.

Plettner E., Slessor K. N. and Winston M. L. (1998) Biosynthesis of mandibular acids in honey bees (*Apis mellifera*): de novo synthesis, route of fatty acid hydroxylation and caste selective ß-oxidation. *Insect Biochem. Molec. Biol.* **28**, 31–42.

Plettner E., Lazar J., Prestwich E. G. and Prestwich G. D. (2000) Discrimination of pheromone enantiomers by two pheromone binding proteins from the gypsy moth *Lymantria dispar. Biochemistry* **39**, 8953–8962.

Prestwich G. D. (1987a) Chemical studies of pheromone reception and catabolism. In *Pheromone Biochemistry*, eds G. D. Prestwich and G. J. Blomquist, pp. 473–527. Academic Press, New York.

Prestwich G. D. (1987b) Chemistry of pheromone and hormone metabolism in insects. *Science* **237**, 999–1006.

Prestwich G. D. and Blomquist G. J. (1987) *Pheromone Biochemistry*, 565 pp. Academic Press, Orlando, Florida.

Raguso R. and Pichersky E. (1995). Floral volatiles of *Clarkia breweri* and *C. concinna* (Onagraceae): recent evolution of floral aroma and moth pollination. *Plant Systematics and Evolution* **194**, 55–67.

Raguso R., Light D. M. and Pichersky E. (1996). Electroantennogram responses of *Hyles lineata* (Sphingidae: Lepidoptera) to floral volatile compounds from *Clarkia breweri* (Onagraceae) and other moth-pollinated flowers. *Journal of Chemical Ecology* **22**, 1735–1766.

Raguso R. A. and Roy B. A. (1998) "Floral" scent production by *Puccinia* rust fungi that mimic flowers. *Molecular Ecology* **7**, 1127–1136.

Raina A. K. and Klun J. A. (1984) Brain factor control of sex pheromone production in the female corn earworm moth. *Science* **225**, 531–533.

Raina A. K., Jaffe H., Kempe T. G., Keim P., Blacher R. W., Fales H. M., Riley C. T., Klun J. A., Ridgway R. L. and Hayes D. K. (1989) Identification of a neuropeptide hormone that regulates sex pheromone production in female moths. *Science* **244**, 796–798.

Raming K., Krieger J. and Breer H. (1989) Molecular cloning of an insect pheromone-binding protein. *FEBS Lett.* **256**, 215–218.

Rau P. and Rau N. (1929) The sex attraction and rhythmic periodicity in giant saturniid moths. *Trans. Acad. Sci St. Louis* **26**, 83–221.

Rogers M. E., Jani M. K., Vogt R. G. (1999) An olfactory-specific glutathione-S-transferase in the sphinx moth *Manduca sexta*. *J. Exp. Biol.* **202**, 1625–1637.

Rogers M. E., Krieger J. and Vogt R. G. (2001) Antennal SNMPs (sensory neuron membrane proteins) of Lepidoptera define a unique family of invertebrate CD36-like proteins. *J. Neurobiol.* **49**, 47–61.

Sandler B. H., Nikonova L., Leal W. S. and Clardy J. (2000) Sexual attraction in the silkworm moth: structure of the pheromone-binding-protein-bombykol complex. *Chem. Biol.* **7**, 143–151.

Schneider D. (1957) Elektrophysiologische Untersuchungen von Chemo- und Mechanorezeptoren der Antenne des Seidenspinners Bombyx mori. *Z. Vergl. Physiol.* **40**, 8–41.

Schneider D. (1969) Insect olfaction: deciphering system for chemical messages. *Science* **163**, 1031–1037.

Schneider D., Lacher V. and Kaissling K.-E. (1964) Die Reaktionsweise und das Reaktionsspektrum von Riechzellen bei *Antheraea pernyi* (Lepidoptera, Saturniidae). *Z. Vergl. Physiol.* **48**, 632–662.

Seybold S. J., Quilici D. R., Tillman J. A., Vanderwel D., Wood D. L. and Blomquist G. J. (1995) *De novo* biosynthesis of the aggregation pheromone components ipsenol and ipsdienol by the pine bark beetles *Ips paraconfusus* Lanier and *Ips pini* (Say) (Coleoptera: Scolytidae). *Proc. Natl. Acad. Sci. USA* **92**, 8393–8397.

Shorey H. H. (1974) Environmental and physiological control of insect sex pheromone behavior. In *Pheromones*, ed. M.C. Birch, pp 62–80. New York, American Elsevier.

Silverstein R. M., Rodin J. O. and Wood D. L. (1966) Sex attractants in frass produced by male *Ips confusus* in ponderosa pine. *Science* **154**, 509–510.

Slifer E. H., Prestage J. J., Beams H. W. (1959) The chemoreceptors and other sense organs on the antennal flagellum of the grasshopper (Orthoptera: Acrididae). *J. Morph.* **105**, 145–191.

Steinbrecht R. A. (1997) Pore structures in insect olfactory sensilla: a review of data and concepts. *Int. J. Insect Morphol. & Embryol.* **26**, 229–245.

Steinbrecht R. A. and Kasang G. (1972) Capture and conveyance of odourmolecules in an insect olfactory receptor. In *Olfaction and Taste IV*, ed. D. Scheeder, pp. 193–199. Wiss. Verl. Ges., Stuttgart.

Steinbrecht R. A. and Müller B. (1971) On the stimulus conducting structures in insect olfactory receptors. *Z. Zellforsch.* **117**, 570–575.

Thompson A. C. and Mitlin N. (1979) Biosynthesis of the sex pheromone of the male boll weevil from monoterpene precursors. *Insect Biochem.* **9**, 293–294.

Tillman J. A., Seybold S. J., Jurenka J. A. and Blomquist G. J. (1999) Insect pheromones – an overview of biosynthesis and endocrine regulation. *Insect Biochem. Mol. Biol.* **29**, 481–514.

Vanderwel D. (1994) Factors affecting pheromone production in beetles. *Arch. Insect Biochem. Physiol.* **25**, 347–362.

Vanderwel D. and Oehlschlager A. C. (1987) Biosynthesis of pheromones and endocrine regulation of pheromone production in Coleoptera. In *Pheromone Biochemistry*, eds G. J. Blomquist and G. D. Prestwich, pp. 175–215. Academic Press, Orlando, FL.

Villet R. H. (1974) Involvement of amino and sylfhydryl groups in olfactory transduction in silkmoths. Nature 248, 707–709.

Villet R. H. (1978) Mechanism of insect sex-pheromone sensory transduction: role of adenyl cyclase. *Comp. Biochem. Physiol.* **61C**, 389–394.

Vogel S. (1983). How much air passes through a silkmoth's antenna? *Journal of Insect Physiology* **29**, 597–602.

Vogt R. G. (1987) The molecular basis of pheromone reception: its influence on behavior. In *Pheromone Biochemistry*, eds G. D. Prestwich and G. J. Blomquist, pp. 385–431. Academic Press, New York.

Vogt R. G., Köehne A. C., Dubnau J. T. and Prestwich G. D. (1989) Expression of pheromone binding proteins during antennal development in the gypsy moth *Lymantria dispar*. *J. Neurosci.* **9**, 3332–3346.

Vogt R. G. and Riddiford L. M. (1981) Pheromone binding and inactivation by moth antennae. *Nature* **293**, 161–163.

Vogt R. G., Riddiford L. M. and Prestwich G. D. (1985) Kinetic properties of a sex pheromone-degrading enzyme: the sensillar esterase of *Antheraea polyphemus*. *Proc. Natl. Acad. Sci. USA* **82**, 8827–8831.

Vogt R. G., Rogers M. E., Franco M. D. and Sun M (2002) A comparative study of odorant binding protein genes: differential expression of the PBP1-GOBP2 gene cluster in *Manduca sexta* (Lepidoptera) and the organization of OBP genes in *Drosophila melanogaster* (Diptera). J. Exp. Biol. **205**, 719–744.

Vosshall L. B. (2000) Olfaction in *Drosophila*. *Curr. Opin. Neurobiol.* **10**, 498–503.

Vosshall L. B. (2001) How the brain sees smells. *Dev. Cell* **1**, 588–590.

Vosshall L. B., Amrein H., Morozov P. S., Rzhetsky A. and Axel R. (1999) A spatial map of olfactory receptor expression in the *Drosophila* antenna. *Cell* **96**, 725–736.

Vosshall L. B., Price J. L., Sehgal A., Saez L. and Young M. W. (1994) Block in nuclear localization of period protein by a second clock mutation, timeless. *Science* **263**, 1606–1609.

Vosshall L. B., Wong A. M. and Axel R. (2000) An olfactory sensory map in the fly brain. *Cell* **102**, 147–159.

Vosshall L. B. and Young M. W. (1995) Circadian rhythms in *Drosophila* can be driven by period expression in a restricted group of central brain cells. *Neuron* **15**, 345–360.

Warr C. G., Clyne P. J., de Bruyne M., Kim J., Carlson J. R. (2001) Olfaction in Drosophila: coding, genetics and e-genetics. *Chem. Senses* **26**, 201–206.

Wieczorek H. and Schweikl H. (1985) Concentrations of cyclic nucleotids and activities of cyclases and phosphodiesterases in an insect chemosensory organ. *Insect Biochem.* **6**, 723–728.

Wojtasek H. and Leal W. S. (1999) Conformational change in the pheromone-binding protein from *Bombyx mori* induced by pH and by interaction with membranes. *J. Biol. Chem.* **274**, 30950–30956.

2

Biology and ultrastructure of sex pheromone-producing tissue

Peter W. K. Ma and Sonny B. Ramaswamy

2.1 Introduction

Since publication of the first edition in 1987 of the book *Pheromone Biochemistry*, studies on explaining the functional ultrastructure of pheromone-producing glands continue to lag behind the significant strides made in explaining pheromone-mediated behavior, chemistry, molecular biology, genetic and hormonal regulation, and sensory processes. But for some recent studies with beetles, where efforts have been initiated to determine the molecular processes in pheromone synthesizing midgut cells using *in situ* hybridization (Hall *et al.*, 2002a, b), the state of our knowledge of the ultrastructure of pheromone-producing cells remains static, with the occasional publication on the ultrastructural details of pheromone-producing cells in yet another species. Actually, ultrastructural studies have been moribund for some time; in concluding their chapter, Percy-Cunningham and MacDonald (1987) stated, "considerable information has accumulated concerning the structure of the cells and cuticle comprising sex pheromone glands, but the studies are far from complete and the biochemical elucidation of the structural–functional interrelationships has barely begun." In reviewing the extant literature since 1987, one can state that their conclusion remains true today.

There have been a couple of recent and excellent reviews of the ultrastructure of exocrine cells in general (Quennedey, 1998) and pheromone-secreting cells in social insects in particular (Billen and Morgan, 1998). This chapter attempts to survey the structural diversity of cells that produce sex pheromone in insects, particularly from a phylogenetic perspective, in part based on advances made since the 1987 chapter, and to describe the current state of knowledge based on recent studies using molecular probes. The chapter is based on descriptions of the morphology of pheromone-secreting cells in Percy-Cunningham and MacDonald (1987) and Quennedey (1998).

2.2 Classification of sex pheromone gland cells

Sex pheromone gland cells can be individual cells on the body, particularly the integument, or clusters of cells forming glandular tissue. These structures are located externally on antennae, the head, thorax, legs, and abdomen (Percy-Cunningham and MacDonald, 1987; Quennedey, 1998). Sex pheromone-producing cells in these structures, are for the most part, modified ectodermal cells, and belong to either the Class I or Class III cell type, in the scheme of Noirot and Quennedey (1974, 1991), particularly based on the route followed by the pheromonal secretions in crossing the cuticular barrier. However, there are at least two different internal abdominal glands involved in the production of sex pheromone; ones involved solely in pheromone production (Levinson *et al.*, 1983), and those whose primary role is some other function, but which secondarily produce sex pheromone (Nardi *et al.*, 2002). Of the latter, one is the ectodermally derived rectal pheromone gland cells characteristic of insects such as tephritid fruit flies (e.g., Little and Cunningham, 1987) and the other is the endodermally derived midgut cells that produce sex pheromone in bark beetles (Nardi *et al.*, 2002). In addition, the mesodermally originated fat body cells have been suggested to produce aggregation pheromone in the boll weevil (Wiygul *et al.*, 1982), although this interpretation remains to be verified.

Contrary to the expectation voiced at the end of the 1987 chapter by Percy-Cunningham and MacDonald, we are yet to decipher the cellular location of the biosynthetic processes, particularly in cells producing complex mixtures. As Quennedey (1998) points out, the low-molecular weight pheromones analyzed by gas chromatography are usually lost during typical preparative procedures needed for ultrastructural studies. What will be required are molecular probes that can be used to localize the biosynthetic molecules *in situ* (e.g. Ferveur *et al.*, 1997; Hall *et al.*, 2002b), as is being done with deciphering the location of nucleic acids and proteins involved in other biosynthetic processes. Alternatively, one can maybe develop antibodies to key enzymes in the biochemical steps or even to the lipoidal intermediates and end products of pheromone biosynthesis, which can then be used to probe the biochemical and subcellular localization of synthetic processes.

The Class I cells occur in several Orders of the Class Insecta. These cells characteristically occur clustered in intersegmental membranes, and are rarely scattered individually (e.g. Percy, 1979). The apical cell membrane of Class I cells abuts the glandular cuticle, and the secretion has to cross this barrier before it is released into the atmosphere (Quennedey, 1998). The Class I cells are characterized by perforated outer epicuticle or cuticulin, which is speculated to allow for enhanced diffusion of secretions. It is possible that the matrix of the cuticle overlaying Class I cells might even contain enzymes that may convert secretions from glandular cells as they are being released. For example, when sex pheromone glands of the spruce budworm moth, *Choristoneura fumiferana*, are examined by analytical methods, one finds monounsaturated 14-carbon acetates, which are hydrolyzed to alcohols and subsequently oxidized by an alcohol oxidase during release to the corresponding aldehydes, the active sex pheromone of this species (Morse and Meighen, 1984). Another modification is the occurrence of extracellular spaces allowing storage of secretions (Quennedey, 1998). The plasma membrane of Class I cells is differentiated both apically and basally, thus increasing the surface area to transfer substances between the hemolymph, glandular cell, and cuticle. Pinocytosis occurs both apically and basally. Class I cells exhibit desmosomes at septate junctions. These cells are characterized by tubular or lamellate form of smooth endoplasmic reticulum, characteristic of non-protein secreting systems. In some species, they may occur in whorls. Sometimes, the endoplasmic reticulum is connected to channels in the microvilli. Occasionally, rough endoplasmic reticulum occurs in close proximity to the well-developed Golgi complex. The abundance, size, and complexity of mitochondria vary, and may be related to differences in types and modes of pheromone secretion. There is also tremendous variability in abundance, shape, and electron density of secretory products, ranging from electron-lucent vesicles containing lipoidal materials to electron-dense secretions containing proteinaceous secretions. In some species lamellated and myeloid secretions are observed. Yet in other species, fibrillar or amorphous secretions are found. In many species, the size of the Class I cells and/or abundance and size of vesicles is correlated with secretory activity (Billen and Morgan, 1998).

The Class II cells, ectodermally derived oenocytes, are not in direct contact with the cuticle, and their secretions pass through surrounding Class I cells (Noirot and Quennedey, 1991). Often, they form aggregates with Class I cells, and may exhibit an extracellular reservoir for storage of secretions (Quennedey, 1998). Class II cells exclusively appear to play a role in species that use hydrocarbon sex pheromones. The oenocytes may be associated with the integument, tracheae, or fat bodies, and are involved in cycling hydrocarbon from pools in the hemolymph to the cuticle (Nardi *et al.*, 1996; Ferveur *et al.*, 1997; Schal *et al.*, 1998a or b). For example, targeted expression of the *transformer* gene in abdominal oenocytes of adult male *Drosophila melanogaster* induced such males to become pheromonally feminized, which produced a female type of hydrocarbon sex pheromone and elicited courtship behavior by other males (Ferveur *et al.*, 1997). The oenocytes in a nitidulid beetle produce the aggregation pheromone, and the tracheal cells have been recruited as conducting ductules (Nardi *et al.*, 1996). The oenocytes in this insect are exceptionally large with significant and deep surface invaginations surrounding the tracheal ductule, which enhances secretion of the pheromone. Uniquely, these glandular cells lack microvilli, but do contain abundant lipid spheres that likely contain pheromone precursors. The size of the secretory cells is correlated with rates of pheromone production. Class II oenocyte cells that form the sternal glands of termites contain numerous electron-lucent or electron-dense vesicles, have large or small mitochondria, and typically lie underneath Class I cells through which the secretions are released across microvilli to the overlying, vacuolated, and sometimes, sponge-like cuticle (Quennedey, 1998).

The Class III cells comprise a group of cells connected to the cuticle by a cuticular duct that serves as a conduit for secretions to be delivered to the outside. In many species, Class III cells are scattered individually or occur in small bundles, often recognized externally by the occurrence of pores in the cuticle through which secretions are deposited on the surface (Ramaswamy *et al.*, 1995). The cuticular duct has an extracellular receiving canal, which is bounded by sponge-like microvilli where secretions can be stored, and a conducting canal, which is surrounded by outer cells (Ramaswamy *et al.*, 1995; Quennedey, 1998). The basal membrane is evaginated (Araujo and Pasteels, 1987), along which surface pinocytosis can occur. Class III cells are also characterized by abundant smooth endoplasmic reticulum, which may be tubular or arranged in whorls. Some may even have rough endoplasmic reticulum. These cells also have abundant mitochondria, often located near the end apparatus formed by the receiving canal, large Golgi complexes, and abundant secretory vesicles that may be electron lucent or dense, lamellated or myeloid, or fibrillar or amorphous. The roles of these cellular organelles in synthesis and release of the pheromonal secretions are yet to be deciphered.

Endodermally derived midgut cells that synthesize pheromone and release into frass to be deposited outside the insect's body have recently been described in male bark beetles (Nardi *et al.*, 2002). In these species, the midgut cells are specialized for digestion and for synthesizing the isoprenoid pheromonal compounds (Hall *et al.*, 2002a, b), and there is no division of labor in the functionality of these cells, i.e., the same cells can be involved in production of digestive enzymes as well as pheromone (Nardi *et al.*, 2002). The cells have abundant apical microvilli and smooth endoplasmic reticulum. These cells appear to secrete by apocrine mechanisms, i.e. via apical extrusions, rather than exocytosis or eccrine mechanism (Nardi *et al.*, 2002). Recently, Hall *et al.* (2002a, b) demonstrated that the gene encoding a key, regulated enzyme, 3-hydroxy-3-methylglutaryl-CoA reductase, for the *de novo* biosynthesis of the bark beetles, *Ips pini* and *Dendroctonus jeffreyi*, aggregation pheromones, ipsdienol and frontalin, respectively, showed high transcript levels in the anterior midgut of male beetles.

2.3 Sex pheromone glands in apterygotes

A review of the literature on pheromones in this group, which comprises mostly very small, soil-dwelling species, including collembolans, thysanurans, proturans, diplurans, and microcoryphians, suggests that the pheromone biology of the springtails is relatively better described, whereas that of the other groups is either unknown or is rudimentary. Surprisingly, nothing is known even in the firebrats and silverfish.

The collembolans use both an aggregation pheromone and an alarm pheromone (Verhoef *et al.*, 1977; Purrington *et al.*, 1991). The former is produced by cells in the posterior pylorus and released in the feces (Verhoef, 1984) and the latter released upon rupture of the body cuticle (Purrington *et al.*, 1991).

The source and identity of the alarm pheromone is unknown; it is a "rapidly vaporizing," hexane-soluble material (Purrington *et al.*, 1991), presumably produced in Class III cells associated with the body wall. The pyloric region of the hindgut of adult male and female collembolans produces aggregation pheromone (Joose and Koelman, 1979; Krool and Bauer, 1987). The posterior pyloric region is characterized by abundant lysosome-like bodies, which are speculated to be involved in pheromone production (Krzysztofowicz and Kuzyk, 1979). The images of the lysosome-like bodies are reminiscent of electron-lucent secretory inclusions in other insects.

Males of the collembolan, *Sinella curviseta*, deposit spermatophores in response to the emission of a putative, volatile sex pheromone produced by females (Waldorf, 1974). The emission of pheromone has been speculated to be related to the molting process (Waldorf, 1974), which suggests that sex pheromone in these species also may well be produced by epidermal cells associated with the integument.

Jacquemin and Bareth (1981) speculated that the numerous bristles found on the anterior edge of the 1st abdominal sternite of male *Campodea chardardi* (Diplura) might be involved in production of a sex pheromone. The bristles, which are hollow and open at the tip, have three sensory-like (sic) cells, a large glandular cell (200 μm $\times$ 20 μm), and envelope cells derived from a trichogen and tormogen cell. The basal plasma membrane is highly invaginated. The glandular cell has numerous smooth endoplasmic reticulum, Golgi bodies, mitochondria, and electron-dense secretory vesicles. The granular secretions from the glandular cell are released into a distal extracellular reservoir that opens into a long canal that traverses the length of the hollow bristle and opens at the tip. The structure of this reservoir and canal is different from that of the end apparatus and conducting canal seen in the Class III cells, but may have a similar origin. Unfortunately, there have been no further studies either of the chemical nature of secretions from these structures or of their behavioral role.

2.4 Sex pheromone glands in the Pterygota-Paleoptera

We did not discover any published literature on the pheromone biology of ephemeropterans and odonates, likely because these two groups are known to use visual signals in courtship behavior (Thornhill and Alcock 1983).

2.5 Sex pheromone glands in the Pterygota-Orthopteroidea

Of the eight orders in this group, no sex pheromones are known in the Grylloblattaria and Embiidina. Of the remaining orders, probably the best studied are the cockroaches and termites, particularly the latter because of the involvement of pheromonal signals in their social structure (Vander Meer *et al.*, 1998).

Thus far no sex pheromone has been described in the Phasmida. Some phasmids produce toxic monoterpenes in typical Class III integumentary glands located behind the head (e.g. Happ *et al.*, 1966). The glands exhibit lipid reserves, carboxylic esterases, phosphatases, alcohol dehydrogenase, and a mevalonate kinase, all of which are suggested to be involved in the production of the toxic compounds (Happ *et al.*, 1966).

Among the Orthoptera, the adult epidermis in *Schistocerca gregaria* (gregarious phase) consists of numerous Class III gland cells that secrete a sexual maturation pheromone under the influence of a putative gonadotropic hormone (Cassier and Delorme-Joulie, 1976a, b, c). The nature of this hormone is unknown, but in the presence of ecdysone it causes morphological and functional changes in the cells, particularly during late 4th and early 5th instar. The active cells have numerous electron-dense vesicles, while the inactive cells lack such vesicles, have numerous microvilli, and appear more vacuolated. Allatectomy and administration of ecdysone stimulate differentiation of glandular cells; in contrast, implantation of corpora allata or application of juvenile hormone (JH) inhibits the differentiation of these sex pheromone-secreting gland cells. Production of the maturation pheromone by Class III dermal scent glands, which trigger reproductive synchrony, appears to be characteristic of many grasshopper species (Hawkes *et al.*, 1987). These Class III cells are characterized by a single end apparatus that has an oblong cavity delimited by the tips of densely packed microvilli. A conducting duct with bulbous thickenings, which is anchored to the cavity by a network of fibrillar material, carries the secretions to the outside.

The sex pheromone gland of female *Poecilocerus pictus*, a pyrogomorphid grasshopper, is a thin sac-like structure that arises as a modification of the tergal epithelium, and is located in the metathoracic segment close to the metanotum, embedded in adipose tissue (Gupta, 1979). The gland comprises a single layer of columnar epithelial cells that rest on a basement membrane, and is internally lined with endocuticle.

Class II oenocytes produce hydrocarbons in grasshoppers (Diehl, 1975); other than its involvement in waterproofing the cuticle and the ootheca, the role of hydrocarbons in the sexual behavior of the grasshopper is not known, unlike in some arctiids, cockroaches, and flies (Schal *et al.,* 1998 a or b). Similarly, while the biochemical processes involved in the cycling of hydrocarbons between oenocytes and target tissues has been described (e.g. Schal *et al.*, 1998), little information is available on such processes in grasshoppers; nor is information available on their ultrastructure.

A pair of globular structures that protrude from the intersegmental membrane between tergites 5 and 6 are displayed during the evening hours by female mantid, *Tarachodes afzelii* (Mantodea). This behavior has been speculated to result in emission of a volatile sex pheromone (Edmunds, 1975). There has been no further work on this or other species of mantids to determine if indeed sex pheromones are involved in their sexual behavior, and on the chemical nature of the secretions and the ultrastructure of the source.

The Blattaria are likely one of the best studied in terms of the different types of sex pheromone-producing cells. This is one group where all three glandular cell types appear to be involved in the production of sex pheromone in different species and in different sexes. The atrial glands located in the genital atrium are the source of Periplanone-B, one of the sex pheromone components of *Periplaneta americana* (Abed *et al.*, 1993). Females adopt a characteristic calling posture exposing the atrial glands, thus releasing sex pheromone. Histologically, the glands have a thick cuticle that overlays clusters of Class I gland cells, whose secretions have been speculated to be delivered to the surface via pores in the cuticle. Interspersed randomly are a few Class III cells, whose role is unknown. Similar atrial glands have been found in *Blatta orientalis* (Abed *et al.*, 1993).

The atrial glands as the source of sex pheromone in *P. americana* has been questioned recently (Yang *et al.*, 1998). When extracts from various female body parts were tested for male responses, hindgut (including both colon and rectum) extracts were most effective. However, when extracts of atrial glands, pygidium, rectum, and colon were tested on male antennae, the colon extract caused the strongest electroantennographic response. Further, gas chromatographic analysis suggested that the colon of a virgin female had highest quantities of Periplanone-A and Periplanone-B. It is unclear whether the colon is the sole sex pheromone source in the female American cockroach or the colon and the atrial glands are both involved.

Virgin females of the German cockroach, *Blattella germanica*, release a volatile sex pheromone that is synthesized by adult female-specific glands on the anterior aspect of the 10th abdominal tergite (Liang and Schal, 1993a, b). The cuticle in this area forms deep depressions with a large number of cuticular pores. The cuticular pores are connected to Class III secretory cells via cuticular ducts surrounded by duct cells. These glandular cells develop in relation to sexual maturation of the female. Teneral females have small secretory cells, contain few secretory vesicles, and little extractable pheromone, but it increases with age. The secretory cells in older females are characterized by a large number of electron-lucent secretory vesicles, abundant rough and smooth endoplasmic reticulum, large nucleus, and a long, convoluted end apparatus that is lined with numerous microvilli. Secretions from the vesicles are released into an extracellular reservoir at the base of microvilli, and then transported to the cuticular surface through the long ducts.

In contrast, the hydrocarbon contact sex pheromone of the German cockroach is synthesized in Class II oenocytes associated with the abdominal sternites (Young *et al.*, 1999; Fan *et al.*, 2002). These large oenocytes, ranging up to 50 μm in diameter in *B. germanica*, have abundant mitochondria and extensive smooth endoplasmic reticulum (Fan *et al.*, 2002). Although the oenocytes are associated with abdominal sternites, the hydrocarbons are released into the hemolymph and loaded, probably across a plasma membrane reticular system, into high-density lipophorin. The lipophorin then likely transports the hydrocarbon to epidermal cells for release onto the cuticle (Fan *et al.*, 2002).

Unlike other cockroaches, males of *Nauphoeta cinerea* produce two successive chemical signals, the sex pheromone and an aphrodisiac (Menon, 1986; Sreng, 1985). Sternal glands produce the sex pheromone, seducin, that attracts the female from a distance. Tergal glands secrete the aphrodisiac, which is licked by females. As the female licks the tergal gland secretions, the male extends his abdomen and clasps her genitalia. The glandular apparatus is well developed, and is composed of five sternal glands and seven tergal glands. The glands appear as a thickening of the epithelium, without significant modification of the external cuticle. The glandular epithelium is made up of several kinds of cells: ordinary epidermal cells, cells with microtubules, Class II cells (oenocytes), and especially Class III glandular units with a secretory cell and a canal cell. The sternal glands secrete primarily volatile substances and fatty acids; in contrast, the tergal glands secrete primarily fatty acids and proteins (Sreng, 1985). Lma-P22 is one of the tergal gland proteins secreted by adult males of *Leucophaea maderae*, and is ingested along with other secretions by females just before copulation (Korchi *et al.*, 1999). Lma-P22 belongs to the lipocalin family that are extracellular proteins that carry hydrophobic compounds and some of them can bind sexual pheromone in vertebrates. Korchi *et al.* (1999) demonstrated that the corresponding mRNA for Lma-P22 is transcribed only in the epidermis of male tergites.

The termites as a group in the Orthopteroidea have been the subject of several studies on pheromone biology, because of the involvement of pheromones in their social stratification. Glands that produce pheromone are located over all regions of the body, in relation to their function, such as labial glands, mandibular glands, frontal glands, and sternal glands (Billen and Morgan, 1998).

Ultrastructurally, these glands include Class I, Class II, and Class III cells, which produce chemical signals in termites with different functions including involvement as trail, alarm, and sex pheromones (Billen and Morgan, 1998; Quennedey, 1998). The trail pheromone, 3, 6, 8-dodecatrien-1-ol, is released from sternal glands and functions as a sex pheromone in a few species of termites (Bordereau *et al.*, 1997; Laduguie *et al.*, 1994). The sternal glands of termites are comprised of Class II oenocytes, contain numerous electron-lucent or electron-dense vesicles, and have large or small mitochondria. They abut Class I cells through which the secretions are released across microvilli to the overlying, vacuolated, and sometimes sponge-like cuticle.

The Dermaptera are another group of insects where sex pheromones have not been described. However, adults and nymphs of the earwig, *Forficula auricularia*, exhibit aggregation behavior in response to aggregation pheromone found in frass and male cuticular lipids (Walker *et al.*, 1993). Whether the aggregation pheromone also enhances contact between males and females and concomitant sexual behavior is unknown. The aggregation pheromone comprises typical normal, monomethyl-, and dimethylalkanes, along with a series of fatty acids and some quinones (Walker *et al.*, 1993). Another study has suggested the tibia to be the source of aggregation pheromone. The ultrastructure of the cells that produce such pheromone is unknown, although based on experience with other species, it is likely that the hydrocarbons in this group are produced by Class II oenocytes.

2.6 Sex pheromone glands in the Pterygota-Hemipteroidea

Of the five orders that comprise this group, no pheromones are known in the lice, the Phthiraptera (Anoplura and Mallophaga). One species belonging to the order Psocoptera uses a female sex pheromone (Betz, 1983), although nothing is known of its identity and source.

Of the remaining orders, probably the best studied, from the perspective of the source of secretions and chemistry, are the true bugs (Hemiptera), particularly the male produced sex pheromones of Pentatomidae (e.g. Aldrich *et al.*, 1984) and female sex pheromones of scales and aphids (e.g. Moreno and Fargerlund; 1975; Galli 1998).

Females of the dictyopharid bug, *Cladodiptera* sp., have a pair of voluminous glandular pouches on the ventral side of the tenth abdominal segment (Bourgoin, 2000). Each gland forms a glandular pouch bearing more than 400 tubercular glandular units. The author speculates that these glands may release a pheromone.

Many hemipterans have metathoracic scent glands (e.g. Oetting and Yonke, 1978), which may produce both allomonal and pheromonal secretions. For example, females of *Alydus eurinus* release an attractant pheromone from their metathoracic scent gland (Aldrich *et al.*, 2000). The internal scent gland apparatus of alydids consists of a ventral metathoracic reservoir and two masses of tubular glands that open to the exterior via an ostiole (Oetting and Yonke, 1978). The scent fluid accumulates as droplets at the ostiole and evaporates from an ostiolar furrow, which has tooth- and ridge-like cuticle and an elevated lobe at the end of the furrow. Gland cells histolyse with age with a reduction in the size of the glandular end apparatus, and produce much less scent fluid. It took 24–36 h for a droplet to be produced by adults, after the 5th ecdysis and after the reservoir had been emptied in response to artificial stimulation. The scent fluid functions as an alarm pheromone (Oetting and Yonke, 1978), and whether the same structures produce the sex pheromone (Aldrich *et al.*, 2000) is unknown.

Scutellarid (Heteroptera) males have integumentary glands that are grouped in one or several pairs of patches on the ventral surface of the abdomen (Carayon, 1984). Each unicellular gland comprises a complex epicuticular formation made from a hollow bristle-like process, the androconia, set into an alveolus. The androconia, which resemble those of lepidopterans, together with the alveolus form a ductule, which in the imago is devoid of a canal cell and is distally blind. The solid secretion appears to be discharged by basal or subapical rupture of the androconia, and acts as a contact pheromone. In some species it is accumulated in friable masses on the patches of androconia. Similar friable secretions from secretory cells with a basal ductule and which occur in paired abdominal patches, accumulate during sexual activity in some pentatomids and lygaeids.

In the predatory spined soldier bug, *Podisus maculiventris*, the dorsal abdominal glands are much smaller in adult females than in males. The female dorsal abdominal gland secretion, a mixture of (E)-2-hexenal, (E)-2-octenal, (E)-2-hexenoic acid, benzaldehyde and nonanal, may be a close-range sexual pheromone (Aldrich *et al.*, 1984), as the dorsal abdominal gland secretion from males is a long-range aggregation pheromone. In a related species, *P. nigrispinus*, the dorsal abdominal glands, which produce the male aggregation pheromone, open between tergites III and IV. These Class III glands consist of a pair of yellowish glands connected by a secretory tubule that opens via a channel into an external opening called an ostiole (Carvalho *et al.*, 1994).

The hind tibiae of oviparous aphids have sex pheromone glands that form plaques (Pettersson, 1971; Crema and Bergamini, 1985; Zeng *et al.*, 1992; Galli, 1998). The glands are large globular cells with a conducting canal, similar to Class III cells, but have a unicellular origin (Pettersson, 1971; Harrington, 1985). The glands undergo organogenesis, maturation, release of sex pheromone, and senility (e.g. Zeng *et al.*, 1992). For example, the sex pheromone glands of the oviparous strawberry aphid *Chaetosiphon folii* are found on the enlarged part of each hind tibia. Each tibia possesses approximately 22 glands, which range in size from 6.54 to 15.71 μm. Externally, the glands are rounded or oval in shape and elevated on the surface of the tibia. The central area of the gland is usually sunken, and the surface of each gland possesses numerous small pores. Several such pores may be aggregated. Internally, every sex pheromone gland is an extremely well-developed, unicellular secretory gland. Early in development of the gland, the cuticle protrudes in the gland-forming area. The glandular cell is larger than the surrounding cells, but the cuticle is still intact and the thickness of the cuticle is the same as that of the surrounding cuticle. The cellular organelles and nuclei are not clearly visible at this stage. Later in development, the cuticle above the glandular cell becomes thinner, and the cell enlarges significantly. The cell develops to form a "half-moon"-shaped protuberance on the surface near the cuticle. However, the secretory canals of the gland have not differentiated and the cells are not fully developed. In a later stage, the cuticle above the glandular cell becomes even thinner. Numerous canals are formed on the epicuticle and nearly connect to the glandular cell, which are large. The nuclei are also very well developed and the cellular organelles are abundant. At this stage, the glandular cells are nearly ready to secrete sex pheromone. However, the membrane of the glandular cell is still intact. The main characteristic of the sex pheromone-releasing stage is that the secretory canals are connected to the well-developed glandular cells and the sex pheromone is secreted and released through the secretion canals. When the females are very old, the glandular cells decline and collapse. The cellular contents disappear and the connections between the cell membrane and canals are completely broken and shrunk. The glands at this stage are completely devoid of sex pheromone.

Tibial plaques of the pea aphid, *Acyrthosiphon pisum*, which produce sex pheromone, are abundant in oviparous aphids but completely absent in males and virginoparous (viviparous) aphids (Crema and Bergamini, 1985). A correlation between the number of tibial plaques and that of viviparous and oviparous ovarioles in the ovaries has been found, suggesting that the morphogenesis of this character may be under ovarian control. Additionally, juvenile hormone induces production of viviparous females that lack sex pheromone-producing tibial plaques (Crema and Bergamini, 1985; Hardie and Lees, 1985).

The common oviduct of the scale insect, *Porphyrophora crithmi* (Homoptera: Margarodidae), has approximately 60 oviducal glands (Foldi, 1986). Each gland is spherical and is linked to the lumen of the oviduct by a long duct, similar to Class III cells. The glands consist of four glandular cells with abundant rough endoplasmic reticulum, mitochondria, and Golgi bodies. Most of these glands produce a mucoproteinaceous secretion released into the oviduct lumen where it is speculated to protect eggs and spermatozoa. Of the 60 glands that are identical in general organization, two differ cytologically. These latter glandular cells have abundant smooth endoplasmic reticulum and appear to produce a pheromone, which is highly attractive to males.

In contrast, in diaspidid scales such as *Aonidiella citrina* and *A. aurantii*, the female sex pheromone glands are found underneath the pygidium (Moreno, 1972; Moreno and Fargerlund, 1975). In these scales, spherical glands appear on either side of the rectum near the anus and vagina after the second molt (Moreno and Fargerlund, 1975). The glands increase in size with age, becoming reniform, and individual cells are discernible. About three to seven days later, the glands release sex pheromone. The glands are typically 70 × 270 μm and are composed of 12 glandular epithelial cells, with large nuclei. The cytoplasm appears to have abundant secretory vesicle-like structures, endoplasmic reticulum, and mitochondria. The glands are connected to the base of the rectum by fragile ducts, and the pheromone is secreted via the anus. In the light micrographs it is hard to decipher how the secretions from the individual cells reach the duct for release. In older virgin females, the glands dissociate, whereas in mated females the pygidium recedes and the glandular cells disintegrate (Moreno and Fargerlund, 1975)

Sternal glands of *Thrips validus* (Thysanoptera) males produce pheromone (Bode, 1978). The secretory cells, which are highly modified Class I cells, have abundant mitochondria, microbodies (*sic*), and agranular, tubular endoplasmic reticulum, the latter enclosed in the numerous apical microvilli. The mitochondrial matrix is probably involved in the elaboration of the secretory pheromonal product, which is extruded and possibly modified by the agranular reticulum, which often lies in contact with mitochondria (Bode, 1978). The secretion is sudanophilic, but not osmiophilic, and the cytoplasm lacks lipid droplets. After storage in the strongly dilated subcuticular space, the secretion is released via epicuticular ductules, which have a diameter of about 120 angstrom.

2.7 Sex pheromone glands in the Pterygota-Neuropteroidea

No neuropteran sex pheromone has been identified to date. However, based mainly on behavioral observations on males, putative sex pheromone glands have been described in this group. In the green lacewing, *Chrysopa perla* (Chrysopidae), the presumed sex pheromone glands are two eversible vesicles that are extruded from the genital aperture during courtship (Wattebled *et al.*, 1978). These vesicles bear sclerotized tubercles, each with a long apical hair and containing a sensory subunit and a glandular subunit. The glandular secretion is released to the outside via a cuticular duct (Wattebled *et al.*, 1978). In contrast, in the spoon wing lacewing, *Palmipenna pilicornis* and *P. aeoleoptera* (Nemopteridae), the male possesses a pair of eversible vesicles that are part of the modified intersegmental membrane between tergites 5 and 6. Each vesicle has an anterior and a posterior portion. When retracted, these vesicles lie beneath the 5th tergite. The folded cuticle of the vesicle is relatively thin compared with the cuticular epithelium elsewhere.

A more intricate sex pheromone gland is found in the male antlion (Myrmeleontidae). In this insect, the Eltringham's organ, which forms a thoracic gland complex (Eltringham, 1932), is a club-like structure that projects from the posterior margin of the hind wings. This club-like structure bears setae, beneath each of which is a specialized epidermal cell that gives off secretions to a subsetal lumen. The organ fits into a pit in the lateral body wall where the thoracic gland opens. The thoracic gland is an extensive hollow tube in the male (Elofsson and Löfqvist, 1974), the wall of which contains glandular and cuticular cells. The glandular cells produce the sex pheromone whose dispersal is aided by the Eltringham's organ (Elofsson and Löfqvist 1974). Studies by Elofsson and Löfqvist (1974) and Wattebled *et al.* (1978) demonstrated that among these three kinds of sex pheromone-producing glands in Neuroptera, the glandular cells belong to Class III exocrine glands that are characterized by the presence of an end apparatus in each gland cell and the glandular secretions are extruded via cuticular ducts connected to the end apparatus that is produced by the duct cells.

The largest order in the Class Insecta, the Coleoptera, has over a quarter of a million described species. Sex pheromone glands have been described in almost every part of the body. The diversity of the location of the sex pheromone gland in Coleoptera can be seen in the scarab beetles. In the scarabeids such as the melolonthine beetle, *Holotrichia parallela*, the sex pheromone gland is located in a ball-shaped sac. Although these authors did not study the ultrastructure of the pheromone gland cells, based on the micrographs of the section through the pheromone gland, we interpret that the columnar epithelium are Class I gland cells, as an obvious end apparatus is not present. In another melolonthine, *Costelytra zealandica*, symbiotic bacterium harbored in the accessory gland produce a phenol that is used by the insect as its own sex attractant (Hoyt *et al.*, 1971; Stringer, 1988). In *Anomala* spp., a ruteline, the sex pheromone is located in the abdominal cuticular epithelium (Tada and Leal, 1997). Interestingly, in another melolonthine, the sex pheromone gland also is located in the abdominal cuticular epithelium (Tada and Leal, 1997).

Pygidial sex pheromone glands are found in many bruchids (Biémont *et al.*, 1992; Ramaswamy *et al.*, 1995; Pierre *et al.*, 1997) and staphylinids (Happ and Happ, 1973). The sex pheromone glands of the female corn rootworm, *Diabrotica virgifera* (Chrysomelidae), are found under either the 7th and 8th tergite or sternite (Lew and Ball, 1978). Similarly, sex pheromone glands in many dermestid species, such as *Trogoderma* spp. (Stanic *et al.*, 1970; Hammack *et al.*, 1973), are found on the 7th and 8th tergite or sternite. In the anobiid, *Lasioderma serricorne*, the sex pheromone gland is located under the 2nd abdominal segment (Levinson *et al.*, 1983).

In the elaterid beetles, a pair of sex pheromone glands is located in the 8th abdominal segment, suspended from the sternite by muscle fibers (Merivee and Erm, 1993). The glandular reservoirs are connected to the intersegmental membrane by thin, winding secretory ducts that are dilated before opening onto the surface, forming "pseudo-ovipositor" pockets. Muscles associated with the spiral ducts may be involved in facilitating pheromone release. The size of the glandular reservoirs varies with season and sexual maturity.

Sex pheromone glands in some beetle species are associated with the midgut, the hindgut, malpighian tubules, fat body, and also the antennae. In the scolytid beetles, *Dendroctonus pseudotsugae*, Zethner-Møller *et al.* (1967) reported that the malpighian tubules in the female are associated with the production of the sex pheromone. They demonstrated that the malpighian tubules in virgin females after feeding are significantly longer than tubules in mated females. However, recent morphological studies by Díaz *et al.* (2000) on the alimentary canal of the *D. frontalis* complex revealed great diversity in the midgut cells and these authors suggested that the midgut in these beetles could be involved in pheromone production. *In situ* hybridization studies conducted by Hall *et al.* (2002) showed high levels of expression of a 3-hydroxy-3-methylglutaryl-CoA reductase transcript, a key regulatory enzyme of the sex pheromone frontalin of the male Jeffrey pine beetle, *Dendroctonus jeffreyi*, in the anterior midgut. Radiotracer studies further confirmed that the anterior midgut is the site of frontalin production in this beetle. Since the malpighian tubules are usually closely associated with the midgut, it is likely this could have contributed to the high biological activity observed in *D. pseudotsugae* (Zethner-Møller and Rudinsky, 1967). In another study on *D. jeffreyi* and *Ips pini*, Nardi *et al.* (2002) demonstrated that midgut cells that produce sex pheromone exhibit ultrastructural characteristics not found in non-pheromone midgut cells, such as the presence of highly ordered arrays of smooth endoplasmic reticulum. These authors concluded that in male *D. jeffreyi* and *I. pini*, most midgut cells can secrete pheromones as well as digestive enzymes. Whether *D. pseudotsugae* is an exception in the scolytid family in the location of their sex pheromone gland requires further clarification.

Bioassay and histological data indicate that sex pheromone is produced in the hindgut of *Trypodendron lineatum* (Schneider and Rudinsky, 1969). Epithelial cells of the hindgut of attractive females are larger and taller, richer in vacuoles, and also possess larger granular nuclei, compared with structures in non-attractive females. The epithelial cells rest on the cuticle, which protrudes into the base of the cells.

In the boll weevil, the male sex pheromone is supposedly produced in the fat body (Wiygul *et al.*, 1982). However, recent studies have questioned the validity of this interpretation (Blomquist and Tittiger, unpublished data) and more work is needed to verify the site of pheromone production.

In cerambycids such as *Hylotrupes bajulus* and *Xylotrechus pyrrhodercus*, the sex pheromone glands occur underneath the prothoracic tergite (Iwabuchi, 1986; Noldt *et al.*, 1995), on which the cuticular openings of secretory units are grouped into "pore fields," unlike the situation in Class III glands of other insects. Further, in *H. bajulus*, many "pore fields" are grouped into a depression called a "pore pit" (Noldt *et al.*, 1995).

Two types of glands on the antenna, the common antennal gland and male-specific antennal gland, are found in cabbage stem flea beetle, *Psylliodes chrysocephala* (Chrysomelidae) (Bartlet *et al.*, 1994). The common antennal glands are synthetically active in both pre- and post-diapause adult males and females, whereas the male-specific antennal glands are active only in post-diapause (reproductively active) males. Behavioral experiments demonstrated that post-diapause males produce secretions that attract females for courtship. The common antennal glands are found in parts of the antennae-bearing sensilla; however, the male-specific glands are found in a glabrous area on flagellar segments 6–10 of the male beetle. The morphology of these glandular cells is typical of Class III glandular cells.

The male tegumentary sex pheromone glands of the leiodid beetle, *Speonomus hydrophilus*, are located on abdominal segments 6–8 and open between segments 8 and 9 (Cazals and Juberthie-Jupeau, 1983). The gland consists of ramified epithelium comprising prismatic cells. The external opening of the gland is decorated with rather intricate cuticular sculpturing, including a porous plate and the mesocuticular cylindrical excrescence and plates. The latter is termed pseudomembranes, and has been suggested to serve the same purpose as an end apparatus in Class III glands. The porous plate consists of an epicuticular layer perforated by tiny pores, and is located at the opening of the gland.

Further demonstrating the diversity of pheromone gland location in the Coleoptera, in a number of sap beetle species including *Carpophilus freemani* (Nitidulidae), the site of production of the aggregation pheromone is located in disk-shaped structures associated with the trachea in the posterior abdomen (Nardi *et al.*, 1996). The disk-like structure occurs only in the abdomen of males. These disks contain fat droplets and are histologically distinct from fat body. These authors suggested that the pheromone is secreted through the spiracle.

A small group of Coleoptera-like, mostly parasitic insects, is the Strepsiptera. The placement of these insects in a totally separate taxonomic group is controversial. A search of the literature did not show any information on their sex pheromone.

2.8 Sex pheromone glands in the Pterygota-Panorpoidea

The mecopterans represent a relatively minor group of insects, but they are relics of a once larger and more widespread order. It is generally believed that a mecopteran ancestor gave rise to the present-day Panorpoid orders, which include the taxonomic groups such as Neuroptera, Mecoptera, Lepidoptera, Trichoptera, Diptera, and Siphonaptera.

Behavioral observations on the courtship behavior of a few species of scorpionflies have established that males release some form of chemical signals that attract females for mating. Two eversible vesicles on the terminal abdominal segments of male *Harpobittacus australis* and *H. nigriceps* release a musty scent that acts as a sex attractant (Bornemissza, 1964, 1966). Similar abdominal vesicles that produce sex pheromone have been demonstrated in other scorpionflies (Thornhill, 1973; Brushwein *et al.*, 1995; Iwasaki, 1996). In *H. australis*, the eversible vesicles consist of three cell types, cuticle secreting cells, the duct cells, and the pheromone-producing cells (Crossley and Waterhouse, 1969), characteristic of Class III glands. These cells also have a moderate amount of mitochondria, smooth and rough endoplasmic reticulum, Golgi apparatus, and vesicles of various electron densities.

The Siphonaptera is another group where a volatile sex pheromone is yet to be demonstrated, although a contact sex pheromone has been demonstrated in preliminary experiments (Ralph Charlton, personal communication). The source and nature of this contact pheromone is yet to be deciphered. Fleas have been speculated to use a certain wavelength of light, CO_2, and visual and thermal stimuli for host localization, as traps that present these kinds of stimuli seem to be effective in trapping fleas (Benton and Lee, 1965; Osbrink and Rust, 1985; Dryden and Broce, 1993).

The Trichoptera is a panorpoid group whose adults are aerial and immatures are aquatic. A glandular structure on the head and thorax of several species of microcaddisflies (Hydroptilidae), called ante-pteral organs and eversible head organ (Roemhild, 1980), has been speculated to produce chemical signals. However, more behavioral and morphological studies are required to establish the pheromone function of these structures. In female hydropsychid and rhyacophilid caddisflies, including *Hydropsyche angustipennis*, *Rhyacophila nubila* and *R. faciata*, the sex pheromone is produced from a paired exocrine gland that is located between the 4th and 5th sternite (Löfstedt *et al.*, 1994). Interestingly, the sex pheromones of these trichopteran species are chemically similar to the eriocraniid moth (Löfstedt *et al.*, 1994). In addition, female eriocraniid moths and the hydropsychid and rhyacophilid caddisflies produce their sex pheromone from apparently homologous glands. Unfortunately, detailed descriptions on the structure of the sex pheromone gland in these insects are unavailable.

Among numerous studies conducted on the structure of sex pheromone glands within the Class Insecta, more than half have been on the Lepidoptera. This is probably due in part to the large number of agricultural pests that are found within this order. Although some of these studies were conducted in males, the majority of the studies have been on females. Percy-Cunningham and MacDonald (1987) provide an excellent account on the structure of the sex pheromone gland in the Lepidoptera, and here we review new information that has appeared since. Insects in this order exhibit great diversity; however, the location of sex pheromone glands in females is remarkably conserved. In the majority of cases, the female sex pheromone gland is localized in the terminal abdominal segments. Often they occur as modified intersegmental membranes found between the terminal abdominal segments (usually between the 8th and 9th abdominal segments). The gross morphology of the modified intersegmental membrane, however, varies. In the simplest form, they are found as dorsal fold, ventral fold, or lateral folds. When the entire intersegmental membrane is glandular, they form the so-called ring gland. In the arctiid moths such as *Callimorpha dominula*, *Tyria jacobeae* and *Utetheisa ornatrix*, however, the dorsal part of the intersegmental membrane is modified into non-eversible tubes (Urbahn, 1913). In the female tiger moth, *Holomelina lamae*, the sex pheromone originates from a pair of internal tubular glands located dorsally in the abdomen. Each tubular structure opens externally to a pore situated on the intersegmental membrane between abdominal segments 8 and 9. The cells lining the tubes are typical Class I cells. Using a related species, *H. aurantiaca*, Schal *et al.* (1998a) showed that the sex pheromone was biosynthesized by Class II oenocytes and transported to the pheromone gland by lipophorin. Based on the observations that the sex pheromone is specifically unloaded at the pheromone gland, these authors suggested that in *Holomelina*, the pheromone gland does not synthesize the pheromone but rather stores and releases the pheromone during calling (Schal *et al.*, 1998a or b).

Perhaps the most unusual female sex pheromone glands in the Lepidoptera are found in the psychid bagworm moth, *Thyridopteryx ephemeraeformis* (Leonhardt *et al.*, 1983; Loeb *et al.*, 1989), the zygaenid moth, *Theresimima (=Ino) ampelophaga*, and the eriocraniid moth *Eriocrania cicatricella* (Zhu *et al.*, 1995). In the psychid bagworm moth, the site of production of sex pheromone is localized in the thorax. The secretory cells secrete sex pheromone on to deciduous hairs on the thorax, and the subsequent shedding of these pheromone-laden hairs attracts the males for mating. In the zygaenid moth, *T. ampelophaga*, the sex pheromone gland is located on the anterior part of the 3rd and 5th abdominal tergites (Hallberg and Subchev, 1997). There are two cell types that make up the pheromone gland, glandular cells and wrapping cells. The glandular cells have a central microvilli-lined cavity, which is directly connected to the lumen of the scale that is directly above the glandular cell. In the eriocraniid moth, *E. cicatricella*, Zhu *et al.* (1995) determined that the female sex pheromone gland is produced from a pair of exocrine glands that are located in the 5th abdominal segment. However, nothing is known of their morphology.

Recently, Jurenka and Subchev (2000) and Subchev and Jurenka (2001) reported the presence of sex pheromone and precursors in the hemolymph of the gypsy moth, *Lymantria dispar* and the noctuid moth *Scoliopteryx libatrix*, respectively. These authors hypothesized a pathway of pheromone production, which suggests that the sex pheromone glands in these insects resemble Class II gland cells. They hypothesize that sex pheromone precursors are synthesized in the oenocytes in the hemolymph and transported to the pheromone gland cells where they are selectively taken up, transformed, and released during calling. Exactly how hemolymph transport and uptake of the precursor occurs in these insects awaits future morphological and biochemical studies.

Most, if not all, of the glandular cells on the intersegmental membrane that forms the sex pheromone glands in Lepidoptera may be classified as Class I glandular cells (Noirot and Quennedey, 1974). Histologically, the modified glandular cells are hypertrophied and they are in direct contact with the overlying cuticle. The hypertrophied epidermal cells exhibit typical ultrastructural characteristics including apical folds or microvilli, smooth endoplasmic reticulum, and lipid spheres. In addition, the overlying cuticle is usually untanned and without large pores.

Unlike its female counterpart, morphology of the sex pheromone glands in male Lepidoptera has received much less attention. The pheromone gland of males can exist as wing glands, brush organs at the ventral junction of the femur and tibia of the hind leg as in the geometriid moth *Semiothisis eleonora* (Percy and Weatherston, 1974), thoracic scent brushes, or as gland cells on the genital appendage such as in *Eldana saccharina* (Atkinson, 1982). The scales associated with the pheromone-producing cells are accessory structures that aid in the dissemination of pheromone. These scales have very intricate surface topography. They are produced by the trichogen cells, which are also responsible for the production of the extracellular cavities. However, Clearwater and Sarafis (1973) pointed out that these trichogen cells could also carry out pheromone biosynthetic function. The glandular cells are simply modified trichogen cells. They exhibit typical morphological characteristics of glandular cells such as hypertrophy, numerous mitochondria, smooth endoplasmic reticulum, and extensive apical and sometime basal membrane folds.

Considered to be one of the largest orders of insects, the order Diptera includes many species that are important agricultural pests as well as species of medical or veterinary importance. To date, sex pheromone has been demonstrated in over a few dozen species of Diptera, both male and female, and many of these pheromones have been isolated and their chemical structures identified. With such a large body of literature on dipteran sex pheromone, one would expect a large amount of information on the structure of the sex pheromone gland. It is surprising that the structure of the sex pheromone gland is described in only a handful of species. The dipteran sex pheromone glands are as widely diverse as those found in Coleoptera. Among the dipterans whose sex pheromone glands have been described, the tephritid fruit fly in the genus *Dacus* has received the most attention. The sex pheromone glands in the male flies of this genus are found in a diverticulum in the posterior rectum. Histology of the presumed sex pheromone gland is described in several species including *D. tryoni*, *D. olea*e, *D. cucurbitae*, and *D. latifrons* (Fletcher, 1969; Economopoulos *et al.*, 1971; Nation, 1981; Little and Cunningham, 1987; Little, 1992). In *D. latifrons*, the diverticulum of the rectum, which is referred to as the "secretory sac" and "reservoir," lacks many of the organelles found in *D. cucurbitae*. Such organelles are morphologically characteristic of sex pheromone-producing glands in other insects (Little, 1992). The significance of differences between *D. cucurbitae* and *D. latifrons* as well as other *Dacus* species remains to be clarified. Nation (1981) examined males and females of species from four genera including *Anastrepha*, *Ceratitis*, *Dacus*, and *Rhagoletis* for the presence of sex pheromone glands. His observations indicated that in some of these species, the salivary glands, the pleural glands, and the anal glands could be sex-specific sex pheromone glands. The exact function of these sex specific glands, however, remains to be described. The sex pheromone gland cells described in all of the tephritids belong to the Class I gland type.

The sex pheromone gland in male *Drosophila grimshawi* is located in the intra-anal lobes at the tip of the abdomen (Hodosh *et al.*, 1979). The intra-anal lobes are characterized by the presence of numerous canaliculi-ducts that empty into the anal region. These authors did not mention the type of gland cells in the anal gland. However, based on the micrographs in the original article, there was no apparent end apparatus in the gland cells, thus we interpret that the gland cells are likely Class I glandular cells. Apparently, secretory products that flow on the surface of the intra-anal lobes are deposited onto the substrate by the finger-like projections on the lobes' surface.

In the psychodid sand fly, *Lutzomyia longipalpis*, the sex pheromone-producing glands occur on the 3rd and 4th abdominal segment of the male insects. The gland consists of large type III secretory cells with prominent end apparatus (Lane and de Souza Bernardes, 1990).

The female sex pheromone gland has been described in three species of Cecidomyiidae midge, *Dasineura brassicae* (Isidoro *et al.*, 1992), *Allocontarinia sorghicola* (Solinas and Isidoro, 1991) and *Mayetiola destructor* (Solinas and Isidoro, 1996). In all three species, the female sex pheromone gland comprises modified epidermal cells located on the intersegmental membrane of the ovipositor between abdominal segments 8 and 9. The glandular epidermis lining the cuticle is typical of Class I gland cells including hypertrophied cells consisting of a single round nucleus, extensive smooth endoplasmic reticulum, numerous mitochondria, and abundant secretory vesicles.

The order Hymenoptera includes many of the social insects. Many social behaviors displayed by members in this group are mediated by pheromones, including such behaviors as nest defense to nest mate recognition. However, relatively little is known of pheromones involved in sexual communication. Perhaps this is due, in part, to the diverse array of exocrine glands found in these insects. These glands may produce secretions affecting many behaviors, thus making the assignment of function more difficult.

Among the social Hymenoptera, one of the best-understood examples is the sex pheromone used by the queen honeybee to attract drones for mating. This pheromone is produced in the mandibular gland that resides in the head capsule (Barbier and Lederer, 1960; Callow and Johnson, 1960; Callow *et al.*, 1964). In honeybees, female tergal gland secretions are known to function as copulatory pheromone (Billen and Morgan, 1998), but the chemical nature of the pheromone and morphology of the glands is unknown. In contrast, the sex pheromone of the bumblebee is produced in labial glands (Stein, 1963; Agren *et al.*, 1979). Sex pheromones in stingless bees are yet to be described. In social wasps, pheromone is involved in mating; however, the exact site of production of this pheromone is yet to be identified (Downing, 1991). Sex pheromone has been demonstrated in many ant species, for example in the pharaoh ant *Monomorium pharaonis* (Hölldobler and Wüst, 1973), where the Dufour's gland and the bursa pouches produce the female sex pheromone. In *Xenomyrmex floridanus*, in contrast, the female sex pheromone is produced in the poison gland (Hölldobler, 1971). Other female ants that use pheromone to attract males during mating include *Harpagoxenus sublaevis* (Buschinger, 1972), *Formica montana* and *F. pergandei* (Kannowski and Johnson, 1969). To date, literature that deals with the morphology of the sex pheromone-producing gland in the formicid ants is lacking.

All the internal exocrine glands found in Hymenoptera, such as the Dufour's gland, mandibular glands, and labial glands, are Class III glands. Three distinct cell types form the glandular tissue: glandular, ductule, and hypodermal cells. Glandular cells are characterized by the presence of well-developed rough and/or smooth endoplasmic reticulum, Golgi apparatus, and microvilli-lined cavities that are surrounded by numerous mitochondria. The ductules are derived from the cavities and are surrounded by the ductule cells. These ductules merge with ductules from other cavities that open into the lumen of the gland. The hypodermal cells, on the other hand, underlie the cuticular walls of the lumen. As indicated above, some hymenopterans such as *Apanteles* spp. do possess epidermal sex pheromone glands, which the authors classify as being Class I glands.

Other than the social and non-social bees, sex pheromones have been identified in sawflies and some parasitic wasps. In the introduced pine sawfly, *Diprion smiles* (Mertins and Coppel, 1972), the female sex pheromone glands are paired structures found on either side of the anterolateral margin of the abdomen. They are connected via a neck-like duct to a slit-like opening in the intertergal membranes II and III. In contrast, the pheromone gland in *Neodiprion sertifer* (Hallberg and Löfqvist, 1981) has been localized in both sexes. However, gross morphological organization of the gland is similar in *D. smiles* and *N. sertifer*.

A sex pheromone has been implicated in the courtship behavior of the ichneumonid wasp, *Campoletis sonorensis* (Vinson, 1972). Although the exact site of production of the sex pheromone is not identified, Vinson suggested that the sex pheromone gland in this insect is associated with the cuticle. In another ichneumonid, *Eriborus terebrans*, one of the sex pheromone components is a hydrocarbon (Shu and Jones, 1993). We speculate that in ichneumonids the pheromone is synthesized in Class II oenocytes, similar to the hydrocarbon secretions produced in other insects (Schal *et al.*, 1998).

In the parasitoid *Apanteles melanoscelus* (Braconidae) (Weseloh, 1980), the sex pheromone gland is located on the 8th abdominal tergite. These cells contain lipid deposits, microvilli, and numerous mitochondria, characteristic of Class I exocrine glands. In another braconid, *A. glomeratus*, the sex pheromone gland is located in the 2nd valvifer (Tagawa, 1977), and there are pores and hairs associated with the glandular region. Judging by the drawing given by the author, it appears that the glandular cells are Class I gland cells. These cells contain lipid deposits, microvilli, and numerous mitochondria.

2.9 Morphological changes associated with pheromone production

The literature is particularly poor in the area of morphological changes associated with pheromone-production. There have been a few reports of changes in gland or cell size or shape, in the types, sizes, and shapes of secretory vesicles, and in the cuticle overlying pheromone-secreting tissue, gland histolysis, hypertrophy, poorly developed organelles, etc. in various species (e.g. Oetting and Yonke, 1978; Agren *et al.*, 1979; Solinas and Isidora, 1991; Zeng *et al.*, 1992; Liang and Schal, 1993a; Nardi *et al.*, 1996; Galli, 1998) in relation to pheromone production. In the Lepidoptera, despite the existence of a large body of literature on the neuroendocrine control of sex pheromone production in moths by pheromone biosynthesis activating neuropeptide (PBAN), fundamental information regarding the action of PBAN, such as changes in the ultrastructure of the pheromone gland cells associated with exposure to the pheromonotropic stimuli, is still unknown. A few recent studies have attempted to address this fundamental question. In *Helicoverpa zea*, Raina *et al.* (2000) conducted detailed chemical analyses of the ovipositor tip of the moth in relation to ultrastructural changes in the pheromone gland cells with and without hormonal manipulations. Other than the usual structural features, such as a large number of mitochondria, rough endoplasmic reticulum, and desmosomes in the intercellular boundaries, they demonstrated that a large number of small vesicles accumulate below the microvilli of the pheromone gland cells. During scotophase, pockets of granular material appeared throughout the cytoplasm as well as in the microvilli. These structural changes coincide with increase in pheromone content of the gland. Using low-temperature scanning electron microscopy, these authors also revealed the presence of small droplets of secretions on the cuticular hairs of the glandular cuticle only during scotophase, the time when high pheromone titers were observed in the pheromone gland cells.

Truncated synthetic Bom-PBAN, ionomycin, and a calcium ionophore stimulate the sex pheromone gland cells to produce bombykol in the silkworm moth *Bombyx mori* (Fónagy *et al.*, 2000). Two important observations were made in this study: (1) The pheromone-producing cells came from the inner layer since they produce bombykol response to the *in vitro* stimuli. (2) The lipid droplets in the pheromone gland cell increase in size and number in response to the *in vitro* stimuli. The *in vitro* observations were later corroborated with *in vivo* experiments where changes in size and number of cytoplasmic lipid droplets were quantified in the pheromone gland (Fónagy *et al.*, 2001). Both the size and number of the lipid droplets increase two days before eclosion. This is followed by fluctuations between photophase and scotophase on the day of emergence through day-3 of adult post-emergence. These changes in size and number of pheromone gland lipid droplets before and post-adult emergence as well as between photophases coincide with the extractable bombykol in the pheromone gland. These changes, however, cease by day-4 of adult post-emergence.

2.10 Developmental and hormonal regulation of pheromone production

A few examples are available in the literature, which have specifically examined the developmental and hormonal regulation of pheromone production. The pheromone gland of the female German cockroach exhibits a clear developmental maturation in relation to sexual maturation of the female (Liang and Schal, 1993). The secretory cells of a newly formed gland in the teneral female are small and contain few secretory vesicles. The amount of extractable pheromone in the gland is low on day-0, but it increases with age and peaks on day-6. The secretory cells in a mature day-6 gland are characterized by a large number of electron-lucid secretory vesicles, abundant rough and smooth endoplasmic reticulum, a large nucleus, and a long, convoluted end apparatus lined with numerous microvilli. The contents of the secretory vesicles are exocytosed into extracellular reservoirs at the base of the microvilli and then transported to the cuticular surface through the long ducts.

Similarly, normal differentiation of the sternal and tergal pheromone glands in male *Nauphoeta cinerea* cockroaches begins before the imaginal molt with the formation of a basic glandular unit of four cells, and ends three to four days after the imaginal molt with the formation of the mature gland with only two cells (Sreng, 1998). Each of the four cells has a specific activity in the developing gland, and apoptosis occurs only after achievement of that purpose. The end apparatus in the mature gland is a vestige of the dead ciliary cell. Decapitation immediately after molting prevents specific and selective apoptosis of two pre-programmed cells, thus indicating that the fate of each cell in the basic glandular unit depends on brain-derived factors. Juvenile hormone III is involved in the differentiation of sternal glands (Sreng *et al.*, 1999).

In the gregarious phase of *Schistocerca gregaria*, final ecdysis allows differentiation of gland cells involved, under exclusive gonadotropic hormone control, in the production of a specific sexual maturation pheromone (Cassier and Delorme-Joulie, 1976c). In crowded females and isolated males and females, gland cells are very rare, widely separated and much less developed, although corpora allata activity and gonadotropic hormone level are more than in crowded males; gland cells do not respond to juvenile hormone. In contrast, in solitary males and gregarious and solitary females, gland cells in the integument are very rare and inactive, or much less active, although corpora allata activity and gonadotropic hormone level are higher than in crowded males. Differences between the presence and activity of tegumental gland cells in crowded and isolated males are dependent upon hormonal conditions at the end of the 4th larval instar and the first two days of the 5th. The exclusive presence of ecdysone favors development of the gregarious phase with numerous gland cells reactive to gonadotropic hormone. The simultaneous presence of ecdysone and juvenile hormone ensures solitary type development with rare and inactive gland cells.

Aphids exhibit vivipary and ovipary, depending on environmental conditions. Oviparous aphids are sex pheromone competent, and females attract males with sex pheromonal signals; in contrast, viviparous females are unable to produce pheromone (Crema and Bergamini, 1985). The pheromone is produced in tibial plaques, which are absent in viviparous females. The occurrence of vivipary and loss of tibial plaques is triggered by juvenile hormone, under conditions that stimulate vivipary.

With a few exceptions, lepidopteran females do not produce sex pheromone until two or three days after adult emergence. Apparently a certain period of time is required for maturation of the sex pheromone gland. In many species that exhibit this delayed development of their sex pheromone gland, the morphology of their "inactive" sex pheromone gland cell is usually not much different than the surrounding non-pheromone-producing cells. Within 24 h of adult emergence, drastic morphological changes occur in the pheromone gland cells. Typically, these include cellular hypertrophy with concomitant development of apical microvilli, rough and smooth endoplasmic reticulum, lipid droplets, and mitochondria. In many cases such as in *H. zea*, these changes occur in relation to increase in pheromone titer (Raina *et al.*, 2000). In the silkworm moth, *B. mori*, the size and number of lipid droplets in the pheromone gland cells are small and few (Fónagy *et al.*, 2001). However, one to two days before eclosion, there is a significant increase in large lipid droplets. After adult emergence, the size and number of the lipid droplets fluctuate during the photophase. However, these changes in the pheromone gland cell cease by day-4. Interestingly, there are extra, large lipid droplets in the pheromone gland cells of 72-h decapitated females and treatment of PBAN restores the formation of small lipid droplets. These authors suggest that the observed changes are related to the storage-pool function of the lipid droplets (Fónagy *et al.*, 2001). The biochemical significance of these changes in relation to the biosynthesis of sex pheromone remains to be seen.

2.11 Conclusions and future research

For years, identification of the sources of sex pheromone usually involved performing behavior assays and chemical analyses, in conjunction with morphological studies. This approach resulted in the unambiguous identification of the sex pheromone glands in many insects. With more information currently available in the literature regarding the biosynthesis of the sex pheromone, the endogenous regulation of their production, the identification of many biosynthetic enzymes, and the availability of sophisticated molecular biology techniques, in addition to traditional techniques such as chemical analyses, behavior assays, and histological techniques, we are now armed with a vast arsenal to locate precisely the site of sex pheromone glands down to the level of individual cells or even to organelles, a level that was extremely difficult, if not impossible, to reach in the past.

Since completion of the sequencing of human genome and the genomic sequence of a few other organisms, including that of insects, it is now possible to obtain the sequences available in the public domain, and develop with relative ease a labeled probe for a pheromone biosynthetic enzyme. *In situ* hybridization then can be performed to locate cells that express this pheromone biosynthetic enzyme. If the endogenous regulation of the biosynthetic enzyme is known, hormonal manipulations can be incorporated into the experiment to observe changes in expression of the biosynthetic enzyme. An example of such an integrative approach has led to the identification that the anterior midgut is the site of synthesis of frontalin in male *D. jeffreyi* (Hall *et al.*, 2002b). Such an approach certainly would allow clarification of the location and function of a number of insect sex pheromone glands whose identification is questionable. Furthermore, this could lead to discovery of even extraordinary locations of insect sex pheromone glands or new sex pheromone production sites.

With developments in cellular and molecular biology, we are tantalizingly close to being able to achieve the expectation voiced by Percy-Cunningham and MacDonald (1987) on the biochemical elucidation of the (ultra)structural–functional interrelationships of pheromone synthesis and release. To this end, advances in microcomputers and their use in instrumentation control, data acquisition, and image analysis in the last one and a half decades have completely changed the face of analytical methods in the laboratory. Technological breakthroughs in liquid chromatography-mass spectrometry (LC-MS) for proteomics research, including micro liquid chromatography (microLC), allow flow rates in the range of nanoliters/min to low microliters/min in splitless gradient mode. With such low flow rates in combination with microcapillary columns, the sensitivity achieved is at least an order of magnitude better than in conventional HPLC. It is theoretically possible to attach an electroantennographic detector (EAD) directly to the microLC, as the minute solvent that goes over the antenna over a long period should not render the antenna inactive. The advantage of such a set-up could allow isolation and analysis of thermolabile pheromones such as that of the German cockroach sex pheromone (Zhang, personal communication). In addition, the sensitivity of the mass spectrometer has improved tremendously in recent years. Although mass spectrometry with quadrupole analyzer remains one of the most popular instruments used by chemical ecologists for pheromone research, availability of hybrid instruments certainly will expand the analytical horizon in this area. These hybrid instruments, including tandem time-of-flight analyzers, tandem quadruple/time-of-flight analyzers, and ion trap/quadruple analyzers in addition to new ionization methods such as matrix-assisted laser desorption/ionization with UV lasers or the more gentle infrared lasers, may become tremendously useful to chemical ecologists. Although many of these new technologies have been developed for protein research, applications of these new technologies in pheromone research are only limited by one's imagination. Already mass spectrometry has been used successfully to analyze peptides in single cells (Garden *et al.*, 1996; Ma *et al.*, 2000). The use of mass spectrometry to analyze individual pheromone gland cells, for example distinguishing pheromone gland cells and surrounding non-pheromone-producing cells based on certain cellular chemical components as markers, is within the reach of the chemical ecologist. Within another 10 to 15 years, we would anticipate a whole new understanding of the (ultra) structural–functional relationships of pheromone biology based on new technologies, both analytical and morphological, used in pheromone research!

References

Abed D., Cheviet P., Farine J. P., Bonnard O., Lequere J. L. and Brossut R. (1993) Calling behaviour of female *Periplaneta americana*: behavioural analysis and identification of the pheromone source. *J. Insect Physiol.* **39**, 709–720.

Agren L., Cederberg B. and Swensson B. G. (1979) Changes with age in ultrastructure and content of male labial glands in some bumblebee species (Hymenoptera, Apidae). *Zoon.* **7**: 1–14.

Aldrich J. R., Blum M. S. and Fales H. M. (1979) Species specific natural products of adult male leaf-footed bugs (Hemiptera: Heteroptera). *J. Chem. Ecol.* **5**, 53–62.

Aldrich J. R., Lusby W. R., Kochansky J. P. and Abrams C. B. (1984) Volatile compounds from the predatory insect *Podisus maculiventris* (Hemiptera: Heteroptera: Pentatomidae) male and female metathoracic scent gland and female dorsal abdominal gland secretions. *J. Chem. Ecol.* **10**, 561–568.

Aldrich J. R., Zhang A. and Oliver J. E. (2000) Attractant pheromone and allomone from the metathoracic scent gland of a broad-headed bug (Hemiptera: Alydidae). *Can. Entomol.* **132**, 915–923.

Araujo J. and Pasteels J. M. (1987) Ultrastructure de la glande défensive d'*Eusphalerum minutum* Kraatz (Coleoptera: Staphylinidae). *Arch. Biol.* **98**, 15–34.

Atkinson P. R. (1982) Structure of the putative pheromone glands of *Eldana saccharina* Walker (Lepidoptera: Pyralidae). *J. Entomol. Soc. S. Africa* **45**, 93–104.

Barbier J. and Lederer E. (1960) Structure chimique de la substance royale de la reine d'abeille. (*Apis mellifica* L.). *C. R. Acad. Sci. Paris* **251**, 1131–1135.

Bartlet E., Isidoro N. and Williams I. (1994) Antennal glands in *Psylliodes chrysocephala*, and their possible role in reproductive behaviour. *Physiol. Entomol.* **19**, 241–250.

Benton A. H. and Lee S. Y. (1965) Sensory reactions of siphonaptera in relation to host finding. *Am. Midl. Nat.* **74**, 119–125.

Betz B. W. (1983) The biology of *Trichadenotecnumc alexanderae* (Psocoptera: Psocidae): duration of biparental and parthenogenetic reproductive abilities. *J. Kansas Entomol. Soc.* **56**, 420–426.

Biémont J.-C., Chaibou, M. and Pouzat J. (1992) Localization and fine structure of the female sex pheromone-producing glands in *Bruchidius atrolineatus (Pic)* (Coleoptera: Bruchidae). *Int. J. Insect Morphol. Embryol.* **21**, 251-262.

Billen J. and Morgan E. D. (1998) Pheromone communication in social insects: sources and secretions. In: *Pheromone Communication in Social Insects: Ants, Wasps, Bees, and Termites.* eds R. K. Vander Meer, M. D. Breed, K. E. Espelie and M. L. Winston, pp. 3–33.

Bode W. (1978) Ultrastructure of the sternal glands in *Thrips validus* (Thysanoptera: Terebrantia). *Zoomorphol.* **90**, 53–66.

Bordereau C., Robert A., Bonnard O. and Lequere J. L. (1997) (3Z, 6Z, 8E)–3, 6, 8-dodecatrien-1-ol. Sex pheromone in a higher fungus-growing termite *Pseudacanthotermes spiniger* (Isoptera : Marotermitidae). *J. Chem. Ecol.* **17**, 2177–2191.

Bornemissza G. F. (1964) Sex attractant of male scorpion flies. *Nature, London* **203**, 786–787.

Bornemisszа G. F. (1966) Observations on the hunting and mating behaviour of two species of scorpionflies (Bittacidae: Mecoptera). *Aust. J. Zool.* **14**, 371–382.

Bourgoin T. (2000) Voluminous paired glands in the tenth abdominal segment in the females of the genus *Cladodiptera* spinola (Hemiptera: Fulgoromorpha: Dictyopharidae). *Ann. Soc. Entomol. Fr.* **36**, 137–142.

Brushwein J. R., Culin J. D. and Hoffman K. M. (1995) Development and reproductive behavior of *Mantispa viridis* Walker (Neuroptera: Mantispidae). *J. Entomol. Sci.* **30**: 99–111.

Buschinger A. (1972). Giftdrüsensekret als sexualpheromon bei der ameise *Harpagoxenus sublaevis. Naturwissenchaften* **59**, 313–314.

Callow R. K. and Johnson N. C. (1960) The chemical constitution and synthesis of queen substance of honeybees (*Apis mellifera* L.). *Bee World* **41**, 152–153.

Callow R. K., Chapman J. R. and Paton P. N. (1964) Pheromones of the honeybee: chemical studies of the mandibular gland secretion of the queen. *J. Apicult. Res.* **3**, 77–89.

Carayon J. (1984) The androconia of some scutellerid (Hemiptera: Heteroptera). *Ann. Soc. Entomol. Fr.* **20**, 113–134.

Carvalho R. S., Vilela E. F., Borges M. and Zanuncio J. C. (1994) Caracterizaçao morfologica da glandula do feromonio do predador *Podisus nigrispinus* (Dallas). *Ann. Soc. Entomol. Br.* **23**, 143–147.

Cassier P. and Delorme-Joulie C. (1976a) Imaginal differentiation of the integument in the desert locust *Schistocerca gregaria.* Part 1: Post imaginal differentiation and its determination. *Arch. Zool. Exp. Gen.* **117**, 95–116.

Cassier P. and Delorme-Joulie C. (1976b) Imaginal differentiation of the integument in the desert locust *Schistocerca gregaria.* Part 2: Development during imaginal molt and its determination in gregarious males. *Ann. Sci. Natur. Zool. et Biol. Anim.* **18**, 295–310.

Cassier P. and Delorme-Joulie C. (1976c) Imaginal differentiation of the integument in the desert locust *Schistocerca gregaria*. Part 3: Phase polymorphism and its hormonal control. *Insectes Sociaux* **23**, 179–198.

Cazals M. and Juberthie-Jupeau L. (1983) Ultrasturcture d'une glande sternale tubuleuse des mâles de *Speonomus hydrophilus* (Coleoptera, Bathysciinae). *Can. J. Zool.* **61**, 673–681.

Clearwater J. R. and Sarafis V. (1973) The secretory cycle of a gland involved in pheromone production in the Noctuid moth, *Pseudaletia separata. J. Insect Physiol.* **19**, 19–28.

Crema R. and Bergamini I. (1985). On the control of a sexual character in the aphid *Acyrthosiphon pisum.* The sex pheromone glands of oviparous females. *Bollett. Zool.* **52**, 359–366.

Crossley A. C. and Waterhouse D. F. (1969) The ultrastructure of a pheromone-secreting gland in the male scorpion-fly *Harpobittacus australis* (Bittacidae: Mecoptera). *Tissuse Cell* **2**, 273–294.

Díaz E., Cisneros R. and Zúñiga G. (2000) Comparative anatomical and histological study of the alimentary canal of the *Dendroctonus frontalis* (Coleoptera: Scolytidae) complex. *Ann. Entomol. Soc. Am.* **93**, 303–311.

Diehl P. A. (1975) Synthesis and release of hydrocarbons by the oenocytes of the desert locust, *Schistocerca gregaria. J. Insect Physiol.* **21**, 1237–1246.

Downing H. A. (1991) The function and evolution of exocrine glands. In *The Social Biology of Wasps*, eds K. G. Ross and R. W. Matthews. Comstock Publ. Ass., Ithaca and London.

Dryden M. W. and Broce A. B. (1993) Development of a trap for collecting newly emerged *Ctenocephalides felis* (Siphonaptera: Pulicidae) in homes. *J. Med. Entomol.* **30**, 901–906.

Economopoulos A. P., Giannakakis A., Tzanakakis M. E. and Voyadjoglou A. V. (1971) Reproductive behavior and physiology of the olive fruit fly. 1. Anatomy of the adult rectum and odors emitted by adults. *Ann. Ent. Soc. Am.* **64**, 1112–1116.

Edmunds M. (1975) Courtship, mating and possible sex pheromones in three species of Mantodea. *Entomol. Monthly Mag.* **111**, 53–57.

Elofsson R. and Löfqvist J. (1974) The Eltringham organ and a new thoracic gland: ultrastructure and presumed pheromone function (Insecta, Myrmeleonidae). *Zool. Scripta* **3**, 31–40.

Eltringham H. (1932) On an extrusible glandular structure in the abdomen of *Mantispa styriaca*, Poda (Neuroptera). *Trans. Entomol. Soc. London* **80**, 103–105.

Fan Y., Zurek L., Dykstra M. J. and Schal C. (2002) Hydrocarbon synthesis by enzymatically dissociated oenocytes of the abdominal integument of the German cockroach, *Blattella germanica. Naturwissenschaften* (in press).

Ferveur J. F., Sarit F., O'Kane C., Sureau G., Greenspan R. J. and Jallon J. M. (1997) Genetic feminization of pheromones and its behavioral consequences in *Drosophila* males. *Science.* **276**, 1555–1558.

Fletcher B. S. (1969) The structure and function of the sex pheromone glands of the male Queensland fruit fly, *Dacus tryoni. J. Insect Physiol.* **15**, 1309–1322.

Foldi I. (1986) Ultrastructure and histochemistry of the oviductal glands of the scale insect *Porphyrophora crithmi* (Homoptera: Margarodidae). *Ann. Soc. Entomol. Fr.* **22**, 145–152.

Fónagy A., Yokoyama N. and Matsumoto S. (2001) Physiological status and change of cytoplasmic lipid droplets in the pheromone-producing cells of the silkmoth, *Bombyx mori* (Lepidoptera, Bombycidae). *Arthr. Str. Funct.* **30**, 113–123.

Fónagy A., Yokoyama N., Okano K., Tatsuki S., Maeda S. and Matsumoto S. (2000).

Pheromone-producing cells in the silkmoth, Bombyx mori: identification and their morphological changes in response to pheromonotropic stimuli. *J. Insect Physiol.* **46**, 735–744.

Galli E. (1998) Mating behaviour in *Tetraneura nigriabdominalis sasaki* (=*akinire* Sasaki) (Hemiptera, Pemphiginae). *Invert. Reprod. Develop.* **34**, 173–176.

Garden R. W., Moroz L. L., Moroz T. P., Shippy S. A. and Sweedler J. V. (1996) Excess salt removal with matrix rinsing: direct peptide profiling of neurons from marine invertebrates using matrix-assisted laser desorption/ionization time-of-flight mass spectrometry. *J. Mass Spectrom.* **31**, 1126–1130.

Gupta B. D. (1979) Sex pheromone of *Poecilocerus pictus* (Acridoidea: Pyrogomorphidae). 2. Histology of female sex pheromone gland. *Folia Morphol. Prague* **27**, 199–202.

Hall G. M., Tittiger C., Andrews G. L., Mastick G. S., Kuenzli M., Luo X., Seybold S. J. and Blomquist G. J. (2002a) Midgut tissue of male pine engraver, *Ips pini*, synthesizes monoterpenoid pheromone component ipsdienol de novo. *Naturwissenschaften* **89**, 79–83.

Hall G. M., Tittiger C., Blomquist G. J., Andrews G. L., Mastick G. S., Barkawi L. S., Bengoa C. and Seybold S. J. (2002b) Male Jeffrey pine beetle, *Dendroctonus jeffreyi*, synthesizes the pheromone component frontalin in anterior midgut tissue. *Insect Biochem. Mol. Biol.* **32** (in press).

Hallberg E. and Löfqvist J. (1981) Morphology and ultrastructure of an intertergal pheromone gland in the abdomen of the pine sawfly *Neodiprion sertifer* (Insecta, Hymenoptera): a potential source of sex pheromones. *Can. J. Zool.* **59**, 47–53.

Hallberg E. and Subchev M. (1997) Unusual location and structure of female pheromone glands in *Theresimima (=Ino) ampelophaga* Bayle-Berelle (Lepidoptera: Zygaenidae). *Int. J. Insect Morphol. Embryol.* **25**, 381–389.

Hammack L., Burkholder W. E. and Ma M. (1973) Sex pheromone localization in females of six *Trogoderma* species (Coleoptera: Dermestidae). *Ann. Entomol. Soc. Am.* **66**, 545–550.

Happ G. M. and Happ C. M. (1973) Fine structure of the pygidial glands of *Bledius mandibularis* (Coleoptera: Staphylinidae). *Tissue & Cell* **5**, 215–231.

Happ G. M., Strandberg J. D. and Happ C. M. (1966) The terpene-producing glands of a phasmid insect: cell morphology and histochemistry. *J. Morphol.* **119**, 143–160.

Hardie J. and Lees A. D. (1985) The induction of normal and teratoid viviparae by a juvenile hormone and kinoprene in 2 species of aphids. *Physiol. Entomol.* **10**, 65–74.

Harrington R. (1985) A comparison of the external morphology of scent plaques on the hind tibiae of oviparous aphids (Homoptera: Aphididae). *Systematic Entomol.* **10**, 135–144.

Hawkes F., Rzepka J. and Gontrand G. (1987) The scent glands of the male south American locust *Schistocerca cancellata.* An electron microscope study. *Tissue & Cell* **19**, 687–704.

Hodosh, R. J., Keough, E. M. and Ringo, J. M. (1979) The morphology of the sex pheromone gland in *Drosophila grimshawi. J. Morphol.* **161**, 177–184.

Hölldobler B. (1971) Sex pheromone in the ant *Xenomyrmex floridanus. J. Insect Physiol.* **17**, 1497–1499.

Hölldobler V. B. and Wüst M. (1973) Ein sexualpheromon bei der pharaoameise *Monomorium pharaonis* (L.). *Z. Tierpsychol.* **32**, 1–9.

Hoyt C. P., Osborne G. O. and Mulcock A. P. (1971) Production of an insect sex attractant by symbiotic bacteria. *Nature* **230**, 472–473.

Isidoro N., Williams I. H., Solinas M. and Martin A. (1992) Mating behaviour and identification of the female sex pheromone gland in the brassica pod midge (*Dasineura brassicae* Winn.: Cecidomyidae, Diptera). *Boll. Ist. Ent.* **47**, 27–48.

Iwabuchi K. (1986) Mating behavior of *Xylotrechus pyrrhoderus* Bates (Coleoptera: Cerambycidae). III. Pheromone secretion by male. *Appl. Entomol. Zool.* **21**, 606–612.

Iwasaki Y. (1996) Hunting and mating behavior in the Japanese hangingfly *Bittacus mastrillii* (Mecoptera: Bittacidae). *Ann. Ent. Soc. Am.* **89**, 869–874.

Jacquemin G. and Bareth C. (1981) Ultrastructure des soies gladulaires et sensorielles du premier sternite abdominal des males de *Campodea chardardi* Condé (Insecta: Diplura): modifications liées a la mue. *Int. J. Insect Morphol. Embryol.* **10**, 463–481.

Joosse E. N. G. and Koelman T. A. C. M. (1979) Evidence for the presence of aggregation pheromones in *Onychiurus armatus* (Collembola): a pest insect in sugar beet. *Entomol. Exp. Appl.* **26**,

Jurenka R. A. and Subchev M. (2000) Identification of cuticular hydrocarbons and alkene precursor to the pheromone in hemolymph of the female gypsy moth, *Lymantria dispar*. *Arch. Insect Biochem. Physiol.* **43**, 108–115.

Kannowski P. B. and Johnson R. L. (1969) Male patrolling behaviour and sex attraction in ants of the genus *Formica*. *Anim. Behav.* **17**, 425–429.

Korchi A., Brossut R., Bouhin H. and Delachambre J. (1999) cDNA cloning of an adult male putative lipocalin specific to tergal gland aphrodisiac secretion in an insect (*Leucophaea maderae*). *FEBS Letters* **449**, 125–128.

Krool S. and Bauer T. (1987) Reproduction development and pheromone secretion in *Heteromurus nitidus* Templeton 1835 (Collembola: Entomobryidae). *Revue d'Ecol. Biol. du Sol.* **24**, 187–196.

Krzysztofowicz A. and Kuzyk S. (1979) Fine structure of the pyloric region and hind gut of *Tetrodontophora bielanensis* (Collembola). *Acta Biol. Acad. Scient. Hungaricae.* **30**, 121–140.

Laduguie N., Robert A., Bonnard O., Vieau F., Lequere J. L., Semon E. and Bordereau C. (1994) Isolation and identification of (3Z, 6Z, 8#0–3, 6, 8-dodecatrien-1-ol in *Reticulitermes santonensis* Feytaud (Isoptera: Rhinotermitidae) – roles of worker trail-following and in alate sex-attraction behaviour. *J. Insect Physiol.* **40**, 781–787.

Lane R. P. and de Souza Bernardes D. (1990) Histology and ultrastructure of pheromone secreting glands in males of the phlebotomine sandfly *Lutzomyia longipalpis*. *Ann. Tropical Med. Parasitol.* **84**, 53–61.

Leonhardt B. A., Neal J. W. Jr., Klun J. A., Schwarz M. and Plimmer J. R. (1983) An unusual lepidopteran sex pheromone system in the bagworm moth. *Science* **219**, 314–316.

Levinson H. Z., Levinson A. R., Kahn G. E. and Schäfer K. (1983) Occurrence of a pheromone-producing gland in female tobacco bettles. *Experientia* **39**, 1095–1097.

Lew A. C. and Ball H. J. (1978). The structure of the apparent pheromone-secreting cells in female *Diabrotica virgifera*. *Ann. Entomol. Soc. Am.* **71**, 685–688.

Liang D., and Schal C. (1993a) Ultrastructure and maturation of a sex pheromone gland in the female German cockroach, *Blattella germanica*. *Tissue & Cell* **25**, 763–776.

Liang D. and Schal C. (1993b) Volatile sex pheromone in the female German cockroach. *Experientia* **49**, 324–328.

Little H. F. (1992) Fine structure of presumed pheromone glands of *Dacus cucurbitae* and *D. latifrons* (Diptera: Tephritidae). *Ann. Ent. Soc. Am.* **85**, 326–330.

Little H. F. and Cunningham R. T. (1987) Sexual dimorphism and presumed pheromone gland in the rectum of *Dacus latifrons* (Diptera: Tephritidae). *Ann. Entomol. Soc. Am.* **80**, 765–767.

Loeb M. J., Neal J. W. J. and Klun J. A. (1989) Modified thoracic epithelium of the bagworm (Lepidoptera: Psychidae): site of pheromone production in adult females. *Ann. Ent. Soc. Am.* **82**, 215–219.

Löfstedt C., Hansson B. S., Petersson E., Valeur P. and Richards A. (1994) Pheromonal secretions from glands on the 5th abdominal sternite of hydropsychid and rhyacophilid caddisflies (Trichoptera). *J. Chem. Ecol.* **20**, 153–170.

Ma P. W. K., Garden R. W., Niermann J. T., O'Connor M., Sweedler J. V. and Roelofs W. L. (2000) Characterizing the Hez-PBAN gene products in neuronal clusters with immunocytochemistry and MALDI MS. *J. Insect Physiol.* **46**, 221–230.

Menon M. (1986) Morphological evidence for a probable secretory site of the male sex pheromones of *Nauphoeta cinerea* (Blattaria: Blaberidae). 2. Electron microscope studies. *J. Morphol.* **187**, 69–80.

Merivee E. and Erm A. (1993) Studies on sex pheromone gland morphology and pheromone components in female elaterid beetles *Agriotes obscurus* L. and *Agriotes lineatus* L. (Coleoptera: Elateridae). *Proc. Estoniann Acad. Sci. Biol.* **42**, 108–117.

Mertins J. W. and Coppel H. C. (1972) Previously undescribed abdominal glands in the female introduced pine sawfly, *Diprion similis* (Hymenoptera: Diprionidae). *Ann. Ent. Soc. Am.* **65**, 33–38.

Moreno D. S. (1972) Location of the site of production of the sex pheromone in the yellow scale and the California red scale. *Ann. Entomol. Soc. Am.* **65**, 1283–1286.

Moreno D. S. and Fargerlund J. (1975) Histology of the sex pheromone glands in the yellow scale and California red scale. *Ann. Entomol. Soc. Am.* **68**, 425–428.

Morse D. and Meighen E. A. (1984) Aldehyde pheromones in Lepidoptera. Evidence for an acetate ester precursor in *Choristoneura fumiferana*. *Science* **226**, 1434–1436.

Nardi J. B., Dowd P. F. and Bartelt R. J. (1996) Fine structure of cells specialized for secretion of aggregation pheromone in a nitidulid beetle *Carpophilus freemani* (Coleoptera: Nitidulidae). *Tissue & Cell* **28**, 43–52.

Nardi J. B., Young A. G., Ujhelyi E., Tittiger C., Lehane M. J. and Blomquist G. J. (2002) Specialization of midgut cells for synthesis of male isoprenoid pheromone components in two scolytid beetles, *Dendroctonus jeffreyi* and *Ips pini*. *Tissue & Cell.* (in press).

Nation J. L. (1981) Sex-specific glands in tephritid fruit flies of the genera *Anastrepha, Ceratitis*, *Dacus* and *Rhagoletis* (Diptera: Tephritidae). *Int. J. Morphol. & Embryol.* **10**, 121–129.

Noirot C. and Quennedey A. (1974) Fine structure of insect epidermal glands. *Ann. Rev. Entomol.* **19**, 61–80.

Noirot C. and Quennedey A. (1991) Glands, gland cells, glandular units: Some comments on terminology and classification. *Ann. Entomol. Soc. Fr. (NS).* **27**, 123–128.

Noldt U., Fettköther R. and Dettner K. (1995) Structure of the sex pheromone-producing prothoracic glands of the male old house borer, *Hylotrupes bajulus* (L.) (Coleoptera: Cerambycidae). *Int. J. Insect Morphol. & Embrol.* **24**, 223–234.

Oetting R. D. and Yonke T. R. (1978) Morphology of the scent gland apparatus of 3 Alydidae. (Hemiptera). *J. Kansas Entomol. Soc.* **51**, 294–306.

Osbrink W. L. A. and Rust M. K. (1985) Cat flea (Siphonaptera: Pulicidae): factors influencing host finding behavior in the laboratory. *Ann. Entomol. Soc. Am.* **78**, 29–34.

Percy J. (1979) Development and ultrastructure of sex pheromone gland cells of the cabbage looper moth, *Trichoplusia ni* (Lepidoptera: Noctuidae). *Can. J. Zool.* **57**, 220–236.

Percy J. and Weatherston J. (1974) Gland structure and pheromone production in insects. In *Pheromones, Frontier of Biology*, ed. M. Birch. North Holland, Amsterdam.

Percy-Cunningham P. and MacDonald J. A. (1987) Biology and ultrastructure of sex pheromone-producing cells. In *Pheromone Biochemistry*, eds G. D. Prestwich and G. J. Blomquist. pp. 27–75. Academic Press, Orlando, FL.

Pettersson J. (1971) An aphid sex attractant. II. Histological, ethological, and comparative studies. *Entomol. Scand.* **2**, 81–93.

Pierre D., Biémont J., Pouzat J., Lextrait P. and Thibeaudeau C. (1997) Location and usltrastructure of sex pheromone glands in female *Callosobruchus maculatus* (Fabricius) (Coleoptera: Bruchidae). *Int. J. Insect Morphol. Embryol.* **25**, 391–404.

Purrington F. F., Kendall P. A., Bater J. E. and Stinner B. R. (1991) Alarm pheromone in a gregarious poduromorph Collembolan (Collembola: Hypogastruridae). *Great Lakes Entomol.* **24**, 75–78.

Quennedey A. (1998) Insect epidermal gland cells: ultrastructure and morphogenesis. In *Microscopic Anatomy of Invertebrates*, Volume 11A: Insecta, pp. 177–207.

Raina A. K., Wergin W. P., Murphy A. C. and Erbe E. F. (2000) Structural organization of the sex pheromone gland in *Helicoverpa zea* in relation to pheromone production and release. *Arthropod Struct. Develop.* **29**, 343–353.

Ramaswamy S. B., Shu S., Monroe W. A. and Mbata G. N. (1995) Ultrastructure and potential role of integumentary glandular cells in adult male and female *Callosobruchus subinnotatus* (Pic) and *C. maculatus* (Fabricius) (Coleoptera: Bruchidae). *Int. J. Insect Morphol. Embryol.* **24**, 51–61.

Roemhild G. (1980) Pheromone glands of Microcaddisflies (Trichoptera: Hydroptilidae). *J. Morphol.* **163**, 9–12.

Schal C., Sevala V. and Cardé R. T. (1998a) Novel and highly specific transport of a volatile sex pheromone by hemolymph lipophorin in moths. *Naturwissenschaften* **85**, 339–342.

Schal C., Sevala V. L., Young H. P. and Bachmann J. A. S. (1998b) Sites of synthesis and transport pathways of insect hydrocarbons: cuticle and ovary as target tissues. *Amer. Zool.* **38**, 382–393.

Schneider I. and Rudinsky J. A. (1969) The site of pheromone production in *Trypodendron lineatum* (Coleoptera: Scolytidae): bioassay and histological studies of the hindgut. *Can. Entomol.* **101**, 1181–1186.

Shu S. and Jones R. L. (1993) Evidence for a multicomponent sex pheromone in *Eriborus terebrans* (Gravenhorst) (Hymenoptera: Ichneumonidae), a larval parasitoid of the European corn borer. *J. Chem. Ecol.* **19**, 2563–2576.

Solinas M. and Isidoro N. (1991) Identification of the female sex pheromone gland in the sorghum midge, *Allocontarinia sorghicola* (Coq.) Solinas (Diptera, Cecidomyiidae). *Redia* **74**, 441–446.

Solinas M. and Isidoro N. (1996) Functional anatomy of female sex pheromone gland of *Mayetiola destructor* Say (Diptera: Cecidomyiidae). *Entomologica Bari* **30**, 43–54.

Sreng L. (1985) Ultrastructure of the glands producing sex pheromones of the male *Nauphoeta cinerea* Insecta: Dictyoptera. *Zoomorphology, Berlin.* **105**, 133–142.

Sreng L. (1998) Apoptosis-inducing brain factors in maturation of an insect sex pheromone gland during differentiation. *Differentiation* **63**, 53–58.

Sreng L., Leoncini I. and Clement J. L. (1999) Regulation of sex pheromone production in the male *Nauphoeta cinerea* cockroach: role of brain extracts, corpora allata (CA), and juvenile hormone (JH). *Arch. Insect Biochem. Physiol.* **40**, 165–172.

Stanic V., Zlotkin E. and Shulov A. (1970) Localization of pheromone excretion in the female of *Trogoderma granarium* (Dermestidae). *Ent. Exp. & Appl.* **13**, 342–351.

Stein G. (1963) Uber den Sexuallockstoff von Hummelmannchen. *Naturwissenschaften* **50**, 305.

Stringer I. A. N. (1988) The female reproductive system of *Costelytra zealandica* (White) (Coleoptera: Scarabaeidae: Melolonthinae). *N. Z. J. Sci.* **15**, 513–533.

Subchev M. and Jurenka R. A. (2001) Sex pheromone levels in pheromone glands and identification of the pheromone and hydrocarbons in the hemolymph of the moth *Scoliopteryx libatrix* L. (Lepidoptera: Noctuidae). *Arch. Insect Biochem. Physiol.* **47**, 35–43.

Suzuki T., Haga K., Kodama S., Watanabe K. and Kuwahara Y. (1988) Secretion of thrips. II. Secretions of three gall-inhabiting thrips (Thysanoptera: Phlaeothripidae). *Appl. Entomol. Zool.* **23**, 291–297.

Tada S. and Leal W. S. (1997) Localization and morphology of the sex pheromone glands in scarab beetles (Coleoptera: Rutelinae, Melolonthinae). *J. Chem. Ecol.* **23**, 903–915.

Tagawa J. (1977) Localization and histology of the female sex pheromone-producing gland in the parasitic wasp, *Apanteles glomeratus. J. Insect Physiol.* **23**, 49–56.

Teerling C. R., Pierce Jr. H. D., Borden J. H. and Gillespie D. R. (1993) Identification and bioactivity of alarm pheromone in the western flower thrips, *Frankliniella occidentalis. J. Chem. Ecol.* **19**, 681–697.

Thornhill A. R. (1973) The morphology and histology of new sex pheromone glands in male scorpionflies, *Panorpa* and *Brachypanorpa* (Mecoptera: Panorpidae and Panorpodidae). *Great Lakes Entomol.* **6**, 47–55.

Thornhill R. and Alcock J. (1983) *The Evolution of Insect Mating Systems.* Harvard University Press, Cambridge, MA.

Urbahn E. (1913) Abdominale Duiftorgans bei weiblichen Schmetterlingen. *Z. Naturwiss.* **50**, 277–355.

Vander Meer R. K., Breed M. D., Espelie K. E. and Winston M. L. (1998) *Pheromone Communication in Social Insects: Ants, Wasps, Bees, and Termites.* Westview Press, Boulder, CO.

Verhoef H. A. (1984) Releaser and primer pheromones in collembola. *J. Insect Physiol.* **30**, 665–670.

Verhoef H. A., Nagelkerke C. J. and Joosse E. N. G. (1977) Aggregation pheromones in Collembola. *J. Insect Physiol.* **23**, 1009–1014.

Vinson S. B. (1972) Courtship behavior and evidence of a sex pheromone in the parasitoid *Campoletis sonorensis* (Hymenoptera: Ichneumonidae). *Enviro. Entomol.* **1**, 409–414.

Waldorf E. S. (1974) Sex pheromone in the springtail, *Sinella curviseta. Environ. Entomol.* **3**, 916–918.

Walker K. A., Jones T. H. and Fell R. D. (1993) Pheromonal basis of aggregation in European earwig, *Forficula auricularia* L. (Dermaptera: Forficulidae). *J. Chem. Ecol.* **19**, 2029–2038.

Wattebled S., Bitsch J. and Rosset A. (1978) Ultrastructure of the pheromone-producing vesicles in males of Chrysopa perla L. (Insecta, Neuroptera). *Cell Tiss. Res.* **194**, 481–496.

Weseloh R. M. (1980) Sex pheromone gland of the gypsy moth parasitoid, *Apanteles melanoscelus*: revaluation and ultrastructural survey. *Ann. Ent. Soc. Am.* **73**, 576–580.

Wiygul G., MacGown M. W., Sikorowski P. P. and Wright J. E. (1982) Localization of pheromone in male boll weevils (*Anthonomus grandis*). *Entomol. Exp. Appl.* **31**, 330–331.

Yang H. T., Chow Y. S., Peng W. K. and Shu E. L. (1998) Evidence for the site of female sex pheromone production in *Periplaneta americana. J. Chem. Ecol.* **24**, 1831–1843.

Young H. P., Bachmann J. A. S., Sevala V. and Schal C. (1999) Site of synthesis, tissue distribution, and lipophorin transport of hydrocarbons in *Blattella germanica* (L.) nymphs. *J. Insect Physiol.* **45**, 305–315.

Zeng R. G., Liu S. K. and Shaefers G. A. (1992) Morphology and histology of the sex pheromone gland of strawberry aphid, *Chaetosiphon fragaefolii* (Cockerell). *Acta Entomol. Sinica.* **35**, 290–293.

Zethner-Møller O. and Rudinsky J. A. (1967) Studies on the site of sex pheromone production in *Dendroctonus pseudotsugae* (Coleoptera: Scolytidae). *Ann. Entomol. Soc. Am.* **60**, 575–582.

Zhu J., Kozlov M. V., Philipp P., Franke W. and Löfstedt C. (1995) Identification of a novel moth sex pheromone in *Eriocrania ciacatricella* (Zett.) (Lepidoptera: Eriocraniidae) and its phylogenetic implications. *J. Chem. Ecol.* **21**, 29–43.

3

Biochemistry of female moth sex pheromones

R. A. Jurenka

3.1 Introduction and historical perspective

Experiments demonstrating the long-range attraction of male moths to female moths were first described in the 1870s and 1880s (Fabre, 1966; Lintner, 1882). The role of volatile chemicals was not realized until later and research continued to define this chemical communication system (see Jacobson, 1972). Many years later, the chemical identification of the first pheromone was made using the sex pheromone glands from thousands of female silkmoths, *Bombyx mori* (Butenandt *et al.*, 1959; Hecker and Butenandt, 1984). Since then, over 1600 species of moths have been investigated regarding their sex pheromones (Arn, 2002). In the previous edition of this book it was reported that the sex pheromones from about 200 species of moths were identified (Bjostad *et al.*, 1987). This dramatic increase in the number of species is primarily due to the relatively simple nature of most of these sex pheromones. Female moth pheromones are relatively simple structures, consisting of a hydrocarbon chain that contains an oxygenated functional group and usually a degree of unsaturation. The functional group can include an ester linkage, alcohols, aldehydes, and epoxides. There are also groups of moths that utilize hydrocarbons as sex pheromones. The structure of the sex pheromones led researchers to speculate that they were fatty acid derived. This provided the key to defining pheromone biosynthetic pathways. These pathways utilize fatty acid enzymes found in normal fatty acid metabolism, but some enzymes have changed through the course of evolution to produce the species-specific pheromone blends. This chapter will detail the research conducted on sex pheromone biosynthesis in female moths since the previous edition of this book (Prestwich and Blomquist, 1987).

The first moth pheromone identified was bombykol (Butenandt *et al.*, 1959), which turned out to be a 16 carbon primary alcohol with two double bonds in the 10 and 12 position (*E*10,*Z*12-hexadecadiene-1-ol, E10,Z12-16:OH[1]). With further research on other moths, it became apparent that most pheromone components were similar in nature. When researchers began utilizing electroantennograms (EAG) and gas-chromatography-EAG (GC-EAG), it became relatively easy to identify further pheromone components (Roelofs, 1984; Struble and Arn, 1984). With the identification of a variety of sex pheromones, the interest turned to determining how the pheromone was produced. Jones and Berger (1978) demonstrated that the pheromone component Z7-12:OAc of the cabbage looper, *Trichoplusia ni*, was produced *de novo* from radiolabeled acetate. The radiolabel was distributed throughout the molecule, demonstrating that it was produced through a fatty acid synthesis pathway. Subsequently, it became important to determine how the various sex pheromone components were produced. A considerable amount of research indicates that most sex pheromones utilized by female moths are composed of a blend of components. This blend is usually produced in precise ratios indicating the importance of double bond position and chain length. Therefore, it became important to determine not only how the pheromone components are produced, but also how the species-specific blend of pheromone components is produced.

3.2 Biosynthetic pathways

Bjostad and Roelofs (1983) were the first to determine correctly how the major pheromone component for a particular moth was biosynthesized. This was done by looking for possible fatty acid intermediates and by monitoring the incorporation of radiolabeled precursors into pheromone components. They showed that glands of the cabbage looper, *Trichoplusia ni*, utilize acetate to produce the common fatty acids octadecanoate and hexadecanoate which undergo Δ11 desaturation to produce Z11-18:acid and Z11-16:acid. But the main pheromone component was Z7-12:OAc, which presumably was made from Z7-12:acid. To demonstrate how the fatty acid precursor Z7-12:acid was produced, [^{3}H-16]-Z11-16:acid was applied to glands and monitored for incorporation. It was incorporated into both Z7-

[1]Shorthand notation for pheromone molecules: for example, (Z)-7-dodecen-1-yl acetate is shortened to Z7-12:OAc where the Z denotes the double bond configuration (Z or E), 7 the double bond position, 12 the number of carbons in the chain, OAc indicates the functional group as an acetate ester. Ald = aldehyde, CoA = coenzyme A, D = deuterium, Epox = epoxide, Hc = hydrocarbon, me = methyl group, OH = alcohol.

12:acid and Z7-12:OAc. They concluded that limited chain shortening of Z11-16:acid could account for this incorporation. The biosynthetic pathways for the other minor components are similar (Figure 3.1). Thus the pheromone components are produced through a fatty acid synthesis pathway involving a Δ11 desaturase and limited chain-shortening enzymes. The proper chain length fatty acid is then reduced and acetylated to form the acetate ester. Subsequent research has demonstrated that similar pathways occur in a wide variety of female moths (Roelofs and Bjostad, 1984; Roelofs and Wolf, 1988).

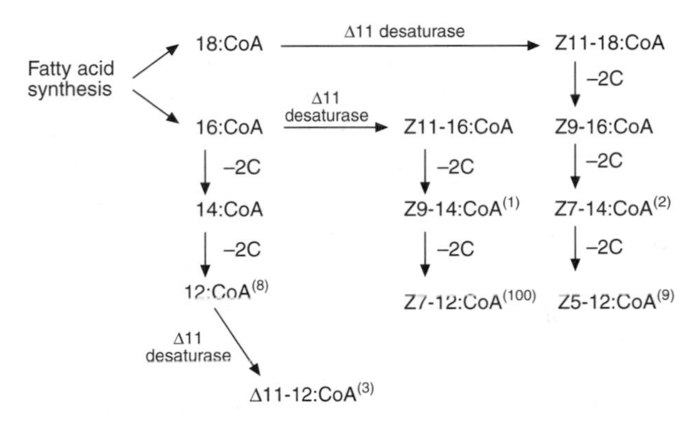

Figure 3.1 Biosynthetic pathways for producing the intermediate CoA derivatives of the pheromone blend of the cabbage looper, *Trichoplusia ni*. The CoA derivatives followed by the superscript number in parentheses are reduced to an alcohol and acetylated to form the acetate esters that make up the pheromone blend. The superscript numbers indicate the approximate ratio of components found in the pheromone gland (Bjostad *et al.*, 1984).

The various chain length fatty acid intermediates found in pheromone biosynthetic pathways could be produced by a mechanism other than limited chain shortening. It was known that a low molecular weight acyl thioester hydrolase interacts with fatty acid synthetase to produce intermediate chain length fatty acids. In the aphid, a thioester hydrolase releases the fatty acid from fatty acid synthetase when the chain length reaches 14 carbons. If the hydrolase is not present, an 18 carbon fatty acid is produced (Ryan *et al.*, 1982). A similar acyl thioester hydrolase is found in the uropygial gland of some ducks and the lactating mammary gland of some mammals (Knudsen *et al.*, 1976; de Renobales *et al.*, 1980). A specific thioester hydrolase was ruled out in the biosynthesis of moth sex pheromones because labeling studies showed that longer chain length fatty acids were incorporated into shorter chain length pheromone components. Also, in most cases the Δ11 desaturase utilized longer chain length precursors that were then chain shortened to produce the right chain length fatty acid (Bjostad and Roelofs, 1981, 1983).

Most female moths produce their pheromone through modifications of fatty acid synthesis pathways, and thus techniques in fatty acid research were utilized. These are outlined in the 1987 *Pheromone Biochemistry* book (Bjostad *et al.*, 1987; Morse and Meighen, 1987b). The main techniques include thin-layer chromatography, GC, and GC/mass spectrometry (MS). The latter technique is the method of choice for determining a biosynthetic pathway. The advantage of using GC/MS is that the label can be explicitly shown to be present in the compound of interest. The labels utilized are the stable isotopes deuterium or carbon-13. Both of these will produce a mass unit that is larger by 1 for each labeled atom present. By monitoring for diagnostic ions that correspond to unlabeled and labeled products, it can be determined with considerable certainty that the label is associated with a particular compound. In addition, the newer GC/MS instruments are more sensitive and relatively easy to operate. A drawback is the expense of such instruments. Illustrations describing how stable isotopes are utilized to elucidate a biochemical pathway are presented using several moths as examples.

3.2.1 Examples of pheromone biosynthetic pathways

3.2.1.1 Almond moth and beet armyworm

The almond moth, *Cadra cautella*, and the beet armyworm, *Spodoptera exigua*, utilize the diene pheromone component Z9,E12-14:OAc and the monoene Z9-14:OAc. These are produced in about a 9:1 ratio of diene:monoene. To determine how these pheromone components are produced, stable isotope precursors were utilized to determine if they became incorporated into the pheromone or into intermediates (Jurenka, 1997). These precursors can be applied in several different ways. The precursor can be injected into the adult female. However, dilution in the hemolymph usually results in very low levels of labeling at the pheromone gland. A more direct approach is to apply the precursor to the cuticular surface of the gland in a solution of dimethyl sulfoxide. This solvent allows for penetration of the fatty acid precursor into the gland. Another technique that can be utilized is an isolated gland incubation. However, it must be determined first that the gland is the site of synthesis for the pheromone. *De novo* biosynthesis of the pheromone can be demonstrated using labeled acetate. For most moths that utilize oxygenated pheromones *de novo*, biosynthesis occurs in the pheromone gland. However, for moths that utilize hydrocarbons, they are biosynthesized in other cells (oenocytes) and transported to the gland (see Section 3.5). Therefore, it must be demonstrated where the pheromone is biosynthesized.

Another technique that is utilized to help ensure that label is incorporated into the pheromone is to apply the precursor at the same time as pheromone biosynthesis activating neuropeptide (PBAN). PBAN is a peptide hormone that regulates pheromone biosynthesis in most, but not all, moths. So, first it must be demonstrated that PBAN regulates pheromone production. In the case of the cabbage looper,

T. ni, it does not, so for each moth it must be demonstrated. In the case of the almond moth and beet armyworm, decapitation, which removes the source of PBAN, after 24 h causes pheromone titers to decline from about 10 ng/gland of the diene to about 0.5 ng/gland (Jurenka, 1997). Administration of exogenous PBAN stimulates pheromone production to normal levels within 3 h. Addition of the labeled precursor directly to the pheromone gland in dimethyl sulfoxide followed immediately by PBAN to decapitated females helped ensure that the label became incorporated into the pheromone.

To demonstrate the biosynthetic pathway, deuterium-labeled precursors were utilized. Previous research with a variety of moths indicated that variations on the scheme presented in Figure 3.1 probably occur. Therefore, a possible biosynthetic pathway was outlined as shown in Figure 3.2A. The pathway starts with fatty acid synthesis to produce 16:CoA. Two different routes could be utilized to produce the monoene and diene. These involve Δ11 desaturation to form Z11-16:CoA or chain shortening to 14:CoA and then Δ9 desaturation to form Z9-14:OAc. To distinguish between these two routes, deuterium-labeled D3-16:acid ([16,16,16-$^2H_3$]hexadecanoic acid) and D3-14:acid (14,14,14-$^2H_3$]tetradecanoic acid) were applied separately to glands, and incorporation into the pheromone was monitored utilizing GC/MS. The data indicated that D3-16:acid became incorporated into the pheromone but not D3-14:acid (Figure 3.3). D3-14:acid apparently was incorporated into E11-14:OAc. This latter compound was tentatively identified based on a retention time with known standards, but still needs to be chemically identified. Therefore, Δ9 desaturation of 14:CoA probably does not occur in pheromone glands of these insects. The possibility still remains for a Δ14 desaturase acting on Z11-16:CoA to produce

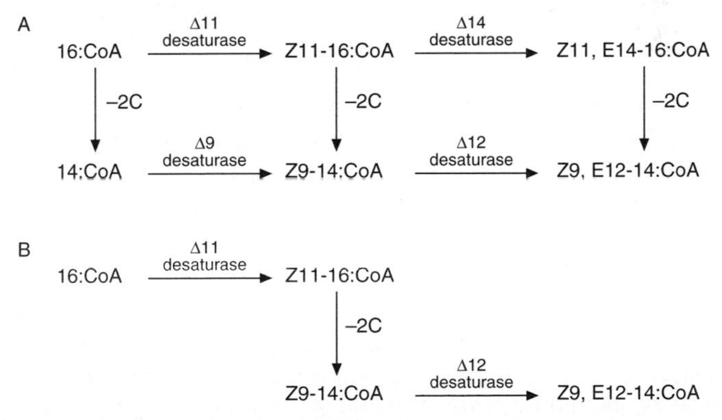

Figure 3.2 A. Possible biosynthetic pathways for producing the sex pheromone components Z9-14:OAc and Z9,E12-14:OAc in the almond moth. The Z9-14:CoA and Z9,E12-14:CoA derivatives are reduced and acetylated to make the acetate esters. B. The biosynthetic pathway as determined by deuterium labeling studies presented in Figure 3.3.

Z11,E14-16:CoA, which is then chain shortened to Z9,E12-14:CoA, or alternatively Z11-16:CoA is chain shortened to Z9-14:CoA and then a Δ12 desaturase produces Z9,E12-14:CoA. Both pathways are possibilities based on the previous identification of a Δ14 desaturase in the Asian corn borer (Zhao *et al.*, 1990) and a Δ12 desaturase in some other insects (de Renobales *et al.*, 1987). To differentiate between the two possible pathways, D9-Z11-16:acid (Z11-[13,13,14,14,15, 15,16,16,16-$^{2}H_{9}$]hexadecenoic acid and D2-Z9-14:acid (Z9-[13,14-$^{2}H_{2}$] tetradecenoic acid) were applied to the glands. The results indicate that D9-Z11-16:acid was chain shortened to Z9-14:acid but was not converted to Z9,E12-14:CoA. This indicates that chain shortening occurs and incorporation of D2-Z9-14:acid into Z9,E12-14:CoA indicates that a Δ12 desaturase is present. The D9-Z11-16:acid was not incorporated into Z9,E12-14:CoA because of an isotope effect. This labeled compound had nine deuteriums on the methyl end where the desaturase enzyme would act, preventing the introduction of a double bond. Combined together, the labeling studies with precursors and intermediates indicate

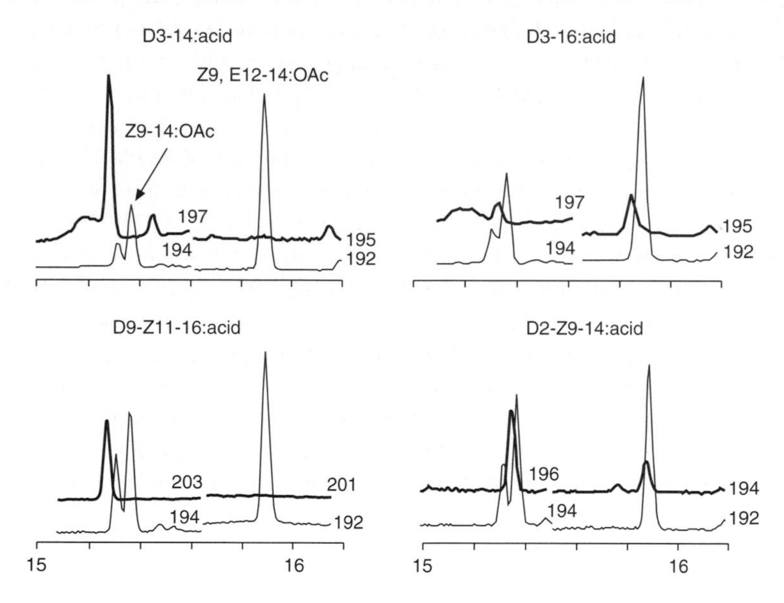

Figure 3.3 Partial GC/MS chromatograms of pheromone gland extracts obtained from female almond moths that were treated with the indicated deuterium-labeled fatty acid. To determine incorporation into the pheromone, single ions were monitored corresponding to (M+)-60 of Z9-14:OAc (197) and Z9,E12-14OAc (195) plus the number of deuteriums present in the precursor. The latter are shown in bold. Control glands were also analyzed to demonstrate that the ions representing labeling were absent (data not shown). The peak eluting just before Z9-14:OAc was tentatively identified as E11-14:OAc.

that female almond moths and beet armyworms utilize a pathway shown in Figure 3.2B.

3.2.1.2 Helicoverpa zea and Helicoverpa assulta

The next example is used to demonstrate how different pathways could produce the same pheromone component. *Helicoverpa zea* and *Helicoverpa assulta* are closely related species that use aldehydes as the major pheromone. *Helicoverpa zea* uses a blend of components with Z11-16:Ald as the major component, and minor components include 16:Ald, Z9-16:Ald, and Z7-16:Ald (Klun *et al.*, 1980). *H. assulta* uses Z9-16:Ald as the major component and Z11-16:Ald as a minor component (Cork *et al.*, 1992; Sugie *et al.*, 1991). The biosynthesis of Z11-16:Ald occurs by Δ11 desaturation of 16:CoA to produce Z11-16:CoA, which is reduced to the aldehyde. This probably occurs in both species, but Z9-16:Ald could be produced by the action of a Δ9 desaturase using 16:CoA as a substrate or by the Δ11 desaturation of 18:CoA to produce Z11-18:CoA that is then chain shortened to Z9-16:CoA. To determine between these two pathways, deuterium-labeled precursors were applied topically to the glands in dimethyl sulfoxide and females injected with PBAN; 1 h later the glands were extracted and analyzed for incorporation using GC/MS (Choi *et al.*, 2002). Figure 3.4 shows the data and biosynthetic pathways.

Incorporation into aldehyde pheromones for both deuterium-labeled D3-16:acid and D3-18:acid were monitored. As expected, D3-16:acid was incorporated into Z11-16:Ald, indicating a Δ11 desaturase in both species. However, in *H. assulta* the major pheromone component Z9-16:Ald is produced by a Δ9 desaturase using 16:CoA as a substrate. D3-18:acid was not incorporated into any of the 16-carbon aldehydes, indicating that chain shortening does not occur in *H. assulta*. In contrast, D3-18:acid was found incorporated into Z9-16:Ald in *H. zea*. This indicates that Z9-16:CoA is probably produced through Δ11 desaturation of 18:CoA, with the resulting Z11-18:CoA chain shortened to Z9-16:CoA. Z7-16:Ald is produced through Δ9 desaturation of 18:CoA to produce Z9-18:CoA which is then chain shortened to Z7-16:CoA. Apparently, the pheromone glands of *H. zea* contain a Δ9 desaturase that uses primarily 18:CoA as a substrate. In contrast, *H. assulta* contains a Δ9 desaturase that uses 16:CoA as a substrate. Both species have a Δ11 desaturase that uses 16:CoA as a substrate. Another difference is that *H. assulta* apparently does not have limited chain-shortening enzymes in pheromone glands (Choi *et al.*, 2002).

The above examples illustrate the use of deuterium labeling to help determine the most likely pathway for biosynthesis of pheromone components. The key components of these pathways are fatty acid biosynthesis, desaturation, chain shortening and specific enzymes to produce a functional group.

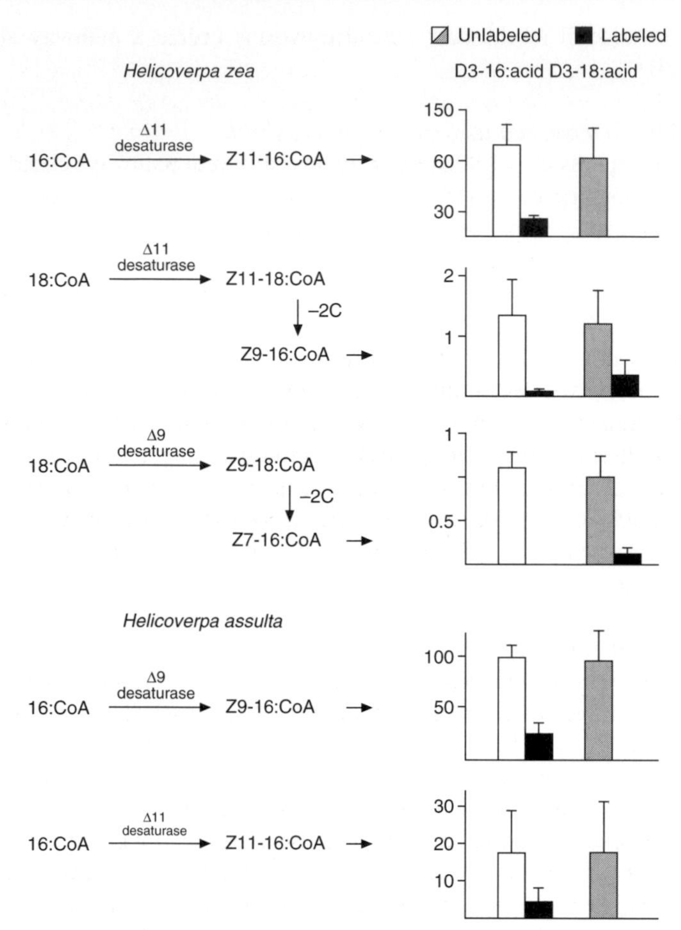

Figure 3.4 Biosynthetic pathways for producing the sex pheromone components of *Helicoverpa zea* and *Helicoverpa assulta*. The CoA derivatives indicated with an arrow are reduced to aldehydes. The unlabeled and labeled aldehyde amounts for each pheromone component are shown in the graphs on the right. The *y*-axis indicates ng/gland for each aldehyde indicated in the biosynthetic pathway. The graphs indicate unlabeled and labeled aldehyde amounts after application of D3-16:acid (left bars) and D3-18:acid (right bars). No label was found in Z7-16:Ald when D3-16:acid was applied to glands of *H. zea*. No label was found in either Z9-16:Ald or Z11-16:Ald when D3-18:acid was applied to glands of *H. assulta*.

3.3 Enzymes involved in pheromone production

3.3.1 Fatty acid synthesis

Saturated fatty acids are produced by the combination of an acetyl-CoA carboxylase and fatty acid synthetase (FAS). Although no direct enzymatic studies have been conducted using pheromone gland cells, these enzymes are presumably similar to enzymes found in other cell types. The acetyl-CoA carboxylase produces malonyl-CoA from acetyl-CoA and bicarbonate using ATP. The malonyl-CoA is used by FAS in the elongation cycle to produce 16:CoA and 18:CoA. In lepidopteran insects, it appears that FAS primarily produces 18:acid followed by 16:acid (Stanley-Samuelson *et al.*, 1988). Labeling studies conducted with acetate indicate that pheromone glands also primarily produce 16:acid and 18:acid saturated products (Bjostad and Roelofs, 1984; Jurenka *et al.*, 1994; Jurenka *et al.*, 1991; Tang *et al.*, 1989). Production of primarily 18:acid is illustrated in studies using the redbanded leafroller (RBLR) moth. Bjostad and Roelofs (1984) applied a topical application of [1-^{14}C]acetate to pheromone glands and then monitored for incorporation by collecting methyl ester fractions from a GC column. They found the highest levels of radioactivity in the 18:acid fraction followed by the 16:acid fraction. Tang *et al.* (1989) observed that decapitation resulted in a decline in pheromone titers and also a decline in 18:acid, whereas 16:acid levels did not change. This indicates that fatty acid synthesis in pheromone glands produces primarily 18:acid. Shorter chain length fatty acids are produced by limited chain shortening.

3.3.2 Chain-shortening enzymes

Insects in general have the ability to shorten long chain fatty acids to specific shorter chain lengths (Stanley-Samuelson *et al.*, 1988). This chain-shortening pathway has not been characterized at the enzymatic level in insects. It presumably is similar to the characterized pathway as it occurs in vertebrates. It is essentially a partial β-oxidation pathway located in peroxisomes (Hashimoto, 1996). The key enzymes involved are an acyl-CoA oxidase (a multifunctional protein containing enoyl-CoA hydratase and 3-hydroxyacyl-CoA dehydrogenase activities) and a 3-oxoacyl-CoA thiolase (Bosch *et al.*, 1992). These enzymes act in concert to chain shorten acyl-CoAs by removing an acetyl group. The evidence for limited chain-shortening enzymes in pheromone glands was originally demonstrated by Bjostad and Roelofs (1983) using the cabbage looper and radiolabeled intermediates, and later demonstrated using the RBLR moth and deuterium-labeled intermediates (Bjostad and Roelofs, 1984). Since then, considerable evidence in a number of moths has accumulated to indicate that limited chain shortening occurs in a variety of pheromone biosynthetic pathways.

In previous studies, radio and stable isotopes were utilized by topical application directly to the pheromone gland *in vivo*. These studies provide evidence for chain shortening because the position of the label was on the terminal methyl carbon, making it difficult for any type of rearrangement to occur. Several studies have also demonstrated limited chain shortening using *in vitro* enzyme assays. The first was conducted using the orange tortrix moth, *Argyrotaenia citrana*, and radiolabeled precursors (Wolf and Roelofs, 1983). In this particular moth, 16:acid was chain shortened to 14:acid. Fatty acid synthesis in moths does not produce 14:acid; therefore, it must be chain shortened from 18:acid to 16:acid. In another study, chain shortening and its inhibition were studied using the Egyptian armyworm, *Spodoptera littoralis* (Rosell *et al.*, 1992). This insect also chain shortens 16:acid to 14:acid, but also, Z11-16:acid is chain shortened to Z9-14:acid.

An *in vitro* enzyme assay was utilized to demonstrate substrate preferences in a study using cabbage looper moths, *T. ni* (Jurenka *et al.*, 1994). This study was prompted by the finding of a mutant line of cabbage loopers that produced a greatly increased amount of Z9-14:OAc (Haynes and Hunt, 1990). The major pheromone component of *T. ni* was found to be Z7-12:OAc (Berger, 1966), with minor components 12:OAc, Z5-12:OAc, Δ11-12:OAc, Z7-14:OAc, and Z9-14:OAc (Bjostad *et al.*, 1984). Increased amounts of Z9-14:OAc indicate that perhaps chain shortening was affected in the mutant cabbage loopers (see Figure 3.1). To determine substrate specificity of chain-shortening enzymes for both normal and mutant cabbage loopers, an *in vitro* enzyme assay was developed utilizing deuterium-labeled acyl-CoA derivatives and NAD as a cofactor (Jurenka *et al.*, 1994). The products were monitored by GC/MS, and in all cases the deuteriums were located on the methyl side of the double bond and would not be lost due to chain shortening. The results indicated that pheromone glands from normal cabbage loopers preferred to chain shorten Z11-16:CoA to Z7-12:CoA. In addition, other substrates were chain shortened to a lesser extent. Chain lengths less than 12 carbons were not observed if monoene fatty-acyl CoAs were used as substrates. This indicates that limited chain shortening stopped when the chain length became 12 carbons. If Z9-14:CoA was utilized, the only product formed was Z7-12:CoA. It appeared that, at most, only two rounds of chain shortening occurred. This is what happens in the normal biosynthesis of the pheromone components as shown in Figure 3.1. Chain shortening of Z11-18:CoA would need three rounds of chain shortening to produce Z5-12:CoA. To achieve this, Z11-18:CoA would be chain shortened to Z9-16:CoA or Z7-14:CoA, and these products would then be chain shortened again by two or one round(s), respectively. Pheromone glands from the mutant cabbage looper apparently have the ability to chain shorten by only one round. Therefore, Z11-16:CoA was chain shortened to Z9-14:CoA. This is the reason why the mutant had higher levels of Z9-14:OAc. Z7-12:OAc was also produced, but at lower levels, because Z9-14:CoA was the starting substrate for one round of chain shortening.

Interestingly, chain shortening has been implicated in the alteration of pheromone ratios in several other species. In a laboratory selection pressure experiment using the RBLR moth, the Z/E ratio of 11-14:OAc could not be changed much from a 92/8 ratio (Roelofs *et al.*, 1986). However, it was found that the ratio of E9-12:OAc/E11-14:OAc could be selected and changed (Sreng *et al.*, 1989). Two lines were selected, one with a low ratio of about 14 percent and one with a high ratio of about 42 percent. By comparing ratios of the 14-/12-carbon pheromone components and Z/E isomers of each chain length it was determined that chain-shortening enzymes were selective for the E isomer (Roelofs and Jurenka, 1996). Therefore, a change in the chain-shortening enzymes may affect the pheromone ratios in this insect. Another example where chain-shortening enzymes may have changed the pheromone components in an insect is the larch budmoth, *Zeiraphera diniana*. The pheromone was identified as E11-14:OAc (Roelofs *et al.*, 1971), but it was later determined that not all larch budmoth males responded to this pheromone (Baltensweiler and Priesner, 1988; Guerin *et al.*, 1984). Some populations used E9-12:OAc as their pheromone. Therefore, apparently, a change in the chain-shortening enzymes helped produce these two populations. Another case where chain shortening may have produced two populations of insects was found in the turnip moth, *Agrotis segetum*. It was found that this moth has two populations with differing pheromone ratios represented by those found in Sweden and Zimbabwe. The Swedish population has a ratio of Z9-14:OAc/Z7-12:OAc/Z5-10:OAc of 29/59/12, whereas the Zimbabwean population has a ratio of 2/20/78. By conducting labeling studies, it was determined that chain-shortening enzymes could be affected to produce the alteration in pheromone ratios (Wu *et al.*, 1998). Apparently the Swedish population has a reduced ability to chain shorten, although a change in reductase activity cannot be ruled out. These studies indicate that alteration in chain-shortening enzymes can have a major effect on pheromone blends. In fact, the combination of desaturases and chain-shortening enzymes can produce many of the possible intermediates that can be converted to identified pheromones (Figure 3.5).

3.3.3 Desaturases

Desaturases introduce a double bond into the fatty acid chain. A variety of desaturases have been described that are involved in the biosynthesis of female moth sex pheromones. The next chapter in this book will describe these enzymes in more detail. The desaturases identified so far include enzymes that act on saturated and monounsaturated substrates. These include Δ5 (Foster and Roelofs, 1996), Δ9 (Löfstedt and Bengtsson, 1988; Martinez *et al.*, 1990), Δ10 (Foster and Roelofs, 1988), Δ11 (Bjostad and Roelofs, 1981; Bjostad and Roelofs, 1983), and Δ14 (Zhao *et al.*, 1990) desaturases that utilize a saturated substrate. The combination of these desaturases along with chain shortening can account for the majority of double bond positions in the various chain length monounsaturated pheromones so far identified (Roelofs and Wolf, 1988). Figure 3.5 illustrates the large number of monounsaturated compounds that can be generated through desaturation and chain shortening. Addition of various functional groups – acetate esters, alcohols, and aldehydes – increases the potential number of pheromone components. Some of the intermediate compounds could be produced in two

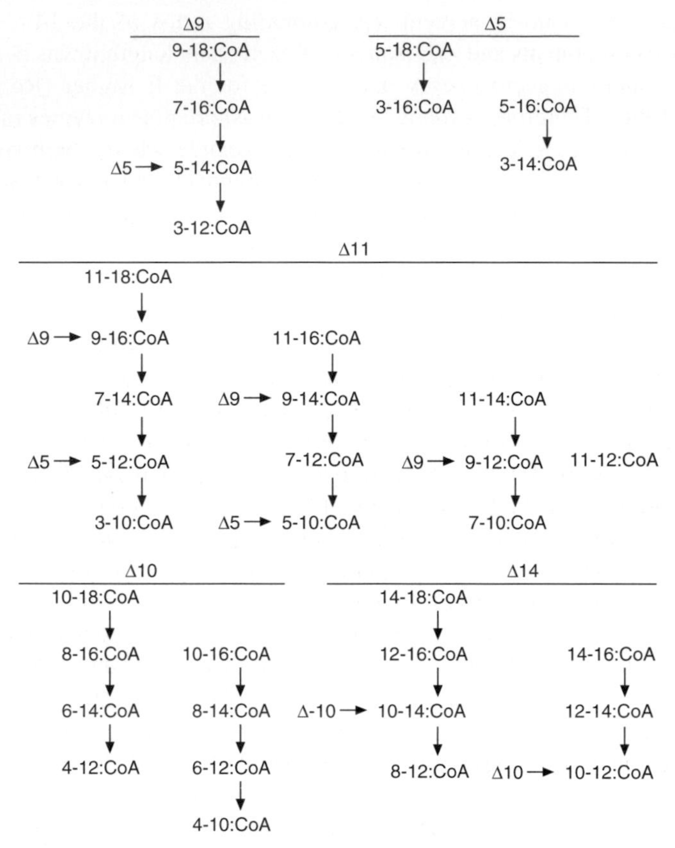

Figure 3.5 Combination of desaturation and chain shortening can produce a variety of monounsaturated acyl-CoA precursors that can be modified to form acetate esters, aldehydes, and alcohols. The number followed by the Δ sign indicates a desaturase that introduces a double bond into the first indicated chain length acyl-CoA. The arrow pointing down indicates limited chain shortening by 2 carbons. The arrow pointing to the right indicates that desaturation could produce the compound found within a chain-shortening pathway. This indicates that certain compounds could be produced in two different ways. Modification of all 16-, 14-, 12-, and 10-carbon acyl-CoA derivatives on the carbonyl carbon can account for the majority of monounsaturated acetate esters, aldehydes, and alcohols identified as sex pheromones.

different ways. For example, within the Δ11 pathways the Δ9 compounds could be produced through Δ9 desaturation as occurs in production of E9-12:OH in the codling moth, *Cydia pomonella* (Löfstedt and Bengtsson, 1988). Production of Z5-14:OAc in the tortricid moth, *Ctenopseustis herana,* occurs by Δ5 desaturation of 14:acid rather than Δ9 desaturation and chain shortening (Foster and Roelofs,

1996). Therefore, although the desaturation and chain-shortening steps occur in a wide variety of moths, the order in which they occur and the type of desaturase must still be determined experimentally.

Some pheromone components are dienes, and these can be produced by either the action of two desaturases or one desaturase and isomerization around the double bond. Some dienes with a 6,9-double bond configuration are produced using linoleic acid (see section 3.5). Desaturases that utilize monounsaturated acyl-CoA substrates include Δ5 (Ono *et al.*, 2002), Δ9 (Martinez *et al.*, 1990), Δ11 (Foster and Roelofs, 1990), Δ12 (Jurenka, 1997), and Δ13 (Arsequell *et al.*, 1990). These can act sequentially to produce the diene. For example, the conjugated diene E9,E11-14:OAc is produced by Δ11 desaturation of 16:CoA followed by chain shortening to E9-14:CoA and another Δ11 desaturation to produce E9,E11-14:CoA (Foster and Roelofs, 1990). The latter is reduced and acetylated to form the acetate ester. Another way to produce conjugated dienes is by the action of one desaturase followed by isomerization. Examples of these include E10,Z12-16:OH (bombykol) (Ando *et al.*, 1988), E8,E10-12:OH (Löfstedt and Bengtsson, 1988), and 10,12-16:Ald (Fang *et al.*, 1995a). It is unclear how this isomerization occurs, but Löfstedt and Bengtsson (1988) suggested that oxidation of the alpha-carbon on either side of the double bond occurs followed by 1,4 elimination of water. This was inferred by observing the loss of one deuterium from the precursor [11,11,11,12,12-D5]E9-12:acid during conversion to E8,E10-12:OH.

The biosynthesis of triene pheromone components has not been extensively investigated. Pheromones with a triene double bond system that is n-3 (3,6,9-) are probably produced from linolenic acid (Millar, 2000). This was demonstrated in the saltmarsh caterpillar, *Estigmene acrea*, and the ruby tiger moth, *Phragmatobia fuliginosa* (Rule and Roelofs, 1989). Moths in the families Geometridae, Arctiidae, and Noctuidae apparently utilize linoleic and linolenic acid as precursors for their pheromones. Most of these pheromones are probably produced by chain elongation and decarboxylation to form hydrocarbons. Oxygen is added across one of the double bonds in the polyunsaturated hydrocarbon to produce an epoxide (Millar, 2000).

3.3.4 Specific enzymes to produce functional group on carbonyl carbon

Once a specific chain length pheromone intermediate is produced, the carbonyl carbon is modified to form a functional group. The majority of oxygenated pheromone components are acetate esters (or other esters), alcohols, and aldehydes. Production of these components requires the reduction of a fatty-acyl precursor. Reduction of fatty-acyl precursors to an alcohol is a two-step reaction requiring a fatty acid reductase and an aldehyde reductase (Morse and Meighen, 1987b). Thus, alcohol formation goes through an aldehyde intermediate. Therefore, aldehydes could be produced by direct reduction of fatty acids. Another route for aldehyde formation is oxidation of alcohols. A cuticular oxidase has been characterized from pheromone glands of *H. zea* and *Manduca sexta* that produce aldehydes as pheromones (Fang *et al.*, 1995b; Teal and Tumlinson, 1988). Due to the non-specific oxidation of alcohols, it is unclear whether or not the oxidase is directly involved in pheromone production. However, if the right alcohol is formed in pheromone glands, then the oxidase will produce the aldehyde. Aldehydes are more reactive, and it could be that the oxidase is located in the cuticle so that the aldehyde is released as soon as it is made (Teal and Tumlinson, 1988). In those insects that utilize both an alcohol and an aldehyde as part of their pheromone, it is unclear how the production of both components would be regulated.

Changes in reductase activity could result in altered pheromone amounts. The lightbrown apple moth, *Epiphyas postvittana*, exhibits a decline in pheromone titers with age that is not due to reduced PBAN levels (Foster and Roelofs, 1994). To determine what is responsible for the decline in titers, different deuterium-labeled precursors were applied to the glands of different aged females. Biosynthesis of the major pheromone component, E11-14:OAc, occurs by chain shortening 16:acid to 14:acid followed by E11 desaturation, reduction, and acetylation (Foster and Roelofs, 1990). Application of deuterium-labeled intermediates indicated that the senescent decline in pheromone titers was due to the reduction in fatty acid reductase activity (Foster and Greenwood, 1997). It is unknown if a decline in reductase enzyme activity is also responsible for observed senescent decline in pheromone titers of other female moths (Foster and Greenwood, 1997).

Production of acetate ester pheromone components utilizes an enzyme called acetyl-CoA:fatty alcohol acetyltransferase that converts a fatty alcohol to an acetate ester (Morse and Meighen, 1987a). Therefore, alcohols could be utilized as substrates for both aldehyde and acetate ester formation. The acetyltransferase utilizes the energy from hydrolysis of acetyl-CoA to make the acetate ester linkage. Other esters could be made using different CoA derivatives. This enzyme has been characterized in several tortricid moths. Morse and Meighen (1987a) first demonstrated its presence in the spruce budworm, *Choristoneura fumiferana*, where it is involved in producing the acetate ester that serves as a precursor to the aldehyde pheromone (Morse and Meighen, 1987b). In this particular moth, the pheromone components are the aldehydes E11-14:Ald and Z11-14:Ald found in a 96/4 ratio. The acetate esters are produced and stored in the gland (Morse and Meighen, 1984). An esterase is also present in the pheromone gland of the spruce budworm to make the conversion from acetate ester to alcohol. It was demonstrated that the acetyltransferase could utilize alcohols with 12–15 carbons and produced the highest activity with Z11-14:OH (Morse and Meighen, 1987a). This is in contrast to the actual E11/Z11 ratio of 96/4 found in the gland. However, it was shown that the desaturase produced primarily the E isomer. The acetyltransferase therefore starts with E11-14:OH, and there is little competition with the Z isomer.

An acetyltransferase that is specific for Z isomers was also found in some other tortricids. The RBLR moth, *Argyrotaenia velutinana*, utilizes a 92/8 ratio of Z/E 11-14:OAc as one of the major pheromone components. The acetyl transferase preferentially converts Z11-14:OH to the acetate ester (Jurenka and Roelofs, 1989). This was demonstrated in an *in vitro* enzyme assay where the ratios of substrates were controlled. A starting ratio of 1/1 showed about a five-fold higher production of Z11-14:OAc over E11-14:OAc. Even a starting ratio of 1/3 Z/E showed about 1.5 times as much Z11-14:OAc produced. These findings were also demonstrated for two additional tortricid species, *Choristoneura rosaceana* and *Platynota idaeusalis*. These results indicate that the family Tortricidae has members that have an acetyltransferase that is specific for the Z isomer of monounsaturated fatty alcohols. In contrast, the European corn borer and cabbage looper moths did not have an acetyltransferase that showed preference (Jurenka and Roelofs, 1989). In addition, no apparent specificity was found in the acetyltransferase in two other studies utilizing noctuid moths (Bestmann *et al.*, 1987; Teal and Tumlinson, 1987). Therefore, this unique acetyltransferase apparently evolved within the Tortricidae.

3.4 Production of specific pheromone blends

Most female moths utilize a blend of components produced in a specific ratio for species-specific pheromone attraction of conspecific males. A major question is how these species-specific ratios of components are produced. Research from several sources indicates that these ratios are produced by the inherent specificity of certain enzymes present in the biosynthetic pathways. The combination of these enzymes acting in concert produces the species-specific pheromone blend. Observations of enzymatic properties of each enzyme in the pathway should yield insights into how the final pheromone blend is produced. Several examples will be illustrated to make this point.

The first example will utilize the cabbage looper, *T. ni*, where the biosynthetic pathway has been known for some time. The biosynthetic pathway is shown in Figure 3.1 with the approximate ratios of pheromone components indicated. The major component is Z7-12:OAc which is produced by Δ11 desaturation of 16:CoA followed by two rounds of chain shortening, reduction, and acetylation. The Δ11 desaturase has been characterized (Wolf and Roelofs, 1986) and the gene isolated (Knipple *et al.*, 1998). These studies indicate that 16:CoA and 18:CoA are substrates, with 16:CoA the preferred substrate. This is in agreement with the intermediate levels found in the gland, with Z11-16:acid being the most abundant monounsaturated fatty acid (Bjostad *et al.*, 1984). The next step in the pathway is limited chain shortening. An *in vitro* enzyme assay indicated that Z11-16:CoA is the preferred substrate for these enzymes (see section 3.3.2) (Jurenka *et al.*,

1994). After chain shortening, the 14-carbon and 12-carbon intermediates are reduced to an alcohol and acetylated. The acetyltransferase enzyme is not specific and will accept a variety of substrates (Jurenka and Roelofs, 1989). Unfortunately, the reductase has not been characterized at the enzymatic level. However, intermediate fatty alcohols are not found in the gland, therefore tight coupling of the reductase and acetyltransferase probably occurs. From these observations, it can be inferred that the final ratio of pheromone components is produced by the specificity found within the Δ11 desaturase and chain-shortening enzymes.

The RBLR moth utilizes a blend of seven acetate esters with the biosynthetic pathway shown in Figure 3.6. This pathway is similar to the one just described, except the Δ11 desaturase starts with 14:CoA as the substrate and produces both Z and E isomers of 11–14:CoA in about a 60/40 ratio (Wolf and Roelofs, 1987). Recent cloning of the Δ11 desaturase from RBLR females indicates that the expressed enzyme produces a ratio of Z/E of about 6/1 (Liu *et al.*, 2002). However, the final ratio of Z11- to E11-14:OAc is 92/8 (Miller and Roelofs, 1980). A selective increase in the Z isomer occurs within the biosynthetic pathway. Timed radiotracer studies using [1-^{14}C]acetate indicated that once 14:CoA was biosynthesized, the subsequent steps of Δ11 desaturation, reduction, and acetylation to the acetate esters occurred in quick succession (Bjostad and Roelofs, 1984). As indicated in section 3.3.4., the acetyl:CoA fatty alcohol acetyltransferase shows specificity for the Z isomer (Jurenka and Roelofs, 1989). Selective acetylation of Z11-14:OH and production of >60 percent Z11-14:CoA indicates that these enzymes have the inherent specificity to produce the 92:8 ratio of the major pheromone components Z11- and E11-14:OAc. Two minor pheromone components are produced by chain shortening Z11- and E11-14:OAc. The ratio of Z9- to E9-

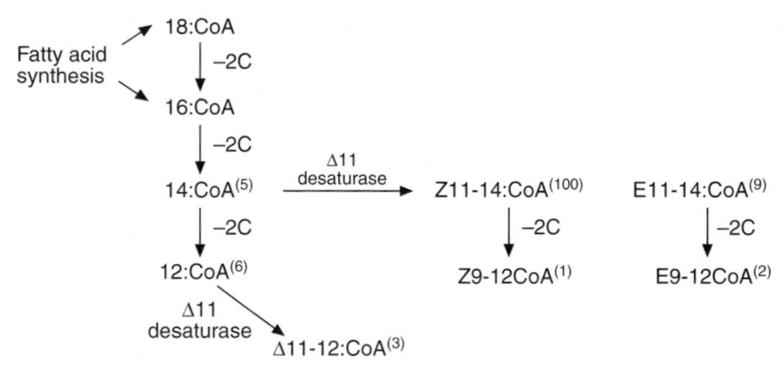

Figure 3.6 Biosynthetic pathways for producing the intermediate CoA derivatives of the pheromone blend of the redbanded leafroller moth, *Argyrotaenia velutinana*. The CoA derivatives followed by a superscript number in parentheses are reduced to an alcohol and acetylated to form the acetate esters that make up the pheromone blend. The superscript numbers indicate the approximate ratio of components found in the pheromone gland (Bjostad *et al.*, 1985).

12:OAc is about 1 to 2. This indicates that the chain-shortening enzymes may prefer E11-14:CoA or that very little Z11-14:CoA is available to chain shorten. To distinguish between these possibilities, the chain-shortening enzymes would need to be characterized *in vitro*. However, a study looking at levels found in two selected lines of RBLR indicates that the chain-shortening enzymes prefer the E isomer (Roelofs and Jurenka, 1996). This combined information indicates that in the RBLR moth pheromone glands the final ratio of pheromone components can be produced through the concerted action of a Δ11 desaturase that produces at least a 60/40 ratio of Z/E intermediate isomers. The final ratio of acetate esters (92/8) is produced through the specificity for the Z isomer by the acetyltransferase. The minor components are produced by specificity in chain shortening.

Another insect that utilizes specific ratios of Z11- and E11-14:OAc is the European corn borer, *Ostrinia nubilalis*. Two strains are known in which one produces a ratio of Z/E of about 97/3 (Z strain) and the other produces an opposite ratio of Z/E of about 1/99 (E strain). Hybridization studies between the two strains indicated that offspring have an acetate ester ratio of Z/E of about 30/70 (Klun and Maini, 1982). The Δ11 desaturase from both strains produced a product with about 30/70 Z/E in an *in vitro* enzyme assay (Wolf and Roelofs, 1987). These results indicate that the final ratio of isomers is produced after the desaturation step. The enzymes that follow the desaturase are a reductase to make an alcohol and an acetyltransferase to produce the acetate esters. An *in vitro* acetyltransferase assay did not find any differences between the two strains or in the hybrids, in that the enzyme from all strains did not prefer one isomer over the other (Jurenka and Roelofs, 1989). This was confirmed in another study in which differences were not found between the strains in the acetylation of labeled alcohols applied to glands *in vivo* (Zhu *et al.*, 1996). However, labeled acids applied to glands *in vivo* were selectively incorporated into the correct pheromone ratio indicating that the reductase shows specificity. The reductase enzyme was not directly assayed, but application of postulated aldehyde intermediates indicated that both strains selectively produced the correct final pheromone ratio (Zhu *et al.*, 1996). Reduction of fatty acids to alcohols involves a fatty acid reductase and an aldehyde reductase (Morse and Meighen, 1987b). The selective conversion of the correct isomer in each strain indicates that the aldehyde reductase has isomeric specificity. Apparently, the fatty acid reductase would also show specificity because aldehydes were not found in the pheromone gland. If the fatty acid reductase was not specific, then some aldehydes of the opposite isomer should be found in the gland and they were not (Zhu *et al.*, 1996). Therefore, the final pheromone ratios produced by females of the European corn borer are made through the action of a Δ11 desaturase that can produce both Z/E isomers in about a 70/30 ratio. The final acetate ester ratio is strain dependent and is produced through the specificity found in the reductase system.

The above three examples illustrate how a species-specific pheromone blend is produced by the concerted action of desaturases, chain-shortening enzymes, and a reductase and an acetyltransferase. The specificity inherent in certain enzymes in the pathway produces the final blend of pheromone components.

Variations in biosynthetic pathways will undoubtedly be found among the wide diversity of moth species. For example, *M. sexta* produces several 16-carbon aldehydes through the combination of desaturation and reduction (Fang *et al.*, 1995a). Correlation of the diene and triene precursor levels in the triacylglycerols and aldehyde pheromone production indicates that the triacylglycerols serve as storage for the precursor fatty acids (Fang *et al.*, 1995c). It appears that fatty acid synthesis and desaturation produce the fatty-acyl precursors on a continual basis with storage occurring in the triacylglycerols. When the aldehyde pheromones are produced during the scotophase, the fatty acid precursors are liberated from the triacylglycerols and reduced to aldehydes. This latter process is apparently under the regulation of PBAN (Fang *et al.*, 1996). A similar process occurs in the eastern spruce budworm, *C. fumiferana*, except acetate esters are used as the storage form of fatty acid precursors to the aldehyde pheromone (Morse and Meighen, 1984). In the light brown apple moth, *Epiphyas postvittana*, fatty acid precursors are also stored in the triacylglycerols and PBAN may release the precursors for formation of the acetate ester pheromone (Foster, 2000).

3.5 Hydrocarbon pheromones

Moths in the families Geometridae, Arctiidae, and some Noctuidae utilize hydrocarbons or epoxides of hydrocarbons as their sex pheromones. Hydrocarbon biosynthesis occurs in oenocyte cells that are associated with either epidermal cells or fat body cells (Wigglesworth, 1970). Once the hydrocarbons are biosynthesized, they are transported to the sex pheromone gland by lipophorin (Schal *et al.*, 1998). The hydrocarbons can be released directly in the case of some moths or they are transformed into epoxides by addition of oxygen across one of the double bonds.

Hydrocarbon biosynthesis occurs through modification of fatty acid metabolic pathways (de Renobales *et al.*, 1991). The enzymes involved include a cytosolic fatty acid synthase for production of straight chain hydrocarbons and a microsomal fatty acid synthase for producing methyl-branched hydrocarbons (see Tillman *et al.*, 1999). Longer chain intermediate fatty acids are produced by a microsomal acyl-CoA elongase (Vaz *et al.*, 1987, 1988). Once the proper chain length is achieved, the acyl-CoA is converted to a hydrocarbon by a decarboxylation reaction involving a cytochrome-P450 (Reed *et al.*, 1994). This conversion is a two-step process involving reduction to an aldehyde intermediate followed by oxidation of the aldehydic carbonyl carbon. The carbonyl carbon is released as carbon dioxide leaving the starting acyl-CoA substrate as a hydrocarbon one carbon shorter. The biosynthesis of methyl-branched hydrocarbons, where the methyl branch is internal, utilizes methylmalonyl-CoA during chain elongation to introduce the methyl group into the growing chain (Juarez *et al.*, 1992). Methyl groups on the number two carbon are usually derived from valine or leucine. The carbons from leucine are incorporated into odd chain length 2-methyl hydrocarbons (Charlton and Roelofs, 1991), whereas valine contributes the carbons for even-chain 2-methyl-branched hydrocarbons (Blailock *et al.*, 1976).

The majority of straight chain hydrocarbons have an odd number of carbons due to the decarboxylation of even chain fatty acid precursors. Sex pheromones that are straight chain hydrocarbons also usually have an odd number of carbons. Most of these are polyunsaturated with double bonds in the n-3 or n-6 positions, indicating that they are derived from linolenic or linoleic acid, respectively (Millar, 2000). Linolenic or linoleic acid would be elongated to 20 or 22 carbons and decarboxylated to form hydrocarbons with 19 or 21 carbons. This has been demonstrated in a couple of arctiid moths (Rule and Roelofs, 1989). A few even chain length hydrocarbon sex pheromones have been identified that also have the n-3 or n-6 double bond configurations, indicating that they are derived from linolenic or linoleic acids. It is not known how these even chain hydrocarbons are formed. Millar (2000) suggested that they could be formed by direct reduction of the carboxyl group without loss of carbon. An alternative pathway could involve chain elongation followed by α-oxidation to an odd chain length fatty acid which is then decarboxylated to an even chain hydrocarbon. In insects, an α-oxidation reaction has been demonstrated in the pheromone glands of the tomato pinworm, *Keiferia lycopersicella*, that uses E4-13:OAc as a sex pheromone. By utilizing uniformly carbon-13-labeled oleic acid (Z9-18:acid) it was determined that five carbons were lost during the transformation to E4-13:OAc (Charlton and Roelofs, unpublished observations). The labeling pattern indicated that α-oxidation of Z9-18:acid produced Z8-17:acid, which was chain shortened by two rounds of β-oxidation with isomerization of the double bond to form E3-13:acid. Whether or not α-oxidation or direct reduction is involved in even-numbered hydrocarbon sex pheromones must await further research.

The formation of hydrocarbon is thought to take place in oenocyte cells (Diehl, 1975; Wigglesworth, 1970). Hydrocarbon sex pheromones are released from sex pheromone glands, which are modified cuticular structures found between abdominal segments 8 and 9 (Percy-Cunningham and MacDonald, 1987). The hydrocarbon must therefore be transported through the hemolymph from the oenocyte cells to the sex pheromone gland. Lipophorin is the protein particle that transports lipids, including hydrocarbons, through the water-based hemolymph (Chino, 1985). When the transport of hydrocarbon sex pheromones in moths was investigated in detail by Schal *et al.* (1998), it was found that a very specific uptake was occurring at pheromone glands. This was demonstrated in the arctiid moths, *Holomelina aurantiaca* and *Holomelina lamae*, both of which use 2me-17:Hc as the main pheromone component. Previous biosynthetic studies showed that leucine injected into the hemocoel labeled the methyl branch but was not incorporated when applied directly to the gland (Charlton and Roelofs, 1991). To demonstrate where biosynthesis occurs, [1-^{14}C]acetate was incubated with isolated sex pheromone glands and isolated abdominal tissue, including epidermal cells and associated oenocytes (Schal *et al.*, 1998). Label was incorporated into the sex pheromone in only the isolated abdominal tissue preparation, indicating that oenocytes are the site of synthesis. Transport through the hemolymph was demonstrated by extraction of hemolymph and analysis by GC. In addition, the lipophorin particle was purified and the sex pheromone was found associated with the purified lipophorin, as were longer chain length hydrocarbons found on the cuticular surface. Other studies also indicate that moth hydrocarbon sex pheromones are found in the hemolymph (Jurenka and Subchev, 2000; Subchev and Jurenka, 2001). Specific uptake by the pheromone gland was demonstrated by GC analysis of pheromone gland extracts. The pheromone glands contained predominately 2me-17:Hc and very little, if any, of the longer chain length hydrocarbons (Schal *et al.*, 1998). This indicates that specific uptake of 2me-17:Hc occurred at the pheromone gland. Further research will decipher how the specific uptake of one hydrocarbon can occur in sex pheromone glands.

A major class of sex pheromones that are derived from hydrocarbons are the polyene monoepoxides (Millar, 2000). These usually have double bonds in the 3,6,9-positions (n-3) or 6,9-positions (n-6), which indicates that they are biosynthesized from linolenic or linoleic acids, respectively. Moths do not have the capability of biosynthesizing these fatty acids, so they must be obtained from the diet (Stanley-Samuelson *et al.*, 1988). Triene and diene hydrocarbons with 17 carbons are probably produced by decarboxylation of linolenic and linoleic acid, respectively. Longer chain length polyenes would be made by chain elongating with malonyl-CoA and then decarboxylation. This was demonstrated for the saltmarsh caterpillar, *Estigmene acrea*, and the ruby tiger moth, *Phragmatobia fuliginosa* (Rule and Roelofs, 1989). Both of these moths utilize 3,6-21:9,10Epox as part of their sex pheromone. It was demonstrated that this compound was produced through chain elongation of linolenic acid followed by decarboxylation and epoxidation. Presumably, the chain elongation and decarboxylation steps occurred in oenocytes; however, it was not shown where in the insect epoxidation took place, because labeled intermediates were injected into female moths (Rule and Roelofs, 1989). Injection of a labeled alkene precursor into female gypsy moth, *Lymantria dispar*, pupae was also used to demonstrate conversion to the sex pheromone, 2me-18:7,8Epox; however, it was not determined where in the insect this conversion was taking place (Kasang and Schneider, 1974).

Evidence that epoxidation takes place in the pheromone gland was shown in a study on the Japanese giant looper, *Ascotis selenaria cretacea*, that uses 6,9-19:3,4Epox as a sex pheromone component (Ando *et al.*, 1997). Deuterium-labeled hydrocarbon precursor, D3-3,6,9-19:Hc, was topically applied to pheromone glands and monitored for conversion to the epoxide (Miyamoto *et al.*, 1999). The hydrocarbon was converted to the epoxide and stimulation occurred when PBAN was injected into decapitated females. This indicates that the epoxidation step takes place in pheromone glands and that it is under the regulation of PBAN. By using a variety of polyene precursors, it was also determined that the monooxygenase regiospecifically attacked the n-3 double bond regardless of chain length or degree of unsaturation. This indicates that the epoxidation enzyme is regiospecific in this insect (Miyamoto *et al.*, 1999).

A current study using the gypsy moth, *L. dispar*, illustrates the overall pathways involved in production of epoxide pheromone components (Figure 3.7). This insect uses disparlure, 2me-18:7,8Epox, as a pheromone component. As indicated above, even chain length 2-methyl-branched hydrocarbons are initiated using valine to supply the carbons for the 2-methyl branch. Injection of deuterium-labeled valine into female gypsy moths resulted in incorporation into both the alkene, 2me-Z7-18:Hc, and 2me-18:6,7Epox (Jurenka and Subchev, unpublished observations). In addition, incubation of isolated abdominal epidermal tissue with deuterium-labeled valine resulted in incorporation into 2me-Z7-18:Hc. This indicates that the oenocyte cells associated with the epidermal tissues biosynthesize 2me-Z7-18:Hc using the carbons of valine to initiate the chain. The double bond is probably introduced by a $\Delta 12$ desaturase. This was demonstrated by incubating isolated abdominal epidermal tissue with D4-18me-19:acid and finding label in 2me-Z7-18:Hc (Jurenka and Fabriás, unpublished observations). Hemolymph transport of 2me-Z7-18:Hc is indicated by the finding of this alkene in the hemolymph (Jurenka and Subchev, 2000). Demonstration that 2me-Z7-18:Hc is converted to the epoxide in the pheromone gland was shown by using D4-2me-Z7-18:Hc and incubation with isolated pheromone glands (Jurenka and Fabriás, unpublished observations). Disparlure is a stereoisomer that has the 7R,8S or (+) configuration. To determine that the monooxygenase made the correct conversion, the resulting deuterium-labeled pheromone will be separated using chiral chromatography (Pu *et al.*, 1999). These results indicate that hydrocarbon pheromones and their epoxides are produced through a pathway outlined in Figure 3.7. The hydrocarbon precursor to the epoxide is produced in oenocytes and then transported to the pheromone gland by lipophorin. If an epoxide is utilized, the epoxidation step takes place in the pheromone gland before the pheromone is released to the environment.

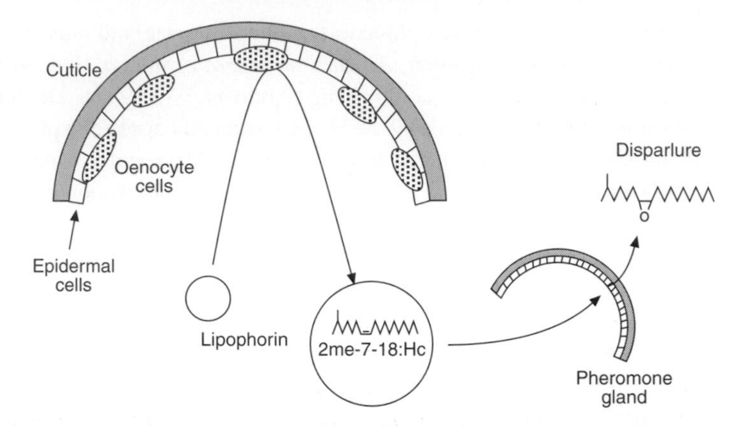

Figure 3.7 Production of the sex pheromone in the gypsy moth, *Lymantria dispar*. The oenocyte cells located in the abdomen biosynthesize the alkene hydrocarbon precursor to the pheromone, 2me-Z7-18:Hc. It is transported through the hemolymph by lipophorin. The alkene is taken up by pheromone gland cells where it is acted upon by an epoxidase to produce the pheromone disparlure, 2me-18:7,8Epox.

3.6 Summary and future directions

A considerable amount of knowledge has accumulated about how pheromone components are produced in female moths since the first pathway was identified some 20 years ago. It appears that most female moths produce their pheromone through modifications of fatty acid biosynthesis pathways. For moths that utilize aldehydes, alcohols, or esters biosynthesis occurs in the pheromone gland. The exceptions are those that utilize linoleic or linolenic acids, which must be obtained from the diet. However, modifications of these fatty acids occur in the gland. For moths that utilize hydrocarbons or epoxides of hydrocarbons, the hydrocarbon is produced in oenocyte cells and then transported to the pheromone gland where the epoxidation step takes place.

The relevant enzymes found within pheromone glands are fatty acid synthesis enzymes including acetyl-CoA carboxylase and fatty acid synthetase to make primarily 18:CoA and 16:CoA. These saturated fatty acids are modified by desaturases and chain-shortening enzymes to produce the correct chain length and degree of unsaturation. There are a variety of desaturases that have been identified to introduce the double bond in the correct position along the carbon chain. Desaturation in combination with chain shortening can produce the wide variety of pheromone precursors found among moths. Modification of the carbonyl carbon by a reductase will produce an alcohol, which is used by some moths as the pheromone. The alcohol can also be oxidized to produce an aldehyde or acted upon by an acetyltransferase to produce an acetate ester (or other ester). Another major group of pheromone components are the hydrocarbons and their epoxides. The hydrocarbons are produced in oenocyte cells, transported to pheromone glands, and oxidation takes place to form a specific epoxide in the gland. Specific ratios of pheromone components produced by a moth are thought to be made by the inherent specificities of the relevant enzymes found in the biosynthetic pathways.

Future research will continue to characterize individual species pheromone biosynthetic pathways. For example, it is unclear how odd chain length oxygenated pheromone components are produced. Characterization of individual enzymes within a biosynthetic pathway will receive more attention. This characterization will undoubtedly be conducted using molecular genetic techniques as is currently being done with the desaturases. Determination of the genes encoding these enzymes will allow for their expression and characterization. Equipped with a better understanding of these complex enzyme systems, we will be able to produce stereochemically pure pheromone components on a competitive basis for expanded use in pest control.

Acknowledgements

Research conducted in the author's lab was supported in part by NSF – INT9813419, USDA-NRI – 2001-35302-10882, and State of Iowa funds.

References

Ando T., Hase T., Arima R. and Uchiyama M. (1988) Biosynthetic pathway of bombykol, the sex pheromone of the female silkworm moth. *Agric. Biol. Chem.* **52**, 473–478.

Ando T., Ohtani K., Yamamoto M., Miyamoto T., Qin X. and Witjaksono (1997) Sex pheromone of Japanese giant looper, *Ascotis selenaria cretacea*: identification and field tests. *J. Chem. Ecol.* **23**, 2413–2423.

Arn H. (2002) The Pherolist. http://www.nysaes.cornell.edu/pheronet/ or http://www-pherolist.slu.se/

Arsequell G., Fabriàs G. and Camps F. (1990) Sex pheromone biosynthesis in the processionary moth *Thaumetopoea pityocampa* by delta-13 desaturation. *Arch. Insect Biochem. Physiol.* **14**, 47–56.

Baltensweiler W. and Priesner E. (1988) A study of pheromone polymorphism in *Zeiraphera diniana* Gn. (Lep., Tortricidae) 3. Specificity of attraction to synthetic pheromone sources by different male response types from two host races. *J. Appl. Entomol.* **106**, 217–231.

Berger R. S. (1966) Isolation, identification and synthesis of the sex attractant of the cabbage looper. *Ann. Entomol. Soc. Am.* **59**, 767–771.

Bestmann H. J., Herrig M. and Attygalle A. B. (1987) Terminal acetylation in pheromone biosynthesis by *Mamestra brassicae* L. (Lepidoptera: Noctuidae). *Experientia* **43**, 1033–1034.

Bjostad L. B., Linn C. E., Du J.-W. and Roelofs W. L. (1984) Identification of new sex pheromone components in *Trichoplusia ni*, predicted from biosynthetic precursors. *J. Chem. Ecol.* **10**, 1309–1323.

Bjostad L. B., Linn C. E., Du J. W. and Roelofs W. L. (1985) Identification of new sex pheromone components in *Trichoplusia ni* and *Argyrotaenia velutinana*, predicted from biosynthetic precursors. In *Semiochemicals: Flavors and Pheromones*, eds T. E. Acree and D. M. Soderlund pp. 223–237. Walter de Gruyter & Co., Berlin.

Bjostad L. B. and Roelofs W. L. (1981) Sex pheromone biosynthesis from radiolabeled fatty acids in the redbanded leafroller moth. *J. Biol. Chem.* **256**, 7936–7940.

Bjostad L. B. and Roelofs W. L. (1983) Sex pheromone biosynthesis in *Trichoplusia ni*: key steps involve delta-11 desaturation and chain-shortening. *Science* **220**, 1387–1389.

Bjostad L. B. and Roelofs W. L. (1984) Biosynthesis of sex pheromone components and glycerolipid precursors from sodium [1-^{14}C]acetate in redbanded leafroller moth. *J. Chem. Ecol.* **10**, 681–691.

Bjostad L. B., Wolf W. A. and Roelofs W. L. (1987) Pheromone biosynthesis in lepidopterans: desaturation and chain shortening. In *Pheromone Biochemistry*, eds G. D. Prestwich and G. J. Blomquist pp. 77–120. Academic Press, Orlando, FL.

Blailock T. T., Blomquist G. J. and Jackson L. L. (1976) Biosynthesis of 2-methylalkanes in the crickets *Nemobius fasciatus* and *Gryllus pennsylvanicus*. *Biochem. Biophys. Res. Commun.* **68**, 841–849.

Bosch H. V. D., Schutgens R. B. H., Wanders R. J. A. and Tager J. M. (1992) Biochemistry of peroxisomes. *Annu. Rev. Biochem.* **61**, 157–197.

Butenandt A., Beckman R., Stamm D. and Hecker E. (1959) Uber den Sexuallockstoff des Seidenspinner *Bombyx mori*, Reidarstellung und Konstitution. *Z. Naturforsch.* **14B**, 283–284.

Charlton R. E. and Roelofs W. L. (1991) Biosynthesis of a volatile, methyl-branched hydrocarbon sex pheromone from leucine by Arctiid moths (Holomelina spp.). *Arch. Insect Biochem. Physiol.* **18**, 81–97.

Chino H. (1985) Lipid transport: biochemistry of hemolymph lipophorin. In *Comprehensive Insect Physiology, Biochemistry and Pharmacology*, Vol. 10, eds G. A. Kerkut and L. I. Gilbert, pp. 115–135. Pergamon, New York.

Choi M.-Y., Han K. S., Boo, K. S. and Jurenka R. (2002) Pheromone biosynthetic pathway in *Helicoverpa zea* and *Helicoverpa assulta*. *Insect Biochem. Mol. Biol.* **32**, 1353–1359.

Cork A., Boo K. S., Dunkelblum E., Hall D. R., Jee-Rajunga K., Kehat M., Jie E. K., Park K. C., Tepgidagarn P. and Xun L. (1992) Female sex pheromone of Oriental tobacco budworm, *Helicoverpa assulta* (Guenee) (Lepidoptera: Noctuidae): identification and field testing. *J. Chem. Ecol.* **18**, 403–418.

Diehl P. A. (1975) Synthesis and release of hydrocarbons by the oenocytes of the desert locust, *Schistocerca gregaria*. *J. Insect Physiol.* **21**, 1237–1246.

de Renobales M., Rogers L. and Kolattukudy P. E. (1980) Involvement of a thioesterase in the production of short-chain fatty acids in the uropygial gland of mallard ducks (*Anas platyrhnchos*). *Arch. Biochem. Biophys.* **205**, 464–471.

de Renobales M., Cripps C., Stanley-Samuelson D. W., Jurenka R. A. and Blomquist G. J. (1987) Biosynthesis of linoleic acid in insects. *Trends Biochem. Science* **12**, 364–366.

de Renobales M., Nelson, D. R. and Blomquist G. J. (1991) Cuticular lipids. In *Physiology of the Insect Epidermis*, eds K. Binnington and A. Retnakaran, pp. 240–251. CSIRO, Melbourne.

Fabre J. H. (1966) The insect world of J. Henri Fabre/in the translation of Alexander Teixeira de Mattos, ed. E. W. Teale p. 333. New York: Dodd, Mead.

Fang N., Teal P. E. A., Doolittle R. E. and Tumlinson J. H. (1995a) Biosynthesis of conjugated olefinic systems in the sex pheromone gland of female tobacco hornworm moths, *Manduca sexta* (L.). *Insect Biochem. Mol. Biol.* **25**, 39–48.

Fang N., Teal P. E. A. and Tumlinson J. H. (1995b) Characterization of oxidase(s) associated with the sex pheromone gland in *Manduca sexta* (L.) females. *Arch. Insect Biochem. Physiol.* **29**, 243–257.

Fang N., Teal P. E. A. and Tumlinson J. H. (1995c) Correlation between glycerolipids and pheromone aldehydes in the sex pheromone gland of female tobacco hornworm moths, *Manduca sexta* (L.). *Arch. Insect Biochem. Physiol.* **30**, 321–336.

Fang N., Teal P. E. A. and Tumlinson J. H. (1996) Effects of decapitation and PBAN injection on amounts of triacylglycerols in the sex pheromone gland of *Manduca sexta* (L.). *Arch. Insect Biochem. Physiol.* **32**, 249–260.

Foster S. P. (2000) Fatty acyl pheromone analogue-containing lipids and their roles in sex pheromone biosynthesis in the lightbrown apple moth, *Epiphyas postvittana* (Walker). *J. Insect Physiol.* **47**, 433–443.

Foster S. P. and Greenwood D. R. (1997) Change in reductase activity is responsible for senescent decline in sex pheromone titre in the lightbrown apple moth, *Epiphyas postvittana* (Walker). *J. Insect Physiol.* **43**, 1093–1100.

Foster S. P. and Roelofs W. L. (1988) Sex pheromone biosynthesis in the leafroller moth *Planotortrix excessana* by Δ10 desaturation. *Arch. Insect Biochem. Physiol.* **8**, 1–9.

Foster S. P. and Roelofs W. L. (1990) Biosynthesis of a monoene and a conjugated diene sex pheromone component of the light brown apple moth by E11-desaturation. *Experientia* **46**, 269–273.

Foster S. P. and Roelofs W. L. (1994) Regulation of pheromone production in virgin and mated females of two tortricid moths. *Arch. Insect Biochem. Physiol.* **25**, 271–285.

Foster S. P. and Roelofs W. L. (1996) Sex pheromone biosynthesis in the tortricid moth, *Ctenopseustis herana* (Felder & Rogenhofer). *Arch. Insect Biochem. Physiol.* **33**, 135–147.

Guerin P. M., Baltensweiler W., Arn H. and Buser H.-R. (1984) Host race pheromone polymorphism in the larch budmoth. *Experientia* **40**, 892–894.

Hashimoto T. (1996) Peroxisomal beta-oxidation: enzymology and molecular biology. *Ann. N. Y. Acad. Sci.* **804**, 86–98.

Haynes K. F. and Hunt R. E. (1990) A mutation in pheromonal communication system of cabbage looper moth, *Trichoplusia ni*. *J. Chem. Ecol.* **16**, 1249–1257.

Hecker E. and Butenandt A. (1984) Bombykol revisited-reflections on a pioneering period and on some of its consequences. In *Techniques in Pheromone Research*, eds H. E. Hummel and T. A. Miller, pp. 1–44. Springer-Verlag, New York.

Jacobson M. (1972) *Insect Sex Pheromones*. Academic Press, New York.

Jones I. F. and Berger R. S. (1978) Incorporation of [1-^{14}C]acetate into *cis*-7-dodecenyl acetate, a sex pheromone in the cabbage looper *Trichoplusia ni*. *Environ. Entomol.* **7**, 666–669.

Juarez P., Chase J. and Blomquist G. J. (1992) A microsomal fatty acid synthetase from the integument of Blattella germanica synthesizes methyl-branched fatty acids, precursors to hydrocarbon and contact sex pheromone. *Arch. Biochem. Biophys.* **293**, 333–341.

Jurenka R. A. (1997) Biosynthetic pathway for producing the sex pheromone component (Z,E)-9,12-tetradecadienyl acetate in moths involves a delta-12 desaturase. *Cell. Mol. Life Sci.* **53**, 501–505.

Jurenka R. A., Haynes K. F., Adlof R. O., Bengtsson M. and Roelofs W. L. (1994) Sex pheromone component ratio in the cabbage looper moth altered by a mutation affecting the fatty acid chain-shortening reactions in the pheromone biosynthetic pathway. *Insect Biochem. Mol. Biol.* **24**, 373–381.

Jurenka R. A., Jacquin E. and Roelofs W. L. (1991) Control of the pheromone biosynthetic pathway in *Helicoverpa zea* by the pheromone biosynthesis activating neuropeptide. *Arch. Insect Biochem. Physiol.* **17**, 81–91.

Jurenka R. A. and Roelofs W. L. (1989) Characterization of the acetyltransferase involved in pheromone biosynthesis in moths: specificity for the Z isomer in Tortricidae. *Insect Biochem.* **19**, 639–644.

Jurenka R. A. and Subchev M. (2000) Identification of cuticular hydrocarbons and the alkene precursor to the pheromone in hemolymph of the female gypsy moth, *Lymantria dispar*. *Arch. Insect Biochem. Physiol.* **43**, 108–115.

Kasang G. and Schneider D. (1974) Biosynthesis of the sex pheromone disparlure by olefin-epoxide conversion. *Naturwissenschaften* **61**, 130–131.

Klun J. A. and Maini S. (1982) Genetic basis of an insect communication system: the European corn borer. *Environ. Entomol.* **11**, 1084–1090.

Klun J. A., Plimmer J. R., Bierl-Leonhardt B. A., Sparks A. N., Primaiani M., Chapman O. L., Lee G. H. and Lepone G. (1980) Sex pheromone chemistry of the female corn earworm moth, *Heliothis zea*. *J. Chem. Ecol.* **6**, 165–175.

Knipple D. C., Rosenfield C., Miller S. J., Liu W., Tang J., Ma P. W. K. and Roelofs W. L. (1998). Cloning and functional expression of a cDNA encoding a pheromone gland-specific acyl-CoA Δ11-desaturase of the cabbage looper moth, *Trichoplusia ni*. *Proc. Natl. Acad. Sci. USA* **95**, 15287–15292.

Knudsen J., Clark S. and Dils R. (1976) Purification and some properties of a medium-chain acyl-thioester hydrolase from lactating-rabbit mammary gland which terminates chain elongation in fatty acid synthesis. *Biochem. J.* **160**, 683–691.

Libertini L. J. and Smith S. (1978) Purification and properties of a thioesterase from lactating rat mammary gland which modifies the product specificity of fatty acid synthetase. *J. Biol. Chem.* **253**, 1393–1401.

Lintner J. A. (1882) *West. N.Y. Hortic. Soc. Proc.* **27**, 52–66.

Liu W., Jiao H., O'Connor M. and Roelofs W. L. (2002) Moth desaturase characterized that produces both Z and E isomers of Δ11-tetradecenoic acids. *Insect Biochem. Mol. Biol.* **32**, 1489–1495.

Löfstedt C. and Bengtsson M. (1988) Sex pheromone biosynthesis of (*E,E*)-8,10-dodecadienol in codling moth *Cydia pomonella* involves *E*9 desaturation. *J. Chem. Ecol.* **14**, 903–915.

Martinez T., Fabriàs G. and Camps F. (1990) Sex pheromone biosynthetic pathway in *Spodoptera littoralis* and its activation by a neurohormone. *J. Biol. Chem.* **265**, 1381–1387.

Millar J. G. (2000) Polyene hydrocarbons and epoxides: a second major class of lepidopteran sex attractant pheromones. *Annu. Rev. Entomol.* **45**, 575–604.

Miller J. R. and Roelofs W. L. (1980) Individual variation in sex pheromone component ratios in two populations of the redbanded leafroller moth, *Argyrotaenia velutinana*. *Envir. Entomol.* **9**, 359–363.

Miyamoto T., Yamamoto M., Ono A., Ohtani K. and Ando T. (1999) Substrate specificity of the epoxidation reaction in sex pheromone biosynthesis of the Japanese giant looper (Lepidoptera: Geometridae). *Insect Biochem. Mol. Biol.* **29**, 63–69.

Morse D. and Meighen E. A. (1984) Aldehyde pheromones in Lepidoptera: evidence for an acetate ester precursor in *Choristoneura fumifernana*. *Science* **226**, 1434–1436.

Morse D. and Meighen E. A. (1987a) Biosynthesis of the acetate ester precursors of the spruce budworm sex pheromone by an acetyl CoA: fatty alcohol acetyltransferase. *Insect Biochem.* **17**, 53–59.

Morse D. and Meighen E. A. (1987b) Pheromone biosynthesis: enzymatic studies in lepidoptera. In *Pheromone Biochemistry*, eds G. D. Prestwich and G. J. Blomquist, pp. 121–158. Academic Press, Orlando, FL.

Ono A., Imai T., Inomata S.-I., Watanabe A. and Ando T. (2002) Biosynthetic pathway for production of a conjugated dienyl sex pheromone of a plusiinae moth, *Thysanoplusia intermixta. Insect Biochem. Mol. Biol.* **32**, 701–708.

Percy-Cunningham J. E. and MacDonald J. A. (1987) Biology and ultrastructure of sex pheromone-producing glands. In *Pheromone Biochemistry*, eds G. D. Prestwich and G. J. Blomquist, pp. 27–75. Academic Press, Orlando, FL.

Prestwich G. D. and Blomquist G. J. (1987) *Pheromone Biochemistry*, p. 565. Academic Press, Orlando, FL.

Pu G.-Q., Yamamoto M., Takeuchi Y., Yamazawa H. and Ando T. (1999) Resolution of epoxydienes by reversed-phase chiral HPLC and its application to stereochemistry assignment of mulberry looper sex pheromone. *J. Chem. Ecol.* **25**, 1151–1162.

Reed J. R., Vanderwel D., Choi S., Pomonis J. G., Reitz R. C. and Blomquist, G. J. (1994) Unusual mechanism of hydrocarbon formation in the housefly: cytochrome P450 converts aldehyde to the sex pheromone component (Z)-9-tricosene and CO_2. *Proc. Natl. Acad. Sci. USA* **91**, 10000–10004.

Roelofs W. L. (1984) Electroantennogram assays: rapid and convenient screening. In *Techniques in Pheromone Research*, eds H. E. Hummel and T. A. Miller, pp. 131–159. Springer-Verlag, New York.

Roelofs W. L. and Bjostad L. B. (1984) Biosynthesis of lepidopteran pheromones. *Bioorg. Chem.* **12**, 279–298.

Roelofs W. L., Cardé R. T., Benz G. and von Salis G. (1971) Sex attractant of the larch bud moth found by electroantennogram method. *Experientia* **27**, 1438.

Roelofs W. L., Du J.-W., Linn C. E., Glover T. J. and Bjostad L. B. (1986) The potential for genetic manipulation of the redbanded leafroller moth sex pheromone blend. In *Evolutionary Genetics of Invertebrate Behavior*, ed. M. D. Heuttel, pp. 263–272. Plenum Press, New York.

Roelofs W. L. and Jurenka R. A. (1996) Biosynthetic enzymes regulating ratios of sex pheromone components in female redbanded leafroller moths. *Bioorganic Med. Chem. Letters* **4**, 461–466.

Roelofs W. L. and Wolf W. A. (1988) Pheromone biosynthesis in Lepidoptera. *J. Chem. Ecology* **14**, 2019–2031.

Rosell G., Hospital S., Camps F. and Guerrero A. (1992) Inhibition of a chain shortening step in the biosynthesis of the sex pheromone of the Egyptian armyworm *Spodoptera littoralis. Insect Biochem. Mol. Biol.* **22**, 679–685.

Rule G. S. and Roelofs W. L. (1989) Biosynthesis of sex pheromone components from linolenic acid in Arctiid moths. *Arch. Insect Biochem. Physiol.* **12**, 89–97.

Ryan R., de Renobales M., Dillwith J., Heisler C. and Blomquist G. (1982) Biosynthesis of myristate in an aphid: involvement of a specific acylthioesterase. *Arch. Biochem. Biophys.* **213**, 26–36.

Schal C., Sevala V. and Cardé R. T. (1998) Novel and highly specific transport of a volatile sex pheromone by hemolymph lipophorin in moths. *Naturwissenschaften* **85**, 339–342.

Sreng I., Glover T. and Roelofs W. (1989) Canalization of the redbanded leafroller moth sex pheromone blend. *Arch. Insect Biochem. Physiol.* **10**, 73–82.

Stanley-Samuelson D. W., Jurenka R. A., Cripps C., Blomquist G. J. and deRenobales M. (1988) Fatty acids in insects: Composition, metabolism and biological significance. *Arch. Insect Biochem. Physiol.* **9**, 1–33.

Struble D. L. and Arn H. (1984) Combined gas chromatography and electroantennogram recording of insect olfactory responses. In *Techniques in Pheromone Research*, eds H. E. Hummel and T. A. Miller, pp. 160–178. Springer-Verlag, New York.

Subchev M. and Jurenka R. A. (2001) Identification of the pheromone in the hemolymph and cuticular hydrocarbons from the moth *Scoliopteryx libatrix* L. (Lepidoptera: Noctuidae). *Arch. Insect Biochem. Physiol.* **47**, 35–43.

Sugie H., Tatsuki S., Nakagaki S., Rao C. B. J. and Yamamoto A. (1991) Identification of the sex pheromone of the oriental tobacco budworm, *Heliothis assulta* (Guenee) (Lepidoptera: Noctuidae). *Applied Entomol. Zool.* **26**, 151–153.

Tang J. D., Charlton R. E., Jurenka R. A., Wolf W. A., Phelan P. L., Sreng L. and Roelofs W. L. (1989) Regulation of pheromone biosynthesis by a brain hormone in two moth species. *Proc. Natl. Acad. Sci. USA* **86**, 1806–1810.

Teal P. E. A. and Tumlinson J. H. (1987) The role of alcohols in pheromone biosynthesis by two noctuid moths that use acetate pheromone components. *Archives of Insect Biochemistry and Physiology* **4**, 261–269.

Teal P. E. A. and Tumlinson J. H. (1988) Properties of cuticular oxidases used for sex pheromone biosynthesis by *Heliothis zea*. *J. Chem. Ecol.* **14**, 2131–2145.

Tillman J. A., Seybold S. J., Jurenka R. A. and Blomquist G. J. (1999) Insect pheromones – an overview of biosynthesis and endocrine regulation. *Insect Biochem. Mol. Biol.* **29**, 481–514.

Vaz A. H., Blomquist G. J., Wakayama E. J. and Reitz R. C. (1987) Characterization of the fatty acyl elongation reactions involved in hydrocarbon biosynthesis in the housefly *Musca domestica*. *Insect Biochem.* **18**, 177–184.

Vaz A. H., Jurenka R. A., Blomquist G. J. and Reitz R. C. (1988) Tissue and chain length specificity of the fatty acyl-CoA elongation system in the American cockroach. *Arch. Biochem. Biophys.* **267**, 551–557.

Wigglesworth V. B. (1970) Structural lipids in the insect cuticle and the function of the oenocytes. *Tissue Cell.* **2**, 155–179.

Wolf W. A. and Roelofs W. L. (1983) A chain-shortening reaction in orange tortrix moth sex pheromone biosynthesis. *Insect Biochem.* **13**, 375–379.

Wolf W. A. and Roelofs W. L. (1986) Properties of the Δ11-desaturase enzyme used in cabbage looper moth sex pheromone biosynthesis. *Arch. Insect Biochem. Physiol.* **3**, 45–52.

Wolf W. A. and Roelofs W. L. (1987) Reinvestigation confirms action of Δ11-desaturase in spruce budworm moth sex pheromone biosynthesis. *J. Chem. Ecol.* **13**, 1019–1027.

Wu W. Q., Zhu J. W., Millar J. and Löfstedt C. (1998) A comparative study of sex pheromone biosynthesis in two strains of the turnip moth, *Agrotis segetum*, producing different ratios of sex pheromone components. *Insect Biochem. Mol. Biol.* **28**, 895–900.

Zhao C., Löfstedt C. and Wang X. (1990) Sex pheromone biosynthesis in the Asian corn borer *Ostrinia furnicalis* (II): biosynthesis of (*E*)- and (*Z*)-12-tetradecenyl acetate involves Δ14 desaturation. *Arch. Insect Biochem. Physiol.* **15**, 57–65.

Zhu J. W., Zhao C. H., Lu F., Bengtsson M. and Löfstedt C. (1996) Reductase specificity and the ratio regulation of E/Z isomers in pheromone biosynthesis of the European corn borer, *Ostrinia nubilalis* (Lepidoptera: Pyralidae). *Insect Biochem. Mol. Biol.* **26**, 171–176.

4

Molecular biological investigations of pheromone desaturases

Douglas C. Knipple and Wendell L. Roelofs

4.1 Introduction and historic context

Sex pheromones in the Lepidoptera are multi-component mixtures consisting mostly of olefinic compounds possessing a terminal aldehyde, alcohol, or acetate moiety. Besides functional group differences, the constituents of lepidopteran sex pheromones vary in hydrocarbon chain length and in the specific number, location, and geometry of double bonds. These chemical structures are formed in biosynthetic pathways involving a limited number of enzymatic steps believed to use fatty-acyl thioesters of coenzyme A (acyl-CoA) as substrates. Key reactions are desaturation, limited β-oxidation, and a small number of terminal functional group modifications (reviewed in Chapter 3).

A general conclusion drawn from the analyses of many moth pheromone biosynthetic pathways is that the diversity of chemical structures used as lepidopteran sex pheromones has arisen by the evolution of novel substrate specificities and catalytic mechanisms of key biosynthetic enzymes, as well as the temporal sequences in which they act. The integral membrane desaturases of lepidopteran sex pheromone biosynthetic pathways have evolved diverse positional and stereochemical specificities, resulting in unsaturated fatty acid products with *Z*9, *E*9, *Z*10, *Z*11, *E*11, *Z*12, *E*12, *Z*14, and *E*14 double bonds, as well as diverse substrate specificities, including the utilization of unsaturated substrates, resulting in multiple unsaturated fatty acids, and of substrates of intermediate (i.e. 12- and 14-carbon) chain length. Thus, the evolution of novel functional properties among the lepidopteran pheromone integral membrane desaturases has contributed substantially to the number of unique unsaturated precursor compounds that are used in the biosynthesis of lepidopteran sex pheromones.

The molecular biological investigations of lepidopteran pheromone desaturases during the last half decade were built upon the conceptual foundations established by biochemical investigations of animal desaturases during the late 1960s and 1970s (Jeffcoat, 1979). Integral membrane desaturases in animals were shown to be the terminal component of the electron transport system of the endoplasmic reticulum that utilizes cytoplasmic NADH and molecular oxygen to effect Δ9 desaturation of CoA esters of long chain fatty acids. The first acyl-CoA desaturase described was a stearoyl-CoA desaturase purified from rat liver (Strittmatter *et al.*, 1974). This early biochemical research showed that NADH-cytochrome b_5 reductase (a flavoprotein) and cytochrome b_5 (a hemoprotein) are integral functional components of a membrane-associated desaturase complex. These studies also indicated that all three enzymatic components of the active desaturase complex are oriented with their catalytic domains on the cytoplasmic side of the endoplasmic reticulum, and that the desaturase component is buried in the microsomal membrane, with only a small portion that contains the catalytic center exposed to the cytoplasm.

In the 1980s, the molecular cloning and sequencing of cDNAs encoding integral membrane desaturases from rat liver (Thiede *et al.*, 1986; Strittmatter *et al.*, 1988) and mouse adipose tissue (Ntambi *et al.*, 1988; Kaestner *et al.*, 1989) revealed the first primary sequences of animal acyl-CoA desaturases, which were highly conserved and composed of more than 60 percent hydrophobic residues, consistent with prior biochemical studies. The subsequent genetically based isolation of the integral membrane Δ9 desaturase-encoding gene, *OLE1*, of the yeast *Saccharomyces cerevisiae* (Stukey *et al.*, 1989, 1990) permitted the comparison of the amino acid sequence of its single long open reading frame with those of the vertebrate acyl-CoA desaturases. Comparisons of the hydropathy plots of the full length rat and yeast desaturases revealed the conservation of four hydrophobic α helices, which are postulated transmembrane domains. The sequence alignment revealed an internal region of 257 amino acids having 36 percent sequence identity and 60 percent sequence similarity (taking into account conservative amino acid substitutions) between the *OLE1* desaturase and a rat liver desaturase (Stukey *et al.*, 1990). This "core" region contains three highly conserved (>70 percent identity) histidine-rich sequence motifs, each with at least ten identical residues. It was postulated that these "histidine boxes" map onto the cytosolic side of the endoplamic reticulum where they participate in iron binding at the catalytic site of the desaturase enzyme.

In another experiment of particular significance it was shown that the yeast *OLE1* desaturase can be functionally replaced *in vivo* by the rat desaturase (Stukey *et al.*, 1990). In this investigation, a desaturase-deficient *ole1* mutant was relieved of its unsaturated fatty acid auxotrophy by transformation with a cDNA encoding a rat liver acyl-CoA desaturase under the control of the yeast *OLE1* promoter. This genetic complementation and the direct confirmation of the production of palmitoleic and oleic acid, in the transformed strain provided compelling evidence for the formation *in vivo* of a functional desaturase complex consisting of the rat Δ9 desaturase and the endogenous yeast cytochrome *b5* and cytochrome *b5* reductase. Subsequent site-directed mutagenesis of the cDNA encoding the rat desaturase demonstrated that mutation of any histidine residue in the three histidine boxes abolished genetic complementation of the *ole1* nutritional defect, a finding consistent with the essential role of these elements in catalytic function (Shanklin *et al.*, 1994). In the same period that the first primary structures of animal and fungal desaturases were being elucidated, investigations of the biochemical properties of Δ11 desaturases present in lepidopteran pheromone glands were being conducted (Wolf and Roelofs, 1986; Rodriguez *et al.*, 1992). These studies showed that the pheromone gland Δ11 desaturases of two different species of moths share a number of biochemical properties with the ubiquitous metabolic integral membrane acyl-CoA Δ9 desaturases of animal cells, including localization in the microsomal membrane fraction, utilization of CoA thioesters of 16- and 18-carbon saturated fatty acids as substrates, inhibition by cyanide, lack of sensitivity to carbon monoxide, and the use of reduced nicotine-adenine dinucleotide cofactor (NADH) as an electron source. The latter investigations provided strong evidence for the homology of pheromone desaturases with the integral membrane desaturases of fungal and vertebrate cells, which had already been characterized at the level of their primary amino acid sequences.

The latter connection provided the conceptual and technical underpinnings of the research on pheromone desaturase structure, function, and evolution that was to ensue during the late 1990s through the present. The following four hypotheses follow directly from the predicted homology of metabolic integral membrane acyl-CoA Δ9 desaturases and lepidopteran pheromone desaturases. First is that some structural features are conserved between metabolic integral membrane Δ9 desaturases and pheromone desaturases, which can provide a technical entry point for the isolation of nucleic acids encoding the latter. A second hypothesis predicts the functional association of the two yeast electron transport proteins with lepidopteran desaturases, which can provide a means for characterizing the functional properties of the latter enzymes. A third hypothesis is that the multiple desaturase activities present in the pheromone glands of many lepidopteran species reflect the existence of an underlying gene family consisting of discrete lineages encoding desaturases with unique functional roles and enzymatic properties. A fourth, and final, hypothesis is that the diverse desaturase catalytic specificities that have been documented in pheromone biosynthetic pathways correspond to specific mutational changes that can be mapped by molecular genetic methods onto their primary amino acid sequences and discrete structural elements. In the remainder of this chapter, we describe specific investigations that address these hypotheses and the picture that is emerging from the exploration of pheromone desaturases.

4.2 Investigations of lepidopteran desaturase-encoding cDNAs

4.2.1 Molecular cloning of a Δ11 desaturase cDNA from *Trichoplusia ni* pheromone gland

In the first effort to clone a lepidopteran pheromone desaturase cDNA, the cabbage looper moth, *T. ni*, was selected as a source of mRNA because research on its pheromone biosynthetic pathways showed that all pheromone components could be produced by Z11-desaturation (Bjostad and Roelofs, 1983; Bjostad *et al.*, 1984, Jurenka *et al.*, 1994) and because developmental progression to functional competency of the pheromone gland, a relatively large eversible sac at the tip of the abdomen, was well characterized (Tang *et al.*, 1991). To isolate pheromone desaturase-encoding cDNA sequences, the pheromone gland RNA was first reverse transcribed to make cDNA. This served as a template in PCR reactions employing degenerate primers designed to amplify cDNA sequences encoding a central region of the desaturase delimited by two of the conserved histidine boxes (HRLWSH and GF/YHNY/FH). The latter sequence motifs are implicated in iron binding and catalytic function (Shanklin *et al.*, 1994), as noted above. In designing these primers, deduced codon bias rules were applied in order to reduce the sequence complexity of the degenerate primers encoding these conserved motifs. To compensate for the lack of completeness of these primers, the annealing phases of the PCR reactions were performed at low temperature and the total number of reaction cycles was increased to 55.

The above homology probing method yielded a 560 base pair product, which was sequenced and found to have a continuous open reading frame encoding an amino acid sequence with 55 percent identity and 72 percent similarity to the central region of the rat stearoyl-CoA Δ9 desaturase and 34 percent identity and 58 percent similarity to the central region of the yeast OLE1 Δ9 desaturase. The amplified cDNA fragment was then labeled and used to probe two independently prepared pheromone gland cDNA libraries. Between 1 and 2 percent of the phage plaques were found to be probe-positive – a remarkably high percentage – and several were isolated and sequenced. Although four sequence polymorphisms were found among the full length cDNAs, the consensus sequence, designated PDesat-TnΔ^{11}Z, contained an open reading frame of 1047 nucleotides encoding a 349 amino acid protein with a predicted molecular mass of 40 240 Da and a pI of 9.12 (Knipple *et al.*, 1998). Alignment of the encoded amino acid sequence with those of the Δ9 desaturases of rat and yeast showed that the highest level of sequence conservation occurred in the region corresponding to the primary 560-

bp PCR product delimited by amino acid positions 86 and 267 of the *T. ni* sequence. Interestingly, one of the amino acids of the upstream histidine box that was part of the target sequence used to design the 5′ degenerate primer differed from the consensus sequence derived from rat and yeast desaturases, with threonine substituted for serine at amino acid position 90. Thus, amplification of the 560 base pair product was the result of the extremely low stringency conditions used in the PCR reaction, allowing mismatch priming to occur.

4.2.2 Functional expression of the *T. ni* Δ11 desaturase in yeast

The next major step was the development of an assay to determine the functional identity of the deduced PDesat-TnΔ11Z desaturase. Initial efforts to develop an *in vitro* reconstitutive biochemical assay combined the recombinant PDesat-TnΔ11Z protein purified from *E. coli* with phospholipids and a biochemical fraction from beef liver containing cytochrome b_5 reductase and cytochrome b_5. These efforts were unsuccessful (Knipple and Roelofs, unpublished).

Subsequently, an assay based on functional expression in the yeast *S. cerevisiae* was tried. Two experimental findings led to the prediction that the unsaturated fatty acid auxotrophy of the desaturase-deficient *ole1* mutant would be relieved by the functional expression of the *T. ni* desaturase, which was predicted to have a regiospecificity for the Δ11 position. The first was the demonstration that the expression of a rat stearoyl-CoA Δ9 desaturase genetically complements the *ole1* unsaturated fatty acid auxotrophy, from which conservation of the functional interactions between the rat desaturases and the yeast electron transport protein components of the desaturase complex was inferred (Stukey *et al.*, 1990). The second was that supplementation of the growth medium with (Z)-11-octadecenoic acid (Z11-18:Acid) also permits growth of the *ole1* mutant in the absence of other unsaturated fatty acids (Knipple and Marsella-Herrick, unpublished). To test this prediction, open reading frames encoding the *T. ni* desaturase sequence variants noted above were inserted into the YEpOLEX expression vector (Knipple *et al.*, 1998) and the plasmids were used to transform the *ole1* mutant L8-14C strain (Stukey *et al.*, 1990). Several clones complemented the host strain's unsaturated fatty acid auxotrophy, whereas several others failed to grow without supplementation. Interestingly, the only consistent amino acid sequence difference between clones that exhibited desaturase activity and those that did not was the presence of lysine at residue 315 in place of glutamic acid (Knipple and Rosenfield, unpublished).

To identify the unsaturated fatty acids produced in the transformants that were able to grow without supplementation with palmitoleic or oleic acid, cells were grown at 30° C in YPD liquid medium to a density of 2×10^7 cells/ml, at which point the cells were pelleted and washed three times with water, and then extracted with methanol/chloroform (1:2). After solvent evaporation, the lipid residue was treated with 0.5 M KOH/methanol, and the resulting fatty acid methyl esters were analyzed by GC/MS. The positions of double bonds in unsaturated fatty acids were confirmed by analysis of the fragmentation products of their dimethyl disulfide adducts (Buser *et al.*, 1983) (Figure 4.1). The analysis unequivocally showed the presence of *Z*11-16 and *Z*11-18 fatty acid methyl esters in approximately the same ratio of the unsaturated fatty acid precursors and products detected in the *T. ni* pheromone gland (Bjostad and Roelofs, 1983; Bjostad *et al.*, 1984; Jurenka *et al.*, 1994).

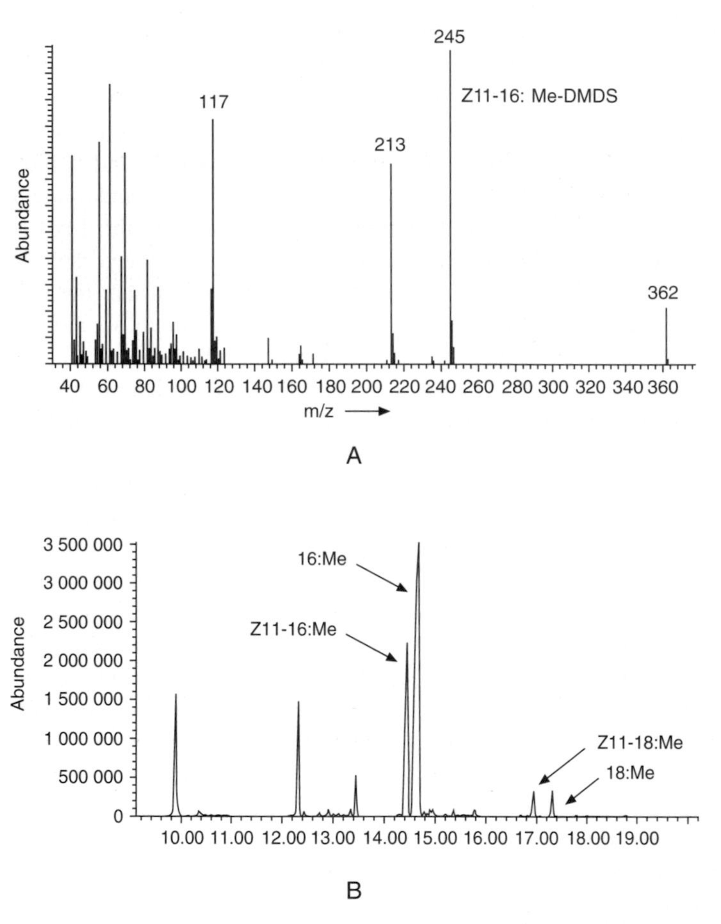

Figure 4.1 GC/MS analysis of methyl esters prepared from a whole cell lipid extract of the YEpOLEX-PDesat-TnD11Z-transformed *ole1* strain of *Saccharomyces cerevisiae*: (A) total ion spectrum of fatty acid methyl esters resolved by capillary GLC; (B) mass spectrum of the degradation products of the DMDS adduct of Z11-16:Me in A. The diagnostic *m/z* values of the DMDS adduct of Z11-16:Me are labeled. (Reproduced with permission from Knipple *et al.*, 1998. © 1998 by The National Academy of Sciences.)

This example illustrates two of the attractive features of using *S. cerevisiae* for functional expression analysis of desaturase-encoding cDNAs. The first is that, since almost all of the endogenous unsaturated fatty acids in wildtype (i.e. *OLE1*) yeast are palmitoleic and oleic acids produced by the OLE1 desaturase, functional expression of complementing cDNAs in the *ole1* null genetic background facilitates the detection of unsaturated fatty acid products of the expressed desaturases by GC analysis. The second is that if the lipids are extracted from the yeast cells during the growth phase when saturated precursors are not rate-limiting, the ratio of the unsaturated fatty acid products detected in the analysis reflects the substrate specificity that is intrinsic to the expressed desaturase (Rosenfield *et al.*, 2001).

4.2.3 Molecular cloning and functional expression of a Δ9 desaturase-encoding cDNA from *T. ni* fat body

A primary functional role of the ubiquitous Δ9 desaturase of animal and fungal cells is to regulate cell membrane fluidity by increasing the amount of unsaturated fatty acids within membranes in response to cold (Tiku *et al.*, 1996; Vigh *et. al.*, 1993), a primitive and highly conserved adaptive process known as homeoviscous adaptation (Hazel and Williams, 1990). Since moths produce palmitoleic and oleic acids that are incorporated into their cell membranes, the existence of Δ9 desaturases that are orthologous to the Δ9 desaturases of animal and fungal cells was predicted. To address this issue, the homology probing method described above was used to isolate a cDNA fragment from *T. ni* fat body mRNA. The amplification product contained an open reading frame with similar levels of sequence identity (63–64 percent) to the core domains of the Δ9 desaturases of rat and the *T. ni* Δ11 desaturase. A full length cDNA, designated TnFBΔ9Ds, encoding a 353 amino acid protein was subsequently obtained by using the RACE (Rapid Amplification of cDNA Ends) procedure (Liu *et al.*, 1999).

An expression construct consisting of the open reading frame of the TnFBΔ9Ds cDNA inserted into the yeast desaturase expression vector YEpOLEX was used to transform the *ole1* strain of *S. cerevisiae* as described above. Many transformant colonies were obtained on medium lacking unsaturated fatty acids, indicating complementation of the *ole1* mutation by the encoded *T. ni* desaturase. GC/MS analysis of the fatty acid methyl esters obtained from the transformants showed that the TnFBΔ9Ds cDNA encoded a Δ9 desaturase that produced oleic acid (Z9-18:Acid) and palmitoleic acid (Z9-16:Acid) (Liu *et al.*, 1999). Quantitation of these unsaturated fatty acids under standard conditions as described above revealed about three times more of the former than the latter (Rosenfield *et al.*, 2001).

Northern blot and RT-PCR analyses showed that the Δ9 desaturase-encoding RNA was present in *T. ni* adult and larval fat body, as well as in pheromone-gland and muscle tissue. Interestingly, no Δ9 desaturase-encoding cDNA was isolated in the effort to obtain desaturase-encoding cDNAs from pheromone glands. The widespread tissue distribution of the Δ9 desaturase RNA suggests that the encoded enzyme probably functions in cold adaptation. The demonstration of two desaturases in *T. ni* with divergent amino acid sequences, different tissue distributions, and distinct enzymatic properties provided the first direct evidence of a desaturase gene duplication in insects and subsequent divergence of functional roles and biochemical properties of the encoded enzymes.

4.2.4 Isolation of multiple desaturase-encoding cDNAs from pheromone glands of *Helicoverpa zea*

The sex pheromones of many moths are blends of constituents derived from unsaturated fatty acid precursors, which, in some instances, are known to reflect the activities of discrete desaturases, as exemplified by *H. zea*, which has pheromone precursors produced by both Δ11 and Δ9 desaturases (Klun *et al.*, 1980; Pope *et al.*, 1984; Teal and Tumlinson, 1986; Jurenka *et al.*, 1991). In a study of desaturase-encoding sequences in the pheromone gland of this species, the homology probing procedure that was used to isolate desaturase-encoding cDNAs of *T. ni* was modified to enhance the recovery of multiple desaturase-encoding cDNAs from pheromone gland RNA (Table 4.1). This resulted in the isolation of three full-length cDNAs and a fourth cDNA fragment encoding distinct amino acid sequences with homology to other previously characterized desaturases (Rosenfield *et al.*, 2001).

The full-length cDNA of one of these encoded a 338 amino acid protein, designated HzPGDs1, with 68 percent sequence identity to the *T. ni* Δ11 desaturase (Knipple *et al.*, 1998) and 52 percent identity to the *T. ni* Δ9 desaturase (Liu *et al.*, 1999). Quantitation of HzPGDs1 RNA levels in pheromone glands of two-day old adult females and in fat bodies of male and female larvae showed that the HzPGDs1 RNA was differentially expressed, with more than 10 000-fold greater abundance in pheromone glands than in larval fat bodies of both sexes. The HzPGDs1 RNA was between five- and ten-fold more abundant in the pheromone gland than two other RNAs encoding Δ9 desaturases (described below), consistent with the role of HzPGDs1 as the desaturase that catalyzes formation of the Z11-16:Acid precursor of the major pheromone component of this species. Yeast *ole1* cells that were transformed with the YEpOLEX plasmid containing the coding region of the HzPGDs1 cDNA were able to grow without supplementation of the growth medium with unsaturated fatty acids. GC analysis of total lipid extracts of transformants showed the production of only Z11-16:Acid (Figure 4.2), which was confirmed by GC/MS analysis of dimethyl disulfide derivatives (not shown). These results are consistent with previous studies of *H. zea* pheromone biosynthetic pathways, which showed abundant Z11-16:Acid pheromone precursor and the absence of Z11-18:Acid (Klun *et al.*, 1980; Pope *et al.*, 1984; Teal and Tumlinson, 1986; Jurenka *et al.*, 1991). The

Table 4.1 Conserved amino acid sequence motifs and encoding nucleotide sequences used to design PCR primers for isolating insect desaturase-encoding cDNAs

Abbreviations: N: A,G,C, or T; K: G or T; R: A or G; Y: C or T.

5′-Primers:

		G A H R* L* W S* (S* = A, T)		Permutations of amino acids R* L* and S*
GAHR-1	5′	GNGCNCAYCGNCTNTGGGC	3′	R1L1A
GAHR-2		GNGCNCAYAGRCTNTGGGC		R2L1A
GAHR-3		GNGCNCAYCGNCTNTGGAC		R1L1T
GAHR-4		GNGCNCAYAGRCTNTGGAC		R2L1T
GAHR-5		GNGCNCAYCGNCTNTGGTC		R1L1S1
GAHR-6		GNGCNCAYAGRCTNTGGTC		R2L1S1
GAHR-7		GNGCNCAYCGNCTNTGGAG		R1L1S2
GAHR-8		GNGCNCAYAGRCTNTGGAG		R2L1S2
GAHR-9		GNGCNCAYCGNTTRTGGGC		R1L2A
GAHR-10		GNGCNCAYAGRTTRTGGGC		R2L2A
GAHR-11		GNGCNCAYCGNTTRTGGAC		R1L2T
GAHR-12		GNGCNCAYAGRTTRTGGAC		R2L2T
GAHR-13		GNGCNCAYCGNTTRTGGTC		R1L2S1
GAHR-14		GNGCNCAYAGRTTRTGGAG		R2L2S2
GAHR-15		GNGCNCAYCGNTTRTGGTC		R1L2S1
GAHR-16		GNGCNCAYAGRTTRTGGAG		R2L2S2

3′-Primer:

		E G F H N Y H	
EGFH	5′	GARGGNTWYCAYAAYTWYCAYC	3′

(Reproduced with permission from Rosenfield *et al.*, 2001. Copyright 2001 by *Insect Biochemistry and Molecular Biology*.)

substrate chain length specificity of the HzPGDs1 desaturase also suggests that the unsaturated precursor of the essential, but minor, *Z*9-16:Ald pheromone component is produced in the pheromone gland by the direct action of a Δ9 desaturase acting on 16:CoA rather than by a Δ11 desaturase acting on 18:CoA and subsequent β-oxidation. However, studies of the biosynthetic pathway with labeled precursors indicate that the latter route to the formation of *Z*9-16:Ald is apparently operational *in vivo* (Choi *et al.*, 2002), as discussed in greater detail below.

The two additional full-length desaturase-encoding cDNAs obtained from the *H. zea* pheromone gland were designated HzPGDs2 and HzFBDs (Rosenfield *et al.*, 2001). The HzPGDs2 cDNA encoded a 353 amino acid protein that had only 64 percent identity to the *T. ni* Δ9 desaturase. Its corresponding RNA was

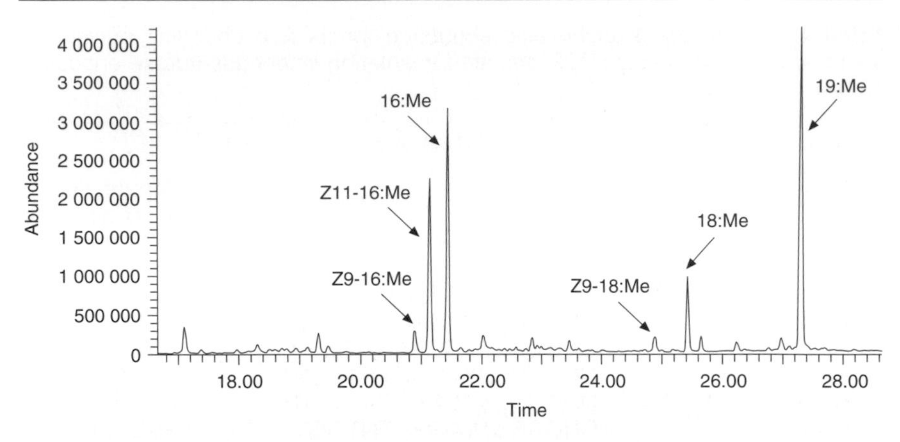

Figure 4.2 GC/MS analysis of methyl esters prepared from a whole cell lipid extract of the YEpOLEX-HzPGDs1-transformed *ole1* strain of *S. cerevisiae*: total ion spectrum of fatty acid methyl esters resolved by capillary GLC. The culture was grown in liquid complete medium (YPD) at 30°C to a density of 2 $\times 10^7$ cells/ml prior to extraction and methylation. The external mass standard nonadecanoic acid methyl ester (19:Me) was added to the cell pellet prior to extraction. Peaks corresponding to Z9-16:Me and Z9-18:Me reflect trace amounts of unsaturated fatty acids in the growth medium. (Reproduced with permission from Rosenfield *et al.*, 2001. © 2001 by *Insect Biochemistry and Molecular Biology*.)

the second most abundant desaturase-encoding RNA in the *H. zea* pheromone gland, occurring at about one-fifth the level of the HzPGDs1 RNA. The HzPGDs2 RNA was 20-fold more abundant in pheromone gland RNA than in RNA from larval fat bodies of both sexes. In the yeast *ole1* functional expression assay, HzPGDs2 was shown to be a Δ9 desaturase that produced about twice as much *Z*9-16:Acid as *Z*9-18:Acid (Figure 4.3). The differential expression in the pheromone gland of the HzPGDs2 RNA and the substrate preference of its encoded enzyme suggests that this desaturase probably plays a role in pheromone biosynthesis and could be responsible for the production of the unsaturated precursor acids of the *Z*9-16:Ald and *Z*7-16:Ald pheromone components, the latter deriving from β-oxidation of *Z*9-18:Acid.

The HzFBDs cDNA isolated from *H. zea* pheromone glands encoded a 354 amino acid protein that is an apparent orthologue of the *T. ni* Δ9 desaturase, with 93 percent sequence identity (Rosenfield *et al.*, 2001). In the yeast functional expression assay, transformants expressing HzFBDs produced a ratio of unsaturated fatty acids similar to that of the *T. ni* Δ9 desaturase, i.e. *Z*9-18:Acid>*Z*9-16:Acid (Rosenfield, You, and Knipple, unpublished). The HzFBDs RNA was moderately abundant in pheromone gland, occurring at about one-eighth the level of the HzPGDs1 RNA and it had roughly equivalent levels in other tissues, similar to

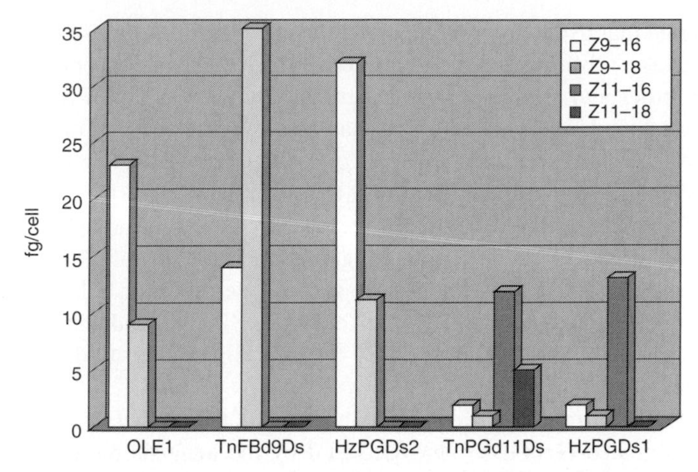

Figure 4.3 Mass quantities of unsaturated fatty acids extracted from *ole1* strains of *S. cerevisiae* transformed with plasmids encoding the wildtype yeast desaturase (OLE1), two moth Δ9 desaturases (TnFBd9Ds and HzPGDs2), and two moth Δ11 desaturases (TnPGd11Ds and HzPGDs1). Growth conditions, extraction procedures, and quantitation were standardized as in Figure 4.2. (Reproduced with permission from Rosenfield *et al.*, 2001. © 2001 by *Insect Biochemistry and Molecular Biology*.)

the broad expression patterns observed for the *T. ni* Δ9 desaturase. The presence of the HzFBDs in the pheromone gland and its demonstrated formation of Δ9 unsaturated fatty acids in yeast suggest that this desaturase also could contribute to the formation of precursors of minor pheromone components. However, the broad tissue distribution of its encoding RNA suggests that this desaturase likely plays a more significant role in homeoviscous adaptation and that the presence of its RNA in the pheromone gland may be largely incidental. The observation that the combined levels of Δ9 desaturase-encoding HzPGDs2 and HzFBDs RNAs are about one-third the level of the Δ11 desaturase-encoding HzPGDs1 RNA in *H. zea* pheromone glands, although pheromone components known or postulated to derive from Δ9 desaturation constitute only a small percentage of the total pheromone blend, suggests that the amount of Δ9 desaturase activity in the gland is regulated at the translational or post-translational levels.

A recent study of the pheromone biosynthetic pathways of *H. zea* and *H. assulta* (Choi *et al.*, 2002) presents some data that are at seeming odds with the above conclusions about the exclusive use of palmitic acid as a substrate of HzPGDs1 and the probable formation of Z9-16:Ald by HzPGDs2. In particular, the latter study reported the detection in *H. zea* pheromone glands of Z11-18:Acid by single ion monitoring GC/MS analysis of DMDS derivatives of total lipid extracts of pheromone glands. Although this finding is inconsistent with the properties of the HzPGDs1 desaturase revealed in the yeast expression system, it is proposed that in the pheromone gland the HzPGDs1 desaturase is able to use stearic acid as a substrate under conditions in which the available pool of palmitic acid is limiting and stearic acid is relatively abundant (Choi *et al.*, 2002). In deuterium-labeling experiments in the same study, the recovery of label in *Z*9-16:Acid from D3-18:Acid applied to *H. zea* pheromone glands was interpreted to suggest that the predominant pathway to this precursor is Δ11 desaturation of 18:Acid followed by β-oxidation. However, the conversion of D3-18:Acid to D3-16:Acid was also shown, indicating that the labeling of *Z*9-16:Acid can result from chain shortening and subsequent Δ9 desaturation. Thus, while this study supports the formation of *Z*9-16:Acid from *Z*11-18:Acid and β-oxidation, it does not preclude the formation of *Z*9-16:Acid directly by Δ9 desaturation of palmitic acid.

A fourth desaturase-homologous sequence isolated from *H. zea* pheromone glands, designated HzPGDs3, was deduced by the analysis of several distinct amplification products of differing lengths corresponding to incompletely processed RNAs. The consensus coding sequence of these fragments was found to be interrupted by two introns and encoded an amino acid sequence that was quite divergent (i.e. 50–58 percent identities) from the core regions of other characterized insect desaturases. It was not possible to obtain full-length cDNAs of this sequence by conventional RACE procedures because of the extremely low abundance of its RNA in pheromone glands, estimated to be six orders of magnitude less than the level of the HzPGDs1 RNA. Such low levels and the prevalence of unspliced transcripts suggest that this sequence is unlikely to code for a functional protein.

The above investigation of desaturase-encoding sequences present in the *H. zea* pheromone gland establishes the functional identities of three desaturases possessing catalytic specificities that, taken together with chain-shortening reactions, can account for the formation of all of the unsaturated fatty acid precursor products predicted by earlier biochemical studies. That the intrinsic properties of these desaturases identified in the yeast functional expression system are not perfectly coincident with the results of a recent *in vivo* study of pheromone biosynthetic precursor acids points to the complex interplay in the pheromone gland of multiple levels of regulation of gene expression, the flux of specific precursor acids, and apparently redundant pathways for minor components. Taken together, these two investigations illustrate the usefulness of combining *in vivo* studies with functional analyses of cloned cDNA sequences and quantitative studies of the steady state levels of the corresponding RNAs in different tissues in order to elucidate the possible functional roles of individual sequences.

4.2.5 Molecular cloning and functional expression of desaturases from a primitive leafroller moth, *Planotortrix octo*, from New Zealand

The greenheaded leafroller moth, *P. octo*, utilizes a Δ10 desaturase in its pheromone biosynthetic pathway (Foster and Roelofs, 1988). A cDNA, designated Pocto-

Z10, was isolated from pheromone glands of this moth by the homology probing method and found to encode a 356 amino acid protein with greatest sequence identity (57–61 percent) to previously described Δ11 desaturases (Hao *et al.*, 2002). A YEpOLEX plasmid containing the open reading frame of this cDNA failed to complement the auxotrophy of the *ole1* mutant. However, when transformants were supplemented with Z11-18:Me in order to obtain satisfactory growth, GC/MS analysis of the total lipid extracts showed that the expressed desaturase produced mainly Z10-16:Acid and a small amount of Z10-14:Acid. Mass spectral analysis of the DMDS adducts showed the fragment ions of *m/z* 231 and 199, which are diagnostic for a Δ10 double bond (Figure 4.4).

A second desaturase-encoding cDNA was isolated from this species and found to encode a 352 amino acid protein that is orthologous to the *H. zea* Δ9 desaturase HzPGDs2 based on its 83 percent sequence identity to the latter vs only 63–64 percent identity to the TnFBDs and HZFBDs desaturases. When this cDNA, designated Pocto-Z9, was functionally expressed in the *ole1* yeast mutant it

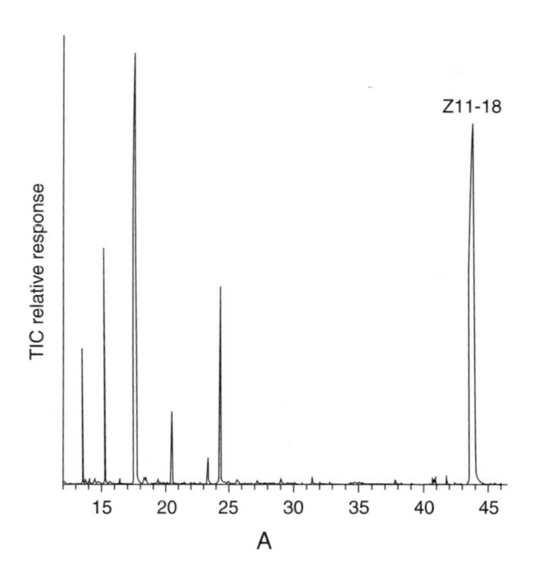

Figure 4.4 GC/MS analysis of DMDS adducts of methyl esters prepared from whole cell lipid extracts of the YEpOLEX-PoctoZ10-transformed *ole1* strain of *S. cerevisiae*: (A) control *ole1* yeast cells supplemented with 0.5 mM Z11-18:Acid; (B) YEpOLEX-PoctoZ10-transformed *ole1* yeast cells supplemented with 0.05 mM Z11-18:Acid; (C) mass spectrum of DMDS adduct of Z10-16:Me in B. GC peaks prior to 25 minutes in A and B correspond to saturated C10-C18 methyl esters. Peaks corresponding to DMDS adducts are found at 44 min for externally added Z11-18:Me and at 38 min in B for Z10-16:Me. (Reproduced with permission from Hao *et al.*, 2002. © 2002 by *Insect Biochemistry and Molecular Biology*.)

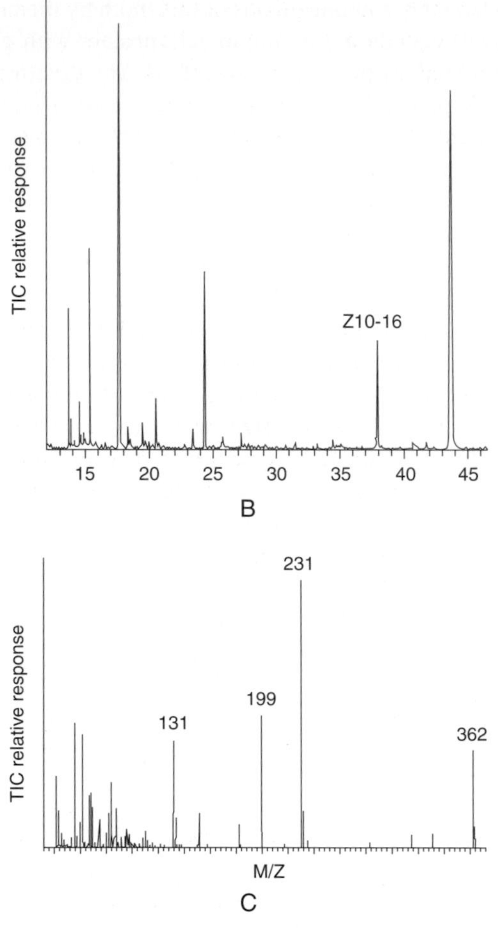

Fig. 4.4 *(Contd)*

relieved the strain's unsaturated fatty acid auxotrophy and produced more palmitoleic acid than oleic acid, like the HzPGDs2 desaturase (Hao *et al.*, 2002).

4.2.6 Characterization of a pheromone desaturase from *Epiphyas postvittana* that produces *E*11 isomers of mono- and diunsaturated fatty acids

Although *E* double bonds are uncommon in unsaturated fatty acids of animals, many have been found among the unsaturated components of lepidopteran pheromones. Early investigations of pheromone biosynthetic pathways showed that *E* isomers are not produced by isomerization of the *Z* double bonds, but rather are produced directly by the catalytic activity of desaturases possessing this stereospecificity (Bjostad and Roelofs, 1981). Biochemical studies of the pheromone biosynthetic pathway of the light brown apple moth, *Epiphyas postvittana*, showed that this species uses such an *E* desaturase in the production of its unsaturated precursor acids (Foster and Roelofs, 1990).

A cDNA designated LBAM-PGE11 was isolated from *E. postvittana* pheromone glands and found to encode a 332 amino acid protein sequence that was the most similar to the Δ11 (*Z*) desaturases of *T. ni* and *H. zea*, with 61–62 percent identity (Liu *et al.*, 2002). This cDNA failed to complement the *ole1* defect. This result suggested that the cDNA could encode an *E* desaturase, since *E* unsaturated fatty acids cannot relieve the unsaturated fatty acid auxotrophy of this strain. The use of a baculovirus/insect cell expression system permitted the functional identity of this cDNA to be determined. GC/MS analysis showed that the encoded desaturase produced both *E*11-14 and *E*11-16 acids. Furthermore, *E*9,*E*11-14:Acid was produced if the cells were provided with the monounsaturated precursor, *E*9-14:Acid (Figure 4.5). The latter compound is produced in the pheromone gland by β-oxidation of the *E*11-16:Acid intermediate (Foster and Roelofs, 1990). Thus, a single desaturase produces all of the *E*-unsaturated fatty acid intermediates identified in the pheromone gland of this species.

Another interesting side observation in the course of this study was that an endogenous Δ9 desaturase of the lepidopteran cell line Sf21, which is derived from *Spodoptera frugiperda*, can also produce conjugated diene products, identified as *Z*9,*E*11-14:Acid and *Z*9,*E*11-16:Acid, when provided with the *E*11-14:Acid and *E*11-16:Acid precursors, respectively. It is of interest to note that the genus *Spodoptera* contains several species that use *Z*9,*E*11-14:Acid as a precursor of the major pheromone component, e.g. *S. littoralis* and *S. litura*. However, the pheromone components of *S. frugiperda* are derived exclusively from monoenes. It is conceivable that a specific Δ9 desaturase in Sf21 cells that catalyzes formation of *Z*9,*E*11-14:Acid is encoded in the *S. frugiperda* genome but is not expressed in that species' pheromone gland, or it may be present in the pheromone gland in the absence of significant amounts of *E*11-14:Acid substrate. Alternatively, the experimental observation could reflect a more universal property of Δ9 desaturases, a lack of specificity resulting from the inability of the hydrophobic substrate-binding pocket of the enzyme to discriminate between saturated fatty acids and unsaturated fatty acids with the *E* double bond conformation, which more closely resembles the extended form of saturated fatty acids than *Z* unsaturated fatty acids.

4.2.7 Characterization of a pheromone desaturase from *Argyrotaenia velutinana* that produces a mixture of *Z*11-14 and *E*11-14 acids

*Z*11-14:Acid and *E*11-14:Acid are the only unsaturated pheromone intermediates of the redbanded leafroller moth, *Argyrotaenia velutinana* (Bjostad and Roelofs, 1981). Stable isotope experiments showed that the mixture of these unsaturated

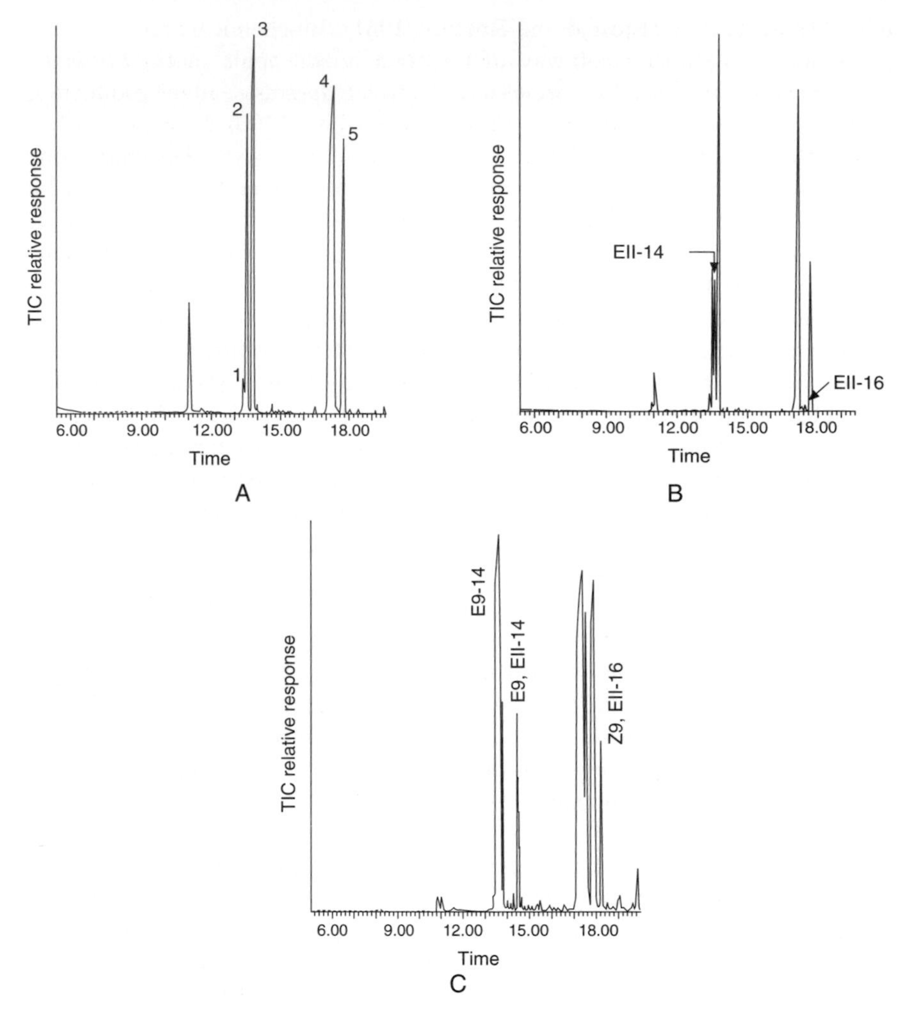

Figure 4.5 GC/MS analysis of methyl esters prepared from whole cell lipid extracts of Sf21 cells transfected with baculovirus vectors encoding the LBAM-PGE11 desaturase: (A) Control Sf21 cells transfected with baculovirus vector-produced acids Z7-14, Z9-14, C14, Z9-16, and C16 corresponding to peaks 1–5, respectively; (B) Sf21 cells transfected with baculovirus encoding the LBAM-PGE11 desaturase exhibit additional peaks corresponding to E11-14 and E11-16 acids (desaturation products catalyzed by the LBAM-PGE11 desaturase); (C) Sf21 cells transfected with baculovirus encoding the LBAM-PGE11 desaturase supplemented with E9-14:Acid exhibit peaks corresponding to E9,E11-14:Acid (desaturation product catalyzed by the LBAM-PGE11 desaturase), in addition to peaks for E11-16:Acid (chain elongation product), and Z9,E11-16:Acid (desaturation product catalyzed by the Z9 desaturase of the Sf21 cells). (Reproduced with permission from Liu *et al.*, 2002. © 1998 by The National Academy of Sciences.)

intermediates is not produced by isomerization, but rather by the direct action of one or more desaturases on myristic acid (14:Acid) (Bjostad and Roelofs, 1986).

A cDNA isolated from *A. velutinana* pheromone glands was found to encode a 330 amino acid protein sequence (Liu *et al.*, 2002) that was the most similar to Δ11 desaturases of *T. ni*, *H. zea*, and *E. postvittana*, with 58 percent, 67 percent, and 75 percent identities, respectively. This cDNA, designated RBLRG-Z/E11, failed to complement the *ole1* defect of *S. cerevisiae*, which led to the use of other expression systems. The cDNA was ligated into the pYES2 vector and the resulting construct was introduced into the INVSc1 strain (Invitrogen). Following the induction of the GAL4 promotor of the pYES2 expression construct and the addition of methyl myristate, GC/MS analysis showed that the cells produced a 6:1 mixture of Z/E11-14:Acids (Figure 4.6). The latter result demonstrates that the mixed geometric isomers of unsaturated fatty acids used as

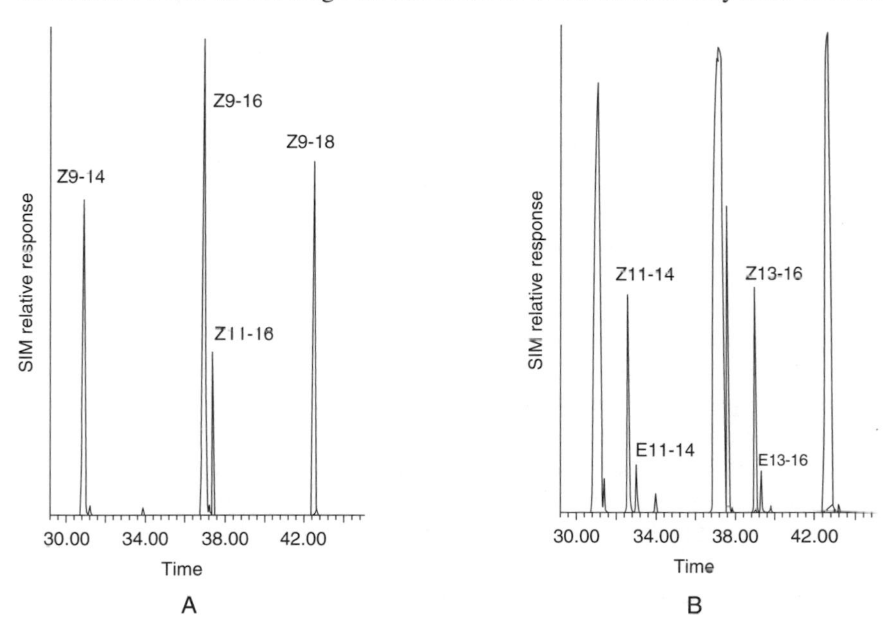

Figure 4.6 GC/MS analysis of DMDS adducts of methyl esters prepared from whole cell lipid extract of pYES2-RBLRG-Z/E11-transformed INVSc1 yeast strain. Single ion monitoring for diagnostic ions 217 and 245 *m/z*: (A) Control INVSc1 yeast cells transformed with vector supplemented with myristic acid (14:Acid) exhibit peaks corresponding to normal Z9-unsaturated fatty acids and Z11-16:Acid produced by chain elongation of Z9-14:Acid; (B) pYES2-RBLRG-Z/E11-transformed INVSc1 yeast strain supplemented with myristic acid exhibits peaks correponding to both geometric isomers of Δ11-14:Acids and their chain elongation products, Δ13-16:Acids. (Reproduced with permission from Liu *et al.*, 2002. © 2002 by *Insect Biochemistry and Molecular Biology*.)

pheromone precursors by this species are produced by the action of a single desaturase. A limitation of this expression system that pertains to the production of 14-carbon unsaturated products is the activity of endogenous chain elongation enzymes that rapidly convert 14-carbon fatty acids into 16-carbon fatty acids. In the present case almost half of the products were converted to Δ13-16:Acids. The baculovirus/SF21 expression system also suffers from this limitation, as well as from the activity of chain-shortening reactions.

4.2.8 Evolution of desaturase sequences in flies and moths

The above examples show that multiple desaturase-encoding sequences are present in the genomes of primitive (Tortricidae) and more derived (Noctuidae) lepidopteran families, consistent with the expansion of a desaturase gene family either early in the lepidopteran lineage or in an ancestor that also gave rise to other insect orders. In order to address the question of when the initial gene duplication events that led to the functional divergence and recruitment of desaturases for pheromone biosynthesis occurred, it is desirable to investigate desaturase-encoding sequences in species from other insect orders, such as the Diptera, where the use of desaturases in chemical signaling pathways has been shown (Tillman *et al.*, 1999). More specifically, flies use unsaturated cuticular hydrocarbons as contact pheromones that specify sex and species recognition (Jallon, 1984; Coyne and Oyama, 1995). The biosynthetic routes to these compounds use fatty acid synthesis, desaturation, chain elongation, and reductive decarboxylation (Blomquist *et al.*, 1987; Pennanec'h *et al.*, 1991; Pennanec'h *et al.*, 1997), and thus share initial steps with lepidopteran pheromone biosynthetic pathways. While more detailed discussions of dipteran pheromones are presented in other chapters of this book, i.e. by Blomquist *et al.* and Jallon *et al.*, we describe below specific aspects of investigations of desaturase-encoding sequences in the genome of the model insect system, *Drosophila melanogaster*, which taken together with the studies of lepidopteran desaturases described here, shed light on the evolution of the desaturase gene family in insects and the establishment of the use of desaturases in mate recognition systems.

In *D. melanogaster*, the genetic locus of a female-specific cuticular hydrocarbon polymorphism was found to contain two tandem desaturase-encoding genes, *desat1* and *desat2*, which were implicated in pheromone biosynthesis (Wicker-Thomas *et al.*, 1997; Coyne *et al.*, 1999; Dallérac *et al.*, 2000). The coding sequences of these two genes encode amino acid sequences with 75 percent similarity. Functional expression of the *desat1* coding sequence in the *ole1* yeast strain showed that it encodes a Δ9 desaturase that produces more palmitoleic acid than oleic acid (Dallérac *et al.*, 2000). Thus, the inferred substrate chain length preference of desat1 (16:0 >18:0) is like that of the HzPGDs2 Δ9 desaturase and its characterized orthologues isolated from *H. assulta*, *P. octo*, *E. postvittana*, and *A. velutinana*, which are the lepidopteran desaturases to which *desat1* has the greatest sequence identity (63–66 percent) (Figure 4.7). Functional expression of the *desat2* coding sequence in the *ole1* yeast strain showed that it encodes a Δ9 desaturase that is specific for the 14-carbon substrate myristic acid (Dallérac *et al.*, 2000). The *desat1* RNA was found in both male and female adult flies, as well as in larvae, whereas *desat2* was found only in adult females. The genetic basis for the observed cuticular hydrocarbon polymorphism of the major component of the female sex pheromone was attributed to the absence of expression of the latter gene in some geographical isolates as a result of a defect in its promotor (Dallérac *et al.*, 2000). A study employing high resolution genetic mapping and DNA sequence analysis of the region containing these two genes confirmed this interpretation and identified a 16-base pair deletion about 150 base pairs upstream of the *desat2* translation start site that specifies the null allele (Takahashi *et al.*, 2001). These results demonstrate that flies and moths both use specific desaturases to synthesize unsaturated fatty acid precursors of their respective sex pheromones, suggesting the commitment of members of the desaturase gene family to this functional role prior to the divergence of flies and moths from a common ancestor more than 250 million years ago.

Interestingly, a BLAST search of the *Drosophila* genome database (FlyBase, 1999) reveals that *desat1* and *desat2* represent only two of the six genes present in the *D. melanogaster* genome that encode desaturase-homologous amino acid sequences (Figure 4.7). At present the functional roles of the other four genes are not known, but their homology relationships to lepidopteran desaturase-homologous sequences were established in a recent study attempting the comprehensive sampling of desaturase-homologous sequence space in the pheromone glands of eight lepidopteran species representing four families (Knipple *et al.*, 2002). In this investigation, between four and nine partial cDNAs encoding additional desaturase-homologous sequences were isolated from each species (for a total of 48 novel sequences) and analyzed along with most of the other desaturase-encoding sequences of flies and moths described above as well as three additional full-length desaturase sequences deduced from cDNAs of *Bombyx mori* (Yoshiga *et al.*, 2000). The results of maximum likelihood analysis (Felsenstein, 1985) identified six statistically supported sequence lineages that contained five of the six *D. melanogaster* desaturase-homologous sequences and 62 of the 69 lepidopteran sequences considered, consistent with the antiquity of the deduced lineages. Interestingly, three of these lineages lacked functional correlates, whereas all of the desaturase sequences to which functional identities have been ascribed were mapped onto the other three, consistent with the data presented in Figure 4.7. Serial analyses of the aggregate data set with different computational tools, including sequence alignment (Altschul *et al.*, 1997), evolutionary trace analysis (Lictarge *et al.*, 1996), hydropathy analysis (Kyte and Doolittle, 1982), and transmembrane prediction analysis (Jones *et al.*, 1994), identified a number of functional class-specific sequence and structural elements (Knipple *et al.*, 2002).

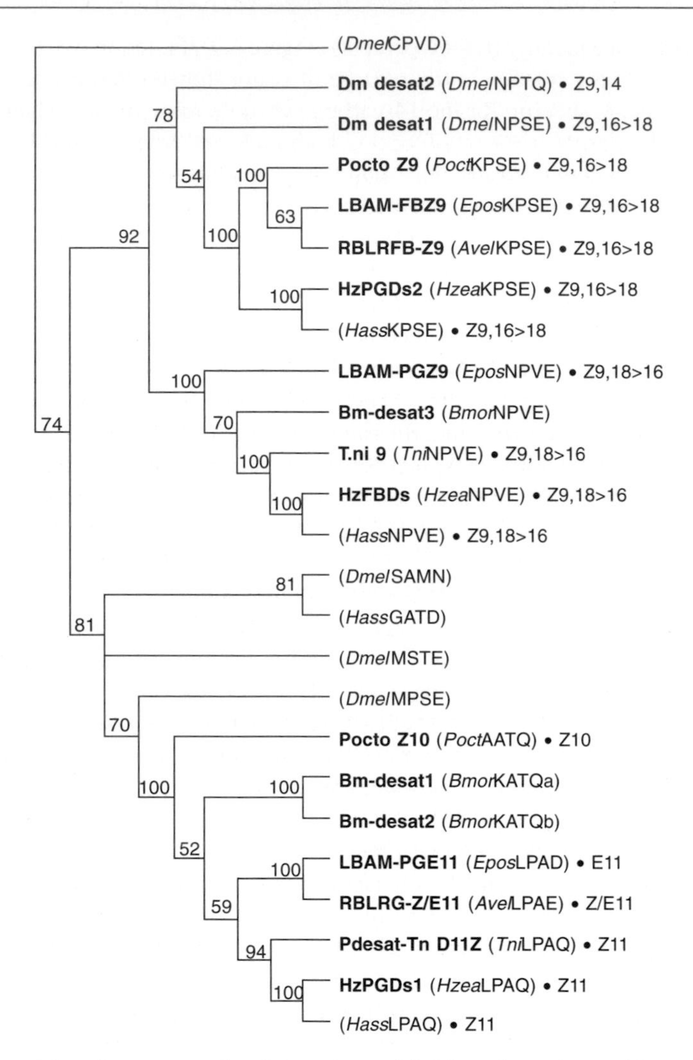

Figure 4.7 Consensus bootstrap tree of full-length amino acid sequences of insect desaturases generated with MacVector 7.0 (Oxford Molecular Limited). Branch points (internal nodes) are retained if they occur in $\geq$50 percent of resampling trees (1000 $\times$ resampling); all other nodes are collapsed. Names in bold are from the published literature; names in parentheses reflect a nomenclature system for insect desaturases (proposed in Knipple *et al.*, 2002), which incorporates an abbreviated biological species name (e.g. *Slit* = *Spodoptera litura*) and an identifier (signature motif) based on a group of four amino acids at positions 165–168. Desaturases that have been functionally expressed are labeled with a filled circle and their functional properties are abbreviated to designate stereo- and regioselectivity and, in the case of the

(Contd)

4.3 Conclusions

In this chapter we describe molecular biological investigations of lepidopteran sex pheromone biosynthetic desaturases, which are implicated in the generation of structural diversity of pheromone components. The strategy for the molecular cloning of cDNAs encoding lepidopteran acyl-CoA desaturases was predicated on the observed similarity of biochemical features of desaturases present in enriched biochemical fractions of lepidopteran pheromone glands with those of previously well-characterized animal and fungal acyl-CoA desaturases. The initial technical entry point for the molecular cloning of lepidopteran desaturase cDNAs used a PCR-based homology probing procedure employing oligonucleotide primer pools that were designed to encode conserved structural motifs of the rat and yeast enzymes. The amplification from *T. ni* pheromone gland cDNA of a cDNA fragment with substantial sequence similarity to the core domains of rat and yeast desaturases and the subsequent isolation of full-length cDNAs encoding desaturase-homologous proteins confirmed the homology of lepidopteran pheromone biosynthetic desaturases with the integral membrane acyl-CoA desaturases of animal and fungal cells (hypothesis one) and provided the basis for testing the conservation of functional associations of the essential elements of the desaturase complex (hypothesis two).

Efforts to test hypothesis two by biochemical reconstitution of a functional desaturase complex *in vitro* proved technically difficult and failed to provide a reproducible assay for functional expression of cloned lepidopteran desaturase sequences. In contrast, an *in vivo* expression system consisting of the yeast *ole1* mutant and YEpOLEX plasmid confirmed hypothesis two and provided a technically facile and robust assay for determining the functional identities of many moth desaturase-encoding cDNAs. Particularly desirable features of this

Figure 4.7 *(Contd)*

Δ9 desaturases, substrate chain length preferences. GenBank Accession numbers of the sequences are as follows: (*Dmel*CPVD) (AE003546.2); **Dm desat2** (*Dmel*NPTQ) (AJ271415.1); **Dm desat1** (*Dmel*NPSE) (U73160); **Pocto Z9** (*Poct*KPSE) (AF268275.2); **LBAM-FBZ9** (*Epos*KPSE) (AY061988); **RBLRFB-Z9** (*Avel*KPSE) (AF243046); **HzPGDs2** (*Hzea*KPSE) (AF272343); (*Hass*KPSE) (AF482906); **LBAM-PGZ9** (*Epos*NPVE) (AF402775); **Bm-desat3** (*Bmor*NPVE) (AF182406); **Tni Δ9** (*Tni*NPVE) (AF038050); **HzFBDs** (*Hzea*NPVE) (AF272345); (*Hass*NPVE) (AF482909); (*Dmel*SAMN) (AAF54920.1); (*Hass*GATD) (AF482905); (*Dmel*MSTE) (AAF57022.1); (*Dmel*MPSE) (AAF57020.1); **Pocto Z10** (*Poct*AATQ) (AY017379); **Bm-desat1** (*Bmor*KATQa) (AF157627); **Bm-desat2** (*Bmor*KATQb) (AF182405); **LBAM-PGE11** (*Epos*LPAD) (AY049741); **RBLRG-Z/E11** (*Avel*LPAE) (AF416738); **PDesat-Tn Δ11Z** (*Tni*LPAQ) (AF035375); **HzPGDs1** (*Hzea*LPAQ) (AF272342); and (*Hass*LPAQ) (AF482908).

functional expression system include the absence of chain-shortening activity under normal growth conditions and the fact that 16- and 18-carbon saturated fatty acid substrates are not limiting during exponential growth under aerobic conditions; thus, the unsaturated fatty acid profiles of the total lipid extracts of complementing transformants reflect the substrate preferences of the expressed desaturases that can make use of these substrates. A limitation of the yeast functional expression system is that many lepidopteran pheromone biosynthetic desaturases use substrates of shorter chain length, or produce unsaturated fatty acid products that do not complement the *ole1* defect. In some instances, it is possible to supplement the growth medium with unsaturated fatty acids that are capable of relieving the *ole1* host strain's auxotrophy and detect the specific products formed by the expressed desaturase against the background of supplemented unsaturated fatty acids. However, a particularly serious limitation of this approach is the very limited tolerance of *S. cerevisiae* for *E* geometric isomers, which cause growth inhibition and arrest of cell division. To circumvent this problem, an alternate functional expression system consisting of the SF9 insect cell line and baculovirus expression vector was employed. While useful to express cDNAs that are predicted to encode desaturases that produce *E* isomers, such as in *E. postvittanna*, a limitation of this system is the presence of the activities of endogenous enzymes of fatty acid metabolism, including desaturation, chain shortening, and chain elongation, which have the effect of obscuring the products of expressed desaturases.

Several studies have identified multiple desaturase-encoding sequences in lepidopteran species, consistent with the existence of a gene family that encodes desaturases with different functional roles (hypothesis three). An unexpected finding was that the number of unique desaturase-encoding transcripts in lepidopteran pheromone glands is much larger than the minimum number required to encode enzymatically distinct desaturases capable of producing the normal cellular unsaturated fatty acids and the known or inferred unsaturated fatty acid precursors of sex pheromone components (Knipple *et al.*, 2002). At present the inference of function of desaturase-encoding cDNAs is based on three distinct approaches: comparative sequence analysis, functional expression to determine the enzymatic properties of the encoded desaturase, and quantitation of RNA in different tissues. The first two approaches are complementary, since closely related sequences tend to encode desaturases that have similar enzymatic properties, as illustrated in Figure 4.7. The use of RNA quantitation to infer function has been used in only a few studies and is based on the following three implicit assumptions. The first is that desaturases encoded by RNAs that are present at high levels in pheromone glands but at much lower levels in other tissues are likely to play a role in sex pheromone biosynthesis. Second, and conversely, RNAs encoding desaturase-homologous sequences that are present at very low levels in pheromone glands are unlikely to play a functional role in pheromone biosynthesis. Third, desaturase-encoding RNAs with broad tissue distribution are likely to function principally in general cellular processes. Of course, while the presence of an RNA in a particular cell or tissue is an essential prerequisite for the functional activity of its encoded enzyme, it is not, in and of itself, proof of a functional role in that tissue.

It is apparent from the limited number of functional expression studies performed to date that the diversity of enzymatic properties of pheromone biosynthetic desaturases with Δ10 and Δ11 regioselectivities map onto a single sequence lineage. However, many questions about desaturase evolution in the Lepidoptera remain to be addressed, such as the biological significance of the two lineages encoding Δ9 desaturases with distinct substrate chain length preferences and the three identified lineages to which no functional attribute has been ascribed. At present, only preliminary results have been obtained towards the identification of specific sequence and structural elements of desaturase regioselectivity, stereoselectivity, and substrate specificity (hypothesis four). Computational analyses identified several functional class-specific sequence and structural elements (Knipple *et al.*, 2002) that provide useful predictions to guide the design and execution of future investigations of desaturase structure and function employing chimeric enzymes and site-directed mutagenesis.

Acknowledgements

The authors' work described in this chapter was supported by grants from the US Department of Agriculture (97-35302-4345 and 2001-35302-09926) and the Environmental Protection Agency/National Science Foundation (BES-9728367) to DCK and from the National Science Foundation (IBN-9514211) to WLR.

References

Altschul S. F., Madden T. L., Schaffer A. A., Zhang J., Zheng Z., Miller W. and Lipman D. J. (1997) Gapped blast and psi-blast: a new generation of protein database search programs. *Nucleic Acids Res.* **25**, 3389–3402.

Bjostad L. B., Linn C. E., Du J.-W. and Roelofs W. L. (1984) Identification of new sex pheromone components in *Trichoplusia ni*, predicted from biosynthetic precursors. *J. Chem. Ecol.* **10**, 1309–1323.

Bjostad L. B. and Roelofs W. L. (1981) Sex pheromone biosynthesis from radiolabeled fatty acids in the red banded leafroller moth. *J. Biol. Chem.* **256**, 7936–7940.

Bjostad L. B. and Roelofs W. L. (1983) Sex pheromone biosynthesis in *Trichoplusia ni*: key steps involve Δ11 desaturation and chain shortening. *Science* **220**, 1387–1389.

Bjostad L. B. and Roelofs W. L. (1986) Sex pheromone biosynthesis in the red banded leafroller moth studied by mass-labeling with stable isotopes and analysis with mass spectrometry. *J. Chem. Ecol.* **12**, 431–450.

Blomquist G. J., Dillwith J. W. and Adams T. S. (1987) Biosynthesis and endocrine regulation of sex pheromone production in Diptera. In: *Pheromone Biochemistry*, eds C. D. Prestwich and G. J. Blomquist, pp. 217–250 Academic Press, London.

Buser H. R, Arn H., Guerin P. and Rauscher S. (1983) Determination of double bond position in mono-unsaturated acetates by mass spectrometry of dimethyl disulfide adducts. *Anal. Chem.* **55**, 818–822.

Choi M.-Y., Han K. S., Boo K. S. and Jurenka R. A. (2002) Pheromone biosynthetic pathways in the moths *Helicoverpa zea* and *Helicoverpa assulta*. *Insect Biochem. Molec. Biol.* **32**, 1353–1359.

Coyne J. A. and Oyama R. (1995) Localisation of pheromonal sexual dimorphism in *Drosophila melanogaster* and its effect on sexual isolation. *Proc. Natl. Acad. Sci. USA* **92**, 9505–9509.

Coyne J. A., Wicker-Thomas C. and Jallon J.-M. (1999) A gene responsible for a cuticular hydrocarbon polymorphism in *Drosophila melanogaster*. *Genet. Res. Camb.* **73**, 189–203.

Dallérac R., Labeur C., Jallon J.-M, Knipple D. C., Roelofs W. L. and Wicker-Thomas, C. (2000) A Δ9 desaturase gene with a different substrate specificity is responsible for the cuticular diene hydrocarbon polymorphism in *Drosophila melanogaster. Proc. Natl. Acad. Sci. USA* **97**, 9449–9454.

Felsenstein J. (1985) Confidence limits on phylogenies: an approach using the bootstrap. *Evolution* **39**, 783–791.

FlyBase (1999) The FlyBase database of the Drosophila genome projects and community literature. *Nucleic Acids Research* **27**, 85–88.

Foster S. P. and Roelofs W. L. (1988) Sex pheromone biosynthesis in the leafroller moth *Planotortrix excessana* by D10 desaturation. *Arch. Insect Biochem. Physiol.* **8**, 1–9.

Foster S. P. and Roelofs, W. L. (1990) Biosynthesis of a monoene and a conjugated diene sex pheromone component of the lightbrown apple moth by Δ11 desaturation. *Experientia* **46**, 269–273.

Hao G., Liu W., O'Connor M. and Roelofs W. L. (2002) Acyl-CoA Z9- and Z10-desaturase genes from a New Zealand leafroller moth species, *Planotortrix octo*. *Insect Biochem. Molec. Biol.* **32**, 961–966.

Hazel J. R. and Williams E. E. (1990) The role of alterations in membrane lipid composition in enabling physiological adaptation of organisms to their physical environment. *Progress in Lipid Research* **29**, 167–227.

Jallon J.-M. (1984) A few chemical words exchanged by *Drosophila* during courtship and mating. *Behaviour Genetics* **14**, 441–478.

Jeffcoat R. (1979) The biosynthesis of unsaturated fatty acids and its control in mammalian liver. *Essays Biochem.* **15**, 1–36.

Jones D. T., Taylor W. R. and Thornton J. M. (1994) A model recognition approach to the prediction of all-helical membrane protein structure and topology. *Biochemistry* **33**, 3038–3049.

Jurenka R. A., Haynes K. F., Adlof R. O., Bengtsson M. and Roelofs W. L. (1994) Sex pheromone component ratio in the cabbage looper moth altered by a mutation affecting the fatty acid chain-shortening reactions in the pheromone bisoynthetic pathway. *Insect Biochem. Molec. Biol.* **24**, 373–381.

Jurenka R. A., Jacquin E. and Roelofs W. L. (1991) Control of the pheromone biosynthetic pathway in *Helicoverpa zea* by the Pheromone Biosynthetic Activating Neuropeptide. *Arch. Insect Biochem. Physiol.* **17**, 81–91.

Kaestner K. H., Ntambi J. M., Kelly T. J. and Lane, M. D. (1989) Differentiation-induced gene expression in 3T3-L1 preadipocytes – a second differentially expressed gene encoding stearoyl-CoA desaturase. *J. Biol. Chem.* **264**, 14755–14761.

Klun J. A., Plimmer J. R., Bierl-Leonhardt B. A., Sparks, A. N., Primiani M., Chapman O. L., Lee G. H. and Lepone, G. (1980) Sex pheromone chemistry of female corn earworm moth, *Heliothis zea*. *J. Chem. Ecol.* **6**, 165–175.

Knipple D. C., Rosenfield C.-L., Miller S. J., Liu W., Tang J. and Ma P. W. K. and Roelofs W. L. (1998) Cloning and functional expression of a cDNA encoding a pheromone gland-specific acyl-CoA Δ11-desaturase of the cabbage looper moth, *Trichoplusia ni*. *Proc. Natl. Acad. Sci. USA* **95**, 15287–15292.

Knipple D. C., Rosenfield C.-L., You K. M. and Jeong S. E. (2002) Evolution of the integral membrane desaturase gene family in moths and flies. *Genetics* **162**, 1737–1752.

Kyte J. and Doolittle R. F. (1982) A simple method for displaying the hydropathic character of a protein. *J. Mol. Biol.* **157**, 105–32.

Lichtarge O., Bourne H. R. and Cohen F. E. (1996) An evolutionary trace method defines binding surfaces common to protein families. *J. Mol. Biol.* **257**, 342–358.

Liu W., Jiao H., Murray N. C., O'Conner M. and Roelofs W. L. (2002) Gene characterized for membrane desaturase that produces (*E*)-11 isomers of mono- and diunsaturated fatty acids. *Proc. Natl. Acad. Sci. USA* **99**, 620–624.

Liu W., Ma P. W. K., Marsella-Herrick P., Rosenfield C.-L., Knipple D. C. and Roelofs W. L. (1999) Cloning and functional expression of a cDNA encoding a metabolic acyl-CoA Δ9-desaturase of the cabbage looper moth, *Trichoplusia ni*. *Insect Biochem. Molec. Biol.* **29**, 435–443.

Ntambi J. M., Buhrow S. A., Kaestner K. H., Christy R. J., Sibley E., Kelly T. J. and Lane M. D. (1988) Differentiation-induced gene expression in 3T3-L1 preadipocytes. *J. Biol. Chem.* **263**, 17291–17300.

Pennanec'h M., Bricard L., Kunesh G. and Jallon J.-M. (1997) Incorporation of fatty acids into cuticular hydrocarbons of male and female *Drosophila melanogaster*. *J. Insect Physiol.* **43**, 1111–1116.

Pennanec'h M., Feveur J. F, Pho D. B. and Jallon J.-M. (1991) Insect fatty acid related pheromones. A review of their biosynthesis, hormonal regulation and genetic control. *Annales de la Société Entomologique de France* **27**, 245–263.

Pope M. M., Gaston L. K. and Baker T. C. (1984) Composition, quantification, and periodicity of sex pheromone volatiles from individual *Heliothis zea* females. *Journal of Insect Physiology* **30**, 943–945.

Rodriguez F., Hallahan D. L., Pickett J. A. and Camps F. (1992) Characterization of the delta-11 palmitoyl-CoA-desaturase from *Spodoptera littoralis* (Lepidoptera, Noctuidae). *Insect Biochem. Molec. Biol.* **22**, 143–148.

Rosenfield C.-L., You K. M., Marsella-Herrick P., Roelofs W. L. and Knipple D. C. (2001) Structural and functional conservation and divergence among acyl-CoA desaturase-encoding genes of two noctuid species, the corn earworm, *Helicoverpa zea*, and the cabbage looper, *Trichoplusia ni*. *Insect Biochem. Molec. Biol.* **31**, 949–964.

Shanklin J., Whittle E. and Fox B. G. (1994) Eight histidine residues are catalytically essential in a membrane associated iron enzyme, stearoyl-CoA desaturase and are conserved in alkane hydroxylase and xylene monooxygenase. *Biochem.* **33**, 12787–12794.

Strittmatter P., Spatz L., Corcoran D., Rogers M. J., Setlow B. and Redline R. (1974) Purification and properties of rat liver microsomal stearoyl coenzyme A desaturase. *Proc. Natl. Acad. Sci. USA* **71**, 4565–4569.

Strittmatter P., Thiede M. A., Hackett C. S. and Ozols J. (1988) Bacterial synthesis of active rat stearoyl-CoA desaturase lacking the 26-residue amino-terminal amino acid sequence. *J. Biol. Chem.* **263**, 2532–2535.

Stukey J. E., McDonough V. M. and Martin C. E. (1989) Isolation and characterization of *OLE1*, a gene affecting fatty acid desaturation from *Saccharomyces cerevisiae*. *J. Biol. Chem.* **264**, 16537–16544.

Stukey J. E., McDonough V. M. and Martin C. E., (1990) The *OLE1* gene of *Saccharomyces cerevisiae* encodes the Δ9 fatty acid desaturase and can be functionally replaced by the rat stearoyl-CoA desaturase gene. *J. Biol. Chem.* **265**, 20144–20149.

Takahashi A., Tsaur S.-C., Coyne J. A., and Wu C. I. (2001) The nucleotide changes governing cuticular hydrocarbon variation and their evolution in *Drosophila melanogaster*. *Proc. Natl. Acad. Sci. USA* **98**, 3920–3925.

Tang J., Wolf W. A., Roelofs W. L. and Knipple D. C. (1991) The development of functionally competent cabbage looper moth sex pheromone glands. *Insect Biochem.* **21**, 573–581.

Teal P. E. A. and Tumlinson J. H. (1986) Terminal steps in pheromone biosynthesis by *Heliothis virescens* and *Heliothis zea*. *J. Chem. Ecol.* **12**, 353–366.

Thiede M. A., Ozols J. and Strittmatter P. (1986) Construction and sequence of cDNA for rat liver stearoyl coenzyme A desaturase. *J. Biol. Chem.* **261**, 13230–13235.

Tiku P. E., Gracey A. Y., Macartney A. I., Beyton R. J. and Crossins A. R. (1996) Cold-induced expression of Δ9-desaturase in carp by transcriptional and posttranslational mechanisms. *Science* **271**, 815–818.

Tillman J. A., Seybold S. J., Jurenka R. A. and Blomquist G. J. (1999) Insect pheromones – an overview of biosynthesis and endocrine regulation. *Insect Biochem.* **29**, 481–514.

Vigh L., Los D. A., Hovath I. and Murata N. (1993) The primary signal in the biological perception of temperature: Pd-catalyzed hydrogenation of membrane lipids stimulated the expression of the *desA* gene in *Synechocystis* PCC6803. *Proc. Natl. Acad. Sci. USA* **90**, 9090–9094.

Wicker-Thomas C., Henriet C. and Dallérac R. (1997) Partial characterization of a fatty acid desaturase gene in *Drosophila melanogaster. Insect Biochem. Molec. Biol.* **27**, 963–972.

Wolf W. A. and Roelofs W. L., (1986) Properties of the Δ11-desaturase enzyme used in cabbage looper moth sex pheromone biosynthesis. *Arch. Insect Biochem. Physiol.* **3**, 45–52.

Yoshiga T., Okano K., Mita K., Shimada T. and Matsomoto S. (2000) cDNA cloning of acyl-CoA desaturase homologs in the silkworm, *Bombyx mori*. *Gene* **246**, 339–345.

5

PBAN regulation of pheromone biosynthesis in female moths

Ada Rafaeli and Russell A. Jurenka

5.1 Introduction

Most moth species utilize sex pheromones to attract a conspecific mate. In most cases it is the female moth that releases the pheromone to attract the male. This is usually done during a specific period of the photoperiod and in most moths it occurs during the scotophase. At this time the male is active and if the pheromone is detected it will fly upwind following the odor plume to its source. Female moths release the sex pheromone as the result of a calling behavior in which the pheromone gland is exposed by extruding the ovipositor tip. The success of this relationship depends in part on the female's ability to regulate pheromone production and the synchronization of these events. In many moth species this synchronization is achieved by neuroendocrine mechanisms present in the female that in turn are influenced by various environmental and physiological events such as temperature, photoperiod, host plants, mating, hormones, neurohormones, and neuromodulators.

It was initially proposed that short-lived adult Lepidoptera that emerge with mature eggs would not exhibit neuroendocrine control of pheromone production; however, in some Lepidoptera that feed as adults, neuroendocrine control over the communication system might occur (Barth, 1965; Barth and Lester, 1973). Conclusive evidence that sex pheromone production was regulated by a neurohormone in a moth was obtained by Raina and Klun (1984) using *Helicoverpa zea*. They demonstrated that neurohormonal activity was found in brain complexes by injecting homogenates of brain-subesophageal ganglion (Br–SEG) complexes into head-ligated female moths and monitoring pheromone titers. The activity was also detected in the hemolymph during the scotophase indicating a hormonal function. The neurohormone was termed pheromone biosynthesis activating neuropeptide (PBAN) (Raina and Menn, 1987). The activity was further characterized to originate from the subesophageal ganglion (SEG) and to be degraded by proteases (Raina *et al.*, 1987). In addition, activity was found in male brain complexes as well as other moth species and the orthopteran, *Blattella germanica*. Many researchers subsequently detected PBAN-like activity in neural tissues of a variety of moths as well as in other insect orders. The peptide was purified from brain-SEG complexes of *H. zea* (Raina *et al.*, 1989) and *Bombyx mori* (Kitamura *et al.*, 1989) and availability of synthetic PBAN facilitated research on characterizing the activity of this neuropeptide.

Although PBAN has been shown to regulate pheromone biosynthesis in a variety of moths, and PBAN-like peptides are probably present in all moths, it is important to note that not all moths utilize PBAN as a pheromonotropic hormone. This is the case for the cabbage looper, *Trichoplusia ni*. These moths have PBAN-like activity in the SEG but it is not regulating pheromone production (Tang *et al.*, 1989). In this moth, the pheromone glands produce relatively high titers of pheromone during both the scotophase and photophase and begin production immediately after adult eclosion. It is the release of pheromone through the calling behavior that is regulated, and not pheromone biosynthesis. The glands become competent to produce pheromone at adult eclosion and this competency is controlled by ecdysteroids during the pupal stage (Tang *et al.*, 1991). However, it was suggested that PBAN may regulate the release of certain pheromone components from the gland during the scotophase (Zhao and Haynes, 1997). Other moths where PBAN apparently does not regulate pheromone production are those that utilize hydrocarbons as sex pheromones. This was demonstrated in *Scoliopteryx libatrix* (Subchev and Jurenka, 2001) and *Antacarsia gemmatalis* (Jeong and Jurenka, unpublished). Therefore, due to the diversity found within the moths, physiological regulation of pheromone production will also be diverse with PBAN regulating pheromone biosynthesis in most, but not all, moths.

In this chapter we will review the current state of knowledge about how pheromone production is regulated in female moths. Discussion of PBAN identification and localization within the nervous system will be followed by how PBAN acts to stimulate pheromone biosynthesis. The final major topic will be a discussion of mediators and inhibitors of PBAN action. A considerable amount of information has accumulated with regard to regulation of pheromone biosynthesis in moths since *Pheromone Biochemistry* (Prestwich and Blomquist, 1987) was first published, and this chapter is not all inclusive. Further information can also be obtained in several reviews (Raina, 1993; Jurenka, 1996; Teal *et al.*, 1996; Rafaeli *et al.*, 1997b; Raina, 1997; Rafaeli, 2002).

5.2 PBAN identification

5.2.1 Pyrokinin/PBAN family of peptides

Once it was established that pheromone biosynthesis was regulated by a peptide produced in the SEG, the next goal was to identify the peptide. In the purification of any biologically active factor, each purification step requires a sensitive bioassay to measure the active material. In the purification of PBAN, the bioassay consisted of head ligated females that were injected with bioactive fractions. After a 1–3 h period of incubation, the pheromone gland was excised and titers of pheromone determined by gas chromatography (GC). The first PBAN was purified and identified from *H. zea* (Raina *et al.*, 1989). Dissection of about 5000 brain–SEG complexes followed by several steps of HPLC purification resulted in a pure peptide that could be sequenced. It was found to be a 33 amino acid peptide with a C-terminal amide (Table 5.1). The peptide was synthesized and was shown to be active in the bioassay in a dose as low as 2 pmol (Raina *et al.*, 1989). In the same year, a PBAN from *B. mori* was purified and sequenced (Kitamura *et al.*, 1989). This peptide also had 33 amino acids with a C-terminal amide and was 82 percent identical to *H. zea* PBAN. A second peptide was subsequently isolated from *B. mori* that had an additional arginine at the N-terminus (Kitamura *et al.*, 1990). It was noted that the N-terminal end of *B. mori* PBAN was identical to the incomplete sequence of melanization and reddish coloration hormone (MRCH) also isolated from *B. mori*. Subsequently, it was shown that *B. mori* PBAN is the same peptide as MRCH (Matsumoto *et al.*, 1990). This was the first indication that these peptides were involved in multiple functions.

We now know that PBAN belongs to a family of peptides that have in common the five C-terminal amino acids (FXPRLamide motif). The first member of this family to be isolated was called leucopyrokinin, based on its ability to stimulate hindgut contraction in the cockroach, *Leucophaea maderae* (Holman *et al.*, 1986). It was termed a pyrokinin because it had a pyroglutamyl N-terminal amino acid. Additional myotropic peptides that belong to this family have since been identified from several insect sources (Table 5.1). An additional function was added to the family when the hormone that induces embryonic diapause in *B. mori* was identified (Imai *et al.*, 1991). Structure-activity studies established FXPRLamide as the minimal sequence required for pheromonotropic activity (Raina and Kempe, 1990, 1992), myotropic activity (Nachman *et al.*, 1986), and induction of embryonic diapause in *B. mori* (Suwan *et al.*, 1994). Cross-reactivity of these peptides was also established for the various functions (Fónagy *et al.*, 1992c; Kuniyoshi *et al.*, 1992b). Therefore, the common C-terminal FXPRLamide defines this family of peptides. In addition, it was demonstrated that a C-terminal β-turn is required for biological activity (Nachman *et al.*, 1991). Structure activity relationships will be discussed in section 5.3.1.

Table 5.1 Amino acid sequences of the pyrokinin/PBAN family of peptides and the species where identified. The FXPRLamide motif is shown in bold. X = S, T, G, or V. Three peptides are shown that have just the SPRLamide or PRLamide ending. Peptides are grouped together based on the primary function for which they were first identified

Function and species		Peptide sequence	Reference
PBAN			
Helicoverpa zea		LSDDMPATPADQEMYRQDPEQIDSRTKY**FSPRL**amide[a]	Raina *et al.* (1989)
Helicoverpa assulta		LSDDMPATPADQEMYRQDPEQIDSRTKY**FSPRL**amide[b]	Choi *et al.* (1998a)
Bombyx mori		LSEDMPATPADQEMYQPDPEEMESRTRY**FSPRL**amide[a]	Kitamura *et al.* (1989)
Lymantria dispar		LADDMPATMADQEVYRPEPEQIDSRNKY**FSPRL**amide[a]	Masler *et al.* (1994)
Agrotis ipsilon		LADDTPATPADQEMYRPDPEQIDSRTKY**FSPRL**amide[b]	Duportets *et al.* (1999)
Mamestra brassicae		LADDMPATPADQEMYRPDPEQIDSRTKY**FSPRL**amide[b]	Jacquin-Joly *et al.* (1998)
Spodoptera littoralis		LADDMPATPADQELYRPDPDQIDSRTKY**FSPRL**amide[b]	Iglesias *et al.* (2002)
Pheromontropic peptides			
Bombyx mori	α	II**FTP**K**L**amide[b]	Kawano *et al.* (1992)
Helicoverpa zea	α	VI**FTP**K**L**amide[b]	Ma *et al.* (1994)
Helicoverpa assulta	α	VI**FTP**K**L**amide[b]	Choi *et al.* (1998a)
Agrotis ipsilon	α	VI**FTP**K**L**amide[b]	Duportets *et al.* (1999)
Mamestra brassicae	α	VI**FTP**K**L**amide[b]	Jacquin-Joly *et al.* (1998)
Spodoptera littoralis	α	VI**FTP**K**L**amide[b]	Iglesias *et al.* (2002)
Bombyx mori	β	SVAKPQTHESLE**FIPRL**amide[b]	Kawano *et al.* (1992)
Helicoverpa zea	β	SLAYDDKSFENVE**FTPRL**amide[b]	Ma *et al.* (1994)
Helicoverpa assulta	β	SLAYDDKSFENVE**FTPRL**amide[b]	Choi *et al.* (1998)
Agrotis ipsilon	β	SLSYEDKMFDNVE**FTPRL**amide[b]	Duportets *et al.* (1999)
Mamestra brassicae	β	SLAYDDKVFENVE**FTPRL**amide[b]	Jacquin-Joly *et al.* (1998)
Pseudaletia separata	β	KLSTDDKVFENVE**FTPRL**amide[b]	Matsumoto *et al.* (1992a)
Spodoptera littoralis	β	SLAYDDKVFENVE**FTPRL**amide[b]	Iglesias *et al.* (2002)
Bombyx mori	γ	TMS**FSPRL**amide[b]	Kawano *et al.* (1992)
Helicoverpa zea	γ	TMN**FSPRL**amide[b]	Ma *et al.* (1994)
Helicoverpa assulta	γ	TMN**FSPRL**amide[b]	Choi *et al.* (1998a)

(Contd)

Agrotis ipsilon	γ	TMN**FSPRL**amide[b]	Duportets *et al.* (1999)
Mamestra brassicae	γ	TMN**FSPRL**amide[b]	Jacquin-Joly *et al.* (1998)
Spodoptera littoralis	γ	TMN**FSPRL**amide[b]	Iglesias *et al.* (2002)
Diapause Hormone			
Bombyx mori		TDMKDESDRGAHSERGALC**FGPRL**amide[a]	Imai *et al.* (1991)
Helicoverpa zea		NDVKDGAASGAHSDRLGLW**FGPRL**amide[b]	Ma *et al.* (1994)
Helicoverpa assulta		NDVKDGAASGAHSDRLGLW**FGPRL**amide[b]	Choi *et al.* (1998a)
Agrotis ipsilon		NDVKDGGADRGAHSDRGGMW**FGPR**lamide[b]	Duportets *et al.* (1999)
Spodoptera littoralis		NEIKDGGSDRGAHSDRAGLW**FGPRL**amide[b]	Iglesias *et al.* (2002)
Pyrokinins, Myotropins			
Locusta migratoria	(I)	GAVPAAQ**FSPRL**amide[a]	Schoofs *et al.* (1990b)
	(II)	EGD**FTPRL**amide[a]	Schoofs *et al.* (1990a)
	(III)	RQQP**FVPRL**amide[a]	Schoofs *et al.* (1992a)
	(IV)	RLHQNGMP**FSPRL**amide[a]	Schoofs *et al.* (1992a)
Locusta migratoria	(I)	pEDSGDGWPQQP**FVPRL**amide[a]	Schoofs *et al.* (1991)
	(II)	pESVPT**FTPRL**amide[a]	Schoofs *et al.* (1993)
Schistocerca gregaria	MT1	GAAPAAQ**FSPRL**amide[a]	Veelaert *et al.* (1997)
	MT2	TSSLFPH**PRL**amide[a]	Veelaert *et al.* (1997)
Periplaneta americana	PK1	HTAG**FIPRL**amide[a]	Predel *et al.* (1997)
	PK2	SPP**FAPRL**amide[a]	Predel *et al.* (1997)
	PK3	LVP**FRPRL**amide[a]	Predel *et al.* (1999)
	PK4	DHLPHDVY**SPRL**amide[a]	Predel *et al.* (1999)
	PK5	GGGGSGETSGMW**FGPRL**amide[a]	Predel *et al.* (1999)
	PK6	SESEVPGMW**FGPRL**amide[a]	Predel and Eckert, (2000)
Periplaneta fuliginosa	PK4	DHLSHDVY**SPRL**amide[a]	Predel and Eckert, (2000)
Leucophaea maderae		pETS**FTPRL**amide[a]	Holman *et al.* (1986)
Drosophila melanogaster		TGPSASSGLW**FGPRL**amide[b]	Choi *et al.* (2001)
		SVP**FKPRL**amide[b]	Choi *et al.* (2001)
Penaeus vannamei (Crustacea)		DFA**FSPRL**amide[a]	Torfs *et al.* (2001)
		ADFA**FNPRL**amide[a]	Torfs *et al.* (2001)

[a]Identified from the amino acid sequence of a purified peptide.
[b]Deduced from the cloned gene sequence.

Pyrokinin/PBAN peptides have been identified primarily from the orders Lepidoptera and Orthoptera. Immunocytochemical evidence discussed in section 5.2.2 indicates that a variety of insects contain pyrokinin/PBAN-like peptides. In addition, it was determined that the white shrimp, *Penaeus vannamei*, has two peptides that can induce myotropic activity (Torfs *et al.*, 2001). This points to the conserved nature of this family of peptides among the arthropods.

In addition to the above indicated peptides, several peptides having a PRL-amide or SPRL-amide C-terminus have been identified. These include three myotropins isolated from Orthoptera (Predel and Eckert, 2000; Predel *et al.*, 1999; Veelaert *et al.*, 1997) and an ecdysis-triggering hormone identified in *Drosophila melanogaster* (Park *et al.*, 1999). It is unknown whether or not these peptides are pheromonotropic. Myomodulin, which has an RLamide C-terminus (PMSMLRLamide), is not pheromonotropic in *H. zea* (Choi and Jurenka, unpublished). The pyrokinin/PBAN family of peptides was shown to accelerate pupariation in two species of fleshflies, with PRLamide as the minimum sequence required for activity (Zdarek *et al.*, 1997, 1998). This finding suggests that puparium formation may involve peptides similar to the ecdysis-triggering hormone of *Drosophila*. In addition, it may also suggest that the receptors for the dipteran ecdysis-triggering hormone and the pyrokinin/PBAN family of peptides are similar.

5.2.2 Immunocytochemical localization of PBAN-like peptides in the CNS

The biological activity of PBAN was first localized to the SEG by dissection of brain-SEG complexes and further testing of the brain and SEG separately (Raina *et al.*, 1987). Upon determination of the PBAN sequence, synthetic peptide was available to make antisera in rabbits. Immunocytochemical localization of PBAN-like-immunoreactivity (ir) in the central nervous system (CNS) of adult moths indicated that several neurons in the SEG contain PBAN-like-ir. These were found as clusters along the ventral midline, one each in the presumptive mandibular, maxillary, and labial neuromeres (Kingan *et al.*, 1992). Tracing of axons indicated that neurons in each neuromere were sending processes along different paths. The mandibular and maxillary neurons have axons that project into the maxillary nerve. This nerve sends fibers to the corpus cardiacum (CC) via nervus corpus cardiaci V (NCC-V). In addition to secretion into the hemolymph through the CC, the NCC-V can be neurosecretory itself, as indicated in a study conducted with *M. sexta* (Davis *et al.*, 1996). The labial neurons send processes to the CC via the NCC-III. In addition, two pairs of maxillary neurons send processes laterally within the SEG that bifurcate. The anterior projections enter the tritocerebrum of the brain and arborize in an area around the esophageal foramen. The posterior projections descend into the paired ventral nerve cord (VNC) and travel its entire length to terminate in the terminal abdominal ganglion (TAG). Arborizations arising from these paired projections were found in each segmental ganglia (Davis *et al.*, 1996). In addition, the segmental ganglia also can have PBAN-like-ir neurons (Davis *et al.*, 1996; Ma and Roelofs, 1995b; Ma *et al.*, 1996). These neurons can send processes anteriorly to the transverse nerve which serves as a neurohemal organ for release of products into the hemolymph.

Subsequent to the above mentioned studies, several reports have indicated similar patterns of neurons containing PBAN-like-ir in other moths (Golubeva *et al.*, 1997; Ichikawa *et al.*, 1995; Ichikawa *et al.*, 1996b). In addition, studies utilizing ELISA or immunoblotting methods have demonstrated the presence of PBAN-like-ir in other moth species (Choi *et al.*, 1998b; Gazit *et al.*, 1992; Jacquin-Joly and Descoins, 1996; Marco *et al.*, 1995). Developmental studies also demonstrate PBAN-like-ir in similar neurons of larvae and pupae. Males also have similar PBAN-like-ir neurons. These findings indicate that the PBAN-like-ir found in males and larvae should be involved in other functions besides regulating pheromone production. PBAN-like-ir has also been demonstrated in other non-lepidopteran insects in apparently homologous cell groups (Ajitha and Muraleedharan, 2001; Choi *et al.*, 2001; Predel and Eckert, 2000; Schoofs *et al.*, 1992b; Shiga *et al.*, 2000; Tips *et al.*, 1993; Jurenka, unpublished observations). Widespread finding of PBAN-like-ir in insects indicates that most antibodies are recognizing the C-terminal ending of this family of peptides. These investigations indicate the widespread distribution of the pyrokinin/PBAN family of peptides in Insecta.

Immunocytochemical evidence has indicated the colocalization of PBAN-like peptides and FMRFamide-like peptides in at least two insects. This was first demonstrated in *H. zea* in which PBAN-like and FMRFamide-like immunoreactivity was colocalized in the same neurons of the SEG (Blackburn *et al.*, 1992). In *D. melanogaster,* several FMRFamides were differentially localized in the CNS with specific antisera (McCormick *et al.*, 1999). Two of these FMRFamides appear to be present in the same neurons that contain PBAN-like peptides (Choi *et al.*, 2001). FMRFamides belong to a large group of peptides that are primarily involved in the regulation of various visceral muscles (Merte and Nichols, 2002). It is interesting that potential myomodulatory peptides are located within the same neuron. The functional physiological significance of this is unknown.

Enzyme-linked immunoabsorbent assays (ELISA) or radioimmunoassays (RIA) have been utilized to measure amounts of PBAN-like-ir in various tissues of female moths (Iglesias *et al.*, 1998; Ma *et al.*, 1996; Marco *et al.*, 1995, 1996; Rafaeli *et al.*, 1991). Consistent with the *in situ* immunocytochemical evidence, the highest amounts of activity were found in the SEG and CC. Lower amounts of activity were found in the thoracic and abdominal ganglia including the TAG. Although some variations were observed in the amount of PBAN-like-ir through the photoperiod, in general PBAN levels were found not to vary much with age. Males were also shown to contain PBAN-like-ir that did not vary with the photoperiod. Several studies also demonstrated that PBAN-like-ir was present in the hemolymph (Iglesias *et al.*, 1998, 1999; Marco *et al.*, 1996). In *M. brassicae* the levels of PBAN-like-ir in the hemolymph were correlated with the time of maximum calling and pheromone production (Iglesias *et al.*, 1999). These findings will be discussed with regard to the physiological mode of action of PBAN (section 5.3.2).

5.2.3 Molecular genetics

The gene encoding PBAN was first characterized from *H. zea* and *B. mori* (Davis *et al.*, 1992; Imai *et al.*, 1991; Kawano *et al.*, 1992; Ma *et al.*, 1994; Sato *et al.*, 1993). The full-length cDNA was found to encode PBAN plus four additional peptide domains with a common C-terminal FXPRL sequence motif including that of the diapause hormone of *B. mori*. Six additional precursor peptides with the common C-termini and sequence homology to those of *H. zea* and *B. mori* have been deduced from cDNA isolated from pheromone glands of *Mamestra brassicae* (Jacquin-Joly and Descoins, 1996), *Helicoverpa assulta* (Choi *et al.*, 1998a), *Agrotis ipsilon* (Duportets *et al.*, 1999), *Spodoptera littoralis* (Iglesias *et al.*, 2002), *Helicoverpa armigera* (Zhang *et al.*, 2001), and *Adoxophyes* sp. (Lee *et al.*, 2001) (Figure 5.1).

Expression of the PBAN mRNA has been investigated in the central nervous system in *B. mori* (Sato *et al.*, 1993, 1994), *H. zea* (Ma *et al.*, 1998), and *H. assulta* (Choi *et al.*, 1998a). In *B. mori*, *in situ* hybridization revealed a total of 12 positive cells which were localized as a cluster at three distinct regions along the ventral midline of the SEG (Sato *et al.*, 1994). Similar clusters of neurons were localized in *H. zea*, and the PBAN encoding gene was shown to be expressed predominantly in the SEG and at relatively low levels in other parts of the CNS (Ma *et al.*, 1998). These findings are similar to the immunocytochemical evidence presented in section 5.2.2.

If the putative processing sites are utilized in the gene sequence, then four additional peptides could be produced (Figure 5.1). Surprisingly, the diapause hormone of *B. mori* was found encoded in the PBAN gene sequence (Kawano *et al.*, 1992). Subsequent studies indicated that the post-translational processed peptides were found in the SEG (Ma *et al.*, 1996; Sato *et al.*, 1993). Surgical removal of the mandibular and maxillary neuronal clusters in the SEG demonstrated that diapause activity was retained, indicating that the labial cluster alone can induce diapause. Surgical removal of the labial neuronal cluster indicated that the mandibular and maxillary clusters are involved in inducing pheromone biosynthesis (Ichikawa *et al.*, 1996b). This study indicates that differential processing of the precursor peptide could occur in each neuronal cluster of the SEG. This was investigated directly in *H. zea* using a combination of immunocytochemistry and matrix assisted laser desorption/ionization mass spectrometry (MALDI MS) (Ma *et al.*, 2000). The MALDI MS data indicated that PBAN was found to a greater extent in the mandibular and maxillary clusters

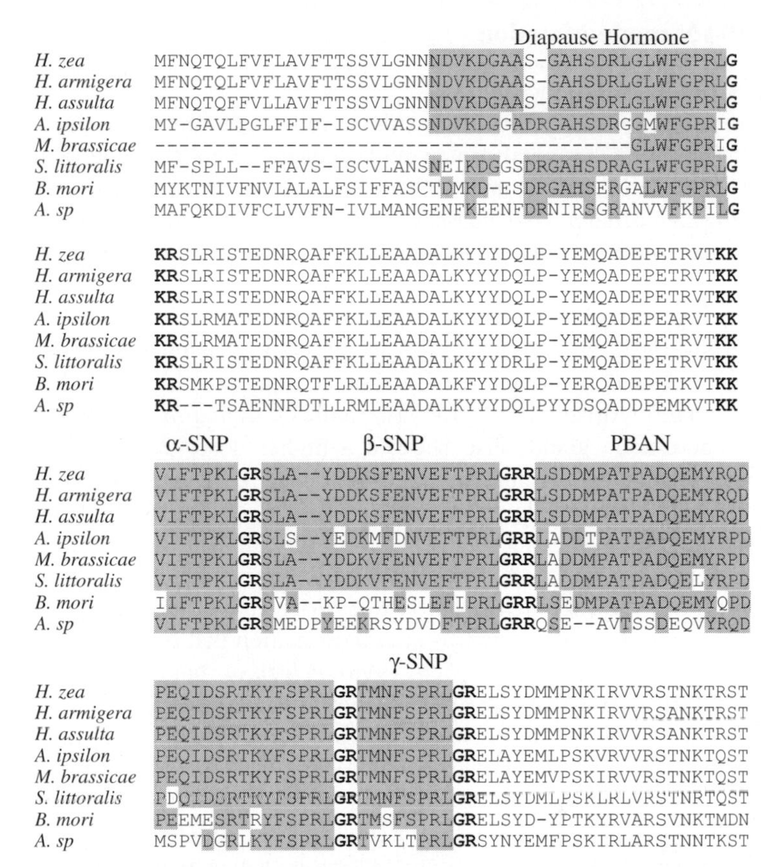

Figure 5.1 Alignment of the deduced amino acid sequences of the diapause hormone/PBAN preprohormone. Putative proteolytic post-translational processing sites are shown in bold with glycine contributing the C-terminal amide. The putatively expressed peptides are shown with conserved amino acids highlighted. GenBank accession number for each protein sequence: *H. zea* – P11159; *H. armigera* – AAL05596; *H. assulta* – AAC64293; *A. ipsilon* – O76818; *M. brassicae* – O45027; *S. littoralis* – AAK84160; *B. mori* – BAA05971; *A.* sp. – AAK72980. References: *H. zea* – Ma *et al.* (1994); *H. armigera* – Zhang *et al.* (2001); *H. assulta* – Choi *et al.* (1998a); *A. ipsilon* – Duportets *et al.* (1999); *M. brassicae* – Jacquin-Joly *et al.* (1998); *S. littoralis* – Iglesias *et al.* (2002); *B. mori* – Imai *et al.* (1991); *A.* sp. – Lee *et al.* (2001). *A.* sp. = *Adoxophyes* sp. SNP = subesophageal neuropeptide.

than in the labial cluster. The other neuropeptides were found in all clusters. In addition, some larger peptide fragments were found indicating alternative processing of the precursor protein. These studies demonstrate that further research is needed to elucidate the processing and functional roles for each of these peptides produced by the PBAN gene in the SEG.

5.3 PBAN mode of action

5.3.1 Structure–activity studies

The minimum sequence essential for stimulation of pheromone production was determined as the C-terminal pentapeptide sequence FXPRLamide (Kuniyoshi *et al.*, 1992a; Raina and Kempe, 1990, 1992). These structure–activity studies were performed using *in vivo* bioassays involving injection into the hemocoel of ligated female moths (Raina and Kempe, 1990). However, 100–1000-fold higher concentrations than the full-length PBAN were required by these minimal sequences. Cyclic-backbone analogs, based on [Arg27-D-Phe30]PBAN27-33NH_2, were also developed, and various PBAN-agonists as well as antagonists to injected PBAN were found to be active although at relatively high concentrations (1 nmol/female) (Altstein *et al.*, 1999a). Studies utilizing *in vitro* bioassays of isolated pheromone glands also showed a higher dose (25–50-fold higher) requirement with the C-terminal fragment peptide (PBAN 28-33) in *H. armigera* (Rafaeli, 1994). Comparable structure–activity studies on the multiple functions of the pyrokinin/PBAN peptide family (Table 5.1) also demonstrated the importance of the C-terminal sequence (Altstein *et al.*, 1996; Matsumoto *et al.*, 1992a 1992b; Nachman *et al.*, 1986; Zdarek *et al.*, 1998). NMR studies performed on the full PBAN sequence showed a type I β-turn conformation that encompassed the C-terminal region (Clark and Prestwich, 1996). A lack of interaction between the C-terminal turn and the rest of the peptide molecule indicated that the C-terminal turn indeed represented the important conformation recognized by the PBAN receptor. Nachman *et al.* (1991) demonstrated significant pheromonotropic activity with a conformationally constrained cyclic pyrokinin/PBAN analogue cyclo-[NTSFTPRL] octapeptide, that retains a β-turn. It can therefore be surmised that the receptors responsible for the different physiological responses by this peptide family do share binding requirements identified by this rigid, well-defined backbone structure. However, the requirement of the full sequence for full activity suggests that other parts of the PBAN sequence may be involved in the interaction with the receptor, and that selective binding capabilities must occur (Nachman *et al.*, 1993).

Indications that the full sequence can impart functional activity was suggested by findings that PBAN can be oxidized on methionine residues and that the oxidized form of PBAN can elicit full biological activity at a lower dose. This was demonstrated for both *B. mori* (Kitamura *et al.*, 1989) and *H. zea* (Raina *et al.*, 1991). However, another study using *Heliothis peltigera* indicated that biological activity was lost upon oxidation of the methionine residues (Gazit *et al.*, 1990). It remains to be determined whether or not the endogenous PBAN found in females is oxidized and if the oxidation does indeed confer greater sensitivity.

5.3.2 Target tissue and receptors

Morphological structural evidence implicated the glandular epithelium between the 8th and 9th abdominal segments of the ovipositor tips as the possible site of pheromone production in females of several species of moths (Aubrey *et al.*, 1983; Percy and Weatherston, 1974; Teal *et al.*, 1983). A study of the structural organization of the sex pheromone gland in *H. zea* demonstrated that pheromone was produced by an almost complete ring of columnar cells situated between the 8th and 9th abdominal segments that correlated to periods of cyclic pheromone production (Raina *et al.*, 2000). Through the development of a sensitive *in vitro* bioassay, studies on *H. armigera* and *H. zea* pioneered the demonstration that brain extracts and synthetic *H. zea* PBAN could stimulate the production of the main pheromone component (Z11-hexadecenal) (Soroker and Rafaeli, 1989; Rafaeli *et al.*, 1990b, 1991, 1993; Rafaeli, 1994; Rafaeli and Gileadi, 1996). The response obtained was specific to the pheromone gland and independent of other abdominal tissues (Rafaeli, 1994; Rafaeli *et al.*, 1997b). Evidence has since accumulated regarding several other species of moths including *B. mori*, *S. littura*, *O. nubilalis*, *Plodia interpunctella*, and *Thaumetopoea pityocampa* (Arima *et al.*, 1991; Fabriàs *et al.*, 1995; Fónagy *et al.*, 1992b; Jurenka, 1996; Ma and Roelofs, 1995a; Rafaeli and Gileadi, 1995a). The PBAN response was delineated to the intersegmental tissues situated between the 8th and 9th abdominal segments of the ovipositor tips. In addition, functional and viable pheromone gland cell clusters were obtained from the intersegmental membrane of *B. mori* using papain enzymatic digestion (Fónagy *et al.*, 2000). These cell clusters produced pheromone (bombykol) in response to applied pheromonotropic peptide indicating a functionally active cell.

Conclusive proof of the direct action of PBAN on the intersegmental membrane, as the tissue responsible for pheromone biosynthesis, awaits demonstration of PBAN binding to specific receptor proteins present in these intersegmental membrane cells. To study the mode of action of PBAN at the receptor level and for subsequent receptor purification, biologically active photoaffinity analogs were synthesized. A biotinylated-PBAN-analogue (N-[N-azido-tetrafluorobenzoyl] biocytinyloxyl-succinimide-PBAN) (Atf-Bct-NHS-PBAN) and a radiolabeled (tritium) photoaffinity label [^{3}H] BzDC-PBAN were used to demonstrate specific PBAN-binding (Elliott *et al.*, 1997; Rafaeli and Gileadi, 1997, 1999). Using crude membrane preparations of the pheromone glands, specific binding of the biotinylated photoaffinity-PBAN ligand to a protein at the 50 kDa range was demonstrated (Rafaeli and Gileadi, 1997, 1999; Rafaeli *et al.*, 2003). Using the [^{3}H] BzDC-PBAN label, two proteins with molecular weights of approximately 100 and 115 kDa were specifically labeled in the supernatant of brain–SEG, VNC and thoracic muscles, and to a lesser extent, in the 100 000xg pellet fractions (Elliott *et al.*, 1997). The signal in pheromone glands prepared from *H. zea* was, however, very weak, which indicates that these proteins may be present in pheromone glands but at a much lower concentration (Jurenka, unpublished). However, neither protein was found in gut or ovarian tissue. The presence of these proteins in the cytoplasm of neural and muscular tissue may indicate that the 100 and 115 kDa proteins are indeed PBAN receptors that have been internalized by receptor-mediated endocytosis following binding to PBAN. Alternatively, they may be binding proteins that are involved in transport of peptides within cells. A binding assay using crude membrane preparations derived from pheromone glands was developed in *H. peltigera*, but only low affinity binding was observed in the μM range (5×10^{-6} M) (Altstein *et al.*, 1999b). High affinity receptors have yet to be characterized for PBAN in pheromone gland cells.

5.3.3 Signal transduction

The signal transduction events that occur after PBAN binds to a receptor have been studied in several model moth species. The main difference found between these species so far is whether or not 3′,5′,cyclic-AMP (cAMP) is used as a second messenger (Figure 5.2). In the case of the heliothines and several others, cAMP is a second messenger. On the other hand, cAMP is thought not to act in pheromone gland cells of *B. mori* and *Ostrinia nubilalis* (Fónagy *et al.*, 1992a; Ma and Roelofs, 1995a). Instead, in these insects, it is thought that an increase in cytosolic calcium directly activates downstream events leading to stimulation of the biosynthetic pathway. It is interesting to note that a difference between *B. mori* and the heliothine moths as discussed in section 5.3.4. is that PBAN is probably affecting a reductase in *B. mori* while in heliothine moths PBAN is probably affecting fatty acid synthesis. Whether or not a connection can be established with regard to utilizing cAMP as a second messenger or not remains to be seen.

Using heliothine moths as a model, it was established that the cellular events, which occur in the pheromone glands as a result of PBAN stimulation, include a calcium influx and cAMP production, both of which act as second messengers (Jurenka *et al.*, 1991c, 1994; Rafaeli and Soroker, 1989, 1994). Extracellular calcium is essential for pheromonotropic activity in all moths studied to date (Fónagy *et al.*, 1992a, 1999; Ma and Roelofs, 1995a; Matsumoto *et al.*, 1995b). Pharmacological compounds such as cAMP analogs, forskolin (an adenylate cyclase activator) and isobutyl methyl xanthine (a phosphodiesterase inhibitor) all stimulated *in vitro* pheromone biosynthesis (Jurenka, 1996; Jurenka *et al.*, 1991c; Rafaeli and Soroker, 1989; Rafaeli *et al.*, 1990a). Stimulation of pheromone biosynthesis by cAMP analogs was shown to be independent of calcium (Rafaeli and Gileadi, 1995b). Sodium fluoride, at concentrations between 1–2 mM, also stimulated pheromonotropic activity thereby indicating that the receptor for PBAN was coupled to a G-protein (Rafaeli and Gileadi, 1996). In addition, intracellular cAMP levels were elevated as a result of PBAN stimulation (Rafaeli and Soroker, 1989; Rafaeli *et al.*, 1990a). Omission of calcium from the incubation medium completely abolished the PBAN-induced increase in intracellular cAMP levels (Soroker and Rafaeli, 1995). The inorganic calcium channel blockers lanthanum, cobalt, nickel and, to a lesser extent, manganese were shown to

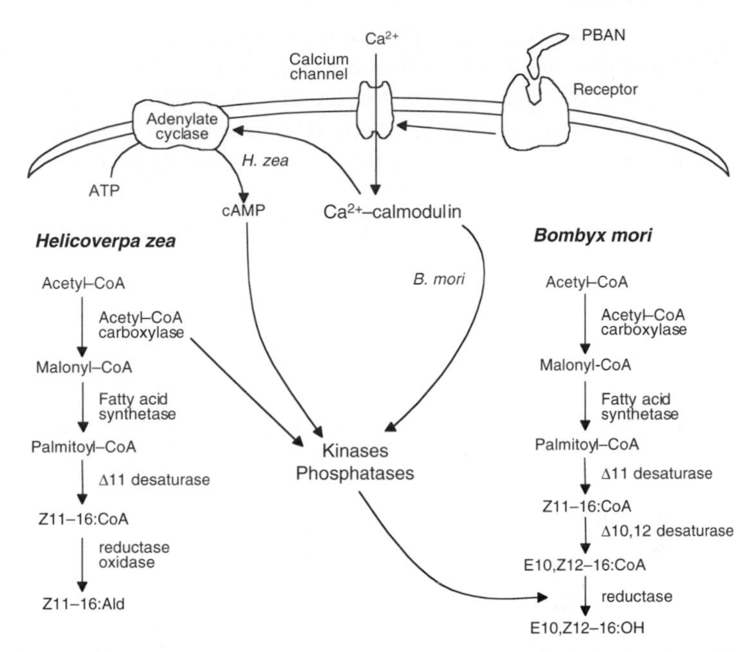

Figure 5.2 Proposed signal transduction mechanisms that stimulate the pheromone biosynthetic pathway in *Helicoverpa zea* and other heliothines as compared with that in *Bombyx mori*. It is proposed that PBAN binds to a receptor present in the cell membrane. Binding to the receptor somehow induces a receptor-activated calcium channel to open causing an influx of extracellular calcium. This calcium binds to calmodulin and in the case of *B. mori* will directly stimulate a phosphatase that will dephosphorylate and activate a reductase in the biosynthetic pathway. This activated reductase will then produce the pheromone bombykol. In *H. zea* and other heliothines like *Helicoverpa armigera* the calcium/calmodulin will activate adenylate cyclase to produce cAMP that will then act through kinases and/or phosphatases to stimulate acetyl-CoA carboxylase in the biosynthetic pathway.

inhibit the production of pheromone in response to PBAN in the presence of extracellular calcium (Jurenka *et al.*, 1991c, 1994). The voltage-gated calcium channel blockers verapamil and nifedipine did not inhibit PBAN stimulation, but the organic calcium channel blocker SKF-96365 (both voltage and receptor-activated calcium channel blocker) was shown to inhibit pheromonotropic stimulation in *H. virescens* (Jurenka, 1996). These results indicate that a receptor-activated calcium channel is involved in the pheromonotropic response. Additionally, calcium release as a result of ionophores (A23187 and ionomycin) duplicated the pheromonotropic effect of PBAN both on pheromone production (Fónagy *et al.*, 1992a; Jurenka *et al.*, 1991c, 1994; Rafaeli and Gileadi, 1996; Soroker and Rafaeli, 1995) and intracellular cAMP synthesis (Rafaeli and Gileadi,

1996; Soroker and Rafaeli, 1995). This confirmed that calcium is involved in pheromone biosynthesis and that there is a "cross-talk" between the two messengers, since the increase in calcium, as a result of the ionophores, induced the increase in cAMP. This reinforces the hypothesis that cAMP stimulation occurs downstream of the calcium influx (Jurenka *et al.*, 1991c; Soroker and Rafaeli, 1995).

It is suggested that free calcium entering the cell binds to calmodulin to form a complex, thereby activating adenylate cyclase and/or phosphoprotein phosphatases. Calmodulin was characterized from pheromone glands of *B. mori* (Iwanaga *et al.*, 1998) and was shown to have an identical amino acid sequence to *Drosophila* calmodulin (Smith *et al.*, 1987). Several inhibitors of calmodulin also were shown to modulate the expected increases observed with PBAN in both isolated glands of *H. armigera* and *B. mori* (Matsumoto *et al.*, 1995b; Soroker and Rafaeli, 1995). These results indicate that calmodulin is involved in the signal transduction cascade, binding calcium after the cell membrane calcium channel is opened.

In *B. mori*, it is suggested that the Ca^{2+}–calmodulin complex directly or indirectly activates a phosphoprotein phosphatase (Matsumoto *et al.*, 1995a). This phosphatase will then activate an acyl-CoA reductase in the biosynthetic pathway (section 5.3.4). Specific inhibitors of calcineurin (phosphoprotein phosphatase 2B) reduced the amount of pheromone when isolated glands were challenged with a pheromonotropic peptide (Fónagy *et al.*, 1999). Two genes encoding calcineurin heterosubunits were identified from the pheromone gland of *B. mori* and were found to be homologous to the catalytic subunit and regulatory subunits of other animal calcineurins (Yoshiga *et al.*, 2002). The calcineurin complex will apparently dephosphorylate an acyl-CoA reductase which catalyzes the formation of bombykol in *B. mori*.

5.3.4 Enzymes affected in the pheromone biosynthetic pathway

Several studies were undertaken to determine how PBAN affects the pheromone biosynthetic pathway. It appears that several different enzymes can be affected, depending on the species of moth. Demonstration of the affected enzyme primarily relies on the following of labeled precursors and intermediates into pheromone in the absence and presence of PBAN. Studies of this nature have so far indicated that PBAN does not affect desaturase activity (Arsequell *et al.*, 1990; Gosalbo *et al.*, 1994). PBAN has been shown to stimulate the reductase that converts an acyl-CoA to an alcohol precursor in several moths, including *B. mori* (Arima *et al.*, 1991; Ozawa *et al.*, 1993), *T. pityocampa* (Gosalbo *et al.*, 1994), *S. littoralis* (Fabriàs *et al.*, 1994; Martinez *et al.*, 1990), and *Manduca sexta* (Fang *et al.*, 1995; Tumlinson *et al.*, 1997). In *Argyrotaenia velutinana* (Tang *et al.*, 1989), *H. zea* (Jurenka *et al.*, 1991b), *Cadra cautella, Spodoptera exigua* (Jurenka, 1997) and *M. brassicae* (Jacquin *et al.*, 1994), it was demonstrated that PBAN controls pheromone biosynthesis by regulating a step during or prior to fatty acid biosynthesis. Circumstantial evidence in *Agrotis segetum* (Zhu *et al.*, 1995) and *H. armigera* (Rafaeli *et al.*, 1990a) also points to the regulation of fatty acid synthesis by PBAN. The rate-limiting step in fatty acid biosynthesis is usually acetyl-CoA carboxylase, but it remains to be determined if this enzyme is up-regulated in response to PBAN treatment. In one study using the moth *Sesamia nonagrioides* it was shown that the acetyltransferase enzyme may be regulated by PBAN (Mas *et al.*, 2000). There appears to be no particular pattern as to which enzyme within the pheromone biosynthetic pathway will be regulated by PBAN. However, in the majority of moths studied it is either the reductase or fatty acid synthesis that is stimulated.

Studies conducted to determine the mode of action of PBAN at the cellular level utilized pharmacological agents to deduce the role of certain proteins. In one study on *B. mori*, phosphatase inhibitors were utilized (Matsumoto *et al.*, 1995a). Both sodium fluoride and p-nitrophenyl phosphate inhibited pheromone production in isolated glands of *B. mori* in response to challenge by a pheromonotropic peptide. This was also demonstrated in *H. zea* and *H. virescens* (Jurenka, 1996). Sodium fluoride (at high concentrations, 10 mM) also inhibited the response to PBAN in *H. armigera* (Rafaeli and Gileadi, 1996). Phosphatase inhibitors were effective if a cAMP analogue was utilized, indicating that the phosphatase is acting downstream of cAMP in the signal cascade. Reductases and acetyl-CoA carboxylase are activated by dephosphorylation in mammalian cells. Therefore, phosphatases may be regulating fatty acid synthesis enzymes in some moths and reductases in other moths.

Several families of moths utilize hydrocarbons and/or their epoxides as sex pheromones (Millar, 2000). The biosynthesis of hydrocarbons occurs in oenocytes associated with fat body or epidermal cells in the abdomen. Hydrocarbons that are used as sex pheromones are then transferred to the sex pheromone gland for release (Schal *et al.*, 1998). It is thought that PBAN does not regulate the production of hydrocarbon sex pheromones as demonstrated in *Scoliopteryx libatrix* (Subchev and Jurenka, 2001). However, PBAN is probably regulating the production of epoxide sex pheromones. This was demonstrated in *Ascotis selenaria cretacea*, where decapitation resulted in pheromone decline that could be restored by injecting PBAN. In addition, it was shown that a deuterium-labeled precursor to the pheromone applied to the pheromone gland was converted to the epoxide only in PBAN-injected females (Miyamoto *et al.*, 1999). Decapitation also decreases the epoxide pheromone titer in the gypsy moth, *Lymantria dispar*, and injection of PBAN can restore pheromone production (Thyagaraja and Raina, 1994). However, decapitation did not decrease the levels of the hydrocarbon precursor in the gypsy moth (Jurenka, unpublished). These findings indicate that PBAN may regulate the epoxidation step in those moths that utilize epoxide pheromones, but not the production of the alkene precursor.

5.4 Mediators and inhibitors of PBAN action

5.4.1 Juvenile hormone

Juvenile hormones, produced by the corpora allata (CA), play an important role in reproductive development of many moth species. However, the reproductive behavior of calling is apparently unaffected by the removal of CA in many species of female moths (Hollander and Yin, 1985; Park and Ramaswamy, 1998; Sasaki and Riddiford, 1984; Tang *et al.*, 1987). However, studies involving the migratory moths *Pseudaletia unipuncta* and *A. ipsilon*, which exhibit a period of reproductive diapause, have shown that JH may be involved in the control of female sex pheromone production and/or its release (Cusson and McNeil, 1989; Gadenne *et al.*, 1993; Picimbon *et al.*, 1995). In both insects, JH was shown to be essential for pheromone biosynthesis and calling behavior, as allatectomized females failed to produce pheromone or call (Cusson *et al.*, 1994; Gadenne *et al.*, 1993; Picimbon *et al.*, 1995). These workers suggested that JH induced these behavioral changes by the induction of PBAN release from the central nervous system and not by controlling PBAN production or priming of PBAN target organs.

However, studies conducted on pharate adults of the non-migratory moth *H. armigera*, showed a clear correlation between CA maturation and the competence of pheromone glands to stimulatory action of PBAN (Fan *et al.*, 1999a). Recently, we have shown that JH influences the appearance of a putative PBAN receptor in pharate adults (Rafaeli *et al.*, 2003). Moreover, JH II challenge in an *in vitro* assay primed pheromone glands of pharate adults to respond to PBAN and induced earlier pheromone production by intact newly emerged females (Fan *et al.*, 1999a). These results indicate that JH is necessary for the primary induction of gland competence to PBAN stimulation, possibly by up-regulating gene expression of proteins required for pheromone production and/or PBAN-receptor proteins. Therefore, PBAN may be regarded as a circadian messenger in this moth.

5.4.2 Bursal factors

Studies conducted to determine how PBAN was affecting pheromone biosynthesis in the moth *A. velutinana* indicated that PBAN could affect fatty acid synthesis or substrate supply for fatty acid synthesis (Tang *et al.*, 1989). Some studies even indicated that another tissue might be affected. To investigate possible roles of other tissues, an isolated abdomen preparation was developed in which tissues could be removed leaving the pheromone gland intact (Jurenka *et al.*, 1991a). Pheromone levels were first reduced by decapitation, and then administration of PBAN to the isolated abdomens stimulated pheromone production. Removal of the VNC did not affect PBAN stimulation; however, removal of the corpus bursae did affect PBAN stimulation. This indicates that the corpus bursae is required for PBAN action. Subsequently a peptide was partially purified from corpus bursae that could stimulate pheromone production (Fabriàs *et al.*, 1992). To date a bursal factor has only been demonstrated in *A. velutinana* and the related tortricids *Choristoneura fumiferana* and *Choristoneura rosaceana* (Delisle *et al.*, 1999). It was not demonstrated in two other tortricids from New Zealand, *Epiphyas postvittana* and *Planotortrix octo* (Foster and Roelofs, 1994). The evolutionary significance for the presence of a bursal factor in some tortricids can only be speculated. Future identification of the bursal factor will contribute greatly to the complete understanding of its role as mediator in the regulation of pheromone biosynthesis in some tortricid moths.

5.4.3 Nervous system

The role of the nervous system in pheromone biosynthesis in moths is not clearly understood. In several moths, including *L. dispar* (Tang *et al.*, 1987; Thyagaraja and Raina, 1994), *H. virescens* (Christensen *et al.*, 1991), *S. littoralis* (Marco *et al.*, 1996), and *M. brassicae* (Iglesias *et al.*, 1998), an intact VNC was reported as necessary for pheromone biosynthesis by intact moths. On the other hand, in many species (even in some of those mentioned above) isolated pheromone glands responded to applied synthetic PBAN in an *in vitro* culture.

Christensen and co-workers (1991, 1992, 1994, 1995) proposed that the neurotransmitter octopamine may be involved as an intermediate messenger during the stimulation of sex pheromone production in *H. virescens*. These workers suggested that octopamine was involved in the regulation of pheromone production, and that PBAN's role lies in the stimulation of octopamine release at nerve endings. However, contradicting results concerning VNC-transection and octopamine-stimulated pheromone production were reported in the same species as well as other moth species (Delisle *et al.*, 1999; Jurenka *et al.*, 1991c; Park and Ramaswamy, 1998; Rafaeli and Gileadi, 1995b; Ramaswamy *et al.*, 1995).

A modulatory role for octopamine was suggested by research conducted on *H. armigera*. Inhibition of pheromone biosynthesis takes place both during the inactive times of the photoperiod as well as after mating. While decapitated females respond to injections of PBAN by producing pheromone for several hours, intact females, injected during the photophase, respond initially, but pheromonotropic activity is reduced after 2 h or less (Rafaeli *et al.*, 1997a). This led to the hypothesis that an endogenous inhibitory feedback for pheromone biosynthesis occurs during the photophase. Studies using octopamine and related biogenic amines, suggested that octopamine may play an important role as neuromodulator of the PBAN-induced pheromonotropic response (Rafaeli and Gileadi, 1995b; Rafaeli *et al.*, 1997a, 1999). Octopamine and several octopaminergic analogues inhibited pheromone production in studies using both *in vitro* and *in vivo* bioassays in two species of moths (Hirashima *et al.*, 2001; Rafaeli and Gileadi, 1995b, 1996; Rafaeli *et al.*, 1997a, 1999). It was determined that octopamine exerted its modulation of the pheromonotropic activity by its interaction with a separate receptor and its inhibitory influence on intracellular cAMP levels (Rafaeli and Gileadi, 1996).

The role of the VNC in pheromone production still requires clarification. It has been suggested that in some moth species (*S. littoralis*) both humoral and neural regulation occurs (Marco *et al.*, 1996). Certainly inhibitory ascending neurons are suggested by the fact that mating suppression is suppressed by VNC transection (section 5.4.4); however, a stimulatory descending neuronal regulation is less defined. Given the diversity of moths, it may not be surprising to find several mechanisms regulating pheromone biosynthesis.

5.4.4 Mating factors

Mating usually results in a behavioral switch in the female from one of receptive to males to one of not receptive and start of ovipositional behaviors. In moths, the switch to being non-receptive usually results in a decrease in pheromone titers for at least one night, or for an extended period of time. The cause of pheromonostasis following mating is usually attributed to the passage of factors, including sperm, from the male to the female during mating. A pheromonostatic peptide (PSP) causing depletion of sex pheromone was found in extracts of male *H. zea* accessory glands (Raina *et al.*, 1990). This peptide was purified from the accessory gland and duplex of males. The purified peptide caused the depletion of sex pheromone within 2 h after injection into females normally producing pheromone during the scotophase (Kingan *et al.*, 1995). It was characterized as a peptide with 57 amino acids containing a single disulfide bridge and a pyroglutamate N-terminus and amidated C-terminus (Kingan *et al.*, 1995). In their study the activity of a synthetic form of PSP was not demonstrated. However, synthetic partial peptides based on the C-terminal were recently shown to have pheromonostatic activity (Eliyahu *et al.*, 2003).

Pheromonostatic activity was observed in other studies conducted on the effect of synthetic *D. melanogaster* sex peptide and a duplex-derived peptide on pheromone biosynthesis in the moth *H. armigera* (Fan *et al.*, 1999b; Fan *et al.*, 2000). It was found that the sex peptide inhibited pheromonotropic activity induced by PBAN in isolated pheromone glands and decapitated females as well as endogenous pheromonotropic activity in females during the scotophase. These findings indicate that factors transferred from the male to the female during mating can influence pheromone production in post-mated females. In moths, these factors need to be characterized. In most cases, it is the presence of sperm or spermatophore in the female following mating that sends a signal via the VNC to initiate post-mating pheromonostasis.

The neural inhibition of pheromone production has been implicated in several moths. In *L. dispar*, it is thought that the presence of sperm in the spermatheca results in a neural signal that is responsible for pheromonostasis (Giebultowicz *et al.*, 1991). Demonstration that a signal sent via the VNC is regulating the release of PBAN was conducted with the lightbrown apple moth, *Epiphyas postvittana* (Foster, 1993). Severing the VNC before mating resulted in pheromone titers that were not different from virgin females, indicating that a VNC signal is required to initiate pheromonostasis. Injection of PBAN into mated females restored pheromone production, indicating that pheromone glands are still capable of producing pheromone. PBAN is still present in the SEG, indicating that the release of PBAN from the CC is affected (Foster, 1993). These results indicate that the neural signal is inhibiting the release of PBAN and the absence of PBAN is causing the post-mating pheromone decline. This scenario has subsequently been demonstrated in several other moths (Ando *et al.*, 1996; Delisle *et al.*, 2000; Delisle and Simard, 2002; Ichikawa *et al.*, 1996a; Jurenka *et al.*, 1993; Ramaswamy *et al.*, 1996). A neurophysiological study conducted with *B. mori* demonstrated that neuronal firing patterns coming from the SEG through NCC-V to the CC were permanently suppressed after mating (Ichikawa, 1998). This latter study confirms that a neuronal signal responsible for releasing PBAN is inhibited following mating. Although the presence of sperm in the spermatheca is partly required for initiating the signal via the VNC other factors may be involved that remain to be determined (Delisle and Simard, 2002).

5.5 Conclusions and future directions

The regulation of pheromone production in female moths has received considerable attention in the last decade since the discovery of PBAN. This neuropeptide apparently regulates pheromone production in most, but not all, moths. It belongs to the pyrokinin/PBAN family of peptides based on functional cross-reactivity with the active sequence FXPRLamide at the C-terminal end. The gene encoding for PBAN has been characterized from several moths and encodes for several pyrokinin/PBAN-like peptides, in addition to PBAN. The diapause hormone of *B. mori* is one of these peptides. Post-translational modification of the propeptide is being studied to try to determine which peptides are involved in regulating pheromone production *in vivo*. *In situ* hybridization studies indicate that this gene is expressed in three groups of cells in the SEG. Studies on the site of PBAN action indicate that pheromone gland cells have a receptor for PBAN that starts the signal cascade of events to trigger the pheromone biosynthetic pathway. Normal activation of pheromone production may involve several interdependent mechanisms, including both neural and humoral modulators. Elucidation of the entire suite of events involved in the physiological regulation of pheromone production in conjunction with environmental factors will keep researchers busy for the next few decades.

Future directions addressing molecular mechanisms involved in pheromone production will undoubtedly lead to the elucidation of receptor proteins involved in both the stimulation and feedback inhibition of pheromone biosynthesis. In addition, such molecular studies will reveal the identity and structure of the key enzymes involved in eliciting the biosynthetic pathway. Such knowledge will contribute greatly to the understanding of insect reproduction and ultimately could lead to drug design for the fertility control of unwanted moth species.

Acknowledgements

Research conducted in the authors' labs was supported in part by a grant from the US–Israel Binational Agricultural Research and Development Fund (IS-2978-98R), by a USDA-NRI grant – 2001-35302-10882 – and State of Iowa funds to R. J. and by an Israel National Academy of Science and Humanities Grant (357/98-2) to A. R.

References

Ajitha V. and Muraleedharan D. (2001) Identification of PBAN-like immunoreactivity in the neuroendocrine system and midgut of *Dysdercus cingulatus* (Heteroptera: Pyrrhocoridae). *Eur. J. Entomol.* **98**, 159–165.

Altstein M., Gazit Y., Ben-Aziz O., Gabay T., Marcus R., Vogel Z. and Barg J. (1996) Induction of cuticular melanization in *Spodoptera littoralis* larvae by PBAN/MRCH: Development of a qualitative bioassay and structure function analysis. *Arch. Insect Biochem. Physiol.* **31**, 355–370.

Altstein M., Ben-Aziz O., Daniel S., Schefler I., Zeltser I. and Gilon C. (1999a) Backbone cyclic peptide antagonists, derived from the insect pheromone biosynthesis activating neuropeptide, inhibit sex pheromone biosynthesis in moths. *J. Biol. Chem.* **274**, 17573–17579.

Altstein M., Gabay T., Ben-Aziz O., Daniel S., Zeltser I. and Gilon C. (1999b) Characterization of a putative pheromone biosynthesis-activating neuropeptide (PBAN) receptor from the pheromone gland of *Heliothis peltigera*. *Invert. Neurosci.* **4**, 33–40.

Ando T., Kasuga K., Yajima Y., Kataoka H. and Suzuki A. (1996) Termination of sex pheromone production in mated females of the silkworm moth. *Arch. Insect Biochem. Physiol.* **31**, 207–218.

Arima R., Takahara K., Kadoshima T., Numazaki F., Ando T., Uchiyama M., Nagasawa H., Kitamura A. and Suzuki A. (1991) Hormonal regulation of pheromone biosynthesis in the silkworm moth, *Bombyx mori* (Lepidoptera: Bombycidae). *Appl. Entomol. Zool.* **26**, 137–147.

Arsequell G., Fabriàs G. and Camps F. (1990) Sex pheromone biosynthesis in the processionary moth *Thaumetopoea pityocampa* by delta-13 desaturation. *Arch. Insect Biochem. Physiol.* **14**, 47–56.

Aubrey J. G., Boudreaux H. B., Grodner M. L. and Hammond A. M. (1983) Sex pheromone-producing cells and their associated cuticle in female *Heliothis zea* and *H. virescens* (Lepidoptera: Noctuidae). *Ann. Entomol. Soc. Am.* **76**, 343–348.

Barth R. H. (1965) Insect mating behavior: endocrine control of a chemical communication system. *Science* **149**, 882–883.

Barth R. H., Lester R. J. (1973) Neurohormonal control of sexual behavior in insects. *Annu. Rev. Entomol.* **18**, 445–472.

Blackburn M. B., Kingan T. G., Raina A. K. and Ma M. C. (1992) Colocalization and differential expression of PBAN- and FMRF-like immunoreactivity in the subesophageal ganglion of *Helicoverpa zea* (Lepidoptera: Noctuidae) during development. *Arch. Insect Biochem. Physiol.* **21**, 225–238.

Choi M. Y., Tanaka M., Kataoka H., Boo K. S. and Tatsuki S. (1998a) Isolation and identification of the cDNA encoding the pheromone biosynthesis activating neuropeptide and additional neuropeptides in the oriental tobacco budworm, *Helicoverpa assulta* (Lepidoptera: Noctuidae). *Insect Biochem. Mol. Biol.* **28**, 759–766.

Choi M. Y., Tatsuki S. and Boo K. S. (1998b) Regulation of sex pheromone biosynthesis in the oriental tobacco budworm, *Helicoverpa assulta* (Lepidoptera: Noctuidae). *J. Insect Physiol.* **44**, 653–658.

Choi M.-Y., Rafaeli A. and Jurenka R. (2001) Pyrokinin/PBAN-like peptides in the central nervous system of *Drosophila melanogaster. Cell Tissue Res.* **306**, 459–465.

Christensen T. A. and Hildebrand J. G. (1995) Neural regulation of sex-pheromone glands in Lepidoptera. *Invert. Neurosci.* **1**, 97–103.

Christensen T. A., Itagaki H., Teal P. E. A., Jasensky R. D., Tumlinson J. H. and Hildebrand J. G. (1991) Innervation and neural regulation of the sex pheromone gland in female *Heliothis* moths. *Proc. Natl. Acad. Sci. USA* **88**, 4971–4975.

Christensen T. A., Lehman H. K., Teal P. E. A., Itagaki H., Tumlinson J. H. and Hildebrand J. G. (1992) Diel changes in the presence and physiological actions of octopamine in the female sex-pheromone glands of heliothine moths. *Insect Biochem. Mol. Biol.* **22**, 841–849.

Christensen T. A., Lashbrook J. M. and Hildebrand, J. G. (1994) Neural activation of the sex-pheromone gland in the moth *Manduca sexta*: real-time measurement of pheromone release. *Physiol. Entomol.* **19**, 265–270.

Clark B. A. and Prestwich G. D. (1996) Evidence for a C-terminal turn in PBAN: an NMR and distance geometry study. *Int. J. Pept. Prot. Res.* **47**, 361–368.

Cusson M. and McNeil J. N. (1989) Involvement of juvenile hormone in the regulation of pheromone release activities in a moth. *Science* **243**, 210–212.

Cusson M., Tobe S. S. and McNeil J. N. (1994) Juvenile hormones – their role in the regulation of the pheromonal communication system of the armyworm moth, *Pseudaletia unipuncta. Arch. Insect Biochem. Physiol.* **25**, 329–345.

Davis M. B., Vakharia V. N., Henry J., Kempe T. G. and Raina A. K. (1992) Molecular cloning of the pheromone biosynthesis-activating neuropeptide in *Helicoverpa zea. Proc. Natl. Acad. Sci. USA* **89**, 142–146.

Davis N. T., Homberg U., Teal P. E. A., Altstein M., Agricola H.-J. and Hildebrand J. G. (1996) Neuroanatomy and immunocytochemistry of the median neuroendocrine cells of the subesophageal ganglion of the tobacco hawkmoth, *Manduca sexta*: immunoreactivities to PBAN and other neuropeptides. *Microscopy Res. Techniq.* **35**, 201–229.

Delisle J., Picimbon J.-F. and Simard J. (1999) Physiological control of pheromone production in *Choristoneura fummiferana* and *C. rosaceana. Arch. Insect Biochem. Physiol.* **42**, 253–265.

Delisle J., Picimbon J.-F. and Simard J. (2000) Regulation of pheromone inhibition in mated females of *Choristoneura fumiferana* and *C. rosaceana. J. Insect Physiol.* **46**, 913–921.

Delisle J. and Simard J. (2002) Factors involved in the post-copulatory neural inhibition of pheromone production in *Choristoneura fumiferana* and *C. rosaceana* females. *J. Insect Physiol.* **48**, 181–188.

Duportets L., Gadenne C. and Couillaud F. (1999) A cDNA, from *Agrotis ipsilon*, that encodes the pheromone biosynthesis activating neuropeptide (PBAN) and other FXPRL peptides. *Peptides* **20**, 899–905.

Eliyahu D., Nagalakmish V., Kubli E. C., Choffat Y., Applebaum S. W. and Rafaeli, A. (2003) Inhibition of pheromone biosynthesis in *Helicoverpa armigera* by pheromonostatic peptides. *J. Insect Physiol.* (in press).

Elliott J. T., Jurenka R. A., Prestwich G. D. and Roelofs W. L. (1997) Identification of soluble binding proteins for an insect neuropeptide. *Biochem. Biophys. Res. Commun.* **238**, 925–930.

Fabriàs G., Jurenka R. A. and Roelofs W. L. (1992) Stimulation of sex pheromone production by proteinaceous extracts of the bursa copulatrix in the redbanded leafroller moth. *Arch. Insect Biochem. Physiol.* **20**, 75–86.

Fabriàs G., Marco M.-P. and Camps F. (1994) Effect of the pheromone biosynthesis activating neuropeptide on sex pheromone biosynthesis in *Spodoptera littoralis* isolated glands. *Arch. Insect Biochem. Physiol.* **27**, 77–87.

Fabriàs G., Barrot M. and Camps F. (1995) Control of the sex pheromone biosynthetic pathway in *Thaumetopoea pityocampa* by the pheromone biosynthesis activating neuropeptide. *Insect Biochem. Mol. Biol.* **25**, 655–660.

Fan Y., Rafaeli A., Gileadi C. and Applebaum S. W. (1999a) Juvenile hormone induction of pheromone gland PBAN-responsiveness in *Helicoverpa armigera* females. *Insect Biochem. Mol. Biol.* **29**, 635–641.

Fan Y., Rafaeli A., Gileadi C., Kubli E. and Applebaum S. W. (1999b) *Drosophilia melanogaster* sex peptide stimulates juvenile hormone synthesis and depresses sex pheromone production in *Helicoverpa armigera*. *J. Insect Physiol.* **45**, 127–133.

Fan Y., Rafaeli A., Moshitzky P., Kubli E., Choffat Y. and Applebaum S. (2000) Common functional elements of *Drosophila melanogaster* seminal peptides involved in reproduction of *Drosophila melanogaster* and *Helicoverpa armigera* females. *Insect Biochem. Mol. Biol.* **30**, 805–812.

Fang N., Teal P. E. A. and Tumlinson J. H. (1995) PBAN regulation of pheromone biosynthesis in female tobacco hornworm moths, *Manduca sexta* (L.). *Arch. Insect Biochem. Physiol.* **29**, 35–44.

Fónagy A., Matsumoto S., Uchhiumi K. and Mitsui T. (1992a) Role of calcium ion and cyclic nucleotides in pheromone production in *Bombyx mori*. *J. Pest. Sci.* **17**, 115–121.

Fónagy A., Matsumoto S., Uchhiumi K., Orikasa C. and Mitsui T. (1992b) Action of pheromone biosynthesis activating neuropeptide on pheromone glands of *Bombyx mori* and *Spodoptera litura*. *J. Pest. Sci.* **17**, 47–54.

Fónagy A., Schoofs L., Matsumoto S., DeLoof A. and Mitsui T. (1992c). Functional cross-reactivities of some locustamyotropins and *Bombyx* pheromone biosynthesis activating neuropeptide. *J. Insect Physiol.* **38**, 651–657.

Fónagy A., Yokoyama N., Ozawa R., Okano K., Tatsuki S., Maeda S. and Matsumoto S. (1999) Involvement of calcineurin in the signal transduction of PBAN in the silkworm, *Bombyx mori* (Lepidoptera). *Comp. Biochem. Physiol. B Biochem. Mol. Biol.* **124**, 51–60.

Fónagy A., Yokoyama N., Okano K., Tatsuki S., Maeda S. and Matsumoto S. (2000) Pheromone-producing cells in the silkmoth, *Bombyx mori*: identification and their morphological changes in response to pheromonotropic stimuli. *J. Insect Physiol.* **46**, 735–744.

Foster S. P. (1993) Neural inactivation of sex pheromone production in mated lightbrown apple moths, *Epiphyas postvittana* (Walker). *J. Insect Physiol.* **39**, 267–273.

Foster S. P. and Roelofs, W. L. (1994) Regulation of pheromone production in virgin and mated females of two tortricid moths. *Arch. Insect Biochem. Physiol.* **25**, 271–285.

Gadenne C., Renou M. and Sreng L. (1993) Hormonal control of pheromone responsiveness in the male black cutworm *Agrotis ipsilon*. *Experientia* **49**, 721–724.

Gazit Y., Dunkelblum E., Benichis M. and Altstein M. (1990) Effect of synthetic PBAN and derived peptides on sex pheromone biosynthesis in *Heliothis peltigera* (Lepidoptera: Noctuidae). *Insect Biochem.* **20**, 853–858.

Gazit Y., Dunkelblum E., Ben-Aziz O. and Altstein M. (1992) Immunochemical and biological analysis of pheromone biosynthesis activating neuropeptide in *Heliothis peltigera*. *Arch. Insect Biochem. Physiol.* **19**, 247–260.

Giebultowicz J. M., Raina A. K., Uebel E. C. and Ridgway R. L. (1991) Two-step regulation of sex-pheromone decline in mated gypsy moth females. *Arch. Insect Biochem. Physiol.* **16**, 95–105.

Golubeva E., Kingan T. G., Blackburn M. B., Masler E. P. and Raina, A. K. (1997) The distribution of PBAN (pheromone biosynthesis activating neuropeptide)-like immunoreactivity in the nervous system of the gypsy moth, *Lymantria dispar*. *Arch. Insect Biochem. Physiol.* **34**, 391–408.

Gosalbo L., Fabriàs G. and Camps F. (1994) Inhibitory effect of 10,11-methylenetetradec-10-enoic acid on a Z9-desaturase in the sex pheromone biosynthesis of *Spodoptera littoralis*. *Arch. Insect Biochem. Physiol.* **26**, 279–286.

Hirashima A., Eiraku T., Watanabe Y., Kuwano E., Taniguchi E. and Eto M. (2001) Identification of novel inhibitors of calling and in vitro [C-14]acetate incorporation by pheromone glands of *Plodia interpunctella*. *Pest Manag. Sci.* **57**, 713–720.

Hollander A. L. and Yin C.-M. (1985) Lack of humoral control in calling and pheromone release by brain, corpora cardiaca, corpora allata and ovaries of the female gypsy moth, *Lymantria dispar* (L.). *J. Insect Physiol.* **31**, 159–163.

Holman G. M., Cook B. J. and Nachman R. J. (1986) Isolation, primary structure and synthesis of a blocked neuropeptide isolated from the cockroach, *Leucophaea maderae*. *Comp. Biochem. Physiol.* **85C**, 219–224.

Ichikawa T. (1998) Activity patterns of neurosecretory cells releasing pheromonotropic neuropeptides in the moth *Bombyx mori*. *Proc. Natl. Acad. Sci. USA* **95**, 4055–4060.

Ichikawa T., Hasegawa K., Shimizu I., Katsuno K., Kataoka H. and Suzuki A. (1995) Structure of neurosecretory cells with immunoreactive diapause hormone and pheromone biosynthesis activating neuropeptide in the silkworm, *Bombyx mori*. *Zool. Sci.* **12**, 703–712.

Ichikawa T., Shiota T. and Kuniyoshi H. (1996a) Neural inactivation of sex pheromone in mated females of the silkworm moth, *Bombyx mori*. *Zool. Sci.* **13**, 27–33.

Ichikawa T., Shiota T., Shimizu I. and Kataoka H. (1996b) Functional differentiation of neurosecretory cells with immunoreactive diapause hormone and pheromone biosynthesis activating neuropeptide of the moth, *Bombyx mori*. *Zool. Sci.* **13**, 21–25.

Iglesias F., Marco M. P., Jacquin-Joly E., Camps F. and Fabriàs G. (1998) Regulation of sex pheromone biosynthesis in two noctuid species, *S. littoralis* and *M. brassicae*, may involve both PBAN and the ventral nerve cord. *Arch. Insect Biochem. Physiol.* **37**, 295–304.

Iglesias F., Jacquin-Joly E., Marco M.-P., Camps F. and Fabriàs G. (1999) Temporal distribution of PBAN-Like immunoreactivity in the hemolymph of *Mamestra brassicae* females in relation to sex pheromone production and calling behavior. *Arch. Insect Biochem. Physiol.* **40**, 80–87.

Iglesias F., Marco P., François M.-C., Camps F., Fabriàs G. and Jacquin-Joly E. (2002) A new member of the PBAN family in *Spodoptera littoralis*: molecular cloning and immunovisualisation in scotophase hemolymph. *Insect Biochem. Mol. Biol.* (in press).

Imai K., Konno T., Nakazawa Y., Komiya T., Isobe M., Koga K., Goto T., Yaginuma T., Sakakibara K. and Hasegawa K., *et al.* (1991) Isolation and structure of diapause hormone of the silkworm, *Bombyx mori. Proc. Japan Acad.* **67(B)**, 98–101.

Iwanaga M., Dohmae N., Fónagy A., Takio K., Kawasaki H., Maeda S. and Matsumoto S. (1998) Isolation and characterization of calmodulin in the pheromone gland of the silkworm, *Bombyx mori. Comp. Biochem. Physiol. B Biochem. Mol. Biol.* **120**, 761–767.

Jacquin E., Jurenka R. A., Ljungberg H., Nagnan P., Löfstedt C., Descoins C. and Roelofs W. L. (1994) Control of sex pheromone biosynthesis in the moth *Mamestra brassicae* by the pheromone biosynthesis activating neuropeptide. *Insect Biochem. Mol. Biol.* **24**, 203–211.

Jacquin-Joly E. and Descoins C. (1996) Identification of PBAN-like peptides in the brain-subesophageal ganglion complex of Lepidoptera using western-blotting. *Insect Biochem. Mol. Biol.* **26**, 209–216.

Jacquin-Joly E., Burnet M., François M., Ammar D., Meillour P. and Descoins C. (1998) cDNA cloning and sequence determination of the pheromone biosynthesis activating neuropeptide of *Mamestra brassicae*: a new member of the PBAN family. *Insect Biochem. Mol. Biol.* **28**, 251–258.

Jurenka R. A. (1996) Signal transduction in the stimulation of sex pheromone biosynthesis in moths. *Arch. Insect Biochem. Physiol.* **33**, 245–258.

Jurenka R. A. (1997) Biosynthetic pathway for producing the sex pheromone component (Z,E)-9,12-tetradecadienyl acetate in moths involves a delta-12 desaturase. *Cell. Mol. Life Sci.* **53**, 501–505.

Jurenka R. A., Fabriàs G. and Roelofs W. L. (1991a) Hormonal control of female sex pheromone biosynthesis in the redbanded leafroller moth, *Argyrotaenia velutinana. Insect Biochem.* **21**, 81–89.

Jurenka R. A., Jacquin E. and Roelofs W. L. (1991b) Control of the pheromone biosynthetic pathway in *Helicoverpa zea* by the pheromone biosynthesis activating neuropeptide. *Arch. Insect Biochem. Physiol.* **17**, 81–91.

Jurenka R. A., Jacquin E. and Roelofs W. L. (1991c) Stimulation of sex pheromone biosynthesis in the moth *Helicoverpa zea*: action of a brain hormone on pheromone glands involves Ca^{2+} and cAMP as second messengers. *Proc. Natl. Acad. Sci. USA.* **88**, 8621–8625.

Jurenka R. A., Fabriàs G., Ramaswamy S. and Roelofs W. L. (1993) Control of sex pheromone biosynthesis in mated redbanded leafroller moths. *Arch. Insect Biochem. Physiol.* **24**, 129–137.

Jurenka R. A., Fabriàs G., DeVoe L. and Roelofs W. L. (1994) Action of PBAN and related peptides on pheromone biosynthesis in isolated pheromone glands of the redbanded leafroller moth, *Argyrotaenia velutinana. Comp. Biochem. Physiol.* **108C**, 153–160.

Kawano T., Kataoka H., Nagasawa H., Isogai A. and Suzuki A. (1992) cDNA cloning and sequence determination of the pheromone biosynthesis activating neuropeptide of the silkworm, *Bombyx mori. Biochem. Biophys. Res. Commun.* **189**, 221–226.

Kingan T. G., Blackburn M. B. and Raina A. K. (1992) The distribution of PBAN immunoreactivity in the central nervous system of the corn earworm moth, *Helicoverpa zea. Cell Tissue Res.* **270**, 229–240.

Kingan T. G., Bodnar W. M. and Hunt D. F. (1995) The loss of female sex pheromone after mating in the corn earworm moth *Helicoverpa zea*: identification of a male pheromonostatic peptide. *Proc. Nat. Acad. Sci. USA* **92**, 5082–5086.

Kitamura A., Nagasawa H., Kataoka H., Inoue T., Matsumoto S., Ando T. and Suzuki A.

(1989) Amino acid sequence of pheromone-biosynthesis-activating neuropeptide (PBAN) of the silkworm, *Bombyx mori*. *Biochem. Biophys. Res. Commun.* **163**, 520–526.

Kitamura A., Nagasawa H., Kataoka H., Ando T. and Suzuki A. (1990) Amino acid sequence of pheromone biosynthesis activating neuropeptide-II (PBAN-II) of the silkmoth, *Bombyx mori*. *Agric. Biol. Chem. Tokyo* **54**, 2495–2497.

Kuniyoshi H., Nagasawa H., Ando T. and Suzuki A. (1992a) N-Terminal modified analogs of C-terminal fragments of PBAN with pheromonotropic activity. *Insect Biochem. Mol. Biol.* **22**, 399–403.

Kuniyoshi H., Nagasawa H., Ando T., Suzuki A., Nachman R. J. and Holman G. M. (1992b) Cross-reactivity between pheromone biosynthesis activating neuropeptide (PBAN) and myotropic pyrokinin insect peptides. *Biosci. Biotech. Biochem.* **56**, 167–168.

Lee J. M., Choi M. Y., Han K. S. and Boo K. S. (2001) Cloning of the cDNA encoding pheromone biosynthesis activating neuropeptide in *Adoxophyes* sp. (Lepidoptera: Tortricidae): a new member of the PBAN family. GenBank Direct submission.

Ma P. W. K., Knipple D. C. and Roelofs W. L. (1994) Structural organization of the *Helicoverpa zea* gene encoding the precursor protein for pheromone biosynthesis-activating neuropeptide and other neuropeptides. *Proc. Natl. Acad. Sci. USA* **91**, 6506–6510.

Ma P. W. K. and Roelofs W. L. (1995a) Calcium involvement in the stimulation of sex pheromone production by PBAN in the European corn borer, *Ostrinia nubilalis* (Lepidoptera: Pyralidae). *Insect Biochem. Mol. Biol.* **25**, 467–473.

Ma P. W. K. and Roelofs W. L. (1995b) Sites of synthesis and release of PBAN-like factor in female European corn borer, *Ostrinia nubilalis*. *J. Insect Physiol.* **41**, 339–350.

Ma P. W. K., Roelofs W. L. and Jurenka R. A. (1996) Characterization of PBAN and PBAN-encoding gene neuropeptides in the central nervous system of the corn earworm moth, *Helicoverpa zea*. *J. Insect Physiol.* **42**, 257–266.

Ma P. W. K., Knipple D. C. and Roelofs W. L. (1998) Expression of a gene that encodes pheromone biosynthesis activating neuropeptide in the central nervous system of corn earworm, *Helicoverpa zea*. *Insect Biochem. Mol. Biol.* **28**, 373–385.

Ma P. W. K., Garden R. W., Niermann J. T., O'Connor M., Sweedler J. V. and Roelofs W. L. (2000) Characterizing the Hez-PBAN gene products in neuronal clusters with immunocytochemistry and MALDI MS. *J. Insect Physiol.* **46**, 221–230.

Marco M.-P., Fabriàs G. and Camps F. (1995) Development of a highly sensitive ELISA for the determination of PBAN and its application to the analysis of hemolymph in *Spodoptera littoralis*. *Arch. Insect Biochem. Physiol.* **30**, 369–382.

Marco M.-P., Fabriàs G., Lázaro G. and Camps F. (1996) Evidence for both humoral and neural regulation of sex pheromone biosynthesis in *Spodoptera littoralis*. *Arch. Insect Biochem. Physiol.* **31**, 157–168.

Martinez T., Fabriàs G. and Camps F. (1990) Sex pheromone biosynthetic pathway in *Spodoptera littoralis* and its activation by a neurohormone. *J. Biol. Chem.* **265**, 1381–1387.

Mas E., Lloria J., Quero C., Camps F. and Fabriàs G. (2000) Control of the biosynthetic pathway of *Sesamia nonagrioides* sex pheromone by the pheromone biosynthesis activating neuropeptide. *Insect Biochem. Mol. Biol.* **30**, 455–459.

Masler E. P., Raina A. K., Wagner R. M. and Kochansky J. P. (1994). Isolation and identification of a pheromonotropic neuropeptide from the brain-suboesophageal ganglion complex of *Lymantria dispar*: a new member of the PBAN family. *Insect Biochem. Mol. Biol.* **24**, 829–836.

Matsumoto S., Kitamura A., Nagasawa H., Kataoka H., Orikasa C., Mitsui T. and Suzuki A. (1990) Functional diversity of a neurohormone produced by the suboesophageal ganglion: molecular identity of melanization and reddish colouration hormone and pheromone biosynthesis activating neuropeptide. *J. Insect Physiol.* **36**, 427–432.

Matsumoto S., Fónagy A., Kurihara M., Uchiumi K., Nagamine T., Chijimatsu M. and Mitsui, T. (1992a) Isolation and primary structure of a novel pheromonotropic neuropeptide structurally related to leucopyrokinin from the armyworm larvae, *Pseudaletia separata. Biochem. Biophys. Res. Commun.* **182**, 534–539.

Matsumoto S., Yamashita O., Fónagy A., Kurihara M., Uchiumi K., Nagamine T. and Mitsui T. (1992b) Functional diversity of a pheromonotropic neuropeptide: induction of cuticular melanization and embryonic diapause in Lepidopteran insects by *Pseudaletia* pheromonotropin. *J. Insect Physiol.* **38**, 847–851.

Matsumoto S., Ozawa R., Uchiumi K., Kurihara M. and Mitsui T. (1995a) Intracellular signal transduction of PBAN action in the common cutworm, *Spodoptera litura*: effects of pharmacological agents on sex pheromone production *in vitro*. *Insect Biochem. Mol. Biol.* **25**, 1055–1059.

Matsumoto S., Ozawa R. A., Nagamine T., Kim G., Uchiumi K., Shono T. and Mitsui T. (1995b) Intracellular transduction in the regulation of pheromone biosynthesis of the silkworm, *Bombyx mori*: suggested involvement of calmodulin and phosphoprotein phosphatase. *Biosci. Biotech. Biochem.* **59**, 560–562.

McCormick J., Lim I. and Nichols R. (1999) Neuropeptide precursor processing detected by triple immunolabeling. *Cell Tissue Res.* **297**, 197–202.

Merte J. and Nichols R. (2002) *Drosophila melanogaster* FMRFamide-containing peptides: redundant or diverse functions? *Peptides* **23**, 209–220.

Millar J. G. (2000) Polyene hydrocarbons and epoxides: a second major class of lepidopteran sex attractant pheromones. *Annu. Rev. Entomol.* **45**, 575–604.

Miyamoto T., Yamamoto M., Ono A., Ohtani K. and Ando T. (1999) Substrate specificity of the epoxidation reaction in sex pheromone biosynthesis of the Japanese giant looper (Lepidoptera: Geometridae). *Insect Biochem. Mol. Biol.* **29**, 63–69.

Nachman R. J., Holman G. M. and Cook B. J. (1986) Active fragments and analogs of the insect neuropeptide leucopyrokinin: structure-function studies. *Biochem. Biophys. Res. Commun.* **137**, 936–942.

Nachman R. J., Roberts V. A., Dyson H. J., Holman G. M. and Tainer J. A. (1991) Active conformation of an insect neuropeptide family. *Proc. Natl. Acad. Sci. USA* **88**, 4518–4522.

Nachman R. J., Kuniyoshi H., Roberts V. A., Holman G. M. and Suzuki A. (1993) Active conformation of the pyrokinin/PBAN neuropeptide family for pheromone biosynthesis in the silkworm. *Biochem. Biophys. Res. Commun.* **193**, 661–666.

Ozawa R. A., Ando T., Nagasawa H., Kataoka H. and Suzuki A. (1993) Reduction of the acyl group: the critical step in bombykol biosynthesis that is regulated *in vitro* by the neuropeptide hormone in the pheromone gland of *Bombyx mori. Biosci. Biotech. Biochem.* **57**, 2144–2147.

Park Y. I. and Ramaswamy S. B. (1998) Role of brain, ventral nerve cord and corpora cardiaca–corpora allata complex in the reproductive behavior of female tobacco budworm (Lepidoptera: Noctuidae). *Ann. Entomol. Soc. Am.* **91**, 329–334.

Park Y., Zitnan D., Gill S. S. and Adams M. E. (1999) Molecular cloning and biological activity of ecdysis-triggering hormones in *Drosophila melanogaster. FEBS Lett.* **463**, 133–138.

Percy J. E. and Weatherston J. (1974) Gland structure and pheromone production in insects. In *Pheromones*, ed. M. C. Birch, pp. 11–34. North Holland Publishing Co., Amsterdam/London.

Picimbon J.-F., Becard J.-M., Sreng L., Clement J.-L. and Gadenne C. (1995) Juvenile hormone stimulates pheromonotropic brain factor release in the female black cutworm, *Agrotis ipsilon*. *J. Insect Physiol.* **41**, 377–382.

Predel R. and Eckert M. (2000) Tagma-specific distribution of FXPRLamides in the nervous system of the American cockroach. *J. Comp. Neurol.* **419**, 352–363.

Predel R., Kellner R., Kaufmann R., Penzlin H. and Gäde G. (1997) Isolation and structural elucidation of two pyrokinins from the retrocerebral complex of the American cockroach. *Peptides* **18**, 61–66.

Predel R., Kellner R., Nachman R. J., Holman G. M., Rapus, J. and Gade G. (1999) Differential distribution of pyrokinin-isoforms in cerebral and abdominal neurohemal organs of the American cockroach. *Insect Biochem. Mol. Biol.* **29**, 139–144.

Prestwich G. D. and Blomquist G. J. (1987) *Pheromone Biochemistry*, p. 565. Academic Press, Orlando, FL.

Rafaeli A. (1994) Pheromonotropic stimulation of moth pheromone gland cultures in vitro. *Arch. Insect Biochem. Physiol.* **25**, 287–299.

Rafaeli A. (2002) Neuroendocrine control of pheromone biosynthesis in moths. *Int. Rev. Cytol.* **213**, 49–91.

Rafaeli A. and Gileadi C. (1995a) Factors affecting pheromone production in the stored product moth, *Plodia interpunctella*: a preliminary study. *J. Stored Prod. Res.* **31**, 243–247.

Rafaeli A. and Gileadi C. (1995b) Modulation of the PBAN-stimulated pheromonotropic activity in *Helicoverpa armigera*. *Insect Biochem. Mol. Biol.* **25**, 827–834.

Rafaeli A. and Gileadi C. (1996) Down regulation of pheromone biosynthesis: cellular mechanisms of pheromonostatic responses. *Insect Biochem. Mol. Biol.* **26**, 797–808.

Rafaeli A. and Gileadi C. (1997) Neuroendocrine control of pheromone production in moths. *Invert. Neurosci.* **3**, 223–229.

Rafaeli A. and Gileadi C. (1999) Synthesis and biological activity of a photoaffinity-biotinylated pheromone-biosynthesis activating neuropeptide (PBAN) analog. *Peptides* **20**, 787–794.

Rafaeli A. and Soroker V. (1989) Cyclic AMP mediation of the hormonal stimulation of ^{14}C-acetate incorporation by *Heliothis armigera* pheromone glands in vitro. *Mol. Cel. Endocrinol.* **65**, 43–48.

Rafaeli A. and Soroker V. (1994) Second messenger interactions in response to PBAN stimulation of pheromone gland cultures. In *Insect Neurochemistry and Neurophysiology, 1993*, eds A. Borkovec and M. J. Loeb, pp. 223–226. CRC Press, Boca Raton, FL.

Rafaeli A., Soroker V., Kamensky B. and Raina A. K. (1990a) Action of pheromone biosynthesis activating neuropeptide on *in vitro* pheromone glands of *Heliothis armigera* females. *J. Insect Physiol.* **36**, 641–646.

Rafaeli A., Soroker V., Klun J. A. and Raina A. K. (1990b) Stimulation of *de novo* pheromone biosynthesis by *in vitro* pheromone glands of *Heliothis* spp. In *Insect Neurochemistry and Neurophysiology 1989*, eds A. B. Borkovec and E. P. Masler pp. 309–312. Humana Press Inc., Clifton Springs, NJ.

Rafaeli A., Hirsch J., Soroker V., Kamensky B. and Raina A. K. (1991) Spatial and temporal distribution of pheromone biosynthesis-activating neuropeptide in *Helicoverpa (Heliothis) armigera* using RIA and *in vitro* bioassay. *Arch. Insect Biochem. Physiol.* **18**, 119–129.

Rafaeli A., Soroker V., Hirsch J., Kamensky B. and Raina A. K. (1993) Influence of photoperiod and age on the competence of pheromone glands and on the distribution of immunoreactive PBAN in *Helicoverpa spp. Arch. Insect Biochem. Physiol.* **22**, 169–180.

Rafaeli A., Gileadi C., Fan Y. and Cao M. (1997a) Physiological mechanisms of pheromonostatic responses: effects of adrenergic agonists and antagonists on moth (*Helicoverpa armigera*) pheromone biosynthesis. *J. Insect Physiol.* **43**, 261–269.

Rafaeli A., Soroker V., Kamensky B., Gileadi C. and Zisman U. (1997b). Physiological and cellular mode of action of pheromone biosynthesis activating neuropeptide (PBAN) in the control of pheromonotropic activity of female moths. In *Insect Pheromone Research: New Directions*, eds R. T. Cardé and A. K. Minks, pp. 74–82. Chapman & Hall, New York.

Rafaeli A., Gileadi C. and Hirashima A. (1999) Identification of novel synthetic octopamine receptor agonists which inhibit moth sex pheromone production. *Pestic. Biochem. Physiol.* **65**, 194–204.

Rafaeli A., Zakarova T., Lapsker Z. and Jurenka R. A. (2003) The identification of an age- and female-specific putative PBAN membrane-receptor protein in pheromone glands of *Helicoverpa armigera:* Possible up regulation by Juvenile Hormone. *Insect Biochem. Mol. Biol.* **33**, 371–380.

Raina A. K. (1993) Neuroendocrine control of sex pheromone biosynthesis in Lepidoptera. *Annu. Rev. Entomol.* **38**, 329–349.

Raina A. K. (1997) Control of pheromone production in moths. In *Insect Pheromone Research: New Directions*, eds R. T. Cardé and A. K. Minks, pp. 21–30. Chapman & Hall, New York.

Raina A. and Kempe T. (1990) A pentapeptide of the C-terminal sequence of PBAN with pheromonotropic activity. *Insect Biochem.* **20**, 849–851.

Raina A. K. and Kempe T. G. (1992) Structure activity studies of PBAN of *Helicoverpa zea* (Lepidoptera: Noctuidae). *Insect Biochem. Mol. Biol.* **22**, 221–225.

Raina A. K. and Klun J. A. (1984) Brain factor control of sex pheromone production in the female corn earworm moth. *Science* **225**, 531–533.

Raina A. K., Menn J. J. (1987) Endocrine regulation of pheromone production in Lepidoptera. In *Pheromone Biochemistry*, eds G. D. Prestwich and G. J. Blomquist, pp. 159–174. Academic Press, Orlando, FL.

Raina A. K., Jaffe H., Klun J. A., Ridgway R. L. and Hayes D. K. (1987) Characteristics of a neurohormone that controls sex pheromone production in *Heliothis zea*. *J. Insect Physiol.* **33**, 809–814.

Raina A. K., Jaffe H., Kempe T. G., Keim P., Blacher R. W., Fales H. M., Riley C. T., Klun J. A., Ridgway R. L. and Hayes D. K. (1989) Identification of a neuropeptide hormone that regulates sex pheromone production in female moths. *Science* **244**, 796–798.

Raina A. K., Bird T. G. and Giebultowicz J. M. (1990) Receptivity terminating factor in *Heliothis zea* and its interaction with PBAN. In *Insect Neurochemistry and Neurophysiology*, eds A. B. Borkovec and E. P. Masler, pp. 299–302. Humana Press Inc., Clifton Springs, NJ.

Raina A. K., Kempe, T. G. and Jaffe H. (1991) Pheromone biosynthesis-activating neuropeptide: Regulation of pheromone production in moths. In *Insect Neuropeptides: Chemistry, Biology and Action*, eds J. J. Menn, T. J. Kelly and E. P. Masler, pp. 100–109. American Chemical Society, Washington, DC.

Raina A., Wergin W., Murphy C. and Erbe E. (2000) Structural organization of the sex pheromone gland in *Helicoverpa zea* in relation to pheromone production and release. *Arthropod Struct. Develop.* **29**, 343–353.

Ramaswamy S. B., Jurenka R. A., Linn C. E. and Roelofs W. L. (1995) Evidence for the presence of a pheromonotropic factor in hemolymph and regulation of sex pheromone production in *Helicoverpa zea*. *J. Insect Physiol.* **41**, 501–508.

Ramaswamy S. B., Qiu Y. and Park Y. I. (1996) Neuronal control of post-coital pheromone production in the moth *Heliothis virescens*. *J. Exp. Zool.* **274**, 255–263.

Sasaki M. and Riddiford L. M. (1984) Regulation of reproductive behaviour and egg maturation in the tobacco hawk moth, *Manduca sexta*. *Physiol. Entomol.* **9**, 315–327.

Sato Y., Oguchi M., Menjo N., Imai K., Saito H., Ikeda M., Isobe M. and Yamashita O. (1993) Precursor polyprotein for multiple neuropeptides secreted from the suboesophageal ganglion of the silkworm *Bombyx mori*: characterization of the cDNA encoding the diapause hormone precursor and identification of additional peptides. *Proc. Natl. Acad. Sci. USA* **90**, 3251–3255.

Sato Y., Ikeda M., Yamashita O. (1994) Neurosecretory cells expressing the gene for common precursor for diapause hormone and pheromone biosynthesis activating neuropeptide in the subesophageal ganglion of the silkworm, *Bombyx mori*. *Gen. Comp. Endocrin.* **96**, 27–36.

Schal C., Sevala V. and Cardé R. T. (1998) Novel and highly specific transport of a volatile sex pheromone by hemolymph lipophorin in moths. *Naturwissenschaften* **85**, 339–342.

Schoofs L., Holman G. M., Hayes T. K., Nachman R. J. and Loof A. D. (1990a). Isolation, identification and synthesis of locustamyotropin II, an additional neuropeptide of *Locusta migratoria*: member of the cephalomyotropic peptide family. *Insect Biochem.* **20**, 479–484.

Schoofs L., Holman G. M., Hayes T. K., Tips A., Vandesande F. and Loof A. D. (1990b) Isolation, identification and synthesis of locustamyotropin (Lom-MT), a novel biologically active insect neuropeptide. *Peptides* **11**, 427–433.

Schoofs L., Holman G. M., Hayes T. K., Nachman R. J. and DeLoof A. (1991) Isolation, primary structure and synthesis of locustapyrokinin: a myotropic peptide of *Locusta migratoria*. *Gen. Comp. Endocrinol.* **81**, 97–104.

Schoofs L., Holman G. M., Hayes T. K., Nachman R. J., Kochansky J. P. and DeLoof A. (1992a) Isolation, identification and synthesis of locustamyotropin III and IV, two additional neuropeptides of *Locusta migratoria*: members of the locustamyotropin peptide family. *Insect Biochem. Mol. Biol.* **22**, 447–452.

Schoofs L., Tips A., Holman G. M., Nachman R. J. and DeLoof A. (1992b) Distribution of locustamyotropin-like immunoreactivity in the nervous system of *Locusta migratoria*. *Regul. Pept.* **37**, 237–254.

Schoofs L., Holman G. M., Nachman R., Proost P., Vandamme J. and Deloof A. (1993) Isolation, identification and synthesis of locustapyrokinin-II from *Locusta migratoria*, another member of the FXPRLamide peptide family. *Comp. Biochem. Physiol. C* **106**, 103–109.

Shiga S., Toyoda I. and Numata H. (2000) Neurons projecting to the retrocerebral complex of the adult blow fly, *Protophormia terraenovae*. *Cell Tissue Res.* **299**, 427–439.

Smith V. L., Doyle K. E., Maune J. F., Munjaal R. P. and Beckingham K. (1987) Structure and sequence of the *Drosophila melanogaster* calmodulin gene. *J. Mol. Biol.* **196**, 471–485.

Soroker V. and Rafaeli A. (1989) In vitro hormonal stimulation of [^{14}C]acetate incorporation by *Heliothis armigera* pheromone glands. *Insect Biochem.* **19**, 1–5.

Soroker V. and Rafaeli A. (1995) Multi-signal transduction of the pheromonotropic response by pheromone gland incubations of *Helicoverpa armigera*. *Insect Biochem. Mol. Biol.* **25**, 1–9.

Subchev M. and Jurenka R. A. (2001) Identification of the pheromone in the hemolymph and cuticular hydrocarbons from the moth *Scoliopteryx libatrix* L. (Lepidoptera: Noctuidae). *Arch. Insect Biochem. Physiol.* **47**, 35–43.

Suwan S., Isobe M., Yamashita O., Minakata H. and Imai K. (1994) Silkworm diapause hormone, structure–activity relationships: indispensable role of C-terminal amide. *Insect Biochem. Mol. Biol.* **24**, 1001–1007.

Tang J. D., Charlton R. E., Cardé R. T. and Yin C.-M. (1987) Effect of allatectomy and ventral nerve cord transection on calling, pheromone emission and pheromone production in *Lymantria dispar*. *J. Insect Physiol.* **33**, 469–476.

Tang J. D., Charlton R. E., Jurenka R. A., Wolf W. A., Phelan P. L. and Sreng L. and Roelofs W. L. (1989) Regulation of pheromone biosynthesis by a brain hormone in two moth species. *Proc. Natl. Acad. Sci. USA* **86**, 1806–1810.

Tang J. D., Wolf W. A., Roelofs W. L. and Knipple D. C. (1991) Development of functionally competent cabbage looper moth sex pheromone glands. *Insect Biochem.* **21**, 573–581.

Teal P. E. A., Carlysle T. C. and Tumlinson J. H. (1983) Epidermal glands in terminal abdominal segments of female *Heliothis*. *Ann. Entomol. Soc. Am.* **76**, 242–247.

Teal P. E. A., Abernathy R. L., Nachman R. J., Fang N., Meredith J. A., Tumlinson J. H. (1996) Pheromone biosynthesis activating neuropeptides: functions and chemistry. *Peptides* **17**, 337–344.

Thyagaraja B. S. and Raina, A. K. (1994) Regulation of pheromone production in the gypsy moth, *Lymantria dispar*, and development of an *in vitro* bioassay. *J. Insect Physiol.* **40**, 969–974.

Tips A., Paemen L., Schoofs L., Ma M. and Blackburn M. (1993) Co-localization of locustamyotropin- and pheromone biosynthesis activating neuropeptide-like immunoreactivity in the central nervous system of five insect species. *Comp. Biochem. Physiol.* **106A**, 195–207.

Torfs P., Nieto J., Cerstiaens A., Boon D., Baggerman G., Poulos C., Waelkens E., Derua R., Calderon J. and De Loof A., *et al.* (2001). Pyrokinin neuropeptides in a crustacean – isolation and identification in the white shrimp *Penaeus vannamei*. *Eur. J. Biochem.* **268**, 149–154.

Tumlinson J. H., Fang N. and Teal P. E. A. (1997) The effect of PBAN on conversion of fatty acyls to pheromone aldehydes in female I. In *Insect Pheromone Research: New Directions*, eds R. T. Cardé and A. K. Minks, pp. 54–55. Chapman & Hall, New York.

Veelaert D., Schoofs L., Verhaert P. and De Loof A. (1997) Identification of two novel peptides from the central nervous system of the desert locust, Schistocerca gregaria. *Biochem. Biophys. Res. Commun.* **241**, 530–534.

Wolfner M. F. (1997) Tokens of love: functions and regulation of *Drosophila* male accessory gland products. *Insect Biochem. Mol. Biol.* **27**, 179–192.

Yoshiga T., Yokoyama N., Imai N., Ohnishi A., Moto K. and Matsumoto S. (2002) cDNA cloning of calcineurin heterosubunits from the pheromone gland of the silkmoth, *Bombyx mori*. *Insect Biochem. Mol. Biol.* **32**, 477–486.

Zdarek J., Nachman R. J. and Hayes T. K. (1997) Insect neuropeptides of the pyrokinin/PBAN family accelerate pupariation in the fleshfly (*Sarcophaga bullata*) larvae. *Annu. NY Acad. Sci.* **814**, 67–72.

Zdarek J., Nachman R. J. and Hayes T. K. (1998) Structure-activity relationships of insect neuropeptides of the pyrokinin/PBAN family and their selective action on pupariation in fleshfly (*Neobelleria bullata*) larvae (Diptera, Sarcophagidae). *European J. Entomol.* **95**, 9–16.

Zhang T., Zhang L., Xu W. and Shen J. (2001) Cloning and characterization of the cDNA of diapause hormone-pheromone biosynthesis activating neuropeptide of *Helicoverpa armigera*. GenBank Direct Submission.

Zhao J. Z. and Haynes K. F. (1997) Does PBAN play an alternative role of controlling pheromone emission in the cabbage looper moth, *Trichoplusia ni* (Hubner) (Lepidoptera: Noctuidae). *J. Insect Physiol.* **43**, 695–700.

Zhu J., Millar J. and Löfstedt C. (1995) Hormonal regulation of sex pheromone biosynthesis in the turnip moth, *Agrotis segetum*. *Arch. Insect Biochem. Physiol.* **30**, 41–59.

6

Biosynthesis and endocrine regulation of pheromone production in the Coleoptera

Steven J. Seybold and Désirée Vanderwel

6.1 Introduction

The beetles (Order Coleoptera) are thought to have arisen as a lineage of animals some 240 to 300 million years ago during the Upper Carboniferous and Permian Periods (Paleozoic Era) when the abundance of plants provided new habitat opportunities, such as spaces within soil and leaf litter (the interstitial zone), rotted wood, and living or moribund plant tissues, including wood (Crowson, 1981; Lawrence and Newton, 1982, 1995; Evans and Bellamy, 1996). The evolution of the strongly sclerotized wing covers or elytra is thought to be one of the many adaptational outcomes to these newly available closed or compact *niches* and, indeed, the possession of the elytra is the namesake of the order (Coleoptera is from the Greek *koleon*, "sheath," and *pteron*, "wing"). Approximately 350 000 species of beetles have been described and estimates of described and undescribed species approach 8 million. Thus, beetles are the largest group of animals in the world, representing one-fifth of all known living organisms and one-fourth of all animals (Evans and Bellamy, 1996). Since 1758, when organisms were first described using binomial nomenclature, an average of four new beetle species have been described each day. Taxonomists have divided this diversity of beetles into 166 families. One family (the weevils, Family Curculionidae) represents nearly one-half of all species of beetles (Lawrence and Newton, 1995).

The extreme biological diversity of the Coleoptera is mirrored in the structural diversity of the sex and aggregation pheromones produced by these insects (Tillman *et al.*, 1999). Intraspecific behavior in the Coleoptera is mediated by complex and diverse structures ranging from acyclic chains (unbranched, branched, and/or variously functionalized) to monocyclic lactones, macrolides, and aromatics, to bicyclic oxygen heterocycles (Vanderwel and Oehlschlager, 1987). These structures may arise through modifications of fatty acid, isoprenoid, and/or amino acid metabolism, and the modifications often include stereospecific formation of the end products. The abundance of chiral centers in the coleopteran pheromone components enhances their complexity and diversity. Some of the most dramatic examples of the interplay between enantiomeric composition and behavior occur in the Coleoptera (reviewed in Seybold, 1993; Mori, 1996; Plarre and Vanderwel, 1999). Although exhaustive lists have not been compiled recently, Bestmann and Vostrowsky (1985) reported 90 pheromone components isolated from 85 species of beetles, whereas Mayer and McLaughlin (1991) reported pheromones from 143 species of beetles (or 0.04 percent of the 350 000 described species). Francke and Schulz (1999) present many more examples of coleopteran pheromone components, but do not give a numerical listing by species or chemical structure.

The biosynthetic pathways and regulatory factors that control coleopteran pheromone production are of interest because they provide the context in which pheromones are made available to beetles in nature. Furthermore, understanding these pathways and their regulation may lead to biotechnological approaches for the syntheses of pheromone components of high chemical and enantiomeric purity for application in pest management programs. The pathways and regulating mechanisms themselves are also potential targets for manipulating populations of pestiferous beetles. When the topic of coleopteran pheromone biosynthesis was reviewed in the predecessor to this book (Vanderwel and Oehlschlager, 1987), the approach was largely bio-organic and predictive. Only one published instance of *de novo* biosynthesis was known. Much has changed since the publication of *Pheromone Biochemistry* (Prestwich and Blomquist, 1987). Strong evidence of *de novo* pheromone biosynthesis is now available from species of Coleoptera from five families and the molecular aspects of the regulation of pheromone biosynthesis are beginning to be understood in one family (Scolytidae, Chapter 7). In this chapter, we review pheromone production and regulation in beetles belonging to seven families: the Silvanidae, Laemophloeidae (formerly Cucujidae), Scarabaeidae, Nitidulidae, Tenebrionidae, Scolytidae, and Curculionidae. Unlike many previous reviews, this overview will focus on experimental results of biosynthetic studies rather than on predicted biochemical transformations based on chemical structures.

6.2 Silvanidae/Laemophloeidae: flat bark beetles

Males in the genera *Oryzaephilus* (family Silvanidae) and *Cryptolestes* (family Laemophloeidae) produce several structurally related macrocyclic lactones, given the trivial name "cucujolides" (Figure 6.1) [the genera *Oryzaephilus* and *Cryptolestes* were formerly in the family Cucujidae, but are now classified in the families Silvanidae and Laemophloeidae, respectively (Lawrence and Newton, 1995)]. Species-specific combinations of these cucujolides function as aggregation pheromones (reviewed in Oehlschlager *et al.*, 1988). Cucujolide I was expected to be of terpenoid origin, due to the characteristic branching pattern and the *E* configuration of the double bonds (Pierce *et al.*, 1984). The other cucujolides were expected to be of fatty acid origin, due to the position (6–7 and/or 9–10 from the terminal carbon) and *Z* geometry of the double bond(s) (Pierce *et al.*, 1984).

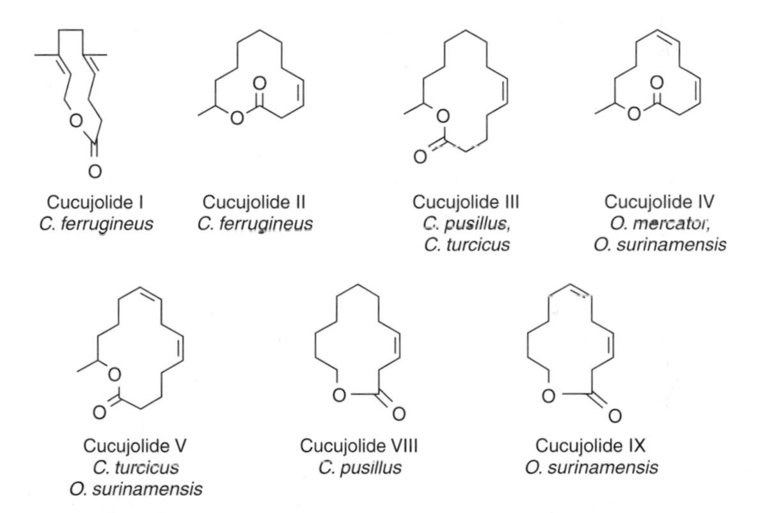

Figure 6.1 Cucujolide pheromone components produced by beetles in the genera *Oryzaephilus* (family Silvanidae) and *Cryptolestes* (family Laemophloeidae): (*E,E*)-4,8-dimethyl-4,8-decadien-10-olide (cucujolide I); (*Z*)-3-dodecen-11-olide (cucujolide II); (*Z*)-5-tetradecen-13-olide (cucujolide III); (*Z,Z*)-3,6-dodecadien-11-olide (cucujolide IV); (*Z,Z*)-5,8-tetradecadien-13-olide (cucujolide V); (*Z*)-3-dodecen-12-olide (cucujolide VIII); and (*Z,Z*)-3,6-dodecadien-12-olide (cucujolide IX) (numbered according to Oehlschlager *et al.*, 1988). Species known to use the compounds as pheromone components are indicated (reviewed in Oehlschlager *et al.*, 1988).

The biosynthesis of the cucujolides was examined in the merchant grain beetle, *O. mercator* Fauvel, and the rusty grain beetle, *C. ferrugineus* (Stephens), using radio- and stable isotope-labeled precursors (Vanderwel *et al.*, 1990, 1992b).

Insects were fed oats impregnated with the precursors of interest. The pheromones emanating from the feeding insects were trapped on a solid adsorbant and analyzed for the incorporation of precursor. The isoprenoid origin of cucujolide I was confirmed by the incorporation of radiolabeled acetate and mevalonate into the pheromone (purified to constant specific activity) (Vanderwel *et al.*, 1990). Deuterium-labeled farnesol was also converted to cucujolide I, as determined by gas chromatography-mass spectrometry (GC-MS) (Vanderwel *et al.*, 1992b): this conversion likely proceeds through oxidative cleavage of the terminal double bond, followed by cyclization (Figure 6.2). Dual labeling studies with ^{2}H and ^{18}O indicated that this cyclization proceeds with retention of the hydroxyl oxygen.

(*E, E*)-Farnesol ^{2}H,^{18}O-labeled

Cucujolide I ^{2}H,^{18}O-labeled

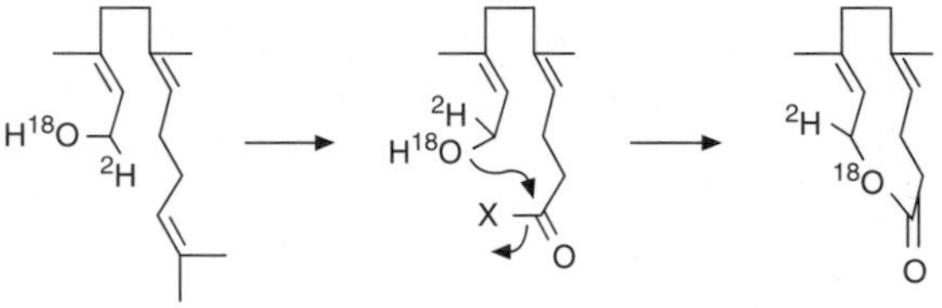

Figure 6.2 Biosynthesis of cucujolide I from (*E,E*)-farnesol. Oxidative cleavage of the terminal double bond would yield a hydroxy acid derivative, which could cyclize to give cucujolide I (X = a leaving group such as CoA). Dual labeling experiments with ^{2}H and ^{18}O indicated that the hydroxyl oxygen of farnesol is retained in the product (Vanderwel *et al.*, 1992b).

The other cucujolides are of fatty acid origin, as evidenced by their production from various radiolabeled fatty acid substrates (Vanderwel *et al.*, 1990). The biosynthesis of cucujolide II (Figure 6.3) is a model for analogous routes to all other cucujolides. Theoretically, the desaturation, hydroxylation, and β-oxidation steps required to convert saturated fatty acids into cucujolide II could occur in any order. However, it appears that the biosynthetic route actually used follows "normal" fatty acid anabolism as closely as possible (Figure 6.3). Insertion of the *Z* double bond (likely through the action of a Δ^{9} desaturase) produces oleic acid as usual; next, chain shortening (possibly through normal β-oxidation) produces the 12-carbon chain. Steps unique to the pheromone biosynthetic pathway occur at the end of the pathway: oxidation at the penultimate (ω-1) carbon to form a hydroxy-acid derivative, and cyclization. Dual labeling studies with ^{2}H and ^{18}O indicated that this conversion proceeds with retention of the hydroxyl oxygen. Not surprisingly, *C. ferrugineus* and *O. mercator* produced cucujolide II of the naturally occurring enantiomeric composition (*S* and *R*, respectively) even when presented with racemic 11-hydroxy-(*Z*)-3-dodecenoic acid, indicating that the cyclization step is enantioselective. The enantioselectivity of the unique ω-1 hydroxylase was not investigated.

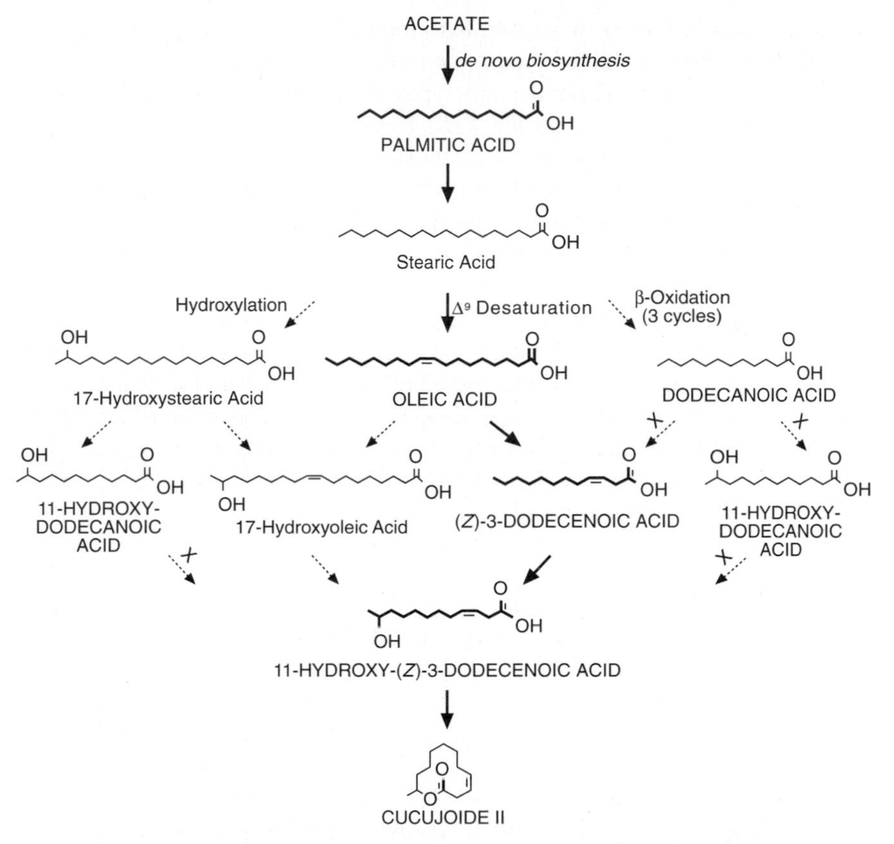

Figure 6.3 Biosynthesis of cucujolide II, showing several possible pathways. Substrates with names entirely in capital letters have been tested for incorporation into cucujolide II (Vanderwel *et al.*, 1990, 1992b). Precursors that were incorporated are shown by solid arrows; those that were not have a cross above the arrow. The actual biosynthetic pathway is shown in heavy arrows, and likely proceeds through activated (e.g. CoA) derivatives of the fatty-acyl intermediates.

The production of the cucujolide aggregation pheromones is very low in the absence of food (Pierce *et al.*, 1984; Oehlschlager *et al.*, 1988). However, rather than feeding serving to provide fatty acid precursors, the effect of feeding is more likely related to hormonal regulation of pheromone biosynthesis. Cucujolides can be produced *de novo* (Vanderwel *et al.*, 1990) and pheromone production is stimulated by the juvenile hormone analogue (JHA) methoprene (Pierce *et al.*, 1986). The activity of the unique ω-1 hydroxylase is apparently at least one of the enzymatic targets subject to hormonal regulation (Vanderwel, 1991). The presentation of 11-hydroxy-(*Z*)-3-dodecenoic acid allows cucujolide II production by *O. mercator* deprived of an adequate food source, but the presentation of (*Z*)-3-dodecenoic acid cannot bypass the point of regulation.

Macrocylic lactones are not the only types of pheromone components used by *Oryzaephilus* species. For example, (3*R*)-1-octen-3-ol (3-octenol) is produced by both sexes of *O. mercator* and *O. surinamensis* L. at a low population density (Pierce *et al.*, 1989). The pheromone is attractive at low doses, but repellent at higher doses. 3-Octenol is also associated with oat volatiles (White *et al.*, 1989), and with the fungi associated with stored grain (Pierce *et al.*, 1991a). This is similar to the situation with scolytid bark beetles, where behaviorally active compounds may be provided not only by the beetles themselves, but also from exogenous sources including microbes and the host (see section 6.6.3.1). This recurring theme may reflect one hypothetical mechanism for evolution of coleopteran pheromones. Certain species of beetles may have evolved to respond positively to volatiles indicative of a favorable environment for reproduction (presence of food, moisture and heat). Incidental production of those same volatiles by the beetles themselves would attract conspecifics, reinforcing the genetic disposition to produce the attractive compounds.

6.2.1 Other silvanid beetles

Other silvanid pheromones have been reported in the literature. For example, Pierce *et al.* (1988) identified the male-produced aggregation pheromone of the square-necked grain beetle, *Cathartus quadricollis* (Guerin-Méneville), as (7*R*, 6*E*)-methylnonen-3-yl acetate, given the trivial name "quadrilure." Both sexes of the foreign grain beetle, *Ahasverus advena* (Waltl) produce the aggregation pheromone 3-octenol (Pierce *et al.*, 1991b). These pheromones appear to be of fatty acid origin, but their biosynthetic pathways remain to be investigated.

6.3 Scarabaeidae: dung and chafer beetles

The female-produced sex pheromones of several different species of scarab beetles have been identified, and they include a wide variety of structures. Although many appear to be fatty acid-derived, isoprenoid- and amino acid-derived compounds are also represented (reviewed in Leal, 1998b, 1999).

The biosyntheses of japonilure and buibuilactone, which are used as components of sex pheromones by several scarab species (particularly those in the subfamily Rutelinae) (Figure 6.4) (Leal, 1999), were investigated in *Anomala cuprea* Hope and *A. osakama* Sawada using deuterium-labeled precursors (Leal *et al.*, 1999). Japonilure and buibuilactone are similar to the cucujolides (section 6.2), in that they appear to be fatty acid derived and their structures involve a lactone ring. Leal *et al.* (1999) confirmed that, like the cucujolides, japonilure and buibuilactone are of fatty acid origin. However, there are significant differences between the two pathways. The hydroxylation steps required to produce japonilure and

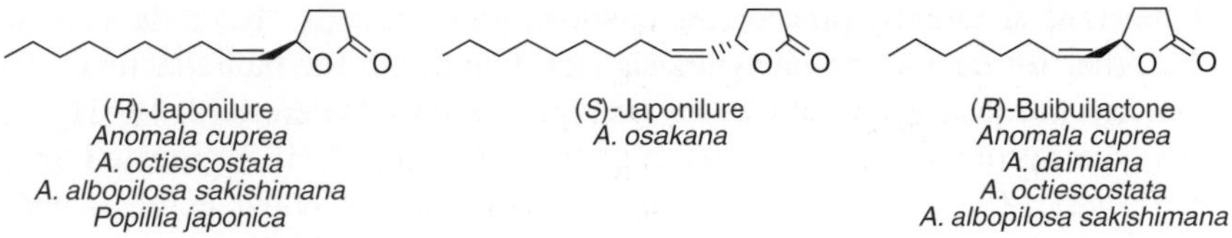

(*R*)-Japonilure
Anomala cuprea
A. octiescostata
A. albopilosa sakishimana
Popillia japonica

(*S*)-Japonilure
A. osakana

(*R*)-Buibuilactone
Anomala cuprea
A. daimiana
A. octiescostata
A. albopilosa sakishimana

Figure 6.4 Lactone pheromone components produced by beetles in the genus *Anomala* (family Scarabaeidae): (*R*)- and (*S*)-(*R,Z*)-5-(–)-(dec-1-enyl)-oxacyclopentan-2-one (japonilure) and (*R,Z*)-5-(–)-(oct-1-enyl)-oxacyclopentan-2-one (buibuilactone). Species known to use the compounds as pheromone components are indicated (reviewed in Leal, 1999).

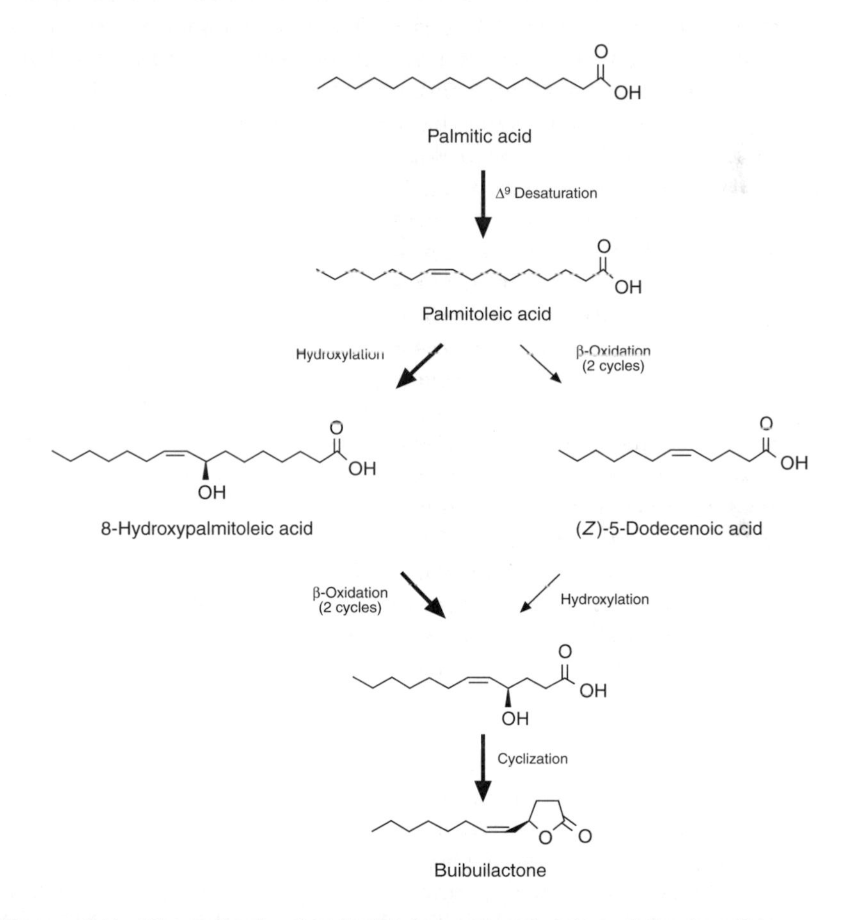

Figure 6.5 Biosynthesis of buibuilactone, showing two of the possible pathways. The actual pathway as elucidated by Leal *et al.* (1999) is shown with heavy arrows.

buibuilactone apparently precede the chain-shortening steps: this is the reverse of the order for cucujolide biosynthesis (see Figure 6.5 for buibuilactone: the biosynthesis of japonilure follows an analogous route). Whereas [9, 10-$^2H_2$]-8-hydroxypalmitoleic acid was efficiently (43 percent) incorporated into buibuilactone, (*Z*)-[5,6-$^2H_2$]-5-dodecenoic acid was not (Leal *et al.*, 1999). Furthermore, *A. cuprea*, which normally produces only (*R*)-buibuilactone, converted the racemic hydroxylated precursor to both (*R*)- and (*S*)-buibuilactone. This suggests that both enantiomers of 8-hydroxypalmitoleic acid are not normally formed *in vivo*: i.e. that the hydroxylation step is highly enantioselective. The mechanism of this unique 8-hydroxylase was investigated using perdeuterated palmitic acid. The buibuilactone produced was fully deuterated, indicating that the hydroxylation does not proceed through preceding desaturation or epoxidation steps, but rather through direct insertion of oxygen. Leal *et al.* (1999) noted that species specificity of sex pheromone production in the Scarabacidae has been achieved by different taxa evolving to specialize in the stereospecificity or regiospecificity of this hydroxylation step.

In contrast to the rutelines, the melolonthine scarabs generally use terpenoid- and amino acid-derived pheromones (reviewed in Leal, 1999). For example, the female large black chafer, *Holotrichia parallela* Motschulsky, produces methyl (2*S*,3*S*)-2-amino-3-methylpentanoate (L-isoleucine methyl ester) as an amino acid-derived sex pheromone (Leal *et al.*, 1992; Leal, 1997). There is no direct evidence that the chafer beetles or any other Coleoptera use the shikimic acid pathway for *de novo* pheromone biosynthesis, but some scarabs and scolytids (see section 6.6.4.2) may convert amino acids such as tyrosine, phenylalanine, or tryptophan to aromatic pheromone components (Leal, 1997, 1999). In another melolonthine species, the female grass grub beetle, *Costelytra zealandica* (White), the phenol sex "pheromone" is produced by symbiotic bacteria (Henzell and Lowe, 1970; Hoyt *et al.* 1971).

Melolonthine scarab beetles in the genera *Holotrichia* and *Phyllophaga* are unusual in that they exhibit calling behavior (Leal *et al.*, 1992, 1993, 1996a). *Holotrichia* beetles exhibit this calling behavior every other evening, soon after sunset (Leal *et al.*, 1993). In *H. parallela*, the titer of the major sex pheromone component (L-isoleucine methyl ester) peaks every 48 hours, corresponding to the unusual "circabidian" periodicity of the calling behavior (Leal *et al.*, 1993). The regulation of sex pheromone production in the Scarobaeidae in general is only poorly understood, but preliminary evidence suggests that hormonal factors besides JH (e.g. a neuropeptide) may be involved (Leal, 1998a).

The site of sex pheromone production has been examined in several species of ruteline and melolonthine beetles through both chemical analyses (extraction of the tissues of interest and analysis for pheromone content by GC-MS) and through histological and morphological studies (Tada and Leal, 1997; Kim and Leal, 1998; reviewed in Leal, 1999). In the three ruteline species examined, the source of the pheromone is apparently female-specific epithelial cells, observed lining the inner surface of the anal plates and two apical sternites. Pheromone is apparently released from these cells to the cuticle surface via fine pores (Tada and Leal, 1997). The source of the sex pheromone of *Holotrichia parallela* was originally described as the ball-shaped abdominal sac exposed during the calling behavior (Tada and Leal, 1997), but was later reported as a layer of cuticular epithelium in the posterior part of the eversible pheromone-producing organ (Kim and Leal, 1998). The yellowish elongate chafer, *Heptophylla picea* Motschulsky, produces the pheromone component (*R*, *Z*)-7,15-hexadecadien-4-olide, which may be derived from stearic acid (Leal *et al.* 1996b; Leal, 2001). In this species pheromone production was localized to epithelial cells lining the inner surface of the anal plates and two apical sternites (Tada and Leal, 1997). Although classified in the Melolonthinae, the biochemistry and biology of pheromone production in *H. picea* bears greater similarity to these processes in the ruteline species (Tada and Leal, 1997).

6.4 Nitidulidae: sap beetles

Within days of being placed on an appropriate food and colonization source, male sap beetles (genus *Carpophilus*) produce aggregation pheromones (reviewed in Bartelt, 1999a). The pheromones are comprised of a series of methyl- or ethyl-branched, conjugated, triene and tetraene hydrocarbons. In all, 22 such hydrocarbons with varying biological activities are produced by the nine species of *Carpophilus* studied (reviewed in Bartelt, 1999a). Similar carbon skeletons have been previously reported in anobiid beetles. For example, the carbon skeleton of stegobinone, produced by the drugstore beetle, *Stegobium paniceum* (L.) (Kuwahara *et al.*, 1978) is identical to that of the *Carpophilus hemipterus* L. pheromone (2*E*,4*E*,6*E*,8*E*)-3,5,7-trimethyl-2,4,6,8-decatetraene (Bartelt *et al.*, 1991) (Figure 6.6). Chuman *et al.* (1983) proposed that the anobiid pheromones

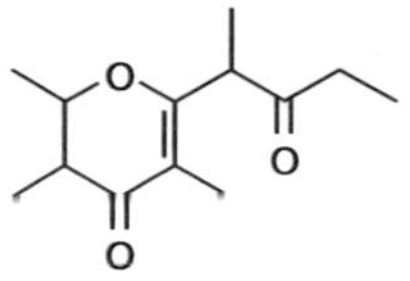

Stegobinone

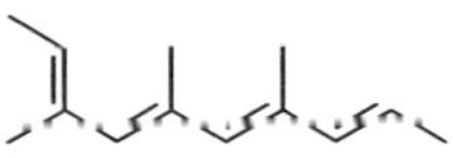

(2*E*,4*E*,6*E*,8*E*)-3,5,7-trimethyl-2,4,6,8-decatetraene

Figure 6.6 Structures of 2,3-dihydro-2,3,5-trimethyl-6-(1′-methyl-2′-oxobutyl)-4*H*-pyran-4-one (stegobinone), pheromone of *Stegobium paniceum* (L.) (Anobiidae) (Kuwahara *et al.*, 1978) and (2*E*,4*E*,6*E*,8*E*)-3,5,7-trimethyl-2,4,6,8-decatetraene, pheromone component of *Carpophilus hemipterus* (L.) (Nitidulidae) (Bartelt *et al.*, 1991). The carbon skeletons are identical, and likely derived through similar pathways (see text).

are of polyketide origin: the branching and oxygenation pattern is consistent with the assembly of the skeleton from acetate and propionate units. Bartelt *et al.* (1990, 1991) proposed that the aggregation pheromones of the nitidulid beetles are likewise assembled through the condensation of small acyl units, including propionate (to introduce the methyl branches) and butyrate (to introduce the ethyl branches). The biosynthetic pathway was not postulated to involve a polyketide intermediate. Rather, it was seen as more likely that the reduction and dehydration steps would proceed in cyclic fashion as during normal fatty acid biosynthesis, leaving the *E*-double bonds (Bartelt *et al.*, 1990, 1991; Petroski *et al.*, 1994). Decarboxylation would yield the hydrocarbon product of appropriate chain length.

The assembly of the carbon skeletons of these unusual hydrocarbons was first studied in *Carpophilus freemani* Dobson, through careful GC-MS and Nuclear Magnetic Resonance (NMR) studies of the incorporation of ^{2}H or ^{13}C-labeled precursors (Petroski *et al.*, 1994). Assembly of the carbon skeleton of the aggregation pheromone of *C. freemani*, (2*E*,4*E*,6*E*)-5-ethyl-3-methyl-2,4,6-nonatriene, involves initiation with acetate; elongation with first propionate (to provide the methyl branch), then butyrate (to provide the ethyl branch); and chain termination with a second butyrate (Figure 6.7). At some point, loss of CO_2 from one of the butyrate units occurs to yield the appropriate hydrocarbon, but Petroski *et al.* (1994) were unable to determine which of the butyrate units loses its carboxyl group. Bartelt and Weisleder (1996) studied the biosynthesis of 15 additional methyl- and/or ethyl-branched, tri- and tetraenes in the related

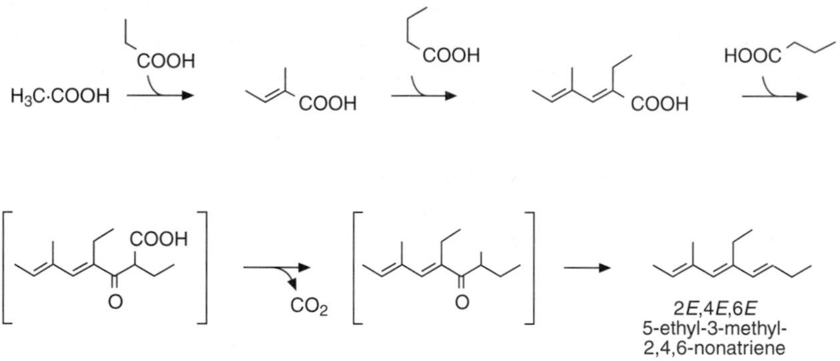

Figure 6.7 Biosynthesis of (2*E*,4*E*,6*E*)-5-ethyl-3-methyl-2,4,6-nonatriene, pheromone component of *Carpophilus freemani* Dobson (Nitidulidae). The pathway is a modification of normal fatty acid biosynthesis, involving initiation with acetate; elongation with first propionate (to provide the methyl branch), then butyrate (to provide the ethyl branch); and chain termination with a second butyrate. The final butyrate to add suffers a loss of CO_2, after adding in a unique head-to-head reaction. Intermediates of the pathway likely occur as activated (e.g. CoA) derivatives (Petroski *et al.*, 1994; Bartelt and Weisleder, 1996).

species *C. davidsoni* Dobson and *C. mutilatus* Erichson. In all cases, the carbon skeletons were assembled from acetate, propionate, and butyrate, in patterns analogous to that of the *C. freemani* pheromone. It was unambiguously determined that the final acyl unit to be added is the one that loses its carboxyl carbon (Figure 6.7). Thus the pathway involves a unique head-to-head assembly of acyl units (Bartelt and Weisleder, 1996; Francke and Schulz, 1999).

It is interesting to speculate on the biosynthetic origin of the complicated blend of hydrocarbons produced by the *Carpophilus* species. Of the 22 methyl- and/or ethyl-branched, tri- and tetraene hydrocarbons identified, only eight have demonstrated behavioral activity (summarized in Bartelt, 1999a). It would seem inefficient for there to be a unique set of enzymes to effect the synthesis of each of these hydrocarbons, since most of them serve no apparent behavioral role. Bartelt and coworkers advanced the hypothesis that all of the methyl- and/or ethyl-branched, tri- and tetraene hydrocarbons produced in a given organism are produced by a single enzyme system: that the variety results from the "imperfect specifity" of this enzyme system for acyl units (Bartelt *et al.*, 1992; Bartelt and Weisleder, 1996; Bartelt, 1999a). This hypothesis can be understood by considering the 11 different tetraenes described from *C. hemipterus* (Table 6.1; Bartelt, 1999a). The most abundant tetraene, designated as 1 in Table 6.1, is biosynthesized through chain initiation with an acetate unit, followed by chain elongation with four propionate units. If tetraene 1 is the "normal" product of the biosynthetic system, the incorrect substitution of one acyl unit would account for the production of tetraenes 2–5. Incorrect substitution of two acyl units would account for the production of tetraenes 6–11. The occurrence of two "errors" would be less likely than the occurrence of one error, and this is reflected in the observed relative abundances of the tetraenes. The occurrence of tetraenes with three or four acyl group substitutions would be even less likely: such tetraenes were not detected. This interesting hypothesis not only provides an explanation for why most of the nitidulid beetles might produce such complicated blends of tri- and tetraene hydrocarbons, but also provides a mechanism for the evolution of new pheromones (Bartelt, 1999a). Interestingly, the actual pheromone for *C. hemipterus* is composed of only the subset of tetraenes (1, 4, 5 and 11) that are formed through the "Ac-Pr-Pr" incorporation pattern of acyl groups (Table 6.1; Bartelt *et al.*, 1992; Petroski and Vaz, 1995; Bartelt, 1999a). Thus, the behavioral activity appears to be correlated with the structural identity (and molecular shape) on one (the chain-initiated) side of the molecules.

The site of hydrocarbon pheromone production in the nitidulid beetles has been identified as very large, round abdominal cells, connected to the tracheal system by fine ducts (Dowd and Bartelt, 1993; Nardi *et al.*, 1996). The secretory cells contain several lipid spheres (presumably pheromone and pheromone precursors), and are dramatically larger in beetles that are actively emitting pheromone (Nardi *et al.*, 1996). Pheromone apparently diffuses from the cells,

Table 6.1 Biosynthesis of male-specific tetraenes of *C. hemipterus* through "imperfect specificity." The simple acyl biosynthetic precursors of each tetraene are indicated; acyl unit substitutions from the main incorporation pattern for tetraene **1** are indicated by an asterisk. Relative abundances and wind tunnel bioassay activities in *C. hemipterus* are also presented. Adapted from Bartelt (1999a) with permission

Acyl unit[a]							
1st	2nd	3rd	4th	5th	Structure	Relative abundance	Bioassay activity
Main pattern of acyl units (tetraene 1):							
Ac	Pr	Pr	Pr	Pr	1	100	Yes
One acyl substitution, relative to tetraene 1:							
Pr*	Pr	Pr	Pr	Pr	2	6	No
Ac	Pr	Bu*	Pr	Pr	3	3	No
Ac	Pr	Pr	Bu*	Pr	4	8	Yes
Ac	Pr	Pr	Pr	Bu*	5	13	Yes
Two acyl substitutions, relative to tetraene 1:							
Pr*	Pr	Bu*	Pr	Pr	6	1	No
Pr*	Pr	Pr	Bu*	Pr	7	1	No
Pr*	Pr	Pr	Pr	Bu*	8	1	No
Ac	Pr	Bu*	Bu*	Pr	9	0.2	No
Ac	Pr	Bu*	Pr	Bu*	10	0.5	No
Ac	Pr	Pr	Bu*	Bu*	11	2	Yes

[a]Ac = acetate; Pr = propionate; Bu = butyrate.

through the ductules, through the tracheal branches, and eventually to the cuticular surface of the male. The common house fly, *Musca domestica* L. (Diptera: Muscidae), also produces its hydrocarbon pheromones in abdominal oenocytes (see Blomquist, Chapter 8, in this volume).

6.5 Tenebrionidae: darkling beetles

6.5.1 Yellow mealworm beetle, *Tenebrio molitor* L.

Several pheromones may be involved in mediating the mating behavior of the yellow mealworm, *Tenebrio molitor* L. (reviewed in Plarre and Vanderwel, 1999), but only one has been identified to date. Tanaka *et al.* (1986, 1989) identified the female-produced male attractant as (4*R*)-(+)-4-methylnonan-1-ol (4-methylnonanol). Females produce the pheromone through a modification of normal fatty acid biosynthesis (Islam *et al.*, 1999; Bacala, 2000). Initiation of the pathway with one unit of propionate results in the uneven number of carbons in the chain; incorporation of another unit of propionate during elongation provides the methyl branch; reduction of the fatty acyl intermediate produces the alcohol pheromone (Figure 6.8).

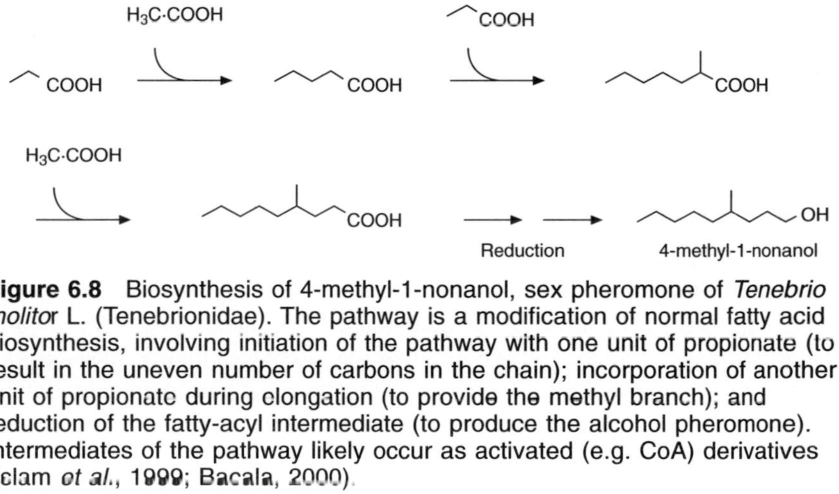

$H_3C{\cdot}COOH$ COOH COOH COOH COOH
$H_3C{\cdot}COOH$ COOH OH
Reduction 4-methyl-1-nonanol

Figure 6.8 Biosynthesis of 4-methyl-1-nonanol, sex pheromone of *Tenebrio molitor* L. (Tenebrionidae). The pathway is a modification of normal fatty acid biosynthesis, involving initiation of the pathway with one unit of propionate (to result in the uneven number of carbons in the chain); incorporation of another unit of propionate during elongation (to provide the methyl branch); and reduction of the fatty-acyl intermediate (to produce the alcohol pheromone). Intermediates of the pathway likely occur as activated (e.g. CoA) derivatives (Islam *et al.*, 1999; Bacala, 2000).

Newly eclosed and immature adult females produce relatively little "sex pheromone," but bioassays indicate that "sex pheromone" production increases as the female matures (Valentine, 1931; Happ and Wheeler, 1969) and that this increase can be stimulated by JH III (Menon, 1970, 1976). Unambiguous interpretation of these results is confounded by the fact that the "sex pheromone" was quantified by bioassay, generally measuring the ability of the female extract to elicit a "copulation release" (CR) response from the males. However, female *T. molitor* produces both 4-methylnonanol, which functions as a male attractant (but does not elicit CR activity at low doses), *and* a CR pheromone (active only in the presence of 4-methylnonanol) (Tanaka *et al.*, 1986; Vanderwel *et al.*, 2003). Both *in vivo* (Islam, 1996) and *in vitro* (Bacala, 2000) radioassays demonstrated that mature females do produce 4-methylnonanol; immature females do not produce 4-methylnonanol; and that immature females can be induced to produce 4-methylnonanol when stimulated by the JHA methoprene. Corpora allata (CA) excised from mature females can produce JH III *in vitro*, whereas those excised from immature females cannot (Judy *et al.* 1975). Thus, as the females mature, the JH titers rise to a level where maturation processes such as ovary development and sex pheromone production are stimulated. The reduction of 4-methylnonanoyl CoA to the alcohol pheromone is at least one of the enzymatic steps affected by JH (Bacala, 2000).

Mating causes a transient (about 12 h) reduction in the activity of female extracts (Griffith *et al.*, 2003). Gas chromatographic (GC) analyses indicate that the quantity of 4-methylnonanol in mature virgin females drops significantly after mating. The biological role of the mating-induced drop in pheromone production by female *T. molitor*, and the mechanism by which it is effected, are not known. Since the last male to mate with the female gains a high level of sperm precedence, it is possible that the responding male induces the inhibition in the female to reduce the possibility of rapid multiple matings between the female and other males (Gage, 1992; Siva-Jothy *et al.*, 1996; Drnevich *et al.*, 2000). In addition to previously reported more physical strategies (e.g. mate guarding, copulatory interruption, and aggression, Gage and Baker, 1991; Palomino *et al.*, 1994), mating induced depression of pheromone production would provide another mechanism to reduce the number of copulatory events with competing males. On the other hand, it may be possible that it is adaptive for the female to reduce the level of sex attractant production for a brief period of time, in order to reduce "harassment" by males so that she can lay some eggs without interruption before entertaining the advances of more males. The female yellow mealworm can lay fertile eggs within 3 h of mating. However, although it may be advantageous for the female to reduce her attractiveness for a transient period, females that do not mate at least every two days do not produce as many eggs as females that mate more regularly (Drnevich *et al.*, 2001). Thus a permanent inhibition of sex pheromone production would not be advantageous to the female.

6.5.2 Flour beetles (*Tribolium* spp.)

Male-produced aggregation pheromones are also present in the *Tribolium* spp. of flour beetles (reviewed in Plarre and Vanderwel, 1999). Many species use 4,8-dimethyldecanal ("tribolure"), and the role of other compounds remains to be clarified. The biosynthesis of 4,8-dimethyldecanal has not been investigated, although it is suspected to be of isoprenoid origin (Vanderwel and Oehlschlager, 1987).

As with many other stored product beetles, the production of aggregation pheromone by *Tribolium* spp. is dependent upon feeding (reviewed in Plarre and Vanderwel, 1999), but the mechanism by which this process is regulated remains to be investigated. In an unusual development, the results of two studies indicate that light can stimulate pheromone production in the red flour beetle, *Tribolium castaneum* Herbst. Male *T. castaneum* maintained under unnaturally long light conditions (24 h light) produced nearly four times as much 4,8-dimethyldecanal per day as those maintained under more normal conditions [16:8 (L:D)] or under total darkness (Hussain, 1994). Bioassay data indicate that the production of the (unidentified) female-produced sex pheromone of the red flour beetle, *T. castaneum* (Herbst), is affected by the time of day: sex pheromone production is low during the scotophase, and peaks during the photophase (Abdel-Kader *et al.*, 1987). Photoperiod has not traditionally been considered a factor affecting pheromone production in the Coleoptera. Clearly, this phenomenon merits further investigation.

6.6 Scolytidae: bark and ambrosia beetles

The first pheromone isolated and identified in the Coleoptera was from a scolytid, the California fivespined ips, *Ips paraconfusus* Lanier. In this case, the multicomponent and synergistic blend of ipsenol, ipsdienol, and *cis*-verbenol is produced by the male and elicits aggregation behavior by both sexes (Silverstein *et al.*, 1966a; Wood *et al.*, 1967, 1968). Given the seminal nature of this work it is not surprising that some of the first studies of pheromone biosynthesis in the Coleoptera involved bark beetles (Hughes, 1974, 1975; Hughes and Renwick, 1977a, b; Fish *et al.*, 1979; Hendry *et al.*, 1980). Our understanding of scolytid pheromone biosynthesis has changed radically during the last decade as the hypothesis that the chemical signals are derived primarily from host monoterpenoids or symbionts has been challenged, tested, and replaced with the new hypothesis that bark beetles play the principal role in regulating and expressing the endogenous synthesis of their pheromones (Seybold and Tittiger, 2003). Recently, this understanding has progressed from the biochemical to the molecular level as key genes related to pheromone biosynthesis via the mevalonate biosynthetic pathway have been cloned and their expression has been studied *in vivo* (see Tittiger, Chapter 7 in this volume). Much has changed since the publication of *Pheromone Biochemistry* (Prestwich and Blomquist, 1987) when the biosynthesis of bark beetle pheromones was reviewed in largely bio-organic terms and only one published case of *de novo* synthesis was known from the entire Coleoptera. In this section we will devote most of our attention to progress that has occurred since 1987 on understanding pheromone biosynthesis in the Scolytidae. Readers are referred to the many other reviews of bark beetle pheromone biosynthesis for items of more historical interest or where greater depth is desired (Borden, 1982,

1985; Wood, 1982; Francke, 1986; Vanderwel and Oehlschlager, 1987; Byers, 1989, 1995; Raffa *et al.*, 1993; Vanderwel, 1994; Francke *et al.*, 1995; Francke and Schulz, 1999; Schlyter and Birgersson, 1999; Tillman *et al.*, 1999; Seybold *et al.*, 2000; Seybold and Tittiger, 2003).

6.6.1 Range of chemical structures

The pheromones described from bark beetles encompass a wide array of chemical structures that includes isoprenoids, fatty acid derivatives, amino acid derivatives, and compounds of unknown origin (Figure 6.9). As such, the Scolytidae are a microcosm of all Coleoptera. The Scolytidae comprise a relatively small family of beetles that is divided into two subfamilies: the Hylesininae (approx. 1200 species) and the Scolytinae (approx. 4600 species) (Wood and Bright, 1992). From the approx. 60 species investigated so far, pheromone components of bark beetles are dominated by hemi- and monoterpenoids (Seybold *et al.*, 2000; Seybold and Tittiger, 2003) and bicyclic acetals (Francke *et al.*, 1996b). Among these, the biosyntheses of only a select few have been investigated experimentally (isoprenoids, Fish *et al.*, 1979; Hendry *et al.*, 1980; Lanne *et al.*, 1989; Gries *et al.*, 1990a; Ivarsson *et al.*, 1993, 1997; Seybold *et al.*, 1995b; Perez *et al.*, 1996; Tillman *et al.*, 1998; Barkawi *et al.*, 2003 and bicyclic acetals, Vanderwel and Oehlschlager, 1992; Vanderwel *et al.*, 1992a; Barkawi *et al.*, 2003). A recent compendium of pheromone components and their application for North American species is available (Skillen *et al.*, 1997), and we reference literature for many scolytid pheromone components from the world fauna (Figure 6.9).

6.6.2 Site of pheromone production in the Scolytidae

With the exception of the European elm bark beetle, *Scolytus multistriatus* (Marsham) (Gore *et al.*, 1977), and the large elm bark beetle, *Scolytus scolytus* (Fabricius) (Gerken and Grüne, 1978), where pheromone production, storage, and release correlate with accessory glands associated with vaginal palpi, accumulation of aggregation pheromone in scolytids has otherwise been generally localized to the alimentary canal, including the Malpighian tubules (Pitman and Vité, 1963; Pitman *et al.*, 1965; Zethner-Møller and Rudinsky, 1967; Borden and Slater, 1969; Borden *et al.*, 1969; Byers, 1983b). Prior to isolation of the aggregation pheromone of *I. paraconfusus*, several investigators used the response of female beetles in a laboratory walking bioassay to localize the attractant to the male hindgut (including the Malpighian tubules) and the fecal pellet that emanated from the hindgut (Borden *et al.*, 1969; Pitman and Vité, 1963; Pitman *et al.*, 1965; Wood *et al.*, 1966). Similar integrated anatomical/behavioral studies were conducted with the Douglas-fir beetle, *Dendroctonus pseudotsugae* Hopkins (Zethner-Møller and Rudinsky, 1967), and the striped ambrosia beetle, *Trypodendron lineatum* (Olivier) (Borden and Slater, 1969). Chemical detection of pheromone components by GC and GC-MS led to more definitive studies of the accumulation of pheromone components in the hemolymph of the mountain pine beetle, *Dendroctonus ponderosae* Hopkins and the red turpentine beetle, *D. valens* Le Conte (Hughes, 1973a) and in various regions of the alimentary canal of *I. paraconfusus* (Byers, 1983b) or throughout the body of *D. pseudotsugae* (Madden *et al.*, 1988). The male spruce engraver, *Pityogenes chalcographus* (L.), produces the fatty acid-derived methyl (2*E*,4*Z*)-decadienoate as part of its aggregation pheromone. Surprisingly this compound accumulates in the head and thorax rather than the abdomen (Birgersson *et al.*, 1990).

Recent work has applied knowledge of the involvement of the mevalonate pathway (see section 6.6.3.2 and Figure 6.10 later in the text) in pheromone

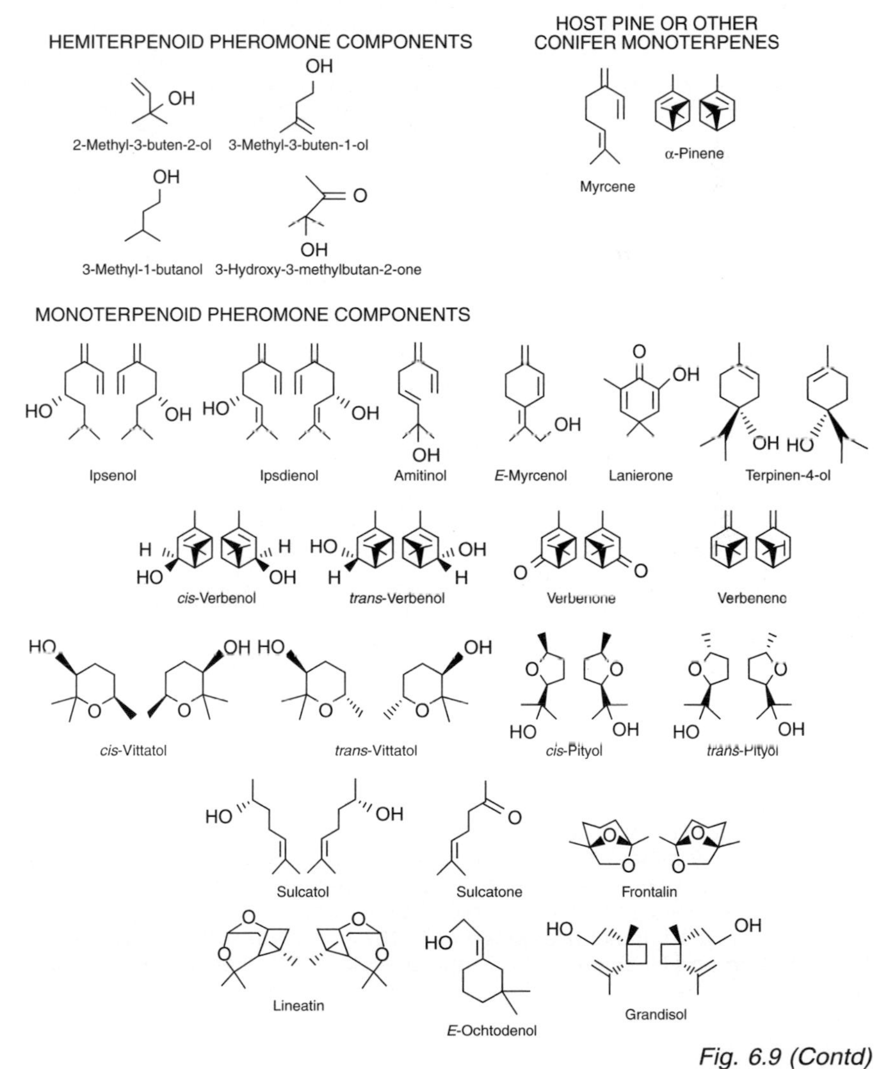

Fig. 6.9 (Contd)

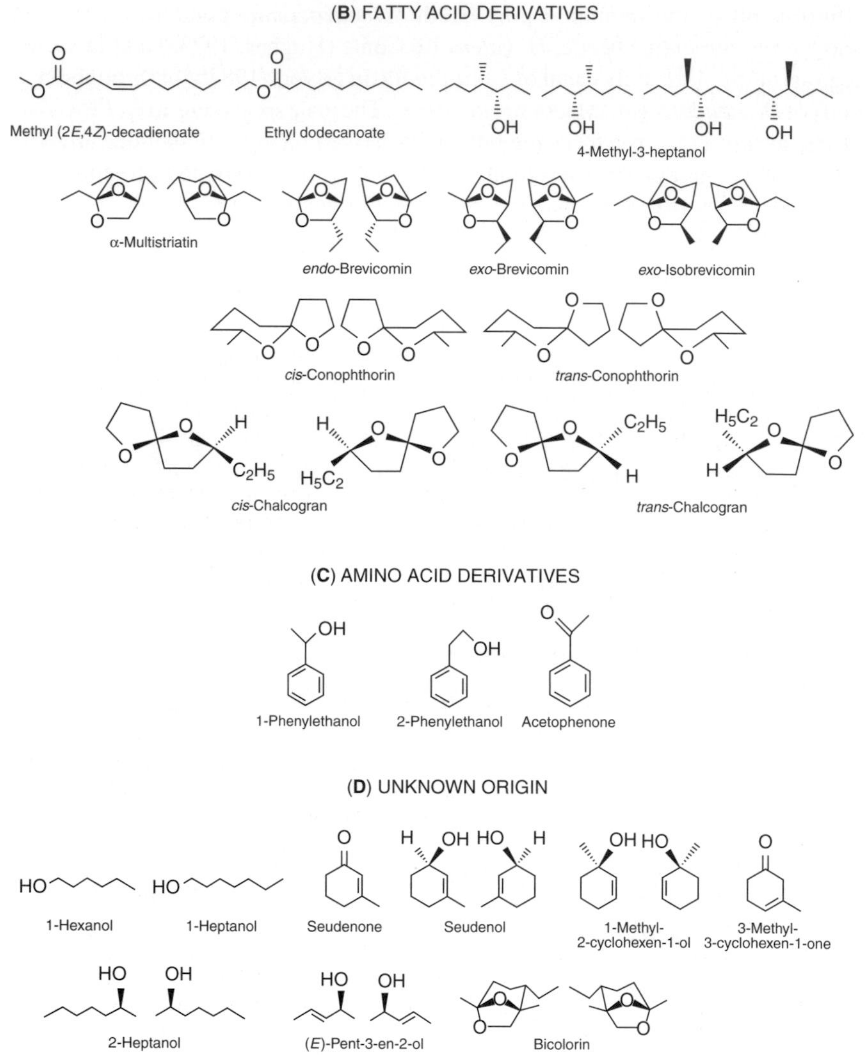

(**B**) FATTY ACID DERIVATIVES

Methyl (2*E*,4*Z*)-decadienoate | Ethyl dodecanoate | 4-Methyl-3-heptanol

α-Multistriatin | *endo*-Brevicomin | *exo*-Brevicomin | *exo*-Isobrevicomin

cis-Conophthorin | *trans*-Conophthorin

cis-Chalcogran | *trans*-Chalcogran

(**C**) AMINO ACID DERIVATIVES

1-Phenylethanol | 2-Phenylethanol | Acetophenone

(**D**) UNKNOWN ORIGIN

1-Hexanol | 1-Heptanol | Seudenone | Seudenol | 1-Methyl-2-cyclohexen-1-ol | 3-Methyl-3-cyclohexen-1-one

2-Heptanol | (*E*)-Pent-3-en-2-ol | Bicolorin

Figure 6.9 Examples of pheromone components of bark beetles (Scolytidae) and ambrosia beetles (Scolytidae and Platypodidae) classified by likely biosynthetic origin (based on Francke and Schulz, 1999). (**A**) *References for identification and/or behavioral activity of isoprenoid pheromone compounds are as follows*: 2-methyl-3-buten-2-ol (Bakke *et al.*, 1977; Giesen *et al.*, 1984; Klimetzek *et al.*, 1989a; Lanne *et al.*, 1989), 3-methyl-3-buten-1-ol (Stoakley *et al.*, 1978; Bowers and Borden, 1990; Bowers *et al.*, 1991; Zhang *et al.*, 2000), 3-methyl-1-butanol (Renwick *et al.*, 1977), 3-hydroxy-3-methylbutan-2-one (Francke and Heeman, 1974; Francke *et al.* 1974), ipsenol and ipsdienol

(Contd)

(Silverstein *et al.*, 1966a; Wood *et al.*, 1967, 1968; Vité *et al.*, 1972a; Harring *et al.*, 1975; Birch *et al.*, 1980; Klimetzek *et al.*, 1989b; Macías-Sámano *et al.*, 1997), amitinol (Silverstein *et al.*, 1966b; Francke *et al.*, 1980), (*E*)-myrcenol (Renwick *et al.*, 1976b; Hunt *et al.*, 1986; Hunt and Borden, 1989b; Byers *et al.*, 1990; Miller *et al.*, 1990), lanierone (Teale *et al.*, 1991, Seybold *et al.*, 1992; Miller *et al.*, 1997), terpinen-4-ol (Kohnle *et al.*, 1985), *cis*- and *trans*-verbenol (Silverstein *et al.*, 1966a; Renwick, 1967; Pitman *et al.*, 1968; Rudinsky *et al.*, 1972; Hughes, 1973b; Francke and Heeman, 1976; Mori, 1976; Plummer *et al.*, 1976; Miller and Lafontaine 1991), verbenone (Renwick, 1967; Kohnle *et al.*, 1992a), verbenene (Gries *et al.*, 1990a, 1992a), *cis*- and *trans*-pityol (Francke *et al.*, 1987; Mori and Puapoomchareon, 1987; Klimetzek *et al.*, 1989a; Dallara *et al.*, 2000), *cis*- and *trans*-vittatol (Klimetzek *et al.*, 1989a), sulcatol (Byrne *et al.*, 1974; Borden *et al.*, 1976, 1980b; Renwick *et al.*, 1977; Borden and McLean 1979; Madden *et al.*, 1988; Klimetzek *et al.*, 1989a; Algarvio *et al.*, 2002), sulcatone (Ryker *et al.*, 1979; Klimetzek *et al.*, 1989a; Francke *et al.*, 1995; Algarvio *et al.*, 2002), frontalin (Kinzer *et al.*, 1969; Stewart *et al.*, 1977; Gries, 1992; Lindgren, 1992; Nishimura and Mori, 1998; Barkawi *et al.*, 2003), lineatin (MacConnell *et al.*, 1977; Borden *et al.*, 1980a; Hoover *et al.*, 2000), (*E*)-ochtodenol (Francke *et al.*, 1989; 1995), and grandisol (Francke *et al.*, 1987, 1989, 1995). (**B**) *References for identification and/or behavioral activity of fatty acid-derived pheromone compounds are as follows*: methyl (2*E*, 4*Z*)-decadienoate (Byers *et al.*, 1988; Birgersson *et al.*, 1990), ethyl dodecanoate (Birgersson *et al.*, 2000), 4-methyl-3-heptanol (Pearce *et al.*, 1975; Mori, 1977; Gore *et al.*, 1977; Plummer *et al.*, 1976; Blight *et al.*, 1978, 1979a, b, c, 1983), α-multistriatin (Mori, 1974; Pearce *et al.*, 1975, 1976; Gore *et al.*, 1977; Lanier *et al.*, 1977; Stewart *et al.*, 1977; Gerken *et al.*, 1978; Elliot *et al.*, 1979; Blight *et al.*, 1980; Birch *et al.*, 1981), *endo*-brevicomin (Rudinsky *et al.*, 1974; Schurig *et al.*, 1983; Gries, 1992; Vanderwel *et al.*, 1992a; Camacho *et al.*, 1993, 1994; Camacho and Borden, 1994), *exo*-brevicomin (Silverstein *et al.*, 1968; Hughes and Renwick, 1977a; Stewart *et al.*, 1977; Francke *et al.*, 1979; Francke, 1981; Ryker and Rudinsky, 1982; Schurig *et al.*, 1983; Borden *et al.*, 1987; Vanderwel and Oehlschalger, 1992; Vanderwel *et al.*, 1992a), *exo*-isobrevicomin (Francke *et al.*, 1996b), *cis*- and *trans*-conophthorin (Francke *et al.*, 1979; Kohnle *et al.*, 1992b; Francke *et al.*, 1995; Birgersson *et al.*, 1995; Pierce *et al.*, 1995; Dallara *et al.*, 2000; Huber *et al.*, 2000; Huber and Borden, 2001), and *cis*- and *trans*-chalcogran (Francke *et al.*, 1977; Birgersson *et al.*, 1990, 2000). (**C**) *References for identification and/or behavioral activity of pheromone compounds derived from amino acids are as follows*: 1-phenylethanol (Renwick *et al.*, 1975; Pureswaran *et al.*, 2000), 2-phenylethanol (Renwick *et al.*, 1976c; Gries *et al.*, 1990b; Ivarsson and Birgersson, 1995; Pureswaran *et al.*, 2000, Zhang *et al.*, 2000), acetophenone (Kohnle *et al.*, 1987; Gries *et al.*, 1992b; Francke *et al.*, 1995; Pureswaran *et al.*, 2000; Barkawi, 2002). (**D**) *References for identification and/or behavioral activity of pheromone compounds of unknown origin are as follows*: 1-hexanol (Renwick *et al.*, 1977; Francke *et al.*, 1995; Birgersson *et al.*, 2000; Algarvio *et al.*, 2002), 1- and 2-heptanol (Renwick and Pitman, 1979; Renwick *et al.*, 1975; Paine *et al.*, 1999; Barkawi, 2002), seudenone, seudenol, 1-methyl-2-cyclohexen-1-ol and 3-methyl-3-cyclohexen-1-one (Vité *et al.*, 1972b; Plummer *et al.*, 1976; Libbey *et al.*, 1976, 1983; Gries, 1992; Lindgren *et al.*, 1992; Francke *et al.*, 1995; Ross and Daterman, 1995, 1998; Borden *et al.*, 1996; Setter and Borden, 1999), (*E*)-pent-3-en-2-ol (Ryker *et al.*, 1979), and bicolorin (Francke *et al.*, 1996a).

biosynthesis in *I. paraconfusus*, the pine engraver, *Ips pini* (Say), and the Jeffrey pine beetle, *Dendroctonus jeffreyi* Hopkins, to localize the site of biosynthesis to endothelial cells of the anterior midgut. Both biochemical and molecular approaches have been used. An *in vitro* bioassay (Ivarsson *et al.*, 1997) of the synthesis of ipsenone and ipsdienone from ^{14}C-acetate in male *I. paraconfusus* demonstrated that of six body regions and organs, radiolabeled product was most abundant in the metathorax (Ivarsson *et al.*, 1998). When fat body, flight muscle, and abdominal tissue were compared in the assay, flight muscle had the highest level of incorporation (perhaps because it surrounds the alimentary canal in the hemocoel). Radiotracer (^{14}C-acetate) incorporation studies with both male *I. pini* and *D. jeffreyi* showed that the pheromone components ipsdienol and frontalin, respectively, were produced in isolated midgut tissue (Hall *et al.*, 2002a, b).

Measurement of the expression of the mevalonate pathway gene coding for 3-hydroxy-3-methylglutaryl CoenzymeA reductase (HMG-CoA reductase = HMG-R, EC 1.1.1.34) (see Tittiger, Chapter 7, in this volume) has also been used quite effectively to more clearly delineate the site of pheromone biosynthesis in scolytids. Since major pheromone components for *I. pini* and *D. jeffreyi* (ipsdienol and frontalin, respectively) are both isoprenoids, strong expression of *HMG-R* should be linked with active sites of biosynthesis of these pheromone components. Following treatment with JH III, HMG-R mRNA accumulated primarily in the male thorax (not head or abdomen) of both *I. paraconfusus* (Ivarsson *et al.*, 1998) and *I. pini* (Hall *et al.*, 2002a). In male *D. jeffreyi*, minor variation in the sectioning method affected the distribution of the signal, but expression of both *HMG-R* and 3-hydroxy-3-methylglutaryl CoenzymeA synthase (HMG-CoA synthase = HMG-S, EC 4.1.3.5) was localized to the junction of the metathorax and abdomen (Tittiger *et al.*, 2000; Hall *et al.*, 2002b). A modified *in situ* hybridization technique using anti-sense RNA for HMG-R has demonstrated that the anterior midgut of pheromone-producing male *I. pini* and *D. jeffreyi* is the site of strongest *HMG-R* expression (Hall *et al.*, 2002a, b). In males of both *I. pini* and *D. jeffreyi*, the JH III-mediated increase in *HMG-R* expression in midgut cells is accompanied by the appearance of large crystalline arrays of smooth endoplasmic reticulum (SER) (Nardi *et al.*, 2002). Since overproduction of SER is common in cholesterol-deprived mammalian cells that produce high levels of HMG-R (Berciano *et al.*, 2000; Chin *et al.*, 1982), pheromone synthesizing midgut cells of bark beetles likely have large amounts of crystalline SER because they are strongly expressing *HMG-R*, and likely have correspondingly elevated HMG-R protein levels.

Research on the site of pheromone synthesis in the Scolytidae using transcript sequences for mevalonate pathway enzymes has led to two major conclusions (Seybold and Tittiger, 2003). First, bark beetles do not have a pheromone gland. Instead, all cells in the midgut appear to be involved in pheromone production because the signals from *in situ* hybridizations are evenly distributed throughout the tissue (Hall *et al.*, 2002a, b), and the SER arrays are observed in all midgut cells (Nardi *et al.*, 2002). Thus, cells in the midgut are actively involved in both digestion and pheromone synthesis and release. Second, *de novo* synthesis of pheromone components in the anterior midgut involves insect tissues, and not symbiotic bacteria. Prokaryotic HMG-Rs are both structurally and mechanistically different from their eukaryotic homologs (Friesen and Rodwell, 1997; Tabernero *et al.*, 1999), so the activity, gene expression, and subcellular changes accompanying *de novo* pheromone production can be attributed solely to scolytid tissues.

6.6.3 Biochemical studies of pheromone production in the scolytidae

6.6.3.1 Microbially assisted biosynthesis, autooxidation of host compounds, and release of sequestered host compounds

Although recent work has focused on the *de novo* synthesis of scolytid aggregation pheromones, there is also substantial evidence for several ecological sources of scolytid semiochemicals ("pheromones"). Brand *et al.* (1975) first demonstrated that a bacterium, *Bacillus cereus* Frankland and Frankland, isolated from *I. paraconfusus* could convert host α-pinene to *cis*- and *trans*-verbenol. Similar conversions by fungi (primarily yeasts) were noted in the southern pine beetle, *Dendroctonus frontalis* Zimmermann (Brand *et al.*, 1976), *D. ponderosae* (Hunt and Borden, 1989a), and the Eurasian spruce engraver, *Ips typographus* (L.) (Leufvén *et al.*, 1984, 1988). We could not locate any reports of the synthesis of hemiterpenoid, acyclic monoterpenoid, or bicyclic acetal "pheromone" components by microorganisms. However, antibiotic treatment and axenic rearing techniques affected production of both cyclic and acyclic monoterpenoid pheromone components by *I. para-confusus* and *D. ponderosae* (Byers and Wood, 1981; Conn *et al.*, 1984; Hunt and Borden, 1989b). Verbenol and verbenone have also been shown to arise through autooxidation of α-pinene (Moore *et al.*, 1956; Hunt *et al.*, 1989), and an enantiomerically enriched starting material results in an enantiomerically enriched oxidized product (Grosman, 1996). Other common conifer monoterpenes (e.g. β-pinene, 3-carene, terpinolene) have not been examined to determine whether they autooxidize to "pheromones".

Sequestration of host compounds for later use as pheromones in their unmodified form appears to be rare in the Scolytidae (Vanderwel and Oehlschlager, 1987). Since pheromone production is often associated with feeding, it is experimentally difficult to distinguish between host compounds that are released from the masticated food or feces, and those that are sequestered by the beetle and released later. Furthermore, depending on the degree to which the beetle assimilates the host compound, the semiochemical may be more aptly referred to as a kairomone rather than as a pheromone (Borden, 1985). For instance, it is possible that *D. pseudotsugae* obtains and sequesters the monoterpene limonene from host Douglas-fir, *Pseudotsuga menziesii* (Mirbel) Franco, oleoresin during feeding. Both sexes of *D. pseudotsugae* release limonene with their respective aggregation pheromone

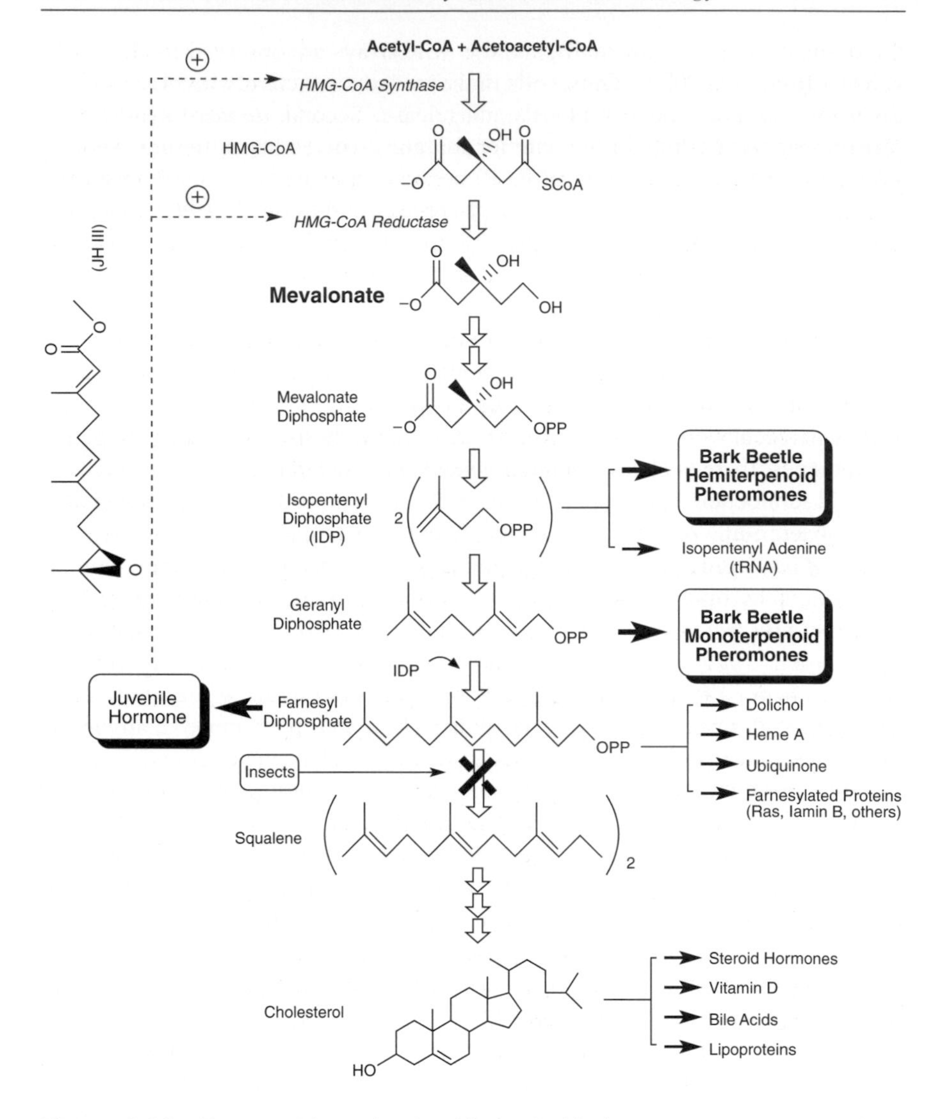

Figure 6.10 *De novo* biosynthesis of isoprenoid pheromone components by bark and ambrosia beetles through the mevalonate biosynthetic pathway. The end products are hemiterpenoid and monoterpenoid pheromone products common throughout the Scolytidae and Platypodidae (Figure 6.9A). The biosynthesis is regulated by juvenile hormone III (JH III), which is a sesquiterpenoid product of the same pathway. The stereochemistry of JH III is indicated as described in Schooley and Baker (1985). Although insects do not biosynthesize sterols *de novo*, they do produce a variety of derivatives of isopentenyl diphosphate, geranyl diphosphate, and farnesyl diphosphate. Figure adapted from Seybold and Tittiger (2003).

components in the presence of acoustic signals from the opposite sex (Rudinsky *et al.*, 1977). Also, the European pine shoot beetle, *Tomicus piniperda* (L.), other *Dendroctonus* spp., and some *Ips* spp. respond at some level to host (*Pinus* spp.) monoterpenes in field studies; however, it is not known whether these monoterpenes are sequestered and released by colonizing beetles (Bedard *et al.*, 1969, 1970; Werner, 1972; Byers *et al.*, 1985; Miller and Borden, 1990a, b; Hobson *et al.*, 1993; Czokajlo and Teale, 1999; Erbilgin and Raffa, 2000). Macías-Sámano *et al.* (1998) demonstrated that the fir engraver, *Scolytus ventralis* LeConte, is attracted to and aggregates on host *Abies* spp. trees in response to host volatiles, which are apparently released when pioneer beetles colonize trees. From an evolutionary perspective, the redundant contributions to pheromone production from endogenous and exogenous sources (microbes, autooxidation, and host) may be a beneficial adaptation for the Scolytidae, but the relative quantities of pheromone from various sources await measurement (Seybold *et al.*, 2000).

6.6.3.2 Endogenous pheromone biosynthesis

During the last decade two themes have emerged related to endogenous pheromone biosynthesis in bark beetles. First, it has been proven conclusively that the mevalonate pathway (Figure 6.10) is central for the *de novo* production of hemiterpenoid and monoterpenoid pheromone compounds by scolytids. Named for its key intermediate, mevalonate [(*R*)-3,5-dihydroxy-3-methylpentanoate], the pathway is the route to an enormous number of biologically important isoprenoids in plants and animals (Cane, 1999). In bark beetles, branched chain amino acids (e.g. leucine and isoleucine) may also serve as carbon sources for this pathway after catabolism to acetyl-CoA (discussed in Barkawi *et al.*, 2003). Francke and Schulz (1999) emphasize that some hemiterpenoids in other insects may be derived directly from these amino acids. In bark beetles, three classes of compounds have been shown to be synthesized via the mevalonate pathway: hemiterpenoids (Lanne *et al.*, 1989); acyclic monoterpenoids (Seybold *et al.*, 1995b; Tillman *et al.*, 1998); and one bicyclic acetal (Barkawi *et al.*, 2003). These syntheses occur in one tribe from each of the two subfamilies of the Scolytidae (Hylesininae-Tomicini and Scolytinae-Ipini). In hindsight, the biosynthesis of isoprenoid pheromones from short-chain metabolic building blocks rather than from host precursors was foreshadowed in *Ips* spp. by numerous studies of the endocrine regulation of pheromone production. Pheromone components readily accumulate in beetle hindgut tissue following treatment with JH III or its analogues (JHA) in the absence of host precursors (Borden *et al.*, 1969; Chen *et al.*, 1988; Hughes and Renwick, 1977a, b; Renwick and Dickens, 1979). Further work that estimated a deficit in the amount of available host precursors (Byers, 1981; Byers and Birgersson, 1990), or that blocked a key enzyme in the *de novo* pathway (Ivarsson *et al.*, 1993), also predicted an inherently scolytid synthesis for these compounds. Presently, there is no evidence that the newly discovered methyl-erythritol-4-

phosphate (MEP) pathway (Rohmer *et al.*, 1996; Rohmer, 1999) is involved in *de novo* pheromone synthesis by bark beetles.

A second theme that has developed recently in relation to endogenous production of pheromones by scolytids is the frequent occurrence of intramolecular cyclizations of unbranched and methyl-branched monounsaturated ketones through their corresponding ketoepoxides (see Figure 6.12 later in text). The key ketoepoxide intermediates are likely formed through the action of P450 enzymes. The capacity to biosynthesize relatively complex bicyclic and tricyclic structures via these reactions is likely to be widespread in the Scolytidae, occurring in species from tribes of both subfamilies [e.g. Hylesininae—Tomicini (Vanderwel *et al.*, 1992a; Vanderwel and Oehlschlager, 1992; Perez *et al.*, 1996; Barkawi *et al.*, 2003) and Scolytinae—Ipini (Perez *et al.*, 1996) and —Dryocoetini (Vanderwel *et al.*, 1992a)]. These reactions even occur in the closely allied weevils (Curculionidae, Perez *et al.*, 1996). This wide distribution suggests that both *de novo* synthesis of isoprenoids via the mevalonate pathway and *de novo* synthesis of bicyclic acetals and related structures are traits that are fundamental in the evolution of the family.

(a) *Biosynthetic experiments.* Experimental approaches to the synthesis of pheromone components in living scolytids or their isolated tissues have involved the application of various unlabeled and isotopically labeled intermediates or precursors. These substrates have been administered *in vivo* through feeding, topical application, exposure in the volatile headspace, or injection. In a few cases, the precursors have been administered *in vitro* by incorporation into the tissue incubation medium. In the vast majority of published accounts, adult beetles were exposed to volatiles of precursors, although in more recent experiments, precursors have been injected through the cuticle. Ultimately these studies have led to the new understanding regarding the frequent occurrence of *de novo* synthesis of pheromone components by bark beetles.

(i) *Experiments with unlabeled host precursors.* In the earliest studies, a *de novo* synthesis was not presumed and unlabeled host monoterpenes of pines (*Pinus* spp.) whose carbon skeletons matched those of the oxidatively-derived behavioral chemicals were used as precursors (Figure 6.11). The bio-organic reactions whereby monoterpenes are metabolized to oxygenated monoterpenoids (allylic alcohols, aldehydes, or ketones) in the Scolytidae have been outlined (Francke and Vité, 1983; Pierce *et al.*, 1987) and these reactions are likely to involve P450 enzymes (White *et al.*, 1979, 1980; Hunt and Smirle, 1988; Feyereisen, 1999). Hughes (1974) and Hughes and Renwick (1977b) first demonstrated that exposure to myrcene in the volatile headspace increased the amounts of ipsenol and ipsdienol in hindgut tissues of male *I. paraconfusus* and related species (Figure 6.11A). Subsequent studies with *I. paraconfusus* established a positive quantitative

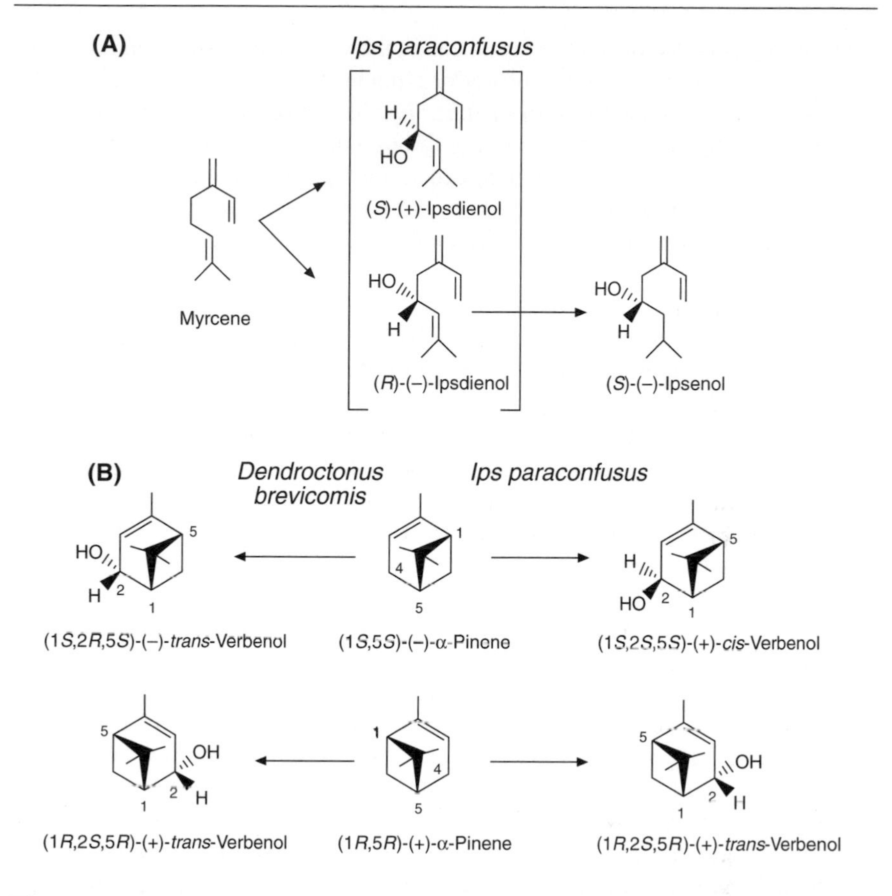

Figure 6.11 Biosyntheses of isoprenoid pheromone components by bark and ambrosia beetles from host conifer monoterpenes. (**A**) Conversion by the male California fivespined ips, *Ips paraconfusus* Lanier (Coleoptera: Scolytidae), of myrcene from the xylem and phloem oleoresin of ponderosa pine, *Pinus ponderosa* Laws., to (4*S*)-(+)-ipsdienol and (4*S*)-(−)-ipsenol, components of the aggregation pheromone (Hendry *et al.*, 1980). (**B**) Conversion by male and female *I. paraconfusus* of (1*S*,5*S*)-(−)-α-pinene (2,6,6-trimethylbicyclo[3.1.1]hept-2-ene) from the xylem and phloem oleoresin of *P. ponderosa* to (1*S*,2*S*,5*S*)-(+)-*cis*-verbenol (*cis*-4,6,6-trimethylbicyclo[3.1.1]hept-3-en-2-ol), an aggregation pheromone synergist and of (1*R*,5*R*)-(+)-α-pinene to (1*R*,2*S*,5*R*)-(+)-*trans*-verbenol (*trans*-4,6,6-trimethylbicyclo[3.1.1]hept-3-en-2-ol), a compound of unknown behavioral activity for *I. paraconfusus*. Male and female western pine beetle, *Dendroctonus brevicomis* LeConte (Coleoptera: Scolytidae), convert (1*S*,5*S*)-(−)-α-pinene to (1*S*,2*R*,5*S*)-(−)-*trans*-verbenol, an aggregation pheromone interruptant and (1*R*,5*R*)-(+)-α-pinene to (1*R*,2*S*,5*R*)-(+)-*trans*-verbenol, a compound of

(Contd)

relationship between the mass of myrcene in the volatile headspace and the mass of ipsenol and ipsdienol in the posterior alimentary canal (Byers *et al.*, 1979). The route of precursor entry from these *in vivo* myrcene exposure studies is unclear, but may have involved passage through the tracheae and tracheoles into the cells and hemolymph of the hemocoel, absorption through the cuticle into the cells and hemolymph, or ingestion and passage into the alimentary canal. Given that much more ipsenol and ipsdienol accumulated in the mid- and hindguts of male *I. paraconfusus* after they had fed on ponderosa pine, *Pinus ponderosa* Laws. Phloem than after exposure to myrcene, Byers (1981) suggested that under "natural conditions" (i.e. during feeding on the host), myrcene entered through the alimentary canal with the food bolus.

In later studies, unlabeled myrcene was delivered in the headspace to male *Dendroctonus* spp. (Renwick *et al.*, 1976b; Byers, 1982; Hunt *et al.*, 1986; Seybold *et al.*, 1992; Barkawi, 2002), to the male Eurasian spruce engraver, *Ips duplicatus* Sahlberg (Ivarsson and Birgersson, 1995), to the male larch engraver, *Ips cembrae* Heer (Renwick and Dickens, 1979), and to male *I. pini* (Lu, 1999). These experiments resulted in the accumulation of ipsdienol and *E*-myrcenol (*Dendroctonus* spp. and *I. duplicatus*) or the accumulation of ipsdienol and ipsenol (*I. cembrae* and *I. pini*) in hindgut tissues. In a creative approach, the quantity of myrcene in host phloem from five species of pines was used as a screening procedure to suggest that *de novo* synthesis of ipsenol and ipsdienol was occurring in male *I. paraconfusus* (Byers and Birgersson, 1990). The enantiomeric composition of the compounds was also largely invariant when *I. paraconfusus* fed on a wider range of hosts, including non-pines (Seybold, 1992). Besides myrcene, unlabeled ipsdienol has been applied topically or as a vapor to *Ips* spp. to confirm its conversion to ipsenol [*I. paraconfusus* and the eastern fivespined ips, *I. grandicollis* (Eichhoff) (Hughes, 1974); *I. cembrae* (Renwick and Dickens, 1979); *I. paraconfusus* and *I. pini* (Lu, 1999)] (see Figure 6.15 later in text).

The conversion of host α-pinene to the diastereomers of verbenol (and the enantiomers of verbenone) has also been studied using unlabeled materials (Figure 6.11B) (Hughes, 1973a, b, 1975; Byers, 1981, 1983a, b; Bridges, 1982; Gries *et al.*, 1990a; Barkawi, 2002). Hughes (1973a, b) exposed adult *Dendroctonus* spp.

Fig. 6.11 (Contd)

unknown behavioral activity (Renwick *et al.*, 1976a and Byers, 1983a). The naturally occurring enantiomeric composition of α-pinene from a distillate of *P. ponderosa* oleoresin in the central Sierra Nevada (habitat of *I. paraconfusus* and *D. brevicomis*) is 95 percent (–) (Hobson *et al.*, 1993). Similar transformations occur in male and female mountain pine beetle, *Dendroctonus ponderosae* Hopkins and male and female Jeffrey pine beetle, *Dendroctonus jeffreyi* Hopkins (Barkawi, 2002). Figure adapted from Tillman *et al.* (1999).

to α-pinene vapors and found that primarily *trans*- and some *cis*-verbenol and verbenone accumulated in hindgut tissue. Topically applied α-pinene was also converted to *trans*-verbenol and other α-pinene derivatives by female *D. ponderosae* and female *D. valens* (Hughes, 1973a). To further establish the precursor/product relationship between α-pinene and the verbenols, Byers (1981) used unlabeled α-pinene to show that the mass of *cis*-verbenol (and to a lesser extent *trans*-verbenol and myrtenol) in hindgut tissue of male *I. paraconfusus* increased with increasing amount of α-pinene in the volatile headspace, and Hughes (1973a) showed a positive quantitative relationship between the duration of exposure to vapors of α-pinene and the amount of *trans*-verbenol present in the hindguts of female western pine beetle, *D. brevicomis* LeConte. In a study that used the enantiomeric composition of α-pinene as a rudimentary labeling technique, Renwick *et al.* (1976a) found that male *I. paraconfusus* produced *cis*-verbenol from the (–)-enantiomer of α-pinene and *trans*-verbenol from the (+)-enantiomer (Figure 6.11B). Similar production of verbenol from α-pinene originating from the host, Norway spruce, *Picea abies*, occurs in *I. typographus* (Klimetzek and Francke, 1980; Lindström *et al.*, 1989). In contrast, both sexes of *D. brevicomis*, *D. jeffreyi*, and *D. ponderosae* convert each enantiomer of α-pinene to the corresponding enantiomers of *trans*-verbenol (Figure 6.11B) (Byers, 1983a; Barkawi, 2002). In *D. jeffreyi* and *D. ponderosae*, the stereospecific conversion to *cis*-verbenol has less fidelity (Barkawi, 2002). Also, in laboratory studies with *D. jeffreyi*, the predominant alkane (heptane) from the host Jeffrey pine, *Pinus jeffreyi* Grev. & Balf., is converted to (2*S*)-(+)-heptanol and 1-heptanol, with the former alcohol present in greater abundance (Barkawi, 2002). 1-Heptanol is an attractant pheromone component for *D. jeffreyi* (Renwick and Pitman, 1979; Paine *et al.*, 1999), but the behavioral activity of 2-heptanol is not clear.

(ii) *Experiments with stable isotopically labeled host and* de novo *precursors.* To provide more convincing evidence of pheromone biosynthetic relationships in the Scolytidae, stable isotopes have been used to label precursors for mass spectrometric (MS) end product detection. ^{2}H-Labeled myrcene, ipsdienol, and ipsdienone have been used to demonstrate that these monoterpenoids are involved in the late stages of the pathway leading to ipsdienol and ipsenol in *I. paraconfusus* (Fish *et al.*, 1979, 1984; Hendry *et al.*, 1980) and *I. pini* (Vanderwel, 1991) (see Figure 6.15 later in this section). Gries *et al.* (1990a) exposed female *D. ponderosae* to volatiles of $^{2}H_{6}$-labeled (–)-α-pinene to prove the syntheses of *trans*-verbenol, verbenene, *p*-mentha-1,5,8-triene, and *p*-cymene from the deuterated precursor.

The biosynthesis of the bicyclic acetals has similarly been investigated using both ^{2}H- and ^{18}O-labeled precursors, and apparently proceeds through the cyclization of the appropriate ketoalkene precursors, via the corresponding ketoepoxide intermediates. For example, after identifying (*Z*)-6-nonen-2-one from volatiles trapped from unfed, male *D. ponderosae*, Vanderwel *et al.* (1992a) demonstrated that [6,7-$^2H_2$]-(*E*)- and (*Z*)-6-nonen-2-one were converted to $^2H_2$-labeled *endo*- and *exo*-brevicomin, respectively, by *D. ponderosae* and by the western balsam bark beetle, *Dryocoetes confusus* Swaine. In both species, the males (the sex typically associated with the pheromone component) produced significantly more pheromone mass per beetle in the presence of precursor, and produced significantly more pheromone mass per beetle than did the females. The direct chromatographic measurement of the enantiomeric composition of labeled *exo*- and *endo*-brevicomin resulting from labeled (*Z*)- and (*E*)-6-nonen-2-one was not reported, but male *D. ponderosae* exposed to unlabeled (*Z*)-6-nonen-2-one produced only the naturally occurring (+)-enantiomer of *exo*-brevicomin, as determined by complexation chromatography (Schurig *et al.*, 1983; Vanderwel *et al.*, 1992a). Taken together, these data provide strong evidence for the involvement of the ketoalkene precursors in the biosynthesis of the brevicomins *in vivo*. In an analogous series of reactions, Perez *et al.* (1996) demonstrated that ^{2}H-labeled 6-methyl-6-hepten-2-one was converted to ^{2}H-frontalin in *D. ponderosae*, in the spruce beetle, *D. rufipennis* (Kirby), in *I. pini*, in *Ips tridens* (Mannerheim), and in the West Indian sugar cane weevil, *Metamasius hemipterus sericeus* (Olivier). The enantiomeric composition of frontalin produced by *D. rufipennis* following exposure to 6-methyl-6-hepten-2-one was 61 to 65 percent-(–) (Perez *et al.*, 1996), which differs substantially from the "pure" (+)-frontalin determined from female *D. rufipennis* that have fed on the phloem of host *Picea engelmannii* Parry (Gries *et al.*, 1988; Gries, 1992).

Through detailed labeling studies, Vanderwel and Oehlschlager (1992) found that the conversion of (*Z*)-6-nonen-2-one into (+)-*exo*-brevicomin by male *D. ponderosae* proceeds through 6,7-epoxynonan-2-one, a keto epoxide intermediate (Figure 6.12A). If this is a general route, the stereochemistry of the bicyclic acetal end products could theoretically be determined by the enzyme-catalyzed formation and/or cyclization of the epoxide. To attempt to distinguish these hypotheses, Vanderwel and Oehlschlager (1992) used knowledge of the stereochemical outcome of the biosynthesis (both in the presence and the absence of substrate) and considered the labeling pattern of the *exo*-brevicomin produced in the presence of various combinations of $^{18}O_2$, $^{18}OH_2$, and (*Z*)-6-nonen-2-one. They found that both the (6*S*,7*R*)- and (6*R*,7*S*)-enantiomers of the *erythro* epoxide derivatives of the ketoalkene are converted to (+)-*exo*-brevicomin. This occurs even when the pheromone is produced in the absence of exogenous precursor (i.e. only from endogenously produced precursor). Thus, in the case of *D. ponderosae* and *exo*-brevicomin synthesis, the epoxidation of the ketoalkene precursor does not control the stereochemistry of the ultimate product. Instead, the enzyme-catalyzed cyclization ensures that only (+)-*exo*-brevicomin is formed.

Although experimental evidence is lacking, the theme of cyclization of unsaturated ketones and related structures is likely to play out in many variations

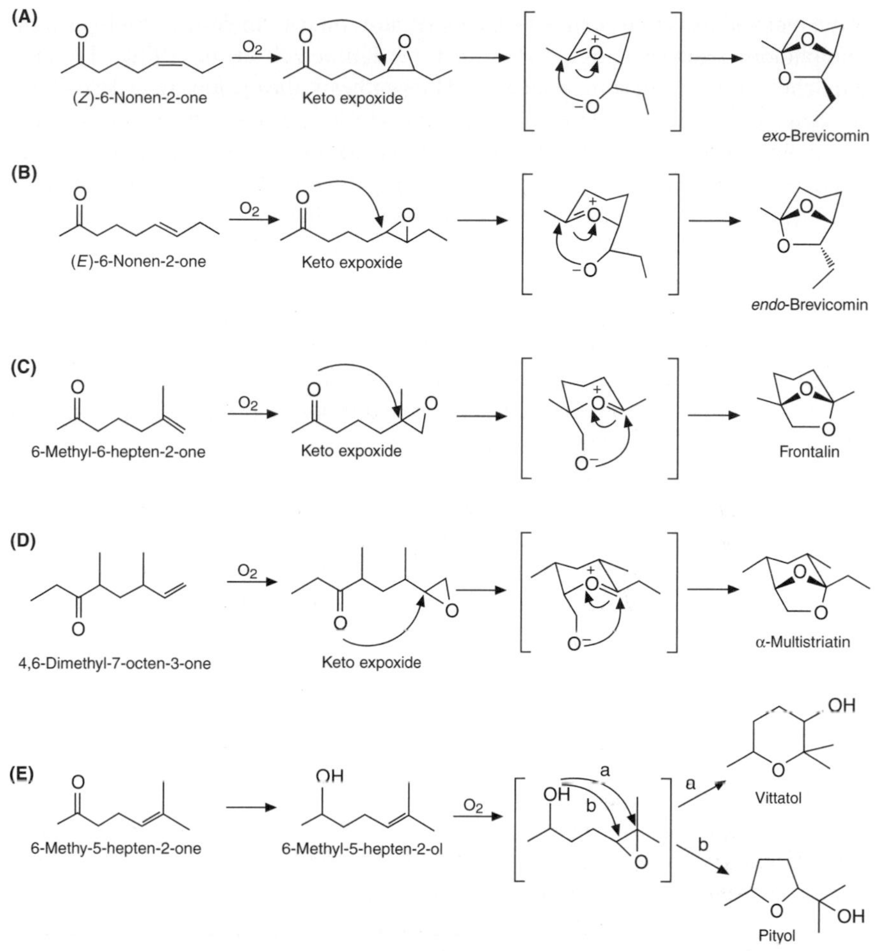

Figure 6.12 Cyclization reactions of acyclic, unbranched and methyl-branched ketones to bicyclic acetals and cyclic alcohols in bark beetles. (**A**) Mountain pine beetle, *Dendroctonus ponderosae*: formation of *exo*-brevicomin [(1*R*,5*S*,7*R*)-(+)-7-ethyl-5-methyl-6,8-dioxabicyclo[3.2.1]octane] from (*Z*)-6-nonen-2-one (Vanderwel and Oehlschlager, 1992; Vanderwel *et al.*, 1992a); (**B**) Western balsam bark beetle, *Dryocoetes confusus*: formation of *endo*-brevicomin [(1*R*,5*S*,7*S*)-7-ethyl-5-methyl-6,8-dioxabicyclo[3.2.1]octane] from (*E*)-6-nonen-2-one (Vanderwel *et al.*, 1992a); (**C**) Spruce beetle, *Dendroctonus rufipennis*: formation of frontalin [(1*S*, 5*R*)-(−)-1,5-dimethyl-6,8-dioxabicyclo[3.2.1]octane] from 6-methyl-6-hepten-2-one (Perez *et al.*, 1996; Francke *et al.*, 1995; Francke and Schulz, 1999); (**D**) European elm bark beetle, *Scolytus multistriatus*: hypothetical formation of α-multistriatin [(1*S*,2*R*,4*S*,5*R*)-(−)-2,4-dimethyl-5-ethyl-6,8-dioxabicyclo[3.2.1]octane] from 4,6-dimethyl-7-octen-3-one (Francke and Schulz, 1999); and E The colored

(Contd)

in the biosyntheses of structures such as vittatol, pityol, lineatin, α-multistriatin, conophthorin, and chalcogran (Figure 6.12; Klimetzek *et al.*, 1989a; Francke and Schulz, 1999). In a somewhat related biosynthesis, the cyclohexene derivatives (seudenone and allied compounds, Figure 6.9D) may be formed through the intramolecular condensation of 2,6-heptadione to yield 3-methylcyclohexenone (seudenone) and then give rise to the corresponding alcohol (seudenol) and also to 1-methyl-2-cyclohexen-1-ol as a product of simple allylic rearrangement (Vanderwel and Oehlschlager, 1987; Francke *et al.*, 1995; Francke and Schulz, 1999). This pathway has not been tested experimentally, but *D. frontalis* can convert 1-methyl-1-cyclohexene to seudenol, seudenone, and 1-methyl-2-cyclohexen-1-ol (Renwick and Hughes, 1975). Pitman and Vité (1974) alluded to "labeled compounds" in reference to seudenol and seudenone produced by *D. pseudotsugae*, but did not elaborate in their methods or results on their labeled precursors or their labeling techniques.

The early stages of the biosynthesis of the ketone precursors to the cyclic acetals and related structures have also not been studied in great detail. The ketones occur frequently among the compounds extracted from beetle tissue or collected from the volatile headspace and this suggests that they might be biogenically related to the pheromone end products. 6-Methyl-6-hepten-2-one has been found in a hindgut extract of females of the eastern larch beetle, *Dendroctonus simplex* LeConte, which produce frontalin (Francke *et al.*, 1995). 6-Methyl-6-hepten-2-one could be derived from 6-methyl-5-hepten-2-one (sulcatone), which has been isolated from *D. pseudotsugae* (Ryker *et al.*, 1979), *D. simplex* (Francke *et al.*, 1995), *D. rufipennis* (Perez *et al.*, 1996), and the colored elm bark beetle, *Pteleobius vittatus* (F.) (Klimetzek *et al.*, 1989a). Nine-carbon ketones such ase (*Z*)-6-nonen-2-one and nonan-2-one have been isolated from *Dryocoetes confusus* and *Dendroctonus ponderosae* (Vanderwel *et al.*, 1992a) and from the European ash bark beetle, *Hylesinus (Leperisinus) varius*

Fig. 6.12 (Contd)

elm bark beetle, *Pteleobius vittatus* (F.): hypothetical formation of vittatol (2,2,6-trimethyl-3-hydroxytetrahydropyran) and pityol [2-(1-hydroxy-1-methylehtyl)-5-methyltetrahydrofuran] from 6-methyl-5-hepten-2-one (sulcatone) and 6-methyl-5-hepten-2-ol (sulcatol) (Klimetzek *et al.*, 1989a; Francke *et al.*, 1995; Francke and Schulz, 1999). For simplicity, stereochemistries of the epoxide intermediates for all products as well as the end products vittatol and pityol are not indicated. Stereochemistries of the bicyclic acetal end products are indicated as the predominant naturally occurring enantiomers. Other bark beetle pheromone components such as lineatin (3,3,7-trimethyl-2,9-dioxatricyclo[3.3.1.0^{4,7}]nonane), conophthorin [(5*R*,7*S*)-7-methyl-1,6-dioxaspiro[4.5]decane], chalcogran [(2*S*,5*R*)-2-ethyl-1,6-dioxaspiro[4.4]nonane], and bicolorin [(1*S*,2*R*,5*R*)-2-ethyl-1,5-dimethyl-6,8-dioxabicyclo[3.2.1]octane] are likely formed through similar mechanisms. Figure courtesy of L. S. Barkawi, University of Minnesota.

(F.) (Francke *et al.*, 1979; Francke, 1981). There are no obvious host sources for these ketones, but sulcatone appears to be synthesized by a fungal symbiont of *D. frontalis* (Brand *et al.*, 1976; Brand and Barras, 1977). The hypothesis that *D. frontalis* might simply cyclize sulcatone from the symbiont to produce frontalin (Brand *et al.*, 1979; Vanderwel and Oehlschalger, 1987) was not supported by experiments by Perez *et al.* (1996), who demonstrated that volatiles of neither sulcatone nor sulcatol were converted to frontalin by male or female *D. rufipennis*. More likely, in a *de novo* synthesis (Figure 6.10), geranyl diphosphate, an intermediate of the mevalonate pathway, might undergo an oxidative cleavage (retroaldol reaction) resulting in endogenous formation of sulcatone, followed by a double bond migration to yield 6-methyl-6-hepten-2-one (Francke, 1986; Vanderwel and Oehlschlager, 1987; Francke and Schulz, 1999). Exogenously presented sulcatone may not be able to access this endogenous biosynthetic system. From recent results on the *de novo* synthesis of frontalin from acetate, mevalonate, and other short-chain precursors, one could infer that the methyl ketone is synthesized *de novo* as well (Barkawi *et al.*, 2003 and below).

(iii) *Experiments with radiolabeled* de novo *precursors.* The *de novo* synthesis of isoprenoid pheromones by scolytids has been established unequivocally through experiments using injected radiolabeled precursors such as ^{14}C-acetate or ^{14}C-mevalonolactone (a cyclic form of mevalonate). The incorporation of radiolabel from these precursors into end products is detected and quantified by liquid scintillation counting. If the end product is isolated from a tissue, whole-body, or headspace extract by preparative chromatography [GC or high performance liquid chromatography (HPLC)] and then derivatized, the comparative GC or HPLC analyses of the labeled end product and its derivative with unlabeled standards provides compelling evidence for the precursor–product relationship (Figures 6.13 and 6.14). To date, no studies have utilized radiolabeling techniques to study the biosynthesis of fatty acid-derived or amino acid-derived pheromones of bark beetles.

In the first application of radiotracers with scolytids, ^{14}C-mevalonolactone (but not ^{14}C-acetate) injected into *I. typographus* (of unknown sex) was incorporated into hindgut and headspace volatile extracts, and the radioactivity was associated with preparative GC fractions that co-eluted with the hemiterpenoid pheromone component, 2-methyl-3-buten-2-ol (Lanne *et al.*, 1989). ^{14}C-Glucose fed to these beetles, either on filter paper or in *Picea abies* logs, did not lead to incorporation of radioactivity into the hemiterpenoid. As a consequence of other *in vivo* studies, radiolabeled products were separated by normal-phase HPLC, and ipsenol, ipsdienol, and amitinol in *I. paraconfusus* and ipsdienol and amitinol in *I. pini* were labeled from ^{14}C-acetate (Seybold *et al.*, 1995b; Tillman *et al.*, 1998) or ^{14}C-mevalonolactone (Tillman *et al.*, 1998; Lu, 1999). The *de novo* syntheses were sex-specific (Figure 6.13A–D), and in *I. paraconfusus* the amount of radioactivity incorporated into ipsenol relative to ipsdienol was approximately

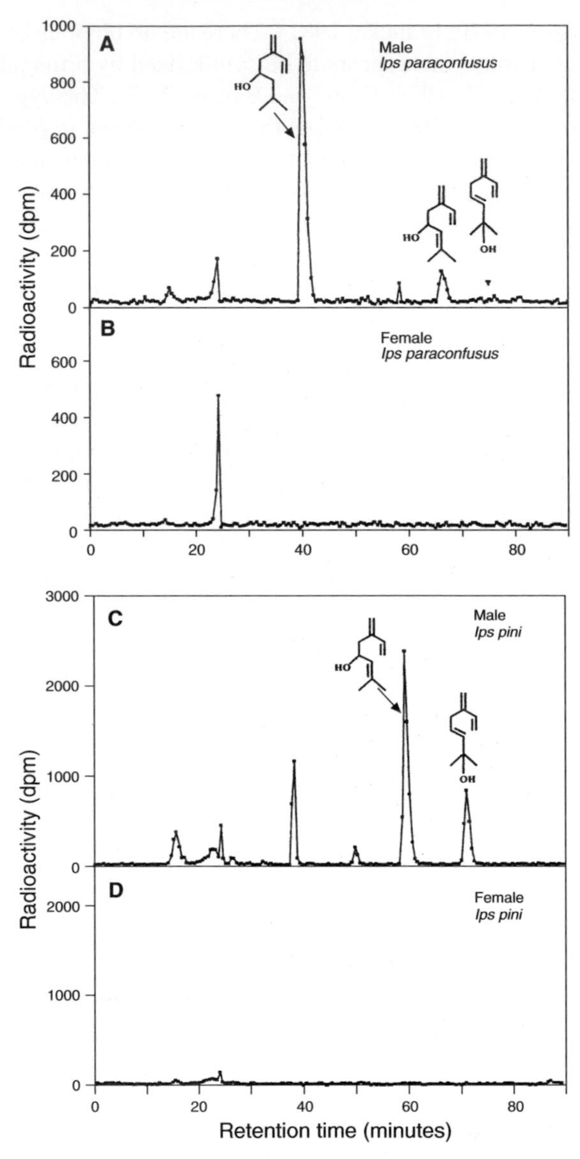

Figure 6.13 Examples of the application of normal-phase, radio-HPLC to the analysis of *de novo* biosynthetic pathways in bark beetles (Scolytidae). Demonstration of sex-specific *de novo* biosynthesis of ipsenol, ipsdienol, and amitinol through radio-HPLC analysis of pentane extracts of Porapak-trapped volatiles from (**A**) male and (**B**) female *Ips paraconfusus* Lanier feeding for 168 h in *Pinus ponderosa* and (**C**) male and (**D**) female *Ips pini* (Say) feeding for 168 h in *Pinus jeffreyi* (Seybold *et al.*, 1995b). Demonstration of sex-specific *de novo* biosynthesis of frontalin through radio-HPLC analysis of pentane extracts of Porapak-trapped volatiles from (**E**) male and (**F**) female

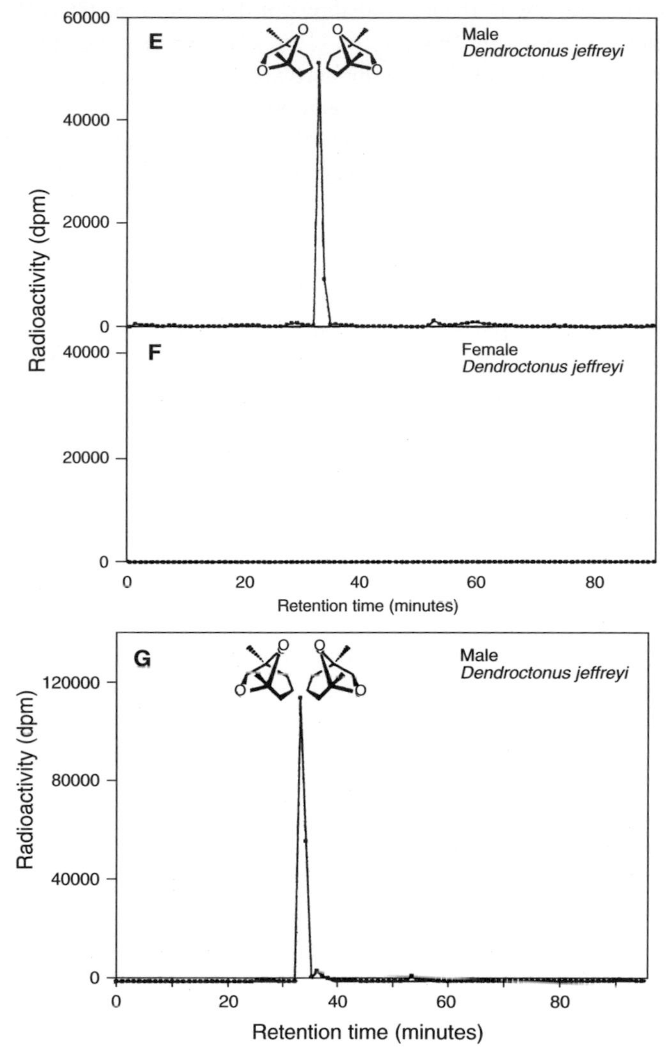

Jeffrey pine beetle, *Dendroctonus jeffreyi* Hopkins, which had been previoucly treated with juvenile hormone III (JH III, 2.2 μg/beetle in acetone) and then placed in an aeration tube for 25 to 30 h. *Ips paraconfusus* and *I. pini* were each injected with 0.2 μCi of sodium [1-^{14}C]acetate prior to placement in cut pine logs and volatile collection, while *D. jeffreyi* were each injected with 3.8 (male) and 3.7 (female) μCi of sodium [1–^{14}C]acetate 6.4 (male) and 10.7 (female) h after JH application. (**G**) The role of the mevalonate pathway in frontalin biosynthesis is supported by the incorporation of radiolabel from [2-^{14}C]mevalonolactone into frontalin by male *D. jeffreyi* (2.2 μg JH III/beetle in acetone, 10 h incubation and volatile collection, 1.1 μCi of [2 ^{14}C] mevalonolactone injected, 20 h volatile collection). Figures adapted from Seybold *et al.* (1995b) and Barkawi (2002).

10 to 1, which agreed with the mass analyses of the sample by GC/FID and with results of mass analyses by Wood *et al.* (1967) during the original isolation of the pheromone components. Following derivatization of the enantiomers of the alcohols to diastereomeric esters, the enantiomeric compositions of the labeled alcohols were also analyzed by normal-phase HPLC (Figure 6.14A–D) and shown to match the natural products (Seybold *et al.*, 1995a, b). ^{14}C-Acetate and ^{14}C-mevalonolactone have also been incorporated into ipsenone (male *I. paraconfusus*) and ipsdienone (male *I. pini*) during an *in vitro* assay of homogenized tissue (Ivarsson *et al.*, 1997). This assay was later used to localize the synthesis of radiolabeled ipsenone to the metathorax of male *I. paraconfusus* (Ivarsson *et al.*, 1998). Isolated midgut tissue from this body region of male *I. pini* converted ^{14}C-acetate to ^{14}C-ipsdienol (Hall *et al.*, 2002a) (see section 6.6.2).

Copious production of frontalin by male *D. jeffreyi* has made it an excellent experimental animal in which to study the biosynthesis of frontalin. When ^{14}C-labeled acetate, mevalonolactone, and isopentenol were injected into male *D. jeffreyi*, radiolabel was present in the HPLC fractions associated with frontalin (Figure 6.13E–G, Barkawi *et al.*, 2003). Incorporation of ^{13}C-acetate into frontalin by this insect was demonstrated unequivocally using isotope ratio monitoring mass spectrometry (IRM-MS). ^{3}H-Leucine can also be used by male *D. jeffreyi* as a substrate for ^{3}H-frontalin synthesis, leading to biosynthetic hypotheses involving fatty acid-like elongation of leucine or catabolism of leucine to acetyl-CoA followed by an isoprenoid synthesis via hydroxy-methylglutaryl-CoA (HMG-CoA) as alternative routes to frontalin in addition to the mevalonate pathway (Barkawi *et al.*, 2003). Finally, isolated midgut tissues from male *D. jeffreyi* also converted ^{14}C-acetate to ^{14}C-frontalin (Hall *et al.*, 2002b).

The late stage reactions of the isoprenoid pathway (those between isopentenyl diphosphate and the pheromone end products) (Figures 6.10 and 6.15) are just beginning to be examined in *Ips* spp. using radiolabeled precursors (Ivarsson *et al.*, 1997, 1998; Seybold *et al.*, 2000; Young *et al.*, 2001; Martin *et al.*, 2003). Activated geraniol (geranyl diphosphate) is expected to be central to this portion of the pathway. Ivarsson and Birgersson (1995) reported small amounts of geraniol from hindgut extracts of male *I. duplicatus*, and the trends in accumulation of geraniol corresponded to treatments expected to stimulate and inhibit pheromone production. Similarly, Wood *et al.* (1966) reported geranyl acetate in extracts of male boring dust (frass) from *I. paraconfusus*, and Francke *et al.* (1995) [(*Pityogenes quadridens* (Hartig), *P. calcaratus* (Eichh.), and *P. bidentatus* (Herbst)] and Zhang *et al.* (2000) [*Ips cembrae*] found geraniol during chemical analyses of hindgut extracts. In the latter case, the relative quantities of geraniol measured from hindgut extracts over time matched those of ipsenol and ipsdienol. Homogenized tissue from male *I. pini* contained geranyl diphosphate in large quantities, indicative of activity of geranyl diphosphate synthase (Young *et al.*, 2001). Surprisingly, when whole-body extracts of *I. pini* from two

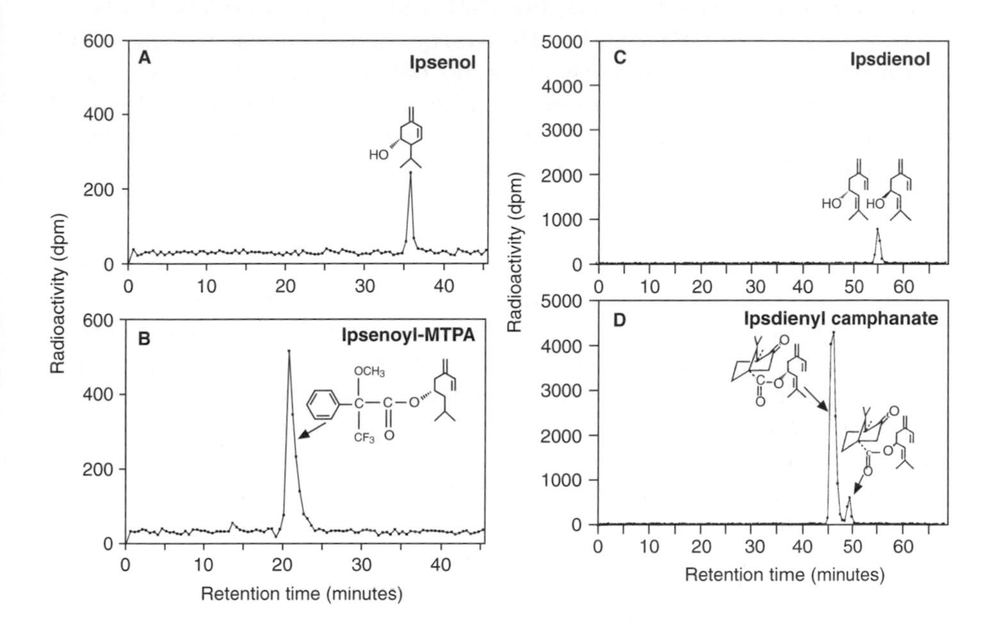

Figure 6.14 Examples of the application of normal-phase, radio HPLC to the analysis of *de novo* biosynthetic pathways in bark beetles (Scolytidae). Derivatization of ^{14}C-labeled ipsenol from *Ips paraconfusus* (**A, B**) and ^{14}C-labeled ipsdienol from *Ips pini* (**C, D**) leads to expected retention time shifts of radioactivity for each compound and derivative. Derivatization of *I. paraconfusus* ipsenol with (2*R*)-(+)-α-methoxy-α-(trifluoromethyl) phenylacetic acid (MTPA) leads to the formation of only one diastereomer [(4*S*)-(–)-ipsenoyl-(2′*R*)-2′-methoxy-2′-phenyl-2′- (trifluoromethyl) phenylacetate] indicating the *de novo* biosynthesis of highly pure (4*S*)-(–)-ipsenol (**B**) Derivatization of *I. pini* ipsdienol with (1*S*)-(–)-camphanic acid leads to the formation of both diastereomers [(4*R*)-(–)- and (4*S*)-(+)-ipsdienyl-(1′*S*)-camphanates] indicating the *de novo* biosynthesis of approximately 90 percent-(4*R*)-(–)-ipsdienol (**D**). Figures adapted from Seybold *et al.* (1995b).

populations were incubated with ^{3}H-geranyl diphosphate, the extracts of males sex-specifically converted the substrate to ^{3}H-myrcene (Martin *et al.*, 2003). This represents the first evidence for a monoterpene synthase and the biosynthesis of a monoterpene (*sensu stricto*) by a metazoan.

In other late-stage reactions, studies of the conversion of myrcene and ketone analogues of ipsenol and ipsdienol to the alcohols have been investigated in *I. paraconfusus* (Fish *et al.*, 1979, 1984; Vanderwel, 1991; Ivarsson and Birgersson,

1995; Ivarsson *et al.*, 1997, 1998; Lu, 1999) and in *I. pini* (Vanderwel, 1991; Ivarsson *et al.*, 1997) (Figure 6.15). The presence of ipsenone and ipsdienone in pheromone extracts from a variety of *Ips* spp. (Francke *et al.*, 1986, 1995 – *Ips sexdentatus* (Boern.) and *Ips cribricollis* (Eich.); Byers and Birgersson, 1990 – *I. paraconfusus*; and Zhang *et al.*, 2000 – *I. cembrae*), from the ipsenol- and ipsdienol-producing fir bark beetle, *Pityokteines elegans* (Swaine) (Macías-Sámano *et al.*, 1997), and from the ipsenol-producing, vine-infesting bark beetle, *Xylocleptes bispinus* Duft. (Klimetzek *et al.*, 1989b) has provided additional evidence for a biogenic relationship between the ketones and the alcohol end products. Isolation of the P450s that are likely to be involved in the transformations of geranyl diphosphate to ipsenol and ipsdienol will help elucidate how *Ips* spp. achieve such remarkable intra- and interspecific control over the enantiomeric composition of the pheromone alcohols (Seybold, 1993; Seybold *et al.*, 1995a). The genomics approach (Seybold and Tittiger, 2003; see Tittiger, Chapter 7, in this volume) will most certainly accelerate our understanding of these P450s. Much work remains to be done to elucidate the "uniquely scolytid" biochemical transformations that promise to be found in the late-stage reactions of the mevalonate pathway.

6.6.4 Studies of the regulation of pheromone production in the Scolytidae

During the earliest studies of pheromone production by male *I. paraconfusus*, both adult male maturity (Wood and Vité, 1961; Borden, 1967; Byers, 1983b) and feeding in host phloem (Wood, 1962; Wood *et al.*, 1966) were noted as regulatory factors (reviewed in Vanderwel, 1994). Later research established that JH III or fenoxycarb (JHA) (Borden *et al.*, 1969; Hughes and Renwick, 1977b; Kiehlmann *et al.*, 1982; Chen *et al.*, 1988) stimulates monoterpenoid pheromone production in *I. paraconfusus*. Further experiments with other *Ips* spp. (Hackstein and Vité, 1978; Renwick and Dickens, 1979; Ivarsson and Birgersson, 1995), *Dendroctonus* spp. (Hughes and Renwick, 1977a; Bridges, 1982; Conn *et al.*, 1984; Barkawi, 2002), *Pityokteines* spp. (Harring, 1978), and *Scolytus* spp. (Blight *et al.*, 1979a) defined a central role for JH III as a regulator for pheromone production throughout the Scolytidae. Although more research is certainly warranted, hormones such as 20-hydroxyecdysone (20-E) and pheromone biosynthesis activating neuropeptide (PBAN), which regulate pheromone production in other insect taxa, do not appear to play a role in regulating bark beetle pheromone production (Ivarsson and Birgersson, 1995). However, several studies have hinted that JH III alone may not be sufficient to explain the regulation of scolytid pheromone biosynthesis (Harring, 1978; Hughes and Renwick, 1977b; Lu, 1999; Tillman *et al.*, 2001). Nearly all research on the regulation of pheromone production has involved isoprenoid pheromone components with only a few studies that have evaluated the regulation of fatty acid- or amino acid-derived pheromone components.

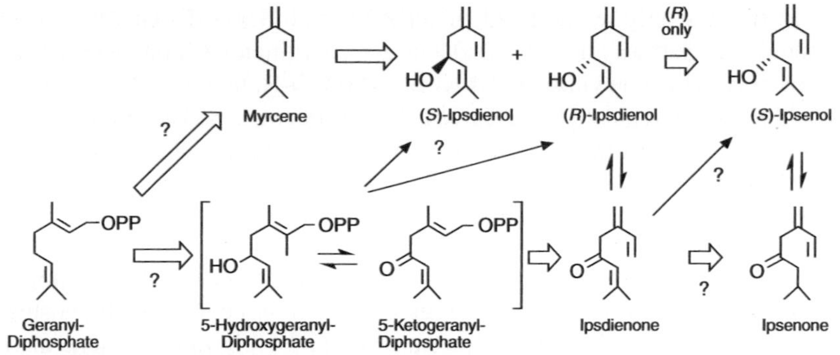

Figure 6.15 Hypothetical alternative late stages of enantiospecific *de novo* biosynthesis of ipsenol and ipsdienol in male *Ips paraconfusus* and *Ips pini*. Biosynthesis may proceed from geranyl diphosphate to myrcene as catalyzed by a sex-specific monoterpene synthase. Terpene synthases (including a myrcene synthase) have been characterized from conifers (Bohlmann *et al.* 1997, 1998). Alternatively, biosynthesis may proceed from geranyl diphosphate to 5-hydroxygeranyl diphosphate (W. Francke, personal communication). Either or both myrcene and 5-hydroxygeranyl diphosphate are then converted to ipsdienol, or 5-hydroxygeranyl diphosphate is converted to ipsdienone via a ketone (5-keto-geranyl diphosphate) intermediate. As outlined in Vanderwel (1991), ipsdienol can be converted directly or via other ketone intermediates to ipsenol. Note that the ultimate enantiomeric composition of ipsdienol may result from the enantioselective insertion of a hydroxyl group at C_5 of geranyl diphosphate or from an enantioselective interconversion of ipsdienol and ipsdienone (modified from Fish *et al.*, 1984; Vanderwel, 1991). For intermediates in this pathway, OPP denotes a diphosphate moiety. Figure adapted from Seybold *et al.* (2000).

6.6.4.1 Regulation of production of isoprenoid pheromones in the Scolytidae

(a) *Major step vs coordinate regulation.* In vertebrates, isoprenoid biosynthesis is negatively regulated by the titer of the primary product, cholesterol (Figure 6.10) (Goldstein and Brown, 1990), and potentially by the titer of farnesol (Meigs and Simoni, 1997). The enzymes HMG-S and HMG-R are central to the regulation in the pathway. Regulation of HMG-R is complex, involving (1) coordinate changes in gene expression; (2) coordinate processing and stability of RNA and protein via feedback inhibition by mevalonate pathway end products (Goldstein and Brown, 1990); and (3) cross-regulation, which occurs when biochemical processes such as cellular energy charge (i.e. ATP level) and cellular oxygen availability modulate levels of HMG-R activity (Hampton *et al.*, 1996). In contrast, in scolytid pheromone biosynthesis, the primary product of the mevalonate pathway is a monoterpenoid (C_{10}) and the pathway is positively regulated by a sesquiterpenoid (C_{16}, JH III) derived from farnesyl diphosphate (Figure 6.10). Ignoring the more complex levels of regulation known from vertebrates and yeast, one hypothetical approach to regulation in the scolytids is that JH III acts only at the level of HMG-S and HMG-R. Alternatively, all enzymes in the pathway may be regulated in a coordinate fashion by JH III, with major regulation exerted through HMG-S and HMG-R. In the Scolytidae, HMG-S would likely be regulated to a lesser extent, reflecting the less prominent role that it has in controlling the mevalonate pathway in vertebrates (Goldstein and Brown, 1990).

(b) *Biochemical studies of regulation through JH III and feeding.* Ivarsson and Birgersson (1995) used the HMG-R inhibitor compactin and the JHA methoprene to provide indirect evidence that JH regulates *de novo* pheromone biosynthesis in male *I. duplicatus*. Using [1-^{14}C]acetate and [2-^{14}C]mevalonolactone *in vivo* and L-[*methyl*-^{3}H]-methionine *in vitro*, Tillman *et al.* (1998) evaluated the relationships among feeding on host (*P. jeffreyi*) phloem, JH biosynthesis, and *de novo* ipsdienol biosynthesis by male *I. pini*. HPLC analysis of corpora allata (CA) extracts demonstrated that the type of JH released by the CA in male *I. pini* is JH III. Furthermore, increasing incorporation of radiolabeled acetate into ipsdienol by male *I. pini* with increasing topical JH III dose demonstrated unequivocally that JH III regulates *de novo* pheromone production (Tillman *et al.*, 1998). However, incorporation of radiolabeled mevalonolactone into ipsdienol by male *I. pini* was not affected by increasing JH III dose, suggesting that JH primarily influences enzymes prior to mevalonate in this pathway (i.e. HMG-S and HMG-R). In contrast, preliminary biochemical studies of some of the late-stage reactions in monoterpenoid biosynthesis suggest that the conversion of ^{3}H-geranyl diphosphate to ^{3}H-myrcene is up-regulated in male *I. pini* by both JH III treatment and feeding on *P. jeffreyi* phloem (Martin *et al.*, 2003).

De novo ipsdienol biosynthesis by male *I. pini* was also stimulated by feeding on host phloem (Tillman *et al.*, 1998) and, thus, feeding may be the initial cue that stimulates the intermediary biosynthesis and release of JH from the CA. This hypothesis was addressed using an *in vitro* assay that compared levels of JH released [likely reflecting new biosynthesis, Feyereisen, 1985] from CA in unfed and previously fed male and female *I. pini*. The rate of JH III release from the CA was significantly higher in male *I. pini* that had fed for 24 hours relative to unfed (24-hour incubated) males, whereas females displayed overall lower rates of JH release and no significant differences between fed and unfed treatment groups at the time points assayed. These radiolabeling studies collectively provide evidence for a behavioral and physiological sequence of events leading to feeding-induced *de novo* pheromone biosynthesis in male *I. pini*: (1) feeding on host phloem; (2) feeding-induced JH III release (i.e. biosynthesis) by the CA; and (3) JH III-stimulated *de novo* ipsdienol biosynthesis.

Topical treatment with JH III also results in the sex-specific accumulation of transcript for HMG-S in *D. jeffreyi* (Tittiger *et al.*, 2000) and for HMG-R in *I. paraconfusus* (Tittiger *et al.*, 1999; Tillman *et al.*, 2001), *I. pini* (Tillman *et al.*, 2001), and *D. jeffreyi* (Tittiger *et al.*, 2003). These effects may result from enhanced mRNA synthesis or stability (or both) (see Tittiger, Chapter 7, in this volume), but the increases in HMG-R transcript abundance in male *D. jeffreyi* are strongly correlated with the production of the isoprenoid frontalin over a range of JH III doses and incubation times (Barkawi, 2002). This type of correlation with ipsenol and ipsdienol production is also evident in *I. paraconfusus* and *I. pini*, though the mass of pheromone produced is attenuated considerably in *I. paraconfusus* (Tillman *et al.*, 2001; see below). Due to the technical difficulty of isolating total or mRNA from phloem-fed bark beetles, the effects of feeding on HMG-S and HMG-R transcription remain to be measured.

It is not clear whether JH III alone is sufficient to up-regulate *de novo* pheromone biosynthesis in all *Ips* spp. Following a series of decapitation and gland (CA and corpora cardiaca (CC)) implantation experiments, Hughes and Renwick (1977b) proposed that juvenile hormone may act indirectly through a brain hormone to stimulate pheromone biosynthesis in male *I. paraconfusus*. To date, no one has conducted enzyme assays of purified HMG-S or HMG-R., However, Tillman *et al.* (2001) monitored the conversion of ^{14}C-HMG-CoA to ^{14}C-mevalonate by crude microsomal preparations from JH III-treated or phloem-fed male *I. paraconfusus* and *I. pini* to establish a measure of HMG-R activity. In this study, the effects of topical application of JH III and feeding were compared with regard to pheromone production, HMG-R activity, and HMG-R transcript levels in both *I. paraconfusus* and *I. pini* (Figure 6.16). The comparative results of these studies indicate that (1) JH III induces 150-fold more of the principal pheromone component of male *I. pini* (ipsdienol) than it does of the principal pheromone component of male *I. paraconfusus* (ipsenol), whereas feeding induces similar amounts of the principal pheromone components in each species; (2) JH III stimulates HMG-R enzyme activity in male *I. pini*, but not in male *I. paraconfusus*, whereas feeding stimulates similar amounts of enzyme activity in each species; and (3) JH III induces similar increases in the transcript for HMG-R in both species (see Tittiger, Chapter 7, in this volume). Lu (1999) repeated this study with decapitated male *I. paraconfusus* and male *I. pini*, largely validated the results of Tillman *et al.* (2001), and showed additionally that decapitation reduced the amount of HMG-R transcript, HMG-R activity, and *de novo* ipsdienol produced by *I. pini*, but only reduced the amount of HMG-R transcript in *I. paraconfusus*. It appears that in *I. paraconfusus*, although JH III alone can up-regulate the transcript for HMG-R, JH III may act in concert with a second feeding-associated factor (hormone) to stimulate ipsdienol and ipsenol biosynthesis through full activity of HMG-R. This could occur through stimulation of HMG-R translation, enhancement of the transcript or protein stability, and/or stimulation of enzyme activity. This second factor does not appear to be necessary in *I. pini*. It is surprising the degree to which the regulation of *de novo* pheromone biosynthesis appears to differ in these two closely related species of Scolytidae. It is clear that one must be cautious when extending information about pheromone regulation from one bark beetle species to another (Tillman *et al.*, 2001).

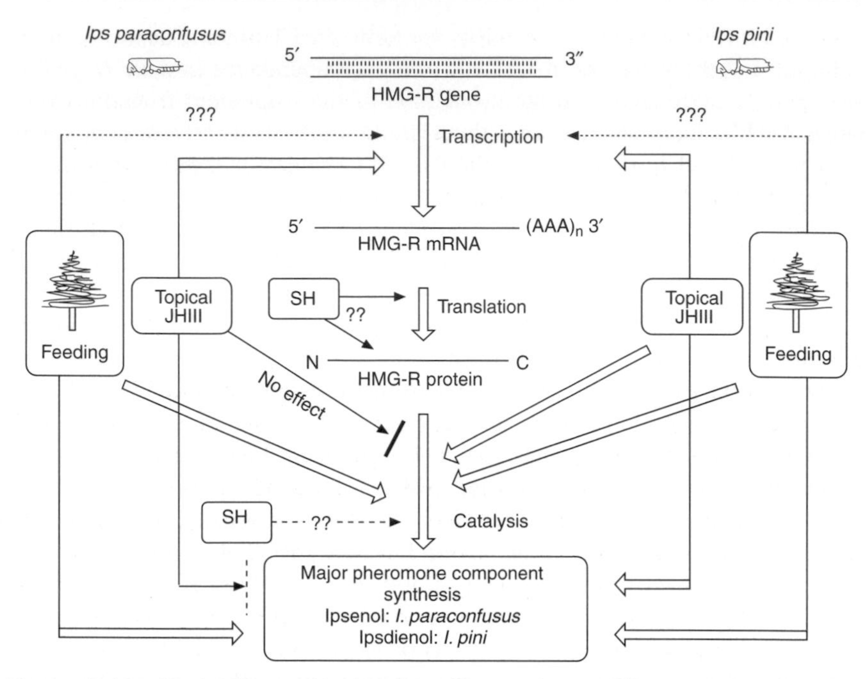

Figure 6.16 Model illustrating interspecific regulatory differences in an early-stage reaction in isoprenoid pheromone biosynthesis between male *Ips paraconfusus* Lanier and *Ips pini* (Say). Feeding on host phloem results in synthesis of the full amount of the major pheromone component and full activity of HMG-R for both species. The impact of feeding on HMG-R transcript levels is yet to be determined. Topical treatment of male *I. pini* with JH III mimics feeding nearly completely in terms of pheromone mass and HMG-R activity. Topical treatment of male *I. paraconfusus* with JH III does not mimic feeding in terms of pheromone mass or HMG-R activity. Topical treatment of both species with JH III results in significantly enhanced levels of HMG-R transcript. One hypothetical explanation for the interspecific difference is that a second hormone (SH) or factor may be associated with the synthesis, stability, and/or activity of HMG-R in *I. paraconfusus.*

6.6.4.2 Regulation of production of non-isoprenoid pheromones in the scolytidae

Our comprehension of how JH III interacts with the biosynthesis of fatty acid- or amino acid-derived scolytid pheromones is limited to what has been learned from several preliminary studies, and in contrast to the isoprenoid pathway there is no information on any endocrine effects on the expression or activity of specific pathway enzymes. Hughes and Renwick (1977a) found that topical application of JH III to newly emerged female *D. brevicomis* resulted in the production of more than 1 μg of *exo*-brevicomin per beetle. Feeding on fresh *P. ponderosa* logs also stimulated *exo*-brevicomin production in females as did feeding on filter paper treated with a methanol extract of *P. ponderosa* phloem. Although no additional studies have been reported, this appears to be an ideal system to investigate the endocrine regulation of fatty acid-derived bicyclic acetal production. In a similar study, Blight *et al.* (1979a) reported that topical treatment of male and female *S. scolytus* with JH III resulted in increased production of both α-multistriatin and *threo*-4-methyl-3-heptanol by males. Female production of either compound was not affected by hormone treatment. The study results are not easy to interpret as the beetles were allowed to feed on elm logs following hormonal treatment.

A bevy of studies have shown that scolytids produce compounds such as toluene and 2-phenylethanol (Figure 6.9C) (Renwick *et al.*, 1976c; Gries *et al.*, 1988, 1990a, b; Ivarsson and Birgersson, 1995; Zhang *et al.*, 2000). Aromatic pheromone components in the Coleoptera could be produced *de novo* by the shikimic acid pathway, but in male *I. pini*, toluene and 2-phenylethanol are clearly derived in axenic beetles from phenylalanine, which is normally available to these beetles in their phloem diet (Gries *et al.*, 1990b). While toluene does not appear to have any pheromonal activity with *I. pini* (Gries *et al.*, 1990b), 2-phenylethanol is weakly attractive to *I. paraconfusus* (Renwick *et al.*, 1976c). Interestingly, the production of 2-phenylethanol by male *I. paraconfusus* and male *I. duplicatus* is stimulated in the absence of phloem feeding by topical treatment with JH III (Hughes and Renwick, 1977b) and the JHA methoprene, respectively (Ivarsson and Birgersson, 1995). Decapitated male *I. paraconfusus* did not produce 2-phenylethanol following treatment with JH III (Hughes and Renwick, 1977b). Given that host phenylalanine is considered to be the sole precursor of 2-phenylethanol in male *I. pini*, it is surprising that host feeding had a negative effect on 2-phenylethanol production in male *I. duplicatus* (Ivarsson and Birgersson, 1995).

6.7 Curculionidae: weevils

The Curculionidae are the largest family of beetles. Although Lawrence and Newton (1995) included the bark beetles (Scolytidae) as a subfamily among the weevils, we are treating the groups separately for historical reasons. The weevils comprise nearly one-half of all species of beetles, and this fascinating group occupies a tremendous range of habitats. Their behavioral responses to pheromones can complicate study of the chemical ecology of weevils. Nonetheless, many workers have begun to elucidate the pheromones of weevils, and Bartelt (1999b) surveyed research on pheromones of 34 species ranging across 11 subfamilies. Most weevil pheromones are long-range, male-produced aggregation pheromones. Cyclic monoterpenoid pheromone components [e.g. grandisol, (*E*)- and (*Z*)-ochtodenol, and the corresponding aldehydes) tend to be produced by weevils in the subfamilies Curculioninae, Cryptorhynchinae, and Pissodinae, whereas straight chain methyl-branched secondary alcohols (likely derived from fatty acids) are characteristic of the Rhynchophorinae. Great progress has been made in pheromone identification for weevils in the latter subfamily, which contains pests of sugar cane as well as of coconut, oil, and ornamental palms (Bartelt, 1999b). The banana weevil, *Cosmopolites sordidus* (Germar), produces the bicyclic acetal sordidin (Mori, 1996; Mori *et al.*, 1996), which is a pheromone component whose biosynthesis is likely analogous to similar end products in the Scolytidae (Figure 6.9B and 6.12). The occurrences of grandisol in *Pityogenes* spp. and *Pityophthorus pityographus* (Ratz.) (both Scolytidae) and of (*E*)-ochtodenol in *Pityogenes* spp. (Francke *et al.*, 1987, 1989, 1995) (Figure 6.9A) also underscore the intimate similarity between the Curculionidae and the Scolytidae. The enantioselective biosyntheses of grandisol and grandisal appear to be significant for the behavior and interspecific discrimination of certain species of weevils (Dickens and Mori, 1989; Hibbard and Webster, 1993).

Biosynthetic studies of pheromone production by the weevils have focused on the boll weevil, *Anthonomus grandis* Boh whose male-produced aggregation pheromone is comprised of (1*R*, 2*S*)-grandisol, (*Z*)-ochtodenol, and (*E*)- and (*Z*)-ochtodenal (Tumlinson *et al.*, 1969). The components can either be formed through *de novo* biosynthesis, or through the derivatization of host plant monoterpenes, such as nerol and geraniol (reviewed in Vanderwel and Oehlschlager, 1987). Pheromone production is stimulated in boll weevils that have been either fed JH III (Hedin *et al.*, 1982) or treated topically with the JHA methoprene (Dickens *et al.*, 1988), or by antennectomy (Dickens *et al.*, 1988). The endocrinology of reproductive behavior has been investigated in female *A. grandis* (Taub-Montemayor and Rankin, 1997; Taub-Montemayor *et al.,* 1997a, b), but unfortunately not in pheromone-producing males. Mating causes a transient inhibition of pheromone production in this species (Dickens and Wiygul, 1987): aggregation pheromone production by males decreased significantly 2–3 days after a single mating, but recovered 1–2 days later. The authors suggest that this effect would help to ensure genetic diversity, since recently mated males would be less attractive. The biochemical mechanism for the mating-induced inhibition of pheromone production has not been investigated. Borden (1967) reported a similar phenomenon with pheromone production in the male scolytid, *I. paraconfusus*, but did not distinguish between simple association of the female with the male in the gallery and mating between the female and male.

Wiygul *et al.* (1982) reported the site of pheromone biosynthesis in male boll weevils as the fat body. To date this is the only insect where the fat body has been proposed as a site for pheromone biosynthesis, and it is an obvious contradiction that in most species of the closely related Scolytidae, isoprenoid pheromone biosynthesis has been reported to occur in the alimentary canal (see section 6.6.2). Wiygul *et al.* (1982) observed that the level of pheromone production measured in fat body (about 720 ng/weevil/day) was much less than the expected value (about 3400 ng/weevil/day).

Preliminary biochemical studies have been performed to elucidate the mechanism for the regulation of pheromone production by JH III. Direct addition of JH III to fat bodies incubated *in vitro* stimulates pheromone biosynthesis, indicating that in this insect, JH may act directly to stimulate pheromone production (Wiygul *et al.*, 1990). Wiygul and Sikorowski (1986, 1991) determined that both staphylococcal enterotoxin B and the heat-stable enterotoxin from *Escherichia coli* (Migula) Castellani and Chalmers reduce pheromone production by fat bodies *in vitro*. The heat-stable toxin is known to increase the level of cyclic guanine monophosphate (cGMP) in the small intestine of mammals (Field *et al.*, 1978; Gianella and Drake, 1979). Although the mechanism of action of the enterotoxins in boll weevils has not been investigated, perhaps their effect on pheromone production in boll weevil fat bodies likewise involves the signal transduction cascade. Certainly greater investigation is warranted in this interesting biochemical system and recent work with *A. grandis* on the isolation of nucleotide sequences for isoprenyl diphosphate synthase and geranyl diphosphate synthase may elucidate the regulation of pheromone production in this economically important species (Nural *et al.*, 2001).

6.8 Summary and future directions

Pheromone biosynthesis has been intensively investigated in several representative species from each of seven families of Coleoptera, particularly the Scolytidae. In these families, the pheromones are derived predominantly through modifications of normal isoprenoid or fatty acid metabolism. Metabolic embellishments to produce unique pheromone components include stereospecific hydroxylation, epoxidation, and/or cyclization reactions; and the stereospecific insertion of methyl or ethyl branches. The hydroxylation and epoxidation reactions will no doubt involve highly specific P450 systems, which remain to be characterized. In the Scolytidae the *de novo* isoprenoid pathways have evolved with an emphasis on hemiterpenoid and monoterpenoid end products. This use of the mevalonate pathway provides an interesting counterexample to plant systems (MEP pathway) and mammalian systems (mevalonate pathway to sterols). However, perhaps because of their evolutionary relationship with higher plants and their tremendous diversity, the Coleoptera provide multiple examples where both *de novo* synthesis and conversion of host pheromone may play a role in isoprenoid precursors biosynthesis (e.g. *Ips* spp, *Dendroctonus* spp., *Anthonomus grandis*). The evolutionary redundancies further include microorganismal synthesis and autooxidation of host compounds as "pheromone" sources in the Coleoptera. Another theme in coleopteran pheromone biosynthesis involves the utilization of amino acids (possibly in some cases host derived) by the Scarabaeidae and the Scolytidae. In certain instances, aromatic amino acids (e.g. phenylalanine) are converted to pheromone components, whereas in other cases alkylated amino acids (e.g. leucine, isoleucine, and valine) are hypothesized precursors for sex pheromone components of the melolonthine scarabs or for the synthesis of hemiterpenoid and monoterpenoid pheromone components by certain scolytids.

In the representative species studied to date, JH III is the predominant endocrine factor regulating pheromone production. However, the hormones involved in the regulation of pheromone biosynthesis in the vast majority of Coleoptera have simply not been investigated. This is particularly true of those that produce fatty acid-derived pheromones. Moreover, studies have in large part focused on the role of the (usually food-induced) stimulation of aggregation pheromone production. There have been relatively few investigations of the factors involved in the regulation of sex pheromones; the effects of physiological factors besides food and maturity (e.g. light, population density, and mating); and the factors involved in the inhibition of pheromone production. The possibility that hormones other than JH III are involved must continue to be explored. The potential roles of 20-hydroxyecdysone and PBAN in the regulation of coleopteran pheromone production bear further investigation. Comparative research between isoprenoid-producing taxa and taxa that produce fatty acid-derived pheromones (e.g. Scarabaeidae, Nitidulidae, Tenebrionidae, and Scolytidae-*Dendroctonus*) would yield particularly valuable insights.

Of all families of Coleoptera, pheromone biosynthesis has been best studied in the Scolytidae. Since the publication of *Pheromone Biochemistry* (Prestwich and Blomquist, 1987), the general biosynthetic routes to two major groups of scolytid pheromones have been elucidated. Studies confirmed expectations that the bicyclic acetals (such as *exo-* and *endo*-brevicomin) can be formed from ketoalkene precursors, likely through ketoepoxide intermediates. In some cases bicyclic acetals are formed *de novo* (e.g. frontalin), but it remains to be demonstrated if this is the general case. In contrast, studies of the biosynthesis of the scolytid hemiterpenoid and monoterpenoid pheromones have yielded exciting and (to many) unexpected results. In the past, the biosynthesis of these pheromones was thought to proceed primarily through the modification of host-derived precursors, with contributions to the semiochemical potpourri by symbiotic microorganisms and autooxidation. However, recent studies unequivocally demonstrated that these isoprenoid pheromones are produced through *de novo* biosynthesis, via the mevalonate pathway. The relative contribution of these routes to overall pheromone production remains to be determined. Despite this progress, there are still a large number of scolytid pheromone components whose biosynthesis is a matter of speculation or for which hypothetical biosyntheses have not been posited (Figure 6.9). Although the techniques for investigating pheromone biosynthesis in beetles have become more sensitive and sophisticated, in many instances the quantities of certain pheromone components produced have limited our attempts to study the biosynthesis or its regulation (e.g. Scolytidae-lanierone, *cis*- and *trans*-verbenol, *cis*- and *trans*-pityol, seudenol and related compounds). In other cases the minute size of the beetles [e.g. *Pityophthorus* or *Pseudopityophthorus* spp. (Scolytidae)] makes experimental manipulation of individual animals practically impossible. Future researchers may embrace some of these microbiochemical and microscopic challenges. Unraveling the biosynthesis of the tricyclic pheromone component lineatin (*Trypodendron* spp., Scolytidae) would represent a true *tour de force* of insect biochemical exploration. In other cases (e.g. *D. brevicomis* and *exo*-brevicomin), relatively high quantities of pheromone produced present unexploited research opportunities for understanding synthesis and regulation. The role of P450s in the stereospecificity of monoterpenoid and bicyclic acetal pheromones of the Scolytidae is also an area that should yield to experimentation in the near future.

The role of JH III in the regulation of pheromone production in the Scolytidae is well documented, and has been extensively studied. In relation to the mevalonate pathway, JH III stimulates the induction of HMG-R at the transcriptional level (see Tittiger, Chapter 7, in this volume). Additional targets of JH III are possible, but remain to be elucidated. Furthermore, JH III may interact with other hormones to regulate pheromone biosynthesis: if so, this could be explored on a molecular level. The mechanism of the JH III-mediated regulation of isoprenoid pheromones outside of the Scolytidae (e.g. in *A. grandis*), and of the regulation of fatty acid-derived pheromones, remains to be investigated.

The diversity of Coleoptera represents a grand opportunity for biochemical and molecular prospecting for unique reactions, regulatory features and gene sequences related to pheromone biosynthesis. Future research in this area will undoubtedly be directed both broadly across all taxa and deeply toward the representative species discussed in this overview. An emphasis on comparative model species at the family, generic, or species levels (e.g. *Ips paraconfusus* and *I. pini*) will continue to elucidate the intricate nuances in synthesis and regulation. It is probable that new techniques that have been recently applied to some species will be soon applied to other species. For example, (1) the application of *in situ* hybridization and eventually immunochemistry to localize cellular sites of synthesis (Hall *et al.*, 2002a, b); (2) any of the PCR-based methods for differential or subtractive screening of nucleic acid libraries to examine life stage-, sex-, or species-related differences linked to pheromone biosynthesis [e.g. differential display (Liang *et al.*, 1993) or representational difference analysis (Lisitsyn *et al.*, 1993; Hubank and Schatz, 1994)]; and (3) genomics-based approaches for broad screening of expressed genes (Seybold and Tittger, 2003; see Tittiger, Chapter 7, in this volume).

Ultimately, just as behavioral chemicals themselves have been extended to pest management, research on pheromone biosynthesis and its regulation may be directed toward application. This might include the culturing of insect tissues or cells, or the transfer of relevant genes into expression systems, for production of behavioral chemicals of high stereochemical purity. Critical here will be developing a thorough understanding the late-stage reactions of all pathways, where, in many cases, stereospecificity or other unique structural attributes are imparted on pheromone components. Perhaps eventually, the isolated genes could be transgenically introduced into microorganisms for areawide treatments, or into agriculturally or silviculturally important plants to produce semiochemicals to disrupt mating, or otherwise interfere with the reproductive biology and host finding of pest beetles.

Abbreviations

Corpora allata = CA, Coenzyme A = CoA, Copulation release = CR, Gas chromatography = GC, Gas chromatography with flame ionization detection = GC-FID, Coupled gas chromatography and mass spectrometry = GC-MS, High pressure liquid chromatography = HPLC, 3-Hydroxy-3-methylglutaryl-CoA = HMG-CoA, HMG-CoA reductase = HMG-R, HMG-CoA synthase = HMG-S, Isotope ratio monitoring mass spectrometry = IRM-MS, Juvenile hormone = JH, JH analog = JHA, Methyl-erythritol-4-phosphate = MEP, Nuclear magnetic resonance = NMR, Pheromone biosynthesis activating neuropeptide = PBAN, Smooth endoplasmic reticulum = SER.

Acknowledgements

Common and scientific names, including higher level taxonomic names, used in this chapter generally follow *Common Names of Insects and Related Organisms 1997* published by the Entomological Society of America. SJS thanks the USDA National Research Initiative Competitive Grants Program (#9302089, #9502551, #9702991, #97-35302-4223, #9802897, #98-35302-6997) for continuous funding of work on the biochemistry and molecular biology of aggregation pheromone production in *Ips* spp. from 1993 to 2000, the National Science Foundation (#IBN-972855 and #IBN-9906530) for support of work on *Dendroctonus* pheromone biosynthesis, and the Human Frontier Science Program (#RGY0382). SJS further acknowledges the roles of D. L. Wood, W. Francke, I. Kubo, and G. J. Blomquist as mentors in scolytid chemical ecology, bioorganic chemistry, and biochemistry and molecular biology, which have been applied to the area of scolytid pheromone biosynthesis. We are grateful to J. A. Tillman and L. S. Barkawi for drafting many of the graphics and for insightful discussions regarding the biosynthesis of beetle pheromones. We also acknowledge the editorial assistance of A. D. Graves. This chapter is a contribution of the Minnesota Agricultural Experiment Station (Project MN-17-070). DV also thanks A. C. Oehlschlager,

H. D. Pierce, Jr, J. H. Borden, and G. J. Blomquist for their roles as teachers, mentors, and sounding boards; R. Bacala for many interesting discussions on insect pheromone biochemistry; and the Natural Sciences and Engineering Research Council (NSERC) for the continuous funding of the work on the biochemistry of pheromone production in beetles from 1992 to the present.

References

Abdel-Kader M. M., Abdu R. M. and Hussien M. A. (1987) Effect of time of day and temperature on sex pheromone production and perception by the rust-red flour beetle, *Tribolium castaneum* (Herbst). *Arab Gulf J. Scient. Res. Agric. Biol. Sci.* **B5**, 147–156.

Algarvio R., Teixeira C., Barata E., Pickett J., Novas P. C. and Figueiredo D. (2002) *Identification of a putative aggregation pheromone from males of* Platypus cylindrus *(Coleoptera: Platypodidae).* Presented at Annu. Meet. Int. Soc. Chem. Ecol., 19th, Hamburg.

Bacala R. (2000) In vitro *studies of sex pheromone biosynthesis in the yellow mealworm beetle,* Tenebrio molitor (*Coleoptera: Tenebrionidae*). MSc thesis. Univ. of Manitoba, 133 pp.

Bakke A., Frøyen P. and Skattebøl L. (1977) Field response of a new pheromonal compound isolated from *Ips typographus*. *Naturwissenschaften* **64**, 98.

Barkawi L. S. (2002) *Biochemical and molecular studies of aggregation pheromones of bark beetles in the genus* Dendroctonus *(Coleoptera: Scolytidae), with special reference to the Jeffrey pine beetle,* Dendroctonus jeffreyi *Hopkins*. PhD thesis. Univ. Nevada, Reno, 193 pp.

Barkawi L. S., Francke W., Blomquist G. J. and Seybold S. J. (2003) Frontalin: *de novo* biosynthesis of an aggregation pheromone component by *Dendroctonus* spp. bark beetles (Coleoptera: Scolytidae). *Insect Biochem. Mol. Biol.* **33**, 773–788.

Bartelt R. J. (1999a) Sap beetles. In *Pheromones of Non-Lepidopteran Insects Associated with Agricultural Plants*, eds J. Hardie and A. K. Minks, pp. 69–89. CABI Publishing, New York.

Bartelt R. J. (1999b) Weevils. In *Pheromones of Non-Lepidopteran Insects Associated with Agricultural Plants*, eds J. Hardie and A. K. Minks, pp. 91–112. CABI Publishing, New York.

Bartelt R. J. and Weisleder D. (1996) Polyketide origin of pheromones of *Carpophilus davidsoni* and *C. mutilatus* (Coleoptera: Nitidulidae). *Bioorg. Med. Chem.* **4**, 429–438.

Bartelt R. J., Dowd P. F., Shorey H. H. and Weisleder D. (1990) Aggregation pheromone of *Carpophilus freemani* (Coleoptera: Nitidulidae): a blend of conjugated triene and tetraene hydrocarbons. *Chemoecology* **1**, 105–113.

Bartelt R. J., Dowd P. F. and Plattner R. D. (1991) Aggregation pheromone of *Carpophilus lugubris*: new pest management tools for the nitidulid beetles. In *Naturally Occurring Pest Bioregulators*, ed P. A. Hedin, pp. 27–40. ACS Symposium Series No. 449, American Chemical Society, Washington, DC.

Bartelt R. J., Weisleder D., Dowd P. F. and Plattner R. D. (1992) Male-specific tetraene and triene hydrocarbons of *Carpophilus hemipterus*: structure and pheromonal activity. *J. Chem. Ecol.* **18**, 379–402.

Bedard W. D., Tilden P. E., Wood D. L., Silverstein R. M., Brownlee R. G. and Rodin J. O. (1969) Western pine beetle: field response to its sex pheromone and a synergistic host terpene, myrcene. *Science* **164**, 1284–1285.

Bedard W. D., Silverstein R. M. and Wood D. L. (1970) Bark beetle phermones. *Science* **167**, 1638–1639.

Berciano M. T., Fernandez R., Pena E., Calle E., Villagra N. T., Rodriguez-Rey J. C. and Lafarga M. (2000) Formation of intranuclear crystalloids and proliferation of the smooth endoplasmic reticulum in Schwann cells induced by tellurium treatment: association with overexpression of HMG CoA reductase and HMG CoA synthase mRNA. *Glia* **29**, 246–259.

Bestmann H. J. and Vostrowsky O. (1985) Pheromones of the Coleoptera. In *Handbook of Natural Pesticides*, eds E. D. Morgan and N. B. Mandava, Vol. IV: Pheromones, Part A, pp. 95–183. CRC Press, Boca Raton, FL.

Birch M. C., Light D. M., Wood D. L., Browne L. E., Silverstein R. M., Bergot B. J., Ohloff G., West J. R. and Young J. C. (1980) Pheromonal attraction and allomonal interruption of *Ips pini* in California by the two enantiomers of ipsdienol. *J. Chem. Ecol.* **6**, 703–717.

Birch M. C., Paine T. D. and Miller J. C. (1981) Effectiveness of pheromone mass-trapping of the smaller European elm bark beetle. *California Agriculture* **35**, 6–7.

Birgersson G., Byers J.A., Bergström G. and Löfqvist J. (1990) Production of pheromone components, chalcogran and methyl (*E,Z*)-2,4-decadienoate, in the spruce engraver *Pityogenes chalcographus*. *J. Insect Physiol.* **36**, 391–395.

Birgersson G., DeBarr G. L., DeGroot P., Dalusky M. J., Pierce H. D., Jr, Borden J. H., Meyer H., Francke W., Espelie K. E. and Berisford C. W. (1995). Pheromones in (The) white pine cone beetle, *Conophthorus coniperda* (Schwarz) (Coleoptera: Scolytidae). *J. Chem. Ecol.* **21**, 143–167.

Birgersson G., Dalusky M. J. and Berisford C. W. (2000) Identification of an aggregation pheromone for *Pityogenes hopkinsi* (Coleoptera: Scolytidae). *Can. Entomol.* **132**, 951–964.

Blight M. M., Wadhams L. J. and Wenham M. J. (1978) Volatiles associated with unmated *Scolytus scolytus* beetles on English elm: differential production of α-multistriatin and 4-methyl-3-heptanol, and their activities in a laboratory bioassay. *Insect Biochem.* **8**, 135–142.

Blight M. M., Wadhams L. J. and Wenham M. J. (1979a) Chemically-mediated behavior in the large elm bark beetle, *Scolytus scolytus*. *Bull. Entomol. Soc. Am.* **25**, 122–124.

Blight M. M., Wadhams L. J. and Wenham M. J. (1979b) The stereoisomeric composition of the 4-methyl-3-heptanol produced by *Scolytus scolytus* and the preparation and biological activity of the four synthetic stereoisomers. *Insect Biochem.* **9**, 525–533.

Blight M. M., Wadhams L. J., Wenham M. J. and King C. J. (1979c) Field attraction of *Scolytus scolytus* (F.) to the enantiomers of 4-methyl-3-heptanol, the major component of the aggregation pheromone. *Forestry* **52**, 83–90.

Blight M. M., Ottridge A. P., Wadhams L. J. and Wenham M. J. (1980) Response of a European population of *Scolytus multistriatus* to the enantiomers of α-multistriatin. *Naturwissenschaften* **67**, 517–518.

Blight M. M., Henderson N. C. and Wadhams L. J. (1983) The identification of 4-methyl-3-heptanone from *Scolytus scolytus* (F.) and *Scolytus multistriatus* (Marsham). Absolute configuration, laboratory bioassay and electrophysiological studies on *S. scolytus*. *Insect Biochem.* **13**, 27–38.

Bohlmann J., Steele C. L. and Croteau R. (1997) Monoterpene synthases from grand fir (*Abies grandis*). *J. Biol. Chem.* **272**, 21784–21792.

Bohlmann J., Meyer-Gauen G. and Croteau R. (1998) Plant terpenoid synthases: molecular biology and phylogenetic analysis. *Proc. Natl. Acad. Sci. USA* **95**, 4126–4133.

Borden J. H. (1967) Factors influencing the response of *Ips confusus* (Coleoptera: Scolytidae) to male attractant. *Can. Entomol.* **99**, 1164–1193.

Borden J. H. (1982) Aggregation pheromones. In *Bark Beetles in North American Conifers*, eds J. B. Mitton and K. B. Sturgeon, pp. 74–139. University of Texas Press, Austin.

Borden J. H. (1985) Aggregation Pheromones. In *Comprehensive Insect Physiology, Biochemistry, and Pharmacology*, eds G. A. Kerkut and L. I. Gilbert, Vol. 9, pp. 257–285. Pergamon Press, Oxford.

Borden J. H. and McLean J. A. (1979) Secondary attraction in *Gnathotrichus retusus* and cross-attraction of *G. sulcatus* (Coleoptera: Scolytidae). *J. Chem. Ecol.* **5**, 79–88.

Borden J. H. and Slater C. E. (1969) Sex pheromone of *Trypodendron lineatum*: production in the female hindgut–Malpighian tubule region. *Ann. Ent. Soc. Am.* **62**, 454–455.

Borden J. H., Nair K. K. and Slater C. E. (1969) Synthetic juvenile hormone: induction of sex pheromone production in *Ips confusus*. *Science* **166**, 1626–1627.

Borden J. H., Chong L., McLean J. A., Slessor K. N. and Mori K. (1976) *Gnathotrichus sulcatus*: synergistic response to enantiomers of the aggregation pheromone sulcatol. *Science* **192**, 894–896.

Borden J. H., Oehlschlager A. C., Chong L., Slessor K. N. and Pierce H. D., Jr, (1980a) Field tests of isomers of lineatin, the aggregation pheromone of *Trypodendron lineatum* (Coleoptera: Scolytidae). *Can. Entomol.* **112**, 107–109.

Borden J. H., Handley J. R., McLean J. A., Silverstein R. M., Chong L., Slessor K. N., Johnston B. D. and Schuler H. R. (1980b) Enantiomer-based specificity in pheromone communication by two sympatric *Gnathotrichus* species (Coleoptera: Scolytidae). *J. Chem. Ecol.* **6**, 445–456.

Borden J. H., Pierce A. M., Pierce H. D., Jr Chong L. J., Stock A. J. and Oehlshlager A. C. (1987) Semiochemicals produced by western balsam bark beetle, *Dryocoetes confusus* Swaine (Coleoptera: Scolytidae). *J. Chem. Ecol.* **13**, 823–836.

Borden J. H., Gries G., Chong L. J., Werner R. A., Holsten E. H., Wieser H., Dixon E. A. and Cerezke H. F. (1996) Regionally-specific bioactivity of two new pheromones for *Dendroctonus rufipennis* (Kirby) (Col., Scolytidae). *J. Appl. Ent.* **120**, 321–326.

Bowers W. W. and Borden J. H. (1990) Evidence for a male-produced aggregation pheromone in the four-eyed spruce bark beetle, *Polygraphus rufipennis* (Kirby) (Col., Scolytidae). *J. Appl. Entomol.* **110,** 292–299.

Bowers W. W., Gries G., Borden J. H. and Pierce H. D., Jr (1991) 3-Methyl-3-buten-1-ol: an aggregation pheromone of the four-eyed spruce bark beetle, *Polygraphus rufipennis* (Coleoptera: Scolytidae). *J. Chem. Ecol.* **17**, 1989–2002.

Brand J. M. and Barras S. J. (1977) The major volatile constituents of a basidiomycete associated with the southern pine beetle. *Lloydia* **40**, 398–400.

Brand J. M., Bracke J. W., Markovetz A. J., Wood D. L. and Browne L. E. (1975) Production of verbenol pheromone by a bacterium isolated from bark beetles. *Nature* **254**, 136–137.

Brand J. M., Bracke J. W., Britton L. N., Markovetz A. J. and Barras S. J. (1976) Bark beetle pheromones: Production of verbenone by a mycangial fungus of *Dendroctonus frontalis*. *J. Chem. Ecol.* **2**, 195–199.

Brand J. M., Young J. C. and Silverstein R. M. (1979) Insect pheromones: a critical review of recent advances in their chemistry, biology, and application. *Progress in the Chemistry of Organic Natural Products* **37**, 1–190.

Bridges J. R. (1982) Effects of juvenile hormone on pheromone synthesis in *Dendroctonus frontalis. Environ. Entomol.* **11**, 417–420.

Byers J. A. (1981) Pheromone biosynthesis in the bark beetle, *Ips paraconfusus*, during feeding or exposure to vapours of host plant precursors. *Insect Biochem.* **11**, 563–569.

Byers J. A. (1982) Male specific conversion of the host plant compound, myrcene, to the pheromone, (+) ipsdienol, in the bark beetle, *Dendroctonus brevicomis. J. Chem. Ecol.* **8**, 363–371.

Byers J. A. (1983a) Bark beetle conversion of a plant compound to a sex-specific inhibitor of pheromone attraction. *Science* **220**, 624–626.

Byers J. A. (1983b) Influence of sex, maturity and host substances on pheromones in the guts of the bark beetles, *Ips paraconfusus* and *Dendroctonus brevicomis*. *J. Insect Physiol.* **29**, 5–13.

Byers J. A. (1989) Chemical ecology of bark beetles. *Experientia* **45**, 271–283.

Byers J. A. (1995) Host-tree chemistry affecting colonization in bark beetles. In *Chemical Ecology of Insects 2*, eds R. T. Cardé and W. J. Bell, pp. 154–213. Chapman & Hall, New York.

Byers J. A. and Birgersson G. (1990) Pheromone production in a bark beetle independent of myrcene precursor in host pine species. *Naturwissenschaften* 77, 385–387.

Byers J. A. and Wood D. L. (1981) Antibiotic-induced inhibition of pheromone synthesis in a bark beetle. *Science* **213**, 763–764.

Byers J. A., Wood D. L., Browne L. E., Fish R. H., Piatek B. and Hendry L. B. (1979) Relationship between a host plant compound, myrcene, and pheromone production in the bark beetle, *Ips paraconfusus*. *J. Insect Physiol.* **25**, 477–482.

Byers J. A., Lanne B. S., Löfqvist J., Schlyter F. and Bergström G. (1985). Olfactory recognition of host-tree susceptibility by pine shoot beetles. *Naturwissenschaften* **72**, 324–326.

Byers J. A., Birgersson G., Löfqvist J. and Bergström G. (1988) Synergistic pheromones and monoterpenes enable aggregation and host recognition by a bark beetle. *Naturwissenschaften* **75**, 153–155.

Byers J. A., Schlyter F., Birgersson G. and Francke W. (1990) *E*-myrcenol in *Ips duplicatus*: An aggregation pheromone component new for bark beetles. *Experientia* **46**, 1209–1211.

Byrne K. J., Swigar A. A., Silverstein R. M., Borden J. H. and Stokkink E. (1974) Sulcatol: population aggregation pheromone in the scolytid beetle, *Gnathotrichus sulcatus*. *J. Insect Physiol.* **20**, 1895–1900.

Camacho A. D. and Borden J. H. (1994) Responses of the western balsam bark beetle, *Dryocoetes confusus* Swaine (Coleoptera: Scolytidae) to host trees baited with enantiospecific blends of *exo*- and *endo*-brevicomin. *Can. Entomol.* **126**, 43–48.

Camacho A. D., Pierce H. D., Jr and Borden J. H. (1993) Geometrical and optical isomerism of pheromones in two sympatric *Dryocoetes* spp. (Coleoptera: Scolytidae), mediates species specificity and response level. *J. Chem. Ecol.* **19**, 2169–2182.

Camacho A. D., Pierce H. D., Jr, and Borden J. H. (1994) Aggregation pheromones in *Dryocoetes affaber* (Mann.) (Coleoptera: Scolytidae): Stereoisomerism and species specificity. *J. Chem. Ecol.* **20**, 111–124.

Cane D. E. (1999) Isoprenoid biosynthesis: overview. In *Comprehensive Natural Products Chemistry*, Vol. 2: Isoprenoids Including Carotenoids and Steroids, eds. D Barton, K. Nakanishi and O. Meth-Cohn, pp. 1–13. Elsevier, Amsterdam.

Chen N. M., Borden J. H. and Pierce H. D., Jr (1988) Effect of juvenile hormone analog, fenoxycarb, on pheromone production by *Ips paraconfusus* (Coleoptera: Scolytidae). *J. Chem. Ecol.* **14**, 1087–1098.

Chin D. J., Luskey K. L., Anderson R. G. W., Faust J. R., Goldstein J. L. and Brown M. S. (1982) Appearance of crystalloid endoplasmic reticulum in compactin-resistant Chinese hamster cells with a 500-fold increase in 3-hydroxy-3-methylglutaryl coenzyme A reductase. *Proc. Natl. Acad. Sci. USA* **79**, 1185–1189.

Chuman T., Mockizuki K., Kato K., Ono M. and Okubo A. (1983) Serricorone and serricorole, new sex pheromone components of the cigarette beetle. *Agric. Biol. Chem.* **47**, 1413–1415.

Conn J. E., Borden J. H., Hunt D. W. A., Holman J., Whitney H. S., Spanier O. J., Pierce H. D., Jr, and Oehlschlager A. C. (1984) Pheromone production by axenically reared

Dendroctonus ponderosae and *Ips parconfusus* (Coleoptera: Scolytidae). *J. Chem. Ecol.* **10**, 281–290.

Crowson, R. A. (1981) *The Biology of the Coleoptera*. Academic Press, London.

Czokajlo D. and Teale S. A. (1999) Synergistic effect of ethanol to α-pinene in primary attraction of the larger pine shoot beetle, *Tomicus piniperda*. *J. Chem. Ecol.* **25**, 1121–1130.

Dallara P. L., Seybold S. J., Meyer H., Tolasch T., Francke W. and Wood D. L. (2000) Semiochemicals from three species of *Pityophthorus* Eichhoff (Coleoptera: Scolytidae) in central coastal California: Identification and field response. *Can. Entomol.* **132**, 889–906.

Dickens J. C. and Mori K. (1989) Receptor chirality and behavioral specificity of the boll weevil, *Anthonomus grandis* Boh. (Coleoptera: Curculionidae), for its pheromone, (+)-grandisol. *J. Chem. Ecol.* **15**, 517–528.

Dickens J. C. and Wiygul G. (1987) Conspecific effects on pheromone production by the boll weevil, *Anthonomus grandis* Boh. (Col., Curculionidae). *J. Appl. Entomol.* **104**, 318–326.

Dickens J. C., McGovern W. L. and Wiygul G. (1988) Effects of antennectomy and a juvenile hormone analog on pheromone production in the boll weevil (Coleoptera: Curculionidae). *J. Entomol. Sci.* **23**, 52–58.

Dowd P. F. and Bartelt R. J. (1993) Aggregation pheromone glands of *Carpophilus freemani* (Coleoptera: Nitidulidae) and gland distribution among other sap beetles. *Ann. Entomol. Soc. Am.* **86**, 464–469.

Drnevich J. M., Hayes E. F. and Rutowski R. L. (2000) Sperm precedence, mating interval, and a novel mechanism of paternity bias in a beetle (*Tenebrio molitor* L.). *Behav. Ecol. Sociobiol.* **48**, 447–451.

Drnevich J. M., Papke R. S., Rauser C. L. and Rutowki R. L. (2001) Material benefits from multiple mating in female mealworm beetles (*Tenebrio molitor* L.). *J. Insect Behav.* **14**, 215–230.

Elliot W. J., Hromnak G., Fried J. and Lanier G. N. (1979) Synthesis of multistriatin enantiomers and their actions on *Scolytus multistriatus* (Coleoptera: Scolytidae). *J. Chem. Ecol.* **5**, 279–287.

Erbilgin N. and Raffa K. F. (2000) Opposing effects of host monoterpenes on responses by two sympatric species of bark beetles to their aggregation pheromones. *J. Chem. Ecol.* **26**, 2527–2548.

Evans A. V. and Bellamy C. L. (1996) *An Inordinate Fondness for Beetles*. Henry Holt & Co., Inc., New York.

Feyereisen R. (1985) Regulation of juvenile hormone titer: synthesis. In *Comprehensive Insect Physiology Biochemistry and Pharmacology*, eds G. A. Kerkut and L. I. Gilbert pp. 391–429. Pergamon Press, Oxford.

Feyereisen R. (1999) Insect P450 enzymes. *Annu. Rev. Entomol.* **44**, 507–533.

Field M., Graf L. H., Laird W. J. and Smith P. L. (1978) Heat-stable enterotoxin of *Escherichia coli*: in vitro effects on guanylate cyclase activity, cyclic GMP concentration, and ion transport in small intestine. *Proc. Natl. Acad. Sci. USA* **75**, 2800–2804.

Fish R. H., Browne L. E., Wood D. L. and Hendry L. B. (1979) Pheromone biosynthetic pathways: conversions of deuterium-labelled ipsdienol with sexual and enantioselectivity in *Ips paraconfusus*. *Tetrahedron Lett.* **17**, 1465–1468.

Fish R. H., Browne L. E. and Bergot B. J. (1984) Pheromone biosynthetic pathways: conversion of ipsdienone to (–)-ipsdienol, a mechanism for enantioselective reduction in the male bark beetle, *Ips paraconfusus*. *J. Chem. Ecol.* **10**, 1057–1064.

Francke W. (1981) Spiroacetale als Pheromone bei Insekten. *Mitt. Dtsch. Ges. Allg. Angew. Ent.* **2**, 248–251.

Francke W. (1986) Convergency and diversity in multicomponent insect pheromones. In *Advances in Invertebrate Reproduction* 4, eds M. Porchet, J.-C. Andries and A. Dhainaut, pp. 327–336. Elsevier Science Publishers, Amsterdam.

Francke W. and Heeman V. (1974) Lockversuche bei *Xyloterus domesticus* L. und *X. lineatus* Oliv. (Coleoptera: Scolytidae) mit 3-hydroxy-3-methylbutan-2-on. *Z. Angew. Entomol.* **75**, 67–72.

Francke W. and Heeman V. (1976) Das Duftstoff-Bouquet des großen waldgärtners *Blastophagus piniperda* L. (Coleoptera: Scolytidae). *Z. Angew. Entomol.* **82**, 117–119.

Francke W. and Schulz S. (1999) Pheromones. In *Natural Products*, Vol. 8 (Including Marine Natural Products, Pheromones, Plant Hormones and Aspects of Ecology, eds D. Barton, K. Nakanishi, and O. Meth-Cohn, pp. 197–261. Elsevier Science Ltd, Oxford.

Francke W. and Vité J. P. (1983) Oxygenated terpenes in pheromone systems of bark beetles. *Z. angew. Entomol.* **96**, 146–156.

Francke W., Heeman V. and Heyns K. (1974) Flüchtige inhaltsstoffe von ambrosiakäfern (Coleoptera: Scolytidae), I. *Z. Naturforsch.* **29c**, 243–245.

Francke W., Heeman V., Gerken B., Renwick J. A. A. and Vité J. P. (1977) 2-Ethyl-1,6-dioxaspiro[4.4]nonane, principal aggregation pheromone of *Pityogenes chalcographus* (L.). *Naturwissenschaften* **64**, 590.

Francke W., Hindorf G. and Reith W. (1979) Alkyl-1,6-dioxaspiro [4.5] decanes — a new class of pheromones. *Naturwissenschaften* **66**, 618–619.

Francke W., Sauerwein P., Vité J. P. and Klimetzek D. (1980) The pheromone bouquet of *Ips amitinus*. *Naturwissenschaften* **67**, 147–148.

Francke W., Pan M.-L., Bartels J., König W. A., Vité J. P., Krawielitzki S., and Kohnle U. (1986) The odor bouquet of three pine engraver beetles (*Ips* spp.). *J. Appl. Ent.* **101**, 453–461.

Francke W., Pan M.-L., König W. A., Mori K., Puapoomachareon P., Heuer H. and Vité J. P. (1987) Identification of pityol and grandisol as pheromone components of the bark beetle, *Pityophthorus pityographus*. *Naturwissenschaften* **74**, 343–345.

Francke W., Bartels J., Krohn S., Schulz S., Baader E., Tengö J., and Schneider D. (1989) Terpenoids from bark beetles, solitary bees and danaine butterflies. *Pure Appl. Chem.* **61**, 539–542.

Francke W., Bartels J., Meyer H., Schröder F., Kohnle U., Baader E. and Vité J. P. (1995) Semiochemicals from bark beetles: New results, remarks, and reflections. *J. Chem. Ecol.* **21**, 1043–1063.

Francke W., Schröder F., Kohnle U. and Simon M. (1996a) Synthesis of (1*S*,2*R*,5*R*)-2-ethyl-1,5-dimethyl-6,8-dioxabicyclo [3.2.1] octane, the aggregation pheromone of male beech bark beetle, *Taphrorychus bicolor* (Col., Scol.). *Liebigs Ann.* **10**, 1523–1527.

Francke W., Schröder F., Phillipp P., Meyer H., Sinnwell V. and Gries G. (1996b) Identification and synthesis of new bicyclic acetals from the mountain pine beetle, *Dendroctonus ponderosae* Hopkins (Coleoptera: Scolytidae). *Bioorg. Med. Chem.* **4**, 363–374.

Friesen J. A. and Rodwell V. W. (1997) Protein engineering of the HMG-CoA reductase of *Pseudomonas mevalonii*. Construction of mutant enzymes whose activity is regulated by phosphorylation and dephosphorylation. *Biochemistry* **36**, 2173–2177.

Gage M. J. G. (1992) Removal of rival sperm copulation in a beetle, *Tenebrio molitor*. *Anim. Behav.* **44**, 587–589.

Gage M. J. G. and Baker R. R. (1991) Ejaculate size varies with socio-sexual situation in an insect. *Ecol. Entomol.* **16**, 331–337.

Gerken B. and Grüne S. (1978) Zur biologischen bedeutung käfereigener duftstoffe des großen ulmensplintkäfers, *Scolytus scolytus* F. (Col. Scolytidae). *Mitt. Dtsh. Ges. Allg. Angew. Ent.* **1**, 38–41.

Gerken B., Grüne S., Vité J. P. and Mori K. (1978) Response of European populations of *Scolytus multistriatus* to isomers of multistriatin. *Naturwissenschaften* **65**, 110–111.

Gianella R. A. and Drake K. W. (1979) Effect of purified *E. coli* heat-stable enterotoxin on intestinal cyclic nucleotide metabolism and fluid secretion. *Infect. Immun.* **24**, 19–23.

Giesen H., Kohnle U., Vité J. P., Pan M.-L. and Francke W. (1984) Das Aggregationspheromon des mediterranen Kiefernborkenkäfers *Ips* (*Orthotomicus*) *erosus*. *Z. angew. Entomol.* **98**, 95–97.

Goldstein J. L. and Brown M. S. (1990) Regulation of the mevalonate pathway. *Nature* **343**, 425–430.

Gore W. E., Pearce G. T., Lanier G. N., Simeone J. B., Silverstein R. M., Peacock J. W. and Cuthbert R.A. (1977) Aggregation attractant of the European elm bark beetle, *Scolytus multistriatus*, production of individual components and related aggregation behavior. *J. Chem. Ecol.* **3**, 429–446.

Gries G. (1992) Ratios of geometrical and optical isomers of pheromones: irrelevant or important in scolytids? *J. Appl. Ent.* **114**, 240–243.

Gries G., Pierce H. D. Jr, Lindgren B. S. and Borden J. H. (1988) New techniques for capturing and analyzing semiochemicals for scolytid beetles (Coleoptera: Scolytidae). *J. Econ. Entomol.* **81**, 1715–1720.

Gries G., Leufvén A., LaFontaine J. P., Pierce H. D. Jr, Borden J. H., Vanderwel D. and Oehlschlager A. C. (1990a) New metabolites of α-pinene produced by the mountain pine beetle, *Dendroctonus ponderosae* (Coleoptera: Scolytidae). *Insect Biochem.* **20**, 365–371.

Gries G., Smirle M. J., Leufvén A., Miller D. R., Borden J. H. and Whitney H. S. (1990b) Conversion of phenylalanine to toluene and 2-phenylethanol by the pine engraver *Ips pini* (Say) (Coleoptera: Scolytidae). *Experientia* **46**, 329–331.

Gries G., Borden J. H., Gries R., LaFontaine J. P., Dixon E. A., Wieser H. and Whitehead A. T. (1992a) 4-Methylene-6,6-dimethylbicyclo[3.1.1]hept-2-ene (verbenene): new aggregation pheromone of the scolytid beetle, *Dendroctonus rufipennis*. *Naturwissenschaften* **79**, 367–368.

Gries G., Gries R., Borden J. H., Pierce H. D. Jr. Johnston B. D. and Oehlschlager A. C. (1992b) 3,7,7-Trimethyl-1,3,5-cycloheptatriene in volatiles of female mountain pine beetles, *Dendroctonus ponderosae*. *Naturwissenschaften* **79**, 27–28.

Griffith O., Vakili R., Currie R. W. and Vanderwel D. (2003) Effects of mating on the sex pheromone of female yellow mealworm beetles (*Tenebrio molitor* L.). *J. Chem. Ecol.* In preparation.

Grosman D. M. (1996) *Southern pine beetle*, Dendroctonus frontalis *Zimmermann (Coleoptera: Scolytidae): quantitative analysis of chiral semiochemicals*. PhD thesis. Viginia Polytechnical Institute.

Hackstein E. and Vité J. P. (1978) Pheromone Biosynthese und Reizkette in der Besiedlung von Fichten durch den Buchdrucker *Ips typographus*. *Mitt. Dtsch. Ges. Allg. Angew. Entomol.* **1**, 185–188.

Hall G. M., Tittiger C., Andrews G. L., Mastick G. S., Kuenzli M., Luo X., Seybold S. J. and Blomquist G. J. (2002a) Midgut tissue of male pine engraver, *Ips pini*, synthesizes monoterpenoid pheromone component ipsdienol *de novo*. *Naturwissenschaften* **89**, 79–83.

Hall G. M., Tittiger C., Blomquist G. J., Andrews G. L., Mastick G. S., Barkawi L. S., Bengoa C. and Seybold S. J. (2002b) Male Jeffrey pine beetle, *Dendroctonus jeffreyi*, synthesizes the pheromone component frontalin in anterior midgut tissue. *Insect Biochem. Mol. Biol.* **32**, 1525–1532.

Hampton R., Dimster-Denk D. and Rine J. (1996) The biology of HMG-CoA reductase: the pros of contra-regulation. *Trends Biol. Sci.* **21**, 140–145.

Happ G. M. and Wheeler J. (1969) Bioassay, preliminary purification, and effect of age, crowding, and mating on the release of sex pheromone by female *Tenebrio molitor*. *Ann. Entomol. Soc. Am.* **62**, 846–851.
Harring C. M. (1978) Aggregation pheromones of the European fir engraver beetles *Pityokteines curvidens, P. spinidens,* and *P. vorontzovi* and the role of juvenile hormone in pheromone biosynthesis. *Z. angew. Entomol.* **85**, 281–317.
Harring C. M., Vité J. P. and Hughes P. R. (1975) "Ipsenol" der populationslockstoff des krummzähnigen tannenborkenkäfers. *Naturwissenschaften* **62**, 488.
Hedin P. A., Lindig O. H. and Wiygul G. (1982) Enhancement of boll weevil *Anthonomus grandis* Boh. (Coleoptera: Curculionidae) pheromone biosynthesis with JH III. *Experientia* **38**, 375–376.
Hendry L. B., Piatek B., Browne L. E., Wood D. L., Byers J. A., Fish R. H., and Hicks R. A. (1980) *In vivo* conversion of a labelled host plant chemical to pheromones of the bark beetle, *Ips paraconfusus*. *Nature* **284**, 485.
Henzell R. F. and Lowe M. D. (1970) Sex attractant of the grass grub beetle. *Science* **168**, 1005–1006.
Hibbard B. E. and Webster F. X. (1993) Enantiomeric composition of grandisol and grandisal produced by *Pissodes strobi* and *P. nemorensis* and their electroantennogram response to pure enantiomers. *J. Chem. Ecol.* **19**, 2129–2141.
Hobson K. R., Wood D. L., Cool L. G., White P. M., Ohtsuka T., Kubo I. and Zavarin E. (1993) Chiral specificity in responses by the bark beetle *Dendroctonus valens* to host kairomones. *J. Chem. Ecol.* **19**, 1837–1846.
Hoover S. E. R., Lindgren B. S., Keeling C. I. and Slessor K. N. (2000) Enantiomer preference of *Trypodendron lineatum* and the effect of pheromone dose and trap length on response to lineatin-baited traps in interior British Columbia. *J. Chem. Ecol.* **26**, 667–677.
Hoyt C. P., Osborne G. O. and Mulcock A. P. (1971) Production of an insect sex attractant by symbiotic bacteria. *Nature* **230**, 472–473.
Hubank M. and Schatz D. G. (1994) Identifying differences in mRNA expression by representational difference analysis of cDNA. *Nuc. Acids Res.* **22**, 5640–5648.
Huber D. W. P. and Borden J. H. (2001) Angiosperm bark volatiles disrupt response of Douglas-fir beetle, *Dendroctonus pseudotsugae*, to attractant-baited traps. *J. Chem. Ecol.* **27**, 217–233.
Huber D. W. P., Borden J. H., Jeans-Williams N. L. and Gries R. (2000) Differential bioactivity of conophthorin on four species of North American bark beetles (Coleoptera: Scolytidae). *Can. Ent.* **132**, 649–653.
Hughes P. R. (1973a) *Dendroctonus*. Production of pheromones and related compounds in response to host monoterpenes. *Z. angew. Entomol.* **73**, 294–312.
Hughes P. R. (1973b) Effect of α-pinene exposure on *trans*-verbenol synthesis in *Dendroctonus ponderosae* Hopk. *Naturwissenschaften* **60**, 261–262.
Hughes P. R. (1974) Myrcene: a precursor of pheromones in *Ips* beetles. *J. Insect Physiol.* **20**, 1271–1275.
Hughes P. R. (1975) Pheromones of *Dendroctonus*: origin of α-pinene oxidation products present in emergent adults. *J. Insect Physiol.* **21**, 687–691.
Hughes P. R. and Renwick J. A. A. (1977a) Hormonal and host factors stimulating pheromone synthesis in female western pine beetles, *Dendroctonus brevicomis*. *Physiol. Entomol.* **2**, 289–292.
Hughes P. R. and Renwick J. A. A. (1977b) Neural and hormonal control of pheromone biosynthesis in the bark beetle, *Ips paraconfusus*. *Physiol. Entomol.* **2**, 117–123.
Hunt D. W. A. and Borden J. H. (1989a) Conversion of verbenols to verbenone by yeasts isolated from *Dendroctonus ponderosae* (Coleoptera: Scolytidae). *J. Chem. Ecol.* **16**, 1385–1397.

Hunt D. W. A. and Borden J. H. (1989b) Terpene alcohol pheromone production by *Dendroctonus ponderosae* and *Ips paraconfusus* (Coleoptera: Scolytidae) in the absence of readily culturable microorganisms. *J. Chem. Ecol.* **15**, 1433–1463.

Hunt D. W. A. and Smirle M. J. (1988) Partial inhibition of pheromone production in *Dendroctonus ponderosae* (Coleoptera: Scolytidae) by polysubstrate monooxygenase inhibitors. *J. Chem. Ecol.* **14**, 529–536.

Hunt D. W. A., Borden J. H., Pierce H. D., Jr, Slessor K. N., King G. G. S. and Czyzewska E. K. (1986) Sex-specific production of ipsdienol and myrcenol by *Dendroctonus ponderosae* (Coleoptera: Scolytidae) exposed to myrcene vapors. *J. Chem. Ecol.* **12**, 1579–1586.

Hunt D. W. A., Borden J. H., Lindgren B. S. and Gries G. (1989) The role of autooxidation of α-pinene in the production of pheromones of *Dendroctonus ponderosae* (Coleoptera: Scolytidae). *Can. J. For. Res.* **19**, 1275–1282.

Hussain A. (1994) *Chemical ecology of* Tribolium castaneum *Herbst (Coleoptera: Tenebrionidae): factors affecting biology and application of pheromone*. PhD thesis. Oregon State University, 118 pp.

Islam N. (1996) *Pheromone biosynthesis and regulation in the yellow mealworm beetle,* Tenebrio molitor. MSc thesis. Univ. of Manitoba, 91 pp.

Islam N., Bacala R., Moore A. and Vanderwel D. (1999) Biosynthesis of 4-methyl-1-nonanol: Female-produced pheromone of the yellow mealworm beetle, *Tenebrio molitor* (Coleoptera: Tenebrionidae). *Insect Biochem. Mol. Biol.* **29**, 201–208.

Ivarsson P. and Birgersson G. (1995) Regulation and biosynthesis of pheromone components in the double spined bark beetle *Ips duplicatus* (Coleoptera: Scolytidae). *J. Insect Physiol.* **41**, 843–849.

Ivarsson P., Schlyter F. and Birgersson G. (1993) Demonstration of *de novo* pheromone biosynthesis in *Ips duplicatus* (Coleoptera: Scolytidae): inhibition of ipsdienol and *E*-myrcenol production by compactin. *Insect Biochem. Molec. Biol.* **23**, 655–662.

Ivarsson P., Blomquist G. J. and Seybold S. J. (1997) *In vitro* production of pheromone intermediates in the bark beetles *Ips pini* (Say) and *I. paraconfusus* Lanier (Coleoptera: Scolytidae). *Naturwissenschaften* **84**, 454–457.

Ivarsson P., Tittiger C., Blomquist C., Borgeson C. E., Seybold S. J., Blomquist G. J. and Högberg H.-E. (1998) Pheromone precursor synthesis is localized in the metathorax of *Ips paraconfusus* Lanier (Coleoptera: Scolytidae). *Naturwissenschaften* **85**, 507–511.

Judy K. J., Schooley D. A., Troetschler R. G., Jennings R. C., Bergot B. J. and Hall M. S. (1975) Juvenile hormone production by corpora allata of *Tenebrio molitor* in vitro. *Life Sciences* **16**, 1059–1066.

Kiehlmann E., Conn J. E. and Borden J. H. (1982) 7-Ethoxy-6-methoxy-2,2-dimethyl-2*H*-1-benzopyran. *Org. Prep. Proc. Int.* **14**, 337.

Kim, J.-Y. and Leal W. S. (1998) The eversible pheromone-producing organ in the Melolonthine beetle, *Holotrichia parallela* (Coleoptera: Scarabaeidae: Melolonthinae). In *Proceedings, Annual Meeting of Japan Society of Applied Entomology*. Nagoya, Japan. March 31-April 2, p. 127.

Kinzer G. W., Fentiman A. F., Jr, Page T. F., Jr, Foltz R. L., Vité J. P. and Pitman G. B. (1969) Bark beetle attractants: identification, synthesis and field bioassay of a new compound isolated from *Dendroctonus*. *Nature* **221**, 477–478.

Klimetzek D. and Francke W. (1980) Relationship between enantiomeric composition of α-pinene in host trees and the production of verbenols in *Ips* species. *Experientia* **36**, 1343–1344.

Klimetzek D., Bartels J. and Francke W. (1989a) Das pheromon-system des bunten ulmenbastkäfers *Pteleobius vittatus* (F.) (Col., Scolytidae). *J. Appl. Ent.* **107**, 518–523.

Klimetzek D., Köhler J., Krohn S. and Francke W. (1989b) Das pheromon-system des waldreben-borkenkäfers *Xylocleptes bispinus* Duft. (Col., Scolytidae). *J. Appl. Ent.* **107**, 304–309.

Kohnle U., Francke W. and Bakke A. (1985) *Polygraphus poligraphus* (L.): response to enantiomers of beetle specific terpene alcohols and a bicyclic ketal. *Z. ang. Ent.* **100**, 5–8.

Kohnle U., Mussong M., Dubbel V. and Francke W. (1987) Acetophenone in the aggregation of the beech bark beetle, *Taphrorychus bicolor* (Col., Scolytidae). *J. Appl. Ent.* **103**, 249–252.

Kohnle U., Densborn S., Duhme D. and Vité J. P. (1992a) Bark beetle attack on host logs reduced by spraying with repellents. *J. Appl. Ent.* **114**, 83–90.

Kohnle U., Densborn S., Kölsch P., Meyer H. and Francke W. (1992b) *E*-7-Methyl-1,6-dioxaspiro[4.5] decane in the chemical communication of European Scolytidae and Nitidulidae (Coleoptera). *J. Appl. Ent.* **114**, 187–192.

Kuwahara Y., Fukami H., Howard R., Ishii S., Matsumura F. and Burkholder W. E. (1978) Chemical studies on the Anobiidae: sex pheromone of the drugstore beetle, *Stegobium paniceum* (L.) (Coleoptera). *Tetrahedron* **34**, 1769–1764.

Lanier G. N., Gore W. E., Pearce G. T., Peacock J. W. and Silverstein R. M. (1977) Response of the European elm bark beetle, *Scolytus multistriatus* (Coleoptera: Scolytidae) to isomers and components of its pheromone. *J. Chem. Ecol.* **3**, 1–8.

Lanne B. S., Ivarsson P., Johnson P., Bergström G. and Wassgren A. B. (1989) Biosynthesis of 2-methyl-3-buten-2-ol, a pheromone component of *Ips typographus* (Coleoptera: Scolytidae). *Insect Biochem.* **19**, 163-168.

Lawrence J. F. and Newton A. F., Jr. (1982) Evolution and classification of beetles. *Ann. Rev. Ecol. Systematics* **13**, 261–290.

Lawrence J. F. and Newton A. F., Jr (1995) Families and subfamilies of Coleoptera (with selected genera, notes, references and data on family group names) In *Biology, Phylogeny, and Classification of Coleoptera. Papers Celebrating the 80th Birthday of Roy A. Crowson*, eds J. Pakaluk, and S. A. Slipinski, pp. 779–1006. Muzeum I Instytut Zoologii PAN, Warsaw.

Leal W. S. (1997) Evolution of sex pheromone communication in plant-feeding scarab beetles. In *Insect Pheromone Research: New Directions*, eds R. T. Cardé, and A. K. Minks, pp. 505–513. Chapman & Hall, New York.

Leal W. S. (1998a) *Biosynthesis of scarab beetle pheromones and neuropeptide regulation.* Presented at Annu. Meet. Ent. Soc. Am., Las Vegas.

Leal W. S. (1998b) Chemical ecology of phytophagous scarab beetles. *Annu. Rev. Entomol.* **43**, 39–61.

Leal W. S. (1999) Scarab beetles. In *Pheromones of Non-Lepidopteran Insects Associated with Agricultural Plants*, eds J. Hardie and A. K. Minks, pp. 51–68. CABI Publishing, New York.

Leal W. S. (2001) Molecules and macromolecules involved in chemical communication of scarab beetles. *Pure Appl. Chem.* **73**, 613–616.

Leal W. S., Matsuyama S., Kuwahara Y., Wakamura S. and Hasegawa M. (1992) An amino acid derivative as the sex pheromone of a scarab beetle. *Naturwissenschaften* **79**, 184–185.

Leal W. S., Sawada M., Matsuyama S., Kuwahara Y. and Hasegawa M. (1993) Unusual periodicity of sex pheromone production in the large black chafer *Holotrichia parallela. J. Chem. Ecol.* **19**, 1381–1391.

Leal W. S., Yadava C. P. S. and Vijayvergia J. N. (1996a) Aggregation of the scarab beetle *Holotrichia consanguinea* in response to female-released pheromone suggests a secondary function hypothesis for semiochemical. *J. Chem. Ecol.* **22**, 1557–1566.

Leal W. S., Kuwahara S., Ono M. and Kubota S. (1996b) (*R,Z*)-7,15-Hexadecadien-4-olide, sex pheromone of the yellowish elongate chafer, *Heptophylla picea*. *Bioorg. Med. Chem.* **4**, 315–321.

Leal W. S., Zarbin P. H. G., Wojtasek H. and Ferreira J. T. (1999) Biosynthesis of scarab beetle pheromones: enantioselective 8-hydroxylation of fatty acids. *Eur. J. Biochem.* **259**, 175–180.

Leufvén A., Bergström G. and Falsen E. (1984) Interconversion of verbenols and verbenone by identified yeasts isolated from the spruce bark beetle *Ips typographus*. *J. Chem. Ecol.* **10**, 1349–1361.

Leufvén A., Bergström G. and Falsen E. (1988) Oxygenated monoterpenes produced by yeasts, isolated from *Ips typographus* (Coleoptera: Scolytidae) and grown in phloem medium. *J. Chem. Ecol.* **14**, 353–361.

Liang P., Averboukh L. and Pardee A. B. (1993) Distribution and cloning of eukaryotic mRNAs by means of differential display: refinements and optimization. *Nuc. Acids Res.* **21**, 3269–3275.

Libbey L. M., Morgan M. E., Putnam T. B. and Rudinsky J. A. (1976) Isomer of antiaggregation pheromone identified from male Douglas-fir beetle: 3–methylcyclohex-3-en-1-one. *J. Insect Physiol.* **22**, 871–873.

Libbey L. M., Oehlschlager A. C. and Ryker L. C. (1983) 1-Methylcyclohex-2-en-1-ol as an aggregation pheromone of *Dendroctonus pseudotsugae*. *J. Chem. Ecol.* **9**, 1533–1541.

Lindgren B. S. (1992) Attraction of Douglas-fir beetle, spruce beetle and a bark beetle predator (Coleoptera: Scolytidae and Cleridae) to enantiomers of frontalin. *J. Entomol. Soc. Brit. Columbia* **89**, 13–17.

Lindgren B. S., Gries G., Pierce H. D., Jr, and Mori, K. (1992) *Dendroctonus pseudotsugae* Hopkins (Coleoptera: Scolytidae): production of and response to enantiomers of 1-methylcyclohex-2-en-1-ol. *J. Chem. Ecol.* **18**, 1201–1208.

Lindström M., Norin T., Birgersson G. and Schlyter F. (1989) Variation of enantiomeric composition of α-pinene in Norway spruce, *Picea abies*, and its influence on production of verbenol isomers by *Ips typographus* in the field. *J. Chem. Ecol.* **15**, 541–548.

Lisitsyn N., Lisitsyn N. and Wigler M. (1993) Cloning the differences between two complex genomes. *Science* 259, 946–951.

Lu F. (1999) *Origin and endocrine regulation of pheromone biosynthesis in the pine bark beetles,* Ips pini *(Say) and* Ips paraconfusus *Lanier (Coleoptera: Scolytidae)*. PhD thesis. Univ. of Reno, Nevada, 152 pp.

MacConnell J. G., Borden J. H., Silverstein R.M. and Stokkink E. (1977) Isolation and tentative identification of lineatin, a pheromone from the frass of *Trypodendron lineatum* (Coleoptera: Scolytidae). *J. Chem. Ecol.* **3**, 549–561.

Macías-Sámano J. E., Borden J. H., Pierce H. D., Jr, Gries, R. and Gries, G. (1997) Aggregation pheromone of *Pityokteines elegans*. *J. Chem. Ecol.* **23**, 1333–1347.

Macías-Sámano J. E., Borden J. H., Gries R., Pierce H. D., Jr, Gries G. and King G. G. S. (1998) Primary attraction of the fir engraver, *Scolytus ventralis*. *J. Chem. Ecol.* **24**, 1049–1075.

Madden J. L., Pierce H. D., Jr, Borden J. H. and Butterfield A. (1988) Sites of production and occurrence of volatiles in Douglas-fir beetle, *Dendroctonus pseudotsugae* Hopkins. *J. Chem. Ecol.* **14**, 1305–1317.

Martin D., Bohlmann J., Gershenzon J., Francke W. and Seybold S. J. (2003) A novel sex-specific and inducible monoterpene synthase activity associated with a pine bark beetle, the pine engraver, *Ips pini*. *Naturwissenschaften* **90**, 173–179.

Mayer M. S. and McLaughlin J. R. (1991) *Handbook of Insect Pheromones and Sex Attractants*. CRC Press, Boca Raton, FL.

Meigs T. E. and Simoni R. D. (1997) Farnesol as a regulator of HMG-CoA reductase degradation: characterization and role of farnesyl pyrophosphatase. *Arch. Biochem. Biophys.* **345**, 1–9.

Menon M. (1970). Hormone-pheromone relationships in the beetle, *Tenebrio molitor*. *J. Insect Physiol.* **16**, 1123–1139.

Menon M. (1976) Hormone-pheromone relationships of male *Tenebrio molitor*. *J. Insect Physiol.* **22**, 1021–1023.

Miller D. R. and Borden J. H. (1990a) ß-Phellandrene: kairomone for pine engraver, *Ips pini* (Say) (Coleoptera: Scolytidae). *J. Chem. Ecol.* **16**, 2519–2531.

Miller D. R. and Borden J. H. (1990b) The use of monoterpenes by *Ips latidens* (LeConte) (Coleoptera: Scolytidae). *Can. Entomol.* **122**, 301–307.

Miller D. R., Gries G. and Borden J. H. (1990) *E*-myrcenol: a new pheromone for the pine engraver, *Ips pini* (Say) (Coleoptera: Scolytidae). *Can. Entomol.* **122**, 401–406.

Miller D. R., Gibson K. E., Raffa K. F., Seybold S. J., Teale S. A. and Wood D. L. (1997) Geographic variation in response of pine engraver, *Ips pini*, and associated species to pheromone, lanierone. *J. Chem. Ecol.* **23**, 2013–2031.

Miller D. R. and Lafontaine J. P. (1991) *cis*-Verbenol: An aggregation pheromone for the mountain pine beetle *Dendroctonus ponderosae* Hopkins (Coleoptera: Scolytidae), *J. Entomol. Soc. Brit. Columbia* **88**, 34–38.

Moore R. H., Golumbic C. and Fisher G. S. (1956) Autoxidation of α-pinene. *J. Am. Chem. Soc.* **78**, 1173–1176.

Mori K. (1974) Synthesis of (1*S*: 2*R*: 4*S*: 5*R*)-(–)-α-multistriatin: the pheromone in the smaller European elm bark beetle, *Scolytus multistriatus*. *Tetrahedron* **32**, 1979–1981.

Mori K. (1976) Synthesis of optically pure (+)-*trans*-verbenol and its antipode, the pheromone of *Dendroctonus* bark beetles. *Agr. Biol. Chem.* **40**, 415–418.

Mori K. (1977) Absolute configuration of (–)-4-methylheptan-3-ol, a pheromone of the smaller European elm bark beetle, as determined by the synthesis of its (3*R*,4*R*)-(+)- and (3*S*,4*R*)-(+)-isomers. *Tetrahedron* **33**, 289–294.

Mori K. (1996) Molecular asymmetry and pheromone science. *Biosci. Biotech. Biochem.* **60**, 1925–1932.

Mori K. and Puapoomchareon P. (1987) Conversion of the enantiomers of sulcatol (6-methyl-5-hepten-2-ol) to the enantiomers of pityol [*trans*-2-(1-hydroxy-1-methylethyl)-5-methyltetrahydrofuran], a male-specific attractant of the bark beetle *Pityophthorus pityographus*. *Liebigs Annalen der Chemie* **3**, 271–272.

Mori K., Nakayama T. and Takikawa H. (1996) Synthesis and absolute configuration of sordidin, the male-produced aggregation pheromone of the banana weevil, *Cosmopolites sordidus*. *Tetrahedron Let.* **37**, 3741–3744.

Nardi J. B., Dowd P. F. and Bartelt R. J. (1996) Fine structure of cells specialized for secretion of aggregation pheromone in a nitidulid beetle *Carpophilus freemani* (Coleoptera: Nitidulidae). *Tissue and Cell* **28**, 43–52.

Nardi J. B., Young A. G., Ujhelyi E., Tittiger C., Lehane M. J. and Blomquist G. J. (2002) Specialization of midgut cells for synthesis of male isoprenoid pheromone components in two scolytid beetles, *Dendroctonus jeffreyi* and *Ips pini*. *Tissue and Cell* **34**, 221–231.

Nishimura, Y. and Mori, K. (1998) A new synthesis of (–)-frontalin, the bark beetle pheromone. *Eur. J. Org. Chem.* **188**, 233–236.

Nural A. H., Tittiger C., Welch W. and Blomquist G. J. (2001) *Isolation and characterization of isoprenyl diphosphate synthase from cotton boll weevil*. Presented at Annu. Meet. Int. Soc. Chem. Ecol. 18th, Lake Tahoe.

Oehlschlager A. C., Pierce A. M., Pierce H. D., Jr, and Borden, J. H. (1988). Chemical communication in cucujid grain beetles. *J. Chem. Ecol.* **14**, 2071–2098.

Paine T. D., Millar J. G., Hanlon C. C. and Hwang J.-S. (1999) Identification of semiochemicals associated with Jeffrey pine beetle, *Dendroctonus jeffreyi*. *J. Chem. Ecol.* **25**, 433–453.

Palomino J. J., Rodriguez M. and Cuerda D. (1994) Comportamiento de cópula y competición de esperma en *Tenebrio molitor*. *Etología*. **4**, 19–26.

Pearce G. T., Gore W. E., Silverstein R. M., Peacock J. W., Cuthbert R. A., Lanier G. N. and Simeone J. B. (1975) Chemical attractants for the smaller European elm bark beetle, *Scolytus multistriatus* (Coleoptera: Scolytidae). *J. Chem. Ecol.* **1**, 115–124.

Pearce G. T., Gore W. E. and Silverstein R. M. (1976) Synthesis and absolute configuration of multistriatin. *J. Org. Chem.* **41**, 2797–2803.

Perez A. L., Gries R., Gries G. and Oehlschlager A. C. (1996) Transformation of presumptive precursors to frontalin and *exo*-brevicomin by bark beetles and the West Indian sugarcane weevil (Coleoptera). *Bioorg. Med. Chem.* **4**, 445–450.

Petroski R. J. and Vaz R. (1995) Insect aggregation pheromone response synergized by "host type" volatiles. Molecular modeling evidence for close proximitiy binding of pheromone and coattractant in *Carpophilus hemipterus* (L.) (Coleoptera: Nitidulidae). In *Computer-Aided Molecular Design*, eds C. H. Reynolds, M. K. Holloway and H. K. Cox, pp. 197–210. ACS Symposium Series 589, American Chemical Society, Washington, DC.

Petroski R. J., Bartelt R. J. and Weisleder D. (1994) Biosynthesis of (2*E*,4*E*,6*E*)-5-ethyl-3-methyl-2,4,6-nonatriene: the aggregation pheromone of *Carpophilus freemani* (Coleoptera: Nitidulidae). *Insect Biochem. Molec. Biol.* **24**, 69–78.

Pierce A. M., Pierce H. D., Jr, Borden J. H. and Oehlschlager A. C. (1986) Enhanced production of aggregation pheromones in four stored-product coleopterans feeding on methoprene-treated oats. *Experientia* **42**, 164–165.

Pierce A. M., Pierce H. D., Jr, Johnston B. D., Oehlschlager A. C. and Borden J. H. (1988) Aggregation pheromone of square necked grain beetle, *Cathartus quadricollis* (GUER). *J. Chem. Ecol.* **14**, 2169–2184.

Pierce A. M., Pierce H. D., Jr, Borden J. H. and Oehlschlager, A. C. (1989) Production dynamics of cucujolide pheromones and identification of 1-octen-3-ol as a new aggregation pheromone for *Oryzaephilus surinamensis* and *O. mercator* (Coleoptera, Cucujidae). *Environ. Entomol.* **18**, 747–755.

Pierce A. M., Pierce H. D., Jr, Oehlschlager A. C. and Borden J. H. (1991a) Fungal volatiles, semiochemicals for stored-product beetles (Coleoptera, Cucujidae). *J. Chem. Ecol.* **17**, 581–597.

Pierce A. M., Pierce H. D., Jr, Oehlschlager A. C. and Borden J. H. (1991b) 1-Octen-3-ol, attractive semiochemical for foreign grain beetle, *Ahasverus advena* (Waltl) (Coleoptera, Cucujidae). *J. Chem. Ecol.* **17**, 567–580.

Pierce H. D., Jr, Pierce A. M., Millar J. G., Wong J. W., Verigin V. G., Oehlschlager A. C. and Borden J. H. (1984) Methodology for isolation and analysis of aggregation pheromones in the genera *Cryptolestes* and *Oryzaephilus* (Coleoptera: Cucujidae). Proc. Third Intern. Working Conf. on Stored-Prod. Ent., Manhattan, KA, pp. 121–137.

Pierce H. D., Jr, Conn J. E., Oehlschlager A. C. and Borden J. H. (1987). Monoterpene metabolism in female mountain pine beetles, *Dendroctonus ponderosae* Hopkins, attacking ponderosa pine. *J. Chem. Ecol.* **13**, 1455–1480.

Pierce H. D., Jr, de Groot P., Borden J. H., Ramaswamy S. and Oehlschlager, A. C. (1995) Pheromones in the red pine cone beetle, *Conophthorus resinosae* Hopkins, and its synonym, *C. banksianae* McPherson (Coleoptera: Scolytidae). *J. Chem. Ecol.* **21**, 169–185.

Pitman G. B. and Vité J. P. (1963) Studies on the pheromone of *Ips confusus* (LeC.). I. Secondary sexual dimorphism in the hindgut epithelium. *Contrib. Boyce Thompson Inst. Plant Res.* **22**, 221–225.

Pitman G. B. and Vité J. P. (1974) Biosynthesis of methylcyclohexenone by male Douglas-fir beetle. *Environ. Entomol.* **3**, 886–887.

Pitman G. B., Kliefoth R. A. and Vité J. P. (1965) Studies on the pheromone of *Ips confusus* (LeConte). II. Further observations on the site of production. *Contrib. Boyce Thompson Inst. Plant Res.* **23**, 13–17.

Pitman G. B., Vité J. P., Kinzer G. W. and Fentiman A. F. (1968) Bark beetle attractants: *trans*-verbenol isolated from *Dendroctonus*. *Nature* **218**, 168–169.

Plarre R. and Vanderwel D. (1999) Stored-product beetles. In *Pheromones of Non-Lepidopteran Insects Associated with Agricultural Plants*, eds J. Hardie and A. K. Minks, pp. 149–198. CABI Publishing, New York.

Plummer E. L., Stewart T. E., Byrne K., Pearce G. T. and Silverstein R. M. (1976) Determination of the enantiomeric composition of several insect pheromone alcohols. *J. Chem. Ecol.* **2**, 307–331.

Prestwich G. D. and Blomquist G. J. (1987) *Pheromone Biochemistry*. Academic Press, Orlando, FL.

Pureswaran D. S., Gries R., Borden J. H. and Pierce Jr, H. D. (2000) Dynamics of pheromone production and communication in the mountain pine beetle, *Dendroctonus ponderosae* Hopkins, and the pine engraver, *Ips pini* (Say) (Coleoptera: Scolytidae). *Chemoecology* **10**, 153–168.

Raffa K. F., Phillips T. W. and Salom S. M. (1993) Strategies and mechanisms of host colonization by bark beetles. In *Beetle-Pathogen Interactions in Conifer Forests*, eds T. D. Schowalter, and G.M. Filip, pp. 103–128. Academic Press, London.

Renwick J. A. A. (1967) Identification of two oxygenated terpenes from the bark beetles *Dendroctonus frontalis* and *Dendroctonus brevicomis*. *Contrib. Boyce Thompson Inst. of Plant Res.* **23**, 355–360.

Renwick J. A. A. and Dickens J. C. (1979) Control of pheromone production in the bark beetle, *Ips cembrae*. *Physiol. Entomol.* **4**, 377–381.

Renwick J. A. A. and Hughes P. R. (1975) Oxidation of unsaturated cyclic hydrocarbons by *Dendroctonus frontalis*. *Insect Biochem.* **5**, 459–463.

Renwick J. A. A. and Pitman G. B. (1979) An attractant isolated from female Jeffrey pine beetles, *Dendroctonus jeffreyi*. *Environ. Entomol.* **8**, 40–41.

Renwick J. A. A., Hughes P. R. and Vité J. P. (1975) The aggregation pheromone system of a *Dendroctonus* bark beetle in Guatemala. *J. Insect Physiol.* **21**, 1097–1100.

Renwick J. A. A., Hughes P. R. and Krull I. S. (1976a) Selective production of *cis*- and *trans*-verbenol from (–)- and (+)-α-pinene by a bark beetle. *Science* **191**, 199–201.

Renwick J. A. A., Hughes P. R., Pitman G. B. and Vité J. P. (1976b) Oxidation products of terpenes identified from *Dendroctonus* and *Ips* bark beetles. *J. Insect Physiol.* **22**, 725–727.

Renwick J. A. A., Pitman G. B. and Vité J. P. (1976c) 2-Phenylethanol isolated from bark beetles. *Naturwissenschaften* **63**, 198.

Renwick J. A. A., Vité J. P. and Billings R. F. (1977) Aggregation pheromones in the ambrosia beetle, *Platypus flavicornis*. *Naturwissenschaften* **64**, 226.

Rohmer M. (1999) A mevalonic-independent route to isopentenyl diphosphate. In *Comprehensive Natural Products Chemistry*, Vol. 2: Isoprenoids including Carotenoids and Steroids, eds D. Barton, K. Nakanishi, and O. Meth-Cohn, pp. 45–67. Elsevier, Amsterdam.

Rohmer M., Seeman M., Horbach S., Bringer-Meyer S. and Sahm H. (1996) Glyceraldehyde 3-phosphate and pyruvate as precursors of isoprenic units in an alternative non-mevalonate pathway for terpenoid biosynthesis. *J. Am. Chem. Soc.* **118**, 2564–2566.

Ross D. W. and Daterman G. E. (1995) Response of *Dendroctonus pseudotsugae* (Coleoptera; Scolytidae) and *Thanasimus undulatus* (Coleoptera: Cleridae) to traps with different semiochemicals. *J. Econ. Entomol.* **88**, 106–111.

Ross D. W. and Daterman G. E. (1998) Pheromone-baited traps for *Dendroctonus pseudotsugae* (Coleoptera: Scolytidae): influence of selected release rates and trap designs. *J. Econ. Entomol.* **91**, 500–506.

Rudinsky J. A., Kinzer G. W., Fentiman A. F., Jr, and Foltz, R. L. (1972) *trans*-Verbenol isolated from Douglas-fir beetle: laboratory and field bioassay in Oregon. *Environ. Entomol.* **1**, 485–488.

Rudinsky J. A., Morgan M. E., Libbey L. M. and Putnam T. B. (1974) Antiaggregative-rivalry pheromone of the mountain pine beetle, and a new arrestant of the southern pine beetle. *Environ. Entomol.* **3**, 90–98.

Rudinsky J. A., Morgan M. E., Libbey L. M. and Putnam T. B. (1977) Limonene released by the scolytid beetle *Dendroctonus pseudotsugae. Z. angew. Entomol.* **82**, 376–380.

Ryker L. C., Libbey L. M. and Rudinsky J. A. (1979) Comparison of volatile compounds and stridulation emitted by the Douglas-fir beetle from Idaho and western Oregon populations. *Environ. Entomol.* **8**, 789–798.

Ryker L. C. and Rudinsky J. A. (1982) Field bioassay of *exo*- and *endo*-brevicomin with *Dendroctonus ponderosae* in lodgepole pine. *J. Chem. Ecol.* **8**, 701–707.

Schlyter F. and Birgersson G. A. (1999) Forest beetles. In *Pheromones of Non-Lepidopteran Insects Associated with Agricultural Plants*, eds J. Hardie and A. K. Minks, pp. 113–148. CABI Publishing, New York.

Schooley D. A. and Baker F. C. (1985) Juvenile hormone biosynthesis. In *Comprehensive Insect Physiology, Biochemistry, and Pharmacology*, eds G. A. Kerkut, and L. I. Gilbert, Vol. 7, pp. 363–389. Pergamon Press, Oxford.

Schurig V., Weber R. A., Nicholsen G. J., Oehlschlager A. C., Pierce H. D., Jr, Borden J. H. and Ryker L. C. (1983) Enantiomer composition of natural *exo*- and *endo*-brevicomin by complexation gas chromatography/selected ion mass spectrometry. *Naturwissenschaften* **70**, 92–93.

Setter R. R. and Borden J. H. (1999) Bioactivity and efficacy of MCOL and seudenol as potential attractive bait components for *Dendroctonus rufipennis* (Coleoptera: Scolytidae). *Can. Ent.* **131**, 251–257.

Seybold S. J. (1992) *The role of chirality in the olfactory-directed aggregation behavior of pine engraver beetles in the genus* Ips *(Coleoptera: Scolytidae)*. PhD thesis. Univ. of Calif., Berkeley, 355 pp.

Seybold S. J. (1993) Role of chirality in olfactory-directed behavior: Aggregation of pine engraver beetles in the genus *Ips* (Coleoptera: Scolytidae). *J. Chem. Ecol.* **19**, 1809–1831.

Seybold S. J. and Tittiger C. (2003) Biochemistry and molecular biology of *de novo* isoprenoid pheromone production in the Scolytidae. *Annu. Rev. Entomol.* **48**, 425–453.

Seybold S. J., Teale S .A., Wood D. L., Zhang A., Webster F. X., Lindahl K. Q., Jr, and Kubo I. (1992). The role of lanierone in the chemical ecology of *Ips pini* (Coleoptera: Scolytidae) in California. *J. Chem. Ecol.* **18**, 2305–2329.

Seybold S. J., Ohtsuka T., Wood D. L. and Kubo I. (1995a) Enantiomeric composition of ipsdienol: A chemotaxonomic character for North American populations of *Ips* spp. in the *pini* subgeneric group (Coleoptera: Scolytidae). *J. Chem. Ecol.* **21**, 995–1016.

Seybold S. J., Quilici D. R., Tillman J. A., Vanderwel D., Wood D. L. and Blomquist G. J. (1995b) *De novo* biosynthesis of the aggregation pheromone components ipsenol and ipsdienol by the pine bark beetles *Ips paraconfusus* Lanier and *Ips pini* (Say) (Coleoptera: Scolytidae). *Proc. Natl. Acad. Sci. USA* **92**, 8393–8397.

Seybold S. J., Bohlmann J. and Raffa K. F. (2000) The biosynthesis of coniferophagous bark beetle pheromones and conifer isoprenoids: evolutionary perspective and synthesis. *Can. Entomol.* **132**, 697–753.

Silverstein R. M., Rodin J. O. and Wood D. L. (1966a) Sex attractants in frass produced by male *Ips confusus* in ponderosa pine. *Science* **154**, 509–510.

Silverstein R. M., Rodin J. O., Wood D. L. and Browne L. E. (1966b) Identification of the two new terpene alcohols from frass produced by *Ips confusus* in ponderosa pine. *Tetrahedron* **22**, 1929–1936.

Silverstein R. M., Brownlee R. G., Bellas T. E., Wood D. L. and Browne L. E. (1968) Brevicomin: principal sex attractant in the frass of the female western pine beetle. *Science* **159**, 889–891.

Siva-Jothy M. T., Blake D. E., Thompson J. and Ryder J. J. (1996) Short- and long-term sperm precedence in the beetle *Tenebrio molitor*: a test of the "adaptive sperm removal" hypothesis. *Physiol. Ent.* **21**, 313–316.

Skillen E. L., Berisford C. W., Camann M. A. and Reardon R. C. (1997) *Semiochemicals of forest and shade tree insects in North America and management applications. FHTET-96-15,* USDA Forest Service, Forest Health Technology Enterprise Team Publication, 182 pp.

Stewart T. E., Plummer E. L., McCandless L. L., West J. R. and Silverstein R. M. (1977) Determination of enantiomer composition of several bicyclic ketal insect pheromone components. *J. Chem. Ecol.* **3**, 27–43.

Stoakley J. T., Bakke A., Renwick J. A. A. and Vité J. P. (1978) The aggregation pheromone system of the larch bark beetle, *Ips cembrae* Heer. *Z. angew. Entomol.* **86**, 174–177.

Tabernero L., Bochar D. A., Rodwell V. W. and Stauffacher C. V. (1999) Substrate-induced closure of the flap domain in the ternary complex structures provides insights into the mechanism of catalysis by 3-hydroxy-3-methylglutaryl-CoA reductase. *Proc. Natl. Acad. Sci. USA* **96**, 7167–7171.

Tada S. and Leal W. S. (1997) Localization and morphology of sex pheromone glands in scarab beetles. *J. Chem. Ecol.* **23**, 903–915.

Tanaka Y., Honda H., Ohsawa K. and Yamamoto I. (1986) A sex attractant of the yellow mealworm, *Tenebrio molitor* L., and its role in the mating behavior. *J. Pesticide Sci.* **11**, 49–55.

Tanaka Y., Honda H., Ohsawa K. and Yamamoto I. (1989) Absolute configuration of 4-methyl-1-nonanol, a sex attractant of the yellow mealworm, *Tenebrio molitor* L. *J. Pesticide Sci.* **14**, 197–202.

Taub-Montemayor T. E. and Rankin M. A. (1997) Regulation of vitellogenin synthesis and uptake in the boll weevil, *Anthonomus grandis. Physiol. Entomol.* **22**, 261–268.

Taub-Montemayor T. E. Dahm K. H., Govindan B. and Rankin M. A. (1997a) Rates of juvenile hormone biosynthesis and degradation during reproductive development and diapause in the boll weevil, *Anthonomus grandis. Physiol. Entomol.* **22**, 269–276.

Taub-Montemayor T. E., Palmer J. O. and Rankin M. A. (1997b) Endocrine regulation of reproduction and diapause in the boll weevil, *Anthonomus grandis* Boheman. *Arch. Insect Biochem. Physiol.* **35**, 455–477.

Teale S. A., Webster F. X., Zhang A. and Lanier G. N. (1991) Lanierone: a new pheromone component from *Ips pini* (Coleoptera: Scolytidae) in New York. *J. Chem. Ecol.* **17**, 1159–1176.

Tillman J. A., Holbrook G. L., Dallara P. L., Schal C., Wood D. L., Blomquist G. J. and Seybold S. J. (1998) Endocrine regulation of *de novo* aggregation pheromone biosynthesis in the pine engraver, *Ips pini* (Say) (Coleoptera: Scolytidae). *Insect Biochem. Molec. Biol.* **28**, 705–715.

Tillman J. A., Seybold S. J., Jurenka R. A. and Blomquist G. J. (1999) Insect pheromones — an overview of biosynthesis and endocrine regulation. *Insect Biochem. Molec. Biol.* **29**, 481–514.

Tillman J. A., Lu F., Donaldson Z., Dwinell S.C., Tittiger C., Hall G. M., Blomquist G. J. and Seybold S. J. (2001) *Biochemical and molecular aspects of the regulation of pheromone biosynthesis in pine bark beetles.* Presented at Annu. Meet. Int. Soc. Chem. Ecol., 18th, Lake Tahoe.

Tittiger C., Blomquist G. J., Ivarsson P., Borgeson C. E. and Seybold S. J. (1999) Juvenile hormone regulation of HMG-R gene expression in the bark beetle, *Ips paraconfusus* (Coleoptera: Scolytidae): implications for male aggregation pheromone biosynthesis. *Cell. Mol. Life Sci.* **55**, 121–127.

Tittiger C., O'Keeffe C., Bengoa C. S., Barkawi L. S., Seybold S. J. and Blomquist G. J. (2000) Isolation and endocrine regulation of an HMG-CoA synthase cDNA from the male Jeffrey pine beetle, *Dendroctonus jeffreyi. Insect Biochem. Molec. Biol.* **30**, 2103–2111.

Tittiger C., Barkawi L. S., Bengoa C. S., Blomquist G. J. and Seybold S. J. (2003) Structure and juvenile hormone-mediated regulation of the HMG-CoA reductase gene from the Jeffrey pine beetle, *Dendroctonus jeffreyi. Mol. Cell. Endocrin.* **199**, 11–21.

Tumlinson J. H., Hardee D. D., Gueldner R. C., Thompson A. C., Hedin P. A., and Minyard J. P. (1969) Sex pheromones produced by the male boll weevil: isolation, identification, and synthesis. *Science* **166**, 1010–1012.

Valentine J. M. (1931) The olfactory sense of the adult mealworm beetle *Tenebrio molitor* (LINN). *J. Exp. Zool.* **58**, 165–227.

Vanderwel, D. (1991) *Pheromone biosynthesis by selected species of grain and bark beetles*. PhD thesis. Simon Fraser Univ. 172 pp.

Vanderwel D. (1994) Factors affecting pheromone production in beetles. *Arch. Insect Biochem. Physiol.* **25**, 347–362.

Vanderwel D. and Oehlschlager A. C. (1987) Biosynthesis of pheromones and endocrine regulation of pheromone production in Coleoptera. In *Pheromone Biochemistry*, eds G. D. Prestwich, and G. J. Blomquist, pp. 175–215. Academic Press, Orlando, FL.

Vanderwel D. and Oehlschlager A. C. (1992) Mechanism of brevicomin biosynthesis from (*Z*)-6-nonen-2-one in a bark beetle. *J. Am. Chem. Soc.* **14**, 5081–5086.

Vanderwel D., Pierce H. D. Jr, Oehlschlager A. C., Borden J. H. and Pierce A. M. (1990) Macrolide (Cucujolide) biosynthesis in the rusty grain beetle, *Cryptolestes ferrugineus* I. *Insect Biochem.* **20**, 567–572.

Vanderwel D., Gries G., Singh S. M., Borden J. H. and Oehlschlager A. C. (1992a) (*E*)- and (*Z*)-6-nonen-2-one: biosynthetic precursor of *endo-* and *exo*-brevicomin in two bark beetles (Coleoptera: Scolytidae). *J. Chem. Ecol.* **18**, 1389–1404.

Vanderwel D., Johnston B. and Oehlschlager A. C. (1992b) Cucujolide biosynthesis in the merchant and rusty grain beetles. *Insect Biochem. Molec. Biol.* **22**, 875–883.

Vanderwel D., Currie R. W., Griffith O. and Hastings C. (2003) Clarification of the roles of the female produced sex pheromones in mediating the mating behaviour of the yellow mealworm, *Tenebrio molitor* L. (Coleoptera: Tenebrionidae). *J. Chem. Ecol.* In preparation.

Vité J. P., Bakke A. and Renwick J. A. A. (1972a) Pheromones in *Ips* (Coleoptera: Scolytidae): occurrence and production. *Can. Ent.* **104**, 1967–1975.

Vité J. P., Pitman G. B., Fentiman A. F., Jr, and Kinzer G. W. (1972b) 3-Methyl-2-cyclohexen-1-ol isolated from *Dendroctonus. Naturwissenschaften* **59**, 469.

Werner R. A. (1972) Aggregation behaviour of the beetle *Ips grandicollis* in response to host-produced attractants. *J. Ins. Physiol.* **18**, 423–437.

White P. R., Chambers J., Walter C. M., Wilkins J. P. G. and Millar J. G. (1989) Saw-toothed grain beetle *Oryzaephilus surinamensis* (L.) (Coleoptera, Silvanidae) collection, identification, and bioassay of attractive volatiles from beetles and oats. *J. Chem. Ecol.* **15**, 999–1013.

White R. A., Jr, Franklin, R. T. and Agosin, M. (1979). Conversion of α-pinene oxide by rat liver and the bark beetle *Dendroctonus terebrans* microsomal fractions. *Pest Biochem. Physiol.* **10**, 233–242.

White R. A., Jr, Agosin M., Franklin R. T. and Webb J. W. (1980) Bark beetle pheromones: evidence for physiological synthesis mechanisms and their ecological implications. *Z. angew. Entomol.* **90**, 255–274.

Wiygul G. and Sikorowski P. P. (1986) The effect of staphylococcal enterotoxin B on pheromone production in fat bodies isolated from male boll weevils. *J. Invertebrate Path.* **47**, 116–119.

Wiygul G., and Sikorowski P. P. (1991). The effect of a heat-stable enterotoxin isolated from *Escherichia coli* on pheromone production in fat bodies isolated from male boll weevils. *Exp. exp. appl.* **60**, 305–308.

Wiygul G., Macgown M. W., Sikorowski P. P. and Wright J. E. (1982) Localization of pheromone in male boll weevils *Anthonomus grandis*. *Exp. Exp. Appl.* **31**, 330–331.

Wiygul G., Dickens J. C. and Smith J. W. (1990) Effect of juvenile hormone III and beta-bisabolol on pheromone production in fat bodies from male boll weevils, *Anthonomus grandis* Boheman (Coleoptera: Curculionidae). *Comp. Biochem. Biophysiol [B]* **95**, 489–491.

Wood D. L. (1962) The attraction created by males of a bark beetle *Ips confusus* (LeConte) attacking ponderosa pine. *Pan. Pac. Entomol.* **38**, 141–145.

Wood D. L. (1982) The role of pheromones, kairomones, and allomones in the host selection and colonization behavior of bark beetles. *Annu. Rev. Entomol.* **27**, 411–446.

Wood D. L. and Vité J. P. (1961) Studies on the host selection behavior of *Ips confusus* (LeConte) (Coleoptera: Scolytidae) attacking *Pinus ponderosa*. *Contrib. Boyce Thompson Inst. of Plant Res.* **21**, 79–96.

Wood D. L., Browne L. E., Silverstein R. M. and Rodin J. O. (1966) Sex pheromones of bark beetles – I. Mass production, bio-assay, source, and isolation of sex pheromone of *Ips confusus* (LeC.). *J. Insect Physiol.* **12**, 523–536.

Wood D. L., Stark R.W., Silverstein R. M. and Rodin J. O. (1967) Unique synergistic effects produced by the principal sex attractant compounds of *Ips confusus* (LeConte) (Coleoptera: Scolytidae). *Nature* **215**, 206.

Wood D. L., Browne L. E., Bedard W. D., Tilden P. E., Silverstein R.M. and Rodin J. O. (1968) Response of *Ips confusus* to synthetic sex pheromones in nature. *Science* **159**, 1373–1374.

Wood S. L. and Bright D. E. (1992) A catalog of Scolytidae and Platypodidae (Coleoptera), Part 2, Taxonomic index, Volumes A and B. *Great Basin Naturalist* **13**.

Young A., Tittiger C., Welch W. and Blomquist G. J. (2001) *Monoterpenoid pheromone biosynthesis: fishing for the elusive geranyl diphosphate synthase in bark beetles.* Presented at Annu. Meet. Int. Soc. Chem. Ecol., 18th, Lake Tahoe.

Zethner-Møller O. and Rudinsky J. A. (1967) Studies on the site of sex pheromone production in *Dendroctonus pseudotsugae* (Coleoptera: Scolytidae). *Ann. Ent. Soc. Am.* **60**, 575–582.

Zhang Q.-H., Birgersson G., Schlyter F. and Chen G.-O. (2000) Pheromone components in the larch bark beetle, *Ips cembrae*, from China: quantitative variation among attack phases and individuals. *J. Chem. Ecol.* **26**, 841–858.

7

Molecular biology of bark beetle pheromone production and endocrine regulation

Claus Tittiger

7.1 Introduction

Until recently, almost all of our knowledge about bark beetle aggregation pheromone biosynthesis and endocrine regulation came from biochemical and behavioral research. Behavioral assays and chemical analyses identified the first pheromone components and their endocrine regulation in *Ips paraconfusus* in the late 1960s (Silverstein *et al.*, 1966; Wood *et al.*, 1967, 1968; Borden *et al.*, 1969). Since then, semiochemicals from dozens of species have been studied using increasingly sophisticated methods (reviewed in Borden, 1985; Seybold *et al.*, 2000). And while our understanding of bark beetle pheromone biosynthesis and regulation has benefited greatly from these approaches, limitations in biochemistry technology left some questions unanswered. Increasingly sophisticated molecular tools have also been available for decades, yet they have been applied to bark beetle pheromone research only within the last five years. Molecular biology offers a complementary perspective to biochemistry, and the overall effect of combining the two approaches is often synergistic. Concentrating on the pine engraver (*Ips pini*), the California fivespined Ips (*I. paraconfusus*), and the Jeffrey pine beetle (*Dendroctonus jeffreyi*), this combined approach has proven very useful for increasing our understanding of bark beetle pheromone production and regulation. Furthermore, molecular biology studies, particularly in species with little established molecular data, invariably provide some evolutionary insights by revealing traits about genome organization and regulation.

7.2 Background

Aggregation pheromones are key elements in bark beetle reproductive strategy (Wood, 1982). The life cycle begins with the invasion of a host tree by a pioneer beetle. For *Ips* spp., the pioneer is a male, while for *Dendroctonus* spp., the pioneer is female (Birch, 1978). Shortly after burrowing through the outer bark, the pioneer produces enormous quantities of volatile aggregation pheromone (Wood and Bushing, 1963). Females and males respond by flying to the tree and burrowing beneath the bark. This "mass attack" is essential for the beetles' reproduction because a healthy tree can produce sufficient toxic resin to drown a few invading insects, but the combined feeding and tunneling of hundreds of beetles in the phloem effectively cuts the flow of water and rapidly kills the tree, leaving the colonizing beetles free to breed beneath the physical protection of the thick bark (Wood, 1982). Even for those species that prefer dead or dying timber, rapid aggregation is needed in order to colonize the wood before it dessicates (Birch, 1978).

The aggregation pheromone of Western *Ips pini* is an enantiomeric blend of 95–98 percent (–) ipsdienol (2-methyl-6-methylene-2,7-octadien-4-ol) (Vité *et al.*, 1972; Birch *et al.*, 1980; Seybold *et al.*, 1995a) and lanierone (2-hydroxy-4,4,6-trimethyl-2,5-cyclohexadien-1-one) (Teale *et al.*, 1991). Ipsenol (2-methyl-6-methylene-7-acten-4-ol) is produced later during the mass attack, possibly as an anti-aggregation signal that "the tree is full" (Birch and Light, 1977; Birch *et al.*, 1977). A sibling species, *I. paraconfusus*, also produces a mixture of ipsdienol and ipsenol as an aggregation pheromone (Silverstein *et al.*, 1966). The aggregation pheromone of *D. jeffreyi* is incompletely understood, though an important semiochemical and likely pheromone component is the bicyclic acetal, frontalin (1,5-dimethyl-6,8-dioxabicyclo(3.2.1)octane) (Paine *et al.*, 1999; Barkawi *et al.*, 2003). Frontalin is produced in abundance by feeding male *D. jeffreyi*, but not by females. Other *D. jeffreyi* pheromone components include *exo*-brevicomin, heptane and heptan-2-ol (Paine *et al.*, 1999; Seybold *et al.*, 2000).

Some early and obvious questions about bark beetle pheromone components concerned their origins: are they synthesized *de novo* from acetate or derived from plant precursor molecules? And are the biochemical reactions performed by insect tissues or symbiotic bacteria? Ipsdienol and ipsenol are clearly monoterpenoid alcohols. Since *de novo* monoterpenoid biosynthesis was unprecedented in the Metazoa before 1995, and monoterpenes are produced by host trees, it seemed logical that monoterpenoid pheromone components were derived from ingested plant precursor molecules (reviewed in Vanderwel and

Oehlschlager, 1987). Ipsdienol, after all, is hydroxylated myrcene, and pine trees make myrcene in abundance (Gershenzon and Croteau, 1991). This argument had an elegant evolutionary component as well, suggesting that bark beetles, during their co-evolution with their host trees, developed detoxification mechanisms to oxidize plant toxins into more hydrophilic and therefore easily excretable forms. The detoxification reaction was subordinated to chemical signaling over time, with different compounds being selectively produced during speciation (Vanderwel and Ohlschlager, 1987). The assumption that pheromone components were derived from plant precursors provided the perspective on which numerous biochemical studies were based, and many provided data in support of this model (reviewed in Seybold *et al.*, 2000). Indeed, despite the recent emphasis on *de novo* biosynthesis (see below), the balance of evidence suggests that some pheromone components, such as *trans*-verbenol, are derived from plant precursors (Renwick *et al.*, 1976).

Other studies challenged the host precursor paradigm, suggesting that some pheromone components may be synthesized *de novo*. Male *Ips* spp. can produce an astonishing amount of ipsdienol and ipsenol (Quilici, Seybold and Blomquist, unpublished observations) and some host trees do not synthesize myrcene in sufficient amounts to support this huge production (Byers, 1981; Byers and Birgersson, 1990). Isoprenoid molecules in animals are synthesized via the mevalonate pathway [formerly the isoprenoid pathway; but with the discovery of an alternative "DOX-P" isoprenoid pathway in prokaryotes and plant plastids (Rohmer, 1999), it is necessary to distinguish between the two] and mevalonate pathway inhibitors attenuate pheromone production in *I. duplicatus* (Ivarsson *et al.*, 1993). Furthermore, judicious use of various radiolabeled precursors in *Ips* spp. established that monoterpenoid pheromone components are synthesized *de novo* via the mevalonate pathway (Figure 7.1) (Seybold *et al.*, 1995b; Ivarsson *et al.*, 1997; Tillman *et al.*, 1998).

In *I. pini,* feeding stimulates JH III production in the *corpora allata*, which in turn stimulates *de novo* pheromone biosynthesis (Tillman *et al.*, 1998). How does JH III regulate the mevalonate pathway in male bark beetles? Guided by extensive literature of mevalonate pathway regulation in vertebrates, research on this question concentrated on HMG-CoA reductase (HMG-R, E.C. 1.1.1.34), the enzyme that catalyzes the first committed, and most tightly controlled, step in the pathway. HMG-R is a textbook example of one of the most highly regulated enzymes known. In mammals, the most important mode of regulation is a negative feedback loop controlled by the principal product of the pathway, cholesterol (reviewed in Goldstein and Brown, 1990; Hampton *et al.*, 1996). Insects, however, are sterol auxotrophs (Karlson, 1970), and insect HMG-R is insensitive to sterol concentrations (Silberkang *et al.*, 1983), so differences in HMG-R regulation between insects and other Metazoa can be expected. Nevertheless, the regulatory role of HMG-R in the mevalonate pathway of other organisms provided a convenient

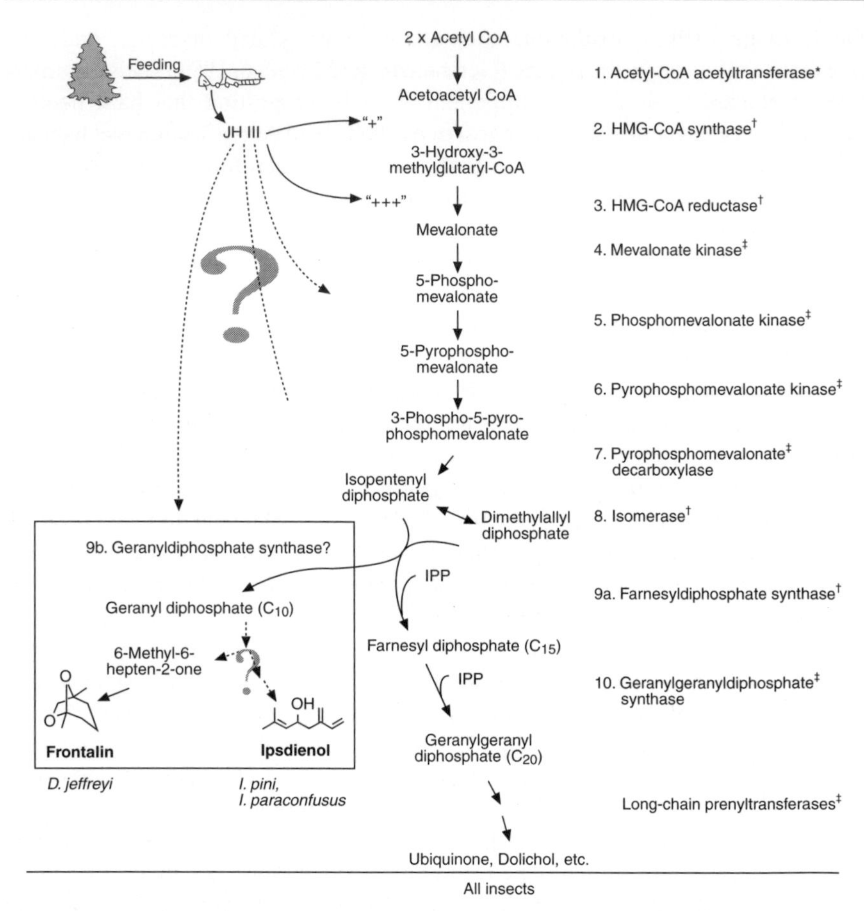

Figure 7.1 Pheromone biosynthesis and regulation in *Ips* spp. and *Dendroctonus jeffreyi*. The mevalonate pathway and associated enzymes are shown to the center-right. †Major regulators of the pathway. *Enzymes for which full or partial cDNAs have been isolated from *I. pini*. ‡uncloned cDNAs. The pheromone biosynthetic pathway (in the box, lower left) branches from the mevalonate pathway at geranyl diphosphate (GPP). Isopentanyl diphosphate (IPP) is added to GPP during consecutive steps in the mevalonate pathway to form longer-chain isoprenoids. Juvenile hormone (JH) III is produced by males upon feeding, and stimulates expression of some mevalonate pathway genes (continuous arrows). Unknown regulatory points and biochemical steps are noted by dashed lines.

starting point for molecular studies in bark beetles. HMG-CoA synthase (HMGS, E.C. 4.1.3.5), which acts in the mevalonate pathway just upstream of HMG-R, was a second early target for molecular investigations because it has a minor role in controlling the mammalian mevalonate pathway (Goldstein and Brown, 1990), and was expected to be regulated similarly to HMG-R in bark beetles.

7.3 Gene isolation

7.3.1 cDNAs

The clear induction of the mevalonate pathway in response to JH III and feeding (Tillman *et al.*, 1998) strongly suggested that HMG-R gene (*HMG-R*) (nomenclature in this chapter follows standard forms: gene names are italicized, while transcripts, cDNAs, and proteins are in normal font) expression levels in pheromone-producing insects would be significantly higher than those in pheromone non-producers. Alignments of metazoan HMG-R proteins found in GenBank revealed numerous highly conserved sections, particularly in the catalytic domain. A highly conserved octapeptide region within the catalytic domain contained three methionine codons and two two-fold degenerate codons, proving an excellent site for a degenerate primer. Located within a kilobase of the 3′ untranslated region (utr), the 32-fold degenerate primer designed to this peptide, "Parahmd," allowed convenient amplification of first-strand cDNA templates by 3′ RACE. Indeed, this primer was used to isolate HMG-R cDNA fragments from *I. paraconfusus* (Tittiger *et al.*, 1999), *I. pini* (Hall *et al.*, 2002a), *Dendroctonus jeffreyi* (Tittiger *et al.*, 2003), and, in a related but separate study, the cotton boll weevil (*Anthonomus grandis*, unpublished observations).

Although the 3′ RACE PCR procedure is not quantitative as we applied it, we expected a qualitative difference when amplifying templates from different sources; that is, HMG-R should have amplified more effectively from pheromone-producing insects (JH III-treated or fed males) compared to non-pheromone producers (females and starved males). Contrary to expectations, the reactions worked best using non-pheromone-producing tissues, even though northern blotting confirmed that HMG-R mRNA levels are lower in pheromone non-producers compared to pheromone producers (see below) (Tittiger *et al.*, 1999; unpublished observations). This suggested that treatment by JH III either changed the chemistry of the cells, making them refractory to RNA isolations, or else HMG-R mRNA was sequestered in such a way that use as a template for cDNA synthesis was inefficient. Indeed, it is frustratingly difficult to reliably isolate RNA from fed bark beetles (Tillman *et al.*, unpublished observations), likely due to the chemical composition of fed gut tissues.

The fragments isolated by 3′ RACE were then used to isolate their corresponding complete cDNAs by a combination of PCR and library screening as necessary. cDNA libraries were constructed for *I. pini* and *D. jeffreyi* and screened either by PCR or by classical plaque hybridizations. For *I. paraconfusus*, a cDNA library was not constructed, so the 5′ end of the HMG-R cDNA was recovered using a combination of RT-PCR (with a degenerate primer near the 5′ end of the open reading frame) and ligation-anchored PCR (LA-PCR; Troutt *et al.*, 1992). The bark beetle HMG-R cDNAs are available in GenBank (accession numbers

AF304440 for *I. pini*, AF071750 for *I. paraconfusus*, and AF159136 for *D. jeffreyi*).

The ~3 kb HMG-R transcripts encode open reading frames (ORFs) whose predicted translation products are highly conserved with respect to other metazoan HMG-Rs. The proteins have an amino-terminal membrane anchor consisting of eight transmembrane domains, followed by a short hydrophilic linker and a C-terminal catalytic domain (reviewed in Hampton *et al.*, 1996). Similar to other HMG-Rs, computer predictions of the insect HMG-R membrane-spanning domains assign a weak probability to the putative seventh transmembrane domain, and this region can even be overlooked depending on the algorithm used (unpublished observations). Elegant substitution studies proved that vertebrate HMG-R has eight membrane-spanning domains in its anchor (Olender and Simoni, 1992), and there is no reason to believe insect HMG-Rs are any different in this respect. The structure of the anchor is important since it senses membrane cholesterol concentrations and regulates HMG-R stability accordingly in mammals (Chun and Simoni, 1992; Kumagai *et al.*, 1995, Sekler and Simoni, 1995). Since insect HMG-R is sterol insensitive (Silberkang *et al.*, 1983), the conserved membrane structure in vertebrates and animals suggests a common role, possibly in subcellular targeting (Gertler *et al.*, 1988) and/or membrane lipid maintenance (Seegmiller *et al.*, 2002).

The start codon for *D. jeffreyi* HMG-R has not been unambiguously determined (Tittiger *et al.*, 2003). Alignments with other HMG-Rs place the start codon at nt 108 on the cDNA. This position is in frame with another start codon upstream at nt 24. It remains unknown which of the two positions, or both, are the legitimate start codon. The context for translation initiation is much stronger for the upstream candidate compared to the downstream position, even though the latter appears to be the correct initiation site. If translation begins at nt 24, the additional 28 amino acids that would precede the start codon predicted from sequence alignments have no identifiably conserved signal sequences or structures. Furthermore, an intron correlating to the 5′ utr in human and hamster HMG-R genes interrupts this area, suggesting that the portion of the cDNA upstream of nt 108 is also a 5′ utr (see below).

7.3.2 HMG-R gene structure

The only genomic HMG-R sequences so far reported from insects are from *D. jeffreyi* (AF159137, AF159138, AH009515), and *Drosophila melanogaster* (FlyBase, 1999). Studies of bark beetle HMG-R gene (*HMG-R*) structures are of interest because of the roles introns play in regulating vertebrate HMG-R activity (Reynolds *et al.*, 1984; Gayen and Peffley, 1995), and although the motivation to study bark beetle *HMG-R* structures stemmed from the desire to understand how the gene is regulated, the results are useful for those involved in debates about how genes evolve. As in other Metazoa, *HMG-R* is a single copy gene in *D.*

jeffreyi and *I. pini* (Tittiger *et al.*, 2003; unpublished results). The *D. jeffreyi* sequence was generated from 11 overlapping PCR products amplified from genomic DNA (gDNA) templates using sequence-specific primers that were designed from the cDNA (Tittiger *et al.*, 2003). Thus, current knowledge of *D. jeffreyi HMG-R* structure is limited to the transcribed portion of the gene and does not include upstream control elements.

A comparison of *D. jeffreyi HMG-R* intron sites with those from Syrian hamster (Reynolds *et al.*, 1984), human (Nakajima *et al.*, 2000) and the fruitfly, as predicted from the *Drosophila* genome project (FlyBase, 1999), may shed light on the function and/or evolution of metazoan *HMG-R*. *D. jeffreyi HMG-R* spans nearly 10 kb and is interrupted by 13 introns (Figure 7.2). Seven of the introns interrupt the 5′ utr and anchor-encoding portions of the gene. With the exception of one, all are located in identical sites compared to the introns in mammalian *HMG-R*s. Intron sites are noticeably less conserved in the linker and catalytic portions of the gene, with only one of the remaining five *D. jeffreyi HMG-R* introns conserved in the mammalian homologs. The asymmetry in the location of conserved sites – with the 5′ utr and anchor domain having seven of eight conserved, while the remainder of the gene has only one of five – suggests functional roles for some introns. This pattern is more difficult to distinguish in *Drosophila HMG-R*, where there are only four introns in total. The reduced

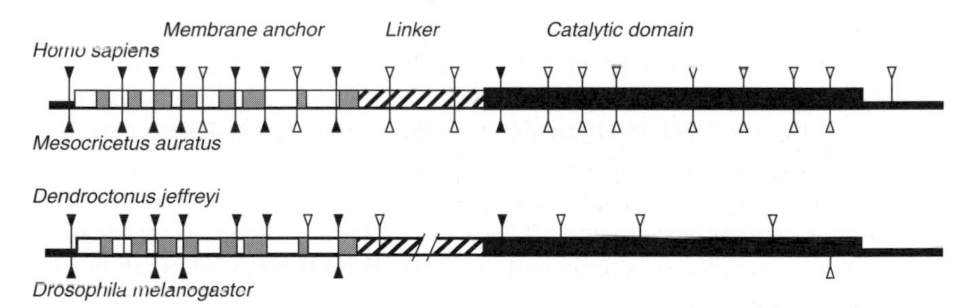

Figure 7.2 Comparison of HMG-R cDNAs from mammals (Syrian hamster, *Mesocricetus auratus* and human) and insects (Jeffrey pine beetle, *Dendroctonus jeffreyi*, and the fruitfly, *Drosophila melanogaster*), showing the relative positions of intron splice sites. Untranslated regions (utrs) are indicated by thin lines and coding regions are indicated by boxes. Boxes representing membrane spanning domains are shaded grey, hydrophilic loops are white, linker domains are striped, and catalytic domains are black. The diagonal slash in the insect cDNAs indicates a gap inserted to maintain the alignment of the catalytic domains. Intron sites are indicated by vertical lines capped with triangles. Black triangles indicate conserved (identical and similar) intron sites in *D. jeffreyi* and mammals, whereas white triangles indicate unconserved intron sites. Figure based on Tittiger *et al.* (2003); reproduced with permission.

number of introns in the *D. melanogaster* gene may reflect the evolutionary drive of that insect towards a compact genome (Petrov and Hartl, 1998).

Introns at the borders of transmembrane (TM) domains and adjacent linkers suggest the evolutionary assembly of the membrane anchor by exon shuffling (Reynolds *et al.*, 1984; Olender and Simoni, 1992). Proteins with at least eight membrane-spanning domains that are involved or implicated in sterol sensing include HMG-R, sterol response element binding protein (SREBP), cleavage-activating protein (SCAP), Patched, and Neimann-Pick type C1 protein. In humans, all have intron sites at or near TM borders (Nakajima *et al.*, 2000). Interestingly, sterol-insensitive polytopic proteins, including insect HMG-R and rhodopsin (Nathans and Hogness, 1983) also show a similar intron distribution pattern. These data suggest that construction of these polytopic proteins arose by exon shuffling (Nakajima *et al.*, 2000), with different proteins possibly acquiring different roles over time (Seegmiller *et al.*, 2002).

In contrast to the possible evolutionary role played by introns in the membrane anchor domain, the intron in the 5′ utr may have a current functional role for insect *HMG-R*s. Human, hamster and beetle genes have an intron located a few tens of bases upstream of the start codon (23, 23, and 25 nt upstream, respectively). Alternative splicing of this intron leads to differential transcript stability in mammals (Reynolds *et al.*, 1984; Gayen and Peffley, 1995). *D. melanogaster* HMG-R mRNA isoforms are developmentally regulated, presumably through alternative splicing (Gertler *et al.*, 1988), and the genomic HMG-R sequence predicted in FlyBase diverges from the published cDNA 50 nt upstream of the start codon. Northern blot experiments infrequently reveal larger than normal HMG-R mRNA signals in *I. pini* (unpublished observations), though it remains unknown if these are due to alternative splicing. Computer analyses of the first intron in *D. jeffreyi HMG-R* predict multiple splice donor and acceptor sites (Tittiger *et al.*, 2003). It is tempting to speculate that similar HMG-R regulation by alternative splicing occurs in bark beetles. To date, however, experiments to determine the presence of alternatively spliced HMG-R mRNA in bark beetles have not been published.

7.3.3 HMG-S

A strategy similar to that used for HMG-R was used to isolate HMG-S cDNA from *D. jeffreyi* (Tittiger *et al.*, 2000; AF166002) and *I. pini* (O'Keeffe, Dhadda and Tittiger, unpublished results). Instead of 3′ RACE, however, degenerate primers based on conserved portions of other insect HMG-S sequences (Martinez-Gonzalez *et al.*, 1993a) were used for RT PCR of first strand cDNA that had been primed with oligo-dT, with the full-length cDNA being isolated through library screening. The 3′ utr of the *D. jeffreyi* HMG-S cDNA has various motifs suggesting that its stability may be modulated (Tittiger *et al.*, 2000). The predicted *D. jeffreyi* HMG-S protein, much like HMG-R, offers no surprises when compared to its homologs in other Metazoa: it is ~58 percent identical to other insectan HMG-Ss, with a well-conserved catalytic domain. Like other insectan HMG-Ss, it is likely a cytosolic protein since there is no recognizable targeting sequence at the N-terminal.

Less information is available for insectan HMG-CoA synthase gene (*HMG-S*) structures. Genomic *HMG-S* sequences from bark beetles have not been published, but southern blotting reveals a single copy *HMG-S* in *D. jeffreyi* (Tittiger *et al.*, 2000), giving an interesting twist in insect genome organization. In vertebrates, separate genes encode mitochondrial and cytosolic forms of HMG-S (Ayté *et al.*, 1990), which are used to produce ketone bodies and isoprenoids, respectively. The first HMG-S study in insects suggested that this precedent applied to insects as well, because two *HMG-S*s were found in the German cockroach, *Blattella germanica* (Buesa *et al.*, 1994). While the single *D. jeffreyi HMG-S* contrasted with the early studies, a single copy *HMG-S* is predicted in the *D. melanogaster* genome (FlyBase, 1999), suggesting that a single copy *HMG-S* may be common in insect genomes (Tittiger *et al.*, 2000). This idea is supported by the finding that one of the *B. germanica HMG-S*s probably evolved from the other by a retrotransposition event (Casals *et al.*, 2001), implying that it is a recent gene duplication and therefore not widely distributed in the Insecta. A survey of fungal mevalonate pathway genes also yielded a single copy *HMG-S* in *Schizosaccharomyces pombe* and *Phycomyces blakesleeanus* (Ruiz-Albert *et al.*, 2002), so perhaps the precedent of genes encoding mitochondrial and cytosolic HMG-Ss is limited to mammals or vertebrates.

A single copy *HMG-S* raises interesting questions about insect metabolism, since insects contain enzymes for ketone body synthesis (Beis *et al.*, 1980, Flybase 1999), but evidently lack a mitochondrial HMG-S. Whether or not insects produce ketone bodies is still questionable (Bailey and Horne, 1972; Shah and Bailey, 1976; Newsholme *et al.*, 1977; Beis *et al.*, 1980; Candy *et al.*, 1997). If they do, the mechanism by which they achieve this without an apparent mitochondrial HMG-S is unknown. One mechanism may involve alternative hnRNA splicing such that exons encoding different signal peptides are selected to produce proteins targeted to different subcellular locations. Alternative splicing of introns in the 5′ utr of hamster cytosolic HMG-S hnRNA is common (Gil *et al.*, 1986), and alternative splicing of the first intron of *HMG-S*, located in the 5′ utr, is predicted in *D. melanogaster* (FlyBase, 1999). However, the various predicted *D. melanogaster* HMG-S mRNA isoforms all have the same apparent open reading frame, so the hypothesis of alternative splicing may not be valid.

7.4 Endocrine regulation

When a male *I. pini* feeds, JH III release (and likely synthesis) rates and titers rise, causing an increase in ipsdienol biosynthesis (Tillman *et al.*, 1998), and unfed (starved) male *I. pini* and *I. paraconfusus* can be induced to synthesize pheromone components if treated with sufficient JH III or JH analogs (Chen *et al.*, 1988; Tillman *et al.*, 1998). Since most ipsdienol biosynthesis is *de novo*, mevalonate pathway enzymes are probably coordinately induced. Major pathway regulators, however, are likely most sensitive to JH III regulation, so it was reasonable to hypothesize that JH III regulated HMG-R gene expression and activity. In particular, we postulated that JH III elevated HMG-R mRNA levels, probably through increased transcription rates, but possibly also due to increased transcript stability. HMG-S would also be regulated, though likely to a lesser extent, reflecting the relatively minor role HMG-S has in controlling the mevalonate pathway (Goldstein and Brown, 1990). Biochemical analyses comparing *I. paraconfusus* and *I. pini* revealed an interesting difference between the two species with respect to pheromone production. Even at peak production, *I. paraconfusus* produces significantly less ipsdienol than *I. pini*, suggesting that the pheromone biosynthetic pathway is not induced as highly in *I. paraconfusus* (Tillman *et al.*, in preparation).

The expected correlation of *HMG-R* expression with pheromone production was notably important for studies with *D. jeffreyi*. The chemical origin of frontalin was unclear from its structure. It was proposed as an isoprenoid derivative (Francke *et al.*, 1995; Perez *et al.*, 1996), but it might also be derived from fatty-acyl precursors (Blomquist, personal communication). Both pathways were theoretically possible, with the additional chance of "cross-talk" between the two (Barkawi *et al.*, 2003). Unfed, mature male *D. jeffreyi* can be induced to synthesize frontalin by topical JH III application. Early work using *I. paraconfusus* was confirming the regulatory role of JH III on *HMG-R* expression in *Ips* beetles (Tittiger *et al.*, 1999), so it was a relatively simple task to do similar studies on *D. jeffreyi.* Following isolation of a *D. jeffreyi* HMG-R cDNA fragment, a preliminary northern blot clearly showed that *HMG-R* was highly expressed in JH III-treated males, but not in females or untreated males (Tittiger *et al.*, 2003). The correlation of *HMG-R* expression with frontalin biosynthesis was the first line of experimental evidence suggesting that frontalin was an isoprenoid, serving to focus biochemical research on the mevalonate pathway. Indeed, radio-tracer studies subsequently confirmed the isoprenoid origin of frontalin, and ruled out the possible involvement of the fatty acid biosynthetic pathway (Hall *et al.*, 2002b; Barkawi *et al.*, 2003).

The effect of JH III on *HMG-R* expression in *I. paraconfusus*, *I. pini* and *D. jeffreyi* was investigated by determining both a dose-response and time course by standard northern blotting. JH III, dissolved in acetone, was applied topically onto the ventral abdomens of unfed adult insects, while control insects were treated with an equivalent volume of acetone. Total or poly $(A)^+$ RNA was then isolated from intact tissues pooled from up to ten individuals for the *Ips* studies (Tittiger *et al.*, 1999; Tillman *et al.*, in preparation). For the *D. jeffreyi* studies, preliminary work localized JH III-induced HMG-R expression to the metathorax/

abdomen region (see below), so isolated thoracic segments pooled from five individuals were used instead of whole insects (Barkawi *et al.*, in preparation; Tittiger *et al.*, 2000, 2003).

The first report on the endocrine regulation of a pheromone biosynthetic gene in an insect was a pilot study of JH III-regulated *HMG-R* expression in *I. paraconfusus* (Tittiger *et al.*, 1999). Unreplicated samples showed that HMG-R mRNA levels increase approximately equally in both male and female beetles following JH III treatment, with levels in males being generally higher than in females (Figure 7.3a). A dose-response curve showed a general increase in HMG-

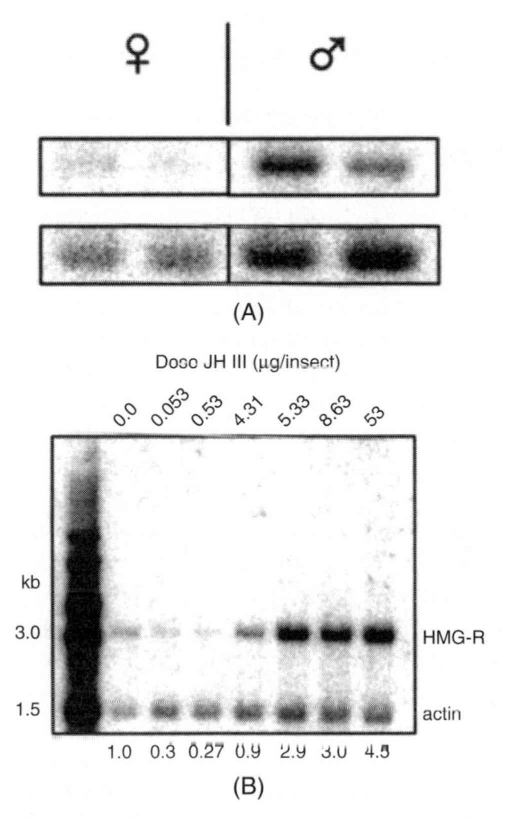

Figure 7.3 Regulation of *HMG-R* expression in male and female *I. paraconfusus*. A JH III-induced HMG-R mRNA levels in females and males following topical application of 5.3 μg JH III. B Northern blot showing the effect of topical JH III application on HMG-R mRNA in males. The doses applied are shown above the image. In both A and B, poly(A)$^+$RNA was isolated 20 h following treatment. The relative levels of HMG-R mRNA (normalized to the actin signal) compared to untreated males is given below each image. Reproduced from Tittiger *et al.* (1999); with permission.

R mRNA with increasing JH III dose, though low doses (<1 μg) evidently depress expression levels (Figure 7.3b). *I. pini HMG-R* expression patterns are similar to *I. paraconfusus*, if more robust (Tillman *et al.*, in preparation). *D. jeffreyi HMG-R* expression has the same general trend, with some differences (Tittiger *et al.*, 2003). *D. jeffreyi* HMG-R mRNA levels increase following topical JH III application in both a dose- and time-dependent manner (Figure 7.4). In contrast to *Ips* spp., sub-μg doses do not depress HMG-R mRNA, and the highest dose attenuated the induction (Figure 7.4c; the effect of a similarly high dose in *I. paraconfusus* has not been published). The amount of maximal induction also varies, from approximately eight to nine-fold for *I. paraconfusus* to up to 30-fold for *D. jeffreyi*. Maximal HMG-R mRNA levels in *D. jeffreyi* are reached within 12 h of treatment. Expression then declines slowly and is still above basal levels after nearly 72 h (Figure 7.4).

The elevation of *HMG-R* expression levels in males is consistent with the requirement for increased flux through the mevalonate pathway during pheromone biosynthesis. In female *I. paraconfusus*, the increase in *HMG-R* expression likely corresponds to sexual maturation, including oogenesis and vitellogenin synthesis and uptake by follicle cells. This response by mevalonate pathway genes in females is precedented in the *B. germanica*, where JH III stimulates oogenesis and elevates HMG-R expression during the gonadotrophic cycle (Martinez-Gonzalez *et al.*, 1993b). *D. jeffreyi* females, however, barely elevate *HMG-R* mRNA above basal levels when treated with JH III (Hall *et al.*, 2002b; Tittiger *et al.*, 2003; Barkawi *et al.*, in preparation). Oogenesis in *D. ponderosae* appears to be triggered by mating, and not by feeding (Reid, 1962), and the requirement for JH III to stimulate HMG-R activity as part of egg production in *D. jeffreyi* has not been shown.

The response of *HMG-S* to JH III has only been investigated in male *D. jeffreyi*. Similar to *HMG-R*, *HMG-S* shows a clear dose response and pulse of elevated expression following JH III application (Figure 7.4) (Tittiger *et al.*, 2000). The relatively modest induction (approximately four fold over basal levels) is consistent with HMG-S having a comparatively minor role in mevalonate pathway regulation (reviewed in Goldstein and Brown, 1990). The generally similar expression patterns of the two genes suggest coordinate regulation of the pheromone biosynthetic pathway. This may be useful to assist in identifying genes involved downstream of the branch away from the mevalonate pathway.

The northern analyses provide tantalizing clues about the regulation of pheromone production at the molecular level. JH III obviously elevates flux through the pheromone biosynthetic pathway by affecting gene expression. At this time, there is no evidence indicating whether JH III acts as a ligand for a specific transcription factor (nuclear receptor) that binds directly to *HMG-R* and *HMG-S* promoters, or if JH III elevates *HMG-R* and *HMG-S* expression indirectly. The time-course studies show a short lag period before mRNA levels begin to

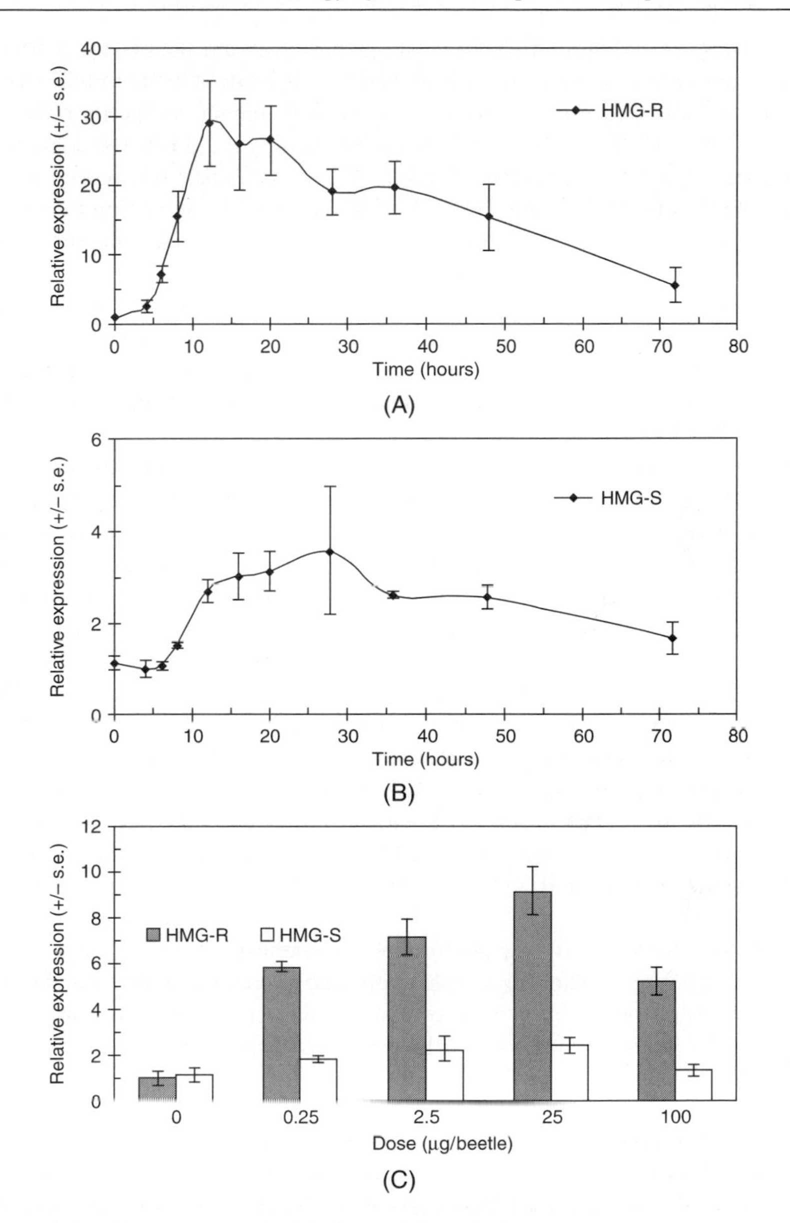

Figure 7.4 Regulation of *HMG-R* and *HMG-S* expression in male *Dendroctonus jeffreyi* by JH III. The time course (A, B) and dose response (C) for each gene in mature (emerged) males was investigated by northern blotting. All values are relative to starved, untreated males. Each point represents the mean +/– standard error of three replicates, five isolated thoraces/sample. Reproduced from Tittiger *et al.* (2000, 2003) with permission.

rise in *D. jeffreyi* (Figure 7.4), suggesting that upstream factors must first be synthesized before those genes are induced. And while it is generally safe to assume that elevated mRNA levels correlate with increased transcription, the possibility that JH III increases *HMG-R* mRNA stability should also be considered. Post-transcriptional regulation of HMG-R by hormones is precedented in vertebrates: estrogen elevates HMG-R mRNA levels in male *Xenopus laevis* without an apparent increase in transcription rates (Chen and Shapiro, 1990). Thus, until nuclear run-off assays or similar studies distinguish between transcription rate and RNA stability, the cause of elevated mRNA levels in JH-treated male bark beetles remains unknown. Similarly, because the mode of JH III-regulated gene expression in any insect is still unknown, it is impossible at this time to determine by what mechanism JH III regulates *HMG-R* and *HMG-S* in bark beetles.

The pulse of mRNA also can be interpreted in different ways, depending on how JH III is released from the cuticle. The decline over time may reflect decreasing JH III dose, since the beetles were only treated once and the JH III titer likely drops below effective dose levels as it is catabolized. On the other hand, if the cuticle acts as a slow-release JH III reservoir, internal JH III levels may be constantly high. If this is so, decreasing expression levels over time may indicate a desensitization to the hormone. The energetic status of the cell may also influence *HMG-R* and *HMG-S* expression levels if pheromone biosynthetic rates depend on the availability of molecular resources. In addition to stimulating pheromone biosynthesis, JH III or JH analogs induce flight muscle degeneration in bark beetles (Borden and Slater, 1968; Sahota and Farris, 1980). It has been suggested that the carbon and energy released from these tissues may be converted to pheromone (e.g. Ivarsson *et al.*, 1998). Thus, the decrease in expression over time in JH III-treated, unfed beetles, may reflect a depletion of precursor molecules.

JH III is sufficient to induce pheromone biosynthesis in starved *I. pini* and *D. jeffreyi*, suggesting a relatively simple endocrine regulation. However, starved *I. paraconfusus* cannot be induced to synthesize pheromone by JH III, despite the fact that *HMG-R* expression rises with JH III treatment. Pheromone biosynthesis in *I. paraconfusus* requires feeding (Seybold *et al.*, 2000; Tillman *et al.*, in preparation). Since *HMG-R* expression is apparently regulated similarly in both species, *I. paraconfusus* must require an additional signal in order to activate pheromone biosynthesis. The necessary factors are likely to act post-transcriptionally or post-translationally on HMG-R. These studies show that extending information from one species to another must be done with caution, as even closely related bark beetles seem to have very different regulatory schema.

Naturally occurring variations in *HMG-R* mRNA levels due to feeding or development remain unknown. Fed bark beetles are refractory to numerous RNA isolation protocols (Tillman and Tittiger, unpublished results), making these types of studies difficult, and a developmental profile of HMG-R gene expression has yet to be done.

7.5 Tissue localization

In most insects, pheromones are synthesized in specialized cells or tissues associated with the epidermis (Tillman *et al.*, 1999). Biochemical analyses traced the localization of Scolytid pheromone accumulation to portions of the alimentary canal, particularly the hindgut (e.g. Borden *et al.*, 1969; Byers, 1983), but the actual tissue source of pheromone components was unknown. Fortunately, the tight correlation of *HMG-R* gene expression with pheromone component biosynthesis meant that hybridization techniques could be used to map the location of pheromone biosynthesis. Northern blots provided the first maps, while *in situ* hybridizations definitively showed which tissues were elevating *HMG-R* mRNA in response to feeding or JH III treatment. As with endocrine regulation studies, the molecular and biochemical data complemented each other.

Northern blots of RNA isolated from approximate head, thorax, and abdomen segments provided a crude map localizing JH III-dependent increases in *HMG-R* mRNA. In all species tested, JH III stimulated *HMG-R* transcript levels mostly in male thoracic tissue. This was shown first with *I. paraconfusus*, where the expression pattern correlated with biochemical mapping of ipsdienone biosynthesis to this region (Ivarsson *et al.*, 1998). A similar result was obtained in *I. pini* (Hall *et al.*, 2002a). The technique was refined in experiments with *D. jeffreyi*, where variations in the way that the animals were dissected affected the distribution of the signal (Hall *et al.*, 2002b). This allowed the map to focus expression of both *HMG-S* and *HMG-R* to the metathorax/abdomen border (Tittiger *et al.*, 2000; Hall *et al.*, 2002b).

Crude mapping by northern blotting could localize the site of *HMG-R* expression, but could not distinguish between tissues, so the question remained: which cells in this area synthesize pheromone components? The alimentary canal was an obvious candidate since it is dermally derived and pheromone components tend to accumulate there. However, biochemical studies in another beetle, *Anthonomus grandis*, had assigned pheromone biosynthesis to the fat body (Wiygul *et al.*, 1982, 1990; Wigul and Sikorowski, 1985), and thc balance of biochemical evidence suggested that flight muscles may synthesize pheromone components in male *I. paraconfusus* (Ivarsson *et al.*, 1998).

For fine mapping, the tissue distribution of expression in *I. pini* and *D. jeffreyi* was determined by *in situ* hybridization. Early attempts to fix *D. jeffreyi* sections to glass substrates were frustrated by the sections' tendency to curl, so an "exposed whole mount" technique was developed to circumvent the problem. This strategy involves sectioning the beetle to a desired depth on a microtome as per a normal *in situ* experiment, but then discarding the sections and performing whole mount hybridizations on the exposed remainder of the carcass (Hall *et al.*, 2002a,b). The striking "beetle-on-the-half-shell" images allow clear differentiation between high- and low-expressing tissues, with the caveat that information from the technique is not quantitative. Subsequent *in situ* hybridizations could then concentrate on more traditional whole mounts of isolated target tissue.

Exposed whole mounts unequivocally show that in pheromone-biosynthetic males, the anterior midgut (ventriculus) specifically and strongly expresses *HMG-R* (Figure 7.5). This area is at the metathorax–abdomen boundary and corroborates the northern blot maps; indeed, it appears that almost all of the HMG-R mRNA on northern blots is contributed by midgut cells. *HMG-R* expression in male midgut tissues correlates tightly with pheromone production: isolated anterior midgut tissues from starved, JH III-treated male *I. pini* and *D. jeffreyi* have elevated *HMG-R* mRNA, as do fed male *I. pini* guts (the correlation with feeding has not been investigated in *D. jeffreyi*). Starved (untreated) male and JH III-treated female *I. pini* midguts do not strongly express HMG-R; neither do they produce pheromone components (Figure 7.5). Radio-tracer incorporation studies confirmed that the pheromone components ipsdienol and frontalin are produced in these tissues by *I. pini* and *D. jeffreyi*, respectively (Hall *et al.*, 2002a, b).

A dramatic change in the subcellular structure of midgut cells accompanies the JH III-mediated increase in *HMG-R* expression (Figure 7.6). Midgut cells from JH III-treated male *I. pini* and *D. jeffreyi* (i.e. pheromone-synthesizing cells) have large amounts of smooth endoplasmic reticulum (SER) ordered in crystalline arrays (Nardi *et al.*, 2002). The arrays are particularly striking in *D. jeffreyi*. Overproduction of SER is common in cholesterol-deprived mammalian cells with abundant HMG-R (Chin *et al.*, 1982; Pathak *et al.*, 1986; Wright & Rine, 1989). In fact, elevated HMG-R protein levels may signal proliferation and crystalloid formation of the SER (Berciano *et al.*, 2000). From this perspective, it is not surprising that pheromone-synthesizing midgut cells have large amounts of crystalline SER since they are strongly expressing *HMG-R*, and therefore likely have correspondingly elevated HMG-R protein levels. This assumes that HMG-R protein localizes to the SER, as is precedented in mammals (e.g. Olender and Simoni, 1992), although the possibility that HMG-R may also distribute to peroxisomes cannot be discounted (Keller *et al.*, 1986). Midgut cells from JH III-treated *I. paraconfusus* have not been analyzed by electron microscopy, though it would be interesting to compare midgut cells from fed, JH III-treated, and untreated males to determine if they have the SER arrays or not. Such information may be useful in determining the point at which the putative secondary factor is required for pheromone production.

Beyond the obvious demonstration that bark beetle pheromones are synthesized in midgut tissues, two other conclusions can be drawn from these studies. First, all the cells in the midgut are involved in pheromone production, since the

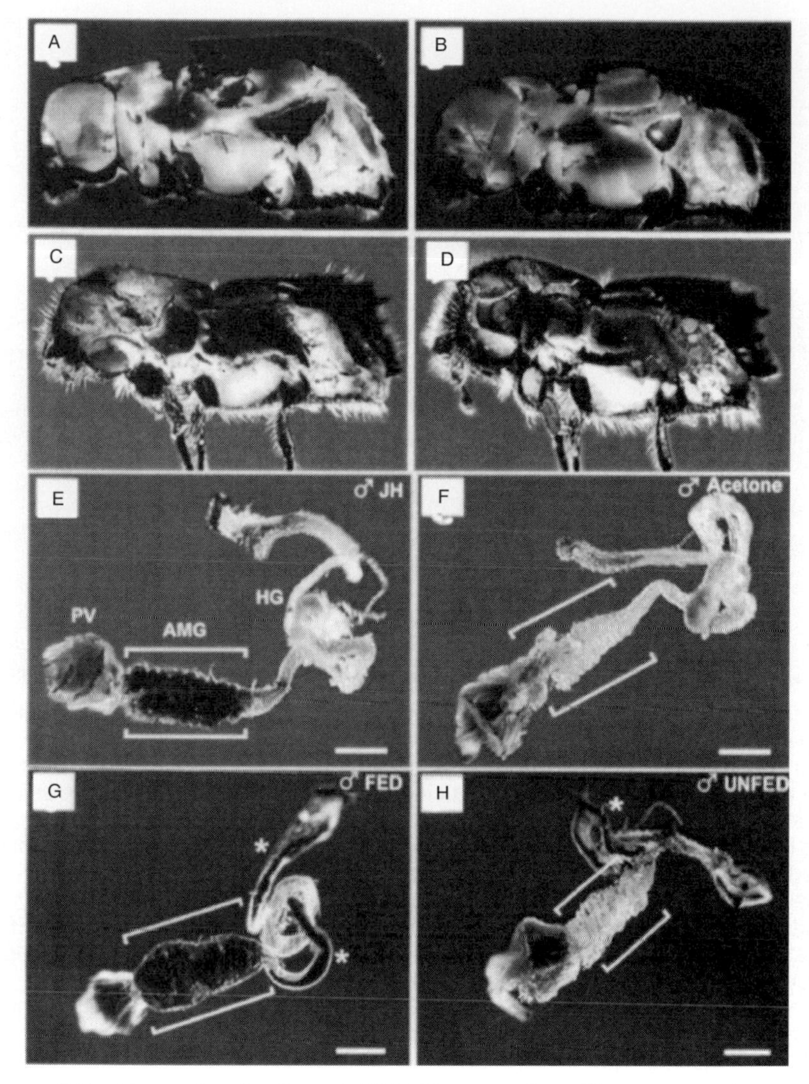

Figure 7.5 Tissue localization of *HMG-R* expression. "Exposed whole mounts" show that HMG-R mRNA is observed in the midgut of JH III-treated male *D. jeffreyi* (A) and *I. pini* (C), but not in untreated insects (B, D). Panels E through H show whole mount hybridizations of isolated *I. pini* alimentary canals. *HMG-R* expression in the anterior midgut (AMG, marked by brackets) correlates with pheromone production in starved, JH III-treated males (E) and fed males (G), while starved and untreated males (F, G), which do not produce monoterpenoid pheromone components, do not strongly express *HMG-R*. Asterisks mark non-specific signal in the hindguts. PV, proventriculus; HG, hindgut. Scale bars = 0.5 mm. Figure modified from Hall *et al.* (2002a, 2002b) with permission.

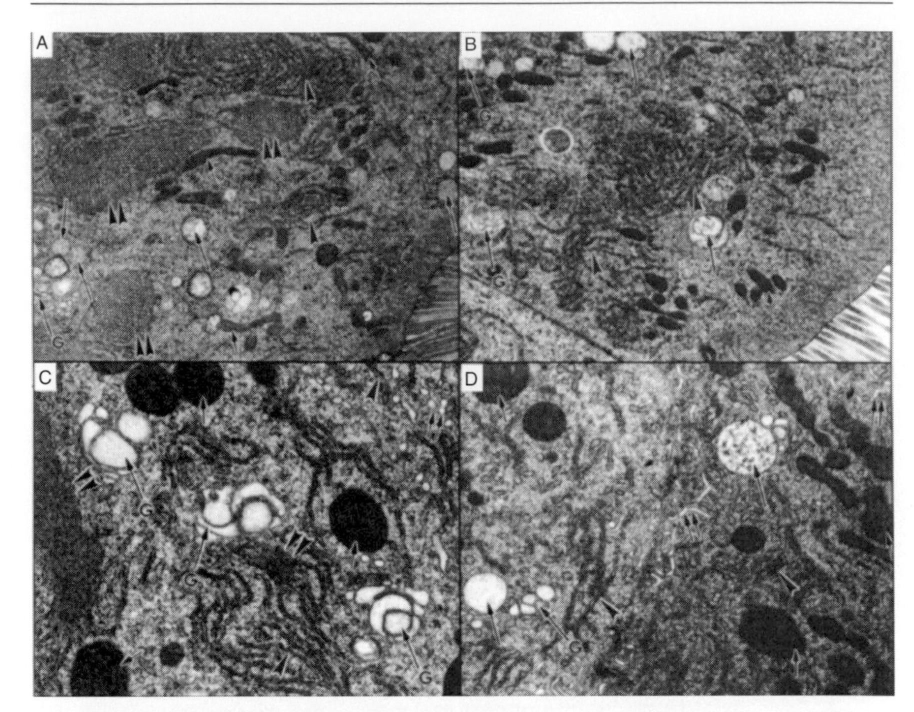

Figure 7.6 Electron microscopy of midgut cells. A Cells from a JH III-treated male *D. jeffreyi* produce frontalin, and contain large whorls and crystalline arrays of smooth endoplasmic reticulum (SER). B The abundant SER clusters are absent in pheromone-non-producing cells from a control (acetone-treated) male *D. jeffreyi.* Microvilli extend from the apical surface of the cells into the gut lumen and are visible in the lower right corners of A and B. C Pheromone-producing cells from a JH III-treated male *I. pini* also have highly ordered arrays of SER, while (D) cells from midguts of JH III-treated female *I. pini,* which do not synthesize pheromone components, do not. In all images, SER arrays are shown by double large arrowheads, single arrowheads mark rough endoplasmic reticulum, long arrows mark vesicles, long arrows with a "G" mark Golgi complex, short single arrows mark mitochondria, and double short arrows mark smooth, tubular endoplasmic reticulum. Figure adapted from Nardi *et al.* (2002) with permission.

signals from *in situ* hybridizations are evenly distributed through the tissue, and the SER arrays are observed in all midgut cells. Thus, there is no regional compartmentalization within the tissue: all cells in the midgut are actively involved in the dual roles of digestion and pheromone synthesis and release (Nardi *et al.*, 2002). Second, pheromone components are produced by insect tissues, and not by symbiotic bacteria. Although there is evidence that pheromone components may be synthesized by symbiotic microorganisms or associated fungi (e.g. Brand *et al.*, 1976; Byers and Wood, 1981), symbionts likely do not contribute to *de novo* pheromone biosynthesis. Prokaryotic HMG-Rs are both structurally and mechanistically different from their eukaryotic homologs (Taberno *et al.*, 1999; Friesen and Rodwell, 1997), so the activity, gene expression, and subcellular changes accompanying bark beetle pheromone production can only be attributed to the beetles' tissues (Hall *et al.*, 2002a, b; Nardi *et al.*, 2002).

7.6 Functional genomics

While classical molecular and biochemical techniques have provided useful insights into the regulation and production of bark beetle pheromone components, two important questions remain: What are the biosynthetic steps between geranyl diphosphate and the final pheromone component (either ipsdienol or frontalin)? And how is JH III acting on pheromone biosynthetic cells?

Answering these questions in bark beetles presents significant challenges. Some of the biochemical steps from geranyl diphosphate to ipsdienol are likely catalyzed by unique enzymes. Similarly, the mode of JH action remains one of entomology's longest-standing mysteries (reviewed in Gilbert *et al.*, 2000), with no clear indication of what kinds of sequences are involved in JH III signaling in bark beetles. The anonymity of relevant proteins involved in pheromone biosynthesis and endocrine regulation limits the effectiveness of identifying and isolating them by classical, homology-based (PCR or library screening) methods. Even if biochemical studies confirm the activities of desired enzymes, the beetles' small size makes enzyme purification – a necessary step to generate antibodies or sequence data useful for DNA isolations – a daunting task. Genetic and/or reverse genetic approaches can identify candidate proteins or genes if a model organism has appropriate physiological traits, but again, these approaches have very limited utility in bark beetles. Bark beetle genetics are very poorly understood (Lanier, 1966; Lanier and Oliver, 1966). And the virtuoso model insect, *D. melanogaster*, with all its enviable genetic tools, is not particularly useful either: fruitflies are not known to synthesize monoterpenoids, and the role of JH in *Drosophila* is complicated by the hormone's interactions with ecdysone (Feyereisen, 1998).

Fortunately, recently developed high-throughput sequencing and hybridization technologies, i.e. functional genomics, make any organism a potential model organism. The approach is the reverse to classical gene isolation and characterization, where candidate sequences are targeted, isolated, and carefully characterized – one or a few at a time. Instead, a genomics study begins with isolating hundreds or thousands of genes indiscriminately and then, through a series of steps including sequence analyses, microarray hybridization studies, and computer modeling, identifies a handful as potential candidates. The candidate genes can then be confirmed through classical functional assays and/or molecular

Petrov D. A. and Hartl D. L. (1998) High rate of DNA loss in the *Drosophila melanogaster* and *Drosophila virilis* species groups. *Mol. Biol. Evol.* **15**, 293–302.

Reid R. W. (1962) Biology of the mountain pine beetle, *Dendroctonus monticolae* Hopkins, in the East Kootney Region of British Columbia: II. Behaviour in the host, fecundity, and internal changes in the female. *Can. Entomol.* **94**, 605–613.

Renwick J. A. A., Hughes P. R. and Krull I. S. (1976) Selective production of *cis*- and *trans*-verbenol from (-)– and (+)-α-pinene by a bark beetle. *Science* **191**, 1735–1740.

Reynolds G. A., Basu S. K., Osborne T. F., Chin D. J., Gil G. and Brown M. S. *et al.* (1984) HMG-CoA reductase: a negatively regulated gene with unusual promoter and 5′ untranslated regions. *Cell* **38**, 275–285.

Robinson G. E. (2002) Honey bee brain EST project. Http://titan.biotec.uiuc.edu/bee/honeybee_project.htm.

Rohmer M. (1999) The discovery of a mevalonate-independent pathway for isoprenoid biosynthesis in bacteria, algae and higher plants. *Nat. Prod. Rep.* **16**, 565–574.

Rondot I., Quennedey B., Courrent A., Lemoine A. and Delachambre J. (1996) Cloning and sequencing of a cDNA encoding a larval–pupal-specific cuticular protein in *Tenebrio molitor* (Insecta, Coleoptera). Developmental expression and effect of a juvenile hormone analogue. *Eur. J. Biochem.* **235**, 138–143.

Rosenfeld J. M. and Osborne T. F. (1998) HLH106, a *Drosophila* sterol regulatory element-binding protein in a natural cholesterol auxotroph. *J. Biol. Chem.* **273**, 16112–16121.

Ruiz-Albert J., Cerdá-Olmedo E. and Corrochano L. M. (2002) Genes for mevalonate biosynthesis in *Phycomyces. Mol. Genet. Genomics* **266**, 768–777.

Sahota T. S. and Farris S. H. (1980) Inhibition of flight muscle degradation by precocene II in the spruce bark beetle, *Dendroctonus rufipennis* (Kirby) (Coleoptera: Scolytidae). *Can. J. Zool.* **58**, 378–381.

Seegmiller A. C, Dobrosotskaya I., Goldstein J. L., Ho Y. K., Brown M. S. and Rawson R. B. (2002) The SREBP pathway in *Drosophila*: regulation by palmitate, not sterols. *Dev. Cell* **2**, 229–238.

Sekler M. S. and Simoni R. D. (1995) Mutation in the lumenal part of the membrane domain of HMG-CoA reductase alters its regulated degradation. *Biochem. Biophys. Res. Com.* **206**, 186–193.

Sevala V. L. and Davey K. G. (1993) Juvenile hormone dependent phosphorylation of a 100 kDa polypeptide is mediated by protein kinase C in the follicle cells of *Rhodnius prolixus. Invert. Reprod. Dev.* **23**, 189–193.

Sevala V. L., Davey K. G. and Prestwich G. D. (1995) Photoaffinity labeling and characterization of a juvenile hormone binding protein in the membranes of follicle cells of *Locusta migratoria. Insect Biochem. Molec. Biol.* **25**, 267–273.

Seybold S. J., Ohtsuka T., Wood D. L. and Kubo I. (1995a) The enantiomeric composition of ipsdienol: a chemotaxinomic character for North American populations of *Ips* spp. in the *pini* subgeneric group (Coleoptera: Scolytidae). *J. Chem. Ecol.* **21**, 995–1016.

Seybold S. J., Quilici D. R., Tillman J. A., Vanderwel D. and Wood D. L. *et al.* (1995b) *De novo* biosynthesis of the aggregation pheromone components ipsenol and ipsdienol by the pine bark beetle, *Ips paraconfusus* Lanier and *Ips pini* (Say) (Coleoptera: Scolytidae). *Proc. Natl. Acad. Sci. USA* **92**, 8393–8397.

Seybold S. J., Bohlmann J. and Raffa K. F. (2000) The biosynthesis of coniferophagous bark beetle pheromones and conifer isoprenoids: evolutionary perspective and synthesis. *Can. Entomol.* **132**, 697–753.

Shah J. and Bailey E. (1976) Enzymes of ketogenesis in the fat body and the thoracic muscle of the adult cockroach. *Insect Biochem.* **6**, 251–254.

Silberkang M., Havel C. M., Friend D. S., McCarthy B. J. and Watson J. A. (1983)

Isoprene synthesis in isolated embryonic *Drosophila* cells: sterol independent eukaryotic cells. *J. Biol. Chem.* **258**, 8503–8511.

Silverstein R. M., Rodin J. O. and Wood D. L. (1966) Sex attractants in frass produced by male *Ips confusus* in ponderosa pine. *Science* **154**, 509–510.

Tabernero L., Bochar D. A., Rodwell V. W. and Stauffacher C. V. (1999) Substrate-induced closure of the flap domain in the ternary complex structures provides insights into the mechanism of catalysis by 3-hydroxy-3-methylglutaryl-CoA reductase. *Proc. Natl. Acad. Sci. USA* **96**, 7167–7171.

Teal S. A., Webster F. X., Zhang A. and Lanier G. N. (1991) Lanierone: a new pheromone component from *Ips pini* (Coleoptera: Scolytidae) in New York. *J. Chem. Ecol.* **17**, 1159–1176.

Tillman J. A., Holbrook G. L., Dallara P. L., Schal C. and Wood D. L. *et al.* (1998) Endocrine regulation of *de novo* aggregation pheromone biosynthesis in the pine engraver, *Ips pini* (Say) (Coleoptera: Scolytidae). *Insect Biochem. Molec. Biol.* **28**, 705–715.

Tillman J. A., Seybold S. J., Jurenka R. A. and Blomquist G. J. (1999) Insect pheromones – an overview of biosynthesis and regulation. *Insect Biochem. Molec. Biol.* **29**: 481–514.

Tittiger C., Blomquist G. J., Ivarsson P., Borgeson C. E. and Seybold S. J. (1999) Juvenile hormone regulation of HMG-R gene expression in the bark beetle *Ips paraconfusus* (Coleoptera: Scolytidae): implications for aggregation pheromone biosynthesis. *Cell. Molec. Life Sci.* **55**, 121–127.

Tittiger C., O'Keeffe C., Bengoa C. S., Barkawi L. S., Seybold S. J. *et al.* 2000. Isolation and endocrine regulation of an HMG-CoA synthase cDNA from the male Jeffrey pine beetle, *Dendroctonus jeffreyi* (Coleoptera: Scolytidae). *Insect Biochem. Molec. Biol.* **30**, 1203–1211.

Tittiger C., Barkawi L. S., Bengoa C. S., Blomquist G. J. and Seybold S. J. (2003) Structure and juvenile hormone-mediated regulation of the HMG-CoA reductase gene from the Jeffreyi pine beetle, *Dendroctonus jeffreyi*. *Mol. Cell. Endocrin.* (in press).

Troutt A. B., McHeyzer-Williams M. G., Pulendran B. and Nossal G. J. (1992) Ligation-anchored PCR: a simple amplification technique with single-sided specificity. *Proc. Natl. Acad. Sci. USA* **89**, 9823–9825.

Vanderwel D. and Oehlschlager A. C. (1987) Biosynthesis of pheromones and endocrine regulation of pheromone production in Coleoptera. In *Pheromone Biochemistry*, eds G. D. Prestwich and G. J. Blomquist, pp. 175–215. Academic Press, Orlando, FL.

Vité, J. P., Bakke A. and Renwick J. A. A. (1972) Pheromones in *Ips* (Coleoptera: Scolytidae): occurrence and production. *Can. Ent.* **104**, 1967–1975.

Wiygul G. and Sikorowski P. P. (1985) The effect of glucose and ATP on sex pheromone production in fat bodies from male boll weevils *Anthonomus grandis* Boheman (Coleoptera: Curculionidae). *Comp. Biochem. Physiol.* **81B**, 1073–1075.

Wiygul G., MacGown M. W., Sikorowski P. P. and Wright J. E. (1982) Localization of pheromone in male boll weevils, *Anthonomus grandis*. *Ent. Exp. & Appl.* **31**, 330–331.

Wiygul, G., Dickens J. C. and Smith J. W. (1990) Effect of juvenile hormone and beta-bisabolol on pheromone production in fat bodies of male boll weevils, *Anthonomus grandis* Boheman (Coleoptera: Curculionidae). *Comp. Biochem. Physiol.* **95B**, 489–491.

Wood D. L. (1982) The role of pheromones, kairomones, and allomones in the host selection and colonization behavior of bark beetles. *Ann. Rev. Entomol.* **27**, 411–446.

Wood D. L. and Bushing R. W. (1963) The olfactory response of *Ips confusus* (LeConte)

(Coleoptera: Scolytidae) to the secondary attraction in the laboratory. *Can. Entomol.* **95**, 1066–1078.

Wood D. L., Stark R. W., Silverstein R. M. and Rodin J. O. (1967) Unique synergistic effects produced by the principal sex attractant compounds of *Ips confusus* (LeConte) (Coleoptera: Scolytidae). *Nature* **215**, 206.

Wood D. L., Browne L. E., Bedard W. D., Tilden P. E. and Silverstein R. M. *et al.* (1968) Response of *Ips confusus* to synthetic sex pheromones in nature. *Science* **159**, 1373–1374.

Wright R. and Rine J. (1989) Transmission electron microscopy and immunocytochemical studies of yeast: analysis of HMG-CoA reductase overproduction by electron microscopy. *Met. Cell. Biol.* **31**, 473–512.

Yamamoto K., Chadarevian A. and Pelligrini M. (1988) Juvenile hormone action mediated in male accessory glands of *Drosophila* by calcium and kinase C. *Science* **239**, 916–919.

Ye J., Rawson R. B., Kumoro R., Chen X. and Dave U. P. *et al.* (2000) ER stress induces cleavage of membrane-bound ATF6 by the same proteases that process SREBPs. *Mol. Cell.* **6**, 1355–1364.

Zhang J., Saleh D. S. and Wyatt G. R. (1996) Juvenile hormone regulation of an insect gene: a specific transcription factor and a DNA response element. *Mol. Cel. Endocrin.* **122**, 15–20.

Zhou S., Zhang J., Hirai M., Chinzei Y. and Kayser H. *et al.* (2002) A locust DNA-binding protein involved in gene regulation by juvenile hormone. *Mol. Cel. Endocrin.* **190**, 177–185.

8

Biosynthesis and ecdysteroid regulation of housefly sex pheromone production

Gary J. Blomquist

8.1 Introduction

Many dipterans use long-chain hydrocarbons as sex pheromones (Table 8.1) (Blomquist *et al.*, 1987, 1993). The hydrocarbon-based dipteran pheromone compounds are present on the cuticle and are structurally related to components in the epicuticular lipids. Thus, these pheromone components are biosynthesized through modifications of the pathways that produce cuticular lipids (Nelson and Blomquist, 1995). Hydrocarbon pheromone biosynthesis has been studied in *Drosophila melanogaster* (Wicker and Jallon, 1995; Pennanec'h *et al.*, 1997; Wicker-Thomas and Jallon, Chapter 9, in this volume), the tsetse fly, *Glossina morsitans morsitans* (Carlson *et al.*, 1978; Langley and Carlson, 1983), and extensively in the common housefly, *Musca domestica*. In *M. domestica*, an ovarian-produced ecdysteroid alters the cuticular lipid composition of maturing females such that hydrocarbon-related pheromone components are produced (Adams *et al.*, 1984a; Blomquist *et al.*, 1987; Tillman *et al.*, 1999).

8.1.1 Insect hydrocarbons and chemical communication

It has become widely recognized over the past several decades that cuticular lipids, especially the hydrocarbons, function in chemical communication in many insect species (Howard, 1993; Nelson and Blomquist, 1995; Blomquist *et al.*, 1998). Semiochemical functions attributed to hydrocarbons include sex attractants

Table 8.1 Cuticular lipids which function in chemical communication in Diptera

Species	Function	Chemicals involved	Reference
Drosophila melanogaster	sex pheromone	(*Z,Z*)-7,11 heptacosadiene	Antony and Jallon (1982)
D. melanogaster Canton-S	homosexual courtship stimulation	(*Z*)-11- and (Z)-13-tritriacontenes	Schaner *et al.* (1989)
D. melanogaster Canton-S	antiaphrodisiac	7,11-heptacosadiene	Scott *et al.* (1988)
D. melanogaster Canton-S	antiaphrodisiac	7-tricosene	Scott (1986)
D. melanogaster Canton-S	antiaphrodisiac	7-pentacosene	Scott and Jackson (1988)
D. americana americana	male aggregation	(*Z*)-9-heneicosene	Bartelt *et al.* (1986)
D. american taxana	male aggregation	(*Z*)-9-heneicosene	Bartelt *et al.* (1986)
D. pallidosa	sex pheromone	(*Z,Z*)-5,7-tritriacontadiene	Nemoto *et al.* (1994)
D. novamexican	male aggregation	(*Z*)-9-heneicosene	Bartelt *et al.* (1986)
D. simulans	sex pheromone	7-tricosene	Jallon (1984)
D. virillis	male aggregation	(*Z*)-10-heneicosene	Bartelt and Jackson (1984)
Fannia canicularis	sex pheromone	(*Z*)-9-pentacosene	Uebel *et al.* (1977)
F. femorallis	sex pheromone	(*Z*)-11-hentriacontene	Uebel *et al.* (1978b)
F. pusio	sex pheromone	(*Z*)-11-hentriacontene	Uebel *et al.* (1978a)
Glossina auteni	contact sex pheromone	15,19-dimethyltritriacontane	Huyton and Langley (1982)
G. morsitans	contact sex pheromone	dimethyl-C37 alkane 15,19,23-trimethylhept triacontane	Carlson *et al.* (1978)
G. morsitans	male antiaphrodisiac	19,23-dimethyltritria cont-1-ene	Carlson and Schlein (1991)
G. pallidepes	contact sex pheromone	13,23-dimethylpenta-triacontane	Carlson *et al.* (1984)
G. tachinoides	contact sex stimulant	11,23-, 13,25-dimethylhepta-triacontane	Carlson *et al.* (1998)

Lycoriella mali	contact sex pheromone	sex pheromone	Kostelc *et al.* (1975)
Lixophage diatraeae	larva position	*n*-alkanes	Thompson *et al.* (1983)
Microdon piperi	chemical mimicry of *Camponotus modoc*	mono- and dimethyl-(*Z*)-4-enes	Howard *et al.* (1990)
Musca autumnalis	sex pheromone	(*Z*)-13- and (*Z*)-14-nonacosene and (Z)-13-heptacosene	Uebel *et al.* (1975b)
M. domestica	sex pheromone	(*Z*)-9-tricosene	Carlson *et al.* (1971)
	arrestant	methylalkanes	Uebel *et al.* (1976)
	sex recognition	C23 epoxide and ketone	Uebel *et al.* (1978a)
			Adams and Holt (1987)
Phormia regina	species recognition	hydrocarbon profile	Stoffolano *et al.* (1997)

enzyme. The results of these studies demonstrated that only minimal change at the 14-position of (*Z*)-9-tricosene is allowed with retention of metabolic activity (Guo *et al.*, 1991).

8.3 Enzymes involved in pheromone biosynthesis

8.3.1 Fatty acid synthase

The soluble and a unique microsomal fatty acid synthase (FAS) which are involved in producing the 18-carbon fatty acyl precursors to hydrocarbons have been purified to homogeneity and characterized (Gu *et al.*, 1997). It appears that the soluble FAS synthesizes the straight chain fatty acids involved in *n*-alkane and alkene formation, whereas the microsomal FAS produces the precursors for the methyl-branched hydrocarbons (Blomquist *et al.* 1995).

8.3.2 Δ9 Desaturase

In the housefly, the ubiquitous Δ9 desaturase is also an important component of pheromone biosynthesis. It has been characterized biochemically (Wang *et al.*, 1982). The observation that [^{13}C]acetate labeled (Z)-9-tricosene equally at both ends of the chain (Dillwith *et al.*, 1982) indicates that 18:0-CoA is synthesized, desaturated and then directly elongated and converted to (Z)-9-tricosene. This is in contrast to diene production in the American cockroach, where the 18:2 precursor is synthesized and stored prior to being elongated and converted to hydrocarbon (Dwyer *et al.*, 1981). The Δ9-desaturase was recently cloned and sequenced in our laboratory (Eigenheer *et al.*, 2002). The complete Δ9 desaturase coding region (1140 bp) was obtained. An amino acid alignment of the deduced housefly Δ9 desaturase amino acid sequence showed an 82.4 percent identity with the amino acid sequence of a Δ9 desaturase from *D. melanogaster* (Wicker-Thomas *et al.*, 1997).

8.3.3 Fatty acyl-CoA elongase

The 18:1-CoA is elongated to fatty acyl-CoAs of 24 to 30+ carbons (Tillman-Wall *et al.*, 1992; Blomquist *et al.*, 1995). Fatty acyl-CoA elongation involves four steps: (1) condensation of a fatty acyl-CoA with malonyl-CoA to form a β-keto fatty acyl-CoA, (2) reduction of the ketone group to an alcohol using NADPH, (3) dehydration and (4) reduction of the unsaturated intermediate with another NADPH (Figure 8.3). In the American cockroach, which produces a C25 alkane and a C27 diene, microsomal preparations elongated 18:0-CoA to C26, and 18:2-CoA to C28:2 (Vaz *et al.*, 1988), suggesting that the chain length of the hydrocarbon products is regulated by the elongation system. The biochemistry of the elongation system has been studied in plants (Wettstein-Knowles, 1995) and studies at the gene level have been performed in both plants (Millar and

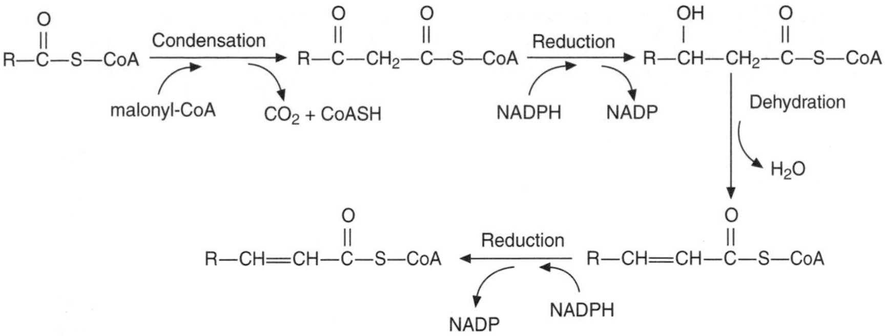

R–C(=O)–S–CoA —Condensation (malonyl-CoA → CO_2 + CoASH)→ R–C(=O)–CH_2–C(=O)–S–CoA —Reduction (NADPH → NADP)→ R–CH(OH)–CH_2–C(=O)–S–CoA

—Dehydration (H_2O)→ R–CH=CH–C(=O)–S–CoA —Reduction (NADPH → NADP)→ R–CH=CH–C(=O)–S–CoA

Figure 8.3 Steps in fatty acyl-CoA elongation: condensation, reduction, dehydration and reduction. The condensation step appears to regulate chain length specificity.

Kunst, 1997) and yeast (Oh *et al.*, 1997). The role of fatty acyl-CoA elongation in regulating the chain length of hydrocarbons in insect pheromone biosynthesis is discussed below (8.4.6).

8.3.4 Decarboxylase: evidence that a cytochrome P450 enzyme is involved

The mechanism of hydrocarbon formation has proven elusive. In an elegant set of experiments in the 1960s and early 1970s, Kolattukudy and coworkers demonstrated that fatty acyl-CoAs were elongated and then converted to hydrocarbon by the loss of the carboxyl group (reviewed in Kolattukudy *et al.*, 1976; Kolattukudy, 1980). In the 1980s and early 1990s, the hypothesis was put forward that the very long chain fatty acyl-CoA was reduced to the aldehyde and then decarbonylated to the hydrocarbon and carbon monoxide, and that this reaction did not require any cofactors or O_2. Evidence for this decarbonylation mechanism was obtained from a plant (Cheesbrough and Kolattukudy, 1984), an algae (Dennis and Kolattukudy, 1991, 1992), a vertebrate (Cheesbrough and Kolattukudy, 1988) and an insect, *Sarcophaga crassipalpas* (Yoder *et al.*, 1992).

In our work on hydrocarbon formation in the housefly, we found that acyl-CoA is reduced to the aldehyde, with the conversion of the aldehyde to hydrocarbon requiring NADPH and molecular oxygen, and that the products were hydrocarbon and carbon dioxide (Figure 8.4) (Reed *et al.*, 1994, 1995). The production of carbon dioxide and not carbon monoxide was determined by radio-GLC (Figure 8.4a, insert). Antibodies to housefly cytochrome P450 and to P450 reductase inhibited hydrocarbon formation in microsomes, as did exposure to CO, and the latter could be partially reversed by white light. GC-MS analyses of specifically deuterated substrates showed that the protons on positions 2,2 and 3,3 of the acyl-CoA were retained during conversion to hydrocarbon, and that the proton

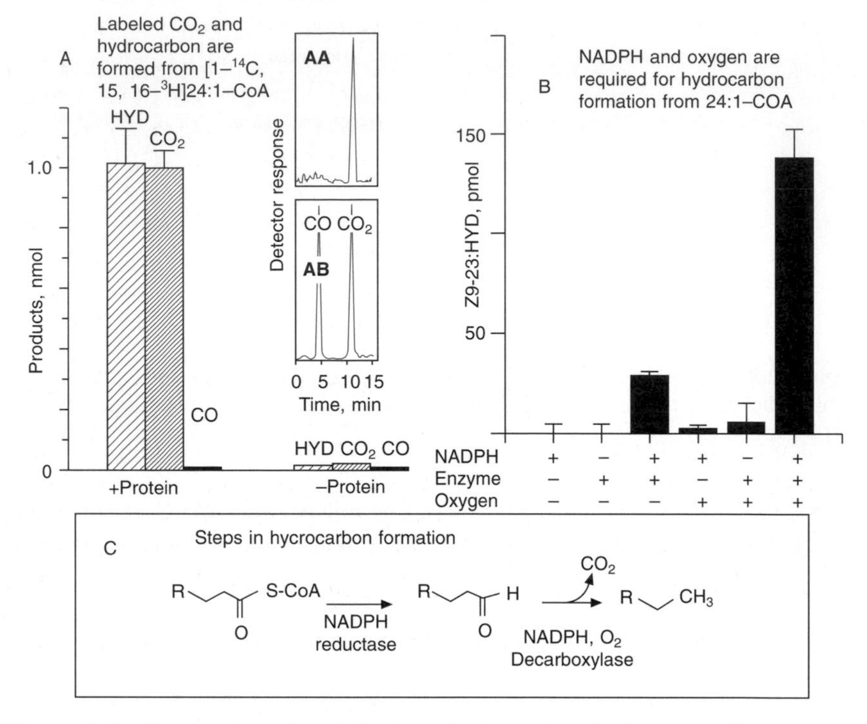

Figure 8.4 Conversion of acyl-CoAs to hydrocarbon in the housefly. Appoximately equal amounts of labeled hydrocarbon and carbon dioxide are produced from the microsomal conversion of [1-^{14}C, 15,16-^{3}H]24:1-CoA to products (A). The inset shows the retention times of labeled CO and CO_2 (A, B) and product from [1-^{14}C]24:1-CoA (AA). NADPH and oxygen are required for microsomal conversion of 24:-CoA to hydrocarbon (B) and (C) shows the steps involved in hydrocarbon formation (data from Reed *et al.*, 1994).

on position 1 of the aldehyde was transferred to the adjacent carbon and retained during hydrocarbon formation. Furthermore, several peroxides could substitute for O_2 and NADPH and support hydrocarbon production. All this evidence strongly supports a cytochrome P450 involvement in hydrocarbon synthesis (Reed *et al.*, 1994).

A mechanism was proposed in which the perferryl iron-oxeme, resulting from heterolytic cleavage of the O–O bond of the iron-peroxy intermediate, abstracts an electron from the C=O double bond of the carbonyl group of the aldehyde. The reduced perferryl attacks the 1-carbon of the aldehyde to form a thiyl-iron-hemiacetal diradical. The latter intermediate can fragment to form an alkyl radical and thiyl-iron-formyl radical. The alkyl radical then abstracts the formyl hydrogen to produce the hydrocarbon and CO_2 (Reed *et al.*, 1995).

We further examined this process in *S. crassipalpis* and five other insect species. In our laboratory, microsomes from all six species required NADPH and O_2 for hydrocarbon production from aldehydes and produced hydrocarbon and CO_2, but not CO (Mpuru *et al.*, 1996). Thus, the mechanism of hydrocarbon formation remains controversial, with evidence obtained favoring both a decarbonylation and a decarboxylation mechanism. The resolution of the problem awaits cloning, expressing and assaying the enzymes involved.

8.3.5 Methyl-branched alkane biosynthesis

Methyl-branched alkanes function as an arrestant in the housefly pheromone (Adams and Holt, 1987; Nelson *et al.*, 1981). They are abundant components of the insect cuticular lipids (Nelson and Blomquist, 1995) where they function in waterproofing (Gibbs, 1998) and serve in chemical communication (Blomquist *et al.*, 1993, 1998).

Methylalkanes are formed by the substitution of methylmalonyl-CoA in place of malonyl-CoA at specific points during chain elongation. Carbon-13 NMR, mass spectrometry and radiochemical studies (Dillwith *et al.*, 1982; Chase *et al.*, 1990) demonstrated that the methylmalonyl-CoA was added during the initial steps of chain elongation in insects by what appears to be a novel microsomal fatty acid synthetase (FAS). A microsomal FAS was first suggested from studies in *Trichoplusia ni* for the formation of methyl-branched very long chain alcohols (de Renobales *et al.*, 1989). In this insect, high rates of methyl-branched very long chain alcohol synthesis were observed in the mid-pupal stages at times when soluble FAS activity was very low or undetectable.

8.3.6 Evidence for a microsomal FAS in methyl-branched hydrocarbon formation

A microsomal FAS was implicated in the biosynthesis of methyl-branched fatty acids and methyl-branched hydrocarbon precursors of the German cockroach contact sex pheromone (Juarez *et al.*, 1992; Gu *et al.*, 1993). A microsomal FAS present in the epidermal tissues of the housefly is responsible for methyl-branched fatty acid production (Blomquist *et al.*, 1994). The housefly microsomal and soluble FASs were purified to homogeneity (Gu *et al.*, 1997) and the microsomal FAS was shown to preferentially use methylmalonyl-CoA in comparison to the soluble FAS. GC-MS analyses showed that the methyl-branching positions of the methyl-branched fatty acids of the housefly were in positions consistent with their role as precursors of the methyl-branched hydrocarbons.

The methylmalonyl-CoA unit, which is the precursor to methyl-branched fatty acids and hydrocarbons, arises from the carbon skeletons of valine and isoleucine, but not succinate (Dillwith *et al.*, 1982). Propionate is also a precursor to methylmalonyl-CoA, and in the course of these studies, a novel pathway for propionate metabolism in insects was discovered. Many insect species, including the housefly, do not contain vitamin B_{12} (Wakayama *et al.*, 1984), and therefore cannot catabolize propionate via methylmalonyl-CoA to succinate. Instead, as first demonstrated in the housefly (Dillwith *et al.*, 1982), insects metabolize propionate to 3-hydroxypropionate and then to acetyl-CoA, with carbons 3 and 2 of propionate becoming carbons 1 and 2 of acetyl-CoA (Halarnkar *et al.*, 1986).

8.4 Endocrine regulation of hydrocarbon pheromone synthesis

8.4.1 Hormonal regulation of pheromone production in insects

The observation that females of certain species have repeated reproductive cycles and that mating occurs only during defined periods of each cycle led to the proposal that pheromone production might be under hormonal control. Studies on a limited number of species demonstrate that pheromone production can be under the regulation of products of the brain, ovary and corpora allata. At least three distinct hormones have been shown to regulate pheromone production in insects (Tillman *et al.*, 1999). In female moths (Lepidoptera) pheromone biosynthesis is often regulated by a 33 or 34 amino acid *p*heromone *b*iosynthesis *a*ctivating *n*europeptide (PBAN) (Raina *et al.*, 1989; Raina, 1993; Jurenka and Rafaeli, Chapter 5, in this volume). PBAN alters enzyme activity at one or more steps during or subsequent to fatty acid synthesis during pheromone production. In some species of Coleoptera (Vanderwel, 1994; Tillman *et al.*, 1998, 1999; Seybold and Vanderwel, Chapter 6, in this volume and Blattodea (Chase *et al.*, 1992; Schal *et al.*, 1997; Schal, Chapter 10, in this volume), juvenile hormone (JH) induces pheromone production. In the female housefly, ovarian-produced ecdysteroids regulate the chain length of cuticular alkenes such that the pheromone component (*Z*)-9-tricosene (muscalure) becomes a major product (described below).

PBAN works through membrane receptors and second messengers to affect the activity of key enzymes in pheromone biosynthesis, with acetyl-CoA carboxylase, fatty acyl-CoA reductase and the delta-11 desaturase potential-regulated enzymes. In contrast, JH and ecdysone exert their action by increasing or decreasing gene transcription. For example, in bark beetles, JH up-regulates 3-hydroxy-3-methylglutaryl-CoA reductase (HMG-R) (Tittiger *et al.*, 1999; Hall *et al.*, 2002) and to a lesser extent HMG-CoA synthase (Tittiger *et al.*, 2000) and geranyl diphosphate synthase (Young *et al.*, 2001) (see Tittiger, Chapter 7, in this volume), allowing an increase in isoprenoid pheromone component synthesis. Evidence, detailed below, points to ecdysteroids regulating pheromone production in the housefly by repressing the synthesis of specific fatty acyl-CoA elongases.

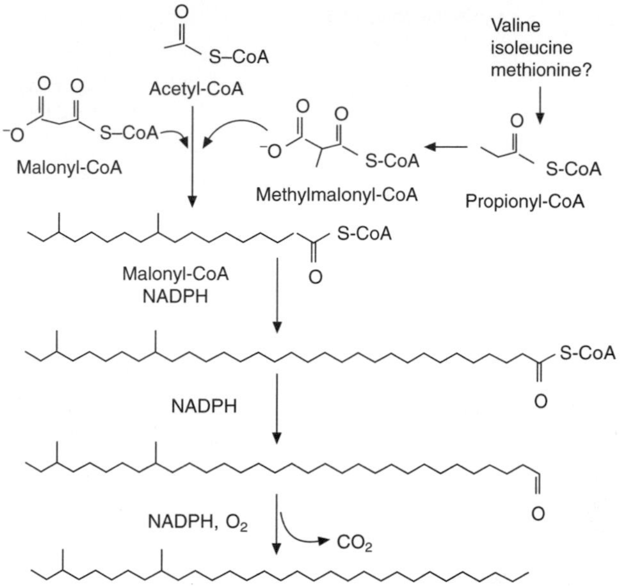

Figure 8.5 Steps in the biosynthesis of methyl-branched hydrocarbons in the housefly.

8.4.2 20-HE induces sex pheromone production in ovariectomized females

Newly emerged female houseflies do not have detectable amounts of the C_{23} sex pheromone components. Sex pheromone production correlates with ovarian development and vitellogenesis. Thus, the C_{23} sex pheromone components first appear when ovaries mature to the early vitellogenic stages (Figure 8.6a, b) and increase in amount until stages 9 and 10 (mature egg) (Dillwith *et al.*, 1983; Mpuru *et al.*, 2001). Females ovariectomized within 6 h of adult emergence did not produce the C_{23} sex pheromone components (Figure 8.6e), whereas control (Figure 8.6c) and allatectomized (Figure 8.6d) females produced abundant amounts of (Z)-9-tricosene (Dillwith *et al.*, 1983). Ovariectomized insects that received ovary implants produced sex pheromone components in direct proportion to ovarian maturation. These data demonstrate that a hormone from the maturing ovary induced sex pheromone production.

Juvenile hormone (JH) regulates both vitellogenesis and pheromone production in some insect species (Tillman *et al.*, 1999). In some Diptera, including the housefly, ovarian-produced ecdysteroids are involved in regulating vitellogenesis (Hagedorn, 1985; Adams *et al.*, 1997) at the transcriptional level (Martin *et al.*, 2001). Because ovariectomy abolished sex pheromone production while alletectomy (which abolishes JH production) had no effect on pheromone production (Blomquist *et al.*, 1992), it was therefore hypothesized that an ecdysteroid, and not JH,

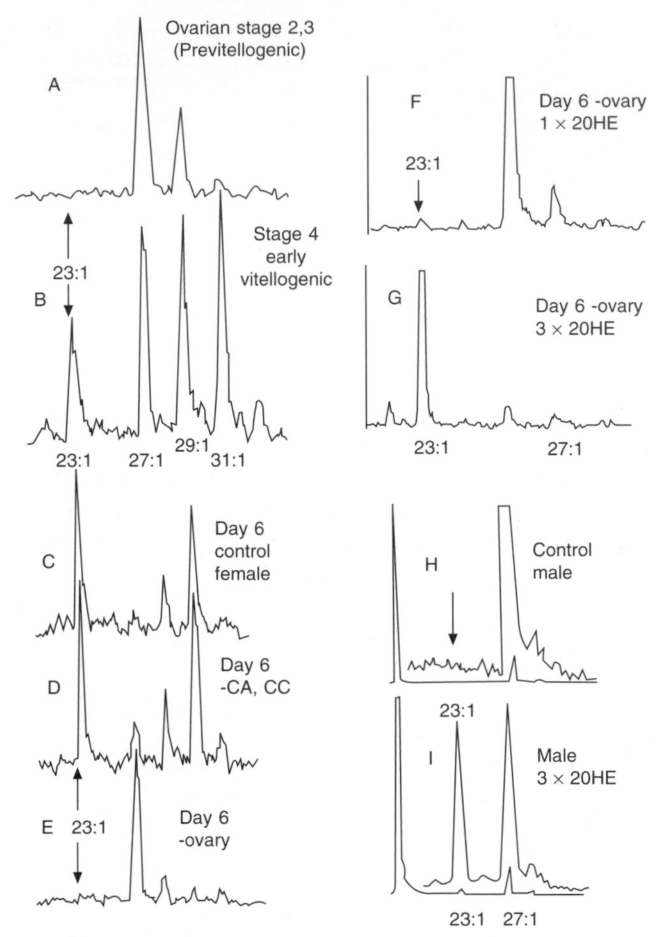

Figure 8.6 Effect of development, surgery and hormone treatments on (Z)-9-tricosene production in male and female houseflies. Radio-GLCs of *in vivo* incorporation of labeled acetate into the alkenes of A ovarian stage 2,3 females, B stage 4 females, C day-6 control female, D day-6 female which had the CA, CC removed within 6 h of emergence, E day-6 female which was ovariectomized within 6 h of emergence, F day-6 female ovariectomized at emergence and treated with one dose of 5 μg 20-HE, G day-6 female ovariectomized at emergence and treated with three doses of 5 μg 20-HE at 24 h intervals, H control male and I males treated with three doses of 5 μg 20-HE at 24 h intervals.

regulated sex pheromone production in the housefly. Injection of 20-hydroxyecdysone (20-HE) at doses as low as 0.5 ng every 6 h induced sex pheromone production in ovariectomized houseflies in a time- and dose-dependent manner (Figure 8.6f, g) (Adams *et al.*, 1984b, 1995). Multiple injections of 20-

HE into ovariectomized insects over several days resulted in as much (Z)-9-tricosene produced as in intact control females. Application of JH or JH analogs alone or in combination with ecdysteroids had no effect on pheromone production, confirming that JH does not have a role in regulating housefly pheromone production (Blomquist *et al.*, 1992).

8.4.3 Induction of female sex pheromone production in male houseflies

Male houseflies normally produce no detectable C_{23} sex pheromone components, but they do produce the same C_{27} and longer alkenes as previtellogenic females (Figure 8.6h). Implantation of ovaries into male houseflies resulted in a change in the chain length specificity of the alkenes such that (Z)-9-tricosene became a major component (Blomquist *et al.*, 1984b, 1987). Likewise, injection of 20-HE into males induces sex pheromone production in a dose-dependent manner (Figure 8.6I). Thus, males possess the biosynthetic capability to produce sex pheromone, but normally do not produce the 20-HE necessary to induce sex pheromone production. This makes male houseflies a very convenient model in which to study the regulation of sex pheromone production, circumventing the need to ovariectomize large numbers of female insects.

8.4.4 What enzymes are affected by 20-HE to induce pheromone production?

The two most likely possibilities to account for the change in the chain length of the alkenes synthesized by the female housefly during production of (Z)-9-tricosene are (1) the chain length specificity of the reductive conversion of acyl-CoAs to alkenes is altered such that 24:1-CoA becomes an efficient substrate or (2) there is a change in the chain length specificity of the fatty acyl-CoA elongation enzymes such that 24:1-CoA is not efficiently elongated, resulting in an accumulation of 24:1-CoA.

8.4.5 Does the specificity of the enzyme that converts 24:1-CoA to hydrocarbon regulate chain lengths of hydrocarbon?

To determine which enzyme activities are affected by 20-HE to regulate the chain length of the alkenes, experiments were performed to examine the chain length specificity of the fatty acyl-CoA reductive conversion of acyl CoAs to alkenes. Day 4 females make abundant amounts of 23:1 hydrocarbon pheromone, whereas day 4 males only make alkenes of 27:1 and longer. Thus, we were quite surprised that microsomal preparations from both males and females of all ages readily converted 24:1-CoA (Figure 8.7) and the 24:1 aldehyde (data not shown) (Reed *et al.*, 1996) to (Z)-9-tricosene. This indicates that 20-HE did not influence this activity (Figure 8.7) (Tillman-Wall *et al.*, 1992; Reed *et al.*, 1995). It also shows that the reductase and decarboxylase have minimal or no chain length preference and are similar in both preparations. Indeed, Reed *et al.* (1996) showed

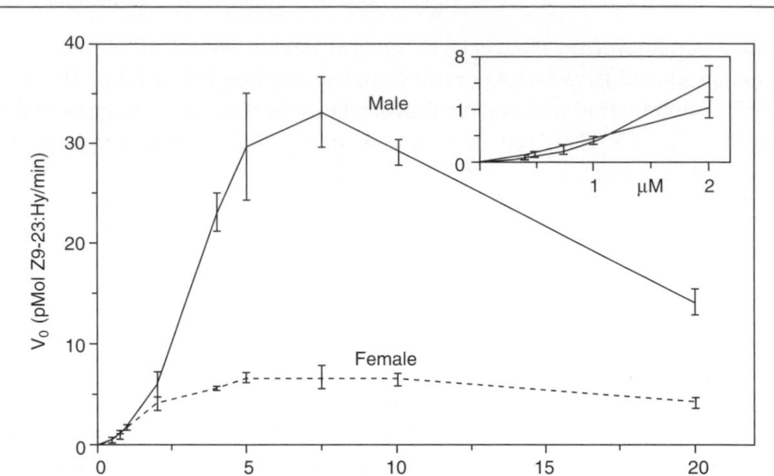

Figure 8.7 Effect of [24:1-CoA] on hydrocarbon production by microsomes prepared from day-4 male and female houseflies.

that both males and females of all ages converted 18-,24- and 28-carbon alkenes to hydrocarbon.

8.4.6 Evidence that the acyl-CoA elongases regulate hydrocarbon chain length and pheromone production

In contrast, microsomes from day-4 females (high ecdysteroid titer and production of (Z)-9-tricosene) did not efficiently elongate either 18:1-CoA or 24:1-CoA beyond 24 carbons, while microsomes from day-4 males or day-1 females (both of which produce alkenes of 27:1 and longer) readily elongated both 18:1-CoA and 24:1-CoA to 28:1-CoA (Figure 8.8A, B) (Tillman-Wall *et al.*, 1992; Blomquist *et al.*, 1995). We envision a process where if the C24:1 to C28:1 elongase condensation activity is present, essentially all of the 24:1-CoA formed is elongated to 26- and 28-carbon unsaturated acyl-CoAs, whereas in the absence of the C24:1 to C28:1 elongase condensation activity, the amount of 24:1-CoA increases and is then converted to 23:1 hydrocarbon. Thus, 20-HE appears to regulate the fatty acyl-CoA elongases and not the enzymatic steps in the conversion of acyl-CoA to hydrocarbon.

8.4.7 Condensation step in fatty acyl-CoA elongation regulates fatty acid chain length

It has been assumed for many years that the first step in the elongase cycle, the condensation step, regulates the chain length of the final product. Direct evidence for this has been obtained in *Arabidopsis*, where the *fatty acid elongation1* (*FAE1*) gene encodes for the condensing enzyme (Millar and Kunst, 1997). The

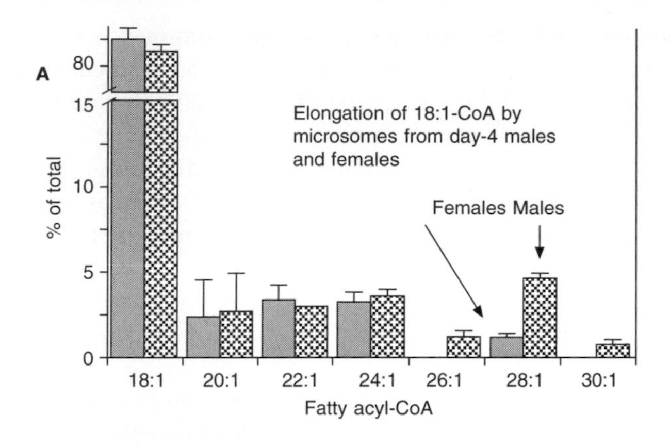

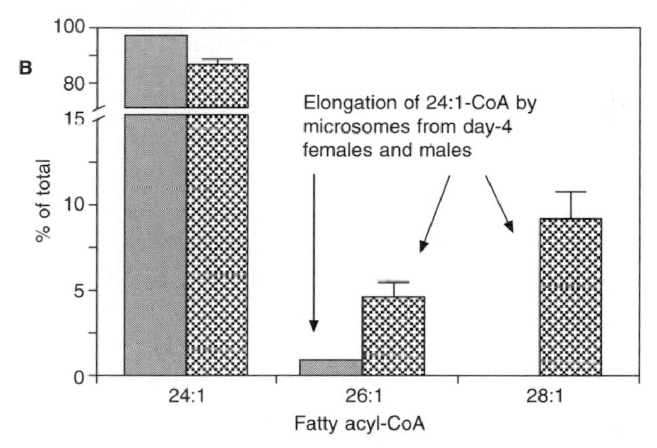

Figure 8.8 Elongation of 18:1-CoA (A) and 24:1-CoA (B) by microsomes prepared from day-4 males and females.

introduction of *FAE1* into yeast and tissues in *Arabidopsis*, where significant amounts of very long chain fatty acids are normally absent, is sufficient for the production of very long chain fatty acids. Likewise, *ELO2* and *ELO3* were isolated from the yeast *Saccharomyces cerevisiae* and were shown to code for the condensation activity in the elongation of 18-carbon fatty acids to 24 and 26 carbons (Oh *et al.*, 1997). Indeed, these sequences were used in BLAST searches to identify at least seven putative elongases from *Drosophila melanogaster* (Wicker-Thomas, personal communication) which facilitated designing primers to isolate the housefly elongases by PCR. We are currently isolating the cDNAs for the housefly elongases and will determine expression patterns and chain length specificities for each of the condensing activities to determine if indeed the repression of an elongase is responsible for pheromone production in the housefly.

8.4.8 Transport of hydrocarbons and sex pheromones

The role of hemolymph in transporting hydrocarbons and hydrocarbon pheromones has only recently become fully appreciated. Older models of hydrocarbon formation showed epidermal-related cells (oenocytes) synthesizing and transporting hydrocarbons directly to the surface of the insect (Hadley, 1984). In the housefly, the role of hemolymph is most clearly seen when (Z)-9-tricosene production is initiated. (Z)-9-Tricosene first accumulates in the hemolymph, and then after a number of hours, is observed on the surface of the insect. Modeling of the process (Mpuru *et al.*, 2001) showed that the delay is surprisingly long, more than 24 h are necessary for transportation from site of synthesis to deposition on the surface of the insect.

In sexually mature females, (Z)-9-tricosene comprised a relatively large fraction of the hydrocarbon of the epicuticle and the hemolymph, but much smaller percentages of the hydrocabons in other tissues, including the ovaries. It appears that certain hydrocarbons were selectively partitioned to certain tissues such as the ovaries, from which pheromone was relatively excluded (Schal *et al.*, 2001). Both KBr gradient ultracentrifugation and specific immunoprecipitation showed that over 90 percent of the hemolymph hydrocarbon was associated with a high-density lipophorin. Lipophorin was composed of two aproproteins under denaturing conditions: apolipophorin I (≈240 kD) and apolipophorin II (≈85 kD) (Schal *et al.*, 2001). The data suggest that lipophorin may play an important role in an active mechanism that selectively deposits certain subsets of hydrocarbon at specific tissues.

8.5 Summary

The regulation of pheromone biosynthesis in *M. domestica* is unique compared to most other insect models studied to date. First, the regulation occurs via an ovarian produced hormone, 20-HE. Other than in *Drosophila melanogaster* (Wicker and Jallon, 1995a), there is no evidence that 20-HE regulates pheromone production in any other insect species. The housefly system is analogous to the JH regulation of pheromone production in the German cockroach *Blattella germanica*, where the same hormone that regulates vitellogenesis (JH in the case of *B. germanica* and ecdysteroids in *M. domestica*) also regulates pheromone production (Schal, Chapter 10, in this volume). Second, the mechanism by which 20-HE regulates pheromone production appears to occur by repressing the synthesis of a specific elongase. PBAN exerts its mode of action by increasing the activity of specific enzymes through second messengers (Rafaeli and Jurenka, Chapter 5, in this volume), and JH regulates pheromone production by up-regulating key isoprenoid pathway enzymes in bark beetles (Seybold and Vanderwel, Chapter 6, in this volume), and, apparently, by inducing the enzyme that hydroxylates a hydrocarbon precursor in a *Blattella germanica* (Schal *et al.*, Chapter 10, in this volume). Third, the regulation of pheromone production involves a modification of the cuticular hydrocarbons in the housefly, and thus pheromone synthesis occurs in epidermal-related tissue, probably oenocytes. Thus, there is no specific gland in the housefly where pheromone is synthesized. The synthesis of hydrocarbon and hydrocarbon-derived pheromones in a variety of taxa appears to follow this pattern, and involves the lipophorin transport of pheromone and pheromone precursors (Schal *et al.*, 1998; Jurenka, Chapter 3, in this volume).

The application of the powerful tools of molecular biology to work on pheromone production in insects, including the housefly, will undoubtedly allow us to obtain a much better understanding of how pheromone biosynthesis is regulated by hormones, and should allow advances in pheromone-based insect control techniques.

Acknowledgements. Research from the author's laboratory was supported in part by Nevada Agricultural Experiment Station, publication #03031275.

References

Adams T. S. and Holt G. G. (1987) Effect of pheromone components when applied to different models on male sexual behavior in the housefly, *Musca domestica. J. Insect Physiol.* **33**, 9–18.

Adams T. S., Holt G. G. and Blomquist G. J. (1984a) Endocrine control of pheromone biosynthesis and mating behavior in the housefly, *Musca domestica.* In *Advances in Invertebrate Reproduction* 3, ed. W. Engels, pp. 441–456. Elsevier, Amsterdam.

Adams T. S., Dillwith J. W. and Blomquist, G. J. (1984b) The role of 20-hydroxyecdysone in housefly sex pheromone biosynthesis. *J. Insect Physiol.* **30**, 287–294.

Adams T. S., Gerst J. W. and Masler E. P. (1997) Regulation of ovarian ecdysteroid production in the housefly, *Musca domestica. Arch. Insect Biochem. Physiol.* **35**, 135–148.

Adams T. S., Nelson D. R. and Blomquist G. J. (1995) Effect of endocrine organs and hormones on (Z)-9-tricosene levels in the internal and external lipids of female house flies, *Musca domestica. J. Insect Physiol.* **41**, 609–615.

Ahmad S., Kirkland K. E. and Blomquist G. J. (1987) Evidence for a sex pheromone metabolizing cytochrome P-450 monooxygenase in the housefly. *Arch. Insect Biochem. Physiol.* **6**, 121–140.

Antony C. and Jallon J.-M. (1982) The chemical basis for sex recognition in *Drosophila melanogaster. J. Insect Physiol.* **28**, 873–880.

Bartelt R. J. and Jackson L. L. (1984) Hydrocarbon component of the *Drosophila virilis* (Diptera: Drosophilidae) aggregation pheromone. *Ann. Entomol. Soc. Am.* **77**, 364–371.

Bartelt R. J., Armold M. T., Schaner A. M. and Jackson L. L. (1986) Comparative analysis of cuticular hydrocarbons of the *Drosophila virilis* group. Comp. *Biochem. Physiol.* **83B**, 731–742.

Blomquist G. J., Dillwith J. W. and Pomonis J. G. (1984a) Sex pheromone of the housefly: metabolism of (Z)-9-tricosene to (Z)-9,10-epoxytricosane and (Z)-14-tricosen-10-one. *Insect Biochem.* **14**, 279–284.

Blomquist G. J., Adams T. S. and Dillwith J. W. (1984b) Induction of female sex pheromone production in male houseflies by ovarian implants or 20-hydroxyecdysone. *J. Insect Physiol.* **30**, 295–302.

Blomquist G. J., Dillwith J. W. and Adams T. S. (1987) Biosynthesis and endocrine regulation of sex pheromone production in Diptera. In *Pheromone Biochemistry*, (ed. G. D. Prestwich and G. J. Blomquist, pp. 217–250. Academic Press, New York.

Blomquist G. J., Adams T. S., Halarnkar P. P., Gu P., Mackay M. E. and Brown L. (1992) Ecdysteroid induction of sex pheromone biosynthesis in the housefly, *Musca domestica.* Are other factors involved? *J. Insect Physiol.* **38**, 309–318.

Blomquist G. J., Guo L., Gu P., Blomquist C., Reitz R. C. and Reed J. R. (1994) Methyl-branched fatty acids and their biosynthesis in the housefly, *Musca domestica* L. (Diptera: Muscidae). *Insect Biochem. Mol. Biol.* **24**, 803–810.

Blomquist G. J., Tillman J. A., Reed J. R., Gu P., Vanderwel D., Choi S. and Reitz R. C. (1995) Regulation of enzymatic activity involved in sex pheromone production in the housefly, *Musca domestica. Insect Biochem. Molec. Biol.* **25**, 751–757.

Blomquist, G. J., Tillman J. A., Mpuru S. and Seybold, S. J. (1998) The cuticle and cuticular hydrocarbons of insects: structure, function and biochemistry. In *Pheromone Communication in Social Insects*, eds R. K. Vander Meer, M. Breed, M. Winston and C. Espelie pp. 34–54. Westview Press, Boulder, CO.

Blomquist G. J., Tillman-Wall J. A., Guo L., Quilici D., Gu P. and Schal C. (1993) Hydrocarbon and hydrocarbon derived sex pheromones in insects: biochemistry and endocrine regulation. In *Insect Lipids: Chemistry and Biology*, eds D. W. Stanley-Samuelson and D. R. Nelson, pp. 317–351. University of Nebraska Press, Lincoln.

Carlson D. A., Mayer M. S., Sillhacek D. L., James J. D., Beroza M. and Bierl B. A. (1971) Sex attractant pheromone of the housefly: isolation, identification and synthesis. *Science* **174**, 76–78.

Carlson D. A., Langley P. A. and Huyton P. (1978) Sex pheromone of the tsetse fly: isolation, identification and synthesis of contact aphrodisiacs. *Science* **201**, 750–753.

Carlson D. A., Nelson D. R., Langley P. A., Coates T. W. and Leegwater-Vander Linden M. E. (1984) Contact sex pheromone in the tsetse fly *Glossina pallidipes* (Austen): identification and synthesis. *J. Chem. Ecol.* **10**, 429–450.

Carlson D. A., Offor I. I., El Messoussi S., Matsyyama K., Mori K., and Jallon J. M. (1998) Sex pheromone of Glossina tachioides: isolation, identification and synthesis. *J. Chem. Ecol.* **24**, 1563–1575.

Carlson D. A. and Schlein Y. (1991) Unusual polymethyl alkenes in tsetse flies acting as abstinon in *Glossina morsitans. J. Chem. Ecol.* **17**, 267–284.

Chase J., Jurenka R. J., Schal C., Halarnkar P. P. and Blomquist G. J. (1990) Biosynthesis of methyl branched hydrocarbons in the German cockroach *Blattella germanica* (L.) (Orthoptera, Blattellidae). *Insect Biochem.* **20**, 149–156.

Chase J., Touhara K., Prestwich, G. D., Schal C. and Blomquist G.J. (1992) Biosynthesis and endocrine control of the production of the German cockroach sex pheromone, 3,11-dimethylnonacosan-2-one. *Proc. Natl. Acad. Sci. USA* **89**, 6050–6054.

Cheesbrough T. M. and Kolattukudy P. E. (1984) Alkane biosynthesis by decarbonylation of aldehydes catalyzed by a particulate preparation from *Pisum sativum. Proc. Natn. Acad. Sci. USA* **81**, 6613–6617.

Cheesbrough T. M. and Kolattukudy P. E. (1988) Microsomal preparations from animal tissue catalyzes release of carbon monoxide from a fatty aldehyde to generate an alkane. *J. Biol. Chem.* **263**, 2738–2743.

de Renobales M., Nelson D. R., Zamboni A. C., Mackay M. E., Dwyer L. A., Theisen M. O. and Blomquist G. J. (1989) Biosynthesis of very long-chain methyl branched alcohols during pupal development in the cabbage looper, *Trichoplusia ni*. *Insect Biochem.* **19**, 209–214.

Dennis M. W. and Kolattukudy P. E. (1991) Alkane biosynthesis by decarbonylation of aldehyde catalyzed by a microsomal preparation from *Botryococcus brauni*. *Arch. Biochem. Biophys.* **287**, 268–275.

Dennis M. W. and Kolattukudy P. E. (1992) A cobalt-porphyrin enzyme converts a fatty aldehyde to a hydrocarbon and CO. *Proc. Natl. Acad. Sci. USA* **89**, 5306–5310.

Dillwith J. W. and Blomquist G. J. (1982) Site of sex pheromone biosynthesis in the female housefly, *Musca domestica*. *Experientia* **38**, 471–473.

Dillwith J. W., Blomquist G. J. and Nelson D. R. (1981) Biosynthesis of the hydrocarbon components of the sex pheromone of the housefly, *Musca domestica* L. *Insect Biochem.* **11**, 247–253.

Dillwith J. W., Nelson J. H., Pomonis J. G., Nelson D. R. and Blomquist G. J. (1982) A ^{13}C-NMR study of methyl-branched hydrocarbon biosynthesis in the housefly. *J. Biol. Chem.* **257**, 11305–11314.

Dillwith J. W., Adams T. S. and Blomquist G. J. (1983) Correlation of housefly sex pheromone production with ovarian development. *J. Insect Physiol.* **29**, 377–386.

Dwyer L. A., de Renobales M. and Blomquist G. (1981) Biosynthesis of (Z,Z)-6,9 heptacosadiene in the American cockroach. *Lipids* **16**, 810–814.

Eigenheer, A. L., Young, S. D., Blomquist, G. J., Borgeson, C. E., Tillman, J. A. and Tittiger, C. (2002) Isolation and molecular characterization of *Musca Domestica* delta-9 desaturase sequences. *Insect Molec. Biol.* **11**, 533–542.

Gibbs A. G. (1998) Water-proofing properties of cuticular lipids. *American Zool.* **38**, 471–482.

Gu P., Welch W. H. and Blomquist G. J. (1993) Methyl-branched fatty acid biosynthesis in the German cockroach, *Blattella germanica*: kinetic studies comparing a microsomal and soluble fatty acid synthetases. *Insect Biochem. Mol. Biol.* **23**, 263–271.

Gu P., Welch W. H., Guo L., Schegg K. M. and Blomquist G. J. (1997) Characterization of a novel microsomal fatty acid synthetase (FAS) compared to a cytosolic FAS in the housefly, *Musca domestica*. *Comparative Biochem. Physiol.* **118B**, 447–456.

Guo L., Latli B., Prestwich G. D. and Blomquist G. J. (1991) Metabolically-blocked analogs of the housefly sex pheromone II. Metabolism studies. *J. Chem Ecol.* **17**, 1769–1782.

Hadley N. F. (1984) Cuticle: biochemistry. In *Biology of the Integument*, eds J. Bereiter-Hahn, A. G. Matoltsy and K. S. Richards, pp. 685–702. Springer-Verlag, Berlin.

Hagedorn H. H. (1985) The role of ecdysteroids in reproduction. In *Comprehensive Insect Physiology, Biochemistry and Pharmacology*, eds G. A. Kerkut and L. I. Gilbert, Vol. 8, pp. 205–262. Pergamon, Oxford.

Halarnkar P. P., Heisler C. R. and Blomquist G. J. (1986) Propionate catabolism in the housefly *Musca domestica* and the termite *Zootermopsis nevadensis*. *Insect Biochem.* **16**, 455–461.

Hall G. M., Tittiger C., Andrews G., Mastick G., Kuenzli M., Luo X., Seybold S. J. and Blomquist G. J. (2002) Male pine engraver beetles, *Ips pini*, synthesize the monoterpenoid pheromone ipsdienol *de novo* in midgut tissue. *Naturwissenschaften* (in press).

Howard R. W. (1993) Cuticular hydrocarbons and chemical communication. In *Insect Lipids: Chemistry, Biochemistry and Biology*, eds D. W. Stanley-Samuelson and D. R. Nelson pp. 177–226. University of Nebraska Press, Lincoln.

Howard R. W., Akre R. D. and Garnett W. B. (1990) Chemical mimicry in an obligate predator of carpenter ants (Hymenoptera: Formicidae). *Ann. Entomol. Soc. Am.* **83**, 607–616.

Huyton P. M. and Langley P. A. (1982) Copulatory behavior of the tsetse flies *Glossina morsitans* and *G. austeni*. *Physiol. Entomol.* **7**, 167–174.
Jallon J.-M. (1984) A few chemical words exchanged by Drosophila during courtship and mating. *Behavior Genetics* **14**, 441–478.
Juarez P., Chase J. and Blomquist G. J. (1992) A microsomal fatty acid synthetase from the integument of *Blattella germanica* synthesizes methyl-branched fatty acids, precursors to hydrocarbon and contact sex pheromone. *Arch. Biochem. Biophys.* **293**, 333–341.
Kolattukudy P. E. (1980) Cutin, suberin and waxes. In *The Biochemistry of Plants: A Comprehensive Treatise*, eds P. K. Stumpf and E. V. Conn, Vol. 4, pp. 571–645. Academic Press, New York.
Kolattukudy P. E., Croteau R. and Buckner J. S. (1976) Biochemistry of plant waxes. In *Chemistry and Biochemistry of Natural Waxes*, ed. P. E. Kolattukudy, pp. 289–347. Elsevier, Amsterdam.
Kostelc J. G., Hendry L. B. and Snetsinger R. J. (1975) A sex pheromone complex of the mushroom infesting sciarid fly, *Lycoriella mali* Fitch. *J. New York Ent. Soc.* **83**, 255–256.
Langley P. A. and Carlson D. A. (1983) Biosynthesis of contact sex pheromone in the female tsetse fly, *Glossina morsitans morsitans* Westwood. *J. Insect Physiol.* **29**, 825–831.
Martin D., Wang S.-F. and Raikhel A. S. (2001) The vitellogenin gene of the mosquito *Aedes aegypti* is a direct target of ecdysteroid receptor. *Mol. Cell. Endocrin.* **173**, 75–86.
Millar A. A. and Kunst L. (1997) Very-long-chain fatty acid biosynthesis is controlled through the expression and specificity of the condensing enzyme. *Plant J.* **12**, 121–131.
Mpuru S., Reed J. R., Reitz R. C. and Blomquist G. J. (1996) Mechanism of hydrocarbon biosynthesis from aldehyde in selected insect species: requirement for O_2 and NADPH and carbonyl group released as CO_2. *Insect Biochem. Molec. Biol.* **2**, 203–208.
Mpuru S., Blomquist G. J., Schal C., Kuenzli M., Dusticier G., Roux M. and Bagneres A.-G. (2001) Effect of age and sex on the production of internal and external hydrocarbons and pheromones in the housefly, *Musca domestica*. *Insect Biochem. Molec. Biol.* **31**, 139–155.
Nelson D. R. and Blomquist G. J. (1995) Insect waxes. In *Waxes: Chemistry, Molecular Biology and Functions*, ed. R. J. Hamilton and W. W. Christie, pp. 1–90. The Oily Press Ltd, England.
Nelson D. R., Dillwith J. W. and Blomquist G. J. (1981) Cuticular hydrocarbons of the housefly, *Musca domestica*. *Insect Biochem.* **11**, 187–197.
Nemoto T., Motomichi D., Oshi K., Matsubay H., Oguma Y., Suzuki T. and Kuwahara Y. (1994) (*Z,Z*)-5,27-tritriacontadiene – major sex pheromone of *Drosophila pallidosa* (Diptera; Drosophilidae). *J. Chem. Ecol.* **20**, 3029–3037.
Noble-Nesbitt, L. P. J. J. (1991) In *Physiology of the Insect Epidermis*, eds K. Binnington and A. Retnakaran, pp. 252–284. CSIRO, East Melbourne, Australia.
Oh C. S., Toke D. A., Mandala S. and Martin C. E. (1997) *ELO2* and *ELO3*, homologues of the *Saccharomyces cerevisiae ELO1* gene, function in fatty acid elongation and are required for sphingolipid formation. *J. Biol. Chem.* **272**, 17376–17384.
Pennanec'h M., Bricard L., Kunesch G., Jallon J. M. (1997) Incorporation of fatty acids into cuticular hydrocarbons of male and female *Drosophila melanogaster*. *J. Insect Physiol.* **43**, 1111–1116.
Raina A. K. (1993) Neuroendocrine control of sex pheromone biosynthesis in Lepidoptera. *Annu. rev. Entomol.* **38**, 329–349.
Raina A. K., Jaffe H., Kempe T. G., Keim P., Blacher R. W., Fales H. M., Riley C. T., Klun J. A., Ridgeway R. L. and Hayes D. K. (1989) Identification of a neuropeptide hormone that regulates sex pheromone production in female moths. *Science* **244**, 796–798.

Reed J. R., Hernandez P., Blomquist G. J., Feyereisen R. and Reitz R. C. (1996) Hydrocarbon biosynthesis in the housefly, *Musca domestica:* substrate specificity and cofactor requirement of P450hyd. *Insect Biochem. Molec. Biol.* **26**, 267–276.

Reed J. R., Quilici D. R., Blomquist G. J. and Reitz R. C. (1995) Proposed mechanism for the cytochrome P450 catalyzed conversion of aldehyde to hydrocarbon in the house fly, *Musca domestica. Biochemistry* **34**, 16221–16227.

Reed J. R., Vanderwel D., Choi S., Pomonis J. G., Reitz R. C. and Blomquist G. J. (1994) Unusual mechanism of hydrocarbon formation in the housefly: cytochrome P450 converts aldehyde to the sex pheromone component (Z)-9-tricosene and CO_2. *Proc. Natl. Acad. Sci. USA* **91**, 10000–10004.

Rogoff W. M., Beltz A. D., Johnson J. O. and Plapp F. W. (1964) A sex pheromone in the housefly, *Musca domestica* L. *J. Insect Physiol.* **10**, 239–246.

Schal C., Ling D. and Blomquist G. J. (1997). Neural and endocrine control of pheromone production and release in cockroaches. In *Insect Pheromone Research: New Directions*, eds R. T. Cardé and A. K. Minks, pp. 3–20. Chapman and Hall, New York.

Schal C., Sevala V. L. and Cardé R. T. (1998) Novel and highly specific transport of a volatile sex pheromone by hemolymph lipophorin in moths. *Naturwissenschaften* **85**, 339–342.

Schal C., Sevala V., Capurro M. D. L., Snyder T. E., Blomquist G. J. and Bagnères A.-G. (2001) Tissue distribution and lipophorin transport of hydrocarbon and sex pheromones in the house fly, *Musca domestica*. 11 pages *J. Insect Sci* **1**, 12. Available online: insectscience.org/1.12.

Schaner A. M., Dixon P. D., Graham K. J. and Jackson L. L. (1989) Components of the courtship-stimulating pheromone blend of young male *Drosophila melanogaster*: (Z)-13-tritriacontene and (Z)-11-tritriacontene. *J. Insect Physiol.* **35**, 341–345.

Scott D. (1986) Sexual mimicry regulates the attractiveness of mated *Drosophila melanogaster* females. *Proc. Natl. Acad. Sci. USA* **83**, 8429–8433.

Scott D. and Jackson L. L. (1988) Interstrain comparison of male-predominant antiaphrodisiacs in *Drosophila melanogaster*. *J. Insect Physiol.* **34**, 863–871.

Scott D., Richmond R. C. and Carlson D. A. (1988) Pheromones exchanged during mating: a mechanism for mate assessment in *Drosophila*. Anim. Behav. **36**, 1164–1173.

Stoffolano J. G., Schauber E., Yin C. M., Tillman J. A. and Blomquist G. J. (1997). Cuticular hydrocarbons and their role in copulatory behavior in *Phormia regina* (Meigen). *J. Insect Physiol.* **43**, 1065–1076.

Thompson A. C., Roth J. P. and King E. G. (1983) Larviposition kairomone of the tachinid *Lixophaga diatraeae*. *Environ. Entomol.* **12**, 1312–1314.

Tillman J. A., Holbrook G. L., Dallara P., Schal C., Wood D. L., Blomquist G. J. and Seybold S. J. (1998) Endocrine regulation of *de novo* aggregation pheromone biosynthesis in the pine engraver, *Ips pini* (Say) (Coleoptera: Scolytidae). *Insect Biochem. Molec. Biol.* **28**, 705–715.

Tillman J. A., Seybold S. J., Jurenka R. A. and Blomquist G. J. (1999) Insect Pheromones – an overview of biosynthesis and endocrine regulation. *Insect Biochem. Molec. Biol.* **29**, 481–514.

Tillman-Wall J. A., Vanderwel D., Kuenzli M. E., Reitz R. C. and Blomquist G. J. (1992) Regulation of sex pheromone biosynthesis in the housefly, *Musca domestica*: relative contribution of the elongation and reductive step. *Arch. Biochem. Biophys.* **299**, 92–99.

Tittiger C., Blomquist G. J., Ivarsson P., Borgeson C. E. and Seybold S. J. (1999) Juvenile hormone regulation of HMG-R gene expression in the bark beetle *Ips paraconfusus*

(Coleoptera: Scolytidae): implications for male aggregation pheromone biosynthesis. *Cell. Mol. Life Sci.* **55**, 121–127.

Tittiger C., O'Keeffe C., Bengoa C. S., Barkawi L., Seybold S. J. and Blomquist G. J. (2000) Isolation and endocrine regulation of HMG-CoA synthase cDNA from the male Jeffrey pine beetle, *Dendroctonus jeffreyi. Insect Biochem. Molec. Biol.* **30**, 1203–1211.

Uebel E. C., Sonnet P. E., Bierl B. A. and Miller R. W. (1975a) Sex pheromone of the stable fly: isolation and preliminary identification of compunds that induce mating strike behavior. *J. Chem. Ecol.* **1**, 377–385.

Uebel E. C., Sonnet P. E., Miller R. W. and Beroza M. (1975b) Sex pheromone of the face fly, *Musca autumnalis* De Geer (Diptera: Muscidae). *J. Chem. Ecol.* **1**, 195–202.

Uebel E. C., Sonnet P. E. and Miller R. W. (1976) House fly sex pheromone: enhancement of mating strike activity by combination of (Z)-9-tricosene with branched saturated hydrocarbons. *Environ. Entomol.* **5**, 905–908.

Uebel E. C., Sonnet P. E., Menzer R. E., Miller R. W. and Lusby W. R. (1977) Mating-stimulant pheromone and cuticular lipid constituents of the little housefly, *Fannia canicularis* (L.). *J. Chem. Ecol.* **3**, 269–278.

Uebel E. C., Schwarz M., Lusby W. R., Miller R. W. and Sonnet P. E. (1978a) Cuticular non-hydrocarbons of the female housefly and their evaluation as mating stimulants. *Lloydia* **41**, 63–67.

Uebel E. C., Schwarz M., Miller R. W. and Menzer R. E. (1978b) Mating stimulant pheromone and cuticular lipid constituents of *Fannia femoralis* (Stein) (Diptera: Muscidae). *J. Chem. Ecol.* **4**, 83–93.

Vanderwel D. (1994) Factors affecting pheromone production in beetles. *Arch. Insect Biochem. Physiol.* **25**, 347–362.

Vaz A. H., Jurenka R. A., Blomquist G. J. and Reitz R. C. (1988) Tissue and chain length specificity of the fatty acyl-CoA elongation system in the American cockroach. *Arch. Biochem. Biophys.* **267**, 551–557.

Wakayama E. J., Dillwith J. W., Howard R. W. and Blomquist G. J. (1984) Vitamin B_{12} levels in selected insects. *Insect Biochem.* **14**, 175–179.

Wang D. L., Dillwith J. W., Ryan R. O., Blomquist G. J. and Reitz R. C. (1982) Characterization of the acyl-CoA desaturase in the housefly, *Musca domestica* L. *Insect Biochem.* **12**, 545–551.

Wettstein-Knowles P. (1995) In *Waxes: Chemistry, Molecular Biology and Functions*, eds R. J. Hamilton and W. W. Christie, pp. The Oily Press Ltd, England.

Wicker C. and Jallon J.-M. (1995a) Influence of ovary and ecdysteroids on pheromone biosynthesis in *Drosophila melanogaster* (Diptera, Drosophilidae). *Eur. J. Entomol.* **92**, 197–202.

Wicker C. and Jallon J.-M. (1995b) Hormonal control of sex pheromone biosynthesis in *Drosophila melanogaster. J. Insect Physiol.* **41**, 65–70.

Wicker-Thomas C., Henriet C. and Dallerac R. (1997) Partial characterization of a fatty acid desaturase gene in *Drosophila melanogaster. Insect Biochem. Mol. Biol.* **11**, 963–972.

Yoder J. A., Denlinger D. L., Dennis M. W. and Kolattukudy P. E. (1992) Enhancement of diapausing flesh fly puparia with additional hydrocarbons and evidence for alkane biosynthesis by a decarbonylation mechanism. *Insect Biochem. Molec. Biol.* **22**, 237–243.

Young A., Tittiger C., Welch W. and Blomquist G. J. (2001) Monoterpenoid pheromone biosynthesis: fishing for the elusive geranyl diphosphate synthase in bark beetles. International Society of Chemical Ecology 18th Annual Meeting. July 7–12, Lake Tahoe. Abstract, p. 158.

9

Genetic studies on pheromone production in *Drosophila*

Jean-Marc Jallon and Claude Wicker-Thomas

9.1 Aphrodisiac compounds in *Drosophila*

The involvement of sex pheromones in *Drosophila* has been postulated since the beginning of studies on sexual isolation mechanisms and courtship (Sturtevant, 1915). Manning (1959) established *D. melanogaster* as a model animal for the study of courtship behavior, which was referenced by Welbergen *et al.* (1987) and Cobb *et al.* (1988). A male wing display is a very conspicuous part of courtship and the wing vibration produces a species-specific pulse song (Burnet and Connolly, 1981; Moulin *et al.*, 2001). Jallon and Hotta (1979) have proposed to use the cumulative amount of wing vibration to describe the intensity of male courtship (a sex appeal parameter). In contrast, other authors use the courtship index, which is the percentage of time during which any of the courtship signals is displayed. Manning (1959) also compared the behavior of sex partners in the sibling species, *D. simulans*, and Spieth (1952) extended the study to comparisons among drosophilids. Cobb *et al.* (1989) compared the ethograms of seven species of the melanogaster subgroup.

A number of studies since 1980 have concentrated on the search for a sex pheromone analogous to Muscalure, the first contact pheromone discovered in mature *Musca domestica* females (Carlson *et al.*, 1971). Other hydrocarbons have been described in various Diptera (Carlson *et al.*, 1978; Blomquist and Jackson, 1979; El Messoussi *et al.*, 1994; Blomquist, Chapter 8, in this volume). In all cases, a cuticular hydrocarbon with low or no volatility and high abundance was necessary, but not sufficient for a response. Its action could be checked only or Caribbean) (Coyne *et al.*, 1999). Two desaturase genes at the same locus, *desat1* and *desat2*, were then shown to control enzyme isoforms, with different specificities, thus leading to ω7 and ω5 unsaturated fatty acids (Dallerac *et al.*, 2000; see Figure 9.1).

9.3 Ontogeny and endocrine control of cuticular hydrocarbons

Cuticular hydrocarbons in mature *D. melanogaster* flies display a marked sexual dimorphism, whereas flies a few hours after imaginal eclosion show strong similarities between sexes. While the mature flies contain hydrocarbon chains with 23–29 carbons, the immature flies have chains with 29–35 carbons including abundant dienes in both sexes (Antony and Jallon, 1981). These dienes are complex mixtures, which have been characterized by epoxidation and analysis of the products by GC-MS with negative ion chemical ionization. Double bonds are more common at carbon position 11 or 13, also found in monoenes, and position 21 or 23 (Péchiné *et al.*, 1988). Moreover, the similarity between both sexes of young *D. melanogaster* was also found within both sexes of young *D. simulans* flies. Among these singular hydrocarbons, the main cuticular compound of mature male or female *D. erecta*, another species of the melanogaster subgroup, is 9,23-tritriacontadiene (Péchiné *et al.*, 1988). Quantities of these compounds did not change much during the first day after eclosion, but then decreased consistently for up to 2–3 days while hydrocarbons characteristic of either mature sex (C23–C29) increased (Figure 9.2).

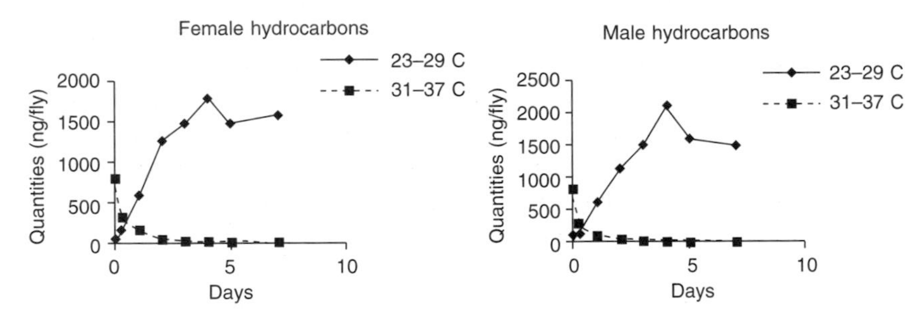

Figure 9.2 Ontogeny of cuticular hydrocarbons in *D. melanogaster* (Canton-S strain). Values (total quantities) are means of four groups of five flies ± SEM.'

The time of appearance and increase differed more with the degree of unsaturation than with sex. In both sexes monoenes appear at 12 h – at least in the wild type Canton-S strain bred at 25°C. Monoene levels then increase in males up to 3 days of age (Arienti *et al.*, unpublished data). In females the ontogeny profile is more complex, as shown on Figure 9.3. The total level of

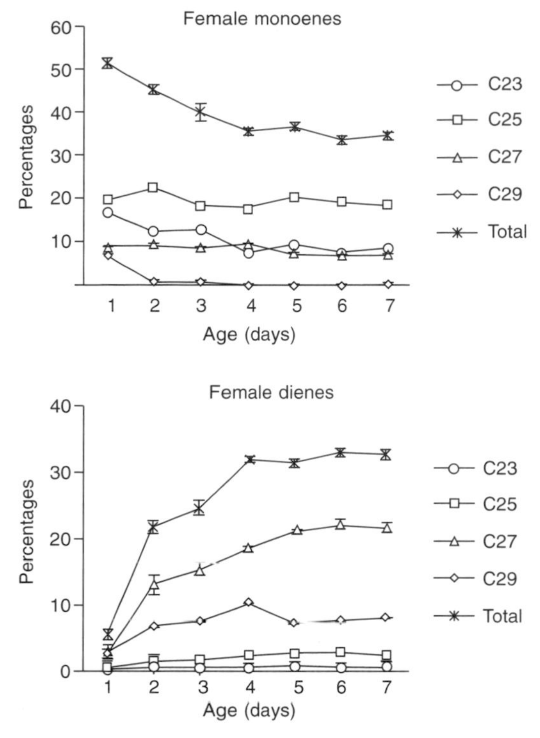

Figure 9.3 Ontogeny of female monoenes and dienes in *D. melanogaster* (Canton-S strain). Values are means of ten flies ± SEM.

unsaturated hydrocarbons increases between 12 and 36 h, after which there are no significant changes or a slight decrease in production. Dienes show homogeneous quantitative changes from 24 h to 3–4 days, irrespective of the chain length being 25, 27 or 29 carbons. It is thus suggested that the maturation of the first desaturation precedes that of the second desaturation. Similar changes with age were observed in African strains like Tai (Arienti *et al.*, unpublished data).

The sexual dimorphism of cuticular hydrocarbons is completed during the first three days after imaginal eclosion. During the same period, important physiological events take place, female oocyte maturation and vitellogenesis. A number of mutations have been described which affect ovarian development or endocrine control. These mutants were used to elucidate a possible hormone control mechanism used to regulate hydrocarbon biosynthesis.

The *apterous* mutation that seems to affect JH metabolism or transport and vitellogenesis markedly delays the replacement of young fly hydrocarbons by sex-specific ones. Three-day-old wild-type flies typically have 1–2 percent of long chain HC, whereas ap^{c} and ap^{56f} flies of the same age have 15–23 percent and 8–11 percent of long chain HC, respectively. A similar result was found in hydrocarbons of the twin *Drosophila* species, *D. melanogaster* and *D. simulans*. This model was tested in later experiments. Chang Yong and Jallon (1986), using radiolabeled acetate, showed that all cuticular hydrocarbons could be synthesized *de novo* by virgin and mature *D. melanogaster* Canton-S males and females. Radioactivity could be found in hydrocarbons 2–3 h after injection, with 2–4 percent of injected radiolabel recovered in hydrocarbons. This global incorporation was maximal 2–3 days after imaginal eclosion and did not vary much between males and females. Parallel radio- and FID-GC studies showed that the radioactivity was equally distributed in all hydrocarbons, linear or branched, and saturated and unsaturated components, regardless of double bond position(s), suggesting acetate is the common precursor, probably involved in the fatty acid biosynthesis steps (Chang Yong and Jallon, 1986).

A relatively high specific activity FAS (fatty acid synthetase) had just been characterized by de Renobales and Blomquist (1984). Later, de Renobales *et al.* (1987) further described this enzyme: it used acetyl-CoA, malonyl-CoA and NADPH as substrates and produced fatty acids with 12–18 carbons (laurate to stearate). At low ionic strength, palmitate was its main product (45 percent), while at high strength, myristate was favored (48 percent). Branched chains are also synthesized by the same FAS (de Renobales *et al.*, 1987). The total fatty acid composition of mature *D. melanogaster* Canton-S flies showed the highest abundance of C16 (21.6 percent 16:0 and 24.6 percent 16:1 in 3-day-old males; 20.5 percent 16:0 and 25.9 percent 16:1 in females of the same age) and the presence of a major position for double bonds at Δ9 (Pennanec'h *et al.*, 1997).

The biosynthesis of cuticular hydrocarbons was further investigated, following topical application of ^{14}C-labeled myristate, palmitate and stearate and ^{3}H-labeled cis-vaccenate. The incorporation levels into unsaturated hydrocarbons were similar in both sexes and markedly depended on the chain length of the saturated precursor, with a maximum yield from myristate; for example, in females, the enrichment in dienes relative to saturated HCs was 2.8 ± 0.1 with C14:0, 1.8 ± 0.1 with C16:0 and 1.0 ± 0.1 with C18:0. Cis-vaccenate only led to unsaturated compounds. With this precursor there was an enhanced incorporation into female monoenes and dienes, maximal in 2–3-day-old females. Cis-vaccenate has been identified among *Drosophila* fatty acids – although in low amounts – along with 17-tetracosenoate (Pennanec'h *et al.*, 1997). All these data supported the elongation-decarboxylation mechanism presented for the *D. melanogaster* hydrocarbon biosynthesis by Jallon (1984). The early steps might be common in both sexes, involving a common desaturase. From FAS production, myristate and palmitate are the main substrates available (de Renobales *et al.*, 1987). As the involved monoenic fatty acids are Δ7, two types of desaturases had to be considered, either a Δ7 using myristate, which would be in better agreement with the labeling experiments described before, or a Δ9 using palmitate, more abundant at low ionic strength. Myristoleate and palmitoleate would then be elongated to produce vaccenate, which has been shown to be an important common precursor for 7-T as well as 7,11-HD (Pennanec'h *et al.*, 1997). Vaccenate is a good candidate as a substrate for the second desaturation present only in *D. melanogaster* mature females.

Using a more rigorous mode of extraction, Pho *et al.* (1996) characterized an internal pool of hydrocarbons, with chain lengths similar to those of the cuticular pool. Studies on the fate of the hydrocarbons synthesized *de novo* after topical application of radiolabeled fatty acid precursors showed a decrease in the internal pool of hydrocarbons with time and a concomitant increase in the cuticular pool. These results suggest that hydrocarbons synthesized at an internal site are transported to the cuticle (Pho *et al.*, 1996). The nature of this internal site was suggested by sex mosaic studies. Studies on gynandromorph allowed Jallon and Hotta (1979) to localize the related sex appeal focus under the abdominal tegument. Later a complete feminization of the cuticular hydrocarbons produced by males was induced by targeted expression of the *transformer* gene in several tissues including oenocytes (Ferveur *et al.*, 1997). Oenocytes are subcuticular abdominal cells of epidermal origin found in segmentally repeated rows (crescent-shaped strands on the tergites, small clusters on sternites). Oenocytes are thus suggested to be the site of at least the last biosynthetic steps that lead from 7-monoenic fatty acids with 23–25 carbons to 7,11-dienoic fatty acids with 27–29 carbons. Their involvement in the production of insect cuticular hydrocarbons has been asserted in previous studies, although it was almost impossible to get oenocytes free of contaminating fat body (Diehl, 1975). Oenocytes have also been implicated in local production of ecdysteroids (Romer, 1991) and might afford a hormonal factor necessary for the production of female pheromones. However, a transporter is needed to bring the synthesized HCs to the cuticle. Such a function has been assigned to the lipoprotein, lipophorin in *Locusta migratoria* (Chino *et al.*, 1977; Katase and Chino, 1982, 1984; Dantuma *et al.*, 1998). Pho *et al.* (1996) purified a similar protein from homogenized *D. melanogaster* adults. It has a native molecular mass of 640 kD and consists of two glycosylated apoproteins of 240 and 75 kD. Gas chromatography and mass spectrometry showed that lipophorins isolated separately from virgin mature male or female flies were associated with specific hydrocarbons that were the same as those found in male and female cuticles, respectively. These lipophorin characteristics and the comparison of the internal and external pool of labeled hydrocarbons thus suggested that hydrocarbons synthesized in internal tissues might be transported to the cuticle by lipophorin (Pho *et al.*, 1996).

9.7 Fatty acid desaturases in *Drosophila*

The recent completion of the sequence of *D. melanogaster* genome (Myers *et al.*, 2000) and its publication in FlyBase (http://flybase.bio.indiana.edu) has facilitated the molecular characterization of new genes.

9.7.1 Desaturase genes in *D. melanogaster*

In *Drosophila melanogaster*, seven putative fatty acid desaturase genes are located on chromosome 3 and one on chromosome 2. The desaturase gene located on chromosome 2, *infertile crescent*, seems to be expressed more specifically in males (FlyBase). Of the seven fatty acid desaturase genes located on chromosome 3, six were found expressed in adults, and two of them (*desat1* and *desat2*) used saturated fatty acids as substrates. One desaturase is not yet completely characterized and the other three appeared to use unsaturated fatty acids rather than saturated fatty acids (R. Dallerac, T. Chertemps, C. Labeur and C. Wicker-Thomas, unpublished data). All these desaturase genes display a high homology at the nucleic acid level.

We will focus on the desaturase locus localized in region 87 B-C, which has been most thoroughly investigated. This locus spans 12 kb and contains the two desaturase genes, *desat1* and *desat2*, separated by about 2 kb. Surprisingly, all of the desaturase Expressed Sequence Tags (ESTs) mapping at 87 B-C locus that have been described in FlyBase came from the *desat1* gene.

9.7.2 *desat1* gene

9.7.2.1 Molecular and functional characterization of desat1

We have used the homology of described rat and yeast Δ9 desaturases to isolate cDNAs encoding a desaturase in *D. melanogaster* Canton-S (Wicker-Thomas *et al.*, 1997). The peptide sequence shows 29 and 43 percent identity with those of yeast and rat, respectively. The *desat1* is expressed in males and females, though at a higher level in females. Similar levels of higher female expression have been also observed in mice, which parallel a higher level of unsaturated fatty acids found in female liver (Lee *et al.*, 1996).

When expressed in a desaturase-deficient yeast, *desat1* open reading frames from 7, 11-HD rich Canton-S and 5, 11-HD rich Tai strains were found to encode a Δ9 desaturase, which preferentially uses palmitate, then stearate, leading to fatty acids preferentially unsaturated in ω7, then ω9 (Dallerac *et al.*, 2000). Since the fatty acids with 16 carbon atoms are the most abundant in *D. melanogaster* (Pennanec'h *et al.*, 1997), a large majority of ω7 fatty acids are produced.

9.7.2.2 Organization of desat1 gene

desat1 has a transcription length of 7992 nucleotides. It contains five exons and four introns. *desat1* ESTs described in Flybase differ in their 5′-end, leading to five different transcripts (T1 to T5). This is due to the presence of five different initiation sites for transcription, resulting in different first exon, but leading to the same open reading frame (Figure 9.7).

Only one EST (LD 19182: 300 nt, GENBANK: AA 539890) has been identified as a T3 transcript. Another one (RH 21795: 617 nt: EMB 583251) has a unique initiation site and corresponds to the T5 transcript (Table 9.1).

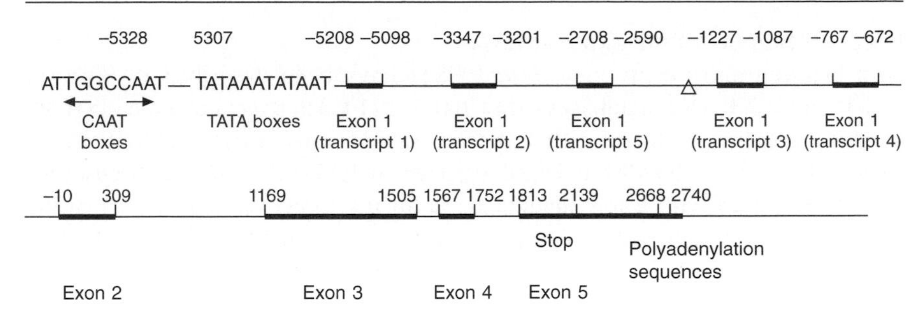

Figure 9.7 Schematic organization of *Drosophila desat1* gene. The two poly A signals at 2668 and 2740 were experimentally determined by RACE-PCR on cDNA from Canton-S and Tai strains. The putative transcription initiation sites were deduced by using the sequences of the ESTs described in databases. The numbers refer to the distances (in bp) from the A of the ATG initiator codon. The putative CAAT, TATA boxes are indicated. A perfect palindromic sequence, AGTTCATCTGAACT, consensus with an ecdysone-response element, is found at –2042. The site of insertion of the *P*-element in the line: *P(w)l(3)S028813* is indicated by Δ; this *P*-line was generated by Deak *et al.* (1997) and the *P*-element mapped at –1706 (Labeur *et al.*, 2002).

Table 9.1 Representation of the five categories of *desat1* transcripts. All transcripts share the same exons 2 to 5

Transcript	Length of the first exon (bp)	Length of the first intron (bp)
T1	111	5088
T2	147	3191
T3	141	1077
T4	96	622
T5	117	2580

This complex organization is very unusual in *Drosophila* and has been also reported for another gene, *fruitless* (*fru*), where a complex set of sex-specific and sex-non-specific transcripts are generated by the use of four promoters (P1–4) and alternative splicing at both the 5′- and 3′-ends of the primary transcripts (Anand *et al.*, 2001). As for *fru* transcripts, *desat1* transcripts might be under the control of different regulatory elements and correspond to different stage or tissue-specific expression and/or specific regulation.

The cDNAs are about 2 kb in length and contain a single open reading frame encoding 383 amino acids, followed by a 600 bp 3′-untranslated sequence. This 3′-untranslated region contains two functional polyadenylation sequences (AAUAAA), which are both used, as demonstrated by RACE-PCR.

9.7.2.3 Analysis of the different ESTs

Two hundred and twenty-nine *desat1* ESTs have been described in Flybase, EMBL or GENBANK databases (May 2003); 221 ESTs have a complete 5′-end, allowing grouping into five different categories of transcripts which only differ in their first exon (Tables 9.1 and 9.2). As shown in Table 9.2, desaturase transcripts have been found with a low abundance in all the tissues tested, except in the head, where they represent a high proportion of the total ESTs (0.2%). The five types of transcripts are not equivalently represented, in particular categories 3 and 5 contain only one EST each. The other three categories of transcripts represent 12 to 73 percent of the ESTs.

In the mouse, three genes have been cloned and characterized (*Scd1*, *Scd2* and *Scd3*, under the control of unique promoters) (Ntambi *et al.*, 1988; Kaestner *et al.*, 1989; Zheng *et al.*, 2001). The mouse desaturase genes are regulated differently in different tissues. For example the *Scd1* expression is constitutive in adipose tissue but inducible in liver in response to a high carbohydrate diet. *Scd2* is induced in adipose tissue, liver and other tissues in response to a high carbohydrate diet and is also constitutively expressed in brain (Kaestner *et al.*, 1989). Both genes are also expressed in mouse skin, leading to *Scd1* mutants exhibiting skin defects such as flaky skin and alopecia (Zheng *et al.*, 1999). *Scd3* expression is restricted to skin (Zheng *et al.*, 2001).

Similar to *Scd2* in mice (Kaestner *et al.*, 1989), *desat1* mRNA is found at higher levels in brain. The role of the desaturase in brain is to supply unsaturated fatty acids necessary for the synthesis of membrane phospholipids, particularly for myelination.

Transcript 2 mRNAs are found most frequently (73 percent of the desaturase ESTs). They are present throughout all developmental stages: embryo, larvae and adult and are predominately expressed in the head (85 percent). The genomic region in close proximity to the T2 initiation site has no apparent TATA or CAAT boxes and might be constitutively expressed in brain, similar to *Scd2* in mice.

Transcript 1 mRNAs represent 11 percent of the *desat* transcripts. They are principally found in Schneider cell culture (Schneider cells originated from embryonic cells which became immortalized) but are less represented in embryos and larvae (10 to 12 percent of the *desat1* transcripts). In the adult they constitute only 3 percent of the *desat* transcripts in head but seem more abundant in testis (two out of two) and ovary (two out of five). In the proximal 5′-non-transcribed region are found two TATA and two CAAT boxes.

Sixteen percent of the *desat* transcripts are transcripts 4. They are particularly abundant in the embryo (36 percent of the *desat1* transcripts), larvae (two out of ten) and constitute only 11 percent of the *desat* transcripts in adult.

A number of 5′-PuGG/TTC/GA-3′ motifs, which are putative ecdysone-response-elements (Antoniewski *et al.*, 1993) are present in the untranslated 5′-region of *desat1* (17 motifs). However, without experimental confirmation, the

Table 9.2 Analysis of the 229 desaturase ESTs described in databases. For 221 ESTs the 5′-sequence began in the first exon and allowed to deduce the initiation of transcription; 6 ESTs had a complete 3′-untranslated sequence which corresponded to the use of the second poly-A signal (at 2740 bp)

Stage or tissue	Library	N (desaturase ESTs)	% of total ESTs	N′ (5′-complete)	Transcripts					Poly-A signal	
					1	2	3	4	5	1	2
Schneider cell culture	SD	19	0.04	19	11	7		1			
	LD	6	0.02	6		1	1	4			
Embryo	RE	27	0.04	27	4	15		8			
	total	33	0.03	33	4	16	1	12			
Larval-early pupal	LP	10	0.11	10	1	5		4			
Adult testis	AT	2	0.01	2	2						
Adult ovary	GM	5	0.04	5	2	1		2			
Adult head	GH	61	0.25	55	1	49		5			5
	RH	99	0.18	97	4	81		11	1		1
	total	160	0.20	152	5	130		16	1		6
Total		226		221	25	159	1	35	1	0	6

significance of these putative response elements remains speculative. A perfect palindromic sequence, composed of inverted repeats of AGTTCA, spaced by two nucleotides, is present at –2042 nt. This motif is a consensus sequence for a putative steroid response element. If these elements were found to be functional, it might explain the higher *desat1* expression in females, where ecdysteroid titers were found higher than in males (Delbecque, Wicker-Thomas and Jallon, unpublished data).

9.7.2.4 Involvement of desat1 gene in pheromone biosynthesis

Several *P*-element insertion lines were mapped in the *desat* locus region. One *P*-element is inserted at –1706 nt from the first ATG in the open reading frame (Figure 9.7). The location, in the first intron of pre-mRNAs transcribed in transcripts 1 and 2 and in the 5′-untranscribed region of transcripts 3 and 4, might affect the transcription of *desat1* and/or the splicing of pre-mRNAs of transcripts 1 and 2.

As shown in Figure 9.8, the insertion of that *P*-element in the *desat1* gene reduced the overall synthesis of unsaturated hydrocarbons in both males and females. The effect was more pronounced in males (–19 percent and –58 percent of the 7-monoenes in males heterozygous and homozygous for the insertion, respectively). In females the effect was only visible in 7,11-dienes (–19 percent) for the heterozygotes and both in monoenes (–71 percent) and dienes (–38 percent) in homozygotes. These effects were principally due to a loss of monoenic fatty acids, which are the substrates both for mono- and di-unsaturated hydrocarbons. The decrease in unsaturated hydrocarbons was compensated by an increase in saturated ones, showing that the effect was at the level of the first desaturation step (Labeur *et al.*, 2002). The significant effect on female dienes might be the consequence of the depletion in monoenes due to the lower transcription of *desat1*. Actually the monoenic substrates are used in different reactions (elongation, desaturation, transport) and the synthesis of saturated, mono- and di-unsaturated hydrocarbons is the result of the relative activities of these different enzymes.

After excision of the transposon, the pheromone phenotype was reversed in 69 percent of the lines and the other excision lines showed generally decreased 7-unsaturated hydrocarbon levels (paralleled with similar increases in linear saturated hydrocarbons). These results clearly show the involvement of *desat1* in the biosynthesis of unsaturated hydrocarbons (Labeur *et al.*, 2002).

9.7.3 *desat2* gene

Females of most *Drosophila* populations have high amounts of 7,11-HD, while those of African and Caribbean populations have high amounts of 5,9-HD (Jallon and Péchiné, 1989). This difference in hydrocarbons has been mapped to a single locus, 87C-D, which is close to the location of *desat1* gene (Coyne *et al.*, 1999). This is due to the second desaturase gene, *desat2*, being located close to *desat1*.

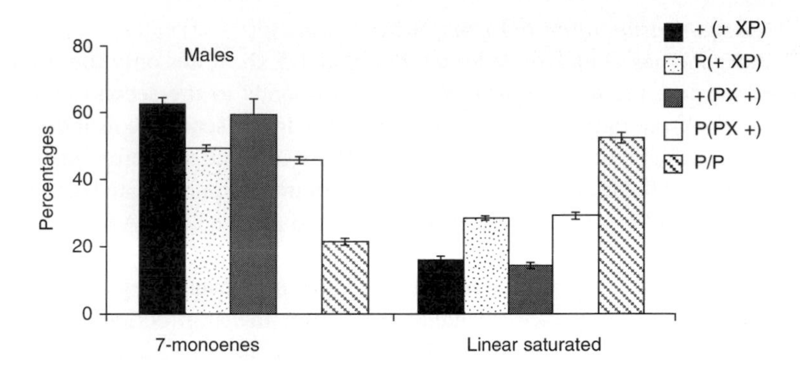

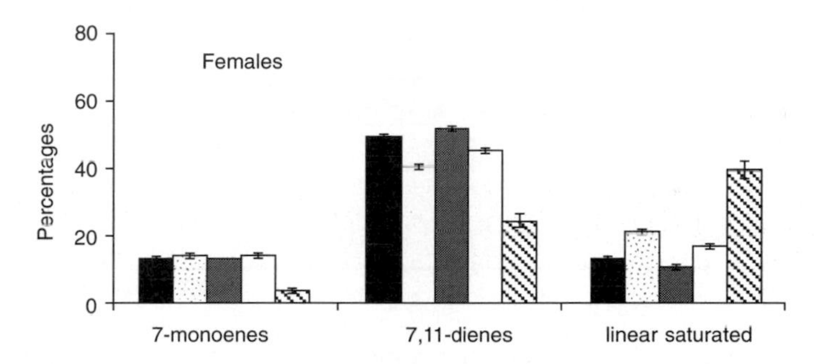

Figure 9.8 7-monoenes, 7,11-dienes and linear saturated hydrocarbons in male and female adult flies of either wild type (+) or heterozygous for the P-element insertion (P) or homozygous for the P-element (P/P). + and P flies were obtained by crossing Canton-S females with P-males (+ XP) or vice versa (PX +).

9.7.3.1 Molecular and functional characterization of desat2

We have cloned and characterized *desat2* in the African Tai strain (Dallerac *et al.*, 2000). At the molecular level, the *desat2* open reading frame encodes a 361 amino acid protein sharing 65 percent identity with *Desat1*. The *Desat2* enzyme was found to be a Δ9 desaturase, with a different substrate specificity from *Desat1*: it favors myristate as a substrate, leading to ω5 fatty acids. This gene seems only to be expressed in female African flies (which have high levels of 5,9-HD) and is not expressed in males of either strains nor in 7,11-HD-rich females (like Canton-S). The strain and sex specificity of *desat2* expression, the substrate specificity of *Desat2* enzyme and the genetic location of 7,11-HD/5,9-HD polymorphism strongly suggest that *desat2* is responsible for the 5,9-HD profile of Tai females.

9.7.3.2 Organization of desat2 gene in Tai strain
The *desat2* gene has a nucleotide length of about 1.5 kb. It has only four exons and three introns; the first exon of *desat2* corresponds to the second exon of *desat1* and the three introns are in similar positions in the sequence as introns 2, 3 and 4 in *desat1* (Figure 9.9). Unlike *desat1*, there is no intron before the open reading frame and therefore there is only one type of transcript generated, although no ESTs for *desat2* have been described yet in FlyBase. The cDNA is about 1.5 kb long and contains a single open reading frame of only 361 amino acids, followed by a short 3′-untranslated sequence. This 3′-untranslated region contains one polyadenylation sequence, located 70 bp after the stop codon, which is functional as demonstrated by RACE-PCR.

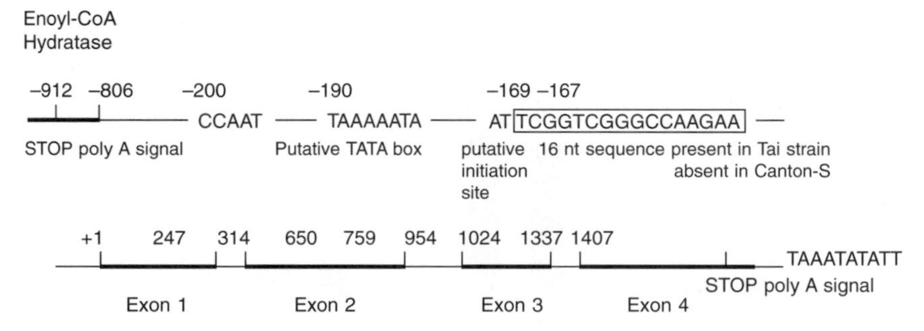

Figure 9.9 5′-flanking region located upstream from the *desat2* open reading frame and organization of *desat2* gene.
Nucleotide position +1 is assigned to the A of the ATG initiator codon. Poly A signals were experimentally determined by RACE-PCR on cDNA from Canton-S and Tai strains (enoyl-CoA hydratase) and Tai (*desat2*). A 16 nt sequence is present in Tai but absent in Canton-S genome. The putative transcription initiation site of *desat2* gene, located at –169, was determined by using the Neural Network Promoter Prediction: Input (http://www.fruitfly.org/seq_tools/promoter.html). A putative TATA box is found 21 nt upstream the putative site of transcription initiation.

The analysis of the 5′-region upstream of the open reading frame shows a putative initiation site located at –169 nt from the ATG. Two putative TATA and CAAT boxes are located at the proximity of this initiation site, but no particular regulatory element could be found.

9.7.3.3 Organization of desat2 *gene in Canton-S strain*
Takahashi *et al.* (2001), through genetic mapping, localized the 7,11-HD/5,9-HD polymorphism to within a 13 kb region just upstream of *desat1*. They then analyzed 43 lines showing 7,11-HD or 5,9-HD phenotype and correlated the 7,11-HD phenotype to a 16 bp deletion upstream of *desat2* gene. It seems possible that this deletion alone abolishes the expression of *desat2*.

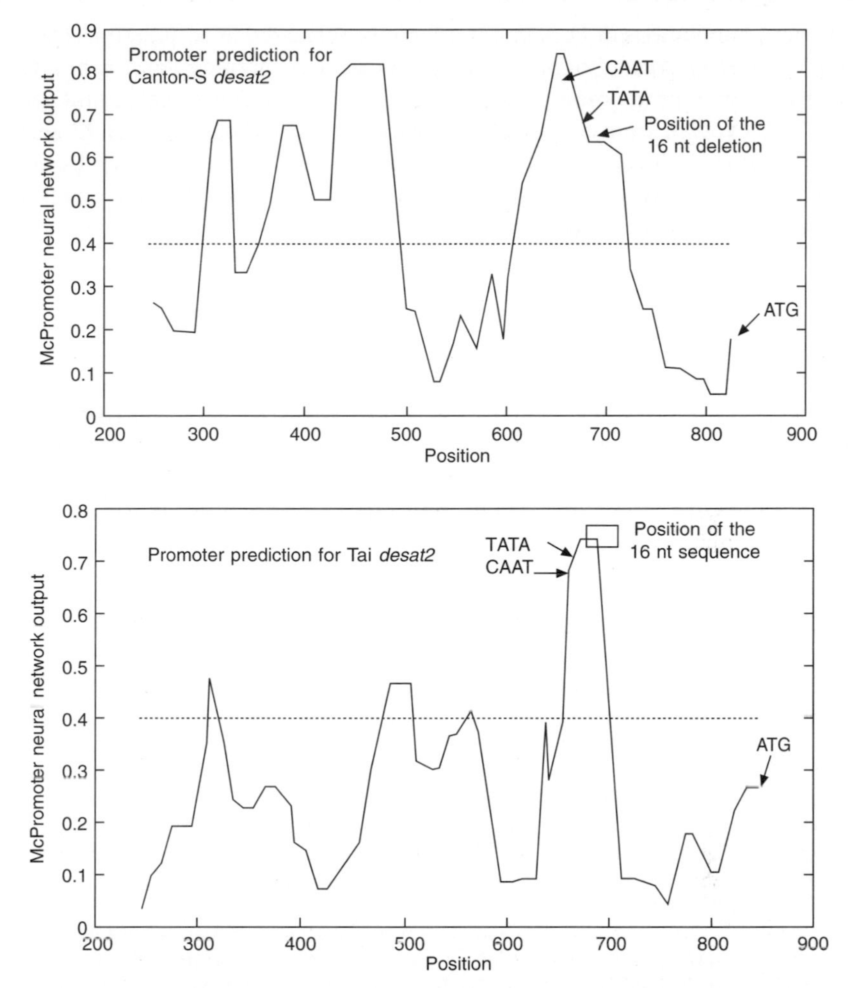

Figure 9.10 Promoter prediction for Canton-S and Tai *desat2*. The untranslated sequences upstream of the ATG of the open reading frame have been analyzed for Canton-S or Tai using McPromoter neural network output. The highest probability for the initiation of transcription for Tai *desat2* is located in the region near the 16 nt sequence which is absent in the Canton-S sequence. In Canton-S, the 16 nt deletion lowers the probability of the initiation of transcription after the two putative CAAT and TATA boxes and might be the cause of the lack of transcription of *desat2* gene in this strain.

In Canton-S strain, *desat2* is not transcribed in either sex (Dallerac *et al.*, 2000), and the same 16 bp deletion is observed in the 5′-region of the gene. According to the promoter prediction shown in Figure 9.10, this deletion results in a lower probability of transcription at the same site and might be responsible for the lack of transcription of *desat2* in Canton-S.

African populations of *D. melanogaster* are considered to be ancestral (David and Capy, 1988). According to the molecular data concerning *desat* genes, the 7,11-HD phenotype in cosmopolitan strains due to a non-functional *desat2* gene must have evolved from a 5,9-HD phenotype due to a functional *desat2* gene. Females with high levels of 7,11-HD have been shown to mate more rapidly than females with high levels of 5,9-HD (Ferveur *et al.*, 1996), which might represent a selective advantage (Ferveur *et al.*, 1996; Roualt *et al.*, 2003).

Conclusion

The identification of enzymes involved in pheromone metabolism, thanks to the characterization of the genome in *Drosophila melanogaster*, has begun. Mutational analysis of these structural genes is a viable approach to the understanding of biosynthetic pathway. The genetics combined with molecular biology will allow greater knowledge and understanding of the regulation of pheromone production and of its evolution in the animal kingdom.

References

Adams T. S. and Holt G. G. (1987) Effect of pheromone components when applied to different models on male sexual behavior in the housefly, *Musca domestica*. *J. Insect Physiol.* **33**, 9–18.

An W. and Wensink P. C. (1995) Integrating sex- and tissue-specific regulation within a single *Drosophila* enhancer. *Genes Dev.* **9**, 256–266.

Anand A., Villella A., Ryner L. C., Carlo T., Goodwin S. F., Song H. J., Gailey D. A., Morales A., Hall J. C., Baker B. S. and Taylor, B. J. (2001) Molecular genetic dissection of the sex-specific and vital functions of the *Drosophila melanogaster* sex determination gene *fruitless*. *Genetics* **158**, 1569–1595.

Antoniewski C., Laval M. and Lepesant J. A. (1993) Structural features critical to the activity of an ecdysone receptor binding site. *Insect Biochem. Mol. Biol.* **23**, 105–14.

Antony C., Davis T. L., Carlson D. A., Péchiné J. M. and Jallon J. M. (1985) Compared behavioral responses of male *Drosophila melanogaster* (Canton S) to natural and synthetic aphrodisiacs. *J. Chem. Ecol.* **11**, 1617–1629.

Antony C. and Jallon J. M. (1981) Evolution des hydrocarbures comportementalement actifs de *Drosophila melanogaster* au cours de la maturation sexuelle. *C. R. Acad. Sci.* **252**, 239–242.

Antony C. and Jallon J. M. (1982) The chemical basis for sex recognition in *Drosophila melanogaster*. *J. Insect Physiol.* **28**, 873–880.

Arthur B. I. Jr, Jallon J. M., Caflisch B., Choffat Y. and Nothiger R. (1998) Sexual behaviour in *Drosophila* is irreversibly programmed during a critical period. *Curr. Biol.* **8**, 1187–1190.

Baker B. S., Burtis K., Goralski T., Mattox W. and Nagoshi R. (1989) Molecular genetic aspects of sex determination in *Drosophila melanogaster*. *Genome* **31**, 638–645.

Bartelt R. J., Schaner A. M. and Jackson L. L. (1985) Cis-vaccenyl acetate as an aggregation pheromone in *Drosophila melanogaster*. *J. Chem. Ecol.* **11**, 1744–1756.

Benamar O. and Jallon J. M. (1983) The production of 7-tricosene, an aphrodisiac of *Drosophila simulans*, is under the control of one gene on the X chromosome (abstr.) *Behav. Genet.* **13**, 533.

Blomquist G. J., Dillwith J. W. and Adams T. S. (1987) Biosynthesis and endocrine regulation of sex pheromone production in Diptera. In *Pheromone Biochemistry*, eds G. D. Prestwitch and G. J. Blomquist, pp. 217–250. Academic Press, London.

Blomquist G. J. and Jackson L. L. (1979) Chemistry and biochemistry of insect waxes. *Prog. Lipid Res.* **17**, 319–345.

Burnet B. and Connolly K. J. (1981) Gene action and the analysis of behaviour. *Br. Med. Bull.* **37**, 107–113.

Burtis K. C. and Baker B. S. (1989) Drosophila *doublesex* gene controls somatic sexual differentiation by producing alternatively spliced mRNAs encoding related sex-specific polypeptides. *Cell* **56**, 997–1010.

Butterworth F. M. (1969) Lipids of *Drosophila*; a new detected lipid in the male. *Science* **163**, 1256–1257.

Carlson D. A., Mayer M. S., Sillhacek D. L., James J. D., Beroza M. and Bierl B. A. (1971) Sex attractant pheromone of the housefly: isolation, identification and synthesis. *Science* **174**, 76–78.

Carlson D. A., Langley P. A. and Huyton P. (1978) Sex pheromone of the tsetse fly: isolation, identification, and synthesis of contact aphrodisiacs. *Science* **201**, 750–753.

Chang Yong T. P. and Jallon J. M. (1986) Synthèse de novo d'hydrocarbures potentiellement aphrodisiaques chez les Drosophiles. *C. r; hebd. Séanc. Acad. Sci.* **303**, 197–202.

Chino H., Downer R. G. H. and Takahashi K. (1977) The role of diacylglycerol-carrying lipoprotein in lipid transport during insect vitellogenesis. *Biochim. Biophys. Acta* **487**, 508–516.

Cline T. W. (1993) The *Drosophila* sex determination signal: how do flies count to two? *Trends Genet.* **9**, 385–390.

Cobb M., Burnet B. and Connolly K. (1988) Sexual isolation and courtship behavior in *Drosophila simulans*, *D. mauritiana*, and their interspecific hybrids. *Behav. Genet.* **18**, 211–225.

Cobb M., Burnett B., Blizard R. and Jallon J. M. (1989) Courtship in *Drosophila sechellia*: its structure, functional aspects, and relationship to those of other members of the *Drosophila melanogaster* species subgroup. *J. Insect Behav.* **2**, 63–89.

Cobb M. and Jallon J. M. (1990) Pheromones, mate recognition and courtship stimulation in the *Drosophila melanogaster* species sub-group. *Anim. Behav.* **39**, 1058–1069.

Coyne J. A. (1996a) Genetics of a difference in male cuticular hydrocarbons between two sibling species, *Drosophila simulans* and *D. sechellia*. *Genetics* **143**, 1689–1698.

Coyne J. A. (1996b) Genetics of differences in pheromonal hydrocarbons between *Drosophila melanogaster* and *D. simulans*. *Genetics* **143**, 353–364.

Coyne J. A. and Charlesworth B. (1997) Genetics of a pheromonal difference affecting sexual isolation between *Drosophila mauritiana* and *D. sechellia*. *Genetics* **145**, 1015–1030.

Coyne J. A., Crittenden A. P. and Mah K. (1994) Genetics of a pheromonal difference contributing to reproductive isolation in *Drosophila*. *Science* **265**, 1461–1464.

Coyne J. A., Wicker-Thomas C. and Jallon J. M. (1999) A gene responsible for a pheromonal polymorphism in *Drosophila melanogaster*. *Genet. Res.* **73**, 189–203.

Dallerac R., Labeur C., Jallon J. M., Knipple D. C., Roelofs W. L. and Wicker-Thomas C. (2000) A Δ9 desaturase gene with a different substrate specificity is responsible for the cuticular diene hydrocarbon polymorphism in *Drosophila melanogaster. Proc. Natl. Acad. Sci. USA* **97**, 9449–9454.

Dantuma N. P., Pijnenburg M. A., Diederen J. H. and Van der Horst D. J. (1998) Multiple interactions between insect lipoproteins and fat body cells: extracellular trapping and endocytic trafficking. *J. Lipid Res.* **39**, 1877–1888.

David J. R. and Capy P. (1988) Genetic variation of *Drosophila melanogaster* natural populations. *Trends Genet.* **4**, 106–111.

Deak P., Omar M. M., Saunders R. D. C., Pal M., Komonyi O., Szidonya J., Maroy P., Zhang Y., Ashburner M., Benos P., Savakis C., Siden-Jiamos I., Louis C., Bolshakov V. N., Kafatos F. C., Madueno E., Modolell J. and Glover, D. M. (1997) *P*-element insertion alleles of essential genes on the third chromosome of *Drosophila melanogaster*: correlation of physical and cytogenetic maps in chromosomal region 86E-87F. *Genetics* **147**, 1697–1722.

de Renobales M. and Blomquist G. J. (1984) Biosynthesis of medium chain fatty acids in *Drosophila melanogaster*. *Arch. Biochem. Biophys.* **228**, 407–414.

de Renobales M., Cripps C., Stanley-Samuelson D. W., Jurenka R. A. and Blomquist G. J. (1987) Biosynthesis of linoleic acid in insects. *Trends Biochem. Sci.* **12**, 364–366.

Diehl P. A. (1975) Synthesis and release of hydrocarbons by oenocytes of the desert locust, *Schistocerca gregaria*. *J. Insect Physiol.* **21**, 1237–1246.

Dillwith J. W., Adams T. S. and Blomquist G. J. (1983) Correlation of housefly sex pheromone production with ovarian development. *J. Insect Physiol.* **29**, 377–386.

El Messoussi S., Wicker C., Arienti M., Carlson D. A. and Jallon J. M. (1994) Hydrocarbons in species recognition in insects. In *Identification and Characterization of Pest Organisms*, ed. D. Hawskworth, pp. 277–287. CABI Press, London.

Ferveur J. F. (1991) Genetic control of pheromones in *Drosophila simulans*. I. *Ngbo*, a locus on the second chromosome. *Genetics* **128**, 293–301.

Ferveur J. F., Cobb M., Boukella H. and Jallon J. M. (1996) World-wide variation in *Drosophila melanogaster* sex pheromone: behavioural effects, genetic basis and potential evolutionary consequences. *Genetica* **97**, 73–80.

Ferveur J. F., Cobb M. and Jallon J. M. (1989) Complex chemical messages in Drosophila. In *Neurobiology of Sensory Systems*, eds R. N. Singh and N. J. Strausfeld. Plenum Publishing Corporation.

Ferveur J. F. and Jallon J. M. (1993a) Genetic control of pheromones in *Drosophila simulans*. II. *kete*, a locus on the X-chromosome. *Genetics* **133**, 561–567.

Ferveur J. F. and Jallon J. M. (1993b) *Nerd*, a locus on chromosome III, affects male reproductive behavior in *Drosophila melanogaster. Naturwissenschaften* **80**, 474–475.

Ferveur J. F. and Jallon J. M. (1996) Genetic control of male cuticular hydrocarbons in *Drosophila melanogaster*. *Genet. Res. Camb.* **67**, 211–218.

Ferveur J. F., Savarit F., O'Kane C. J., Sureau G., Greenspan R. J. and Jallon J. M. (1997) Genetic feminization of pheromones and its behavioral consequences in *Drosophila* males. *Science* **276**, 1555–1558.

Ferveur J. F. and Sureau G. (1996) Simultaneous influence on male courtship of stimulatory and inhibitory pheromones produced by live sex mosaic *Drosophila melanogaster*. *Proc. R. Soc.* **B263**, 967–973.

Finley K. D., Taylor B. J., Milstein M. and McKeown M. (1997) *Dissatisfaction*, a gene involved in sex-specific behavior and neural development of *Drosophila melanogaster*. *Proc. Natl. Acad. Sci. USA* **94**, 913–918.

Jallon J. M. (1984) A few chemical words exchanged by *Drosophila* during courtship and mating. *Behav. Genet.* **14**, 441–478.

Jallon J. M., Antony C. and Benamar O. (1981) Un anti-aphrodisiaque produit par les mâles de *Drosophila melanogaster* et transféré aux femelles lors de la copulation. *C. R. Acad. Sci. Paris* **292**, 1147–1149.

Jallon J. M., Antony C., Chang Yong T. P. and Maniar S. (1986) Genetic factors controlling the production of aphrodosiac substances in *Drosophila*. In *Advances in Invertebrate Reproduction, vol. 4* eds M. Porchet J. C. Andries and A. Dhainaut, pp. 445–452. Amsterdam, Elsevier.

Jallon J. M. and David J. R. (1987) Variations in cuticular hydrocarbons among the eight species of the *Drosophila melanogaster* subgroup. *Evolution* **4**, 294–302.

Jallon J. M. and Hotta Y. (1979) Genetic and behavioral studies of female sex appeal in *Drosophila*. *Behav. Genet.* **9**, 257–275.

Jallon J. M., Laugé G., Orssaud L. and Antony C. (1988) Female pheromones in *Drosophila melanogaster* are controlled by the *doublesex* locus. *Genet. Res. Camb.* **51**, 17–22.

Jallon J. M. and Péchiné J. M. (1989) Une autre race de chimique *Drosophila melanogaster* en Afrique. *C. R. Acad. Sci.* **309**, 1551–1556.

Kaestner K. H., Ntambi J. M., Kelly T. J. Jr, and Lane M. D. (1989) Differentiation-induced gene expression in 3T3-L1 preadipocytes. A second differentially expressed gene encoding stearoyl-CoA desaturase. *J. Biol. Chem.* **264**, 14755–14761.

Katase H. and Chino H. (1982) Transport of hydrocarbons by the lipophorin of insect hemolymph. *Biochim. Biophys. Acta* **710**, 341–348.

Katase H. and Chino H. (1984) Transport of hydrocarbons by hemolymph lipophorin in *Locusta migratoria*. *Insect Biochem.* **14**, 1–6.

Labeur C., Dallerac R. and Wicker-Thomas C. (2002) Involvement of *desat1* gene in the control of *Drosophila melanogaster* pheromone biosynthesis. *Genetica* **114**, 269–274.

Lee K. N., Pariza M. W. and Ntambi J. M. (1996) Differential expression of hepatic stearoyl-CoA desaturase gene 1 in male and female mice. *Bioch. Biophys. Acta* **1304**, 85–88.

Manning A. (1959) The sexual behaviour of two sibling *Drosophila* species. *Behaviour* **15**, 123–145.

McKeown M., Belote J. M. and Boggs R. T. (1998) Ectopic expression of the female *transformer* gene product leads to female differentiation of chromosomally male *Drosophila*. *Cell* **53**, 887–895.

McRobert S. P. and Tompkins L. (1985) The effect of *transformer*, *doublesex* and *intersex* mutations on the sexual behavior of *Drosophila melanogaster*. *Genetics* **111**, 89–96.

Moulin B., Rybak F., Aubin T. and Jallon J. M. (2001) Compared ontogenesis of courtship song components of males from the sibling species, *D. melanogaster* and *D. simulans*. *Behav. Genet.* **31**, 299–308.

Myers E. W., Sutton G. G., Delcher A. L., Dew I. M., Fasulo D. P., Flanigan M. J., Kravitz S. A., Mobarry C. M., Reinert K. H., Remington K. A., Anson E. L., Bolanos R. A., Chou H. H., Jordan C. M., Halpern A. L., Lonardi S., Beasley E. M., Brandon R. C., Chen L., Dunn P. J., Lai Z., Liang Y., Nusskern D. R., Zhan M., Zhang Q., Zheng X., Rubin G. M., Adams M. D. and Venter J. C. (2000) A whole-genome assembly of *Drosophila*. *Science* **287**, 2196–2204.

Nemoto T., Doi M., Oshio K., Matsubayashi H., Oguma Y., Suzuki T., Kuwahara Y. (1994) (Z,Z)-5,27-tritriacontadiene: major sex pheromone of *Drosophila pallidosa* (Diptera; Drosophilidae). *J. Chem. Ecol.* **20**, 3029–3037.

Ntambi J. M., Buhrow S. A., Kaestner K. H., Christy R. J., Sibley E., Kelly T. J. Jr, and Lane M. D. (1988) Differentiation-induced gene expression in 3T3-L1 preadipocytes. *J. Biol. Chem.* **263**, 17291–17300.

Oguma Y., Nemoto T. and Kuwahara Y. (1992a) (Z)-11-Pentacosene is the major sex pheromone component in *Drosophila virilis* (Diptera). *Chemoecology* **3**, 60–64.

Oguma Y., Nemoto T. and Kuwahara Y. (1992b) A sex pheromone study of a fruit fly *Drosophila virilis* Sturtevant (Diptera: Drosophilidae): additive effect on cuticular alkadienes to major sex pheromone. *Appl. Entomol. Zool.* **27**, 499–505.

Oguma Y., Jallon J. M., Tomaru M. and Matsubayashi H. (1996) Courtship behavior and sexual isolation between *Drosophila auraria* and *D. triauraria* in darkness and light. *J. Evol. Biol.* **9**, 803–815.

Péchiné J. M., Antony C. and Jallon J. M. (1988) Precise characterization of cuticular compounds in young *Drosophila* by mass spectrometry. *J. Chem. Ecol.* **14**, 1071–1085.

Pennanec'h M., Ferveur J. F., Pho D. B. and Jallon J. M. (1991) Insect fatty acid related pheromones: a review of their biosynthesis, hormonal regulation and genetic control. *Ann. Soc. Ent. Fr.* **27**, 245–263.

Pennanec'h M., Bricard L., Kunesh G. and Jallon J. M. (1997) Incorporation of fatty acids into cuticular hydrocarbons of male and female *Drosophila melanogaster*. *J. Insect Physiol.* **43**, 1111–1116.

Pho D. B., Pennanec'h M. and Jallon J. M. (1996) Purification of adult *Drosophila melanogaster* lipophorin and its role in hydrocarbon transport. *Arch. Insect Biochem. Physiol.* **31**, 289–303.

Roelofs W. and Wolf W. A. (1988) Pheromone biosynthesis in Lepidoptera. *J. Chem. Ecol.* **14**, 2019–2031.

Romer F. (1991) The oenocytes of insects: differentiation, changes during molting, and their possible involvement in the secretion of molting hormone. In *Recent Advances in Comparative Arthropod Morphology, Physiology and Development*, ed. A. P. Gupta, Vol. 3, pp. 542–566. Rutgers University Press, New Brunswick, NJ.

Rouault J., Capy P. and Jallon J. M. (2001) Variations of male cuticular hydrocarbons with geoclimatic variables: an adaptative mechanism in *Drosophila melanogaster*? *Genetica* **110**, 117–130.

Rouault J., Marican C., Wicker-Thomas C. and Jallon J. M. (2003) Relations between cuticular hydrocarbon polymorphism, resistance against dessication and breeding temperature: a model for their evolution in *D. melanogaster* and *D. simulans*. *Genetica* (in press).

Ryner L. C., Goodwin S. F., Castrillon D. H., Anand A., Villella A., Baker B. S., Hall J. C., Taylor B. J. and Wasserman S. A. (1996) Control of male sexual behavior and sexual orientation in *Drosophila* by the *fruitless* gene. *Cell* **87**, 1079–1089.

Savarit F. and Ferveur J. F. (2002) Genetic study of the production of sexually dimorphic cuticular hydrocarbons in relation with the sex-determination gene *transformer* in *Drosophila melanogaster*. *Genet. Res.* **79**, 23–40.

Savarit F., Sureau G., Cobb M. and Ferveur J. F. (1999) Genetic elimination of known pheromones reveals the fundamental chemical basis of mating and isolation in *Drosophila*. *Proc. Natl. Acad. Sci. USA* **96**, 9015–9020.

Scott D. (1994) Genetic variation for female mate discrimination in *Drosophila melanogaster*. *Evolution* **48**, 112–121.

Scott D. and Jackson L. L. (1988) Interstrain comparison of male predominant antiaphrodisiacs in *Drosophila melanogaster*. *J. Insect Physiol.* **34**, 863–871.

Scott D. and Richmond R. C. (1988) A genetic analysis of male-predominant pheromones of *Drosophila melanogaster*. *Genetics* **119**, 639–646.

Spieth T. H. (1952) Mating behavior within the genus *Drosophila* (Diptera). *Bull. Am. Museum Nat. His.* **99**, 395–474.

Steinmann-Zwicky M., Amrein H. and Nothiger R. (1990) Genetic control of sex determination in *Drosophila*. *Adv. Genet.* **27**, 189–237.

Sturtevant A. H. (1915) Experiments of sex recognition and the problem of sexual selection in *Drosophila*. *J. Anim. Behav.* **5**, 351–366.

Sureau G. and Ferveur J. F. (1999) Co-adaptation of pheromone production and behavioural responses in *Drosophila melanogaster* males. *Genet. Res. Camb.* **74**, 129–137.

Takahashi A., Tsaur S.C., Coyne J.A. and Wu C.I. (2001) The nucleotide changes governing cuticular hydrocarbon variation and their evolution in *Drosophila melanogaster. Proc. Natl. Acad. Sci. USA* **98**, 3920–3925.

Tompkins L. and McRobert S. P. (1989) Regulation of behavioral and pheromonal aspects of sex determination in *Drosophila melanogaster* by the *Sex-lethal* gene. *Genetics* **123**, 535–541.

Tompkins L. and McRobert S. P. (1995) Behavioral and pheromonal phenotypes associated with expression of loss-of-function mutations in the *Sex-lethal* gene of *Drosophila melanogaster*. *J. Neurogenet.* **9**, 219–226.

Waterbury J. A., Jackson L. L. and Schedl P. (1999) Analysis of the *doublesex* female protein in *Drosophila melanogaster*: role in sexual differentiation and behavior and dependence on *intersex*. *Genetics* **152**, 1653–1667.

Welbergen P. van Dijken F. R. and Scharloo W. (1987) Collation of the courtship behaviour of the sympatric species *Drosophila melanogaster* and *Drosophila simulans*. *Behaviour* **101**, 253–274.

Wicker C. and Jallon J. M. (1995a) Influence of ovary and ecdysteroids on pheromone biosynthesis in *Drosophila melanogaster* (Diptera: Drosophilidae). *Eur. J. Entomol.* **92**, 197–202.

Wicker C. and Jallon J. M. (1995b) Hormonal control of sex pheromone biosynthesis in *Drosophila melanogaster*. *J. Insect Physiol.* **41**, 65–70.

Wicker-Thomas C., Henriet C. and Dallerac R. (1997) Partial characterization of a fatty acid desaturase gene in *Drosophila melanogaster*. *Insect Biochem. Molec. Biol.* **27**, 963–972.

Yamamoto D., Jallon J. M. and Komatsu A. (1997) Genetic dissection of sexual behavior in *Drosophila melanogaster*. *Ann. Rev. Entomol.* **42**, 551–585.

Zawitowski S. and Richmond R. C. (1986) Inhibition of courtship and mating of *Drosophila melanogaster* by the male-produced lipid, cis-vaccenyl acetate. *J. Insect Physiol.* **32**, 189–192.

Zheng Y., Eilertsen K. J., Ge L., Zhang L., Sundberg J. P., Prouty S. M., Stenn K. S. and Parimoo S. (1999) *Scd1* is expressed in sebaceous glands and is disrupted in the *asebia* mouse. *Nat. Genet.* **23**, 268–270.

Zheng Y., Prouty S. M., Harmon A., Sundberg J. P., Stenn K. S. and Parimoo S. (2001) *Scd3* – a novel gene of the stearoyl-CoA desaturase family with restricted expression in skin. *Genomics* **71**, 182–191.

10

Regulation of pheromone biosynthesis, transport, and emission in cockroaches

Coby Schal, Yongliang Fan, and Gary J. Blomquist

10.1 Introduction

Chemical signaling is fundamental for the survival and reproduction of most animals. Chemical cues allow animals to appraise their environment, to detect food, toxins, prey, predators and pathogens, to identify kin, and to evaluate and bias mate-choice decisions. Many insects depend on sex pheromones for the identification of reproductive partners and the onset of reproductive physiology and mating behavior. However, the prevalence and importance of chemical signaling in various insect groups depends on the natural history of the insects. Cockroaches are the quintessential chemical communicators. In contrast to the closely related grasshoppers and crickets (Orthoptera), which rely on sound as their primary modality in communication, cockroaches use olfactory and tactile signals in their social behavior. Most of the 4000+ described species of cockroaches are nocturnal. Many use long-range volatile pheromones in mate-finding and cuticular contact pheromones in the final recognition process. Short-range volatile pheromones emitted by the male, coupled with nuptial tergal secretions, facilitate proper alignment of the pair prior to copulation. Pheromones also mediate intrasexual conflicts, especially when males establish dominance hierarchies and territories, in parent–offspring communication, stage and population recognition, trail-following behavior, and as epidiectic pheromones that mediate dispersion behavior. In this chapter we discuss only sex pheromones that are used in mate-finding and recognition. First, we describe reproductive modes in cockroaches as they relate to pheromone production. We then present a conceptual framework for neuroendocrine regulation of cockroach reproductive biology. Last, we discuss pheromone production in several of the best studied species, reviewing for each what is known about the tissues, biochemical pathways, transport routes, and neuroendocrine regulation of pheromone production. Cockroach sex pheromones in general were last reviewed by Gemeno and Schal (2003) and regulation of pheromone production in cockroaches was last reviewed by Schal and Smith (1990), Tillman *et al.* (1999) and Blomquist *et al.* (2003).

10.2 Reproduction and pheromones in cockroaches

Blattodea (Order Dictyoptera) is divided into five families, three of which – Blattidae, Blattellidae, and Blaberidae – contain most of the cockroach species (McKittrick, 1964) and all the species for which sex pheromones have been identified to date. The other two families, Cryptocercidae and Polyphagidae, are poorly known. This highly diverse group of insects exhibits a variety of reproductive strategies, including obligatory and facultative parthenogenesis, oviparity, ovoviviparity, and viviparity (Roth, 1970). Their social organization ranges from solitary individuals to genetically related families with monogamous parents. They live in temperate as well as tropical habitats, deserts, caves and hollow trees, bromeliad pools, in the nests of birds and social insects, and in association with human-built structures, sewers, and dump sites (Schal *et al.*, 1984).

Oviparous females oviposit their eggs into a protective egg case (ootheca), which is either deposited soon after being produced or carried by the mother until the nymphs hatch, depending on the species. In ovoviviparous and viviparous species the egg case is reduced, the embryos are incubated within a brood sac or uterus, and live nymphs hatch from the mother (parturition). The reproductive cycle in cockroaches is regulated by several lipid and peptide hormones, but juvenile hormone III (JH III), a C_{16} sesquiterpenoid, is a critical adult gonadotropic hormone. JH III stimulates the fat body to produce vitellogenin – a yolk protein precursor – and the oocytes to endocytose vitellogenin, it stimulates the production of oothecal proteins by accessory glands, and it elevates food intake in vitellogenic females. As will become apparent in this chapter, behavioral and physiological events related to mate-finding and sexual receptivity are also regulated in a coordinated manner by this vital hormone. Generally, the female cockroach undergoes sexual maturation for 1–3 days after the imaginal molt, she feeds and drinks extensively as her corpora allata (CA) synthesize and secrete JH III, and the fat body responds by synthesizing massive amounts of vitellogenin. In response to JH III the virgin female also produces and releases sex pheromones, and mates. Her basal oocytes continue to synchronously take up yolk protein precursors and grow, and as hemolymph ecdysteroids peak the JH III titer dramatically declines, the oocytes become chorionated, they are ovulated, fertilized, and the eggs packaged into an ootheca. Most oviparous cockroaches carry the ootheca for just one to a few days, but ovoviviparous species retract it into the uterus and incubate the embryos for several weeks or months. During this gestation period, the JH III titer in the hemolymph is low, the gravid female feeds little and only sporadically, and she remains sexually unreceptive. As the neonates synchronously hatch from the egg case, a second gonotrophic cycle ensues. Depending on the species, an adult female can undergo ~6–30 such cycles. In connection with pheromone production, sexual receptivity may be permanently suppressed after mating, or it may reappear during the vitellogenic stage of each successive reproductive cycle.

Cockroaches have long served as a model system for studies of invertebrate biochemistry, endocrinology, and neurobiology (Tobe and Stay, 1985; Scharrer, 1987; Huber *et al.*, 1990). Therefore, a wealth of information is available about endocrine regulation of reproduction, which provides a useful foundation for studies of the neural and endocrine bases of pheromone production and emission. The cockroach system has other advantages. First, both volatile and contact pheromones have been characterized chemically and behaviorally in some species. Second, their large size facilitates biochemical and physiological investigations. And finally, cockroaches are major urban pests and cockroach infestations spread disease and serve as a major source of allergens that can cause asthma. It is therefore conceivable that understanding when, where, and how sex pheromones are produced will help shape new approaches on environmentally compatible pest control.

10.3 Regulatory mechanisms of pheromone production and release

Regulation of sex pheromone production and emission can occur through several major regulatory mechanisms, but it is likely that several of these mechanisms operate in concert in most insects:

1 Organogenesis of sex pheromone-producing glands occurs during pre-imaginal development. In the adult stage, a period of glandular differentiation is followed by alternating phases of sexual receptivity followed by sexual inactivity, generally during gestation. In some insects, pheromone secretory cells hypertrophy and gain competence for the production of pheromone during periods of sexual receptivity and regress during stages of sexual inactivity.
2 Availability of biosynthetic precursors from food, as in some danaid butterflies and arctiid moths, which modify plant pyrrolizidine alkaloids into volatile derivatives that are used as male pheromones (e.g. Schneider *et al.*, 1975). In

insects that produce large amounts of cuticular lipids, food availability (e.g. malonyl-CoA) may control the flux of carbon into the pheromone biosynthetic pathway. A deficiency of specific methyl branch donors, for example, may impede production of methyl-branched hydrocarbons.

3 Endocrine mediators, including specific pheromonotropins (e.g. pheromone biosynthesis activating neuropeptide [PBAN]) (reviewed in Raina, 1993; Nagasawa *et al.*, 1994; Rafaeli, 2002; Rafaeli and Jurenka, Chapter 5, in this volume), JH (reviewed in Tillman *et al.*, 1999), and ecdysteroids (reviewed in Blomquist *et al.*, 1993; Blomquist, Chapter 8, in this volume), can regulate pheromone biosynthesis in different insect species.
4 Neuronal signals that descend from the central nervous system (CNS) or ascend through the ventral nerve cord (VNC) can modulate sex pheromone production and/or its release.
5 Chemical signaling can be regulated at the level of calling – the emission of attractant sex pheromones during specific behaviors. In some insects, pheromone production continues uninterrupted in the adult stage, and mate-finding is regulated only through controlled release of the pheromone (e.g. Charlton and Roelofs, 1991; Schal *et al.*, 1998a).

10.4 Barth's hypothesis revisited

In 1965 Barth proposed a hypothesis relating neuroendocrine control of pheromone production to life history strategies of insects. Specifically, neuroendocrine control was predicted for insects with a long-lived adult stage and with multiple reproductive cycles interrupted by periods during which sexual receptivity and mating are not appropriate or not even possible anatomically. Conversely, in insects that eclose with mature oocytes, and live for only a few days as adults, Barth (1965) predicted that pheromone signaling would be part of the adult metamorphic process and not subject to neuroendocrine control. Early work on long-lived cockroaches and beetles and short-lived moths lent support to this hypothesis. However, the discovery of PBAN in a number of lepidopteran insects (see Rafaeli, 2002; Rafaeli and Jurenka, Chapter 5, in this volume) called for a reassessment of this hypothesis. Yet, two fundamental elements of Barth's hypothesis were overlooked in a rush to dismantle this rather useful conceptual framework. First, Barth clearly stated that "in those [Lepidoptera] species which do feed as adults, oocytes maturation requires the participation of JH and perhaps neuroendocrine factors as well. So far as we know the control of mating behavior in Lepidoptera which feed as adults has not been investigated but, . . . the existence of neuroendocrine regulating mechanisms for mating behavior would not be surprising. In fact we might venture to predict that some form of neuroendocrine regulation of mating behavior is likely in the majority of insects which feed as adults and which depend on JH or neuroendocrine factors for oocytes maturation" (Barth and Lester, 1973). Second, the hypothesis relates the existence of neuroendocrine control of pheromone production to the type of reproductive cycle exhibited. It predicts that neuroendocrine control would also be found in short-lived moths that exhibit periods of sexual inactivity. Delaying reproductive activity until after migration is completed (reviewed in Dingle, 2002) or a suitable host plant is located (e.g. Raina *et al.*, 1997) are examples of such a strategy.

We propose a reconsideration of Barth's hypothesis, taking into account the coordination of reproductive developmental processes with mating-related events. In long-lived insects, such as cockroaches, pheromone production is regulated synchronously with other reproductive processes by the same hormone, usually JH. Cellular remodeling of the pheromone glands plays a prominent role in this group of insects, resulting in a slow stimulation of pheromone production. The cessation of pheromone production after mating is also slow, and precise control of pheromone signaling, therefore, is not at the level of pheromone production, but rather at the behavioral level through pheromone emission during calling. Conversely, in short-lived moths rapid modulation of rate-limiting enzymes in the pheromone biosynthetic pathway is much more prominent than developmental processes, and pheromone biosynthesis is turned on or off in coordination with activity cycles (day versus night) and sexual receptivity (virgin versus mated). Control of sexual signaling occurs at the level of pheromone production as well as emission, but these two events are usually regulated by different factors. Thus, both groups of insects exhibit neuroendocrine control of pheromone production. In cockroaches, pheromone production is coordinated with the gonotrophic cycle and the major gonadotropic hormone – JH – has been recruited to control both by acting at several target tissues. In most moths, on the other hand, reproduction and pheromone production are regulated by different hormones. But here also, the hormones that control pheromone production (e.g. PBAN) also affect other target tissues as myotropins, melanization agents, and diapause and pupariation factors. An interesting departure from the moth model occurs in migratory moth species in which reproduction is delayed by migration (low levels of JH production and sexual inactivity), and pheromone production and its release are JH dependent (Cusson *et al.*, 1994; Dingle, 2002). All these observations are consistent with our interpretation of Barth's model.

10.5 Anatomical sites of pheromone production and calling behavior

10.5.1 Volatile pheromones

Although early reports considered various organs in the head, thorax, and abdomen as possible sites of pheromone production in cockroaches, most of the more recent research implicates specialized abdominal glands. Nevertheless, the biosynthetic tissue may be dorsal (tergal), ventral (sternal), genital (e.g. atrial glands), or associated with the digestive tract. Periplanones-A and -B, the sex pheromone components of *Periplaneta americana* (L.) (American cockroach, Blattidae), are a case in point. They are highly volatile and readily adsorb to females, their feces, and the cage in which virgin females are kept. This, coupled with the extraordinary behavioral sensitivity of males to periplanones, seriously confounded early behavioral bioassays of female extracts. Indeed, the exact location of the pheromone in the abdomen has yet to be resolved. There is general agreement that the greatest amount of pheromone activity is in the midgut (Takahashi *et al.*, 1976; Persoons *et al.*, 1979), and that it can be recovered from feces. However, female calling involves opening the genital vestibulum (=atrium; posterior genital pouch where the egg case is formed – terminology of McKittrick, 1964), without excretion of feces (Abed *et al.*, 1993b), and this glandular tissue (atrial glands) in fact attracts males (Seelinger, 1984). GC-MS analysis of extracts from the alimentary canal, the last two abdominal segments (which contain the atrial glands), and the anterior part of the abdomen confirmed that the atrial glands contain most of the periplanone-B, up to 60 ng (Abed *et al.*, 1993b). In *Blatta orientalis* L. (Oriental cockroach), a closely related species, the atrial glands also elicited the strongest male responses and of various tissues tested, only the atrial glands stimulated any males to court by raising their wings (Abed *et al.*, 1993a). The epithelium of the atrial gland consists of class-1 glandular cells, in which the secretion passes directly to the cuticle and not through a duct. However, Yang *et al.* (1998) extracted several abdominal tissues of *P. americana*, including the atrial glands, and concluded that both periplanone-A and periplanone-B were most abundant in the colon (0.34 and 8.31 ng, respectively), and that this tissue elicited the strongest electroantennogram (EAG) responses in males. Moreover, feces extracted from the colon elicited strong sexual responses in males, showing that fecal material contained pheromone before it could contact the atrial glands. Unfortunately, in these two species, as in other cockroaches that produce volatile sex pheromones, no definitive proof of the sites of pheromone biosynthesis is available through isotope tracing of pheromone precursors in isolated tissues.

The volatile sex pheromone of female *Blattella germanica* L. (German cockroach, Blattellidae) is produced in the 10th abdominal tergite, termed the pygidium. Because this pheromone has not been chemically identified, this conclusion is based upon surgical manipulations followed by bioassays with both live females and female extracts. In olfactometers, fewer males were attracted to the pygidium of males than to that of virgin females (Abed *et al.*, 1993c), and in two-choice assays females whose pygidium was ablated attracted only 13 percent of the males, whereas sham-operated females attracted 87 percent of the males (Liang and Schal, 1993a). Clear tergal modifications can be seen on the anterior of the female pygidium but not in nymphs or males. The pheromone gland consists of three groups of cuticular depressions, one medial and two lateral, and each group contains numerous cuticular depressions, within which are 1–32 orifices (Liang and Schal, 1993c; Tokro *et al.*, 1993). Each orifice is ~0.5 μm in diameter and leads to a duct that penetrates through the cuticle and inserts deeply within a large secretory cell located in the modified epithelium beneath the cuticle. This class-3 secretory unit consists of a large secretory cell and a much smaller duct cell. Within the secretory cell a complex end-apparatus is elaborated which consists of numerous long microvilli. The secretory cells of mature, pheromone-producing virgin females are characterized by abundant mitochondria, SER, RER, a large nucleus, and numerous secretory vesicles. The latter, presumably containing pheromone, are discharged into the end-apparatus and up through the duct to the cuticular surface. Large amounts of secreted material can be found within the ducts and in the cuticular depressions (Liang and Schal, 1993c). The volatile pheromone of *B. germanica* is emitted while the female calls, a behavior during which the wings are elevated, abdomen lowered, and the genital vestibulum of the female occasionally exposed (Liang and Schal, 1993b). Calling females have been shown to release more pheromone and they are more likely to mate than non-calling females.

Similar pheromone glands are found in other cockroaches that produce volatile sex pheromones, but the glandular pores are usually more uniformly distributed on the cuticular surface and do not form the specialized regions evident in *B. germanica*. While calling, female *Supella longipalpa* (F.) (brownbanded cockroach, Blattellidae) release a volatile pheromone, 5-(2*R*′,4*R*′-dimethylheptyl)-3-methyl-2*H*-pyran-2-one (supellapyrone), that elicits sexual responses in males (Charlton *et al.*, 1993; Leal *et al.*, 1995). Although the genital vestibulum is exposed intermittently during calling, as in *Periplaneta* and *Blattella*, behavioral, electrophysiological, and morphological studies have localized the sex pheromone to the 4th and 5th tergites of virgin females (Schal *et al.*, 1992). The tergal glands of *S. longipalpa* females are composed of multiple class-3 secretory units, each leading through a long unbranched duct to a single cuticular pore, as in *Blattella*. Cuticular pores occur on all tergites, but their density is highest on the lateral margins of the 4th and 5th tergites.

Parcoblatta lata (Brunner) and *Parcoblatta caudelli* Hebard (Blattellidae) are forest-dwelling wood cockroaches endemic to North America. Sexually receptive females call, during which they release long-range sex pheromones to which males are attracted (Gemeno *et al.*, 2003). Based on a correlation between the density of cuticular pores that overlie class-3 glandular cells and behavioral and EAG analyses of tissue extracts, as in *S. longipalpa*, it was concluded that the sex pheromone is produced only in tergites 1–7 (Gemeno *et al.*, 2003). However, it remains to be demonstrated unequivocally that these cells produce pheromone, as in other species, because similar integumental glands are found elsewhere on the body, and they have been shown to serve multiple functions, including pheromone production (Noirot and Quennedey, 1974).

Volatile sex pheromones are also released by male cockroaches, usually from tergal or sternal glands. Two types of behaviors are evident in males: (1) calling behavior, analogous to female calling, occurs in some species the absence of females; and (2) courtship behavior, during which tergal glands are exposed and release short-range attractants. The best studied example is *Nauphoeta cinerea* (Olivier) (lobster cockroach, Blaberidae). Glands on sternites 3–7 are exposed during calling, releasing a volatile male pheromone (Sreng, 1979a). The sternal glands are made up of four categories of cells including the class-3 glandular units seen in other species. Orifices of the class-3 cells are found only on the anterior zone of each sternite, but no specialized cuticular structures are evident (Sreng, 1979b, 1984, 1985). The sternal glands produce a mixture of the four components 3-hydroxy-2-butanone, 2-methylthiazolidine, 4-ethyl-2-methoxyphenol, and 2-methyl-2-thiazoline (Sreng, 1990; Sirugue *et al.*, 1992). Similar glands are also present in the closely related *Rhyparobia* (=*Leucophaea*) *maderae* (F.) (Madeira cockroach), but on sternites 2–7 (Sreng, 1984).

The tergal glands of *N. cinerea* also exhibit sexual dimorphism with more orifices on tergites 1 and 2 in males than in females (Brossut and Roth, 1977). These glands also lack cuticular modifications, other than individual cuticular pores, and are composed of five categories of cells, including class-3 glandular units, and three of the four compounds found in the sternal glands are also present in the tergal glands, but at much lower amounts. Presumably, these components of the pheromone are released during courtship, as the male raises his wings, and they stimulate the female to mount the male and feed from his tergal secretions. This wing-raising stance is common to male cockroaches of other species.

In some species the male tergal glands form specialized regions on the cuticular surface and produce a blend of close-range attractants, phagostimulants, and nutrients that the male deploys to place the female in a pre-copulatory position. For example, the tergal glands of *B. germanica* consist of traverse cuticular depressions on the 7th and 8th tergites, connected to numerous class-3 secretory cells (Sreng and Quennedey, 1976; Brossut and Roth, 1977), as in the female's pygidial gland (above). The tergal secretions include a number of volatile compounds, mainly carboxylic acids (>92 percent), but none have been tested for behavioral effects on females (Brossut *et al.*, 1975). Because males and nymphs also feed upon the exposed tergal secretions of courting males, these secretions appear to serve as non-specific food attractants.

The non-volatile fraction of the tergal secretion of male *B. germanica* contains lipids, proteins, and carbohydrates (Brossut *et al.*, 1975), but their identity and role(s) in courtship behavior have not been investigated. Recently, however, Nojima *et al.* (1999, 2002) identified several male compounds that serve as phagostimulants, including a mixture of oligosaccharides (D-maltose, its analogs D-maltotriose and D-maltotetraose, four oligoglucosyl trehaloses with $(1 \rightarrow 4)$-α-glycosidic linkages), and two trisaccharides. Activity of an artificial blend of these sugars is significantly enhanced by a polar lipid fraction that contains a mixture of phosphatidylethanolamines and phosphatidylcholines (lecithins) (Kugimiya *et al.*, 2002).

10.5.2 Contact pheromones

Cuticular contact pheromones mediate species and sex recognition and, in most cases, they function as courtship-inducing pheromones. Although they are probably present in most cockroach species, sex-specific contact pheromones have been identified in only two species, *B. germanica* females and *N. cinerea* males. They are thought to be distributed throughout the epicuticular surface and are perceived by means of antennal contact and with the mouth parts. Male-produced contact pheromones may function in male–male recognition and to inhibit courtship in other males (Fukui and Takahashi, 1983).

The best studied contact pheromone, indeed the only cockroach pheromone that has been investigated with isotope tracing, is the blend of hydrocarbon-derived ketones of *B. germanica* females that elicits courtship responses in males. The female German cockroach produces a pheromone blend that includes (3*S*,11*S*)-dimethylnonacosan-2-one as the major component, and three related compounds [29-hydroxy-(3*S*,11*S*)-dimethylnonacosan-2-one, 29-oxo-(3,11)-dimethylnonacosan-2-one, and 3,11-dimethylheptacosan-2-one] are also behaviorally active (Nishida *et al.*, 1974, 1976; Nishida and Fukami, 1983; Schal *et al.*, 1990b). Other putative pheromone components with a 3,11-dimethyl branching pattern are present on the female cuticular surface, including C_{31} and C_{33} methyl ketones (Schal, unpublished results), but their biological activity has not been confirmed. Critical to studies on the source of this pheromone was understanding its biosynthetic pathway (section 10.7.2.1, Figure 10.1) and evidence that a fatty acid synthetase and methyl-branched fatty acid precursors could be isolated from the abdominal epidermis but not from the fat body (Juarez *et al.*, 1992). Gu *et al.* (1995) incubated various tissues with [1–^{14}C]propionate, which labels methylmalonyl-CoA and methyl-branched hydrocarbons (Figure 10.1), and concluded that only the integument underlying abdominal sternites and tergites produced methyl branched hydrocarbons and the methyl ketone pheromone.

Oenocytes have been shown to biosynthesize hydrocarbons in several insect species in which the oenocytes are within the hemocoel and could be readily separated from other tissues (Diehl, 1973, 1975; Romer, 1980, 1991). These cells are characteristically very large, among the largest somatic cells in insects, and are rich in mitochondria and SER, as are steroidal cells in mammalian systems, suggesting participation in lipid synthesis (Rinterknecht and Matz, 1983). In the German and American cockroaches, however, the oenocytes are

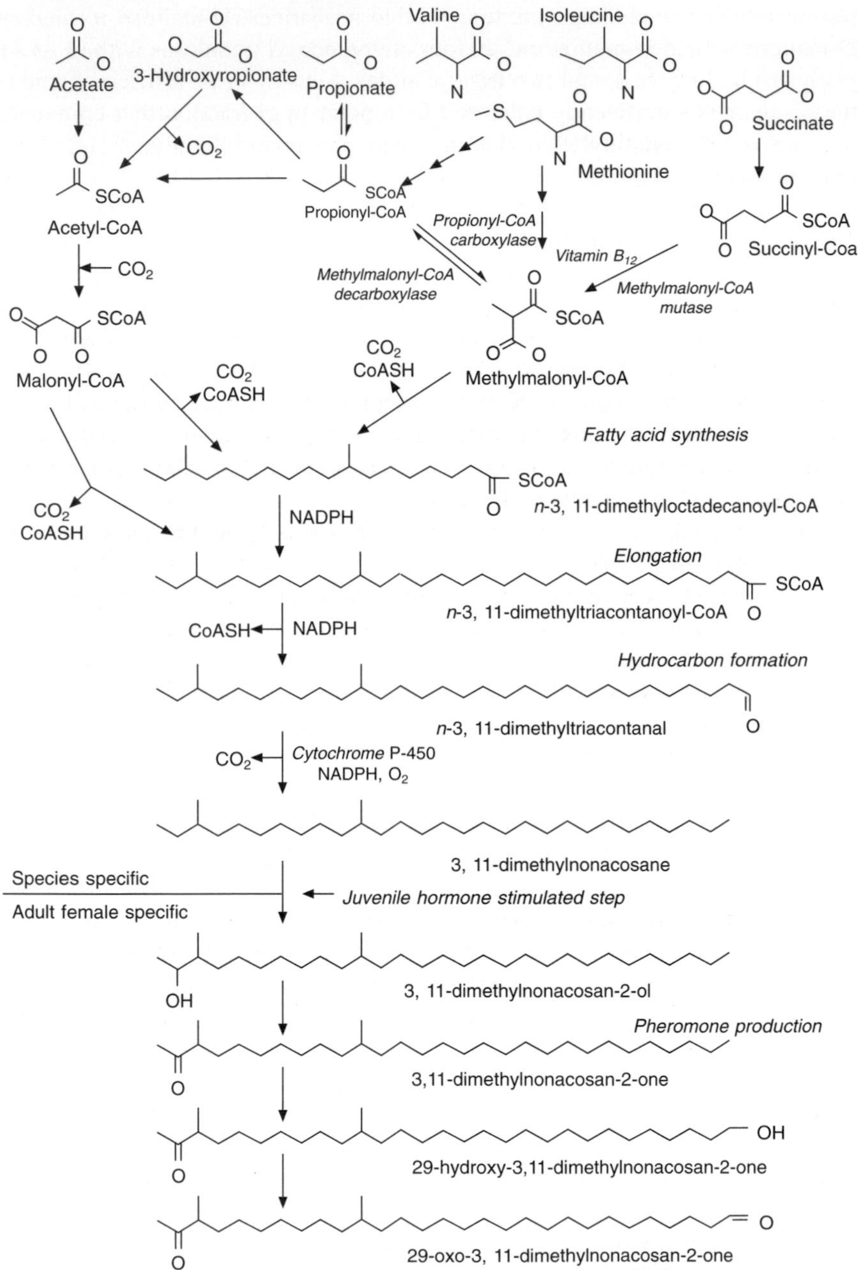

Figure 10.1 Metabolic pathways for the biosynthesis of 3,11-dimethylnonacosan-2-one and related sex pheromone components of the female contact sex pheromone of the German cockroach, *Blattella germanica*.

localized within the abdominal integument, separated from the hemocoel by a basal lamina (Kramer and Wigglesworth, 1950; Liang and Schal, 1993c). As a consequence, the definitive localization of hydrocarbon biosynthesis in these insects has been hampered by the inability to separate the integument into its component cell types. Recently, Fan *et al.* (2003) used enzymatic digestion of the basal lamina and the extracellular matrix to dissociate the integument underlying the sternites of female *B. germanica* into a cell suspension. Based upon ultrastructural details, two cell fractions were obtained: an "epidermal cell-enriched" sample that contained ~85 percent epidermal cells (<15 μm) and only 15 percent oenocytes (>30 μm), and an "oenocyte-enriched" sample that contained 60 percent oenocytes. Each cell suspension was then further separated into fractions in a Percoll gradient. In both gradients there was broad overlap between the two cell types, but nonetheless, epidermal cells predominated in more buoyant fractions, while most oenocytes were in fractions of higher density. Hydrocarbon synthesis was then measured in each fraction by incorporation of [1-^{14}C]propionate into methyl-branched hydrocarbons. Only very large cells produced hydrocarbons, whereas the much larger population of small cells did not (Fan *et al.*, 2003). These results demonstrate, for the first time using direct biochemical evidence, that hydrocarbons are produced by oenocytes and not by other cell types in an insect in which oenocytes are confined to the abdominal integument. They support the elegant genetic approach of Ferveur *et al.* (1997) showing that targeted expression of the *transformer* gene in oenocytes of male *Drosophila melanogaster* resulted in feminization of his hydrocarbon pheromone mixture.

10.6 Pheromone production regulated by gland development and cellular plasticity

Developmental regulation of pheromone biosynthesis must be universal among insects because organogenesis of a sex-specific gland occurs only in the pre-imaginal stage. However, this mechanism has received meager attention from researchers. In short-lived insects that emerge as reproductively competent adults it is expected that functional competence of the gland for pheromone production might be regulated by the same factors that control pheromone gland development (e.g. ecdysteroids – see Tang *et al.*, 1991, for example). In insects whose pheromone gland acquires functional competence during an imaginal maturation period, developmental regulation might involve factors that also control adult reproductive readiness. Examples include many insects that mature sexually for several days after eclosion, and adult insects that enter diapause or migrate before the onset of reproduction. Competence for the production of pheromone, once attained, may be maintained for the life of the insect. Little is known of the cellular processes that occur during the stimulation of pheromone production and its suppression after mating, even in well-studied moths. Recent studies of the silkmoth *Bombyx mori* suggest that lipid droplets within pheromone secreting cells increase in size in the pre-adult stage (Fónagy *et al.*, 2000). In the young adult female, the lipid droplets declined in size but increased in number, and these changes could be prevented by decapitation and stimulated by PBAN. Fónagy *et al.* (2001) also showed that these changes followed a diel pattern in relation to pheromone production. In mid-photophase, when females produced pheromone and called, small lipid droplets predominated, whereas large lipid droplets accumulated at other times. Mated females, which ceased pheromone production, also accumulated large droplets. Together, these results suggest that the lipid droplets contain pheromone precursors that undergo daily and physiological cycles of accumulation and depletion (Fónagy *et al.*, 2001). Indeed, Matsumoto *et al.* (2002) confirmed that triacylglycerols were a major component of the lipid spheres, and the pheromone precursor Δ10,12-hexadecadienoate predominated. It is likely that other ultrastructural changes in pheromone gland cells follow these diel and physiological cycles. Likewise, it is expected that such changes will be discovered in insects that exhibit cycles of sexual receptivity, including some short-lived moths.

In long-lived insects, in which reproduction is interrupted by long periods of sexual inactivity, developmental regulation of the sex pheromone gland can result in alternating cycles of acquisition and subsequent loss of competence through maturation and retrogression of the glandular machinery, respectively. In cockroaches, pheromone production is controlled by both developmental differentiation of gland cells before the imaginal molt and by cyclic maturational changes in the gland that coincide with the ovarian cycle.

The best studied examples of differentiation and maturation of sex-specific pheromone glands are the tergal and sternal glands of *N. cinerea* and *B. germanica* males. Both possess class-3 glandular units composed of two cells – a secretory cell and a duct cell (section 10.5.1) (Noirot and Quennedey, 1974). But after apolysis and before the imaginal molt the immature gland contains four concentric cells, including also an enveloping cell and a ciliary cell in addition to the adult cells (Sreng, 1985, 1998; Sreng and Quennedey, 1976). During several days after the adult molt the gland matures by undergoing apoptosis (programmed cell death): the ciliary cell gives rise to a part of the microvillar end-apparatus, then dies, whereas the enveloping cell forms an upper portion of the duct, then it too dies (Sreng, 1998). Although the factors regulating sex pheromone gland differentiation in the pre-adult stage are not known, it is thought that differentiation proceeds in response to a low JH III titer throughout the last stadium and a rise in ecdysteroids late in the intermolt period.

Regulation of the maturational stage of pheromone gland differentiation in the adult is better understood. Before day-5 the immature sternal glands of *N. cinerea* males contain numerous pycnotic nuclei (from the dying enveloping and ciliary cells) and produce little pheromone (Sreng, 1998; Sreng *et al.*, 1999). By day-5, however, each glandular unit consists of only a glandular cell and a duct cell, and its pheromone content increases significantly. Sreng (1998) proposed that brain factors mediate these maturational changes. He showed that decapitation or allatectomy (removal of the CA) of *N. cinerea* males before a critical point during maturation completely blocked the apoptotic process. Nevertheless, pheromone accumulated in the sternal gland of decapitated or allatectomized males, albeit at a slower rate, suggesting that pheromone production is less dependent on brain factors than is cellular differentiation. Moreover, because pheromone release was probably hindered in the undifferentiated glands, further accumulation of pheromone would result. The brain factor was shown not to be PBAN-like neuropeptides because injection of brain extracts or synthetic PBAN did not restore gland differentiation or stimulate greater pheromone production (Sreng *et al.*, 1999). JH III administered to allatectomized males, on the other hand, restored both cellular differentiation of the gland and pheromone production.

Female *B. germanica* employ similar class-3 glands to produce a volatile sex pheromone that is yet to be identified. Ultrastructural studies and behavioral and electrophysiological assays have shown that, as in males, the pheromone gland undergoes cellular maturation during the period of sexual maturation and attainment of sexual receptivity (Abed *et al.*, 1993c; Liang and Schal, 1993c; Torko *et al.*, 1993; Schal *et al.*, 1996). The secretory cells of newly formed glands in the imaginal female are small (~5–10 μm in diameter) and they contain few, primarily electron-lucid secretory vesicles, few short microvilli around the end-apparatus, and little material within the end-apparatus. The amount of behaviorally and EAG-active material in the gland is low after the imaginal molt (day-0) but it increases with age and peaks on day-6, corresponding to the physiological stage when females become sexually receptive and begin to emit the pheromone (Liang and Schal, 1993c; Schal and Chiang, 1995). In tandem, the glandular epithelium thickens and the size of pheromone-secreting cells increases; mature day-6 glands are characterized by large secretory cells (about three-fold larger than on day-0) containing a large number of spherical or oval electron-lucid and electron-dense secretory vesicles, an end-apparatus lined with numerous long microvilli, and a large spherical nucleus (Abed *et al.*, 1993c; Liang and Schal, 1993c). Often, large amounts of secreted material can be found inside the cuticular depressions into which the ducts empty.

The mature pheromone gland of *B. germanica* undergoes a cyclical pattern of cellular hypertrophy and retrogression in relation to successive reproductive cycles. Both the thickness of the glandular tissue and the size of each secretory cell decline markedly after mating, when pheromone production is no longer needed, secretory vesicles become small and sparsely distributed, and the gland remains atrophied during gestation (Abed *et al.*, 1993; Liang and Schal, 1993c; Torko *et al.*, 1993; Schal *et al.*, 1996). EAG and behavioral assays confirmed that the pheromone content of the gland declines after mating and remains low throughout gestation (Abed *et al.*, 1993; Liang and Schal, 1993c; Schal *et al.*, 1996). But as a new vitellogenic cycle begins after the infertile egg case of virgins or the viable egg case of mated females is deposited, the pheromone gland undergoes rapid regrowth and proliferation of cellular machinery, the epithelium thickens again, and extracts of active glands from both virgin and mated females in the second reproductive cycle are EAG and behaviorally active. This pattern has been related to the pattern of JH III production (and the titer of JH III in the hemolymph) (Liang and Schal, 1993c; Schal *et al.*, 1996), but no experimental manipulations of hormone titers have been conducted to verify the hypothesis that JH III controls the cellular plasticity of the pheromone gland. Interestingly, this appears to be the only known case in insects where pheromone production is modulated in relation to the reproductive cycle (and JH), independently of the mating status of the female. The pattern of chemical signaling in *B. germanica* females is thus controlled primarily through regulated pheromone emission during calling, not through regulation of pheromone production. Calling behavior, too, is modulated by the titer of JH III (section 10.8.1).

To our knowledge, this pattern of cellular plasticity has not been described in any other insect species. It is tempting to speculate that this mechanism of regulating pheromone production is unusual and has evolved to conserve cell energy in insects that require long-term arrestment of pheromone synthesis during periods of sexual inactivity (e.g. during a long gestation period, diapause, migration). However, the CA – the endocrine glands that produce JH – also exhibit cycles of growth and retrogression of CA cells and these changes play a major role in the regulation of JH III biosynthesis during reproductive cycles in both *Blattella* and *Supella* females (Chiang *et al.*, 1991; Chiang and Schal, 1994). Female *Supella* retain the egg case for less than one day, and thus exhibit rapid reproductive cycles. It is therefore plausible that developmental regulation of pheromone biosynthesis is not restricted to species with long periods of gestation. In *Supella*, a major function of developmental changes in the pheromone gland might be in the long-term suppression of pheromone production after mating, rather than in tracking changes in JH III titers at each ovarian cycle as in *Blattella* (Smith and Schal, 1990a). Comparative studies of oviparous cockroaches – with rapid, short reproductive cycles – and ovoviviparous cockroaches – with slow, protracted reproductive cycles – would be enlightening in this regard.

10.7 Pheromone production regulated by juvenile hormone

The previous section highlighted the pivotal role that JH plays not only in pheromone gland differentiation, but also in its cyclic maturation in relation to the reproductive cycle. Most aspects of the female reproductive cycle in cockroaches are coordinately controlled by JH. Removal of the CA abolishes vitellogenin synthesis, its uptake by oocytes, and in some species, both pheromone production and sexual receptivity are suppressed by this operation. Barth and Lester (1973) and Schal and Smith (1990) reviewed the early literature on endocrine regulation of pheromone production in cockroaches.

10.7.1 Volatile pheromones

Unfortunately, no studies are available on the regulation of volatile pheromone production using analytical or radiochemical approaches. The most detailed studies have used behavioral and EAG responses of males as estimates of the relative amount of pheromone in females or their pheromone glands. The best studied species are *B. germanica* and *S. longipalpa*. Virgin *Supella* females initiate pheromone production 4 days after the imaginal molt (Smith and Schal, 1990a), in relation to increasing JH III biosynthesis by the CA (Smith *et al.*, 1989). Likewise, *Blattella* females begin to accumulate pheromone in the pygidial glands in relation to increasing rates of JH III synthesis (Liang and Schal, 1993c). Ablation of the CA of newly emerged adult females prevented pheromone production in both species, and pheromone production could be restored by reimplantation of active CA or by treatment with JH III or JH analogues. Interestingly, although growth of the vitellogenic oocytes is controlled by and highly correlated with JH III biosynthetic rates, direct or even intermediary involvement of the ovaries in regulating calling and pheromone production in both species was excluded by ovariectomies (Smith and Schal, 1990a).

It is not known whether JH exerts its pheromonotropic effects directly on secretory cells of the pheromone gland, or whether it acts indirectly by stimulating the synthesis and/or release of pheromonotropic neuropeptides. Notably, although cockroach brain extracts could induce PBAN-like pheromonotropic activity in moth pheromone glands (Raina *et al.*, 1989), injections of brain extracts of cockroaches or synthetic *Hez*-PBAN failed to induce pheromone production in allatectomized *Nauphoeta* males (Sreng *et al.*, 1999) or in *Supella* females (Schal, unpublished results). Moreover, lack of pheromone production in mated females that periodically produce large amounts of JH III suggests that JH plays a "permissive" role (Schal *et al.*, 1997). That is, its presence is *required* for pheromone to be produced, but even when the JH III titer is high pheromone production can be suppressed by neural or humoral pheromonostatic factors.

10.7.2 Contact pheromones

B. germanica has served as an important model for delineating the regulation of non-volatile cuticular pheromones. In this section we review findings on the biochemical pathways and the role of JH in facilitating the production of (3*S*,11*S*)-dimethylnonacosan-2-one, the major component of the courtship-inducing female pheromone (see reviews in Schal *et al.*, 1991, 1996; Blomquist *et al.*, 1993; Tillman *et al.*, 1999).

10.7.2.1 Biosynthetic pathway

Central to investigations of the biosynthetic pathway and regulation of the contact pheromone of *B. germanica* was the observation that the major cuticular hydrocarbon in all life stages of this species is an isomeric mixture of 3,7-, 3,9- and 3,11-dimethylnonacosane (Jurenka *et al.*, 1989). The presence of only the 3,11-isomer in the cuticular dimethyl ketone fraction and only in adult females prompted Jurenka *et al.* (1989) to suggest that production of the pheromone might result from the female-specific oxidation of its hydrocarbon analog. This scheme follows the well-established conversion of hydrocarbons to methyl ketone and epoxide pheromones in the housefly (Blomquist *et al.*, 1984; Ahmad *et al.*, 1987).

This model has since been validated with several independent approaches. Biochemical studies on the biosynthesis of methyl-branched alkanes showed that the methyl branches are added during the early stages of chain elongation (Chase *et al.*, 1990). Using carbon-13 labeling and NMR analyses Chase *et al.* (1990) showed that carbons 1 and 2 of acetate are incorporated as the chain initiator, and that the carbon skeleton of propionate serves as the methyl branch donor (Figure 10.1). Further, propionate and succinate labeled the hydrocarbons and the methyl ketone pheromone, as did the amino acids valine, isoleucine and methionine, all of which can be metabolized to propionate. NMR studies confirmed that these substrates were metabolized to methylmalonyl-CoA for incorporation into the methyl branch unit of hydrocarbons (Chase *et al.*, 1990), as in the housefly (Dillwith *et al.*, 1982; Halarnkar *et al.*, 1987), the American cockroach (Halarnkar *et al.*, 1985), the cabbage looper moth (de Renobales and Blomquist, 1983), and the termite *Zootermopsis* (Chu and Blomquist, 1980).

Methyl-branched fatty acids are intermediates in branched alkane biosynthesis (Juarez *et al.*, 1992). Thus, [1-^{14}C]propionate labeled methyl-branched fatty acids of 16–20 carbons, but did not label straight chain-saturated and monounsaturated fatty acids (Chase *et al.*, 1990).

Chase *et al.* (1992) examined the hypothesis that the methyl ketone sex pheromone arises from the insertion of an oxygen into the preformed 3,11-dimethyl alkane. When high-specific activity, tritiated,3,11-dimethyl nonacosane was topically applied on the cuticle of *B. germanica* females, it readily penetrated the cockroach and radioactivity from the alkane was detected in both 3,11-dimethylnonacosan-2-ol and 3,11-dimethylnonacosan-2-one. When tritiated 3,11-dimethylnonacosan-2-ol was applied to the cuticle it was readily and highly efficiently converted to the corresponding methyl ketone pheromone. But, surprisingly, the dimethyl ketone pheromone was derived from the corresponding alcohol not only in females, as expected, but also in males. These results suggest that the sex pheromone of *B. germanica* arises via a female-specific hydroxylation of 3,11-dimethylnonacosane and a subsequent non-sex-specific oxidation, probably involving a polysubstrate monooxygenase system, to the (3*S*,11*S*)-dimethylnonacosan-2-one pheromone (Figure 10.1). Chase *et al.* (1992) also suggested that a similar hydroxylation and subsequent oxidation at the 29-position of 3,11-dimethylnonacosan-2-one might give rise to 29-hydroxy- and 29-oxo-(3,11)-dimethylnonacosan-2-one, the other components of the contact pheromone blend, but this hypothesis has yet to be tested. It is quite likely, as well, that the same mechanism converts the C_{27} dimethyl alkane to the corresponding methyl ketone pheromone, and perhaps its 27-hydroxy- and 27-oxo- analogues.

10.7.2.2 Juvenile hormone-mediated contact pheromone production

JH plays an essential regulatory role in the metabolism of (3*S*,11*S*)-dimethylnonacosane to (3*S*,11*S*)-dimethylnonacosan-2-one, probably by increasing the activity of a female-specific polysubstrate monooxygenase. This subject has been reviewed in relation to pheromone biosynthesis in other insects (Schal *et al.*, 1996; Blomquist *et al.*, 1993; Tillman *et al.*, 1999). The pattern of synthesis and accumulation of pheromone on the female's cuticle suggested the involvement of JH. Incorporation of radiolabel from injected [1-^{14}C]propionate into the methyl-branched sex pheromone is low in previtellogenic and in gravid females, when the JH III titer is also low, and high in vitellogenic females, when the JH III titer is greatly elevated (Schal *et al.*, 1994; Sevala *et al.*, 1999). Likewise, the amount of pheromone on the cuticle increases in relation to the vitellogenic cycle and the titer of JH III (Schal *et al.*, 1990a). But most convincing is the experimental evidence from allatectomized females, females treated with JH analogs, and from dietary manipulations that influence both hormone titer and pheromone production (Schal *et al.*, 1990a, 1991, 1994; Chase *et al.*, 1992). Females treated so as to reduce their JH III titer (allatectomy, anti-allatal drugs such as precocene, starvation, or implantation of an artificial egg case into the genital vestibulum, which inhibits JH biosynthesis) produced less pheromone (Schal *et al.*, 1990a, 1991, 1994). But pheromone production was greatly stimulated by treatments with JH III or with JH analogs. Because only the hydroxylation of 3,11-dimethylnonacosane to 3,11-dimethylnonacosan-2-ol is regulated in a sex-specific manner, it appears that this step is under JH III control (Figure 10.1) (Chase *et al.*, 1992).

All life stages of the German cockroach produce 3,11-dimethylnonacosane, suggesting that production of a sex-specific contact pheromone may be less dependent on differentiation of sexually dimorphic pheromone glands, as is the case for volatile pheromones, and more so on the endocrine milieu of the adult female. Normally, adult male cockroaches produce much less JH III than do females, and males have a much lower titer of JH III in the hemolymph (Piulachs *et al.*, 1992; reviewed in: Tobe and Stay, 1985; Wyatt and Davey, 1996). However, when newly emerged males were exposed to filter papers treated with the JH mimic hydroprene they exhibited a six-fold elevation in female pheromone on their cuticle (Schal, 1988). Although substantial, this limited stimulation indicates either a limited capacity to express the putative female-specific polysubstrate monooxigenase or a loss of JH receptor sites in the adult male oenocytes. Nevertheless, the parallels with estrogen induction of vitellogenin synthesis in the male liver of oviparous vertebrates, and JH induction of vitellogenin synthesis in male cockroaches (e.g. Mundall *et al.*, 1983) are striking. The emerging model suggests, at a minimum, that the endocrine milieu of the vitellogenic female contributes significantly to the production of female-specific contact pheromones in cockroaches.

Yet another feature distinguishes JH control of contact pheromone production from control of volatile pheromone production in cockroaches. Although based solely on behavioral data, lack of JH III in females completely suppresses production of attractant pheromones, as for instance in *S. longipalpa* (section 10.7.1). The contact pheromone of *B. germanica*, on the other hand, is still biosynthesized in allatectomized females, albeit at a lower rate, probably because regulation of contact sex pheromone production operates at several levels, including the regulated production of its 3,11-dimethylnonacosane precursor (section 10.7.2.3).

10.7.2.3 Pheromone production regulated by food availability

There are numerous examples in insects of sequestration or metabolism of plant compounds for their own chemical defenses (Eisner and Meinwald, Chapter 12, in this volume). For sex pheromones, on the other hand, *de novo* biosynthesis is much more common than the sequestration of pheromones or their precursors from plant compounds. Nevertheless, some pheromones, especially those which are produced in large quantities (micrograms), require large amounts of food-derived biosynthetic substrate, and a deficiency of building blocks may significantly interfere with pheromone production. Production of the methyl ketone contact pheromone in the female German cockroach depends on the biosynthesis of considerable amounts of the parent methyl-branched hydrocarbon, production of which is regulated to a large extent by availability of food, not JH (Schal *et al.*, 1994). All life stages of *B. germanica* produce and secrete onto the cuticle large amounts of 3,11-dimethylnonacosane. In both nymphs and adults, surges in hydrocarbon synthesis occur during feeding stages that follow each molt, but hydrocarbon synthesis declines dramatically when nymphs cease to feed in preparation for the next molt or when adult females reduce their food intake at the end of the vitellogenic phase (Schal *et al.*, 1994, 1996; Young *et al.*, 1997, 1999a). Therefore, hydrocarbon production is high in the adult female when the ecdysteroid titer is low and it wanes well before any appreciable decline in the JH III titer, suggesting that neither of these two hormones directly regulates hydrocarbon biosynthesis. Rather, hydrocarbon production in both nymphs and adults appears to change only in relation to food intake. Young *et al.* (1999a) showed a clear causal relationship between these two processes in last instar *B. germanica*: starved nymphs ceased to produce hydrocarbons, whereas refed nymphs resumed hydrocarbon biosynthesis. Interestingly, this pattern was reflected in short-term *in vitro* culture of tergites in a nutrient-rich medium, suggesting that starvation reduced not only the availability of substrates for hydrocarbon biosynthesis but also the *capacity* of oenocytes to produce hydrocarbons.

A complex interaction between food intake and JH III titer controls contact pheromone production in adult female *B. germanica*. In normal females, food consumption facilitates both the production of hydrocarbons by the oenocytes and JH III biosynthesis by the CA (Schal *et al*., 1993; Osorio *et al*., 1998; Holbrook *et al*., 2000). A high JH III titer, in turn, stimulates the conversion of the hydrocarbon to contact pheromone (Figure 10.1, section 10.7.2.2). At ovulation and throughout the ensuing 3-week gestation period, food intake is minimal and the JH III titer generally remains at or near undetectable levels; much less hydrocarbon and pheromone are produced. Production of all three lipids resumes after parturition when a new vitellogenic and feeding cycle commences.

Considerable insight into this system has been gained from studies of allatectomized females. Young allatectomized females ate less and therefore synthesized less hydrocarbons during the first few days of adult life (Schal *et al*., 1994). However, for several interrelated reasons, large amounts of hydrocarbons accumulated over time in the hemolymph of allatectomized females. First, in the absence of JH III females did not provision their eggs, and therefore vast amounts of nutrients accumulated in the hemolymph. Second, food intake was not suppressed in older allatectomized females because this normally requires the presence of an egg case, resulting in a further build-up of nutrients in the hemolymph. Third, the relatively low rate of hydrocarbon biosynthesis in older allatectomized females far exceeded that of gravid females of the same age. And finally, large amounts of hydrocarbons are normally deposited into the vitellogenic oocytes (Fan *et al*., 2002), but because ovarian uptake of hydrocarbons was impeded in allatectomized females, the hemolymph titer of hydrocarbons increased well beyond normal levels (Schal *et al*., 1994). Fan *et al*. (2002) demonstrated that in normal females, as the oocytes matured and hydrocarbon uptake was reduced, a higher fraction of hemolymph hydrocarbons (section 10.8.2.3) was directed to the epicuticular surface. Thus, as allatectomized females accumulated more hydrocarbons in their hemolymph, the amount of cuticular pheromones also increased. In fact, the amount of 3,11-dimethylnonacosan-2-one increased beyond normal levels in both the hemolymph and the cuticle, suggesting that excess 3,11-dimethylnonacosane was metabolized to pheromone.

These patterns suggest that under normal conditions, feeding in adult females is modulated in a stage-specific manner, regulating the amount of 3,11-dimethylnonacosane that is available for pheromone production. Early in the vitellogenic cycle JH mediates the metabolism of 3,11-dimethylnonacosane to 3,11-dimethylnonacosan-2-one and other pheromone components. Later in the reproductive cycle, however, the oocytes sequester large amounts of hydrocarbons and much less pheromone is produced. This system illustrates the complex interactions among multiple mechanisms that regulate the biosynthesis of lipids that serve multiple functions. Hydrocarbons serve not only as pheromone precursors, but also as water repellents on the cuticular surface and as significant maternal provisions to progeny.

10.8 Transport and emission of pheromones

10.8.1 Volatile pheromones

Little is known of the cellular processes that deliver volatile pheromones from secretory cells to the cuticular surface, even in the intensively researched Lepidoptera. Based on results from ultrastructural studies of the moth *Heliothis virescens*, Raina *et al.* (2000) recently speculated that secretory cells of the pheromone gland somehow deliver pheromone or its precursors to hollow cuticular hairs. During calling behavior the female exposes the gland and also the cuticular hairs that exude pheromone droplets. Raina *et al.* (2000) further posit that as the female retracts the ovipositor more pheromone is "squeezed" onto the exposed surface, thus recharging the cuticular hairs.

In cockroaches, electron microscopy studies often show accumulation of secretion in the end-apparatus, ducts, and around the cuticular pores of class-3 exocrine glands in both males and females. Whether these are carrier proteins that deliver hydrophobic pheromones to the surface, or other excretions, is not known. Recent studies in *Leucophaea maderae* have identified and sequenced an epicuticular protein, Lma-p54, that is expressed specifically in the tergites and sternites of adult males and females, but not in nymphs (Cornette *et al.* in 2002). The sequence of this protein is closely related to aspartic proteases, but because it appears to be enzymatically inactive the authors speculate that it serves as a ligand-binding protein. Cornette *et al.* (2002) further hypothesize that Lma-p54, alone or together with a ligand, serves in sexual recognition. Other ligand-binding proteins, namely the lipocalins Lma-p22 and Lma-p18, have been isolated only from male tergal secretions of *L. maderae* (Korchi *et al.*, 1999; Cornette *et al.*, 2001). These exciting findings suggest that carrier proteins might be involved in transport of volatile pheromones to the cuticle but functional studies will be needed to verify this hypothesis. It is worth noting that Bla g4, an allergen that was suggested to serve a similar role in *B. germanica*, is expressed in male reproductive tissues and appears to play no role in pheromone transport (Schal, unpublished results).

Attractant sex pheromones are usually emitted while the female or male cockroach performs a species-specific calling behavior, as in most lepidopterans. In females, calling is generally characterized by elevated tegmina and wings, a recurved abdomen, and occasional exposure of the genital vestibulum (see Gemeno and Schal, 2003; Gemeno *et al.*, 2003). Although it has been shown for several species that pheromones are released during this behavior, the precise connection between calling and pheromone emission remains unknown. In some species, calling exposes pheromone gland openings that are normally obscured by the wings or by more anterior tergites or sternites (in most males, e.g. *B. germanica*, *S. longipalpa*, *N. cinerea*). In other species, however, including females of *B. germanica* and *S. longipalpa*, and especially wingless females of *Parcoblatta*, the tergal openings of the pheromone glands do not seem to become appreciatively more exposed during calling. Moreover, in *Supella* and *Parcoblatta* there is no reservoir or storage space for pheromone on the cuticular surface, suggesting that pheromone is emitted "on demand" only when needed. In these species, calling behavior is expressed exclusively by sexually receptive females and volatiles collected from calling females are much more attractive to males than volatiles collected from non-calling females of the same age (Smith and Schal, 1990a; Liang and Schal, 1993b; Gemeno *et al.*, 2003). This suggests that the postures employed by calling females must have a direct effect on the release of the pheromone by the individual glands. Both pheromone emission and calling behavior are physiologically regulated, but the exact mechanisms by which calling behavior effects the transfer of pheromone from secretory cells to the exterior of the body is unclear. It is worth noting, in this context, that the pheromone glands are extensively innervated (Liang and Schal, 1993c) and it is possible that the same neuronal signals that control calling also control pheromone release.

JH III regulates calling behavior in both *B. germanica* and *S. longipalpa*. In both species, transection of the nerves connecting the CA to the brain, an operation that significantly accelerates the rate of JH III biosynthesis by the CA (Gadot *et al.*, 1989; Schal *et al.*, 1993), also hastens the age when calling first occurs (Smith and Schal, 1990a; Liang and Schal, 1994). In *B. germanica*, the central role of JH has been confirmed by ablation of the CA and with rescue experiments with a JH analogue (Liang and Schal, 1994). In this species, JH also is required for females to become sexually receptive and accept courting males (Schal and Chiang, 1995).

10.8.2 Contact pheromones

The view that newly synthesized integumental lipids, including contact pheromones, pass directly from oenocytes to the epicuticle has recently been critically re-examined. An alternative model has gained support (Figure 10.2): a lipoprotein-mediated indirect transport pathway plays a major, if not exclusive, role in delivering lipids to the cuticular surface. This hypothesis is supported by the following observations (reviewed in Schal *et al.*, 1998b). First, the hydrocarbons in the hemolymph of several insect species are similar to the hydrocarbon profile of the respective cuticle (e.g. Chino and Downer, 1982; Katase and Chino, 1984; reviewed in Schal *et al.*, 1998b). Second, in some insects the male and female

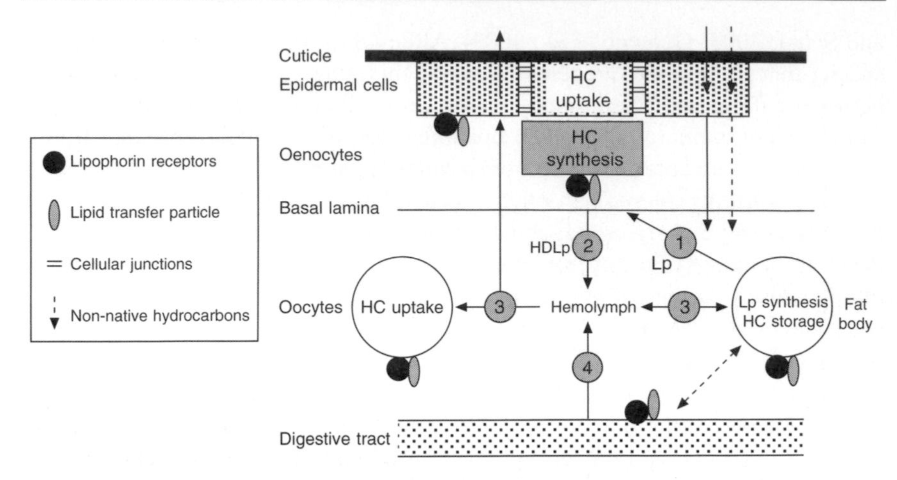

Figure 10.2 A conceptual model depicting transport of hydrocarbons (HC) and hydrocarbon metabolites from the oenocytes that produce them to the cuticular surface, fat body, oocytes, and digestive tract. Arrows represent directions of movement of lipophorin (Lp) and hydrocarbons. Lipophorin is biosynthesized in the fat body, but it is not known whether nascent Lp is loaded with hydrocarbons before it is released into the hemolymph (1) Epidermal cells and oenocytes are enclosed within a basal lamina. Oenocytes biosynthesize hydrocarbons, which are then loaded into HDLp (2) HDLp delivers hydrocarbons to the cuticular surface, oocytes, and fat body (3) After a molt, some exuvium hydrocarbons can also be reclaimed through the digestive tract (4) Non-native hydrocarbons can enter through the cuticle or the digestive tract, and removed through fat body metabolism, as well as through the digestive system. The involvement of lipid transfer particle (LTP) in uptake or unloading of hydrocarbons in any of these tissues remains unknown. It is also not known if HDLp enters target tissues (epidermis, fat body) and how hydrocarbons traverse the cuticle. The recognition of non-native hydrocarbons by the transport system and target tissues has not been studied. Based in part on a similar scheme for diacylglycerol transport (Arrese *et al.*, 2000; Canavoso *et al.*, 2001).

hemolymph carries different hydrocarbons, some of which serve as sex pheromones, and the composition of hemolymph is determined by the types of sex-specific hydrocarbons biosynthesized by the insect (e.g. cockroach – Gu *et al.*, 1995; *Drosophila* – Pho *et al.*, 1996; houseflies – Mpuru *et al.*, 2001; moths – Schal *et al.*, 1998a; Jurenka and Subchev, 2000). Third, in insects with oenocytes localized within the hemocoel (see section 10.5.2) a hemolymph transport pathway is clearly required to deliver lipids to the cuticular surface. However, in all species, hydrocarbons also need to be delivered to integument that does not synthesize hydrocarbons as well as to internal tissues, such as the ovaries (Fan *et al.*, 2002), a process that requires hemolymph transport regardless of where the oenocytes are located.

10.8.2.1 Hydrocarbon transport in hemolymph: role of lipophorin

Hydrocarbons and contact pheromones can be transported from biosynthetic tissues (e.g. abdominal integument) to tissues that do not synthesize them (e.g. head, wings, legs) either by physical translocation on the cuticular surface or by transport through hemolymph. To test the hypothesis that hemolymph is required for transport of lipids to tissues that do not synthesize them Gu *et al.* (1995) severed the veins that enter the fore-wings. The amount of hydrocarbons that appeared on the wings was significantly lower than on intact fore-wings of the same insects. Because the wings do not biosynthesize hydrocarbons, these results show that hydrocarbon transport from the abdominal oenocytes to various other body parts is hemolymph mediated.

It turns out that transport of hydrophobic ligands, including hydrocarbons and their semiochemical metabolites, requires plasma lipoproteins. The seminal work of Chino and co-workers showed that hydrocarbons associate with an M_r ~600 kDa high-density lipophorin (HDLp). In most insects HDLp is characterized by two constituent apoproteins, apoLp-I and apoLp-II, ~240 kDa and ~70 kDa, respectively (Chino *et al.*, 1981; Chino, 1985; Kanost *et al.*, 1990; Law *et al.*, 1992; Van der Horst *et al.*, 1993; Blacklock and Ryan, 1994; Soulages and Wells, 1994; Arrese *et al.*, 2000; Ryan and Van der Horst, 2000; Canavoso *et al.*, 2001). HDLp serves multiple transport functions. Its shell consists of the two apoproteins, phospholipids, and cholesterol, while the core contains non-polar lipids. Flying insects generally have large amounts of diacylglycerol which Lp delivers from the midgut to the fat body for storage, and to flight muscles for utilization as metabolic fuel. An important feature of Lp is its dynamic nature; it can range from a low-density particle (LDLp), resulting from an exchangeable association of apoLp-III (~17–20 kDa) and diacylglycerol with HDLp, to a very high-density complex (VHDLp) that is depleted of lipids. Neither of the latter forms has been isolated from cockroaches.

HDLp also carries cholesterol and carotenoids, and it has specific, high-affinity binding sites for JH III in Coleoptera, Isoptera, Diptera, Hymenoptera, and Dictyoptera (Trowell, 1992; Sevala *et al.*, 1997). In addition to its recognized central roles in transporting hydrophobic ligands, Lp is also a plasma coagulogen, involved in clotting of insect hemolymph (Ferstl *et al.*, 1988) and it appears to mediate immune responses, presumably by modulating cell adhesion responses of hemocytes (Coodin and Caveney, 1992; Dettloff *et al.*, 2001) and formation of lipopolysaccharide–Lp complexes which constitute an anti-bacterial defense system (Kato *et al.*, 1994). The multiplicity of its functions and its use by different developmental stages suggest that both the concentration of HDLp in hemolymph and its lipid composition are probably related to the stage-specific function that it serves. For example, the hemolymph concentration of HDLp, which serves as a JH-binding protein, is unrelated to the JH III titer, but more correlated with the titer of the much more abundant hemolymph hydrocarbons (Sevala *et al.*, 1999).

In the cockroaches *P. americana* and *B. germanica* hydrocarbons are synthesized by the abdominal epidermis, and virtually all newly synthesized hydrocarbons that enter the hemolymph are bound to HDLp (Figure 10.2) (Chino *et al.*, 1981; Gu *et al.*, 1995). In both species, hydrocarbons constitute up to 50 percent of the total lipids in HDLp and they are qualitatively the same as the cuticular hydrocarbons. Moreover, in both species, newly synthesized hydrocarbons can only be transferred from the integument to an incubation medium if HDLp is present, and other hemolymph lipoproteins, such as vitellogenin, cannot mediate this transfer (Katase and Chino, 1982, 1984; Chino and Downer, 1982; Fan *et al.*, 2002a). Abundance of epicuticular hydrocarbons on body parts that do not synthesize hydrocarbons (e.g. wings, legs) lends further support to the idea that HDLp serves as the sole vehicle for transport of species-specific hydrocarbons from the oenocytes to the cuticle. However, while Lp must deliver hydrocarbons to integument that does not synthesize hydrocarbons, its role in delivering hydrocarbons and pheromones to integumental tissue that synthesizes hydrocarbons (i.e. sternites and tergites) is not clear.

10.8.2.2 Loading of hydrocarbons into lipophorin

Newly synthesized hydrocarbons are loaded from oenocytes into HDLp and delivered to the cuticle, oocytes, and to the fat body for storage (Figure 10.2). In non-feeding stages that do not synthesize hydrocarbons, as, for example, late in the last stadium, the fat body appears to load hydrocarbons into HDLp for delivery to the cuticle (Young and Schal, 1997; Young *et al.*, 1999b). In newly ecdysed insects, hydrocarbons reclaimed from the ingested exuvium are loaded into HDLp from the midgut (Young and Schal, unpublished results). Thus, at least three tissues take part in loading HDLp with hydrocarbons – the oenocytes, midgut, and fat body.

The mechanisms by which hydrocarbons are taken up by HDLp are poorly understood. In some insects a very high-density lipid transfer particle (LTP) plays a crucial role in mediating diacylglycerol transfer from the midgut to HDLp (Canavoso and Wells, 2001). LTP also carries hydrocarbons (Blacklock and Ryan, 1994), and Takeuchi and Chino (1993) suggested that it may catalyze hydrocarbon transport in the American cockroach. Hydrocarbons comprise 40 percent of the lipids in LTP of *P. americana*, while only 28 percent of the lipids in HDLp are hydrocarbons; nevertheless, LTP carries much less total hydrocarbons than does HDLp (Takeuchi and Chino, 1993). Although the capacity of LTP to transfer hydrocarbons between Lp particles was clearly demonstrated in the American cockroach (Takeuchi and Chino, 1993), results from *in vitro* experiments with purified HDLp of *P. americana* and *B. germanica* have shown that HDLp accepts hydrocarbons from oenocytes without involvement of LTP (Katase and Chino, 1982, 1984; Fan *et al.*, 2002). In fact, purified HDLp (without LTP) is no less efficient in hydrocarbon transport than hemolymph, which also contains

LTP. Although these results suggest that LTP might not be required for the exchange of hydrocarbons, two confounding features of this system have yet to be investigated. First, it is possible that despite extensive washing of tissues and cells, LTP could remain bound to tissues. Second, LTP might be produced by either oenocytes or epidermal cells. Van Heusden and Law (1989) showed that although diacylglycerol could be transferred *in vitro* from fat body to purified *M. sexta* HDLp, pre-incubation of fat body with LTP antibody reduced the amount of diacylglycerol loaded onto HDLp. Canavoso and Wells (2001) also significantly reduced diacylglycerol transfer from labeled midgut sacs to Lp by pretreatment with LTP antibody; when LTP was added, diacylglycerol exchange was restored. Of course, it is also possible that hydrocarbon uptake by HDLp does not require LTP. In *M. sexta*, for example, cholesterol transfer occurs by a mass action mechanism, without involvement of LTP. Pretreatment of the midgut with anti-LTP IgG did not inhibit the transfer of cholesterol from midgut to HDLp, showing that the uptake of cholesterol is fundamentally different from diacylglycerol uptake (Yun *et al*., 2002).

Whatever the mechanism(s) of hydrocarbon and contact pheromone uptake by HDLp from various tissues, it appears that oenocyte-containing tissues can exchange their hydrocarbons with both native and heterospecific HDLp, indicating that the uptake of hydrocarbons into Lp lacks species specificity (Katase and Chino, 1984; Fan and Schal, unpublished results). Interestingly, this appears to be the case for diacylglycerol loading as well (Chino and Kitazawa, 1981).

10.8.2.3 Hydrocarbon exchange from lipophorin into tissues

Unlike vertebrate lipoproteins, insect Lp is generally considered to be a reusable particle that shuttles lipids among tissues without entering cells (Chino and Kitazawa, 1981; Van Heusden *et al*., 1991). Although delivery of hydrocarbons and pheromones to cells has not been studied in any insect, this process clearly involves hemolymph HDLp. When *B. germanica* sternites, which synthesize hydrocarbons, or pronotum, which does not synthesize hydrocarbons, were co-incubated with hydrocarbon-prelabeled sternites, radiolabeled hydrocarbons were transferred into the unlabeled tissue only when HDLp was present in the incubation medium (Schal, unpublished results). The accumulated evidence, though insufficient, supports the idea that the hydrocarbons and contact pheromone components of *B. germanica* are produced by oenocytes within the abdominal integument, carried by HDLp, and deposited in the cuticle (Figure 10.2). Whether epidermal cells are involved in the deposition process, and if so, how, remains to be determined.

The best evidence of hydrocarbon transfer into cells is with oocytes. Oocyte development in cockroaches, as in most oviparous animals, is characterized by a dramatic growth period during which all maternally derived macromolecules are sequestered. Vitellogenin uptake by the oocyte has been extensively studied through successive stages of the gonadotrophic cycle, they experience various checkpoints that determine whether they should carry on or resorb their oocytes. Reproduction in oviparous species with rapid reproductive cycles (e.g. *Supella*) can also be arrested by external sensory and internal physiological signals. Cockroaches appear to have evolved a tight link between the gonadotrophic cycle and glandular maturation (both CA and pheromone glands). Thus, even short 2–3-day periods of sexual inactivity result in a concomitant regression of the cellular machinery within these glands. This is especially clear in ovoviviparous females that experience long periods of sexual inactivity. Such periods are followed in cockroaches by a proliferation of the cellular machinery involved in pheromone production. Because cockroaches are long-lived, many such cycles can occur within pheromone secreting cells. The factors that regulate these cellular cycles remain to be elucidated.

5 *Role of JH III in pheromone production and calling behavior*. In addition to its fundamental role as a gonadotropic hormone, in many cockroaches JH III plays a pivotal permissive role in pheromone production, calling behavior, and sexual receptivity of the female. Without it, these components of mate-finding are not expressed or are terminated. In most such examples, however, JH appears to exert its effects as a priming hormone – its effects require a long time to become evident and JH appears to prepare cells to respond to it or to other regulatory agents (see Wyatt and Davey, 1996 for discussion). To date, the only example of a possible regulatory (i.e. "releaser") action of JH on pheromone production in cockroaches is in *B. germanica*, where JH III induces the metabolism of 3,11-dimethylnonacosane to 3,11-dimethyl-nonacosan-2-ol, a precursor of the contact pheromone. How this is accomplished and what enzymes catalyze this change remains unknown.

   It is imperative to demonstrate whether JH III operates directly on pheromone glands or whether it facilitates the activity of other pheromonotropic regulators. Studies of mated females, which produce JH III but not pheromone, have made it abundantly clear that unknown regulatory elements downstream of the action of JH must be involved. These factors probably interact with inhibitory signals from the terminal abdominal ganglion that ascend the VNC and control CNS activity. A concerted effort is needed to identify neuropeptides and other factors that activate and terminate pheromone production, emission, and sexual receptivity.

6 *Transport of pheromones*. Two major routes for translocation of pheromones have been considered in this chapter: (a) from the secretory cell directly through the cuticle overlying it; and (b) indirectly through the hemolymph. Cockroaches share with even the most studied lepidopterans an almost complete lack of information on the former pathway. Transport of hydrocarbons and contact sex pheromones, on the other hand, has been extensively studied in cockroaches, commencing with the work of Chino and colleagues. It has

become clear that HDLp serves as pheromone carrier not only in cockroaches, but also in termites, flies, and moths. However, details of this pathway remain unknown. How are pheromone components loaded into HDLp? How are they imported into cells? Are Lp receptors involved? Does LTP catalyze these transfers? And finally, how are different types of hydrocarbons sorted among various tissues? Resolution of these questions in cockroaches will certainly further our understanding of similar mechanisms in other insects because it appears that in some moths as well hydrocarbon pheromones are transported by lipophorin (Schal *et al.*, 1998a).

7 *Role and mechanics of calling behavior.* Calling behavior in female cockroaches is associated with the release of volatile sex pheromones. However, we have only a rudimentary understanding of how the female calling posture results in the release of pheromone and what role atrial glands play in calling and courtship. Although the orifices of class-3 pheromone glands are always exposed, only when the female calls is the pheromone emitted. The relationship between the motor patterns that characterize calling behavior and the release of pheromone needs to be studied. Do these movements facilitate transport of the pheromone products along the ducts? What is the function of exposing the genital vestibulum during calling in females whose pheromone appears to be produced only by the tergites?

8 *Role of neural directives.* Insertion of the spermatophore in the genital chamber has a multiplicity of effects on the female's endocrinology and behavior: it stimulates JH production and inhibits pheromone production, calling, and sexual receptivity. Sperm, likewise, inhibits further sexual behaviors, while the ootheca also inhibits JH biosynthesis. The prevalent model is that these actions are conveyed to the CNS through the VNC. But it is not known whether mechanoreceptors transduce this information, how, and what events transpire within the CNS to prevent the expression of sexual behaviors.

9 *Cockroaches as integrative models.* Cockroaches have served as early models for investigations of mechanisms that regulate pheromone production. Research on ovoviviparous blaberids, over four decades ago, and more recent studies with the blattellids *B. germanica* and *S. longipalpa* have shown that in cockroaches, several mechanisms are integrated to promote or suppress production and emission of pheromones. Almost all physiological and behavioral aspects of female reproduction are coordinately regulated and paced by JH. Food intake and mating also intervene in the regulation of pheromone production and emission in females, in part by modulating the production of JH. Mechanoreceptive signals from the spermatophore and sperm appear to play roles in inhibiting pheromone production and emission. And pheromone-producing glands undergo cycles of competence for pheromone production. Because they are long-lived and reproduce cyclically, cockroaches will remain an important model system for integrative studies of regulation of pheromone production.

Schal C. and Bell W. J. (1985) Calling behavior in female cockroaches (Dictyoptera, Blattaria) *J. Kansas Entomol. Soc.* **58**, 261–268.

Schal C. and Chiang A.-S. (1995). Hormonal control of sexual receptivity in cockroaches. *Experientia* **51**, 994–998.

Schal C. and Smith A. F. (1990) Neuroendocrine regulation of pheromone production in cockroaches. In *Cockroaches as Models for Neurobiology: Applications in Biomedical Research*, eds I. Huber E. P. Masler and B. R. Rao, pp. 179–200. CRC Press, Boca Raton.

Schal C., Gautier J.-Y. and Bell W. J. (1984) Behavioural ecology of cockroaches. *Biol. Rev.* **59**, 209–254.

Schal C., Burns E. L. and Blomquist G. J. (1990a) Endocrine regulation of female contact sex pheromone production in the German cockroach, *Blattella germanica. Physiol. Entomol.* **15**, 81–91.

Schal C., Burns E. L., Jurenka R. A. and Blomquist G. J. (1990b) A new component of the female sex pheromone of *Blattella germanica* (L.) (Dictyoptera: Blattellidae) and interaction with other pheromone components. *J. Chem. Ecol.* **16**, 1997–2008.

Schal C., Burns E. L., Gadot M., Chase J. and Blomquist G. J. (1991) Biochemistry and regulation of pheromone production in *Blattella germanica* (L.) (Dictyoptera, Blattellidae). *Insect Biochem.* **21**, 73–79.

Schal C., Liang D., Hazarika L. K., Charlton R. E. and Roelofs W. L. (1992) Site of pheromone production in female *Supella longipalpa* (Dictyoptera: Blattellidae): behavioral, electrophysiological, and morphological evidence. *Ann. Entomol. Soc. Am.* **85**, 605–611.

Schal C., Chiang A-S., Burns E. L., Gadot M. and Cooper R. A. (1993). Role of the brain in juvenile hormone synthesis and oocyte development: effects of dietary protein in the cockroach *Blattella germanica* (L.). *J. Insect Physiol.* **39**, 303–313.

Schal C., Gu X., Burns E. L. and Blomquist G. J. (1994) Patterns of biosynthesis and accumulation of hydrocarbons and contact sex pheromone in the female German cockroach, *Blattella germanica. Arch. Insect Biochem. Physiol.* **25**, 375–391.

Schal C., Liang D. and Blomquist G. J. (1996) Neural and endocrine control of pheromone production and release in cockroaches. In *Insect Pheromone Research: New Directions*, eds R. T. Cardé and A. K. Minks pp. 3–20. Chapman and Hall, New York.

Schal C., Holbrook G. L., Bachmann J. A. S. and Sevala V. L. (1997) Reproductive biology of the German cockroach, *Blattella germanica*: juvenile hormone as a pleiotropic master regulator. *Arch. Insect Biochem. Physiol.* **35**, 405–426.

Schal C., Sevala V. and Cardé R. T. (1998a) Novel and highly specific transport of a volatile sex pheromone by hemolymph lipophorin in moths. *Naturwissenschaften* **85**, 339–342.

Schal C., Sevala V. L., Young H. P. and Bachmann J. A. S. (1998b) Synthesis and transport of hydrocarbons: cuticle and ovary as target tissues. *Amer. Zool.* **38**, 382–393.

Scharrer B. (1987) Insects as models in neuroendocrine research. *Annu. Rev. Entomol.* **32**, 1–16.

Schneider D., Boppré M., Schneider H., Thompson W. R., Boriack C. J., Petty R. L. and Meinwald J. (1975) A pheromone precursor and its uptake in male *Danaus* butterflies. *J. Comp. Physiol. A* **97**, 245–256.

Seelinger G. (1984) Sex-specific activity patterns in *Periplaneta americana* and their relation to mate-finding. *Z. Tierpsychol.* **65**, 309–326.

Sevala V. L., Bachmann J. A. S. and Schal C. (1997) Lipophorin: a hemolymph juvenile hormone binding protein in the German cockroach, *Blattella germanica. Insect Biochem. Mol. Biol.* **27**, 663–670.

Sevala V., Shu S., Ramaswamy S. B. and Schal C. (1999) Lipophorin of female *Blattella germanica* (L.): characterization and relation to hemolymph titers of juvenile hormone and hydrocarbons. *J. Insect Physiol.* **45**, 431–441.

Sirugue D., Bonnard O., Le Quere J. L., Farine J.-P. and Brossut R. (1992) 2-Methylthiazolidine and 4-ethylguaiacol, male sex pheromone components of the cockroach *Nauphoeta cinerea* (Dictyoptera, Blaberidae): a reinvestigation. *J. Chem. Ecol.* **18**, 2261–2276.

Smith A. F. and Schal C. (1990a) Corpus allatum control of sex pheromone production and calling in the female brown-banded cockroach, *Supella longipalpa* (F.) (Dictyoptera: Blattellidae). *J. Insect Physiol.* **36**, 251–257.

Smith A. F. and Schal C. (1990b) The physiological basis for the termination of pheromone-releasing behaviour in the female brown-banded cockroach, *Supella longipalpa* (F.) (Dictyoptera: Blattellidae). *J. Insect Physiol.* **36**, 369–373.

Smith A. F., Yagi K., Tobe S. S. and Schal C. (1989) *In vitro* juvenile hormone biosynthesis in adult virgin and mated female brown-banded cockroaches, *Supella longipalpa*. *J. Insect Physiol.* **35**, 781–785.

Soulages J. L. and Wells M. A. (1994) Lipophorin: the structure of an insect lipoprotein and its role in lipid transport in insects. *Adv. Protein. Chem.* **45**, 371–415.

Sreng L. (1979a) Pheromones and sexual behaviour in *Nauphoeta cinerea* (Olivier) (Insecta; Dictyoptera). *Cr. Acad. Sci. D. Nat.* **289**, 687–690.

Sreng L. (1979b) Ultrastructure and chemistry of the tergal gland secretion of the male of *Blattella germanica* (L.) (Dictyoptera: Blattelidae). *Intl. J. Insect Morphol.* **8**, 213–227.

Sreng L. (1984) Morphology of the sternal and tergal glands producing the sexual pheromones and the aphrodisiacs among the cockroaches of the subfamily Oxyhaloinae. *J. Morphol.* **182**, 279–294.

Sreng L. (1985) Ultrastructure of the glands producing sex pheromones of the male *Nauphoeta cinerea* (Insecta, Dictyoptera). *Zoomorphology* **105**, 133–142.

Sreng L. (1990) Seducin, male sex pheromone of the cockroach *Nauphoeta cinerea*: isolation, identification, and bioassay. *J. Chem. Ecol.* **16**, 2899–2912.

Sreng L. (1998) Apostosis-inducing brain factors in maturation of an insect sex pheromone gland during differentiation. *Differentiation* **63**, 53–58.

Sreng L. and Quennedey A. (1976) Role of a temporary ciliary structure in the morphogenesis of insect glands. An electron microscope study of the tergal glands of male *Blattella germanica* L. (Dictyoptera, Blattellidac). *J. Ultrastructure Res.* **56**, 78–95.

Sreng L., Leoncini I. and Clement J. L. (1999) Regulation of sex pheromone production in the male *Nauphoeta cinerea* cockroach: role of brain extracts, corpora allata (CA), and juvenile hormone (JH). *Arch. Insect Biochem. Physiol.* **40**, 165–172.

Still W. C. (1979) (±)-Periplanone-B. Total synthesis and structure of the sex excitant pheromone of the American cockroach. *J. Amer. Chem. Soc.* **101**, 2493–2495.

Sun J., Hiraoka T., Dittmer N. T., Cho K. and Raikhel A. S. (2000) Lipophorin as a yolk protein precursor in the mosquito, *Aedes aegypti*. *Insect Biochem. Mol. Biol.* **30**, 1161–1171.

Takahashi S., Kitamura C. and Waku Y. (1976) Site of the sex pheromone production in the American cockroach, *Periplaneta americana* L. *Appl. Entomol. Zool.* **11**, 215–221.

Takeuchi N. and Chino H. (1993) Lipid transfer particle in the hemolymph of the American cockroach: evidence for its capacity to transfer hydrocarbons between lipophorin particles. *J. Lipid Res.* **34**, 543–551.

Tang J. D., Wolf W. A., Roelofs W. L. and Knipple D. C. (1991) Development of functionally competent cabbage looper moth sex pheromone glands. *Insect Biochem.* **21**, 573–581.

Tillman J. A., Seybold S. J., Jurenka R. A. and Blomquist G. J. (1999) Insect pheromones – an overview of biosynthesis and endocrine regulation. *Insect Biochem. Mol. Biol.* **29**, 481–514.

Tobe S. S. and Stay B. (1985) Structure and regulation of the corpus allatum. *Adv. Insect Physiol.* **18**, 305–432.

Tokro P. G., Brossut R. and Sreng L. (1993) Studies on the sex pheromone of female *Blattella germanica* L. *Insect Sci. Appl.* **14**, 115–126.

Trowell S. C. (1992) High affinity juvenile hormone carrier proteins in the hemolymph of insects. *Comp. Biochem. Physiol. B* **103**, 795–808.

Van der Horst D. J., Weers P. M. M. and Marrewijk W. J. A. (1993) Lipoproteins and lipid transport. In *Insect Lipids: Chemistry, Biochemistry and Biology*, eds D. W. Stanley-Samuelson and D. R. Nelson, pp. 1–24. University of Nebraska Press, Lincoln, Neb.

Van Heusden M. C. and Law J. H. (1989) An insect lipid transfer particle promotes lipid loading from fat body to lipoprotein. *J. Lipid Res.* **32**, 1789–1794.

Van Heusden M. C., Van der Horst D. J., Kawooya J. K. and Law J. H. (1991) *In vivo* and *in vitro* loading of lipid by artificially lipid-depleted lipophorins: evidence for the role of lipophorin as a reusable lipid shuttle. *J. Biol. Chem.* **264**, 17287–17292.

Wolfner M. F. (1997) Tokens of love: functions and regulation of *Drosophila* male accessory gland products. *Insect Biochem. Mol. Biol.* **27**, 179–192.

Wolfner M. F. (2002) The gifts that keep on giving: physiological functions and evolutionary dynamics of male seminal proteins in *Drosophila. Heredity* **88**, 85–93.

Wyatt G. R. and Davey K. G. (1996) Cellular and molecular actions of juvenile hormone. II. Roles of juvenile hormone in adult insects. *Adv. Insect Physiol.* **26**, 1–155.

Yang H.-T., Chow Y.-S., Peng W.-K. and Hsu E.-L. (1998) Evidence for the site of female sex pheromone production in *Periplaneta americana. J. Chem. Ecol.* **24**, 1831–1843.

Young H. P. and Schal C. (1997) Cuticular hydrocarbon synthesis in relation to feeding and developmental stage in *Blattella germanica* (L.) (Dictyoptera: Blattellidae) nymphs. *Ann. Entomol. Soc. Am.* **90**, 655–663.

Young H. P., Bachmann J. A. S. and Schal C. (1999a) Food intake in *Blattella germanica* (L.) nymphs affects hydrocarbon synthesis and its allocation in adults between epicuticle and reproduction. *Arch. Insect Biochem. Physiol.* **41**, 214–224.

Young H. P., Bachmann J. A. S., Sevala V. and Schal C. (1999b) Site of synthesis, tissue distribution, and lipophorin transport of hydrocarbons in *Blattella germanica* (L.) nymphs. *J. Insect Physiol.* **45**, 305–315.

Yun H. K., Jouni Z. E. and Wells M. A. (2002) Characterization of cholesterol transport from midgut to fat body in *Manduca sexta* larvae. *Insect Biochem. Molec. Biol.* **32**, 1151–1158.

11

Pheromone biosynthesis in social insects

Gary J. Blomquist and Ralph W. Howard

11.1 Introduction

In their 1990 review of chemical communication in insect communities (with special attention to social insects), Ali and Morgan (1990) stated the following: "The study of insect pheromones has grown from nothing in three decades to be now far too large to be reviewed comprehensively." Twelve years later we can confidently say that the same statement is true for just the pheromones of social insects. The last major review of social insect pheromones was the monograph edited by Vander Meer *et al.* (1998), and the papers therein provide a solid introduction to the field and should be consulted for the cogent reviews provided by the major researchers in this area. Even a cursory examination of the field will reveal that the structural diversity of social insect pheromones is as great as in non-social insects. Furthermore, social insects frequently show multiple uses of one or more pheromones in different behavioral interactions, which greatly complicates the task of teasing out structure–function relationships. Complicating the matter further is that many of these pheromones are biosynthesized, stored and released from a multiplicity of glandular complexes (over 53 described so far – Billen and Morgan, 1998) and often more than one chemical is produced from a given gland. We will focus in this chapter on some of the major advances made in social insect pheromones since 1998 with particular attention to biosynthetic aspects which are probably the least well known part of this field.

A summary of some of the recent papers on the pheromones of social insects is found in Table 11.1. A critical discussion of these papers is not possible in this

Table 11.1 Summary of some recent papers on the pheromones of social insects

Semiochemical role	Taxon	Chemical class	Reference
Alarm pheromones	ant	alcohols, aldehydes, ketones, esters, terpenes	Hughes *et al.* (2001)
	bee	alcohols, esters	Wager and Breed (2000)
	wasp	6 spiroketals + 36 misc. compounds	Dani *et al.* (2000)
	termite	monoterpenes, sesquiterpenes	Reinhard and Clément (2002)
Brood pheromones	bee	fatty acid esters	Pankiw and Page (2001)
Mimicry of species-specific pheromones	ant	cuticular hydrocarbons	Johnson *et al.* (2001)
	ant	Dufour gland extracts	Mori *et al.* (2000)
	ant: spider	cuticular hydrocarbons	Allan *et al.* (2002)
	wasp	cuticular hydrocarbons	Turillazzi *et al.* (2000)
Nestmate recognition	ant	cuticular hydrocarbons	Wagner *et al.* (2000); Lahav *et al.* (2001); Astruc *et al.* (2001); Boulay *et al.* (2000); Lenoir *et al.* (2001); Chen and Nonacs (2000); Liang and Silverman (2000)
	ant	cuticular hydrocarbons, alcohols, ketones	Hernández *et al.* (2002)
	bee	cuticular hydrocarbons	Arnold *et al.* (2000)
	wasp	cuticular hydrocarbons	Ruther *et al.* (2002); Panek *et al.* (2001)
	wasp	cuticular hydrocarbons, alcohols	Zanetti *et al.* (2001)
	wasp	none (cuticular hydrocarbons)	Pickett *et al.* (2000)
Phagostimulus pheromone	termite	1,4-dihydroxybenzene	Reinhard *et al.* (2002)
Primer pheromone	ant	none identified	Vander Meer and Alonso (2002)
Queen pheromones	ant	alkaloids, lactones, pyrones, sesquiterpenes, pentadecene	Cruz-López *et al.* (2001)
	ant	cuticular hydrocarbons	Tentschert *et al.* (2001)

	bee	fatty acids	Moritz *et al.* (2000); Pankiw *et al.* (2000); Moritz *et al.* (2002); Ledoux *et al.* (2001)
	bee	esters from Dufour's gland	Katzav-Gozansky *et al.* (2001); Katzav Gozansky *et al.* (2002); Oldroyd *et al.* 2002
Recruitment/Trail pheromones	ant	pyrazines; unidentified; 6-methylsalicylate	Hölldobler *et al.* (2001); Liefke *et al.* (2001); Gobin *et al.* (2001); Tentschert *et al.* (2000); Kohl *et al.* (2000); Heredia and Detrain (2000); Hillery and Fell (2000)
	termite	(Z)-dodec-3-en-1-ol; unidentified	Peppuy *et al.* (2001a); Peppuy *et al.* (2001b); Cornelius and Brand (2001); Reinhard and Kaib 2001; Gessner and Leuthold (2001); Affolter and Leuthold (2000)
Food-related scent marks	bees	hydrocarbons; 2-heptanone	Goulson *et al.* (2000); Stout and Goulson (2001)
Social status pheromones	ants	hydrocarbons	Cuvillier-Hot *et al.* (2001); Wagner *et al.* (2001);
	wasps	hydrocarbons	Sledge *et al.* (2001)
Territorial marking pheromone	ants	unidentified cloacal gland components	Wenseleers *et al.* (2002)
Unknown pheromonal functions	ants: Dufour gland	hydrocarbons; aldehydes; acetates; ketones; esters; farnesenes	Gökçen *et al.* (2002); Maile *et al.* (2000)
	ants: venom gland	4-methyl-3-heptanol; esters; long-chain alcohols	Maile *et al.* (2000)
	bees: Dufour	all-trans-geranylgeranyl acetate;	Cruz-López *et al.* (2001)

(Contd)

Table 11.1 *(Contd)*

Semiochemical role	Taxon	Chemical class	Reference
	gland	cyclic ketals; mono-, sesqui- and diterpene compounds; acetates; other oxygenated compounds	
	bees: labial gland	hydrocarbons	Katzav-Gozansky *et al.* (2001)
	bees: mandibular gland	hydroxy-acids; others	Simon *et al.* (2001); Cruz-López *et al.* (2002)

chapter, but several notable areas are worth highlighting. At least two new glands have been discovered in ants: one between abdominal sternites 6 and 7 in *Cylindromyrmex whymperi*, which produces secretions that function in mass recruitment (Gobin *et al.*, 2001), and antennal glands in the fire ant *Solenopsis invicta* of unknown function (Isidoro *et al.*, 2000). The chemistry and biology of Dufour glands in the social Hymenoptera continue to be an active area of research reflecting the complexity of the biosynthetic and functional aspect of these glands. Two papers serve as exemplars of this complexity. Gökçen *et al.* (2002) describe over 112 compounds from ants in the *Cataglyphis bicolor* group, including hydrocarbons, alcohols, aldehydes, acetates, ketones, isobutyrates, and long-chain esters with both straight-chain and branched components. Cruz-López *et al.* (2001) described a complex mixture of over 40 components from the stingless bees *Nannotrigona testaceicornis* that included the diterpene ester all-trans-geranylgeranyl acetate and a complex mixture of cyclic ketals, mono, sesqui- and diterpene compounds as well as acetates and various other oxygenated compounds. Clearly, any attempts at unraveling the biosynthetic and functional aspects of glandular products of this complexity will face daunting challenges.

New findings on the chemistry and biology of cuticular hydrocarbons in social insects continue to grow exponentially. Considerable attention in recent years has gone into understanding the role of these compounds in various levels of recognition processes (from the individual to colony and population level). Recent papers dealing with the roles of individual classes of hydrocarbons (Dani *et al.*, 2001) and individual hydrocarbons themselves (Ruther *et al.*, 2002) promise to open significant new understanding in this complex area. Additional studies have addressed the issue of which environmental or physiological processes might be important in explaining the variation in cuticular hydrocarbons known to occur in their various semiochemical roles (Cuvillier-Hot *et al.*, 2001; Liu *et al.*, 2001; Sledge *et al.*, 2001; Tentschert *et al.*, 2001; Wagner *et al.*, 2001). These various factors, such as age, sex, ovarian activity, reproductive status and social status, have important implications for studies on biosynthesis and physiological regulation. Examples of the use of cuticular hydrocarbons in mimicry systems also continue to grow (Allan *et al.*, 2002). Finally, an intriguing paper by Goulson *et al.* (2000), reported that tarsal scent marks (hydrocarbons in the low C20 range) left by foraging bumblebees were perceived and responded to by other bumblebees at concentrations of 10^{-12} ng or greater! The nature of receptors capable of responding to levels of hydrocarbon this low (approaching single molecules) remains unknown.

11.2 Biosynthesis of pheromones in social insects

Pheromone biosynthesis in the social insects has not been looked at in as much detail as the pheromones of other insect groups, perhaps in part due to the complex mixtures of chemicals and the low amounts often present. Many of the pheromone components are probably derived from the chain elongation or chain shortening of fatty acids or from the isoprenoid pathway, but in most cases the details are not known. Chain elongation of appropriate fatty acids followed by the loss of the carboxyl carbon leads to the formation of hydrocarbons, which function widely in chemical communication in the social insects, and some aspects of their biosynthesis have been studied. In an elegant set of experiments, the biosynthesis of the honeybee queen mandibular pheromone has been shown to occur by chain shortening after hydroxylation of a fatty acid, and this process is reviewed herein. There appears to be no work on the biosynthesis of any of the isoprenoid derived pheromone components. One can speculate on the biosynthesis of some components based on their similarity to other compounds, whereas for still others, the biosynthesis is completely unknown.

11.3 Hydrocarbons

Insects, particularly social insects, use long-chain hydrocarbons in chemical communication (Howard, 1993). The hydrocarbons are used as attractants, sex recognition cues, arrestants, species recognition cues and nestmate recognition cues, among others. In some cases, individual components are recognized to alter behavior, whereas in other cases, especially for species and nestmate recognition cues, it appears the mixture of hydrocarbon components is important. Features of hydrocarbons that are important for chemical communication include the length of the carbon chain, the number and positions of methyl-branch groups and the number and positions of double bonds. While we don't yet have a clear picture of how insects regulate and control these parameters, we are gaining a better understanding of the biosynthesis of n-, methyl-branched and unsaturated hydrocarbons that are used in chemical communication. The next section will review our understanding of hydrocarbon biosynthesis and transport, with an emphasis on the work done in social insects.

The biosynthesis of hydrocarbons occurs by the microsomal elongation of straight chain, methyl-branched and unsaturated fatty acids to produce very long-chain fatty acyl-CoAs (Figure 11.1). The very long chain fatty acids are then reduced to aldehydes and converted to hydrocarbon by loss of the carboxyl carbon. The mechanism of hydrocarbon formation has been controversial. Kolattukudy and coworkers have reported that for a plant, an algae, a vertebrate and an insect, the aliphatic aldehyde is decarbonylated to the hydrocarbon and carbon monoxide, and that this process does not require cofactors (Cheesbrough and Kolattukudy, 1984; 1988; Dennis and Kolattukudy, 1991, 1992; Yoder *et al.*, 1992). In contrast, the Blomquist laboratory has presented evidence that the aldehyde is converted to hydrocarbon and carbon dioxide in a process that

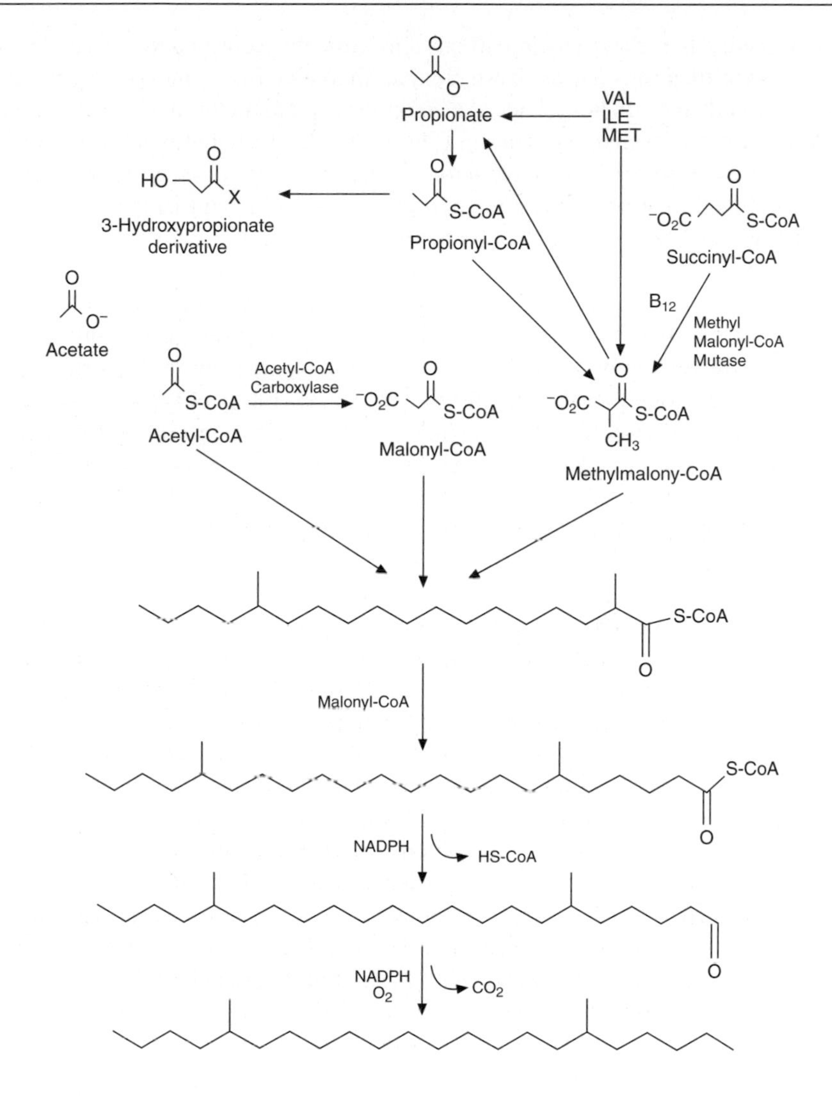

Figure 11.1 Biosynthetic pathways leading to methyl-branched alkanes in insects.

requires NADPH and molecular oxygen. This has been shown to occur in a variety of insects including the termite *Zootermopsis nevadensis* (Reed *et al.*, 1994, 1995; Mpuru *et al.*, 1996). The resolution of this controversy over the mechanism of hydrocarbon formation must await the molecular characterization of the enzymes involved.

The methyl branches in hydrocarbons arise from the incorporation of a malonyl-CoA at specific points during chain elongation to give rise to mono-, di-, tri- and tetramethylalkanes. In houseflies and cockroaches, the methylmalonyl-CoA unit is inserted early in chain elongation (Dwyer *et al.*, 1981; Dillwith *et al.*, 1982) and it arises from the carbon skeleton of valine (Dillwith *et al.*, 1982; Chase *et al.*, 1990). In contrast, the labeled amino acids valine and isoleucine did not efficiently label methyl-branched hydrocarbons in termites (Guo *et al.*, 1991). In the dampwood termite, *Zootermopsis nevadensis*, the pattern of incorporation of [1,4-^{14}C]succinate and [2,3-^{3}H]succinate was consistent with succinate serving as the carbon source for methylmalonyl-CoA (Chu and Blomquist, 1980). Direct evidence for succinate serving as the carbon source for methyl-branched hydrocarbons was obtained by monitoring the incorporation of [2,3-^{13}C]succinate into the methyl-branching carbon and tertiary carbon of methyl-branched hydrocarbons in *Z. nevadensis* (Blomquist *et al.*, 1980). Carbon-13 NMR analysis showed that [1,4-^{13}C]succinate labeled carbon 6 of 5-methylheneicosane and positions 6 and 18 of 5,17- dimethylheneicosane, indicating that the methylmalonyl-CoA units were incorporated as the third unit to form 5-methylheneicosane and the third and ninth units to form 5,17-dimethylheneicosane (Guo *et al.*, 1991). Termites are unique among insects in that they have abundant levels of vitamin B12, especially in their gut tracts (Wakayama *et al.*, 1984). Analysis of organic acids after the *in vivo* metabolism of [2,3-^{14}C]succinate showed that succinate was converted to propionate and methylmalonate. Labeled succinate injected into the hemolymph was readily taken up by the gut tract. Isolated gut tissue efficiently converted succinate to acetate and propionate, both of which were released into the incubation media. Mitochondria from termite tissue (minus gut tract) converted succinate to methylmalonate and propionate only in the presence of malonic acid, an inhibitor of succinate dehydrogenase. These data are interpreted to indicate that while termite mitochondria are able to convert succinate to propionate and methylmalonate, most of the propionate used for methyl-branched hydrocarbon biosynthesis is produced by gut tract microorganisms. The propionate is then presumably transported through the hemolymph to epidermal cells for use in methyl-branched hydrocarbon biosynthesis.

The regulation of chain length and the number and positions of double bonds and methyl groups has not been studied in social insects. However, work describing the regulation of the chain length of alkenes in the housefly strongly points to the importance of the condensing activity of the acyl-CoA elongase enzymes in regulating the chain length of products, with the reduction of the acyl-CoA to aldehyde and decarboxylation reactions showing little chain length specificity. No information is available in any system as to how the number and positions of the methyl groups are regulated.

11.4 Biosynthesis of hydrocarbons by inquilines of social insects

A large number of alien insects have managed to invade and colonize the nests of social insects, many of them doing so by mimicking to a lessor or greater extent the cuticular lipids of their hosts (Dettner and Liepert, 1994; Howard and Akre, 1995; Lenoir *et al.*, 2001). Although the literature in this area is voluminous, the vast majority of it deals with description of the chemicals involved, the extent of the qualitative and quantitative correspondence in hydrocarbons between inquiline and host, and behavioral and chemical ecological interactions among the inquilines and the hosts, with only scant knowledge of biosynthetic processes. Two hypotheses have been set forth as to how the inquilines come by their mimicking chemicals: one by *de novo* biosynthesis, and the other by acquiring them from their hosts by some sort of mechanical transfer process. It is now clear that both processes operate and the major question is the relative importance of each process for each given inquiline–host pairing.

Direct evidence for an inquiline biosynthesizing cuticular hydrocarbons that closely mimicked that of their hosts was first reported for the staphylinid beetle *Trichopsenius frosti*, a highly integrated, host-specific inquiline associated with the subterranean termite *Reticulitermes flavipes* (Howard *et al.*, 1980). Not only did this beetle possess all of the hydrocarbons found on the cuticle of *R. flavipes* and no other hydrocarbons, but also radiolabelling experiments with [1-^{14}C]acetate indicated that all hydrocarbon classes incorporated substantial amounts of the label, thus suggesting that the beetle made essentially all of its own hydrocarbons and did not depend on having to acquire them from the termite hosts. A second sympatric species of *Reticulitermes (R. virginicus*) has three host-specific staphylinid inquilines associated with it, and Howard *et al.* (1982) showed that these three beetles *(Trichopsenius depressus*, *Xenistusa hexagonalis*, and *Philotermes howardi*) also possessed exactly the same cuticular hydrocarbons (and no others) as found on *R. virginicus* (which has its own species-specific hydrocarbons profile). Experiments with [1-^{14}C]acetate in *X. hexagonalis* indicated that it too biosynthesized all of its own cuticular hydrocarbons. Radiolabelling experiments were not done with the other two species of inquilines because sufficient insects were not available (it should be noted that inquilines frequently occur at the ratio of about one inquiline per 50 000 or more hosts).

To date, only two other sets of inquilines have been directly examined for their ability to biosynthesize suites of cuticular hydrocarbons that mimic that of their hosts. Larvae of the syrphid fly *Microdon albicomatus* are obligate predators of the ant *Myrmica incompleta*, and have qualitatively identical hydrocarbons as host pupae (but different relative abundances). As with the beetles above, [1-^{14}C]acetate was directly incorporated into the fly cuticular hydrocarbons in the same proportions as the various classes of hydrocarbons present (Howard *et al.*, 1990). The other study was by Akino *et al.* (1999) and involved the larvae of the caterpillar *Maculinea rebeli*, an inquiline of *Myrmica schencki*. In this system, the caterpillar larvae first biosynthesize hydrocarbons that mimic those of their host ant brood, and later acquire from the ants additional hydrocarbons that enhance the mimicry of their host colony's odor.

11.5 Transport of hydrocarbons

Another step in the regulation of the hydrocarbons that appear on the surface of insects could occur during the transport of hydrocarbons to the surface. A number of studies show that hydrocarbons are synthesized by oenocytes, whose location varies widely among insects (Romer, 1991) and can be associated with epidermal cells or the fat body. Nevertheless, even in insects whose oenocytes are associated with the epidermal cells, it appears that hydrocarbons are first transported to the hemolymph where they are carried by lipophorin (Schal *et al.*, 1998). In social insects, hydrocarbon profiles serve as species-, colony-, and caste-specific recognition cues. Thus social insects must closely regulate their cuticular profiles and exocrine secretions (Howard, 1993; Clément and Bagnères, 1998). The mechanism of hydrocarbon transport has been addressed in the termite *Zootermopsis nevadensis* (Sevela *et al.*, 2000). The hemolymph of all castes, including workers, soldiers, nymphs, and female and male alates, contained large amounts of hydrocarbon. All hemolymph samples had nearly identical hydrocarbon profiles and the hydrocarbon was associated with lipophorin. However, the hemolymph profile was quite different from the caste-specific cuticular profile. These data support the hypothesis that lipophorin transports hydrocarbons from the site of synthesis to the site of deposition on the integument. In many social insects, different castes have different proportions of the same hydrocarbon and lipophorin appears to play an important role in regulation of externalization and internalization of specific hydrocarbon components, and therefore in the attainment of caste-specific chemical profiles (Sevela *et al.*, 2000). In several Lepidoptera, hydrocarbons are transported by the hemolymph, and the process is regulated such that the shorter-chain pheromone and pheromone precursor hydrocarbons are sequestered by the pheromone gland, whereas the longer-chain cuticular hydrocarbons appear elsewhere on the cuticle (Schal *et al.*, Chapter 10, and Jurenka, Chapter 3, in this volume).

In the desert ant, *Cataglyphis niger*, both the hemolymph and crop contain the same hydrocarbons that are found in the postpharyngeal gland (Soroker and Hefetz, 2000). Biosynthetic studies showed that hydrocarbon was synthesized by fat body tissue (perhaps oenocytes associated with fat body?) and not by postpharyngeal glands. This suggests that hydrocarbons are transported from the site of synthesis by hemolymph, presumably bound to lipophorin, to both the cuticle and postpharyngeal glands, and subsequently to the crop (Soroker and Hefetz, 2000).

In several ant species, the postpharyngeal gland serves as a "gestalt" organ where hydrocarbons from different sources accumulate, mix and are then distributed between insects by trophallaxis and/or allogrooming (Soroker *et al.*, 1994; Meskali *et al.*, 1995). In *Camponotus fellah*, exchange of hydrocarbons among individuals was necessary to maintain similar hydrocarbon profiles (Boulay *et al.*, 2000). Ants separated by isolation periods of 20–40 days had a greater change in hydrocarbon profile and made residents intolerant towards introduced nestmates. Thus in ants there must be both transport within individuals (by lipophorin) and between individuals (by trophallaxis and allogrooming) to obtain the appropriate hydrocarbon profiles.

11.6 Honeybee queen mandibular pheromone

The best understood of the sexual pheromones of social insects is the queen substance of honeybees. Interestingly, the queen substance used for "queen control" inside the nest is also the substance used by virgin queens to attract drones for mating. Callow and Johnston (1960) and Barbier and Lederer (1960) identified 9-keto-2(E)-decenoic acid ([E]-9-oxodec-2-enoic acid) (9-ODA) as major components of the queen mandibular glands. 9-Hydroxy-2(E)-decenoic acid is also present (Callow *et al.*, 1964) and together both attract drones. Additional components of the queen retinue pheromone have recently been identified (Keeling *et al.*, 2003).

In an elegant set of experiments, Plettner *et al.* (1996, 1998) elucidated the biosynthetic pathways for the honeybee queen mandibular pheromone (QMP). The QMP contains major components 9-hydroxy- and 9-keto-2(E)-decenoic acids. The workers produce compounds that are structurally similar, but don't have the hydroxy and keto groups on the n-2 positions and include 10-hydroxy-2(E)-decenoic acid and the corresponding diacid (Figure 11.2). Plettner *et al.* (1996, 1998) used carbon-13 and deuterated precursors to demonstrate: (1) the *de novo* synthesis of stearic acid in worker mandibular glands, (2) the hydroxylation of stearic acid at the n- (workers) and n-1 (queens) positions, (3) chain shortening through β-oxidation to the 10 and 8 carbon hydroxy acids, and (4) oxidation of n- and n-1 hydroxy groups to give diacids and 9 keto-2(E)-decenoic acid, respectively. Stearic acid was shown to be the main precursor of the pheromone molecules as it was converted to C10 hydroxy acids and diacids more efficiently than either 16 or 14 carbon fatty acids.

There were at least two possibilities in the conversion of the stearic acid to the C10 hydroxyacids: they could be first hydroxylated and then chain shortened, or they could be first chain shortened and then hydroxylated. That the former occurs was demonstrated by two lines of evidence: (1) mandibular glands chain shortened 18-hydroxy and 17-hydroxy stearic acid to the C10 acids and (2) both n-1 and n-2 hydroxyacids with more than 10 carbons accumulated when

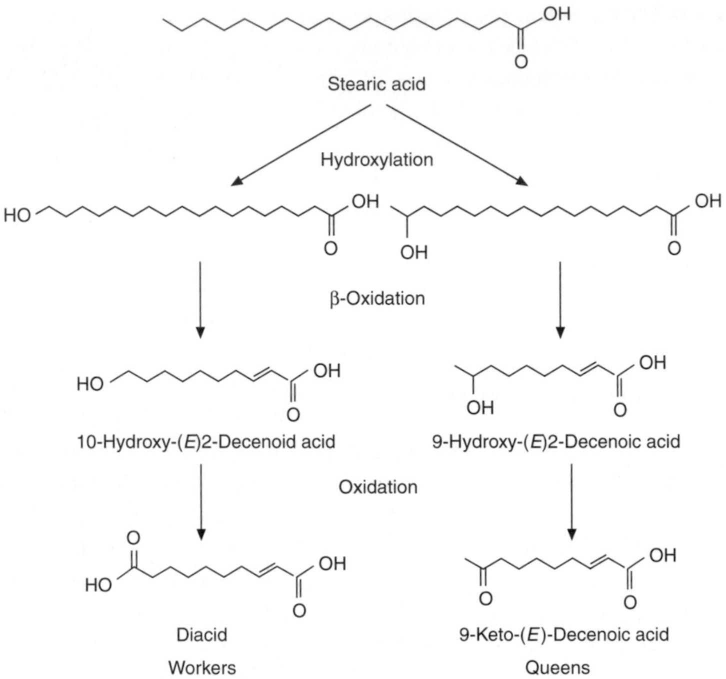

OH
O
Stearic acid
Hydroxylation
HO OH O
OH OH O
β-Oxidation
HO OH O
10-Hydroxy-(*E*)2-Decenoid acid
OH OH O
9-Hydroxy-(*E*)2-Decenoic acid
Oxidation
O HO OH O
Diacid
Workers
OH O O
9-Keto-(*E*)-Decenoic acid
Queens

Figure 11.2 Biosynthetic pathways leading to honeybee worker and queen mandibular gland pheromones. Reprinted with permission from Plettner E., *et al.* (1966), © 1966 American Association for the Advancement of Science.

2-fluorostearic acid was used as an inhibitor of β-oxidation (Plettner *et al.*, 1996). Furthermore, using deuterated substrates, they showed that a direct hydroxylation occurred, and not a desaturation followed by hydration. Worker bees are able to elongate hydroxy acids to the next higher two carbon homolog and are able to reduce hydroxy-2-(E)-decenoic acids to the corresponding hydroxydecanoic acids, transformations that were not observed in queens.

11.7 Future directions

Billen and Morgan (1998) show structures of 125 trail, alarm, sex and queen recognition pheromones. In some cases it is clear that components could arise from chain elongating or shortening of fatty acids, a common theme in pheromone biosynthesis, or from modification of isoprenoid derived intermediates, but in most cases no work has been done on the biosynthesis of these compounds. In other insect groups, PBAN, juvenile hormone and ecdysteroids regulate pheromone production. To our knowledge, the endocrine regulation of pheromone production has not been explored in any social insect, and it is an area that begs for attention.

Acknowledgements

Contribution of the Nevada Agricultural Experiment Station, MS # 03031276.

References

Affolter J. and Leuthold R. H. (2000) Quantitative and qualitative aspects of trail pheromones in *Macrotermes subhyalinus* (Isoptera, Termitidae). *Insectes Soc.* **47**, 256–262.

Akino T., Knapp J. J., Thomas J. A. and Elmes G. W. (1999) Chemical mimicry and host specificity in the butterfly *Maculinea rebeli*, a social parasite of *Myrmica* ant colonies. *Proc. R. Soc. London Ser. B*, **266**, 1419–1426.

Ali M. F. and Morgan E. D. (1990) Chemical communication in insect communities: a guide to insect pheromones with special emphasis on social insects. *Biol. Rev.* **65**, 227–247.

Allan R. A., Capon R. J., Brown W. V. and Elgar M. A. (2002) Mimicry of host cuticular hydrocarbons by salticid spider *Cosmophasis bitaeniata* that preys on larvae of tree ants *Oecophylla smaragdina*. *J. Chem. Ecol.* **28**, 835–848.

Arnold G., Quenet B. and Masson C. (2000) Influence of social environment on genetically based subfamily signature in the honeybee. *J. Chem. Ecol.* **26**, 2321–2333.

Astruc C., Malosse C. and Errard C. (2001) Lack of intraspecific aggression in the ant *Tetramorium bicarinatum*: a chemical hypothesis. *J. Chem. Ecol.* **27**, 1229–1248.

Barbier J. and Lederer E. (1960) Structure chemique de la substance royale de la reine d'abeille (*Apis mellifica* L.). *C. R. Acad. Sci. Paris* **251**, 1131–1135.

Billen J. and Morgan E. D. (1998) Pheromone communication in social insects: sources and secretions. In *Pheromone Communication in Social Insects*, eds R. K. Vander Meer, M. D. Breed and M. L. Winston and E. K. Espelie pp. 3–33, Westview Press, Boulder, CO.

Blomquist G. J., Chu A. J., Nelson J. H. and Pomonis J. G. (1980) Incorporation of [2,3-13C]-succinate into methyl-branched alkanes in a termite. *Arch. Biochem. Biophys.* **204**, 648–650.

Boulay R., Hefetz A., Soroker V. and Lenoir A. (2000) *Camponotus fellah* colony integration: worker individuality necessitates frequent hydrocarbon exchanges. *Animal Behavior* **59**, 13: 1127–1133.

Callow R. K. and Johnston N. C. (1960) The chemical constitution and synthesis of queen substance of honeybees (*Apis mellifera* L.). *Bee World* **41**, 152–153.

Callow R. K., Chapman J. R. and Paton P. N. (1964) Pheromones of the honeybee: chemical studies of the mandibular gland secretion of the queen. *J. Apicult. Res.* **3**, 77–89.

Chase J., Jurenka R. A., Schal C., Halarnkar P. P. and Blomquist G. J. (1990) Biosynthesis of methyl branched hydrocarbons in the German Cockroach *Blatella germanica* (L.) (Orthoptera, Blatellidae). *Insect Biochem.* **20**, 149–156.

Cheesbrough T. M. and Kolattukudy P. E. (1984) Alkane biosynthesis by decarbonylation of aldehydes catalyzed by a particulate preparation from *Pisum sativum*. *Proc. Natl. Acad. Sci. USA* **81**, 6613–6617.

Cheesbrough T. M. and Kolattukudy P. E. (1988) Microsomal preparations from animal tissue catalyzes release of carbon monoxide from a fatty aldehyde to generate an alkane. *J. Biol. Chem.* **263**, 2738–2743.

Chen J. S. C. and Nonacs P. (2000) Nestmate recognition and intraspecific aggression based on environmental cues in Argentine ants (Hymenoptera: Formicidae). *Ann. Entomol. Soc. Am.* **93**, 1333–1337.

Chu A. J. and Blomquist G. J. (1980) Biosynthesis of hydrocarbons in insects: succinate is a precursor of the methyl branched alkanes. *Arch. Biochem. Biophys.* **201**, 304–312.

Clément J.-L. and Bagnères A.-G. (1998) Nestmate recognition in termites. pp. 126–155, In eds R. K. Vander Meer, M. Breed, M. Winston and K. Espelie, *Pheromone Communication in Social Insects: Ants, Wasps, Bees, and Termites*, Westview Press, Boulder, CO.

Cornelius M. L. and Brand J. M. (2001) Trail-following behavior of *Coptotermes formosanus* and *Reticulitermes flavipes* (Isoptera; Rhinotermitidae): is there a species-specific response? *Environ. Entomol.* **30**, 457–465.

Cruz-López L. Patricio E. F. L. R. A. and Morgan E. D. (2001a) Secretions of stingless bees: the Dufour gland of *Nannotrigona testaceicornis*. *J. Chem. Ecol.* **27**, 69–80.

Cruz-López L., Rojas J. C., De L Cruz-Cordero, R. and Morgan, E. D. (2001b) Behavioral and chemical analysis of venom gland secretion of queens of the ant *Solenopsis geminata*. *J. Chem. Ecol.* **27**, 2437–2445.

Cruz-López L., Patricio E. F. L. R. A., Maile R. and Morgan E. D. (2002) Secretions of stingless bees: cephalic secretions of two *Frieseomelitta* species. *J. Insect Physiol.* **48**, 453–458.

Cuvillier-Hot V., Cobb M., Malosse C. and Peeters C. (2001) Sex, age and ovarian activity affect cuticular hydrocarbons in *Diacamma ceylonese*, a queenless ant. *J. Insect Physiol.* **47**, 485–493.

Dani F. R., Jeanne R. L., Clarke S. R., Jones G. R., Morgan E. D., Francke W. and Turillazzi S. (2000) Chemical characterization of the alarm pheromone in the venom of *Polybia occidentalis* and of volatiles from the venom of *P. sericea*. *Physiol. Entomol.* **25**, 363–369.

Dani F. R., Jones G. R., Destri S., Spencer S. H. and Turillazzi S. (2001) Deciphering the recognition signature within the cuticular chemical profile of paper wasps. *Animal Behaviour* **62**, 165–171.

Dennis M. W. and Kolattukudy P. E. (1991) Alkane biosynthesis by decarbonylation of aldehyde catalyzed by a microsomal preparation from *Botryococcus brauni*. *Arch. Biochem. Biophys.* **287**, 268–275.

Dennis M. W. and Kolattukudy P. E. (1992) A cobalt-porphyrin enzyme converts a fatty aldehyde to a hydrocarbon and CO. *Proc. Natl. Acad. Sci. USA* **89**, 5306–5310.

Dettner K. and Liepert C. (1994) Chemical mimicry and camouflage. *Annu. Rev. Entomol.* **39**, 129–154.

Dillwith J. W., Nelson J. H., Pomonis J. G., Nelson D. R. and Blomquist G. J. (1982) A 13 C-NMR study of methyl-branched hydrocarbon biosynthesis in the housefly. *J. Biol. Chem.* **257**: 11305–11314.

Dwyer L. A., Blomquist G. J., Nelson J. H. and Pomonis J. G. (1981) A 13 C-NMR study of the biosynthesis of 3-methylpentacosane in the American cockroach. *Biochim. Biophys. Acta* **663**, 536–566.

Gessner S. and Leuthold R. H. (2001) Caste-specificity of pheromone trails in the termite *Macrotermes bellicosus*. *Insectes Soc.* **48**, 238–244.

Gobin B. Rüppell, Hartmann A., Jungnickel, Morgan E. D. and Billen J. (2001) A new type of exocrine gland and its function in mass recruitment in the ant *Cylindromyrmex whymperi* (Formicidae, Cerapachyinae). *Naturwissenschaften* **88**, 395–399.

Gökçen O. Q., Morgan E. D., Dani F. R., Agosti D. and Wehner R. (2002) Dufour gland contents of the *Cataglyphis bicolor* group. *J. Chem. Ecol.* **28**, 71–87.

Goulson D., Stout J. C., Langley J. and Hughes W. O. H. (2000) Identity and function of scent marks deposited by foraging bumblebees. *J. Chem. Ecol.* **26**, 2897–2911.

Guo L., Quilici D. R, Chase J. and Blomquist G. J. (1991) Gut tract microorganisms supply the precursors for methyl-branched hydrocarbon biosynthesis in the termite, *Zootermopsis nevadensis*. *Insect Biochem.* **21**, 327–333.

Heredia A. and Detrain C. (2000) Worker size polymorphism and ethological role of sting associated glands in the harvester ant *Messor barbarus*. *Insectes Soc.* **47**, 383–389.

Hernández J. V., López H. and Jaffe K. (2002) Nestmate recognition signals of the leaf-cutting ant *Atta laevigata*. *J. Insect Physiol.* **48**, 287–295.

Hillery A. E. and Fell R. D. (2000) Chemistry and behavioral significance of rectal and accessory gland contents in *Camponotus pennsylvanicus* (Hymenoptera: formicidae). *Ann. Entomol. Soc. Am.* **93**, 1294–1299.

Hölldobler B., Morgan E. D., Oldham N. J. and Liebig J. (2001) Recruitment pheromone in the harvester ant genus *Pogonomyrmex*. *J. Insect Physiology*. **47**, 369–374.

Howard R. W. (1993) Cuticular hydrocarbons and chemical communication. *Insect Lipids: Chemistry, Biochemistry and Biology*. In eds D. W. Stanley-Samuelson and D. R. Nelson, pp. 176–226. University of Nevada Press, Lincoln.

Howard R. W. and Akre R. D. (1995) Propaganda, crypsis and slave-making. *Chemical Ecology of Insects 2*, eds R. T. Cardé and W. J. Bell, pp. 364–424. Chapman and Hall, New York.

Howard R. W., McDaniel C. A. and Blomquist G. J. (1980) Chemical mimicry as an integrating mechanism: cuticular hydrocarbons of a termitophile and its host. *Science* **210**, 431–433.

Howard R. W., McDaniel C. A. and Blomquist G. J. (1982) Chemical mimicry as an integrating mechanism for three termitophiles associated with *Reticulitermes virginicus* (Banks). *Psyche* **89**, 157–167.

Howard R. W., Stanley-Samuelson D. W. and Akre R. D. (1990) Biosynthesis and chemical mimicry of cuticular hydrocarbons from the obligate predator, *Microdon albicomatus* Novak (Diptera: Syrphidae) and its ant prey, *Myrmica incompleta* Provancher (Hymenoptera: Formicidae). *J. Kansas Entomol. Soc.* **63**, 437–443.

Hughes W. O. H., Howse P. E. and Goulson D. (2001) Mandibular gland chemistry of grass-cutting ants: species, caste and colony variation. *J. Chem. Ecol.* **27**, 109–124.

Isidoro N., Romani R., Velasquez D., Renthal R., Bin F. and Vinson S. B. (2000) Antennal glands in queen and worker of the fire ant, *Solenopsis invicta* Buren: first report in female social Aculeata (Hymenoptera, formicidae). *Insectes Soc.* **47**, 236–240.

Johnson C. A., Vander Meer R. K. and Lavine B. (2001) Changes in the cuticular hydrocarbon profile of the slave-maker ant queen, *Polyergus breviceps* Emery, after killing a *Formica* host queen (Hymenoptera: Formicidae). *J. Chem. Ecol.* **27**, 1787–1804.

Katzav-Gozansky T., Soroker V. and Hefetz A. (2002) Evolution of worker sterility in honey bees: egg-laying workers express queen-like secretion in Dufour's gland. *Behav. Ecol. Sociobiol.* **51**, 588–589.

Katzav-Gozansky T., Soroker V., Ibarra F., Francke W. and Hefetz A. (2001a) Dufour's gland secretion of the queen honeybee *(Apis mellifera)*: an egg discriminator or a queen signal? *Behav. Ecol. Sociobiol.* **51**, 76–86.

Katzav-Gozansky T., Soroker V., Ionescu A., Robinson G. E. and Hefetz A. (2001b) Task-related chemical analysis of labial gland volatile secretion in worker honeybees *(Apis mellifera ligustica)*. *J. Chem. Ecol.* **27**, 919–926.

Keeling, C. I., Slessor, K. N., Higo, H. A. and Winston, M. L. (2003) New components of the honey bee (*Apis mollifera* L.) Queen retinue pheromone. *Proc. Natl. Acad. Sci. USA* **100**, 4486–4491.

Kohl E., Hölldobler B. and Bestmann H.-J. (2000) A trail pheromone component of the ant *Mayriella overbecki* Viehmeyer (Formicidae: Myrmicinae). *Naturwissenschaften* **87**, 320–322.

Lahav S., Soroker V., Vander Meer R. K. and Hefetz A. (2001) Segregation of colony odor in the desert ant *Cataglyphis niger*. *J. Chem. Ecol.* **27**, 927–943.

Ledoux M. N., Winston M. L., Higo H., Keeling C. I., Slessor K. N. and LeConte Y. (2001) Queen and pheromonal factors influencing comb construction by simulated honey bee *(Apis mellifera* L.) swarms. *Insectes Soc.* **48**, 14–20.

Lenoir A., Cuisset D. and Hefetz A. (2001a) Effects of social isolation on hydrocarbon pattern and nestmate recognition in the ant *Aphaenogaster senilis* (Hymenoptera, Formicidae). *Insectes Soc.* **48**, 101–109.

Lenoir A., D'Ettore P. D., Errard C. and Hefetz A. (2001b) Chemical ecology and social parasitism in ants. *Annu. Rev. Entomol.* **46**, 573–599.

Liang D. and Silverman J. (2000) "You are what you eat": diet modifies cuticular hydrocarbons and nestmate recognition in the Argentine ant, *Linepithema humile*. *Naturwissenschaften* **87**, 412–416.

Liefke C., Hölldobler B. and Maschwitz U. (2001) Recruitment behavior in the ant genus *Polyrachis* (Hymenoptera, Formicidae). *J. Insect Behav.* **14**, 637–657.

Liu Z. B., Bagnères A. G., Yamane S., Wang Q. C. and Kojima J. (2001) Intra-colony, intercolony and seasonal variations of cuticular hydrocarbon profiles in *Formica japonica* (Hymenoptera, Formicidae). *Insectes Soc.* **48**, 342–346.

Maile R., Jungnickel H., Morgan E. D., Ito F. and Billen J. (2000) Secretion of venom and Dufour glands in the ant *Leptogenys diminuta*. *J. Chem. Ecol.* **26**, 2497–2506.

Meskali M., Bonavita-Cougourdan A., Provost E., Bagnères A.-G., Dusticier G. and Clément J. L. (1995) Mechanism underlying cuticular hydrocarbon homogeneity in the ant *Camponotus vagus* (Scop.) (Hymenoptera: Formicidae): role of post pharyngeal glands. *J. Chem. Ecol.* **21**, 1127–1148.

Mori A., Grasso D. A., Visicchio R. and Le Moli F. (2000) Colony founding in *Polyergus rufescens*: the role of Dufour's gland. *Insectes Soc.* **47**, 7–10.

Moritz R. F. A., Crewe R. M. and Hepburn H. R. (2002) Queen avoidance and mandibular gland secretion of honeybee workers *(Apis mellifera* L.). *Insectes Soc.* **49**: 86–91.

Moritz R. F. A., Simon U. E. and Crewe R. M. (2000) Pheromonal contests between honeybee workers *(Apis mellifera capensis)*. *Naturwissenschaften* **87**, 395–397.

Mpuru S., Reed J. R., Reitz R. C. and Blomquist G. J. (1996) Mechanism of hydrocarbon biosynthesis from aldehyde in selected insect species: requirement for O_2 and NADPH and carbonyl group released as CO_2. *Insect Biochem. Molec. Biol.* **2**: 203–208.

Oldroyd B. P., Ratnieks F. L. W. and Wossler T. C. (2002) Egg-marking pheromones in honeybees *Apis mellifera*. *Behav. Ecol. Sociobiol.* **51**, 590–591.

Panek L. M., Gamboa G. J. and Espelie K. E. (2001) The effect of a wasp's age on its cuticular hydrocarbon profile and its tolerance by nestmate and non-nestmate conspecifics (*Polistes fuscatus*, Hymenoptera: Vespidae). *Ethology* **107**, 55–63.

Pankiw T. and Page R. E., Jr (2001) Brood pheromone modulates honeybee *(Apis mellifera* L.) sucrose response thresholds. *Behav. Ecol. Sociobiol.* **49**, 206–213.

Pankiw T., Winston M. L., Fondrk M. K. and Slessor K. N. (2000) Selection on worker honeybee responses to queen pheromone (*Apis mellifera* L.). *Naturwissenschaften* **87**, 487–490.

Peppuy A., Robert A., Sémon E., Bonnard O., Son N. T. and Bordereau C. (2001a) Species specificity of trail pheromones of fungus-growing termites from northern Vietnam. *Insectes Soc.* **48**, 245–250.

Peppuy A., Robert A., Semon E., Ginies C., Lettere M., Bonnard O. and Bordereau C. (2001b) (Z)-dodec-3-en-1-ol, a novel termite trail pheromone identified after solid phase microextraction from *Macrotermes annandalei*. *J. Insect Physiol.* **47**, 445–453.

Pickett K. M., McHenry A. and Wenzel J. W. (2000) Nestmate recognition in the absence of a pheromone. *Insectes Soc.* **47**, 212–219.

Plettner E., Slessor K. N., Winston M. L. and Oliver J. E. (1996) Caste-selective pheromone biosynthesis in honeybees. *Science* **271**, 1851–1853.

Plettner E., Slessor K. N. and Winston M. L. (1998) Biosynthesis of mandibular acids in honey bees (*Apis mellifera*): de novo synthesis, route of fatty acid hydroxylation and caste selective oxidation. *Insect Biochem. Molec. Biol.* **28**, 31–42.

Reed J. R., Quilici D. R., Blomquist G. J. and Reitz R. C. (1995) Proposed mechanism for the cytochrome P450 catalyzed conversion of aldehyde to hydrocarbon in the house fly, *Musca domestica*. *Biochemistry* **34**, 16221–16227.

Reed J. R., Vanderwel D., Choi S., Pomonis J. G., Reitz R. C. and Blomquist G. J. (1994) Unusual mechanism of hydrocarbon formation in the housefly: cytochrome P450 converts aldehyde to the sex pheromone component (Z)-9-tricosene and CO_2. *Proc. Natl. Acad. Sci. USA* **91**, 10000–10004.

Reinhard J. and Clément J.-L. (2002) Alarm reaction of European *Reticulitermes* termites to soldier head capsule volatiles (Isoptera, Rhinotermitidae). *J. Insect Behav.* **15**, 95–107.

Reinhard J. and Kaib M. (2001) Trail communication during foraging and recruitment in the subterranean termite *Reticulitermes santonensis* De Feytaud (Isoptera: Rhinotermitidae). *J. Insect Behav.* **14**, 157–171.

Reinhard J., Lacey M. J., Ibarra F., Schroeder F. C., Kaib M. and Lenz M. (2002) Hydroquinone: a general phagostimulating pheromone in termites. *J. Chem. Ecol.* **28**, 1–14.

Romer F. (1991) The oenocytes of insects: differentiation, changes during molting, and their possible involvement in the secretion of molting hormone. *Morphogenetic Hormones of Arthropods; Roles in Histogenesis, Organogenesis, and Morphogenesis*. In ed. A. P. Gupta, pp. 542–566. Rutgers University Press, New Brunswick, New Jersey.

Ruther J., Sieben S. and Schricker B. (2002) Nestmate recognition in social wasps: manipulation of hydrocarbon profiles induces aggression in the European hornet. *Naturwissenschaften* **89**, 111–114.

Schal C., Sevala V. L., Young H. and Bachman J. A. S. (1998) Synthesis and transport of hydrocarbons: cuticle and ovary as target tissue. *Am. Zool.* **38**, 382–393.

Sevela V. L., Bagnères A.-G., Kuenzli M., Blomquist G. J. and Schal C. (2000) Cuticular hydrocarbons of the dampwood termite, *Zootermopsis nevadensis*: caste differences and role of lipophorin in transport of hydrocarbons and hydrocarbon metabolites. *J. Chem. Ecol.* **26**, 765–789.

Simon U. E., Moritz R. F. A. and Crewe R. M. (2001) The ontogenetic pattern of mandibular gland components in queenless worker bees (*Apis mellifera capensis* Esch.). *J. Insect Physiol.* **47**, 735–738.

Sledge M. F., Boscaro F. and Turillazzi S. (2001) Cuticular hydrocarbons and reproductive status in the social wasp *Polistes dominulus*. *Behav. Ecol. Sociobiol.* **49**, 401–409.

Soroker V. and Hefetz A. (2000) Hydrocarbon site of synthesis and circulation in the desert ant *Cataglyphis niger*. *J. Insect Physiol.* **46**, 1097–1102.

Soroker V., Vienne C., Nowbahari E., and Hefetz A. (1994) The postpharyngeal gland as a “gestalt” organ for nestmate recognition in the ant *Cataglyphis niger*. *Naturwissenschaften* **81**, 510–513.

Stout J. C. and Goulson D. (2001) The use of conspecific and interspecific scent marks by foraging bumblebees and honeybees. *Animal Behaviour* **62**, 183–189.

Tentschert J., Bestmann H.-J., Hölldobler B. and Heinze J. (2000) 2,3-Dimethyl-5-(2-methylpropyl)pyrazine, a trail pheromone component of *Eutetramorium mocquerysi* Emery (1899) (Hymenoptera: Formicidae). *Naturwissenschaften* **87**, 377–380.

Tentschert J., Kolmer K., Hölldobler B., Bestmann H.-J., Delabie J. H. C. and Heinze J. (2001) Chemical profiles, division of labor and social status in *Pachycondyla* queens (Hymenoptera: Formicidae). *Naturwissenschaften* **88**, 175–178.

Turillazzi S., Sledge M. F., Dani F. R., Cervo R., Massolo A. and Fondelli L. (2000) Social hackers: integration in the host recognition system by a paper wasp social parasite. *Naturwissenschaften* **87**, 172–176.

Vander Meer R. K. and Alonso L. E. (2002) Queen primer pheromone affects conspecific fire ant (*Solenopsis invicta*) aggression. *Behav. Ecol. Sociobiol.* **51**, 122–130.

Vander Meer R. K., Breed M. D., Espelie K. E. and Winston M. L. (eds) (1998) *Pheromone Communication in Social Insects. Ants, Wasps, Bees, and Termites*. Westview Press, Boulder, CO.

Wager B. R. and Breed M. D. (2000) Does honey bee sting alarm pheromone give orientation information to defensive bees? *Ann. Entomol. Soc. Am.* **93**, 1329–1332.

Wagner D., Tissot M. and Gordon D. (2001) Task-related environment alters the cuticular hydrocarbon composition of harvester ants. *J. Chem. Ecol.* **27**, 1805–1819.

Wagner D., Tissot M., Cuevas W. and Gordon D. M. (2000) Harvester ants utilize cuticular hydrocarbons in nestmate recognition. *J. Chem. Ecol.* **26**, 2245–2257.

Wakayama E. J., Dillwith J. W., Howard R. W. and Blomquist G. J. (1984) Vitamin B12 levels in selected insects. *Insect Biochem.* **14**, 175–179.

Wenseleers T., Billen J. and Hefetz A. (2002) Territorial marking in the desert ant *Cataglyphis niger*: does it pay to play Bourgeois? *J. Insect Behav.* **15**, 85–93.

Yoder J. A., Denlinger D. L., Dennis M. W. and Kolattukudy P. E. (1992) Enhancement of diapausing flesh fly puparia with additional hydrocarbons and evidence for alkane biosynthesis by a decarbonylation mechanism. *Insect Biochem. Molec. Biol.* **22**, 237–243.

Zanetti P., Dani F. R., Destri S., Fanelli D., Massolo A., Moneti G., Pieraccini G. and Turillazzi S. (2001) Nestmate recognition in *Parischnogaster striatula* (Hymenoptera: Stenogastrinae), visual and olfactory recognition cues. *J. Insect Physiol.* **47**, 1013–1020.

12

Alkaloid-derived pheromones and sexual selection in Lepidoptera

Thomas Eisner and Jerrold Meinwald

12.1 Introduction

Courtship in insects may involve more than the copulatory act itself. Following the initial phases of the behavior, in which mate recognition, attraction, and localization are at play, the prospective partners do not necessarily proceed at once to mate, but first may show elaborate precopulatory interactions, obviously communicative in nature. What it is that the sexes "say" to one another in that context, and why they should even "bother" to communicate once they have achieved the proximity necessary for copulation, is often a mystery.

Our primary purpose in this chapter is to describe work, largely from our laboratories, that has led us to believe that, in certain insects at least, "foreplay" is basically a sexually selective process, involving assessment by the female of certain male traits that are a measure of an eventual benefit to the offspring. The foreplay is in the nature of a pheromone-mediated dialog, and the insects are certain butterflies and moths. Our collaborators in these studies are former graduate students, undergraduates, and postdoctoral fellows, instrumental not only in doing much of the research but in generating and refining some of the ideas. Much work related to these studies has been the subject of reviews by others (Ackery and Vane-Wright, 1984; Boppré, 1984, 1986, 1990; Edgar, 1984; Brown and Trigo, 1995; Hartmann and Ober, 2000; Rothschild *et al.*, 1979; Trigo *et al.*, 1996; von Nickisch-Rosenegk and Wink, 1993; Weller *et al.*, 1999), of which that by Nishida (2002) is particularly insightful. Our treatment of the subject here is admittedly limited in that it focuses primarily on species that we ourselves have studied and on the concepts that have driven our work. It is also intended to provide some notion of how these concepts took form over the course of time.

12.2 Danaidone: first characterization of a hairpencil secretion (*Lycorea ceres*)

What kindled our interest in this area of research was the seminal paper by Brower *et al.* (1965) on the courtship of the queen butterfly, *Danaus gilippus*, and the motion picture that these investigators made of this behavior. Their data showed clearly that the two brush-like structures, or "hairpencils," that the males ordinarily kept tucked away in their abdomen are in fact everted and splayed during courtship, and brushed against the female prior to copulation. Glandular in nature, the hairpencils seemed to function as an "aphrodisiac" device that effected its action chemically. But the nature of the presumed pheromone and its precise communicative significance remained unknown.

No butterfly pheromone had previously been characterized, let alone assayed for activity, and we were intrigued by the prospects of doing so. We knew that hairpencils occurred in other danaine butterflies (family Nymphalidae; subfamily Danainae) besides the queen butterfly. Through the help of Jocelyn Crane, Lincoln Brower, and others, we were able to secure a source of *Lycorea ceres* (Figure 12.1A), a beautiful tropical danaine with particularly large hairpencils (Figure 12.1B). Extirpation of the brushes proved easy, and we proceeded to excise dozens of them for chemical extraction, partly from individuals shipped to us live from Trinidad. Three major compounds were characterized from these extracts, a crystalline, nitrogenous ketone, for which we could establish structure **I**, and two aliphatic esters, **II** and **III** (Meinwald *et al.*, 1966; Meinwald and Meinwald, 1966). The relative inaccessibility of *Lycorea*, plus the fact that the animal courts in dense tropical forest, made possibilities for bioassaying the compounds dim.

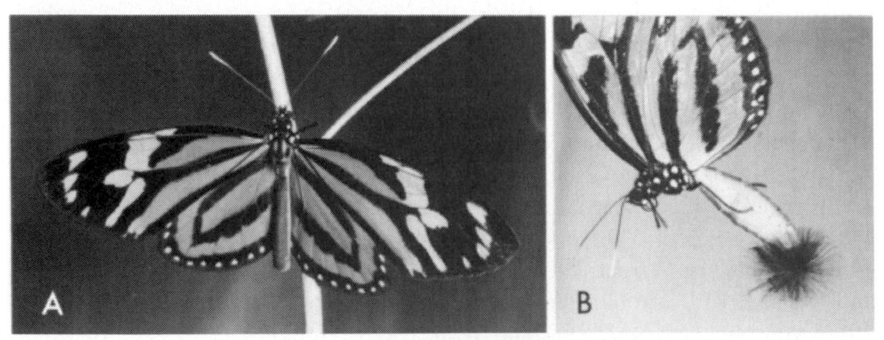

Figure 12.1 A *Lycorea ceres* from Trinidad. B Male of same, with hairpencils everted. (Photos by T. Eisner, from Meinwald *et al.*, 1966, © AAAS).

Formulas

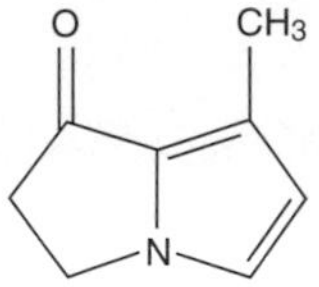

O CH_3

N

I

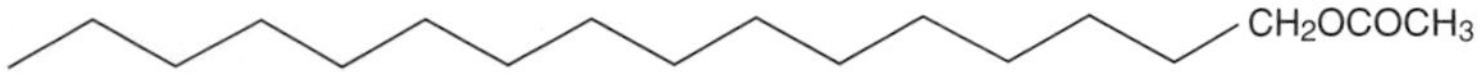

CH_2OCOCH_3

II

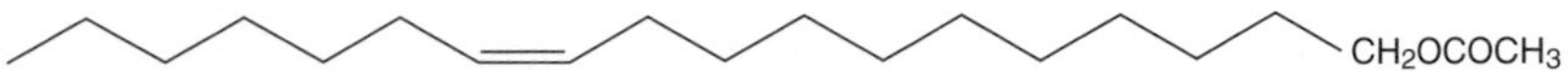

CH_2OCOCH_3

III

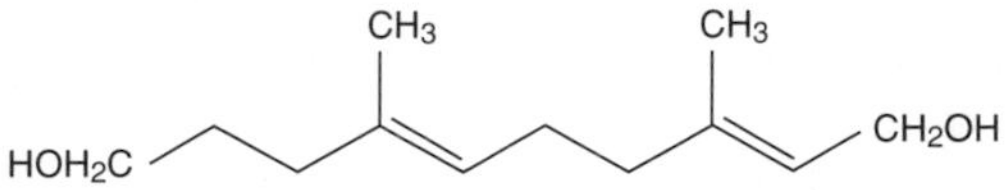

CH_3 CH_3

HOH_2C CH_2OH

IV

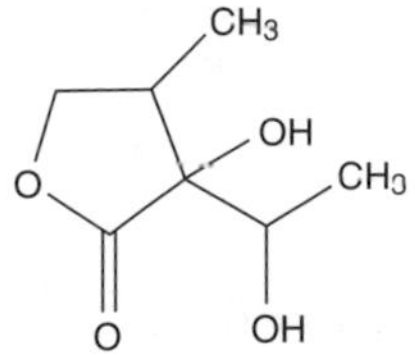

CH_3

OH

O CH_3

O OH

V

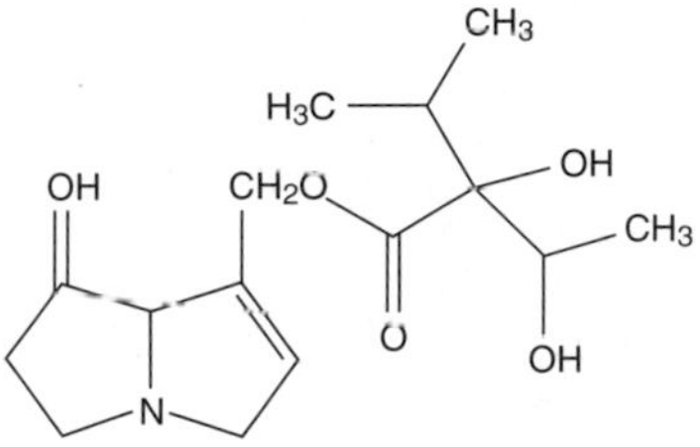

CH_3

H_3C

OH CH_2O OH

CH_3

N O OH

VI

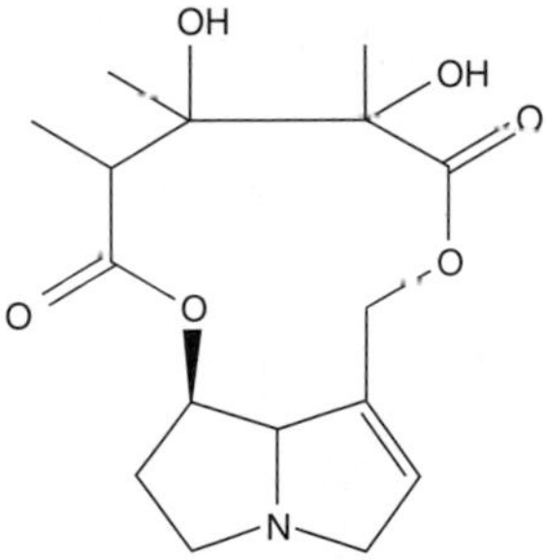

OH

OH

O

O O

O

N

VII

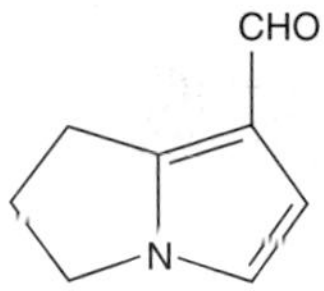

CHO

N

VIII

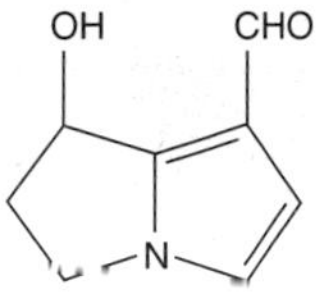

OH CHO

N

IX

(Contd)

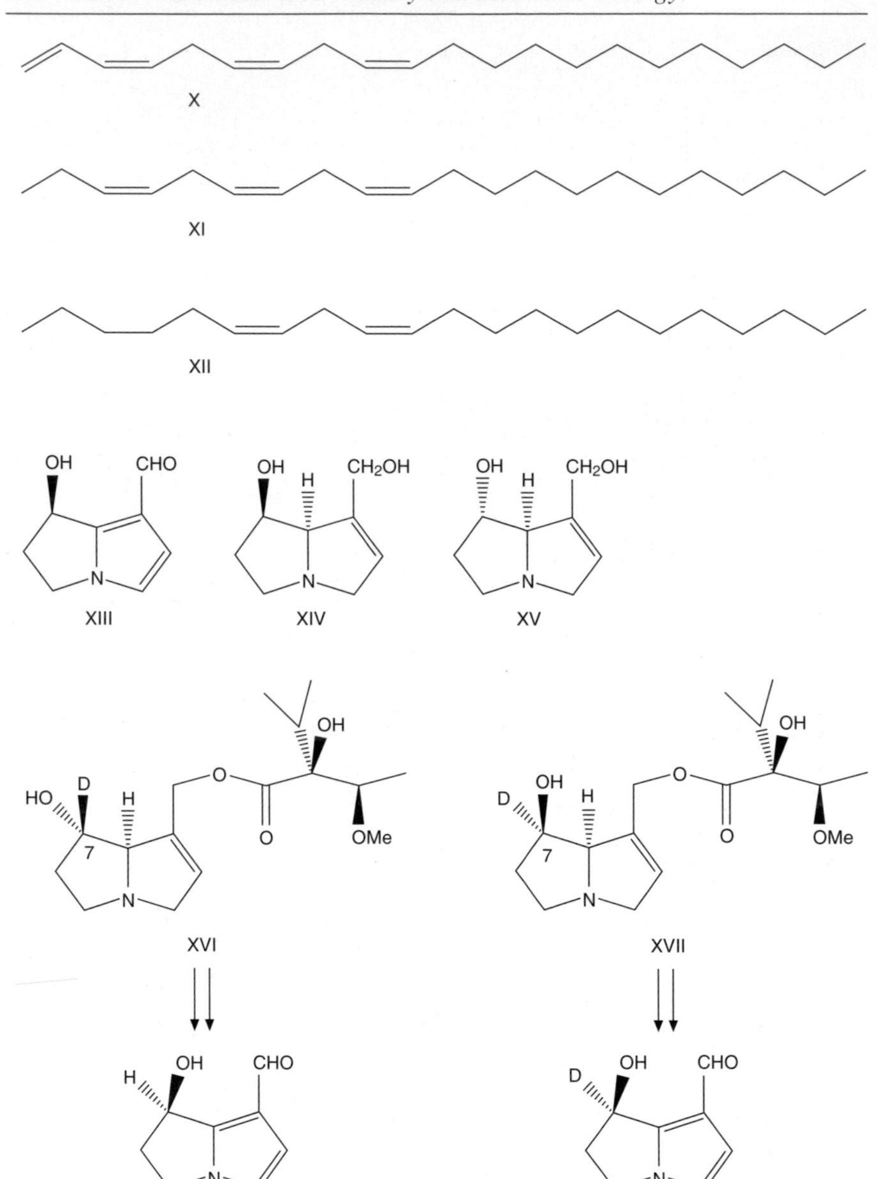
X
XI
XII
OH
CHO
N
XIII
OH
H
CH₂OH
N
XIV
OH
H
CH₂OH
N
XV
HO
D
H
7
N
O
O
OH
OMe
XVI
D
OH
H
7
N
O
O
OH
OMe
XVII
H
OH
CHO
N
XIII
D
OH
CHO
N
XVIII

We were struck by one concomitant of our finding, however, and speculated at the time about its possible significance. The pyrrolizidine **I**, since named "danaidone," bore no resemblance to any previously characterized natural products except certain alkaloids, the so-called senecio or pyrrolizidine alkaloids (PAs), present not in animals but in a diversity of plants. This led us to suggest that the *Lycorea* compound might be obtained or derived from a food source (Meinwald *et al.*, 1966; Meinwald and Meinwald, 1966). We proceeded to look into the question experimentally, but were thwarted by happenstance. *Lycorea* male collection had proven relatively easy for our suppliers in Trinidad because the butterflies could be lured in numbers to senescent branches of a plant known locally as "fedegoso," a species of *Heliotropium* (Boraginaceae). We secured extracts of these plants with the intent of looking into the chemistry, but the material proved so allergenic to one of our associates, the late James W. Wheeler, that we discontinued the project. We turned instead to a study of the queen butterfly in Florida.

12.3 Danaidone: proven pheromonal function (*Danaus gilippus*)

The hairpencils of *Danaus gilippus* (Figures 12.2A and 12.2C) turned out to be chemically similar to those of *Lycorea*. While they lacked the esters, and had instead the viscous terpenoid alcohol **IV**, they too were laden with danaidone (**I**) (Meinwald *et al.*, 1969). Structurally they were notable in that their bristles were densely beset with tiny cuticular pellets, the hairpencil "dust" (Figure 12.2D). Our hypothesis had it that the pellets acted as a carrier for the pheromonal secretion of the hairpencils and that they were transferred to the surface of the female during precopulatory "hairpencilling". Transfer of pellets was readily demonstrated. Virgin females, recaptured after they had been courted by males within a large experimental cage that we had built for our purposes in Florida, could be shown on surface examination with powerful epi-illumination optics to bear pellets on their antennae, where principal chemoreceptors could be expected to be located.

Through an unforeseen circumstance we were able to prove that danaidone is indeed a pheromone. Male *Danaus* that we had raised indoors in cages had proved singularly unsuccessful in courtship. They pursued females normally and hairpencilled them but vis-à-vis wild males were only about 20 percent as likely to be accepted by a female. Such laboratory-raised males were found to be virtually devoid of danaidone. We did not initially know the reason for the deficiency but did find that the males were not irreversibly impotent. By subsidizing them with danaidone, added to their hairpencils either as native secretion from wild males or as synthesized material (dissolved in synthetic **IV** or in mineral oil), one could restore their potency. The danaidone, we concluded, was not only

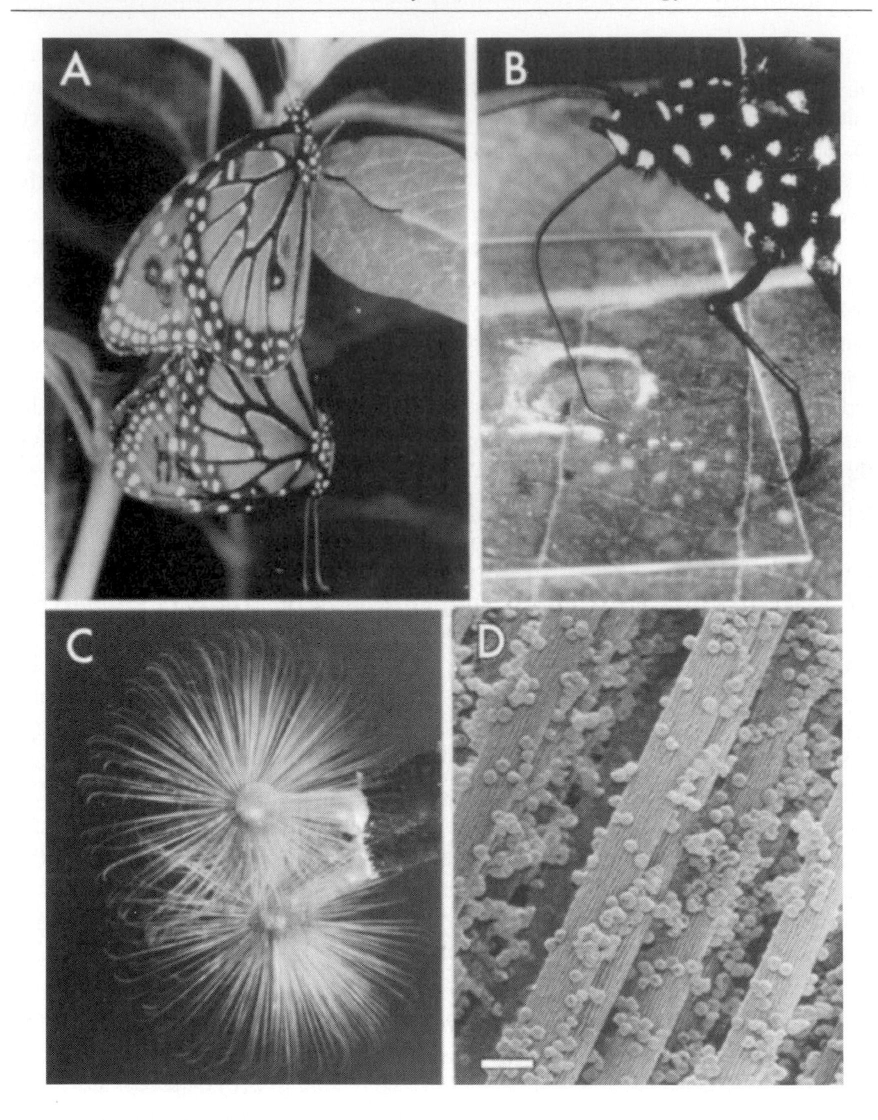

Figure 12.2 *Danaus gilippus*, the queen butterfly, from Florida. A Mating pair, the female is hanging downward, coupled to the abdomen of the perched male. B Male, feeding on monocrotaline (*N*-oxide) crystals. The animal has liquefied some of the crystals with regurgitated fluid and is imbibing the solution. C Everted hairpencils of male. D Scanning electronmicrograph of bristles of a hairpencil, showing the cuticular pellets that are transferred to the surface of the female during courtship and are carriers of the pheromone (danaidone, **I**). Bar = 10 μm (C From Pliske and Eisner, 1969, © AAAS.)

the principal communicative component of the secretion but, possibly the only one. The terpene alcohol seemed to act primarily as carrier and as glue for the dust pellets (Pliske and Eisner, 1969). Independent work by Schneider and Seibt (1969) showed the antenna of *Danaus* to be electrophysiologically sensitive to danaidone, but only minimally sensitive to the terpene alcohol, and Myers and Brower (1969) showed in behavioral experiments that certain antennal zones in the female are essential for chemical signal reception in courtship.

Much information has been added since to our knowledge of the courtship behavior in danaine butterflies. First, it is becoming clear that danaidone and closely related pyrrolizidines are very generally present in the hairpencils of these insects, having been found in a number of genera beside *Danaus* and *Lycorea* (references in Ackery and Vane-Wright, 1984, and Nishida, 2002). Accompanying compounds are also present, but these are variable and have been identified in only a few species (e.g. Meinwald *et al.*, 1974; Nishida, 2002; Petty *et al.*, 1977).

Second, and perhaps most interesting, the pyrrolizidines do indeed appear to be derived from dietary PAs (Edgar *et al.*, 1973; Edgar, 1984; Schneider *et al.*, 1975). Adult male danaines visit PA-containing plants (including species of *Heliotropium*) and routinely feed on the fluid excrescences that ooze from senescent parts of such plants (they may even scratch the plants to induce excrescence; Boppré, 1984). Feeding on these fluids, or on PA sources that can be offered to them as laboratory alternatives (Figure 12.2B), is essential if they are to produce hairpencil pyrrolizidine, hence the lack of danaidone in our laboratory-reared *Danaus* males and their consequent lack of success in courtship. Additional factors complicate the story. Subsidiary glandular structures on the wings of the males (wing pouches, scent patches) present in some (e.g. *Danaus*, *Amauris*), but not all, danaines (e.g. *Lycorea*), which the males periodically wipe or otherwise bring into contact with the hairpencils, may contribute in important ways to the derivation of pyrrolizidine pheromone from alkaloid precursor, but the precise basis of interaction of these glandular structures remains to be clarified (Seibt *et al.*, 1972; Boppré *et al.*, 1978). Also complicating the story is the fact that some danaines obtain their PAs as larvae from their foodplants, rather than by adult procurement from PA-containing plants. Such is the case in *Idea leuconoe,* which sequesters macrocyclic PAs from *Parsonia laevigata* (Apocynaceae), and derives two hairpencil components from the PAs, danaidone and viridifloric-β-lactone. Interestingly, in this species, the PAs in the foodplant also serve as ovipositon stimulants (Nishida, 2002; Nishida *et al.*, 1991, 1996).

One species, the monarch butterfly *(Danaus plexippus)*, anomalous by virtue of its migratory habits, is exceptional also in that it visits PA-containing plants and sequesters PAs, but fails to produce a pheromonal pyrrolizidine (Meinwald *et al.*, 1968; Pliske, 1975a; Edgar *et al.*, 1971, 1976a). Such an exception does not obscure the central fact that danaines, as a group, show male sequestration of PAs and production, by degradation of these PAs, of pheromonal pyrrolizidines. The pheromones – if generalization from the single proven case (Pliske and Eisner, 1969) of the queen butterfly is justified – serve the males as a critical key to success in mating.

Further relevant findings have been made in butterflies related to the danaines, the so-called ithomiines (family Nymphalidae; subfamily Ithomiinae) (Brown, 1984; Trigo and Brown, 1990; Trigo *et al.*, 1996). Male ithomiines have tufts of hair on the costal margins of the hindwings, which are "aired" behaviorally, as if for scent dissemination, under various conditions (aggregation, male–male interaction, female attraction) not all clearly defined or understood (Pliske, 1975b; Haber, 1978). Ithomiines also visit PA-containing plants, and chemical analyses of the wing glands showed presence of a lactone (**V**), which, by virtue of its structural similarity to the acid moiety of certain PAs (such as lycopsamine, **VI**), is believed to be a derivative of these plant products (Edgar *et al.*, 1976b).

12.4 Pyrrolizidine alkaloids: proven defensive role (*Utetheisa ornatrix*)

Pyrrolizidine alkaloids (PAs) are secondary metabolites, and as such, can be toxic, certainly to vertebrates (Bull *et al.*, 1968; Mattocks, 1972), and could therefore serve in plants for protection against at least some herbivores. Sequestration of PAs by insects could thus be viewed as an adaptive strategy on the insects' part for arming themselves with prefabricated, and hence relatively "low-cost," defenses. Since sequestration of PAs is by no means restricted to adult danaines, but can potentially occur in any of the diverse insects known to feed on PA-containing plants as larvae or adults, it seemed that the strategy might be fairly widespread. Proof that PA sequestration actually conveys a defensive advantage, however, was lacking. We came upon the opportunity to provide such proof in a fortuitous way.

Work from our laboratories of some years back had shown that Lepidoptera are protected against entanglement in spider webs, by their investiture of scales. Instead of sticking to webs as "naked" insects typically do, they simply lose scales to points of contact with an orb and flutter free (Eisner *et al.*, 1964). The site in Florida where we made these observations was the habitat of *Utetheisa ornatrix*, a well-protected aposematic arctiid moth which, like its congeners (Rothschild, 1972, 1973, 1979), feeds as a larva on PA-containing food plants (Fabaceae of the genus *Crotalaria*) (Bull *et al.*, 1968) (see Figure 12.5B). *Utetheisa*, we noted, makes no effort to struggle loose in a spider web, but simply folds its wings and remains quiescent. The spider pounces upon it, but on contact-inspection immediately pulls it from the web or, as we observed many times with our chief experimental spider, *Nephila clavipes*, frees it by severance of the entangling threads. The *Utetheisa* invariably survives uninjured.

In the laboratory we succeeded in rearing *Utetheisa* on a semisynthetic diet based on pinto beans, totally devoid of PAs (Miller *et al.*, 1976). Such moths, although visually and in every other respect indistinguishable from *Crotalaria*-fed counterparts, proved palatable to *Nephila* (Figures 12.3 and 12.4). Further experiments provided proof that the PAs themselves are deterrent to *Nephila*. Mealworms with a topical additive of monocrotaline (**VII**, the principal PA of *Crotalaria spectabilis*, one of *Utetheisa*'s food plants) proved substantially less acceptable to the spider than untreated controls (Eisner, 1980, 1982) (Figure 12.4). PA sequestration, it seemed, could be adaptive to any insect capable of withstanding systemic incorporation of the compounds, including, of course, danaine butterflies, in which the PAs could be viewed as an adult supplement of the defensive cardenolides incorporated by these insects from the milkweed plants (Asclepidaceae) they eat as larvae (Eisner, 1980; Ackery and Vane-Wright, 1984; Boppré, 1984). More recently, it has been shown that PAs may play a protective role vis-à-vis spiders in ithomiine butterflies as well (Brown, 1984; Vasconcellos-Neto and Lewinsohn, 1984; Orr *et al.*, 1996). Moreover, PAs can

Figure 12.3 *Utetheisa ornatrix* that were offered to the orb-weaving spider *Nephila clavipes*. The specimen on the right, rejected intact, was raised on one of its normal, pyrrolizidine alkaloid-containing food plants (*Crotalaria mucronata*). The one on the left, raised on an artificial diet devoid of alkaloid, was eaten. See also Figure 12.4. (From Eisner, 1982. © American Institute of Biological Science.)

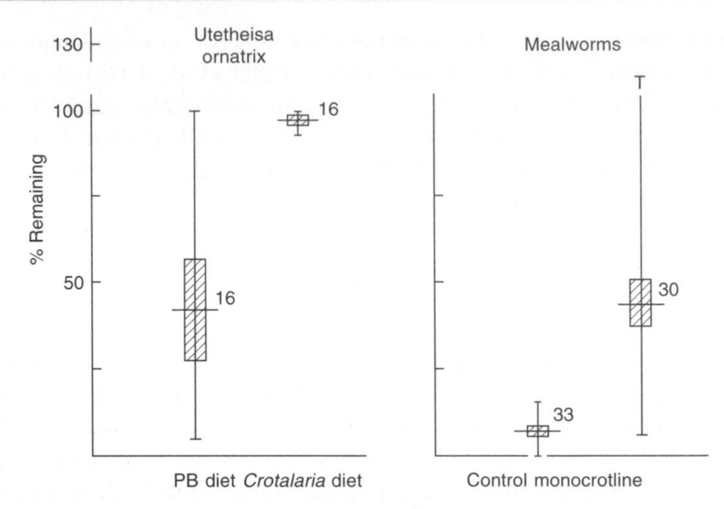

Figure 12.4 Percent of prey item (*Utetheisa*, mealworms) remaining following attack by the orb-weaving spider *Nephila clavipes*. Items were placed into individual webs with forceps (mealworms) or flipped from vials (*Utetheisa*). The field-collected *Utetheisa* can be expected to have fed as larvae on normal pyrrolizidine alkaloid-containing food plants (*Crotalaria* sp.). The control *Utetheisa* were laboratory reared on alkaloid-free [pinto bean (PB)-based] diet. The experimental mealworms were treated by topical addition of 200 μg monocrotaline (free base); controls were alkaloid free. With *Utetheisa*, the percent remaining was calculated from weighings of each moth before and after the test (all field-collected *Utetheisa* survived the tests without noticeable injury; the pinto bean-fed *Utetheisa* were all killed and partly to almost totally eaten (see Figure 12.3). With the mealworms, the percent remaining was calculated from final dry weight relative to mean dry weight of a sample of 10 mealworms. Sample sizes are given by numbers in parentheses. [From T. Eisner, W. Conner, K. Hicks and D. Aneshansley (data first presented in Eisner and Meinwald, 1987).]

offer protection to earlier developmental stages of a lepidopteran. Thus, in *Utetheisa*, where the PAs are sequestered from the food plant in the larval stage, the larvae are protected against wolf spiders by the alkaloids (Eisner and Eisner, 1991). And, as will be noted below, PAs can serve for protection of the eggs. The data are scant, however, as pointed out by Nishida (2002), with regard to protection by acquired PAs against vertebrate enemies.

12.5 Hydroxydanaidal: pheromonal indicator of systemic alkaloid load (*Utetheisa ornatrix*)

A paper by Culvenor and Edgar (1972), reporting the presence of two pyrrolizidines closely related to danaidone, danaidal (**VIII**) and hydroxydanaidal (**IX**), from the coremata (analogs of hairpencils) of Australian *Utetheisa*, set us looking into the courtship of *U. ornatrix*. The question was whether this moth also produced a corematal pyrrolizidine, and whether it used the substance for sexual communicative purposes as danaines do. The potential for parallel was striking, since these aldehydes seemed also to be derived from PAs, albeit from PAs ingested by larvae rather than adults.

Courtship in *Utetheisa* turned out to proceed in two stages (Conner *et al.*, 1980, 1981). The initial stage, in which the sexes are brought together, and which in danaines and other butterflies involves visual pursuit by day, is mediated by a sex attractant in *Utetheisa*. The female broadcasts the pheromone after sunset while stationary, luring males from downwind. Principal components of the attractant, which is released from a pair of tubular abdominal glands, are three C_{21} unsaturated hydrocarbons, **X**, **XI,** and **XII** (Conner *et al.*, 1980; Huang *et al.*, 1983; Jain *et al.*, 1983). An incidental finding – providing a first demonstration of temporal patterning in an aerial pheromonal signal – was that the secretion is released discontinuously, in short pulses (Conner *et al.*, 1980), which could enable the female to increase the range of attraction of the pheromone (Dusenbery, 1989). Female *Utetheisa* raised on a pinto bean diet had normal titers of attractant, indicating that production of the hydrocarbon is, as was to be expected, independent of PA intake (Conner *et al.*, 1980).

Utetheisa uses its coremata at close range during what we used to call the "seductant" phase of the behavior but prefer now to call the sexually selective phase. Once the male has reached the female, he flutters around her and repeatedly thrusts his abdomen against her, simultaneously everting the coremata. The eversions are momentary and involve direct strokings of the female (Figures 12.5A and 12.5C). Chemical analyses showed the coremata to contain hydroxydanaidal, but only in moths reared on *Crotalaria.* Such moths proved consistently successful in courtship, unlike individuals with excised hairpencils or those raised on pinto bean diet which were shown to be lacking in hydroxydanaidal. Further tests showed that hydroxydanaidal induced the wing-raising response in females that is the usual prelude to copulation. This stimulatory effect was maximal with the particular stereoisomeric form, the *R*(–) configuration (**XIII**), in which hydroxidanaidal occurs in the coremata (Conner *et al.*, 1981). The pheromonal role of hydroxydanaidal seemed established, as was the apparent dependence of hydroxydanaidal production on dietary PAs.

The nagging question was whether the derivation of a sexual pheromone from phytotoxin, a phenomenon now demonstrated for two disparate phyletic lineages, was to be viewed strictly as a manifestation of metabolic expediency or whether it could be justified on entirely different adaptive grounds. What we proposed is that the pheromonal pyrrolizidines could function, in both *Utetheisa* and *Danaus*, as a chemical yardstick by which the female gauges the PA load of her suitor. By exercising such assessment, and by favoring males of higher PA content, the

Figure 12.5 *Utetheisa ornatrix.* A Male (above) stroking its everted coremata against female during courtship. B Larva feeding on seed pod of its natural, pyrrolizidine alkaloid-containing food plants (*Crotalaria spectabilis*). C Scanning electronmicrograph of abdominal tip of male, showing coremata in everted (left) and retracted condition. [Bar = 1 mm A and C from Eisner, 1980].

females could be selecting for males adept at PA sequestration, a trait that could be heritable. In *Utetheisa* the adeptness could manifest itself in improved larval competitive ability for *Crotalaria* seeds, the parts of the plant richest in PAs (Culvenor and Smith, 1957; Sawhney *et al.*, 1967), favored not only by fellow larvae but other herbivores as well (in *Crotalaria mucronata*, for example, larvae of another moth, *Etiella zinkenella*, compete with *Utetheisa* for the seed pods). In *Danaus*, the favored male would be one with proven ability to locate and sequester PA as an adult.

We recognized, when we advanced the hypothesis (Eisner and Conner, in Eisner, 1980), that it was contingent on a number of premises, including major uncertainties. Is the PA content in male *Utetheisa* a reflection of the amount of PA ingested, and does the hydroxydanaidal titer provide a measure of the male's alkaloid content?

While many details remain unknown, some basic facts have been established.

We learned that adult *Utetheisa* differ substantially in their PA content in nature, and that their PA load is higher when they are feeding on seed-bearing *Crotalaria* than on immature plants (Conner *et al.*, 1990). Moreover, we have found that PAs are potently phagostimulatory, and that the larvae are literally driven to eat by the alkaloids (Bogner and Eisner, 1991, 1992). Further, we established that, under laboratory conditions, the amount of alkaloid incorporated systemically by the larvae is a function of the seed content (and, by inference, the PA content) of their diet. In addition, and this was particularly telling, we showed that there was a proportionality between the amount of hydroxydanaidal in the hairpencils and the amount of PA stored systemically by the male (Dussourd *et al.*, 1991). The female, in other words, could indeed gauge the male's alkaloid content by the intensity of his corematal scent. Her antennae, in line with expectation, bear sensilla sharply attuned to *R*(–) hydroxydanaidal (Grant *et al.*, 1989).

Interesting in this connection were data from other hydroxydanaidal-producing arctiid moths (*Creatonotos* spp.) in which the coremata vary dramatically in size, as a function of systemic PA load (Schneider *et al.*, 1982; Boppré and Schneider, 1986, 1989; Schulz *et al.*, 1993). Whether effectiveness of the organs in courtship is also a function of size, and whether sequestrative ability finds expression through this morphogenetic effect, remains unclear. In *Creatonotos*, where the males seemingly show lek behavior, the hairpencils may also mediate aspects of female attraction and male–male interaction (Schneider, 1983). Such may be the case also in other arctiid moths (Willis and Birch, 1982).

Courtship in arctiids, as in danaines, both sizeable taxonomic groups, is subject to considerable adaptive variation (Krasnoff and Dussourd, 1989; Krasnoff and Roelofs, 1989; Weller *et al.*, 1999). *Utetheisa* and *Danaus*, therefore, while illustrative of striking behavioral convergence, cannot each be taken to typify the courtship strategy of its group. We do feel, however, that the notion that close-range precopulatory interactions in insects are sexually selective in character, and that they involve assessment by the female, through indirect, measurable characters manifest by the male, of traits (chemical, acoustic, tactile, or visual) indicative of fitness, is worth pondering and could be broadly applicable. The courtship of that most discrepant of all danaine butterflies, the monarch, might itself be interpretable in such light. Male monarchs produce no pyrrolizidine pheromone (Meinwald *et al.*, 1968) and may entirely forego use of their rudimentary hairpencils in courtship. They seize females forcibly and bring them down for mating after capture in midair (Pliske, 1975a), an approach that has been likened to rape (Rothschild, 1978). But might not the female be assessing the male for vigor when she is in his aerial grasp, and by so doing put him to the test for a trait that is eminently adaptive in the context of migration?

Evidence was eventually to show that our interpretation of the sexual selective strategy of *Utetheisa* and the queen butterfly was too simplistic.

12.6 Pyrrolizidine alkaloids: parental transmission to egg (*Utetheisa ornatrix* and *Danaus gilippus*)

When a *Utetheisa* female raised on a pinto bean diet is mated with a male raised on *Crotalaria*, she lays eggs endowed with PA, indicating that she must have obtained a nuptial gift of PA from the male which she transmitted to the eggs. The experiment was not a fortuitous one but was deliberately designed to look into the possibility of seminal transfer of PA from male to female in mating (Dussourd *et al.*, 1988). Within the eggs, moreover, the PA acts protectively. Coccinellid beetles, for example, which are relatively reluctant to consume eggs containing PA, find PA-free eggs (from parents both raised on pinto bean diet) significantly more palatable (Dussourd *et al.*, 1988). The female, if herself raised on *Crotalaria*, provides a subsidy of her own PA to the eggs, complementing the amount supplied by the male. The male's contribution amounts to less than the female's contribution, but it suffices, in itself, to decrease the acceptability of the eggs to coccinellids. Ants and chrysopid larvae have been shown to reject *Utetheisa* eggs as well, when these contain PAs (Eisner *et al.*, 2000; Hare and Eisner, 1993).

The paternal provisioning of eggs puts the sexual selective strategy of *Utetheisa* in a new light. We were forced to review our interpretation of the corematal message. Rather than serving exclusively for proclamation of the genetic capacity to compete in the larval race for PA acquisition, hydroxydanaidal could serve for advertisement of the magnitude of the male's alkaloidal gift. The two traits could be linked. We predicted that in males there should be proportionality not only of pheromonal pyrrolizidine to body PA content, but also of PA content to quantity of PA transmitted to the female, and we found both relationships to hold (Dussourd *et al.* 1991).

It seemed obvious that there might be a comparable egg-endowment mechanism in *D. gilippus*. Males of the queen butterfly can be made to readily sequester PA in the laboratory. Crystalline monocrotaline *N*-oxide that is offered to them by the hundreds of micrograms is first liquefied with regurgitated fluid and then imbibed. The males transfer the chemical largely to the reproductive accessory glands, and then at mating to the female. She, in turn, passes it on to the eggs (Dussourd *et al.*, 1989).The parallel for danaines was established and appears to hold for other members of the subfamily as well (Nishida, 2002).

The story could doubtless be complicated by variation on the basic theme. In danaines, too, for example, there could be maternal contribution of PAs to the eggs, since both sexes sequester PAs as adults in some species, and, in exceptional cases even as larvae from the food plants (Edgar, 1984; Nishida, 2002). There are arctiids, in turn, and also species of the related family Ctenuchidae, that sequester PAs as adults in the manner of danaines, by visitation of PA-containing plants (Pliske, 1975c; Goss, 1979). In one species of ctenuchid, *Cisseps fulvicollis*, we have shown that PA procured by the adult male is transmitted through the female to the eggs (Dussourd *et al.*, 1984; Dussourd, 1986).

As regards receipt of PA from the male, there is also the interesting fact that the female herself, rather than only the eggs, may benefit from the acquisition. Female *Utetheisa* devoid of alkaloid (raised on pinto bean diet), if mated with a PA-laden male, become unacceptable to wolf spiders. The effect takes hold promptly and endures: females are unacceptable to spiders from the moment they uncouple from the male and they remain unacceptable as they age. Chemical data showed that the female allocates the received PA quickly to all body parts (González *et al.*, 1999a; Rossini *et al.*, 2001).

Although the evidence is relatively scant, it appears likely that generally, in lepidopterans that sequester PAs, the alkaloids find themselves transmitted in part, at least, to the eggs. This may be so irrespective of whether the PAs are procured by one parent or both (Eisner *et al.*, 2002; Nishida, 2002). The evidence is indirect but strong, for instance, that in the ithomiines the eggs are endowed, as they are in the queen butterfly, with paternally procured PAs (Brown, 1984).

12.7 Sperm selection (*Utetheisa ornatrix*)

Utetheisa females are promiscuous, raising the question whether the individual male has assurance of fathering offspring when he mates. Is there sperm mixing in multiply mated female *Utetheisa* or do the sperm of some males "win out" at the expense of others?

By use of enzymatic markers we were able to show that in twice-mated female *Utetheisa* the progeny is almost exclusively sired by the larger of the two males (LaMunyon and Eisner, 1993). Variables such as duration of copulation, mating order, or between-mating interval were not determinants of mating success.

Furthermore, it seems that the female herself may control the mechanism by which one set of sperm is favored over the other. If females are anesthetized, so that they are presumably prevented from activating the musculature by which they ordinarily convey the sperm, the normal routing of sperm is inhibited. The sperm themselves are not inactivated by the anesthesia, indicating that it is not on their own that they ordinarily sort themselves out.

We have evidence that the female assesses male size indirectly, by gauging the size of the spermatophore, which in *Utetheisa* is proportional in size to body size (LaMunyon and Eisner, 1994). If large males are experimentally caused to produce inordinately small spermatophores (as a consequence of having mated shortly beforehand with another female), their sperm will lose out to that of a smaller male, whose spermatophore is then the relatively larger one.

The female's strategy is an interesting one. By accepting multiple males she is able to accrue multiple alkaloidal gifts, which are available to her as a sum total for investment in self-defense and defense of eggs. She does not, however, use sperm indiscriminately, but favors that of larger males. Larger males, it turns out, are also richer in PA content (LaMunyon and Eisner, 1993), indicating that it is they which as larvae might have been the more successful competitors for PA. A further point concerns the presence of nutrient in the spermatophore. The *Utetheisa* male transmits upwards of 10 percent of his body mass with the spermatophore, as a result of which the female is enabled to increase her egg output by 15 percent (LaMunyon, 1997). Multiple mating can therefore lead to substantial nutrient gain and to a major increase in fecundity. Earlier data suggested that female *Utetheisa* mate on average four to five times over their life span of a few weeks (Pease, 1968). However, females in a Florida population that we recently monitored were shown to have an average lifetime mating incidence of 11, determined by counts of the spermatophore remnants (the colla) in the bursa (Eisner, unpublished).

A point worth noting is that by exercising postcopulatory sperm assessment, the female is given the option of taking corrective action relative to the favoring of mates. If, on a given evening, she accepted a male of moderate size and PA content, she can still discriminate genetically against such a male, by utilizing the sperm of a larger, more richly PA-endowed male that she accepted on a subsequent night. But the earlier mating is canceled in the genetic sense only, since the female does not also forego use of the earlier male's nutritive and alkaloidal gifts.

Sperm selection also provides females with the option of checking on potential "liars," males which might have misrepresented themselves in courtship by producing hydroxydanaidal at exaggerated levels relative to their PA content. Unless such males also are able to produce exaggerated-sized spermatophores, they are likely to be "found out" when they are put to the test in sperm selection.

12.8 Heritability of the sexually selected trait (*Utetheisa ornatrix*)

The fact that, in *Utetheisa,* male hydroxydanaidal titer correlates with systemic PA content, and systemic PA correlates with body mass (as well as with spermatophore mass), means that the female, on the two occasions that she exercises mate choice, favors larger males. Body size is at issue in the precopulatory context when she chooses on the basis of hydroxydanaidal titer, and it is at stake after mating when she makes her choice on the basis of spermatophore mass.

We showed experimentally that hydroxydanaidal is the only basis of measure by which the female gauges male size during precopulatory interaction. Females failed to differentiate between males that differed in body mass or PA content if the males lacked hydroxydanaidal, but they chose between males that were size-matched and PA-free, if one of the males was given an experimental subsidy of hydroxydanaidal (Iyengar *et al.*, 2001). We showed further that females can differentiate between males containing different quantities of hydroxydanaidal, and not just, as we had shown previously (Conner *et al.*, 1981), between males that either possessed, or totally lacked hydroxydanaidal (Iyengar *et al.*, 2001).

A critical question concerned the nature of the benefits derived by the female as a consequence of her choosiness. What does she gain by selecting larger males? Is it strictly the phenotypic benefits, received in the form of increased quantities of PA and nutrient? Or does mating with a large male provide her with larger sons and daughters, sons more likely to be successful in courtship, and daughters bound to be more fecund (egg production is a function of female size in *Utetheisa*; LaMunyon, 1997)? In other words, is body mass a heritable trait in *Utetheisa,* and does the female receive genetic benefits as a consequence of her selectivity?

We demonstrated that body mass is indeed heritable for both sexes in *Utetheisa* (Iyengar and Eisner, 1999a), and that by favoring large males females do obtain genetic benefits for the offspring. Choosy females do have sons more likely to be accepted as mates and daughters that produce more eggs (Iyengar and Eisner, 1999b).

12.9 Heritability of the female's preference for the sexually selected trait (*Utetheisa ornatrix*)

A question that inevitably arises is whether the female's choosiness, that is, her predilection for large males, is itself subject to variation and heritable. Evidence recently obtained (Iyengar *et al.*, 2002) indicates that it is. Specifically, the evidence shows that the female's mating preference is inherited via the father rather than the mother, which in Lepidoptera has special significance, given that these insects have a reverse genetic architecture (*ZZ/ZW*), where males are homogametic (*ZZ*) and females heterogametic (*ZW*). The fact that the preference genes are received by the female from the father only, permits assignment of these genes to the Z sex chromosome, which is strictly inherited paternally by daughters. The evidence indicates further that the preferred male trait and the female's preference for that trait are correlated, as females with larger fathers have a stronger preference for larger males. These findings were predicted by a theory recently proposed – the protected invasion theory (Reeve, 1993; Reeve and Shellman-Reeve, 1997; Reeve and Pfennig, 2002) – which asserts that male homogametic sex chromosome systems, such as those found in Lepidoptera, are conducive to the evolution of exaggerated male traits through sexual selection. Specifically, the theory predicts that, because female preference alleles arising on the *Z* chromosome are transmitted to all sons that have the father's attractive trait (rather than to only some of the sons), such alleles are likely to be subject

nearly enough is known about the chemistry of male accessory glands, composition of spermatophores, and defensive substances in insect eggs. Nor is enough known about the numerous pheromone glands that are "aired" by male insects prior to mating, or even orally sampled by females, and about the chemical relationship of the products of such glands to substances that a male might transfer to the female at mating. The transferred substances, moreover, could be nutritive rather than defensive (e.g. Boggs and Gilbert, 1979; Greenfield, 1982; Marshall, 1982; Schal and Bell, 1982), in which case the males might be assessed by females for magnitude of intended nutritive rather than defensive gift. The area of inquiry, quite clearly, is wide open (Gwynne and Morris, 1983; Rutowski, 1982; Thornhill and Alcock, 1983; Trivers, 1985), and, to the extent that it may link the field of pheromone chemistry to that of sexual selection opened by Charles Darwin over a century ago, may prove fruitful.

Acknowledgements

Our studies on this general subject have been supported by the National Institutes of Health (Grants AI02908 and GM53830), Hatch Funds and fellowship stipends from the Johnson & Johnson Corporation (AG). Maria Eisner was most helpful in the preparation of the illustrations and took the scanning electron micrograph in Figure 12.2D. We are greatly indebted to the staff of the Archbold Biological Station, Lake Placid, Florida, where many of our studies on lepidopteran courtship were carried out.

References

Ackery P. R. and Vane-Wright R. I. (1984) *Milkweed butterflies, Their Cladistics and Biology*. Cornell Univ. Press, Ithaca, New York.

Bell T. W. and Meinwald J. (1986) Pheromones of two arctiid moths (*Creatonotos transiens* and *C. gangis*): chiral components from both sexes and achiral female components. *J. Chem. Ecol.* **12**, 385–409.

Bell T. W., Boppré M., Schneider D. and Meinwald J. (1984) Stereochemical course of pheromone biosynthesis in the arctiid moth, *Creatonotos transiens*. *Experientia* **40**, 713–714.

Boggs C. L. and Gilbert L. E. (1979) Male contribution to egg production in butterflies: evidence for transfer of nutrient at mating. *Science* **206**, 83–84.

Bogner F. X. (1996) Interspecific advantage results in intraspecific disadvantage: chemical protection versus cannibalism in *Utetheisa ornatrix* (Lepidoptera: Arctiidae). *J. Chem. Ecol.* **22**, 1439–1451.

Bogner F. and Eisner T. (1991) Chemical basis of egg cannibalism in a caterpillar (*Utetheisa ornatrix*). *J. Chem. Ecol.* **17**, 2063–2075.

Bogner F. and Eisner T. (1992) Chemical basis of pupal cannibalism in a caterpillar (*Utetheisa ornatrix*). *Experientia* **48**, 97–102.

Boppré M., (1984) Chemically mediated interaction of butterflies. In *The Biology of Butterflies*, eds R. I. Vane-Wright and P. R. Ackery, pp. 259–275. Academic Press, New York.
Boppré M. (1986) Insects pharmacophagously utilizing defensive plant chemicals (pyrrolizidine alkaloids). *Naturwissenschaften* **73**, 17–26.
Boppré M. (1990) Lepidoptera and pyrrolizidine alkaloids: exemplification of complexity in chemical ecology. *J. Chem. Ecol.* **16**, 165–185.
Boppré M. and Schneider D. (1986) Pyrrolizidine alkaloids quantitatively regulate both scent organ morphogenesis and pheromone biosynthesis in male *Creatonotos* moths (Lepidoptera: Arctiidae). *J. Comp. Physiol. A.* **157**, 569–577.
Boppré M. and Schneider D. (1989) The biology of *Creatonotos* (Lepidoptera: Arctiidae) with special reference to the androconial system. *Zool. J. Linnean Soc.* **96**, 339–356.
Boppré M., Petty R. L., Schneider D. and Meinwald J. (1978) Behaviorally mediated contacts between scent organs: Another prerequisite for pheromone production in *Danaus chrysippus* males. *J. Comp. Physiol.* **126**, 97–103.
Brattsten L. B. (1979) Biochemical defense mechanisms in herbivores against plant allelochemicals. In *Herbivores, Their Interaction with Secondary Plant Metabolites*, eds G. A. Rosenthal and D. H. Janzen, pp. 199–270. Academic Press, New York.
Brattsten L. B. (1992) Metabolic defenses against plant allelochemicals. In *Herbivores: Their Interactions with Secondary Plant Metabolites*, 2nd edn, Vol. 2, eds G. A. Rosenthal and M. R. Berenbaum, pp. 175–242. Academic Press, New York.
Brower L. P., Brower J. V. Z. and Cranston F. P. (1965) Courtship behavior of the queen butterfly, *Danaus gilippus berenice*. *Zoologica* **50**, 1–39.
Brown K. S. (1984) Adult-obtained pyrrolizidine alkaloids defend ithomiine butterflies against a spider predator. *Nature (London)* **309**, 707–709.
Brown, K. S. and Trigo, J. R. (1995). The ecological activity of alkaloids. *The Alkaloids* **47**, 227–354.
Bull L. B., Culvenor C. C. J. and Dick A. T. (1968) *The Pyrrolizidine Alkaloids*. North Holland, New York.
Conner W. E., Eisner T., Vander Meer R. K., Guerrero A., Ghiringelli D. and Meinwald J. (1980) Sex attractant of an arctiid moth (*Utetheisa ornatrix*): a pulsed chemical signal. *Behav. Ecol. Sociobiol.* **7**, 55–63.
Conner W. E., Eisner T., Vander Meer R. K., Guerrero A. and Meinwald J. (1981) Precopulatory sexual interaction in an arctiid moth (*Utetheisa ornatrix*): role of a pheromone derived from dietary alkaloids. *Behav. Ecol. Sociobiol.* **9**, 227–235.
Conner W. E., Roach B., Benedict E., Meinwald J. and Eisner. T. (1990) Courtship pheromone production and body size as correlates of larval diet in males of the arctiid moth, *Utetheisa ornatrix*. *J. Chem. Ecol.* **16**, 543–552.
Conner W. E., Boada R., Schroeder F. C., González A., Meinwald J. and Eisner. T. (2000) Chemical defense: Bestowal of a nuptial alkaloidal garment by a male moth upon its mate. *Proc. Natl. Acad. Sci. USA* **97**, 14406–14411.
Culvenor C. C. J. and Edgar J. A. (1972) Dihydropyrrolizine secretions associated with coremata of *Utetheisa* moths (family Arctiidae). *Experientia* **28**, 627–628.
Culvenor, C. C. J. and Smith L. W. (1957) The alkaloids of *Crotalaria spectabilis* Roth. *Aust. J. Chem.* **10**, 474–479.
Dusenbery, D. B. (1989) Calculated effect of pulsed pheromone release on range of attraction. *J. Chem, Ecol.* **15**, 971–977.
Dussourd D. (1986) *Adaptations of insect herbivores to plant defenses*. PhD thesis, Cornell University, Ithaca, New York.

Dussourd D., Ubik K., Resch J. F., Meinwald J. and Eisner T. (1984) Egg protection by parental investment of plant alkaloids in Lepidoptera. Abstract. *17th Int'l Cong. Entomol., Hamburg*, 840.

Dussourd D. E., Harvis C. A., Meinwald J. and Eisner T. (1989) Paternal allocation of sequestered plant pyrrolizidine alkaloid to eggs in the danaine butterfly, *Danaus gilippus*. *Experientia* **45**, 896–898.

Dussourd D. E., Harvis C. A., Meinwald J. and Eisner T. (1991) Pheromonal advertisement of a nuptial gift by a male moth *Utetheisa ornatrix*. *Proc. Natl. Acad. Sci. USA* **88**, 9224–9227.

Dussourd D. E., Ubik K., Harvis C., Resch J., Meinwald J. and Eisner T. (1988) Biparental defensive endowment of eggs with acquired plant alkaloid in the moth *Utetheisa ornatrix*. *Proc. Natl. Acad. Sci., USA* **85**, 5992–5996.

Edgar J. A. (1984) Parsonsieae: ancestral larval food plants of the Danainae and Ithomiinae. In *The Biology of Butterflies*, eds R. I. Vane-Wright and P. R. Ackery, pp. 91–93. Academic Press, New York.

Edgar J. A., Culvenor C. C. J. and Smith L. W. (1971) Dihydropyrrolizine derivatives in hairpencil secretion of danaid butterflies. *Experientia* **27**, 761–762.

Edgar J. A., Culvenor C. C. J. and Robinson G. S. (1973) Hairpencil dihydropyrrolizines of Danainae from the New Hebrides. *J. Aust. Entomol. Soc.* **12**, 144–150.

Edgar J. A., Cockrum P. A. and Frahn J. L. (1976a) Pyrrolizidine alkaloids in *Danaus plexippus* and *Danaus chrysippus*. *Experientia* **32**, 1535–1537.

Edgar J. A., Culvenor C. C. J. and Pliske T. E. (1976b) Isolation of a lactone, structurally related to the esterifying acids of the pyrrolizidine alkaloids, from the costal fringes of male Ithomiinae. *J. Chem. Ecol.* **2**, 263–270.

Eisner T. (1980) Chemistry, defense, and survival: case studies and selected topics. In *Insect Biology and the Future*, eds M. Locke and D. S. Smith, pp. 847–878. Academic Press, New York.

Eisner T. (1982) For love of nature: exploration and discovery at biological field stations. *BioScience* **32**, 321–326.

Eisner T. and Eisner M. (1991) Unpalatability of the pyrrolizidine alkaloid containing moth, *Utetheisa ornatrix,* and its larva, to wolf spiders. *Psyche* **98**, 111–118.

Eisner T., Alsop R. and Ettershank G. (1964) Adhesiveness of spider silk. *Science* **146**, 1058–1061.

Eisner T. and Meinwald J. (1987) Alkaloid-derived pheromones and sexual selection in Lepidoptera. In *Pheromone Biochemistry*, eds G. D. Prestwich and G. J. Blomquist, pp. 251–269. Academic Press, Orlando.

Eisner T., Smedley S. R., Young D. K., Eisner M., Roach B. and Meinwald J. (1996a). Chemical basis of courtship in a beetle (*Neopyrochroa flabellata*): cantharidin as "nuptial gift." *Proc. Natl. Acad. Sci. USA* **93**, 6499–6503.

Eisner T., Smedley S. R., Young D. K., Eisner M., Roach B. and Meinwald J. (1996b) Chemical basis of courtship in a beetle (*Neopyrochroa flabellata*): cantharidin as precopulatory "enticing" agent. *Proc. Natl. Acad. Sci. USA* **93**, 6494–6498.

Eisner T., Goetz M. A., Hill D. E., Smedley S. R. and Meinwald J. (1997) Firefly "femmes fatales" acquire defensive steroids (lucibufagins) from their firefly prey. *Proc. Natl. Acad. Sci. USA* **94**, 9723–9727.

Eisner T., Eisner M., Rossini C., Iyengar V. K., Roach B. L., Benedikt E. and Meinwald J. (2000) Chemical defense against predation in an insect egg. *Proc. Natl. Acad. Sci. USA* **97**, 1634–1639.

Eisner T., Rossini C., Gonzàlez A., Iyengar V. K., Seigler M. V. S. and Smedley S. R. (2002) Paternal investment in egg defense. In *Chemoecology of Insect Eggs and Egg Deposition*, eds M. Hilker and T. Meiners, pp. 91–116, Blackwell Verlag GmbH, Berlin.

González A., Rossini C., Eisner M. and Eisner T. (1999a) Sexually transmitted chemical defense in a moth (*Utetheisa ornatrix*) *Proc. Natl. Acad. Sci. USA* **96**, 5570–5574.
González A., Hare J. F. and Eisner T. (1999b) Chemical egg defense in *Photuris* firefly "femmes fatales." *Chemoecology* **9**, 177–185.
Goss G. J. (1979) The interaction between moths and plants containing pyrrolizidine alkaloids. *Environ. Entomol.* **8**, 487–493.
Grant A. J., O'Connell R. J. and Eisner T. (1989) Pheromone-mediated sexual selection in the moth *Utetheisa ornatrix*: olfactory receptor neurons responsive to a male-produced pheromone. *J. Ins. Behav.* **2**, 371–386.
Greenfield M. D. (1982) The question of parental investment in Lepidoptera: male-contributed proteins in *Plodia interpunctella. Int. J. Invert. Reprod.* **5**, 323–330.
Gwynne D. T. and Morris G. K. (1983) *Orthopteran Mating System.* Westview Press, Boulder, CO.
Haber W. A. (1978) *Evolutionary ecology of tropical mimetic butterflies.* PhD thesis, University of Minnesota, St Paul.
Hare J. F. and Eisner T. (1993) Pyrrolizidine alkaloid deters ant predators of *Utetheisa ornatrix* eggs: effects of alkaloid concentration, oxidation state, and prior exposure of ants to alkaloid-laden prey. *Oecologia* **96**, 9–18.
Hare J. F. and Eisner T. (1996) Cannibalistic caterpillars (*Utetheisa ornatrix*) fail to differentiate between eggs on the basis of kinship. *Psyche* **102**, 27–33.
Hartman T. and Ober D. (2000) Biosynthesis and metabolism of pyrrolizidine alkaloids in plants and specialized insect herbivores. *Topics in Current Chemistry* **209**, 208–243.
Huang W., Pulaski S. P. and Meinwald J. (1983) Synthesis of highly unsaturated insect pheromones: (Z,Z,Z)-1,3,6,9-heneicosatetraene and (Z,Z,Z)-1,3,6,9-nonadecatetraene. *J. Org. Chem.* **48**, 2270–2274.
Iyengar V. K. and Eisner T. (1999a) Heritability of body mass, a sexually selected trait, in an arctiid moth (*Utetheisa ornatrix*). *Proc. Natl. Acad. Sci. USA* **96,** 9169–9171.
Iyengar, V. K., and Eisner, T. (1999b). Female choice increases offspring fitness in an arctiid moth (*Utetheisa ornatrix*). *Proc. Natl. Acad. Sci. USA* **96**, 15013–15016.
Iyengar V. K., Reeve H. K. and Eisner T. (2002) Paternal inheritance of a female moth's mating preference. *Nature* (in press).
Iyengar V. K., Rossini C. and Eisner T. (2001) Precopulatory assessment of male quality in an arctiid moth (*Utetheisa ornatrix*): hydroxydanaidal is the only criterion of choice. *Behav. Ecol. Sociobiol.* **49**, 283–288.
Jain S., Dussourd D., Conner W. E., Eisner T., Guerrero A. and Meinwald J. (1983) Polyene pheromone components from an arctiid moth (*Utetheisa ornatrix*): characterization and synthesis. *J. Organic Chem.* **48**, 2266–2270.
Krasnoff S. B. and Dussourd D. E. (1989) Dihydropyrrolizine attractants for arctiid moths that visit plants containing pyrrolizidine alkaloids. *J. Chem. Ecol.* **15**, 47–60.
Krasnoff S. B. and Roelofs W. L. (1989) Quantitative and qualitative effects of larval diet on male scent secretions of *Estigmene acrea, Phragmatobia fuliginosa,* and *Pyrrharctia Isabella* (Lepidoptera: Arctiidae). *J. Chem. Ecol.* **15**, 1077–1094.
LaMunyon C. W. (1997) Increased fecundity, as a function of multiple mating, in an arctiid moth, *Utetheisa ornatrix. Ecol. Entomol.* **22**, 69–73.
LaMunyon C. W. and Eisner T. (1993) Post copulatory sexual selection in an arctiid moth (*Utetheisa ornatrix*). *Proc. Natl. Acad. Sci. USA* **90**, 4689–4692.
LaMunyon C. W. and Eisner T. (1994) Spermatophore size as determinant of paternity in an arctiid moth (*Utetheisa ornatrix*). *Proc. Natl. Acad. Sci. USA*. **91**, 7081–7084.
Marshall L. D. (1982) Male nutrient investment in Lepidoptera: what nutrients should males invest? *Am. Natural.* **120**, 273–279.

Mattocks A. R. (1972) Toxicity and metabolism of *Senecio* alkaloids. In *Phytochemical Ecology*, ed. J. B. Harborne pp. 179–200. Academic Press, New York.

Meinwald J. and Meinwald Y. C. (1966) Structure and synthesis of the major components in the hairpencil secretion of a male butterfly, *Lycorea ceres ceres* (Cramer). *J. Am. Chem. Soc.* **88**, 1305–1310.

Meinwald J., Meinwald Y. C., Wheeler J. W., Eisner T. and Brower L. P. (1966) Major components in the exocrine secretion of a male butterfly (*Lycorea*). *Science* **151**, 583–585.

Meinwald J., Chalmers A. M., Pliske T. E. and Eisner T. (1968) Pheromones. III. Identification of *trans*, *trans*-10-hydroxy-3,7-dimethyl-2,6-decadienoic acid as a major component in "hairpencil" secretion of the male monarch butterfly. *Tetrahedron Letters* **1968**, 4893–4896.

Meinwald J., Meinwald Y. C. and Mazzocchi P. H. (1969) Sex pheromone of the queen butterfly: chemistry. *Science*, **164**, 1174–1175.

Meinwald J., Boriack C .J., Schneider D., Boppré M., Wood D. F. and Eisner T. (1974) Volatile ketones in the hairpencil secretion of danaid butterflies (*Amauris* and *Danaus*). *Experientia* **30**, 721–722.

Miller J. R., Baker T. C., Cardé R. T. and Roelofs W. L. (1976) Reinvestigation of oak leaf roller sex pheromone components and the hypothesis that they vary with diet. *Science* **192**, 140–142.

Myers J. and Brower L. P. (1969) A behavioral analysis of the courtship pheromone receptors of the queen butterfly, *Danaus gilippus berenice*. *J. Insect Physiol.* **15**, 2117–2130.

Nakanishi, K. (1991). *A Wandering Natural Products Chemist*, pp. 91–94. American Chemical Society, Washington, DC.

Nishida R. (2002) Sequestration of defensive substances from plants by Lepidoptera. *Ann. Rev. Entomol.* **47**, 57–92.

Nishida R., Kim C. S., Fukami H. and Irie R. (1991) Ideamine *N*-oxides: pyrrolizidine alkaloids sequestered by the danaine butterfly, *Idea leuconoe*. *Agric. Biol. Chem.*, **55**, 1787–1792.

Nishida R., Schulz S., Kim C. S., Fukami H., Kuwahara Y., Honda K. and Hayashi N. (1996) Male sex pheromone of a giant danaine butterfly, *Idea leuconoe*. *J. Chem. Ecol.* **22**, 949–972.

Orr A. B., Trigo J. R., Witte L. and Hartmann T. (1996) Sequestration of pyrrolizidine alkaloids by larvae of *Tellervo zoilus* (Lepidoptera: Ithomiinae) and their role in the chemical protection of adult against the spider *Nephila maculata* (Araneidae). *Chemoecol.* **7**, 68–73.

Pease R. W. (1968) The evolutionary and biological significance of multiple pairing in Lepidoptera. *J. Lepidopterists' Soc.* **22**, 197–209.

Petty R. L., Boppré M., Schneider D. and Meinwald J. (1977) Identification and localization of volatile hairpencil components in male *Amauris ochlea* butterflies. *Experientia* **33**, 1324–1326.

Pliske T. E. (1975a) Courtship behavior of the monarch butterfly, *Danaus plexippus*. *Ann. Entomol. Soc. Am.* **68**, 143–151.

Pliske T. E. (1975b) Courtship behavior and use of chemical communication by males of certain species of ithomiine butterflies. *Ann. Entomol. Soc. Am.* **68**, 935–942.

Pliske T. E. (1975c) Attraction of Lepidoptera to plants containing pyrrolizidine alkaloids. *Environ. Entomol.* **4**, 455–473B.

Pliske T. E. and Eisner T. (1969) Sex pheromones of the queen butterfly: biology. *Science* **164**, 1170–1172.

Reeve H. K. (1993) Haplodiploidy, eusociality and absence of male parental and alloparental care in Hymenoptera: a unifying genetic hypothesis distinct from kin selection theory. *Phil. Trans. R. Soc. Lond. B* **342**, 335–352.

Reeve H. K. and Shellman-Reeve J. S. (1997) The general protected invasion theory: sex biases in parental and alloparental care. *Evol. Ecol.* **11**, 357–370.

Reeve H. K. and Pfenning D. W. (2002) Genetic basis for showy males. *Nature* (in review).

Rossini C., González A. and Eisner T. (2001) Fate of an alkaloidal nuptial gift in the moth *Utetheisa ornatrix*: systemic allocation for defense of self by the receiving female. *J. Insect Physiol.* **47**, 639–647.

Rothschild M. (1972) Some observations on the relationship between plants, toxic insects and birds. In *Phytochemical Ecology*, ed. J. B. Harborne, pp. 2–12. Academic Press, New York.

Rothschild M. (1973) Secondary plant substances and warning colouration in insects. In *Insect/Plant Relationships*, ed. H. F. van Emden, pp. 59–83. Blackwell, London.

Rothschild M. (1978) Hell's angels. *Antenna* **2**, 38–39.

Rothschild M. (1979) Mimicry, butterflies and plants. *Symb. Bot. Upsal.* **22**, 82–99.

Rothschild M., Aplin R. T., Cockrum P. A., Edgar J. A., Fairweather P. and Lees R. (1979) Pyrrolizidine alkaloids in arctiid moths (Lep.) with a discussion on host plant relationships and the role of these secondary plant substances in the Arctiidae. *Biol. J. Linnean Soc.* **12**, 305–326.

Rutowski R. L. (1982) Male choice and lepidopteran mating behavior. *Florida Entomol.* **65**, 72–82.

Sawhney R. S., Girotra R. N., Atal C. K., Culvenor C. C. J. and Smith L. W. (1967) Phytochemical studies on genus *Crotalaria*: Part VII. Major alkaloids of *C. mucronata, C. brevifolia, C. laburnifolia. Indian J. Chem.* **5**, 655–656.

Schal C. and Bell W. J. (1982) Ecological correlates of parental investment of mates in a tropical cockroach. *Science*, **218**, 170–173.

Schlatter C., Waldner E. E. and Schmid H. (1968) Zur Biosynthese des Cantharidins. *Experientia* **24**, 994–995.

Schneider D. (1983) Kommunikation durch chemische Signale bei Insekten: Alte und neue Beispiele von Lepidopteren. *Verhandl. Dtsch. Zool. Gesellsch.* **1983**, 5–16.

Schneider D. and Seibt U. (1969) Sex pheromone of the queen butterfly: electroantennogram responses. *Science* **164**, 1173–1174.

Schneider D., Boppré M., Schneider H., Thompson W. R., Boriack C. J., Petty R. L. and Meinwald J. (1975) A pheromone precursor and its uptake in male *Danaus* butterflies. *J. Comp. Physiol.* **97**, 245–256.

Schneider D., Boppré M., Zweig J., Horsley S. B., Bell T. W., Meinwald J., Hansen K. and Diehl E. W. (1982) Scent organ development in *Creatonotus* moths: regulation by pyrrolizidine alkaloids. *Science* **215**, 1264–1265.

Schulz S., Francke W., Boppré M., Eisner T. and Meinwald J. (1993) Insect pheromone biosynthesis: stereochemical pathway of hydroxydanaidal production from alkaloidal precursors in *Creatonotos transiens* (Lepidoptera: Arctiidae). *Proc. Natl. Acad. Sci. USA* **90**, 6834–6838.

Seibt U., Schneider D. and Eisner T. (1972) Duftpinsel, Flügeltaschen und Balz des Tagfalters *Danaus chrysippus. Z. Tierpsychol.* **31**, 513–530.

Sierra J. R., Woggon W.-D. and Schmid H. (1976) Transfer of cantharidin during copulation from the adult male to the female *Lytta vesicatoria* (Spanish flies). *Experientia* **32**, 142–144.

Smedley S. R. and Eisner T. (1995) Sodium uptake by "puddling" in a moth. *Science*. **270**, 1816–1818.

Smedley S. R. and Eisner T. (1996) Sodium: a male moth's gift to its offspring. *Proc. Natl. Acad. Sci. USA* **93**, 809–813.

Thornhill R. and Alcock J. (1983) *The Evolution of Insect Mating Systems*. Harvard Univ. Press, Cambridge, MA.

Trigo J. R. and Brown K. S. (1990) Variation of pyrrolizidine alkaloids in Ithomiinae: a comparative study between species feeding on Apocynaceae and Solanaceae. *Chemoecol.* **1**, 22–29.

Trigo J. R., Brown K. S., Jr, Witte L., Hartmann T., Ernst L. and Barata L. E. S. (1996) Pyrrolizidine alkaloids: different acquisition and use patterns in Apocynaceae and Solanaceae feeding ithomiine butterflies (Lepidoptera: Nymphalidae). *Biol. J. Linnean Soc.* **58**, 99–123.

Trivers R. L. (1985) *Social Evolution*. Benjamin Cummings, Menlo Park, CA. von Nickisch-Rosenegk E. and Wink M. (1993) Sequestration of pyrrolizidine alkaloids in several arctiid moths (Lepidoptera: Arctiidae). *J. Chem. Ecol.* **19**, 1889–1903.

Vasconcellos-Neto J. and Lewinsohn T. M. (1984) Discrimination and release of unpalatable butterflies by *Nephila clavipes*, a neotropical orb-weaving spider. *Ecol. Entomol.* **9**, 337–344.

Weller S. J., Jacobson N. L. and Conner W. E. (1999) The evolution of chemical defenses and mating systems in tiger moths (Lepidoptera: Arctiidae). *Biol. J. Linn. Soc.* **68**, 557–578.

Willis M. A. and Birch M. C. (1982) Male lek formation and female calling in a population of the arctiid moth *Estigmene acrea*. *Science* **218**, 168–170.

Part 2

Pheromone Detection

13

The biochemistry of odor detection and its future prospects

Lawrence J. Zwiebel

13.1 Introduction

These are exciting times in the field of chemosensory reception in general and olfaction in particular. In the decade since the landmark identification of a novel class of candidate odorant receptors (ORs) in rats (Buck and Axel, 1991), we have seen an explosion of similar studies involving other vertebrate as well as several insect species. In addition to an ever-increasing wealth of behavioral and physiological studies, insect systems provide arguably the most robust experimental system for the study of olfaction as well as a profound demonstration of the universal conservation of olfactory signal transduction mechanisms.

As is the case for all sensory pathways, the capacity to perceive and respond to olfactory cues (odorants) is the combined result of events that take place in both peripheral and central processing centers. These steps, which will be discussed in detail below, begin with the molecular transduction of chemical signals in the form of odorants into electrical activity by olfactory receptor neurons (ORNs) in the periphery whose axonal projections form characteristic synaptic connections with elements of the central nervous system (CNS). Within the CNS, complex patterns of olfactory signals are integrated and otherwise processed to afford recognition and ultimately, the behavioral responses to the insect's chemical environment. Within the context of pheromone recognition these responses would likely be centered on various elements of the insect's reproductive cycle.

13.2 The biochemistry and neurobiology of odor detection

While vertebrates and invertebrates differ in many aspects of the overall morphology of their olfactory apparatus, they share a number of fundamental mechanisms through which they are able to sense chemical environments. At the onset, an important distinction between vertebrate and insect olfactory systems is the segregation of insect ORNs and their support cells into distinct olfactory hairs (sensilla) on both antennal and maxillary palp structures (Stocker, 1994). In vertebrates, the detection of an odorant is mediated by specialized ORNs that are clustered together within the main olfactory epithelium of the nasal cavity or the vomeronasal organ (VNO). The first synaptic relay of ORNs, both in insects and vertebrates, occurs in anatomically similar neuropil structures of the central nervous system (CNS) known as glomeruli, that in vertebrates are found within the olfactory bulb (OB) and range in number from approximately 1000 (rat) to 5000 (dog) (Hildebrand and Shepherd, 1997). In insects, antennal ORN axons project either ipsilaterally or bilaterally to glomeruli located in antennal lobes (AL) (Stocker, 1994). The number of glomeruli in arthropods range from as low as 20–25 in *Aedes aegypti* (Anton, 1996) to approximately 300 in a crustacean olfactory system (Blaustein *et al*., 1993).

In addition, many insect behaviors and olfactory responses may be controlled under the broad umbrella of circadian rhythms. Flight activity (Jones *et al*., 1967), sugar feeding (Haddow *et al*., 1961), oviposition (Haddow and Ssenkubuge, 1962) and biting (Boorman, 1961) all occur with 24-h cycles of activity in *Anopheles gambiae* and in the yellow-fever vector mosquitoes *A. aegypti* and *Anopheles farauti* (Taylor, 1969). The peaks of activity occur around dusk ("lights off," in the laboratory) and dawn ("lights on," in the laboratory). This "V"-shaped diurnal (crepuscular) pattern of flight activity and biting have also been documented in the tsetse fly, *Glossina morsitans* (Brady, 1975), as well as in several other important disease vector insects. More recently, electrophysiological analysis of *Drosophila melanogaster* reveals circadian rhythmicity of odorant-stimulated antennal activity (Krishnan *et al*., 1999). Because all the above mentioned activities rely on olfaction, it is plausible to suggest that some of the genes that play a role in olfactory signal transduction may be under circadian control at the molecular level. Indeed, several genes encoding elements of the olfactory signal transduction pathway have recently been shown to cycle in abundance in *D. melanogaster* using microarray-based screens (McDonald and Rosbash, 2001) (Claridge-Chang *et al*., 2001).

In all systems studied thus far, olfactory-stimulated signal transduction is mediated by G-protein-coupled receptor (GPCR) pathways involving the synthesis of second messengers such as cyclic AMP (cAMP) and/or inositol 1,4,5-triphosphate (IP_3) (Boekhoff *et al*., 1994; Reed, 1992). Moreover, several proteins that are involved in fundamental aspects of olfactory signal transduction have been identified in both vertebrate and invertebrate systems. These include components of the biochemical machinery necessary for chemo-electric signal transduction, such as olfactory-specific G-proteins, adenylate cyclase, cyclic nucleotide and IP_3-gated channels, as well as other inositol carrier proteins (reviewed in Hildebrand and Shepherd, 1997).

There is also a broad class of water-soluble proteins found in olfactory mucosa and sensilla that have been associated with a vital role in the olfactory process. These secreted proteins are known as odorant binding proteins (OBPs) and have been hypothesized to act as odorant carriers to facilitate the solubilization and raise the effective concentration of hydrophobic odorants as a prelude to receptor binding. OBPs were first defined based on the ability to directly bind known odorants in both vertebrate (Pevsner *et al.*, 1985) and insect (Vogt and Riddiford, 1981) systems. Indeed, in the latter case, a subset of OBPs is expressed in male-specific, pheromone-sensitive hairs and have been shown to bind pheromones *in vitro*. Thus, they are classified as pheromone-binding proteins (PBPs) (Vogt and Riddiford, 1981). In addition, moths have been shown to express a second group of PBP-like proteins associated with general odorant-sensitive neurons, found in both male and female antenna structures, and are known as general odorant-binding proteins (GOBPs) (Vogt *et al.*, 1991). There are conflicting reports of several classes of OBPs identified in vertebrates, one of which has been shown to bind a broad array of ligands with no apparent specificity, as well as specific OBPs capable of recognizing and binding separate odorants (Dear *et al.*, 1991). Several putative members of the OBP/PBP family of olfactory proteins have been isolated from several insects including *D. melanogaster* (Pikielny *et al.*, 1994), *Apis mellifera* (Briand *et al.*, 2001) and true bugs (Dickens *et al.*, 1998).

Interestingly, in *Drosophila*, mutations in one candidate OBP gene, *lush*, display defects in ethanol sensitivity (Kim *et al.*, 1998) and another member of this family of proteins has recently been linked to the regulation of complex social behaviors in the fire ant, *Solenpsis invicta* (Krieger and Ross, 2002). While the precise functions of OBPs remain elusive, this latter study considerably raises the significance of these proteins in the overall process of olfactory signal transduction. No doubt similar findings will soon emerge in other systems that together will serve to underscore the often overlooked relevance of this family of highly expressed olfactory proteins. One hypothesis that is particularly attractive has OBPs serving not only as carrier or shuttle proteins that are responsible for bringing odorant ligands in proximity to ORs but, more importantly, OBPs, as a consequence of their differential affinity for particular odorants would be able to add to the overall complexity of the olfactory repertoire. In this model the overall representation of the olfactory sensitivity of an insect would reflect the multiplicative binding affinities of both ORs and OBPs.

In addition to ligand (agonist)-based activation, an important component of sensory perception is the cessation or reduction of signaling that is directly related to the phenomenon of adaptation, whereby a progressively weaker response is generated to repeated or persistent stimuli. At the molecular level, this is characterized by a process known as desensitization, which has been observed in all chemosensory systems, ranging from bacterial chemotaxis to neural transmission in humans, and can vary from complete termination of signaling, as seen in visual and olfactory systems, to graded attenuation of agonist potency in other systems (Dohlman *et al.*, 1991). Desensitization of GPCR-mediated signal transduction is mediated through an impairment of the receptor's ability to activate its corresponding G-protein and is carried out principally through the combined activity of two classes of proteins: G-protein-coupled serine/threonine receptor kinases (GRKs) and arrestins (reviewed in Freedman and Lefkowitz, 1996).

While phosphorylation and slow desensitization of specific intracellular residues on GPCRs can occur by more generalized second messenger-induced kinases, such as cAMP-dependent protein kinase A (PKA) and protein kinase C (PKC), GRKs phosphorylate only the agonist-bound (activated) form of GPCRs and are responsible for rapid receptor-specific desensitization (Inglese *et al.*, 1993). Phosphorylation by GRKs serves to promote the binding of arrestin proteins, which further uncouple GPCRs from the G-protein-based signaling cascade (Pippig *et al.*, 1993). Furthermore, while the role of GRKs and arrestins in desensitization pathways has been well established, it is likely that they are also intimately involved in GPCR internalization (sequestration), which is an integral component of GPCR resensitization (Ferguson *et al.*, 1996). More recently, visual arrestins have been shown also to be expressed and function in olfactory signal transduction pathways in *D. melanogaster* and *A. gambiae* (Merrill *et al.*, 2002). Moreover, in contrast to the large number of ORs and OBPs that are present in both these insects, dual-functional arrestins are encoded by at most only three genes, making them particularly attractive targets for reducing the olfactory sensitivity of insects of medical and economic importance.

GPCRs represent the largest superfamily of proteins currently known, with more than 5000 members (Gether, 2000), and most contain seven transmembrane-spanning regions of 20–25 amino acids (Strader *et al.*, 1994). These proteins link ligands and downstream effectors, not only by transmitting the signal, but also by amplifying and integrating other cellular signals (Dohlman *et al.*, 1991). As members of the GPCR superfamily, ORs are postulated to operate through a signal transduction pathway with features identical to other GPCRs and with specific components unique to olfactory tissue, such as G_{olf} (a G_s-like protein), adenylate cyclase III and a cAMP-gated channel (Pilpel *et al.*, 1998). Following the discovery of the first ORs in rats, other candidate receptors have been cloned from humans, dogs, cows, zebrafish, mice, pigs and several other vertebrates (reviewed in Mombaerts, 1999).

The molecular characterization of genes that encode candidate ORs was brought to the forefront when Buck and Axel (Buck and Axel, 1991) used a polymerase chain reaction (PCR)-based approach to clone the first odorant receptors (ORs) from rat olfactory epithelium. The family of rat ORs was subsequently estimated to contain between 70 and 200 members, displaying between 40 percent and 80 percent amino acid identity (Mombaerts, 1999). Moreover, in several other vertebrate systems, a large family of genes has been identified that encodes the putative ORs; there are approximately 100 OR genes in fish, 900 OR genes in humans, and 1300 ORs in mice (Ngai *et al.*, 1993; Troemel *et al.*, 1995; Zhang and Firestein, 2002). Apart from containing a conserved serpentine architecture comprising the seven transmembrane domains characteristic of GPCRs, these putative ORs display striking divergence. They contain highly variable regions within the third, fourth and fifth transmembrane domains that presumably correspond to domains responsible for odorant recognition and binding.

In order to establish whether individual members of this class of GPCRs actually function as ORs, several experimental approaches have been used for a very limited number of candidate vertebrate ORs. Most notably, heterologous expression of putative rat ORs in Sf9 cells led to odorant-specific generation of IP_3 (Raming *et al.*, 1993), while adenovirus-mediated overexpression of the rat I7 OR in rat olfactory epithelium led to an increased sensitivity to C7–C10-saturated aliphatic aldehydes (Zhao *et al.*, 1998). In a notable extension of these studies, viral overexpression in olfactory epithelium has been used to detail the molecular receptive range of I7 OR (Araneda *et al.*, 2000). These studies along with others which used a novel approach that combined calcium imaging and reverse genetics to examine the olfactory sensitivity of a cluster of mouse ORNs (Malnic *et al.*, 1999) provide compelling evidence that ORs recognize multiple odorants spanning a range of molecular features. Conversely, in the latter studies individual odorants were recognized by multiple ORNs each of which was shown to express a unique OR protein. Taken together, these experiments suggest that in these mammalian systems olfactory sensitivity employs a combinatorial code wherein different odorants stimulate unique sets of ORs and ORNs.

Of course, it must be stressed that the vast majority of candidate ORs remain orphan receptors and that it is prudent to avoid overly sweeping generalizations based on a very limited number of studies. Owing to a host of technical problems ranging from achieving correct protein targeting to the plasma membrane to lack of adequate expression levels, ORs have been for the most part refractory to functional studies. However, the fact that the bulk of ORs must therefore remain classified as orphan receptors should not be taken as an indication that their status as ORs is tenuous, but rather as a sign that we are still in the infancy of the field of OR characterization. As discussed below, there is every indication this will continue to mature and incorporate both standard and novel techniques in the overall objective of understanding the functional characteristics of olfactory signal transduction.

The first invertebrate organism in which candidate ORs were identified was

Caenorhabiditis elegans. These genes were identified through a bioinformatics-based screen of the complete *C. elegans* genomic sequence for potential signaling molecules (Troemel *et al.*, 1995). Like the vertebrate ORs, the *C. elegans* ORs are seven transmembrane GPCRs, although the *C. elegans* ORs bear almost no similarity to the vertebrate ORs. Moreover, within *C. elegans*, there is only between 10 percent and 48 percent identity among the ORs, indicating that they are much more divergent than the vertebrate ORs. In another departure from the vertebrate model of ORs in which the data are consistent with one type of receptor expressed per neuron (Pilpel *et al.*, 1998), GFP-fusion expression patterns of the *C. elegans* ORs reveal that more than one type of receptor is expressed in a given sensory neuron (Troemel *et al.*, 1995). Of course this result is hardly surprising in light of the over 800 candidate ORs that have currently been identified in *C. elegans* (Robertson, 2001), an organism that has only 32 chemosensory neurons. Importantly, mutants in one putative *C. elegans* OR (*odr-10*) display defects in chemosensory responses to diacetyl, strongly suggesting that this gene encodes the diacetyl OR (Sengupta *et al.*, 1996).

Because olfaction is mediated by GPCRs in both vertebrates and at least one invertebrate, it was assumed that insects would also utilize these proteins in olfactory signal transduction. Indeed, after a considerable period of investigation using a variety of approaches, a large family of candidate ORs was recently identified in *D. melanogaster* (Clyne *et al.*, 1999; Vosshall *et al.*, 1999; Gao and Chess, 1999). In the first of these studies, putative *D. melanogaster* ORs (DORs) were identified using a novel computer algorithm that searched for diagnostic features of the GPCR superfamily, including hydropathy, polarity and weighted amino acid composition of the predicted protein (Kim *et al.*, 2000). This approach was unique in that, rather than relying on putative local or global homology (which might miss a divergent member of a particular family), it utilized physicochemical properties from known transmembrane proteins. The structures that were ultimately identified using these strategies are members of a highly divergent family of receptors, displaying between 10 percent and 75 percent identity and bearing no significant homology to any other GPCR family (Smith, 1999).

Importantly, *in situ* hybridization with DOR antisense probes demonstrated that the majority of these genes (almost 30 percent of putative DORs were undetectable in any tissues) were selectively and stereotypically expressed in olfactory sensory neurons of the fly (Clyne *et al.*, 1999; Vosshall *et al.*, 1999; Gao and Chess, 1999; Vosshall *et al.*, 2000). Indeed, consistent with its presumed role as an OR, antisera generated against a C-terminal peptide of DOR43b labeled selective ORN cell bodies and dendritic processes in *D. melanogaster* antennae (Elmore and Smith, 2001). In contrast to the case in *C. elegans* and reminiscent of mammalian ORNs, studies using two-color (double labeling) *in situ* hybridization suggest that with one notable caveat, *D. melanogaster* ORNs are likely to express a single DOR gene (Vosshall *et al.*, 2000). The apparent exception to this rule of a single OR gene/ORN is DOR83b (originally designated A45), which is expressed throughout the antennal and maxillary palp ORNs of *D. melanogaster.* This has led to the hypothesis that its function may be independent of odorant binding (Vosshall *et al.*, 2000). Indeed, there is compelling evidence for the existence of GPCR oligomeric structures from a wide range of systems (as reviewed in Bouvier, 2001). One particularly relevant example of the importance of GPCR dimeric complexes is the metabotropic $GABA_B$ receptor, where coexpression of GBR1 and GBR2 isoforms is required for the formation of a functional GABA receptor on the cell surface (Kaupmann *et al.*, 1998; Jones *et al.*, 1998; White *et al.*, 1998).

Furthermore, by taking advantage of the wealth of experimental tools available in *D. melanogaster*, DOR promoter elements have been effectively used to drive neuronal reporters to elegantly describe ORN synapses to both contralateral and ipsilateral glomeruli within the antennal lobe (Vosshall *et al.*, 2000). More recently, another family of novel chemosensory receptors has been described in *D. melanogaster* that are likely to encode both olfactory and gustatory receptors (Clyne *et al.*, 2000; Scott *et al.*, 2001). More recently, important inroads in the functional characterization of candidate ORs have been made through the heterologous expression of DOR43a in Xenopus oocytes (Wetzel *et al.*, 2001) or overexpression in *D. melanogaster* (Storkuhl and Kettler, 2001) which resulted in increased sensitivity in both contexts to the identical set of four odorants. Taken together, these studies provide compelling evidence that DOR43a is indeed acting as an OR protein. It is hoped that functional studies of additional DORs as well as candidate ORs from other species will be carried out in the future.

In the first report of insect ORs outside of the model insect system *D. melanogaster*, my laboratory, as part of a collaborative effort with the Robertson and Carlson groups, has recently identified a similar family of candidate ORs from *A. gambiae* (AgORs) (Fox *et al.*, 2001) which, as predicted, are selectively expressed in olfactory tissues of the mosquito. In addition, the completion of the *Anopheles gambiae* genome project has allowed us (at present) to identify at least 79 members of the AgOR family in this important vector mosquito (Hill *et al.*, 2002). Furthermore, in the course of our initial characterization of the AgOR family, we have demonstrated that, as is the case for several OR families in other organisms, several groups of AgORs are clustered tightly together in the Anopheles genome and, more importantly, display little or no homology to any *Drosophila* OR genes suggestive of expansion of this gene family in Anopheles and perhaps other mosquitos (Hill *et al.*, 2002). Even more intriguing is the observation that one candidate OR from *A. gambiae* denoted as AgOr7, displays an extraordinarily high degree of conservation relative to Dor 83b, the one DOR gene that displays broad expression across most, if not all *Drosophila* ORNs. This homology along with the observation that AgOR7 shows a similarly broad pattern of expression

(R. J. Pitts and L. J. Zwiebel, unpublished observation) supports the hypotheses that these genes are orthologs and, as discussed above, may act to facilitate the formation of OR receptor heterodimeric complexes.

Another compelling observation is that at least five of the AgORs characterized thus far displays female-specific expression, a feature that may be especially relevant for disease transmission that is only carried out by female mosquitoes during the course of taking the bloodmeals required for the completion of their gonotrophic cycle. In a series of experiments that bears directly on the role of AgORs in host preference behaviors, we have also shown that several of these genes are all dramatically downregulated following blood feeding by adult females. Interestingly, the small subset of AgORs thus far examined also displayed variable rates of recovery from this inhibition.

While there is precedent in *C. elegans* for the activity-dependent regulation of chemosensory receptors (Peckol *et al.*, 2001), these observations are especially compelling in light of behavioral and electrophysiological studies in *A. gambiae* that showed an inhibition of host-seeking response and olfactory sensitivity to human odorants following blood feeding (Takken *et al.*, 2001). These latter experiments parallel similar data in another mosquito vector, *A. aegypti*, in which host seeking is inhibited following a blood meal and not fully restored until after the completion of the gonotrophic cycle, as indicated by the depositing of eggs (Klowden and Blackmer, 1987). In this system, the inhibition of host seeking is likely to result from two distinct mechanisms: the first occurs following a blood meal and is caused by distension of the midgut (Klowden and Lea, 1979) while the second occurs later in the reproductive cycle where a humoral factor is synthesized by the ovaries and reduces the sensitivity of a subset of lactic acid receptor cells (Davis, 1984).

It is tempting to speculate that AgORs might provide part of the molecular switch that underlies this dramatic behavioral transition. Because so many mosquito behaviors are olfactory mediated, insight into the olfactory signal transduction cascade in this system could dramatically enhance our understanding of mosquito behaviors as well as strengthen our understanding of insect olfaction in general.

13.3 Central and higher order olfactory pathways

Considerable evidence concerning the central olfactory pathways of insects is available and may be used to draw a reasonable hypothesis as to the organization and physiology (Hansson and Anton, 2000). In these model dipteran and lepidopteran systems, electrical activity in the form of action potentials that arise in response to the signal transduction events initiated by the olfactory stimulus is directed along ORN axons to the AL where synaptic connections are made that mediate signaling to higher centers within the insect CNS (Hildebrand and

Shepherd, 1997). In most insect systems, antennal ORNs project solely to the ipsilateral AL, although in flies several bilaterally projecting ORN afferents have been described (Stocker, 1994).

Within the AL, the ORN axonal projections are directed to spherical clusters of neuropil known as glomeruli that are separated from each other by glial cells; glomeruli are sites of synaptic connections between ORN axons and interneurons of the CNS (Tolbert and Hildebrand, 1981). Considerable evidence from several insect systems suggests that each individual ORN projects to a single glomerulus within the AL. Furthermore, in *D. melanogaster,* where the molecular cloning of individual OR genes (see below) has facilitated the labeling of specific classes of ORNs, it has been shown that the relationship between a particular class of ORN and specific AL glomeruli is stereotypically conserved between *D. melanogaster* individuals (de Bruyne *et al.*, 2001; Vosshall *et al.*, 1999). This observation mirrors similar findings in vertebrate model systems (Buck and Axel, 1991) and suggests that OR expression and identity may play an essential role in the development of the neuronal connections within the AL and other regions of the CNS. These speculations must be tempered, however, with the recognition that in *Drosophila*, it appears likely that ORs are not expressed until after the relevant synaptic connections have been established (Vosshall *et al.*, 2000).

The number of AL glomeruli found varies from species to species. In *A. aegypti*, the single vector insect examined in this regard to date, approximately 32 glomeruli has been identified (Bausenwein and Nick, 1998). Most other insects typically contain between 40 and 160 individual glomeruli (Hansson and Anton, 2000), while greater than 1000 glomeruli have been described in the locust (Ernst *et al.*, 1977).

A glomerulus within the AL can be viewed as a "cluster of neuronal dendrites organized in relation to a set of input fibers" (Shepherd, 1990). There are three types of AL interneurons that are juxtaposed within glomeruli with ORN axons: intrinsic AL local neurons (LN), projection neurons (PN) that form output pathways to higher order CNS structures and centrifugal neurons (CN) that surround the AL. In most insect species almost all AL neurons have the cell bodies in clusters located at the AL periphery, although there is considerable variation between species in the location and orientation of these clusters (Anton and Homberg, 1999). Moreover, there is considerable evidence from both vertebrate and invertebrate systems that a significant degree of neuronal convergence takes place in the AL and is a hallmark of olfactory signaling pathways. By comparing quantitative estimates of such events between several vertebrate and invertebrate species one can gain insight into the general pattern of neuronal convergence within the olfactory glomeruli. Indeed, in one such study Boeckh *et al.* (1990) estimate that anywhere from 10^5 to 10^8 ORN axons converge onto 10 to 10^3 glomeruli from which, in turn, 1 to 10^2 PNs convey information to higher CNS structures.

In several insect systems PN have been observed to have both uni- as well as multiglomerular arborizations, and their axons exit the AL via any of several antennoglomerular tracts where they make connections to different regions of the lateral protocerebrum (Hansson and Anton, 2000). Of these, the most prominent is the calyces of the corpora pedunculata or mushroom bodies (MBs) that consist of large numbers of parallel Kenyon cell fibers. The MBs are assumed to be the principal site of higher-order processing and storage of chemosensory information (Heisenberg, 1989).

13.4 Olfactory coding

There is good evidence from several lepidopteran systems and *D. melanogaster* that ORN axonal projections are reorganized upon entry into the AL from somatotopic (position-specific) to odotopic (odor-specific) groupings (Christensen *et al.*, 1996; Vosshall *et al.*, 2000). In addition to synaptic contact with ORN, PNs have been shown also to receive indirect signals from ORNs via LN connections in two well-studied non-vector insects, *Manduca sexta* (Sun *et al.*, 1997) and *Periplaneta americana* (Distler and Boeckh, 1997). While similar studies have yet to be carried out for a vector insect, the current thinking in the field is that such relationships between ORNs, LNs and PNs will prove to be broadly true throughout insects.

In addition to the glomerular organization of the AL already discussed, the male AL of several species of Lepidoptera, Dictyoptera and Hymenoptera insects contain an enlarged, sexually dimorphic glomerular structure (Rospars, 1988). This structure is known as the macroglomerular complex (MGC) in moths and bees and the macroglomerulus (MG) in cockroaches; these structures exclusively receive input from ORNs that are specifically sensitive to female sex pheromones (Boeckh and Boeckh, 1979; Christensen and Hildebrand, 1987; Matsumoto and Hildebrand, 1981). As such, the MGC and MG constitute the first described example of a "labeled-line code" where odorant specific (in this case, pheromone) ORNs are directly connected to odotopically specialized CNS structures. Interestingly, in the one similar study carried out to date in a vector insect no obvious sexual size dimorphism (apart from a slight exaggeration of AL size in females) was observed in the primary olfactory processing centers of *A. aegypti* (Anton, 1996).

Another important example of a labeled line within the olfactory system is found in the yellow fever vector, *A. aegypti,* where electrophysiological studies have identified CO_2-detecting ORNs in specific club-shaped sensilla basiconica (sometimes referred to as palpal pegs) on the distal-most (fourth) segment of the maxillary palp (Grant and O'Connell, 1996). Using a combination of anterograde track tracings with HRP and induced degeneration studies, all maxillary palp

ORN afferents were mapped to a distinct glomerular grouping in the ventroposterior region of the ipsilateral AL which are distinct from other AL glomeruli that receive antennal innervation (Distler and Boeckh, 1997 a or b). Similar patterns of "reserved" glomeruli or labeled lines for CO_2 input from the labial palps have been reported in three lepidopteran species (Bogner *et al.*, 1986; Kent *et al.*, 1986; Lee and Altner, 1986). In *A. aegypti*, it should be stressed that because the maxillary palps are also sensitive to other odorants such as n-heptane, acetone and amyl acetate (Kellogg, 1970) and as some of these diverse ORNs are co-localized in palpal pegs along with CO_2 ORNs (McIver, 1982), these studies suggest these "reserved" glomeruli would be expected to process signals from a diverse collection of maxillary palp ORNs.

The molecular cloning of a large number of candidate odorant receptors in *D. melanogaster* (DORs) (Clyne *et al.*, 1999; Vosshall *et al.*, 1999) has facilitated the use of an extremely robust genetic and experimental insect model system to examine the mechanisms that underlie olfactory coding within the AL. In an elegant series of genetic studies, Vosshall *et al.* (2000) used transgenic flies in which specific DOR promoters were used to drive expression of either β-galactosidase (LacZ) or a carboxy-terminal fusion of green fluorescent protein (GFP) and neuronal synaptobrevin (nsyb-GFP); the expression of these reporters allowed direct visualization of both ORN cell bodies as well as the terminal axonal projections in the AL. In these studies, ORNs expressing a particular DOR gene were mapped to one or two spatially invariant glomeruli within the AL. The establishment of such a conserved topographic map of DOR projectives onto the AL is consistent with earlier efforts that used 2-deoxyglucose mapping to map metabolic activity in the AL in response to specific olfactory stimuli and which demonstrated that different odorants elicit distinct patterns of glomerular activity (Rodrigues, 1988). In addition to the labeled-line pheromone tracing studies mentioned above, similar conclusions were also generated using calcium imaging techniques in the honeybee (Galizia *et al.*, 1999; Sachse *et al.*, 1999). These studies, along with parallel efforts that employed functional tracing, genetic, and imaging methods to examine the organization of the mammalian olfactory system reviewed by Buck (2000), reinforce the labeled-line code and suggest that the mechanisms that underlie odor discrimination are largely conserved between insects and mammals.

At first glance, labeled-line coding of olfactory signals may seem in contrast to the "ensemble or across-fiber code" (Shepherd, 1985) where complex mixtures of odorants or even individual odorant components are perceived as patterns of activity across an ensemble of neurons and AL glomeruli. However, recent experiments examining odor coding of individual ORNs in *Drosophila* and mammalian olfactory systems demonstrate that individual ORNs are capable of a wide spectrum of responses. In the fly, a particular odor can excite one neuron while inhibiting another, and a particular neuron can be excited by one odor and inhibited by another as well as exhibiting different modes of termination kinetics when stimulated by alternative odorants (de Bruyne *et al.*, 1999, 2001). In mammals one ORN type can recognize multiple odorants (with differing degrees of affinity) and, conversely, one odorant can be recognized by multiple ORNs (Malnic *et al.*, 1999). These studies raise the possibility of combinatorial receptor codes for odorants that are sensitive to both the nature and concentration of any particular stimuli. Taken together and given the fact that outside of laboratory studies, vector and non-vector insects are much more likely to experience complex odor cues made up of many components rather than purified individual odorants, these studies reinforce the ensemble code hypothesis where the diversity of ORN response to particular odorants would create a complex pattern of glomeruli response in the AL. There is every likelihood that neither the labeled-line nor ensemble codes will prove to be universally applicable but rather each modality will be useful for describing specific olfactory pathways.

In parallel to the spatial pattern of AL activity, temporal considerations of odor delivery are also expected to play significant roles in establishing olfactory responses. In biological situations, because of variations such as wind speed and turbulence, odorants are likely to be delivered to an insect's olfactory apparatus in discrete packets rather than in continuous plumes. Indeed, in the one preliminary study available for a vector insect, *A. aegypti* displayed more robust olfactory responses to dose-dependent pulses of CO_2 compared to continuous streams of stimulant (Grant and O'Connell, 1996). Considerably more work has been carried out with several lepidopteran species and with locust. In wind tunnel studies with pheromone stimulation of male moths, pulsed odor plumes were shown to be required for efficient upwind flight towards the odor source, while a homogeneous cloud of pheromone resulted in a cross-wind flight with no resulting upwind movement (Baker and Vicker, 1996). In an exciting study that used simultaneous recording from PNs of the MGC along with a broad field recording of the ipsilateral antennae (electroantennogram, EAG) in response to odor plumes generated in wind tunnels, it was shown that glomerular output is tightly coupled to the temporal pattern of antennal responses (Vickers *et al.*, 2001). In real time recordings as moths moved through a changing odor plume (and hence olfactory input) PN activity was shown in this work to be modulated within a millisecond timeframe relative to the fluctuating olfactory stimulus that was generated in the odor plume. The strict correspondence between AL output and olfactory stimulus is somewhat in contrast to several reports that use immobilized locusts (Laurent *et al.*, 1996; Wehr and Laurent, 1996) and honeybees (Stopfer *et al.*, 1997) to suggest that different odors activate specific groups of PNs in the AL. Importantly, however, in locust and honeybee, the activity of the PNs were constrained by broad network oscillations and that the coding of each odor is represented by a different temporal pattern of activity across the PN ensemble.

It remains to be seen how these two viewpoints will sort out with regard to the question of odor representations within the AL. It is possible that alterations in the spatio-temporal pattern of stimulus presentation (as found in a biological context or a laboratory wind tunnel) may account for much of the differences in these studies. Nevertheless, it is clear that a considerable degree of processing occurs between the G-protein-mediated events of signal transduction that take place in ORN dendrites and the coordinate neuronal activity that subsequently follows in the AL and, in turn, in the higher-order structures such as the MBs of the protocerebrum. In this manner, sensory information is coupled to both behavioral responses as well as those of learning and memory, although we are only just beginning to acquire an understanding of the molecular, cellular and system level events that comprise these pathways.

13.5 What the future holds

With great progress come great challenges. The prospect for the future of research into olfactory processes in general, as well as insect pheromone-related pathways in particular, is especially bright. First and foremost, given the recent advances (described above) in examining the molecular genetics of olfaction in *D. melanogaster* and *A. gambiae*, the stage is set for extending these studies to other insects. This is especially compelling for the many insect systems where a wealth of pheromone-related olfactory studies have already been carried out. Of particular interest in this regard would be the identification of OR and other olfactory genes from several lepidopteran systems where the precise chemical nature of pheromones has already been well characterized.

Based upon a considerable degree of interspecific homology, there is every reason to expect that the identification of several classes of olfactory-active genes will be forthcoming. These include, but are not restricted to, olfactory arrestins, G-proteins, effector and biotransformation enzymes as well as the large family of putative odorant-binding proteins However, given the critical role that genomics have played in the cloning of DOR and AgOR genes, it is reasonable to expect that, in the absence of concerted genome sequencing programs, the isolation of other insect OR genes remains an uncertain undertaking without a firm prospect of success. That much being said, this cautionary note should be tempered by our observation that several DOR and AgOR amino acid sequences exhibit a significant and uncharacteristic degree of similarity. The hope is that a subset of insect OR genes may display sufficient homology to undertake the design of effective oligonucleotide primers for interspecific PCR-based cloning projects or to facilitate cross-hybridization studies.

Apart from the identification of additional insect olfactory genes, there are still numerous questions at the level of the peripheral signaling pathways that remain unanswered. A precise functional characterization of the ligand specificity for the large number of currently identified orphan insect ORs is arguably a top priority for many research programs. There is every reason to expect that this knowledge will bring a greater appreciation of how olfactory coding is set up within both the peripheral and central nervous systems. Considering what is known from the sparse number of studies that have been carried out with mammalian ORs, the expectation is that there will be some degree of combinatorial cross-talk between ORs that are sensitive to similarly structured odorant ligands. In addition, we can expect many of the uncertainties regarding OR receptor structure and processing to be elaborated in the not too distant future. Furthermore, as discussed above, with an eye towards other, more understood, vertebrate GPCR-based pathways, it seems reasonable to expect that insect ORs will also function as heteromeric complexes with the logical co-receptor being members of the DOr83b/AgOr7 family of proteins.

Another important question in olfactory signaling that will no doubt be addressed centers on understanding the currently enigmatic role of OBPs in olfactory signal transduction pathways of insects. As discussed above, the high levels of OBP expression within insect olfactory tissue, the several intriguing studies which suggesting that OBPs are responsible for directly mediating olfactory driven behaviors, and evidence that OBPs are capable of binding odorants with a high degree of specificity suggest that these proteins play a critical role in the olfactory processes of insects.

Because of its abundance of bioinformatic, genetic and molecular resources, *Drosophila* is likely to continue to act as the vanguard insect system for much of these future studies. Considerable progress has been made in developing similar tools for other insect species as evidenced by the recent completion of the genomic sequence of *A. gambiae,* as well as the onset of smaller-scale genomics efforts for other disease vector mosquitoes. Advances in establishing effective protocols for germline transformation in Anopheline mosquitoes (Catteruccia *et al.*, 2000; Grossman *et al.*, 2001) and other insects (Handler, 2001) set the stage for extending the analysis of insect olfactory pathways beyond academic model systems such as *D. melanogaster* to agricultural pests and disease vectors of critical importance to global public health. Moreover, the emergence of a variety of robust gene silencing strategies (Fire, 1999) such as RNA interference (RNAi) opens the door for the application of forward genetics-based approaches to the study of olfactory processes in insect systems that otherwise lack sufficient genetic resources.

Given the wealth of breakthrough-level work that has taken place over the past several years (in part detailed above), there seems little doubt that our overall understanding of the chemoreception has entered a robust period of growth and maturation. At first glance, it should be noted that the primacy of the conservation of general principles of olfactory signal transduction has now been extended over several representative species of both vertebrates and invertebrate lineages. The importance of this is underscored not only in terms of its validation of the current paradigm, but more importantly, in that it firmly lays the foundation for future work extending these efforts to a still wider range of systems.

While there can be no doubt that a large portion of the landmark studies carried out thus far could not have been undertaken if not for the abundance of biochemical, genetic, molecular and neurological tools that are available to olfaction researchers using model biological systems, one must at the same time be wary of the myopic view that may inadvertently arise when exclusively utilizing a small number of experimental systems. The very act of extending these analyses into novel biological systems is a strong countermeasure to this tendency and provides a meaningful level of contextual diversity through which olfactory processes may be viewed. Furthermore, by conducting this work in insects of medical and economic importance, such as agricultural pests and disease vectors where olfaction underlies several behaviors critical for host/crop selection and overall vectorial capacity, chemosensory research may also help to advance efforts to reduce the catastrophic impact of these insects. Indeed, in this context, an ideal marriage can be forged between pure research into the mechanisms and evolution of this ancient and complex sensory process and applied studies that could significantly impact public health and economic development on a global scale.

References

Anton S. (1996) Central olfactory pathways in mosquitoes and other insects. *Ciba Found. Symp.* **200**, 184–192; discussion 192–196, 226–232.

Anton S. and Homberg, U. (1999) Antennal lobe structure. In *Insect Olfaction*, ed. B. Hansson, Springer-Verlag, Berlin: pp. 97–124.

Araneda R. C., Kini A. D. and Firestein S. (2000) The molecular receptive range of an odorant receptor. *Nat. Neurosci.* **3**, 1248–1255.

Baker T. and Vicker N. (1996) Pheromone-mediated flight in moths. In *Insect Pheromone Research: New Directions*, eds R. Cardé and A. Minks, pp. 248–264. Chapman Hall, New York.

Bausenwein B. and Nick P. (1998) Three dimensional reconstruction of the antennal lobe in the mosquito, Aedes aegypti. In *New Neuroethology on the Move*, eds R. Wehner and N. Elsner, p. 386. Thieme Stuttgart.

Blaustein D. N., Simmons R. B., Burgess M. F., Derby C. D., Nishikawa M. and Olson K. S. (1993). Ultrastructural localization of 5'AMP odorant receptor sites on the dendrites of olfactory receptor neurons of the spiny lobster. *J. Neurosci.* **13**, 2821–2828.

Boeckh J. and Boeckh V. (1979) Threshold and odor specificity of pheromone-sensitive neurons in the deutocerebrum of *Antheraea pernyi* and *A. polyphemus* (Saturnidae). *J. Comp. Physiol.* **132**, 235–242.

Boeckh J., Distler P., Ernst K., Hosl M. and Malun D. (1990) Olfactory bulb and antennal lobe. In *Chemosensory Information Processing*, ed. D. Schild, pp. 201–228. Springer Heidelberg, Germany.

Hill C. A., Fox A. N., Pitts R. J., Kent L. B., Tan P. L., Chrystal M. A., Cravchik A., Collins F. H., Robertson H. M. and Zwiebel L. J. (2002) G protein-coupled receptors in *Anopheles gambiae*. *Science* **298**, 176–178.

Inglese J., Freedman N. J., Koch W. J. and Lefkowitz R. J. (1993) Structure and mechanism of the G protein-coupled receptor kinases. *J. Biol. Chem.* **268**, 23735–23738.

Jones M. D., Hill M. and Hope A. M. (1967) The circadian flight activity of the mosquito *Anopheles gambiae*: phase setting by the light regime. *J. Exp. Biol.* **47**, 503–511.

Jones K. A., Borowsky B., Tamm J. A., Craig D. A., Durkin M. M., Dai M., Yao W. J., Johnson M., Gunwaldsen C., Huang L. Y., Tang C., Shen Q., Salon J. A., Morse K., Laz T., Smith K. E., Nagarathnam D., Noble S. A., Branchek T. A. and Gerald C. (1998). GABA(B) receptors function as a heteromeric assembly of the subunits GABA(B)R1 and GABA(B)R2. *Nature* **396**, 674–679.

Kaupmann K., Malitschek B., Schuler V., Heid J., Froestl W., Beck P., Mosbacher J., Bischoff S., Kulik A., Shigemoto R., Karschin A. and Bettler B. (1998) GABA(B)-receptor subtypes assemble into functional heteromeric complexes. *Nature* **396**, 683–687.

Kellogg, F. E. (1970) Water vapour and carbon dioxide receptors in *Aedes aegyptii*. *J. Insect Physiol.* **16**, 99–108.

Kent K. S., Harrow I. D., Quartararo P. and Hildebrand J. G. (1986) An accessory olfactory pathway in Lepidoptera: the labial pit organ and its central projections in *Manduca sexta* and certain other sphinx moths and silk moths. *Cell Tissue Res.* **245**, 237–245.

Kim M. S., Repp A. and Smith D. P. (1998) LUSH odorant-binding protein mediates chemosensory responses to alcohols in *Drosophila melanogaster*. *Genetics* **150**, 711–721.

Kim J., Moriyama E. N., Warr C. G., Clyne P. J. and Carlson J. R. (2000) Identification of novel multi-transmembrane proteins from genomic databases using quasi-periodic structural properties. *Bioinformatics* **16**, 767–775.

Klowden M. J. and Blackmer J. L. (1987) Humoral control of pre-oviposition behaviour in the mosquito, *Aedes aegypti*. *J. Insect. Physiol.* **33**, 689–692.

Klowden, M. J., and Lea, A. O. (1979). Abdominal distension terminates subsequent host-seeking behavior of *Aedes aegypti* following a blood meal. *J. Insect. Physiol.* **25**, 583–585.

Krieger M. J. and Ross K. G. (2002) Identification of a major gene regulating complex social behavior. *Science* **295**, 328–232.

Krishnan B., Dryer S. E. and Hardin P. E. (1999) Circadian rhythms in olfactory responses of *Drosophila melanogaster*. *Nature* **400**, 375–378.

Laurent G., Wehr M. and Davidowitz H. (1996) Temporal representations of odors in an olfactory network. *J. Neurosci.* **16**, 3837–3847.

Lee J. K. and Altner H. (1986) Primary sensory projections of the labial palp-pit organ of *Pieris rapae* L. (Lepidoptera: Pieridae). *Int. J. Insect. Morph. Embryol.* **15**, 439–448.

Malnic B., Hirono J., Sato T. and Buck L. B. (1999) Combinatorial receptor codes for odors. *Cell* **96**, 713–723.

Matsumoto S. and Hildebrand J. (1981) Olfactory mechanisms in the moth *Manduca sexta*: response characteristics and morphology of central neurons in the antennal lobes. *Proc. R. Soc. London Ser. B* **213**, 249–277.

McDonald M. J. and Rosbash M. (2001) Microarray analysis and organization of circadian gene expression in Drosophila. *Cell* **107**, 567–578.

McIver, S. B. (1982). Sensilla mosquitoes (Diptera: Culicidae). *J. Med. Entomol.* **19**, 489–535.

Merrill C. E., Riesgo-Escovar J., Pitts R. J., Kafatos F. C., Carlson J. R. and Zwiebel L.

J. (2002) Visual arrestins in olfactory pathways of Drosophila and the malaria vector mosquito Anophelesgambiae. *Proc. Natl. Acad. Sci. USA* **99**, 1633–1638.

Mombaerts P. (1999) Molecular biology of odorant receptors in vertebrates. *Annu. Rev. Neurosci.* **22**, 487–509.

Ngai J., Dowling M. M., Buck L., Axel R. and Chess A. (1993) The family of genes encoding odorant receptors in the channel catfish. *Cell* **72**, 657–666.

Peckol E. L., Troemel E. R. and Bargmann C. I. (2001) Sensory experience and sensory activity regulate chemosensory receptor gene expression in *Caenorhabditis elegans*. PNAS **98**, 11032–11038.

Pevsner, J., Trifiletti, R., Strittmatter, S. M., and Synder, S. H. (1985). Isolation and characterization of an olfactory receptor protein for odorant pyrazines. *Proc. Natl. Acad. Sci. USA* **82**, 3050–3054.

Pikielny C. W., Hasan G., Rouyer F. and Rosbash M. (1994). Members of a Family of drosophila putative odorant-binding proteins are expressed in different subsets of olfactory hairs. *Neuron* **12**, 35–49.

Pilpel Y., Sosinsky A. and Lancet D. (1998) Molecular biology of olfactory receptors. *Essays in Biochemistry* **33**, 93–104.

Pippig S., Andexinger S., Daniel K., Puzicha M., Caron M. G., Lefkowitz R. J. and Lohse M. J. (1993) Overexpression of beta-arrestin and beta-adrenergic receptor kinase augment desensitization of beta 2-adrenergic receptors. *J. Biol. Chem.* **268**, 3201–3208.

Raming K., Krieger J., Strotmann J., Boekhoff I., Kubick S., Baumstark C. and Breer H. (1993) Cloning and expression of odorant receptors. *Nature* **361**, 353–356.

Reed, R. (1992). Signalling pathways in odorant detection. *Neuron* **8**, 205–209.

Robertson H. M. (2001) Updating the str and srj (stl) families of chemoreceptors in Caenorhabditis nematodes reveals frequent gene movement within and between chromosomes. *Chem. Senses* **26**, 151–159.

Rodrigues, V. (1988). Spatial coding of olfactory information in the antennal lobe of *Drosophila melanogaster*. *Brain. Res.* **453**, 299–307.

Rospars J. P. (1988) Structure and development of the insect antennodeutocerebral system. *Int. J. Insect. Morphol. Embryol.* **17**, 243–294.

Sachse S., Rappert A. and Galizia C. G. (1999) The spatial representation of chemical structures in the antennal lobe of honeybees: steps towards the olfactory code. *Eur. J. Neurosci* **11**, 3970–3982.

Scott K., Brady J. R., Cravchick A., Morozov P., Rzhetsky A., Zuker C. and Axel R. (2001) A chemosensory gene family encoding candidate gustatory and olfactory receptors in *Drosophila*. *Cell* **104**, 661–673.

Sengupta P., Chou J. H. and Bargmann C. I. (1996) odr-10 encodes a seven transmembrane domain olfactory receptor required for responses to the odorant diacetyl. *Cell* **84**, 899–909.

Shepherd G. (1985) Are there labeled lines in the olfactory pathway? In *Taste, Olfaction and the Central Nervous System*, ed. D. Pfaff, pp. 307–321. Rockefeller Press, New York.

Shepherd G. M. (1990) Contribution toward a theory of olfaction. In *Frank Allison Linville's R. H. Wright Lectures on Olfactory Research*, ed. C. K, Simon Fraser University Burnaby, BC, Canada, pp. 61–109.

Smith D. P. (1999) Drosophila odor receptors revealed. *Neuron* **22**, 203–204.

Stocker R. F. (1994) The organization of the chemosensory system in *Drosophila melanogaster*: a review. *Cell & Tissue Research* **275**, 3–26.

Stopfer M., Bhagavan S., Smith B. H. and Laurent G. (1997) Impaired odour discrimination on desynchronization of odour-encoding neural assemblies. *Nature* **390**, 70–74.

Storkuhl K. F. and Kettler R. (2001) Functional analysis of an olfactory receptor in *Drosophila melanogaster*. *PNAS* **98**, 9381–9385.

Strader C. D., Fong T. M., Tota M. R. and Underwood D. (1994) Structure and function of G protein-coupled receptors. *Annu. Rev. Biochem.* **63**, 101–132.

Sun X. J., Tolbert L. P. and Hildebrand J. G. (1997) Synaptic organization of the uniglomerular projection neurons of the antennal lobe of the moth *Manduca sexta*: a laser scanning confocal and electron microscopic study. *J. Comp. Neurol.* **379**, 2–20.

Takken W., van Loon J. J. A. and Adam W. (2001) Inhibition of host-seeking response and olfactory responsiveness in *Anopheles gambiae* following blood feeding. *J. Insect. Physiol.* **47**, 303–310.

Taylor, B. (1969). Geographical range and circadian rhythms. *Nature* **222**, 296–297.

Tolbert L. and Hildebrand J. (1981) Organization and synaptic ultrastructure of glomeruli in the antennal lobes of the moth *Manduca sexta*: a study using thin sections and freeze-structure. *Phil. Trans. R. Soc. London Ser B* **213**, 279–301.

Troemel E. R., Chou J. H., Dwyer N. D., Colbert H. A. and Bargmann C. I. (1995). Divergent seven transmembrane receptors are candidate chemosensory receptors in C. elegans. *Cell* **83**, 207–218.

Vickers N. J., Christensen T. A., Baker T. C. and Hildebrand J. G. (2001) Odour-plume dynamics influence the brain's olfactory code. *Nature* **410**, 466–470.

Vogt R. G. and Riddiford L. M. (1981) Pheremone binding and inactivation by moth antennae. *Nature* **293**, 161–163.

Vogt R. G., Rybczynski R. and Lerner M. R. (1991) Molecular cloning and sequencing of general odorant-binding proteins GOBP1 and GOBP2 from the tobacco hawk moth *Manduca sexta*: comparisons with other insect OBPs and their signal peptides. *J. Neurosci.* **11**, 2972–2984.

Vosshall L. B., Amrein H., Morozov P. S., Rzhetsky A. and Axel, R. (1999). A spatial map of olfactory receptor expression in the Drosophila antenna. *Cell* **96**, 725–736.

Vosshall L. B., Wong A. M. and Axel R. (2000) An olfactory sensory map in the fly brain. *Cell* **102**, 147–159.

Wehr M. and Laurent G. (1996) Odour encoding by temporal sequences of firing in oscillating neural assemblies. *Nature* **384**, 162–166.

Wetzel C. H., Behrendt H., Gisselmann G., Storkuhl K. F., Hovemann B. and Hatt H. (2001) Functional expression and characterization of a *Drosophila* odorant receptor in a heterologous cell system. *PNAS* **98**, 9377–9380.

White J. H., Wise A., Main M. J., Green A., Fraser N. J., Disney G. H., Barnes A. A., Emson P., Foord S. M. and Marshall F. H. (1998). Heterodimerization is required for the formation of a functional GABA(B) receptor. *Nature* **396**, 679–682.

Zhang X. and Firestein S. (2002) The olfactory receptor gene superfamily of the mouse. *Nat. Neurosci.* **5**, 124–33.

Zhao H., Ivic L., Otaki J. M., Hashimoto M., Mikoshiba K. and Firestein S. (1998) Functional expression of a mammalian odorant receptor. *Science* **279**, 237–242.

14

Biochemical diversity of odor detection: OBPs, ODEs and SNMPs

Richard G. Vogt

14.1 Introduction

Insects decode their olfactory environment using olfactory sensilla with diverse odor-sensitive phenotypes. Olfactory sensilla are traditionally classified by their external morphology; for example, Table 14.1 lists the types and numbers of olfactory sensilla on the antennae of *Manduca sexta* and *Drosophila melanogaster*. These morphological types can be further subdivided based on physiological responses to specific odors (e.g. de Bruyne *et al*., 2001; Kalinová *et al*., 2001; Shields and Hildebrand 2001a,b). And underlying these morphological and physiological phenotypes are diverse genes expressing proteins that instruct cells to take on specific morphologies, proteins that self-assemble into cuticles of a specific form, and that detect and process odor signals.

The biochemistry of odor detection involves at least three types of protein: odor receptors (ORs); odorant binding proteins (OBPs); and odor degrading enzymes (ODEs). ORs are expressed by olfactory receptor neurons (ORNs) and localized in the membranes of the ciliated dendrites (Figure 14.1). The result of detection is translated into neuronal electrical activity by transductory proteins. But while transductory proteins are more or less common for all olfactory neurons, differential expression of ORs, OBPs and ODEs allows the neurons to detect specific odor molecules.

Efforts to understand the biochemistry of odor detection have certainly been driven by a search for ORs, but along the way, this search yielded OBPs and ODEs. In the early 1980s there were few laboratories studying the biochemistry

Table 14.1 Olfactory sensilla types and numbers

***Manduca sexta* (male) (Lee and Strausfeld, 1990)**					***Drosophila melanogaster* (male/female) (Shanbhag *et al.* 1999)**		
	Sensilla per annulus	Neurons per annulus	Sensilla per antennae @ 80 annuli per antennae	Neurons per antennae @ 80 annuli per antennae		Sensilla per antenna	Neurons per antenna
S. trichodea type I (long)	834	1664	66 720	133 120	S. trichodea (3 types)	117/113	219/205
S. trichoidea type II (short)	736	1921	58 880	153 680	S. intermedia (2 types)	12/21	28/49
S. basiconica type I	166	390	13 280	31 200	S. basiconica (7 types)	193/229	490/564
S. basiconica type II	372	833	29 760	66 640			
S. chaetica type-I	8	40	640	3200			
S. chaetica type-II	9	9	720	720			
S. coeloconica type-I	22	110	1760	8800	S. coeloconica (2 types)	57/54	147/130
S. coeloconica type-II	3	9					
	2150	4976	172 000	398 080		419/457	974/1038* 1306/1316**

*Numbers based on estimated sensilla numbers. **Numbers based on counts of axon profiles in nerve.

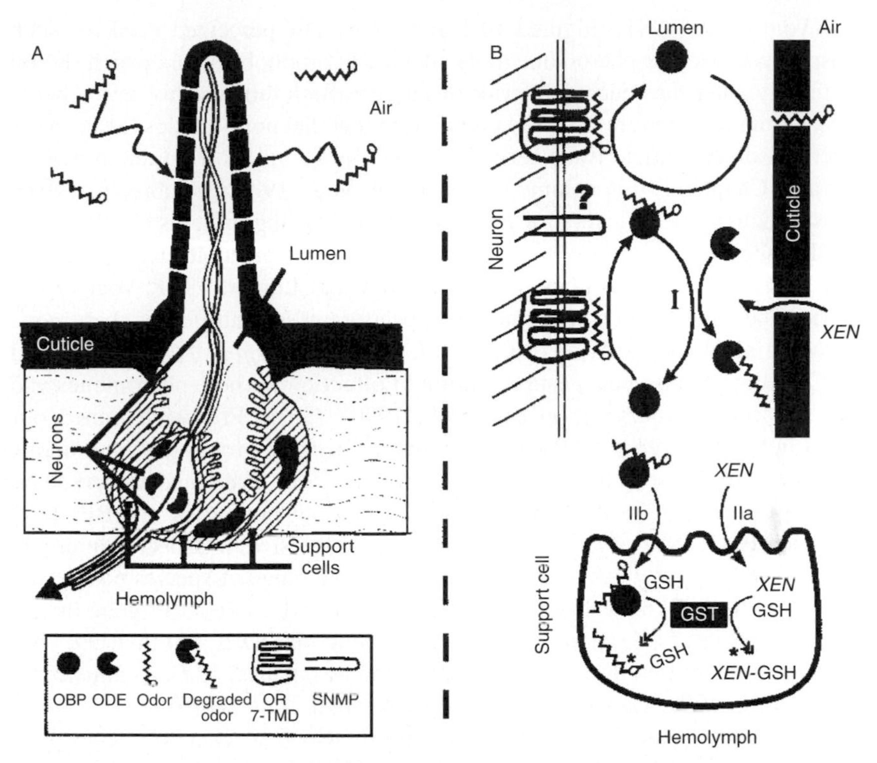

Figure 14.1 Schematic of olfactory sensillum and a generalized biochemical pathway of odor reception. A An olfactory sensillum includes 2–3 neurons surrounded by 3 support cells; olfactory dendrites/cilia project up the fluid filled lumen of a cuticular hair. The sensillum lumen is isolated from hemolymph by a cellular barrier. Modified from Steinbrecht (1969); see Steinbrecht (1999) for more details. B Hydrophobic odor molecules enter the aqueous sensillum lumen via pores penetrating the cuticular hair wall. Hydrophilic OBPs are proposed to bind and transport odors to receptor proteins located in the neuronal membranes. ODEs (pathway I) in the sensellum lumen are proposed to degrade these odor molecules. Cytoplasm of support cells contain xenobiotic inactivating enzymes, such as glutathione-S-transferase (GST) (pathway IIa) which may also serve to inactivate odor molecules (pathway IIb). Interactions between OBPs and ORs and the function of SNMP are unclear. Modified from Rogers *et al.* (1999).

of odor detection in insects. In 1981 we reported the first identification of a pheromone binding protein (PBP) and a sensillar esterase (SE); both appeared uniquely expressed in male antenna of the silk moth *Antheraea polyphemus* and were present in the extracellular fluid of the sensillum lumen (Vogt and Riddiford, 1981). In 1985 we proposed a scheme in which pheromone molecules first bound to PBPs for transport to ORs and were subsequently rapidly degraded by

SEs (Vogt *et al.*, 1985) (Figures 14.1 and 14.7). The perceived need for such transport was because pheromone molecules are hydrophobic odors which should not readily enter the aqueous interior of the sensillum lumen. This new scheme was a dramatic departure from the prevailing view that pore-tubules (anatomical structures of the cuticle wall) served as conduits for odors from air to neuron (see also Chapter 1, this volume; reviewed by Vogt, 1987; Steinbrecht, 1997). Our subsequent identification of general odorant binding proteins (GOBP1 and GOBP2) suggested this new scheme might be extended to other olfactory modalities such as the detection of plant volatiles (Vogt and Lerner, 1989; Vogt *et al.*, 1991a). The identification of an antennal-specific but sex-indifferent pheromone degrading aldehyde oxidase (AOX) in *M. sexta* and *A. polyphemus* suggested that ODEs might also be a general feature of olfaction for both pheromones and plant volatiles (Rybczynski *et al.*, 1989, 1990). Glenn Prestwich joined our efforts in the mid-1980s synthesizing valuable chemical probes and pheromone analogues (e.g. Prestwich *et al.*, 1984; Vogt *et al.*, 1985, 1988) and embarked on his own efforts to elucidate the pheromone binding properties of PBPs (e.g. Prestwich 1985, 1987). Many other labs became interested in OBPs beginning in the late 1980s, greatly expanding knowledge of the range of species possessing these proteins and knowledge of the size and diversity of the OBP gene family in single species. Little more work has been done on ODEs, but in 1999, ORs were finally identified from the sequenced genome database of *D. melanogaster* (Clyne *et al.*, 1999; Gao and Chess, 1999; Vosshall *et al.*, 1999).

Insects present an exceptional landscape in which to examine genes with roles in processing environmental information. More than half the known animal species are insects, somewhere between the 800 000 named species and another 800 000 to 30 million undiscovered species (Erwin, 1982; Freeman and Herron, 1998; Novotny *et al.*, 2002). Insects are organized into 29 extant Orders, 25 of which belong to the division Neoptera (Figure 14.2); the Neoptera account for about 98 percent of known insect species, and the 11 Orders of the neopterous Endopterygota (insects with complete metamorphosis, also called holometabolous insects) account for about 88 percent of these species (Borror *et al.*, 1989; Kristensen, 1991). Fossil evidence suggests the Neoptera emerged and its extant Orders diverged around 300 million years ago (Carboniferous Era) (Labandeira and Sepkoski, 1993). The lineages of neopterous Orders were well established before the ascendancy of Angiosperms (about 130 million years ago, Jurassic Era), but their association with plants doubtless drove the enormous species expansion of many neopterous Orders. Each of the major insect lineages possesses unique characteristics of life history, development and physiology, and some of these phylogenetically distinct characteristics are clearly supported by unique lineages of gene families. Insects clearly offer an extraordinary opportunity to explore the molecular/genetic evolutionary history of olfactory diversity.

Today we know that OBPs comprise a multigene family that is present

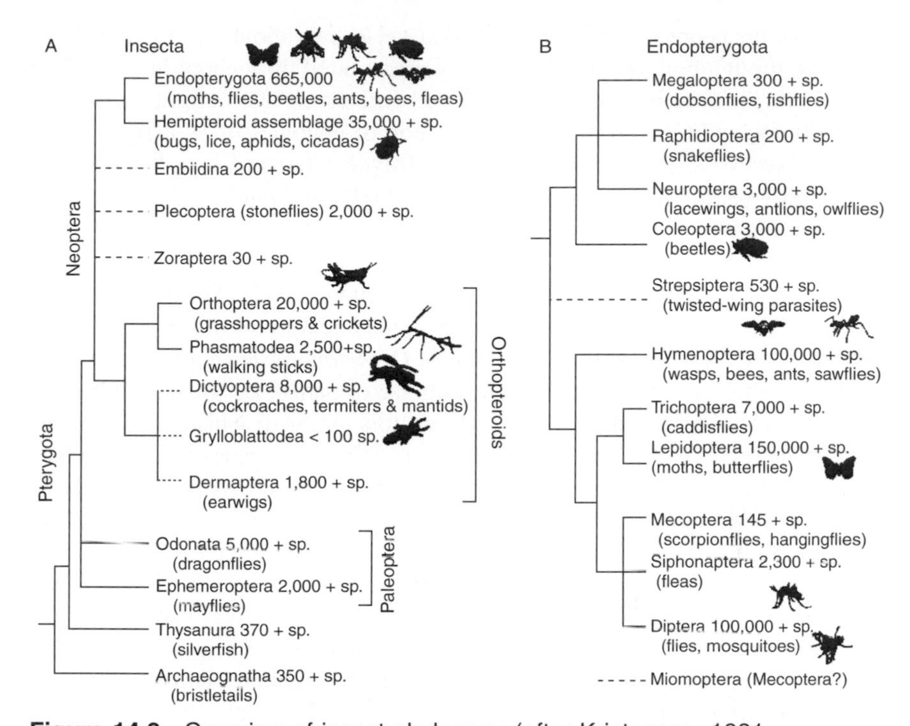

Figure 14.2 Overview of insect phylogeny (after Kristensen, 1981; Boudreaux, 1979; Hennig, 1981). In A, the representation of a monophyletic orthopteroid group is debatable (Maddison and Maddison, 1998); this group was suggested by Boudreaux (1979) and Hennig (1981), but was not included by Kristensen (1991) due in part to uncertainties regarding the relationships of the Plecoptera. The monophyletic division of Endopterygota (holometabolous insects) (Kristensen, 1991; Whiting *et al.*, 1997) is supported by most authors in part because of the unique development in this group; all members undergo a complete metamorphosis from the non-reproductive larval stage to the reproductive adult stage. A shared common ancestor for the Endopterygota and Hemipteroid lineages is suggested by morphological (e.g. Hennig, 1981; Kristensen, 1991; Whiting *et al.*, 1997) and molecular data (e.g. Whiting *et al.*, 1997). In B, two supported endopterygote lineages are shown, one including the Coleoptera and a second including Hymenoptera, Diptera and Lepidoptera. The fossil record suggests that the orthopteroids are more ancient than most endopterygote Orders (late Carboniferous vs early Permian); among the Endopterygota, the Coleoptera (Permian) are more ancient than the Diptera/Lepidoptera lineage, and the Diptera (Triassic) are more ancient than the Lepidoptera (Jurassic) (Kukalová-Peck, 1991). The majority of molecular- or biochemical-based olfactory research has been done on a subset of these orders (noted by the insect symbols). Numbers of named species in each lineage are shown, taken from various entries in *The Insects of Australia* (Achterberg *et al.*, 1991); these numbers are generally viewed as gross underestimates, but accurately reflect the relative success of the respective lineages.

throughout the neopterous insects. Less is as yet known about the range of ODEs and ORs, though both are unquestionably represented throughout insects given the central role chemodetection plays for all organisms. Our current knowledge is rapidly expanding with the characterization of genomes of *D. melanogaster* (Diptera; see Vosshall, Chapter 19 and de Bruyne, Chapter 23, in this volume) *Anopheles gambiae* (Diptera; e.g. Fox *et al.*, 2002; Hill *et al.*, 2002; Vogt, 2002) and *Heliothis viresens* (Lepidoptera; e.g. Krieger *et al.* 2002); and in the United States, the National Science Foundation has recently funded research on the genomes of several more insect species (see Couzin, 2002). The genome of *D. melanogaster* has given us a reasonably complete picture of the size of the OR and OBP gene families in a single species. *D. melanogaster* contains about 60 OR genes which express in either the antennae or palps (see Vosshall, Chapter, 19, in this volume), about 60 gustatory receptors, a few of which are expressed in antennae (Scott *et al.*, 2001), and about 50 OBP related genes, many of which express in the antennae and palps, but some of which express in non-olfactory tissues (Galindo and Smith, 2001; Graham and Davies 2002; Hekmat-Scafe *et al.*, 2002; Vogt *et al.*, 2002). Similar numbers of these genes have been identified in the recently published *A. gambiae* genome (e.g. Fox *et al.*, 2002; Hill *et al.*, 2002; Biessmann *et al.*, 2002; Vogt, 2002). Experiments suggest that ORs and OBPs differentially express among distinct populations of sensilla (Voshall *et al.*, 2000; Shanbhag *et al.*, 2001). Such implied combinatorial expression of olfactory genes presumably underlies the odor specificities of olfactory sensilla.

There are significant questions regarding the exact roles of OBPs and ODEs, and the interactions of these proteins with the ORs (see Figure 14.7). One concern has been whether the association between odor and OBP is so stable that the odor cannot be released from OBP in a physiologically relevant time, within the 150–500 msec observed for moths altering their behavior in response to changed pheromone concentration (Baker and Haynes, 1987; Baker and Vogt, 1988). This concern has led recent queries into whether there might be secondary mechanisms which orchestrate the release of odor from OBP in the near-region of the OR (see Leal, Chapter 15 and Plettner, Chapter 16, in this volume). One protein that might contribute to such release is called SNMP; SNMP is a very abundant protein located in the receptor membrane of ORNs, and has been characterized in several lepidopteran species (Rogers *et al.*, 1997, 2001a, b; Krieger *et al.*, 2002). SNMPs may have some other function, but whatever their exact role, these proteins have properties that strongly suggest they also have central actions in odor detection.

This chapter reviews aspects of OBPs, ODEs and SNMPs in the context of ORs and odor detection. Other chapters in this book focus on ORs (see Zwiebel, Chapter 13, Vosshall, Chapter 19, Krieger and Breer, Chapter 20 and de Bruyne Chapter 23, in this volume). Too often, studies have focused on individual proteins, investigating mechanisms in a non-integrative manner. Much of this was doubtless due to the lack of data on ORs and thus we were unable to include these central components in any studies of OBPs and ODEs. This time is past. The opportunities before us include investigations of a biochemical network that orchestrates odor detection and odor-based behavior, of the divergence and diversification of these network components through gene duplication and evolution, and of the gene regulatory mechanisms that insure appropriate selection and expression of unique combinations of these gene products and thereby establish unique functional phenotypes for different sensilla. Here indeed is a rich and behaviorally relevant genetic landscape with a strong sensory, ecological and evolutionary context that is ripe for exploration.

14.2 Odorant binding proteins – OBPs

Times are always changing. Until recently OBPs were defined by several criteria including their unique expression in antennae and demonstrated presence in the lumen of olfactory sensilla, their demonstrated ability to bind odors, and their size, sequence similarity and inclusion of six cysteins. Not all criteria were always satisfied, but enough were to argue that described proteins were indeed OBPs and likely functioned in odor detection. Genomics has changed this. Now "OBP" refers to a rather large family of genes, perhaps 51 in *D. melanogaster* (e.g. Hekmat-Scafe *et al.*, 2002) which share structural properties but not necessarily biological properties; many of the genes we are now calling OBPs presumably have no function at all in chemodetection, let alone olfactory-based behavior. Such is the makeup of large families.

OBPs comprise a multigene family that includes a group of small hydrophilic proteins which bind and transport small hydrophobic ligands. Higher vertebrates and insects both have OBPs, though the sequences and structures are clearly different (e.g. Tegoni *et al.*, 2000; Pelosi, 2001) indicating that mammalian and insect OBPs are independently derived (i.e. unrelated). Direct cloning/sequencing has identified 13 OBPs in the moth *Manduca sexta*, and a combination of direct cloning/sequencing and genome analysis has identified about 50 OBP-related genes in the fruitfly *D. melanogaster* and a similar number in the malaria mosquito *A. gambiae* (Table 14.2). OBPs have been identified in a range of species that suggests they are present at least throughout the neopterous insects, or in at least about 98 percent of all insect species, and consequently at least 50 percent of all named and currently extant eukaryotic species (see Vogt *et al.*, 1999).

The first identified insect OBP was the pheromone binding protein (PBP) of the silk moth *A. polyphemus* (Vogt and Riddiford, 1981). This 14 kDa protein appeared to be specific to the male antenna, was perhaps the most abundant soluble protein in the antenna, was located in the aqueous extracellular fluid that bathed the pheromone sensitive neurons, and could bind sex pheromone. The concentration of *Apol*PBP within the sensillum fluid was estimated to be about

Table 14.2 OBP-related proteins with published sequences available 8/12/02

Order	Species	Name	Accession #	Reference
Lepidoptera	*Agrotis segetum*	*Aseg*-PBP	AH007940	LaForest *et al.* (1999)
(moths)	Noctuoidea; Noctuidae			
	Antheraea pernyi	*Aper*-PBP1 **Sn Pi**	X96773	Raming *et al.* (1990)
	Bombycoidea;	*Aper*-PBP2 **Pi**	X96860	Krieger *et al.* (1991)
	Saturniidae	*Aper*-PBP3 **Pi**	AJ277265	Maida *et al.* (2000)
		Aper-GOBP1	Y10970	Mameli *et al.* (1997)
		Aper-GOBP2 **Sn**	X96772	Breer *et al.* (1990)
		Aper-ABPX	AJ002519	Krieger *et al.* (1997)
	Antheraea	*Apol*-PBP1 **Sx,v**	X17559	Raming *et al.* (1989)
	polyphemus	**Px,h,i,a,e**	AJ277266	Maida *et al.* (2000)
	Bombycoidea;	*Apol*-PBP2 **Pi**	AJ277267	,,
	Saturniidae	*Apol*-PBP3 **Pi**	P34169	Vogt *et al.* (1991a)
		Apol-GOBP2 **Sv, Px**	(peptide)	
	Argyrotaenia velutinana	*Avel*-PBP	AF177641	Willet, (2000)
	Tortricoidea; Tortricidae			
	Bombyx mori	*Bmor*-PBP **Sv,n,p**	X94987	Krieger *et al.* (1996)
	Bombycoidea;	**Po,t**	X94988	,,
	Bombycidae	*Bmor*-GOBP1 **Sp**	X94989	,,
		Bmor-GOBP2 **Sv,n,p**	X94990	,,
		Bmor-ABPX		
	Choristoneura fumiferana	*Cfum*-PBP	AF177643	Willet, (2000a)
	Tortricoidea; Tortricidae			
	Epiphyas postvittana	*Epos*-PBP1	AF416588	Newcomb *et al.* (2002)
	Tortricoidea;	*Epos*-PBP2	AF411459	,,
	Tortricidae	*Epos*-GOBP2	AF411460	,,
	Galleria mellonella	Sericotropin (non-	L41640	Filippov *et al.* (1995)
	Pyraloidea; Pyralidae	olfactory)		
	Heliothis virescens	*Hvir*-PBP **Se′**	X96861	Krieger *et al.* (1993)

Noctuoidea;	*Hvir*-GOBP1	X96862	,,
Noctuidae	*Hvir*-GOBP2 **Se′**	X96863	,,
	Hvir-ABPX	AJ002518	Krieger *et al.* (1997)
Helicoverpa zea	*Hzea*-PBP	AF090191	Callahan *et al.* (2000)
Noctuoidea; Noctuidae			
Helicoverpa	*Harm*-PBP **Se′**	AJ278992	Wang and Guo, (2000)
armigera	*Harm*-GOBP1	AY049739	Wang *et al.* (2001)
Noctuoidea;	*Harm*-GOBP2 **Se′**	AJ278991	Wang and Guo, (2000)
Noctuidae			
Hyalophora cecropia	*Hcec*-PBP	P34175 (peptide)	Vogt *et al.* (1991a)
Bombycoidea;	*Hcec*-GOBP2	P34172 (peptide)	,,
Saturniidae			
Lymantria dispar	*Ldis*-PBP1 **Sa′, Pa′,s**	AF007867	Merritt *et al.* (1998)
Noctuoidea;	*Ldis*-PBP2 **Sa′, Pa′,s**	AF007868	,,
Lymantriidae	*Ldis*-GOBP2	P34173 (peptide)	Vogt *et al.* (1991a)
Mamestra brassicae	*Mbra*-PBP1 **Pe,b**	AF051143	Maïbèche-Coisné *et al.* (1998a)
Noctuoidea;	*Mbra*-PBP2 **Pe,b**	AF051142	,,
Noctuidae	*Mbra*-PBP4	AF461143	
	Mbra-GOBP2	AF051144	Cristiani *et al.* (2001) Maïbèche-Coisné *et al.* (1998b)
Manduca sexta	*Msex*-PBP1	M21798	Györgyi *et al.* (1988)
Sphingidea;	*Msex*-PBP1(rev)	AF323972	Vogt *et al.* (2002)
Sphingidae	**Sy,w,z**	AF117589	Robertson *et al.* (1999)
	Msex-PBP2	AF117581	,,
	Msex-PBP3	M73797	Vogt *et al.* (1991b)
	Msex-GOBP1 **Sy,z**	AF323972	Vogt *et al.* (2002)
	Msex-GOBP2 **Sy,z Pj**	AF117577	Robertson *et al.* (1999)
	Msex-ABPX	AF117591	,,
	Msex-ABP1	AF393491	Robertson *et al.* (2001)
	Msex-ABP2	AF393488	,,

(Contd)

Table 14.2 *(Contd)*

Order	Species	Name	Accession #	Reference
		Msex-ABP3	AF393490	Robertson *et al.* (2001)
		Msex-ABP4	AF393498	,,
		Msex-ABP5	AF393499	,,
		Msex-ABP6	AF393500	,,
		Msex-ABP7		
	Orgyia pseudosugata Noctuoidea; Lymantriidae	*Opse*-PBP	P34178 (peptide)	Vogt *et al.* (1991a)
	Pectinophora gossypiella Gelechioidea; Gelechiidae	*Pgos*-PBP	AF177656	Willet, (2000b)
	Spodoptera exigua Noctuoidea; Noctuidae	*Spexi*-GOBP2	AJ294808	Guirong and Yuyuan (2000)
	Synanthedon exitiosa Sesioidea; Sesiidae	*Sexi*-PBP	AF177660	Willet, (2000b)
	Yponomeuta cagnagellus Yponomeutoidea; Yponomeutidae	*Ycag-PBP*	AF177661	Willet, (2000b)
Coleoptera (beetles)	*Anomala cuprea* Scarabaeiformia; Scarabaeidae; Rutelinae	*Acup*-PBP1-	AB040980	Nikonov *et al*, (2002)
		Acup-PBP2	AB040141	
	Anomala octiescostata Scarabaeidae; Rutelinae	*Aoct*-PBP1 **Pq**	AB040143	Nikonov *et al*, (2002)
		Aoct-PBP2	AB040981	
	Anomala osakana Scarabaeidae; Rutelinae	*Aosa*-PBP1 **Pc′**	AB040985	Wojtasek *et al.* (1998)
		Aosa-PBP2	AF031492	Peng and Leal, (2001)
	Exomala orientalis Scarabaeidae; Rutelinae	*Eori*-PBP1	AB040144	Peng and Leal, (2001)
		Eori-PBP2	AB040986	
	Holotrichia parallela Scarabaeidae; Melolonthinae	*Hpar*-OBP1	AB026554	Deyu and Leal (2002)
		Hpar-OBP2	AB026556	

	Phyllopertha diversa	*Pdiv*-OBP1 **Pd′**	AB026552	Wojtasek *et al.* (1999b)
	Scarabaeidae; Rutelinae	*Pdiv*-OBP2 **Pd′**	AB026553	
	Popillia japonica	*Pjap*-PRP1 **Pc′**	AF031491	Wojtasek *et al.* (1998)
	Scarabaeidae; Rutelinae			
	Rhynchophorus palmarum	*Rpal*-OBP2	**AF139912**	Jacquin-Joly *et al.* (1999)
	Cucujiformia; Phytophaga;			
	Curculionidae;			
	Dryophthorinae			
	Tenebrio molitor	*Tmol*-B1	M97916	Paesen and Happ, (1995)
	Cucujiformia;	*Tmol*-B2	M97917	
	Tenebrionidae	*Tmol*-THP12 (non-olfactory)	U24237	Graham *et al.* (2001)
Diptera (flies)	*Drosophila*	*Dmel*-99D	CG7584	b,c,d
	melanogaste	*Dmel*-99C	CG12665	b,c,d
	Brachycera;	*Dmel*-99B **Sk**	CG7592	a,b,c,d
	Muscomorpha;	*Dmel*-99A **NSk**	CG18111	a,b,c,d
	Ephydroidea;	Dmel-93A*	CG17284	d
	Drosophilidae	Dmel-85A*	CG11732	d
		Dmel-84A **Sk**	PBPRP4	Pikielny *et at.*, (1994)
		Dmel-83F	CG15583	a,c,d
	Drosophila OBP	*Dmel*-83E **NSk**	CG15583	a,b,c,d
	names are based on	*Dmel*-83D	CG15582	a,c,d
	cytological map	*Dmel*-83C **Sk**	CG15582	a,c,d
	locations (number);	*Dmel*-83B **Sk,l,u**	OSE	McKenna *et al.* (1994)
	letters distinguish	*Dmel*-83A **Sk,l,u**	OSF	,, and Pikielny *et al.*
	between genes within	*Dmel*-76A **Sk,u**	LUSH	(1994)
	that region. "*"	*Dmel*-69A **Sk**	PBPRP1	Kim *et al.* (1998)
	indicates a divergent	Dmel-58A	CG13517	Pikielny *et al.* (1994)
	group of presumptive	Dmel-58B*	CG13518	d
	OBPs with uniquely	Dmel-58C*	CG13524	d

(Contd)

Table 14.2 *(Contd)*

Order	Species	Name	Accession #	Reference
	large number of	Dmel-58D*	CG13519	d
	cysteins, reported by	*Dmel*-57E **Sk**	CG13429	d
	Hekmat-Scafe *et al.*	*Dmel*-57D **Sk**	AF457149	a,b,c,d
	(2002).	*Dmel*-57C **Sk**	CG13421	a,c,d
		Dmel-57B **Sk**	AF457147	a,b,c,d
	CG numbers are gene	*Dmel*-57A **Sk**	AF457148	a,c,d
	product identifiers.	*Dmel*-56I	AF457143	a,c,d
	AF numbers include	*Dmel*-56H **Sk**	CG13874	a,c,d
	nucleotide sequences.	*Dmel*-56G **Sk**	CG13873	a,b,c,d
	Names (e.g. OSE) are	*Dmel*-56F	AF457146	a,b,c,d
	formal gene	*Dmel*-56E **Sk**	CG8462	a,b,c,d
	identifiers.	*Dmel*-56D **Sk**	CG11218	a,b,c,d
		Dmel-56C **Sk**	CG15129a	a,b,c,d
	In Reference column,	*Dmel*-56B **Sk**	CG15129b	a,b,c,d
	lower case letters	*Dmel*-56A **NSk**	CG11797	a,c,d
	indicate source	*Dmel*-51A	AF457145	a,b,c,d
	publications: "a"	Dmel-50A*	??	c,d
	Galindo and Smith	Dmel-50B*	CG13940	d
	2001; "b" Vogt *et al.*	Dmel-50C*	??	d
	2002; "c" Graham and	Dmel-50D*	??	d
	Davies, 2002; "d"	Dmel-50E*	CG13939	d
	Hekmat-Scafe	Dmel-49A*	CG8769	d
	et al. 2002.	Dmel-47B*	CG13208	d
		Dmel-47A	CG12944	d
		Dmel-46A*	CG12905	a,b,c,d
		Dmel-44A	CG2297	d
		Dmel-28A **Sk,r,u**	PBPRP5	b,c,d
		Dmel-22A	AF457144	Pikielny *et al.* (1994)
		Dmel-19D **NSk**	PBPRP2	b,c

		Dmel-19C **Sk**	CG15457	Pikielny *et al.* (1994)
		Dmel-19B **Sk**	CG1670	a,b,c,d
		Dmel-19A **Sk**	CG11748	a,b,c,d
		Dmel-18A **NSk**	CG15883	a,b,c,d
		Dmel-8A	CG12665	c,d
				b,c,d
	Phormia regina	*Preg*-CSRBP	S78710	Ozaki *et al.* (1995)
	Brachycera;	(taste)		
	Muscomorpha;			
	Oestroidea; Calliphoridae			
	Ceratitis capitata	*Ccap*-OBPRP	AJ252076	Christophides *et al.* (2000)
	(medfly)	*Ccap*-OBPRP	AJ252077	
	Brachycera;	serum protein	Y08954	
	Muscomorpha;	serum protein	Y19146	
	Tephritoidea; Tephritidae			
Diptera	*Anopheles gambiae*	*Agam*-OBP1	AF393487	Robertson *et al.* (2001)
(mosquitoes)	Nematocera;	*Agam*-OBP2	AF393485	,,
	Culicoidea	*Agam*-OBP-1(OS-F)*	AF437884	Biessmann *et al.* (2002)
		Agam-OBP-2*	AF437885	,,
		Agam-OBP-3(OS-E)*	AF437886	,,
		Agam-OBP-4*	AF437887	,,
		Agam-OBP-5	AF437888	,,
		Agam-OBP-6*	AF437889	,,
		Agam-OBP-7*	AF437890	,,
		Agam-OBPRP-1*	EAA00498	Vogt, (2002)*
		Agam-OBPRP-2*	EAA00779	,,
		Agam-OBPRP-3*	EAA00788	,,
		Agam-OBPRP-4*	EAA00801	,,
		Agam-OBPRP-5*	EAA01392	,,
		Agam-OBPRP-6*	EAA01491	,,

(Contd)

Table 14.2 *(Contd)*

Order	Species	Name	Accession #	Reference
		Agam-OBPRP-7*	EAA03447	,,
		Agam-OBPRP-8*	EAA03742	,,
		Agam-OBPRP-9*	EAA03745	,,
		Agam-OBPRP-10*	EAA06799	,,
		Agam-OBPRP-11*	EAA06803	,,
		Agam-OBPRP-12*	EAA07741	,,
		Agam-OBPRP-13*	EAA07997	,,
		Agam-OBPRP-14*	EAA09324	,,
		Agam-OBPRP-15*	EAA12996	,,
		Agam-OBPRP-16*	EAA14622	,,
	Aedes aegypti	*Aaeg*-OBPRP	AY062131	Bohbot and Vogt, (2001)
	Nematocera; Culicoidea	*Cqui*-OBPRP	AF468212	Ishida *et al.* (2002)
	Culex quinquefasciatus			
	Nematocera; Culicoidea			
Hymenoptera	*Apis mellifera*	*Amel*-ASP1 **Sf Pf**	AF166496	Danty *et al.* (1998, 1999)
	(honeybee)	*Amel*-ASP2 **Sc Pg**	AF166497	Danty *et al.* (1997)
	Apoidea; Apidae	*Amel*-ASP4	AF393495	Robertson *et al.* (2001)
		Amel-ASP5	AF393497	,,
		Amel-ASP6*	AF393496	,, (*also OBP8, Forret and Maleszka, (2001); AF339140)
	Solenopsis sp.	*S. amblychila*	AF427889	Krieger and Ross, (2002)
	(fireant*)	*S. aurea*	AF427890	
	Formicidae;	*S. geminata*	AF427905	
	Myrmicinae;	*S. globularia l.*	AF427906	
		S. interrupta	AF427891	
		*S. invicta** Gp-9	AF459414	
		S. macdonaghi	AF427901	
		S. quinquecuspis	AF427902	

		S. richteri	AF427904	
		S. saevissima	AF427892	
Hemiptera (True Bugs)	*Lygus lineolaris* Miridae; Mirini	*Llin*-LAP **Sb′**	AF091118	Vogt *et al.* (1999)
Dictyoptera Blattaria	*Leucophaea maderae* (cockroach) Blattaria; Blaberoidea; Blaberidae	*Lmad*-PBP	AY116618	Riviere *et al.* (2003)
Dictyoptera Isoptera	*Zootermopsis nevadensis* (termite) Isoptera; Termopsidae; Zootermopsis	*Znev*-OBP1 *Znev*-OBP2 *Znev*-OBP3		Ishida *et al.* (2002)

S = demonstration of sensilla localization; NS = expressed but not associated with sensilla (*Drosophila*); P = demonstration of ligand binding; a–e′ = appropriate references (below).

a. Bette *et al.* (2002)
b. Bohbot *et al.* (1998)
c. Briand *et al.* (2001b)
e. Campanacci *et al.* (2001a)
f. Danty *et al.* (1999) (IS)
g. Danty *et al.* (1997) (IS)
h. Du *et al.* (1994)
i. Du and Prestwich, (1995)
j. Feng and Prestwich, (1997)
k. Galindo and Smith, (2001) (T)
l. Hekmat-Scafe *et al.* (1997) (AB)
m. Jacquin-Joly *et al.* (2000)
n. Laue and Steinbrecht, (1997) (AB)
o. Leal, (2000)
p. Maida *et al.* (1999) (AB)
q. Nikonov *et al*, (2002)
r. Park *et al.* (2000) (AB)
s. Plettner *et al.* (2000)
t. Sandler *et al.* (2000)
u. Shanbhag *et al.* (2001a) (AB)
v. Steinbrecht *et al.* (1995) (AB)
w. Steinbrecht, (1996) (AB)
x. Vogt and Riddiford, (1981) (D)
y. Vogt *et al.* (1991a) (D)
z. Vogt *et al.* (2002) (IS)
a′. Vogt *et al.* (1989) (D, AB')
b′ Vogt *et al.* (1999)
c′. Wojtasek *et al.* (1998)
d′. Wojtasek *et al.* (1999)
e′. Zhang *et al.* (2001) (AB)

AB = imunoreactive with antisera generated to *A. polyphemus* PBP or GOBP2 or *B. mori* GOBP1 for moths, or to indicated protein for *Drosophila* (AB′ is antisera to *L. dispar* PBP1 and PBP2); IS = *in situ* hybridization. D = direct purification from isolated sensilla.

10 mM (Klein, 1987). PBPs were subsequently identified by tissue specificity and N-terminal sequence from the Lepidoptera *A. polyphemus*, *Hyalophora cecropia*, *Bombyx mori*, *Lymantria dispar*, *M. sexta* and *Orgyia pseudosugata* (Vogt, 1987; Vogt *et al.*, 1989; 1991a); the first full-length OBP sequences were obtained for *M. sexta* PBP1 (Györgyi *et al.*, 1988) and *A. polyphemus* PBP1 (Raming *et al.*, 1989). The identification of two PBPs in *L. dispar* was the first indication of multiple PBP genes within a single species (Vogt *et al.*, 1989). The identification and cloning of the general odorant binding proteins GOBP1 and GOBP2 (Vogt and Lerner 1989; Breer *et al.*, 1990; Vogt *et al.*, 1991a, b) were the first indication that OBPs were members of a multigene family. These data on lepidopteran OBPs provided a basis for identifying the first OBPs in *D. melanogaster* (OS-E and OS-F, McKenna *et al.*, 1994; PBPRP1-5, Pikielny *et al.*, 1994; LUSH, Kim *et al.*, 1998). Collectively, the OBPs described above might be viewed as the "gold standard" OBPs against which all others are compared.

14.2.1 The OBP gene family

Table 14.2 lists the majority of sequenced insect OBPs available at the time of writing (August 2002), including one from the cockroach (Riviere *et al.*, 2003) and three from termite (Ishida *et al.*, 2002b). Those OBPs which have actually been shown to bind ligands are marked, as are those which have been localized to chemosensory sensilla. The rest of the proteins were identified as OBPs solely based on their sequence similarity to the "gold standard proteins," including the presence of six cysteins (although the "six cystein rule" has not been strictly followed). The main tool for such identifications has been the BLAST program which compares a sequence of interest with sequences stored in a database. BLAST analyses return matches along with a numerical significance value (e-value) which reflects the probability that a match is due to random chance; an e-value of <0.05 is considered significant (Karlin and Altschul, 1990).

Published analyses of the fully sequenced *D. melanogaster* genome have included 25, 34, 38 and 51 genes as candidate members of the OBP family, the numbers depending largely on the criteria applied (Vogt *et al.*, 2002; Galindo and Smith 2001; Graham and Davies, 2002; Hekmat-Scafe *et al.*, 2002). This approach, currently being applied to the *A. gambiae* genome, is identifying genes as OBPs which are quite divergent from the "gold standard" OBPs mentioned above. Twenty-five *Agam*OBPs are listed in Table 14.2, but nearly 60 sequences have recently been submitted as candidate OBP genes based on a thorough review of the *A. gambiae* genome. Both the *D. melanogaster* and *A. gambiae* OBP genes are distributed throughout the chromosomes, although many OBP-related genes reside in clusters. Such distribution suggests that the gene family has expanded by gene duplication and become distributed by translocation events; the evolutionary history of these genes is often evident in the conservation of specific exon/intron boundaries (e.g. Hekmat-Scafe *et al.*, 2002; Vogt *et al.*, 2002; Vogt, 2002).

Not all OBPs are involved in odor detection. The genes included in Table 14.2 include serum proteins from the medfly *Ceratitis capitata* (Y08954 and Y19146) and beetle *Tenebrio molitor* (THP12); accessory gland proteins from *T. molitor* (B1 and B2), and the *Drosophila* protein PBPRP2 which is localized in non-chemosensory tissues or spaces (Park *et al.*, 2000). In a recent study by Galindo and Smith (2001), the expression of 34 candidate *D. melanogaster* OBPs was characterized using an approach (GAL4/UAS) where a presumptive regulatory region of an OBP gene was allowed to drive expression of a reporter gene in a transgenic animal. Nine of these OBP-regulatory constructs expressed only in "chemosensory tissue" (sensilla of antennae and palps), four expressed only in "gustatory tissue" (sensilla of legs and wings), nine expressed in both tissues, five expressed in non-sensillar tissue, and seven did not express at all, perhaps due to inappropriate choice of regulatory region or the genes in question being pseudogenes (Galindo and Smith, 2001). Clearly, though many OBP genes are assumed to express proteins which have an important role in odor detection, assigned membership in this gene family does not necessarily imply an olfactory or even chemosensory function, although it should imply common ancestry (i.e. the genes all derived from some common ancestor). The OBP gene family was presumably derived through duplication of a gene with a function that was likely not chemosensory, and the family has expanded through subsequent gene duplications. Alleles of these duplicated OBP genes accumulated, and evolutionary selection acted on these alleles leading to changes in the properties and functions of the predominant alleles in the species. Speciation events allowed allelic selection to proceed in manners supporting unique olfactory behaviors (life histories) of individual species. Depending on the function (non-chemosensory or chemosensory) of the founder gene(s), subsequent lineages diverged to either chemosensory or non-chemosensory functions. The genes currently listed as OBPs of *D. melanogaster* and *A. gambiaea* include sequences with 4–12 cysteins, differing significantly from the six cysteins of the gold standard OBPs, differences which no doubt influence the function of these proteins.

14.2.2 Properties of OBPs

In general, OBPs are small, water soluble, extracellular proteins around 14 kDa and that range between 120 and 150 amino acids long. OBPs presumably bind small ligands, and if involved in odor detection, those ligands are presumably odor molecules. OBPs are expressed with leader sequences which are removed during secretion. Most OBPs contain six cysteins in a certain asymmetric pattern which has become a diagnostic feature of OBPs; these cysteins form disulfide bonds which stabilize the functionally relevant three-dimensional structure of the protein (Leal *et al.*, 1999; Briand *et al.* 2001a). OBPs involved in odor detection are expressed in sensilla support cells and secreted by these cells into the aqueous sensilla lumen.

Binding of odor to OBP has been demonstrated for a number of pheromone–PBP systems (e.g. Vogt and Riddiford, 1981; Vogt *et al.*, 1989; Wojtasek *et al.*, 1998, 1999; Maïbèche-Coisné *et al.*, 1998a; Jacquin-Joly *et al.*, 2000, 2001; Nikonov *et al.*, 2002). Studies comparing multiple PBPs in the gypsy moth *L. dispar* and the silk moth *A. pernyi* have shown selective and differential binding between PBPs and pheromone components (Vogt *et al.*, 1989; Du *et al.*, 1994; Du and Prestwich, 1995; Plettner *et al.*, 2000; Kowcun *et al.*, 2001; Bette *et al.*, 2002). Binding constants between pheromone and PBP have been estimated as low as 60 nM (Kaissling, 1986) but more typically in the range 0.5–2 μM (Vogt and Riddiford, 1986; Du and Prestwich 1995; Kowcun *et al.*, 2001; Campanacci *et al.*, 2001a; Bette *et al.*, 2002). Kds of 20–30 μM have also been determined for certain compounds and conditions (Bette *et al.*, 2002; Kowcun *et al.*, 2001); the Kd for a PBP of *Mamestra brassicae* was estimated at 0.2 μM (Campanacci *et al.*, 2001a). Pheromone–PBP interactions are strongly influenced by pH (Wojtasek *et al.*, 1998; Kowcun *et al.*, 2001), leading to suggestions that lower pH environments in the microenvironment near the neuronal membrane might destabilize the pheromone–PBP complex to release pheromone to receptors (Wojtasek *et al.*, 1998; Horst *et al.*, 2001). These studies are discussed in greater detail in the chapters by Leal and Plettner. For OBPs not involved in pheromone detection, odor binding has only been demonstrated from ApolGOBP2 (Vogt and Riddiford, 1981) and MsexGOBP2 (Feng and Prestwich, 1997). There is no odor binding data for any of the dipteran OBPs. However, there is behavioral-genetic evidence that the OBP LUSH protein of *D. melanogaster* is required for the correct behavioral response to ethanol (Kim *et al.*, 1998; Kim and Smith, 2001).

The structure of two OBP-related proteins has been determined: PBP1 of the moth *Bombyx mori* (*Bmor*PBP1; Sandler *et al.*, 2000) and THP12 of the beetle *Tenebrio molitor* (Rothemund *et al.*, 1999; Graham *et al.*, 2001). *Bmor*PBP1 is known to be located within pheromone-sensitive sensilla (Steinbrecht, 1998) and to bind sex pheromone (Wojtasek and Leal, 1999b; Leal, 2000). The protein includes six alpha-helix segments and a hydrophobic binding pocket formed by four antiparallel helices, and may function as a dimer (Sandler *et al.*, 2000). *Bmor*PBP1 is stabilized through disulfide bridges between Cysteins 1-3, 2-5, 4-6 (Leal *et al.*, 1999), a disulfide bridge pattern has also been observed for the honeybee OBP ASP2 (Briand *et al.*, 2001a). THP12 is a hemolymph protein; while the protein is speculated to be a ligand-binding protein, the ligand(s) it binds is unknown; this protein also contains six alpha-helix segments and a presumed binding pocket (Rothemund *et al.*, 1999; Graham *et al.*, 2000). *Bmor*PBP1 and THP12 are quite different in sequence, so the conservation of structure between such highly divergent OBPs suggests this overall structure may be conserved among all insect OBPs. This assumed conservation allows one to use the structurally based ligand binding information for *Bmor*PBP1 to model the odor binding pockets of other OBPs, including the prediction of specific amino acids that may be involved in odor binding (e.g. Peng and Leal, 2001).

14.2.3 Evolutionary genomics of OBPs

Structural relationships between different OBPs can be explored to some extent using phylogenetic analysis tools (Figure 14.3). Such analyses compare amino acid sequences emphasizing similarities and differences. Amino acid sequences are aligned using programs such as ClustalX (Thompson *et al*., 1997), and trees are constructed using programs such as PAUP (Swofford, 2000) or MEGA (Kumar *et al*., 2001) and methods such as neighbor joining (Saitou and Nei, 1987) or maximum likelihood (Nei and Kumar, 2000) available within these programs. The results of such analyses, when applied to proteins rather then nucleic acids, usually do not imply any evolutionary history, but rather suggest properties that certain proteins may share. Other criteria must be considered to infer evolutionary relatedness and history such as a demonstrated positive selection (e.g. significant non-synonymous differences in codon nucleotides) or conservation of gene structure (e.g. exon/intron boundaries, relative positions in chromosomes, identification of neighboring genes; Vogt *et al*., 2002; Vogt, 2002).

The sequence similarities of most OBPs listed in Table 14.2 are summarized in the neighbor-joining tree illustrated in Figure 14.3. Overall, these OBPs are highly divergent, indicated by the long lengths of the branches. Several branches include multiple taxa which define specific similarity groups, and such groupings may suggest related functions. One such group is the Lepidoptera specific PBP/GOBP gene family. Other similarity groups include a group of ant OBPs with a reputed role in governing the social behavior of *Solenopsis* sp. (fire ants and their relatives; Ross and Keller, 1998; Ross, 1997; Bourke, 2002; Keller and Parker, 2002; Krieger and Ross, 2002), two groups that include the *Drosophila* OBPs LUSH and OS-E/OS-F, and a group that includes the lepidopteran ABPX proteins. Both the LUSH and ABPX groups include genes belonging to multiple insect orders, a feature that may suggest that these proteins have a functional role common to or of central importance to diverse insects. However, most branches do not include OBPs from multiple Orders, suggesting that selection acting on these genes is strongly biased by features unique to the Orders, Families or genera represented.

The PBPs and GOBPs of Lepidoptera represent a group that is not present in other insect Orders, but these PBPs and GOBPs are clearly only a subgroup of lepidopteran OBPs. Their absence from Diptera (a sister lineage) and the relatively recent emergence of Lepidoptera (implied by the fossil record) suggest that the PBP/GOBP genes may have arisen within Lepidoptera, or at least within the lepidopteran/trichopteran lineage (Figure 14.2B). PBP, GOBP1 and GOBP2 genes comprise three distinct subgroups; amino acid sequence identities between PBPs and GOBPs are consistently between 25 and 30 percent, and between GOBP1s

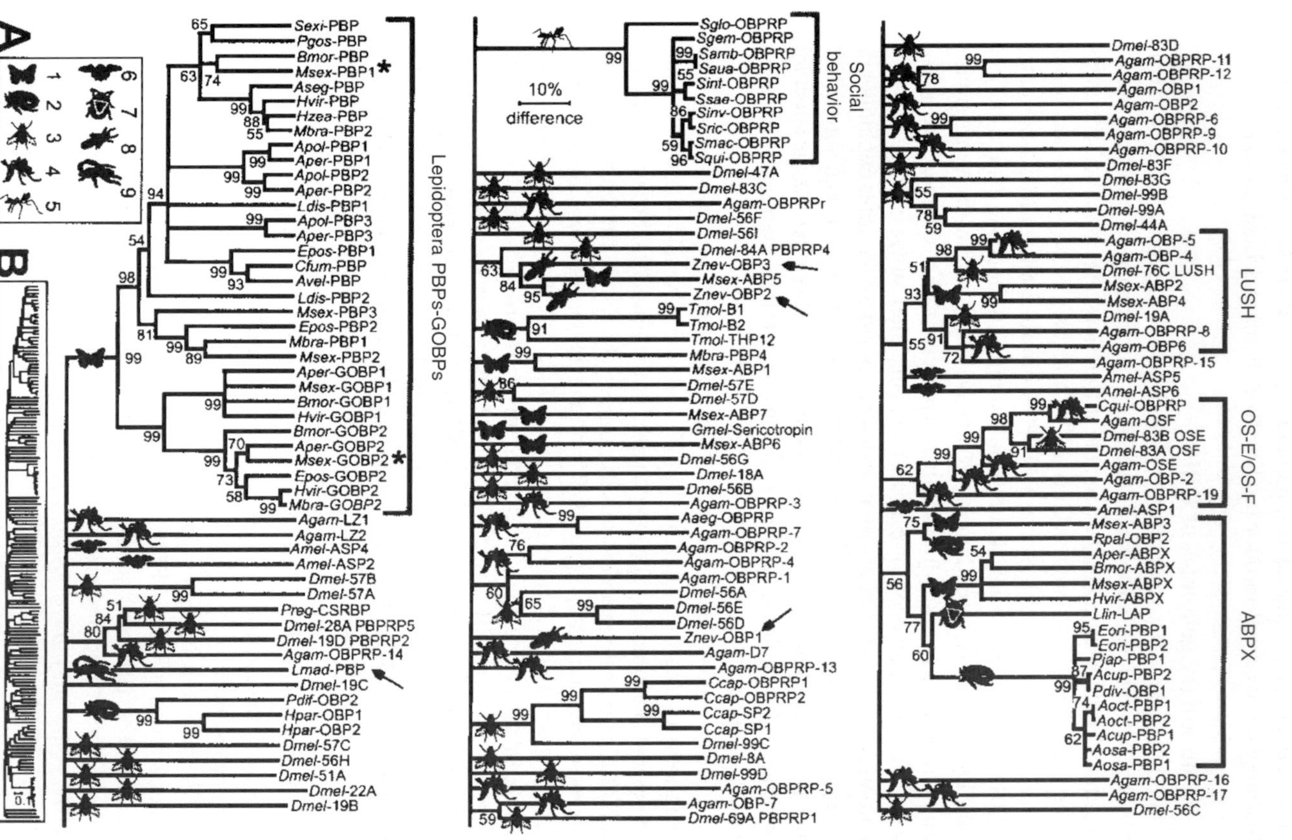

Figure 14.3 A neighbor-joining tree of most OBPs listed in Table 14.2; bootstrap support is based on 1000 replicates and only branches with >50% support are shown. A single tree is shown, broken into thirds to fit the page. Insert A is a key to the taxa. Insert B is the uncollapsed tree (taxa not shown) to illustrate full branch lengths (percent sequence differences). A size bar indicating 10% sequence difference is shown. Named clusters (e.g. ABPX) are referred to in the text.

and GOBP2 consistently about 50 percent. GOBP1s and GOBP2s are highly conserved (around 90 percent identical within each group) while PBPs are highly divergent (20–95 percent identical). Also, multiple PBPs have been identified in individual species while only single genes are yet known for GOBP1s and GOBP2s, suggesting that gene duplications are being retained for PBPs but not GOBPs and that selection is acting very differently on these subgroups as might be expected for systems supporting reproduction (PBPs) vs general maintenance (i.e. feeding, GOBPs).

The genes encoding PBP2 and GOBP2 of *M. sexta* (*Msex*PBP1 and *Msex*GOBP2) are adjacent to one another, with coding regions separated by about 2000 bp, and both genes have identical exon/intron structures (Vogt *et al.*, 2002a). Proximity and conserved gene structure argue strongly that PBP1 and GOBP2 derived from a gene duplication, yet the two proteins have very different sequences and temporal and spatial expression patterns. *Msex*PBP1 and *Msex*GOBP2 express in different types of sensilla (see Figure 14.4); *Msex*PBP1 expresses only in adult moths while *Msex*GOBP2 expresses in both larvae and adults, but the expression of both is regulated by ecdysteroid hormones (Vogt *et al.*, 1993, 2002). If PBP1 and GOBP2 did derive from a gene duplication, then presumably GOBP1 derived from a subsequent duplication event involving GOBP2. The origins of multiple PBPs are less clear. Multiple PBPs may have derived from subsequent duplications involving PBP1, or the PBP1/GOBP2 pair may have derived from a duplication involving one of these other PBPs. PBP diversification may have been driven by reproductive selective pressures within species, genera or families (Merrit *et al.*, 1998); however, this does not appear to be the case with the GOBPs which are both highly conserved and broadly distributed among the lepidopteran families. An in-depth comparative study of the PBPs and GOBPs across the Lepidoptera might reveal much of the olfactory evolution of this insect Order.

OS-E and OS-F of *Drosophila* are also presumably derived from a gene duplication based on the neighboring relationship of their genes (separated by about 1000 bp), similar exon/intron structure, and similar sequence. However, unlike *Msex*PBP1/*Msex*GOBP2, OS-E and OS-F coexpress in olfactory sensilla of the *Drosophila* antenna. Orthologs of OS-E and OS-F have been tentatively identified in the malaria mosquito *A. gambiae*, suggesting that the OS-E/OS-F gene duplication preceded the divergence of flies and mosquitoes within the Diptera (Vogt, 2002). A gene with significant sequence similarity to OS-E and OS-F has also been identified in the mosquito *Culex quinquefasciatus* (Ishida *et al.*, 2002); the sequence conservation of these presumed OS-E/OS-F orthologs among different Dipteran groups suggests their singular importance in the olfactory biology of these species. Unlike in *Drosophila*, the presumptive *A. gambiae* OS-E/OS-F genes are separated by about 30 million nucleotides suggesting a chromosomal translocation event has occurred; it will be interesting to learn whether this separation has also resulted in a divergence in expression/function between these *A. gambiae* genes.

The PBP/GOBP genes of Lepidoptera and OS-E/OS-F genes of Diptera are just two stories of duplication and expansion of olfactory genes. A look at the distribution of ORs and OBP and SNMP homologues on the *Drosophila* chromosomes reveals a rich evolutionary history of gene duplication and translocation in the context of sensory biology and behavioral selection (see Figure 14.8). Increased access to genomes permits the identification gene orthologs and the following of gene change across insect Families and Orders, and will allow us to hypothesize the function in one species based on the characterized function of a gene ortholog in another.

14.2.4 Allelic variation of PBPs

Several efforts have examined allelic variation of PBPs between distinct populations. We characterized electrophoretic variants (possible multiple alleles) of both *Apol*PBP1 and *Apol*SE between populations of *A. polyphemus* from New York (Long Island) and Wisconsin; however, this apparent allelic variation appeared high within, but similar between, each population and nothing definitive could be concluded regarding how this variation might affect behavior (Vogt, 1987; Vogt and Prestwich, 1988). LaForest and colleagues (1999) characterized PBP genes from individuals of two populations of the noctuid moth *Agrotis segetum* which use the same pheromone components but in different ratios; however, only neutral sequence variation was observed in the PBP sequences and again nothing definitive could be concluded regarding how such variation might affect pheromone perception and behavior. Similar results were obtained from studies of distinct populations of the European corn borer *Ostrinia nubilalis* (Willett and Harrison, 1999a, b). Positive selection was observed for PBPs of species using different pheromone structures, but it was not clear that this selection resulted from the changes in pheromone components (Willet, 2000a, b). A limitation of all three studies is that each characterized only one PBP; closer examination has consistently demonstrated that lepidopteran species have several PBPs (Table 14.2). It is likely that multiple PBPs derived from relatively recent gene duplications which occurred within specific lepidopteran lineages rather than at the base of the lepidopteran lineage (Merritt *et al.*, 1998). If this is true, then it is also possible that in certain species PBP genes did not duplicate or that duplicated genes have become lost, but either is difficult to know in the absence of a full genome sequence. If positive selection has occurred among the PBPs, it may be discernible only by analyzing the larger repertoire of PBP genes within a species.

14.2.5 Differential expression of OBPs

The differential expression of OBPs within specific classes of sensilla, along with the demonstrated ability of OBPs to bind pheromone and non-pheromone odors, are strong evidence that OBPs play an important role in odor detection. Differential expression was first noted for the PBPs and GOBPs of Lepidoptera, where PBPs were present in much higher levels in males vs females in contrast to GOBPs which were present equally in both sexes (Vogt *et al.*, 1991a). The asymmetric association of PBPs with male antennae was consistent with their role in pheromone detection because only males were reported to detect pheromone. The equal distribution of GOBP1 and GOBP2 between males and females was consistent with the detection of general odorants (plant volatiles, non-pheromones) by both sexes and led to the names given to these two OBP classes. We now know that PBPs are not exclusively expressed in male antennae (e.g. Callahan *et al.*, 2000), although this feature seems stronger in some Families (e.g. Noctuidae) than in others (e.g. Saturniidae and Sphingidae). In *M. sexta* (Sphingidae), *Msex*PBP1 expresses in pheromone-sensitive long trichoid sensilla of male antennae as well as a small number of discrete but otherwise uncharacterized sensilla in female antennae; *Msex*GOBP2 expresses in basiconic sensilla of both male and female antennae, and in olfactory sensilla of larval antennae and palps (Vogt *et al.*, 2002; Figure 14.4).

The distribution of PBPs, GOBP1, GOBP2 and ABPX proteins have been characterized in several moth species using immunohistochemistry and electronmicroscopy; PBP proteins are present in the lumen of long trichoid sensilla, GOBP2 and ABPX proteins are present in the lumen of subsets of basiconic sensilla and GOBP1 can be present in both trichoid and basiconic sensilla (e.g. Steinbrecht *et al.*, 1995; Steinbrecht, 1996; Laue and Steinbrecht, 1977; Maida *et al.*, 1999; Zhang *et al.*, 2001). This broad expression of GOBP1 in both basiconic and trichoid sensilla was also observed in *M. sexta* (Vogt *et al.*, 2002). A similar pattern has been described in Diptera. In *Drosophila*, OS-E, OS-F and LUSH coexpress in trichoid sensilla, OS-E and OS-F but not LUSH coexpress in intermediate sensilla, and PBPRP5 expresses in a subset of basiconic sensilla (Hekmat-Scafe *et al.*, 1997; Park *et al.*, 2000; Shanbhag *et al.*, 1999, 2001a,b). Galindo and Smith (2001) tested the expression patterns of 34 presumptive OBP promoters in *Drosophila*; nine drove expression in olfactory tissue alone (antenna and palps) while another nine drove expression in both olfactory and gustatory (wings and legs) tissues. These studies show that OBPs differentially express in sensilla with diverse odor specificities. Several studies have shown that OBPs interact with odor molecules differentially (Vogt *et al.*, 1989; Du *et al.*, 1994; Prestwich *et al.*, 1995; Feng and Prestwich, 1997; Plettner *et al.*, 2000; Kowcun *et al.*, 2001; Bette *et al.*, 2002). Taken together, these observations suggest that subtypes of OBP, differentially associating with subtypes of olfactory sensilla, support the unique specificities of the sensilla in which they are expressed.

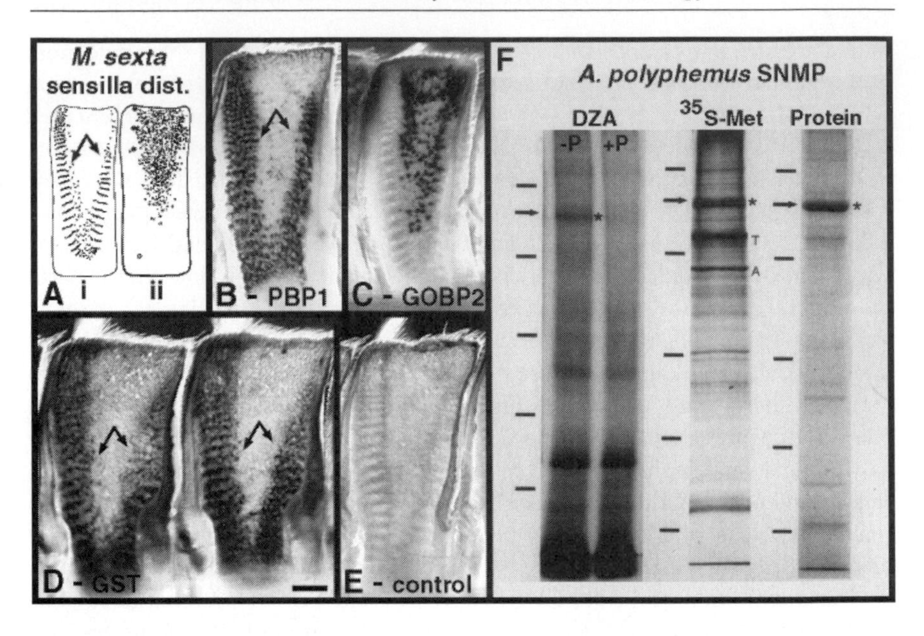

Figure 14.4 *In situ* hybridizations of *Msex*PBP1, *Msex*GOBP2 and *Msex*GSTolf in adult male *M. sexta* antenna (A–E) and visualizations of sensory membrane proteins of *A. polyphemus*, including *Apol*SNMP1 (F). **Expression of *M. sexta* PBP1, GOBP2 and GST.** A shows the non-overlapping distribution of pheromone-sensitive long trichoid sensilla (i) vs other olfactory sensilla (after Lee and Strausfeld, 1990). B and C show the non-overlapping expression of *Msex*PBP1 and *Msex*PBP2, and D shows the PBP-like expression of *Msex*GSTolf (Rogers *et al.*, 1999; Vogt *et al.*, 2002a). E shows a control *in situ* using an non-insect probe. Arrows point to pheromone-sensitive sensilla. *Scale bars* 100 μm. Modified from Rogers *et al.* (1999). **Visualizations of *A. polyphemus* SNMP1.** F shows three experiments showing proteins of the receptor membranes of olfactory neurons of male *A. polyphemus*, visualized by tritiated photoaffinity analogue of sex-pheromone, ^{35}S-methionine, or the protein stain Coomassie blue. "DZA": dendrite membranes were photolabeled with [^{3}H]-DZA in the absence (–P) and presence (+P) of excess non-radioactive pheromone; asterisk marks SHMP69 thought to be a candidate pheromone receptor (Vogt *et al.*, 1988). "^{35}S-Met": dendrite membrane proteins were examined after animals were injected with "^{35}S-Met", revealing a prominent band near 69 kDa (asterisk) as well as tubulin (T) and actin (A) (Vogt, unpublished; for details of method see Vogt *et al.*, 1989). "Protein": membranes were isolated from olfactory sensilla isolated from 800 male antennae of animals attracted to female moths releasing sex pheromone. One-tenth of this preparation was analyzed revealing an abundant protein near 69 kDa (asterisk); this band (*Apol*SNMP1) was electroblotted and sequenced (Rogers *et al.*, 1997). Arrows identify the 69 kDa MW marker. "DZA" and "^{35}S-Met" are fluorograms while "Protein" is Coomassie blue stain.

14.3 Sensory appendage proteins – SAPs

Another group of proteins, small and water soluble but unrelated to the OBPs, may nevertheless have a function similar to OBPs in insect olfaction. Referred to as OS-D (olfactory specific-D) (McKenna *et al.*, 1994), CSP (chemosensory protein) (e.g. Mameli *et al.*, 1996; Maïbèche-Coisné *et al.*, 1997; Angeli *et al.*, 1999; Marchese *et al.*, 2000; Picimbon *et al.*, 2000) or SAP (sensory appendage protein) (Robertson *et al.*, 1999), these proteins have been identified in several orthopteroid and endopterigote species suggesting they are distributed throughout the Neoptera (see Figure 14.2). A single insect species may have multiple SAP genes, but these are highly conserved. SAPs are also highly conserved between species as evolutionarily distant as walking stick (phasmid), cockroach, and moth (e.g. Mameli *et al.*, 1996; Picimbon and Leal, 1999; Jacquin-Joly *et al.*, 2001); this is in stark contrast to OBP genes which are highly divergent both within and between species. SAPs are expressed in antennal and non-antennal regions, but have been visualized within the lumen of olfactory sensilla and are capable of binding odorants (Jacquin-Joly *et al.*, 2001; Bohbot *et al.*, 1998). The structure of a moth SAP has been determined in both free and ligand-bound states (Campanacci *et al.*, 2001b; Lartigue *et al.*, 2002; Mosbah *et al.*, 2002). The function of these proteins remains less clear than that of the OBPs, though one suggestion is that the SAPs, showing less ligand specificity than the OBPs, might prevent pheromone from becoming lost to neuronal membranes (Lartigue *et al.*, 2002). The SAP/CSP family is discussed more fully by Nagnan-Le Meillour and Jacquin-Joly, Chapter 17 and Picimbon, Chapter 18, in this volume.

14.4 Odor degrading enzymes – ODEs

Signal termination plays a critical role in all chemically mediated biological processes, and this is no less so in odor detection. The process of pheromone degradation has been studied for some years, at least as far back as Kasang (1971) in *B. mori* and Ferkovich *et al.* (1973a, b) in *Trichoplusia ni*. Since then, a few pheromone degrading enzymes have been identified and characterized in detail and the general principle of pheromone degradation has become well established. Yet surprisingly little work is being done to investigate ODEs, in contrast with current efforts on ORs and OBPs. One reason to expand efforts on ODEs is their potential in insect control. If it is true that odors are perceived as precise mixtures, then the targeted inhibition of the ODE for a specific component should alter the blend ratio within a sensillum resulting in misperception of the odor. Of the three protein classes (ORs, OBPs and ODEs), ODEs may be the least specific and thus the more generally targetable protein for behavioral inhibition.

ODEs are enzymes selectively evolved to degrade odor molecules (see Prestwich, 1987). However, to call an enzyme an ODE it is not enough to merely demonstrate the ability of an enzyme to degrade an odor; that enzyme has to be shown to reside in a space that is relevant to odor detection. This principle should apply to OBPs and ORs as well. In general, those ODEs studied have been ones which attack specific functional groups, such as acetate esters, aldehydes, alcohols, ketones and epoxides. ODEs attacking such different functional groups presumably belong to different gene families. The 55 kDa antennal specific esterase (*Apol*SE, see below) that degrades the acetate–ester pheromone component of the silkmoth *A. polyphemus* (e.g. Vogt and Riddiford, 1981; Vogt *et al.*, 1985) is most certainly a member of some gene family that encodes insect esterases. Similarly, the 150 kDa antennal-specific aldehyde oxidases (AOXs) that degrade the aldehyde components of the *A. polyphemus* and *M. sexta* pheromones (e.g. Rybczynski *et al.*, 1989, 1990) must belong to a distinct gene family of insect AOXs. Little is known about ODE genes, except for the antennal-specific glutathione-S-transferase (*Msex*GSTolf; Rogers *et al.*, 1999; see below).

The following are discussions of four categories of ODEs, based on their location in the animal. To some degree this has functional significance, distinguishing between ODEs of the sensillum lumen (soluble or membrane bound), support cell cytosol and body surface.

14.4.1 Category 1. Soluble extracellular ODEs of the sensillum lumen

The early studies of the fate of pheromone within the antenna were the first biochemical studies of insect odor detection. At the time, much seemed known about pheromone detection in moths: the pheromone of *B. mori* had been identified and radiolabeled (Butenandt *et al.*, 1959; Kasang, 1968); pheromone detection had been characterized by electrophysiology in *B. mori* (e.g. Schneider, 1969); the ultrastructure of pheromone sensitive sensilla had been described for *B. mori* (e.g. Steinbrecht and Müller, 1971). Adam and Delbrück (1968) had recently used the *B. mori* pheromone system to justify their "reduction in dimension" hypothesis for ligand-receptor targeting. Steinbrecht and Müller (1971) had suggested that pore-tubules might serve as conductive structures for the transport of pheromone molecules from the sensillum surface to receptors of the olfactory neurons (the first suggestion that OBPs performed this transport was made in 1985 by Vogt *et al.*). With this background, a series of studies were performed where radiolabled bombykol was applied to intact male *B. mori* (Bombycidae) antennae and the products analyzed following solvent extraction; these studies observed the general degradation of the alcohol bombykol to its acid metabolite, but identified no specific enzymes or enzyme activities (Kasang 1971, 1973; Kasang and Kaissling 1972; Kasang and Weiss 1974; Kasang *et al.*, 1989). About the same time an independent series of studies examined degradation of the *T. ni* (Noctuidae) pheromone from the active acetate ester to the inactive alcohol metabolite; these studies demonstrated esterase activities in the antenna but did not identify any antennal specific enzymes (Ferkovich *et al.*, 1973a, b,

1980, 1982; Mayer, 1975; Taylor *et al.*, 1981). Although no specific enzymes were identified, these efforts drew attention to the issues of pheromone degradation as an important component of the pheromone detection process.

14.4.1.1 Antennal specific esterase

The antennal-specific sensilla esterase (SE) of *A. polyphemus* (*Apol*SE) was the first identified pheromone degrading enzyme (Vogt and Riddiford, 1981). The activity of *Apol*SE was visualized on non-denaturing PAGE, and was shown to be specific to the male antennae and to be located in the sensillar fluid (Vogt and Riddiford, 1981). *Apol*SE molecular mass was estimated at about 55 kDa (Klein, 1987). *A. polyphyemus* was reported to have a two-component pheromone consisting of 9:1 ratio of acetate:aldehyde (Kochansky *et al.*, 1975); *Apol*SE was shown to degrade the acetate component. *Apol*SE was subsequently purified and subjected to kinetic analysis (Vogt *et al.*, 1985). We established a spectrophotometric assay using alpha- and beta-napthyl acetate as substrates, and a radioactive-TLC assay using tritiated pheromone. *Apol*SE strongly preferred beta- over alpha-napthyl acetate, was inhibited by the volatile trifluoroketone 1,1,1-trifluoro-2-tetradecanone (IC_{50} = 5 nM), and, most importantly, degraded the sex pheromone with unexpected aggression. Making adjustments for the concentrations and volumes within a sensillum lumen, we conservatively estimated the *in vivo* half-life of pheromone to be 15 msec in the presence of this enzyme. The observation of rapid pheromone degradation prompted us to propose a new model for pheromone reception, in which OBPs served as pheromone transporters (in place of pore-tubules), and enzymes degraded pheromone molecules to rapidly terminate these odor signals within the sensillum (see Figures 14.1 and 14.7).

*Apol*SE was used to assess interactions between pheromone and PBP (Vogt and Riddiford, 1986a). *Apol*SE and *Apol*PBP1 were purified from antennae, and the ability of a fixed concentration of *Apol*SE to degrade pheromone was examined in various concentrations of PBP1 and pheromone. *Apol*SE activity was unaffected under conditions where PBP1 concentrations exceeded pheromone concentrations by 1000 times, suggesting the PBP did not provide general protection from *Apol*SE, and was not binding pheromone irreversibly. Reduced activity was observed at all pheromone concentrations when PBP1 exceeded 1 μM; we suggested that this transition was a consequence of protection during binding site-occupancy and that the Kd of the pheromone–PBP interaction must be around 1 μM. The activity of *Apol*SE below 1 μM PBP suggested to us that pheromone–PBP binding was either very slow or very ephemeral (binding and release occurred rapidly). These conclusions are currently being challenged by views that the pheromone–PBP complex may be quite stable (see earlier discussions).

Since this manuscript was submitted, the *Apol*SE has reportedly been cloned, using PCR primers designed to conserved regions of known insect esterase enzymes (Ishida and Leal, 2002), indicating that *Apol*SE is indeed a member of the larger class of insect esterases. *Apol*SE mRNA was uniquely expressed in male antennae and encoded a 60 kDa protein.

14.4.1.2 Antennal-specific aldehyde oxidase

An antennal-specific aldehyde oxidase (AOX) of *M. sexta* (*Msex*AOX) was the next identified pheromone-degrading enzyme (Rybczynski *et al.*, 1989). The activity of *Msex*AOX was visualized on non-denaturing PAGE, and was shown to be antennal specific but present in sensilla of both male and female antennae. *Msex*AOX was observed as a dimer with a combined estimated molecular mass of 295 kDa. *M. sexta* uses a multicomponent pheromone consisting exclusively of aldehydes including bombykal (Starratt *et al.*, 1979; Tumlinson *et al.*, 1989, 1994); *Msex*AOX was shown to degrade bombykal to its carboxylic acid. Both TLC and spectrophotometric assays were established and a variety of substrates and inhibitors were characterized. Making adjustments for the concentrations and volumes within a sensillum lumen, the *in vivo* half-life of pheromone was estimated at 0.6 msec in the presence of this enzyme (Rybczynski *et al.*, 1989).

Many species use pheromone components with chemically diverse functional groups, and thus presumably require multiple ODEs. The pheromone of *A. polyphemus* is a blend of acetate and aldehyde components (Kochansky *et al.*, 1975), and that of *B. mori* a blend of alcohol and aldehyde (Kasang *et al.*, 1978). Both *A. polyphemus* and *B. mori* have antennal-specific AOXs; both proteins were identified in antennal extracts, and *Apol*AOX was identified in extracts of isolated pheromone sensilla of male antennae (Rybczynski *et al.*, 1990). *Apol*AOX and *Bmor*AOX are both high MW proteins, similar to *Msex*AOX, and are present in male and female antennae. Both *Apol*AOX and *Bmor*AOX were shown to degrade bombykal (Rybczynski *et al.*, 1990). *B. mori* antennae thus contain both alcohol oxidase or dehydrogenase activity (Kasang *et al.*, 1989) and antennal-specific AOX activities, and *A. polyphemus* sensilla contain both antennal-specific SE and AOX activities. *Apol*SE is only found in male antennae, while *Apol*AOX is abundant in both male and female antennae, suggesting that *Apol*SE was selected to function specifically in pheromone degradation while *Apol*AOX has the multiple function of inactivating both pheromone and plant volatile odorants (Rybczynski *et al.*, 1990).

Tasayco and Prestwich (1990a, b, c) studied AOX and aldehyde dehydrogenase (ALDH) activities in the antennae of noctuid moths *Heliothis verescins*, *Heliothis subflexa* and *Helicoverpa zia*. Antennal-specific AOXs were observed in both males and females; MWs were similar to those of *M. sexta*, *A. polyphemus* and *B. mori*. The noctuid enzymes also converted aldehyde pheromone components to their corresponding carboxylic acid without the aid of additional cofactors. Antennnal-specific ALDH enzymes were detected through radiolabeling with a substrate analogue, but the location of the enzymes within the antenna was not determined. These ALDH enzymes required the cofactor NAD(P^+). *Apol*SE and

*Apol*AOX, as well as the other AOXs described above, function without the assistance of any cofactors, enabling them to act autonomously in the extracellular environment of the sensillum lumen where access to cellular metabolites might be limited. The ALDH requirement for cofactors may suggest these enzymes are located in the cytoplasm of support cells where cofactors would be available. Antennal ALDH may serve as a biotransformation enzyme for pheromones and possibly xenobiotics in a manner similar to the antennal specific GST of *M. sexta* (Rogers *et al.*, 1999; see below).

Several studies have reported slow rates of pheromone degradation on antennae (Kasang, 1971, 1974; Kasang and Kaissling, 1974; Kanaujia and Kaissling, 1985; Klun *et al.*, 1991, 1996, 1998). These studies typically applied excessively high (non-physiological) concentrations of pheromone to antennae and subsequently extracted metabolites by incubating treated intact antennae in solvent, and analyzing the resulting solvent extract. It is difficult to interpret these results as no efforts were made to identify the tissue location of observed enzymatic activity. Enzymes capable of degrading pheromone are present in the cuticle, the cytoplasm of cells and the hemolymph in addition to the lumen of sensilla; observed activities may have been from any of these tissue spaces. We have suggested that observed slow degradation rates may be due to significant portions of pheromone migrating into the cuticle matrix rather than directly entering sensilla, and that pheromone thus entering the cuticle matrix might exit and encounter enzymes at slow rates and thus be degraded at slow rates (Vogt *et al.*, 1985; Vogt, 1987).

14.4.2 Category 2. Membrane-bound ODEs

Evolution recruits what is available. While it might seem most efficient for ODEs to be soluble and evenly distributed throughout the sensilla fluid, certain classes of enzymes may only be available in membrane-bound forms. Such may be the case with epoxide hydrolases (EH). The epoxide pheromone disparlure can be degraded by EH activity that is antennal specific and associated with isolated sensilla; this activity is not water soluble suggesting that it might instead be associated with neuronal or support cell membranes (Vogt, 1987; Prestwich, 1987; Graham and Prestwich, 1992, 1994). A membrane-bound ODE that degrades purinergic odor molecules has been characterized in lobster (Carr *et al.*, 1990; Trapido-Rosenthal *et al.*, 1987, 1990; Grünert and Ache, 1988; Gleeson *et al.*, 1992). Lobsters, arthropods by all rights, smell many things, including the purines ATP and ADP; the ratio of these molecules is thought to inform the lobster of how recently its food was alive (Zimmer-Faust *et al.*, 1988). Certain olfactory neurons are stimulated by ATP and ADP, and the membranes of these neurons contain enzymes which degrade these purinergic signals (Trapido-Rosenthal *et al.*, 1987, 1990; Gleeson *et al.*, 1992). These enzymes are localized to membrane regions at the base of the sensillum, with their activities oriented towards the lymph cavity.

14.4.3 Category 3. Cytosolic ODEs: multifunctional ODEs inactivating odors and xenobiotics

Olfactory and respiratory tissues of animals are constantly under attack by toxic chemicals from the environment. These xenobiotic compounds have presumably driven the enrichment in such tissues of biotransformation enzymes such as glutathionine-S-transferases (GSTs), cytochrome P-450 oxygenases and UDP-glucuronosyltransferases, and various dehydrogenases and oxidases (for reviews see Dahl, 1988; Mannervik and Danielson, 1988; Clark, 1990; Feyereisen, 1999). Many of these enzymes can also attack odor molecules, and this has led to suggestions in mammals that biotransformation molecules may function in odor signal termination (Nef *et al.*, 1989; Burchell, 1991; Lazard *et al.*, 1990, 1991; Ding *et al.*, 1991; Ben-Arie *et al.*, 1993).

Several biotransformation enzymes have been identified in insects. These enzymes have features suggesting they serve a dual function of attacking both xenobiotics and odor molecules. One of these, *Msex*GSTolf, is an antennal-specific GST in *M. sexta* capable of transforming aldehyde odorants (Rogers *et al.*, 1999). GSTs often function as detoxification enzymes, complexing xenobiotics to glutathione and thereby rendering them harmless (Yu, 1983, 1984, 1989; Fournier *et al.*, 1992; Snyder, 1995). Though present in male and female antennae, in male antennae *Msex*GSTolf is restricted to cells underlying the pheromone-sensitive trichoid sensilla (Figure 14.4a); this restricted expression to pheromone sensilla suggests that *Msex*GSTolf plays a role in pheromone inactivation (Rogers *et al.*, 1999). *Msex*GSTolf mRNA encodes a 219 amino acid protein which lacks an encoded leader sequence; the absence of a leader sequence suggests that GSTolf is not extracellular but rather is retained in the cytoplasm of the cells expressing it. Pheromone molecules entering these cells, perhaps carried by PBPs, would be complexed and inactivated by *Msex*GSTolf and thus prevented from later stimulating the neurons (Figures 14.1 and 14.7). *Msex*GSTolf shares significant similarity with an antennal GST from *B. mor*i (AJ006502; Krieger); *Msex*GSTolf and *Bmor*GSTolf are distinct from other insect GSTs and thus may define a unique class of multifunctional biotransformation enzymes (Rogers *et al.*, 1999).

Antennae of the beetle *Phyllopertha diversa* contain a P450 enzyme that may also serve the dual function of inactivating xenobiotics and pheromone molecules; the enzyme was shown to be male specific and to degrade sex pheromone (Wojtasek and Leal, 1999a). Discussions earlier in this section have already described noctuid ALDHs that require cofactors (Tasayco and Prestwich, 1990a, b). This requirement for cofactors in enzymes such as GSTs, P450s and ALDHs could be difficult to satisfy in the extracellular space of the sensilla lumen and may alone suggest a cytosolic location of these enzymes and possible roles in biotransformation depending on tissue location.

A cytosolic location of an ODE may seem incompatible with its function;

once an odor/pheromone passes out of the sensillum lumen and enters a cytoplasm it would seem effectively inactivated and out of reach from the perspective of the ORs. Perhaps this is not so. The restricted expression of *Msex*GSTolf to pheromone sensilla may be the consequence of strong evolutionary selection acting on pheromone systems to maintain low signal backgrounds. Cytosolic signal degradation would ensure that pheromone molecules entering the support cells could not later diffuse back into the sensillum lymph or activate ORs located in the nearby cell bodies of the olfactory neurons.

14.4.4 Category 4. Cuticular ODEs: surface catabolism of background noise

Airborne pheromone and other odors are hydrophobic and tend to adsorb onto the waxy surface of the insect cuticle. Body surfaces thus can collect odors and become sources of background noise if these odors are later released. Degradation of these surface-bound odor molecules might significantly reduce such signal noise.

Pheromone-degrading esterase activity associates with the scales covering the outer surface of the silkmoth *A. polyphemus* (Vogt and Riddiford, 1986b). Isolated scales were loaded into small cartridges through which volatilized radioactive pheromone was blown ([H^3]6E,11Z-HDA); alcohol metabolites were subsequently extracted indicating that the scales degraded adsorbed pheromone. Isolated scales were suspended in detergent solution containing either alpha-napthyl acetate or pheromone; both compounds were rapidly degraded. The degradative activity was apparently cross linked to the scale cuticle as removal of the scales reversibly removed the enzymatic activity. Boiling scales had only a partial effect on diminishing enzyme activity, suggesting that the presumed enzyme might be structurally stabilized by cross-linking to the cuticle. Activity was inhibited by trifluoroketone 1,1,1-trifluoro-2-tetradecanone and *O*-ethyl *S*-phenyl phosphoramidothiolate (EPPAT), the same compounds that inhibited the pheromone degrading *Apol*SE (Vogt *et al.*, 1985). Scales isolated from different body surfaces of *A. polyphemus* had somewhat different activities, and scales taken from other Lepidoptera had little or no activity compared to *A. polyphemus* suggesting the activity was species specific for the *A. polyphemus* pheromone. These data suggested the activity represented an essentially solid state and fairly indestructible enzyme selected to degrade sex pheromone adsorbed to the body surface, and that the *A. polyphemus* activity was tuned to the *A. polyphemus* pheromone (assuming the other species tested had their own scale esterases). This enzyme may function to reduce the noise that adsorbed pheromone might create if later released at an inappropriate moment.

Surface catabolism of pheromones has been observed anecdotally but rarely studied with passion. Several studies have shown that homogenates or sonicates of general body parts of moths will degrade pheromone (Kasang, 1971; Mayer,

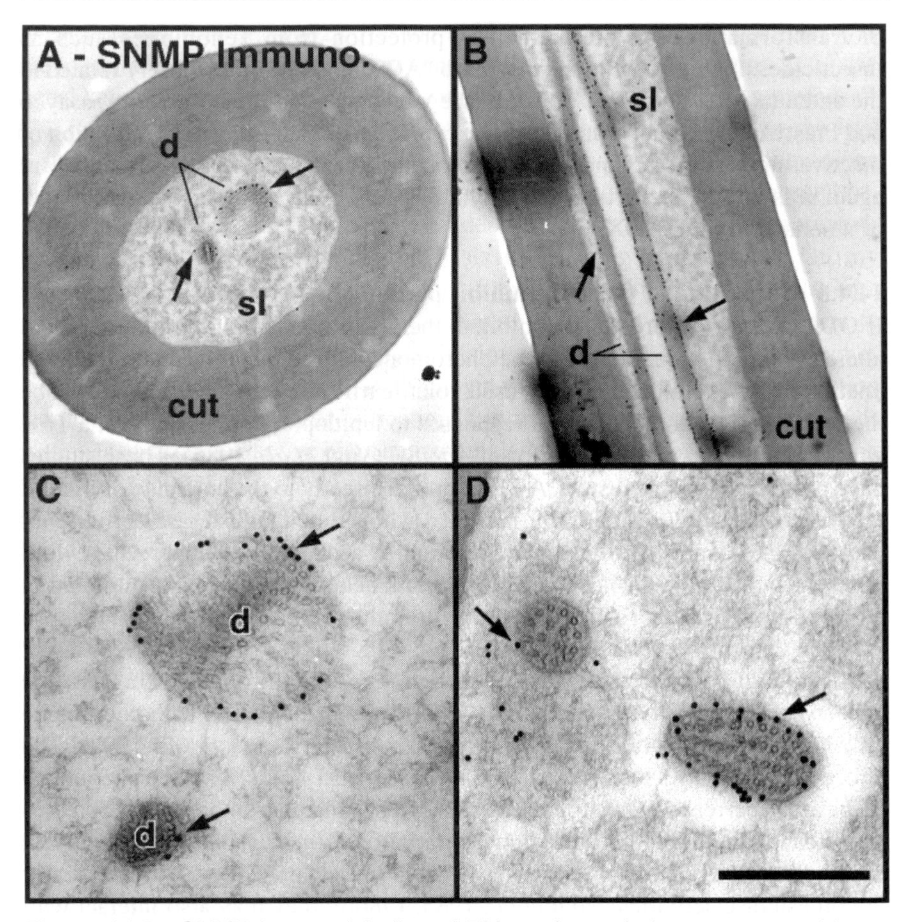

Figure 14.5 SNMP immunolabeling of EM sections of pheromone-sensitive trichoid sensilla. SNMP antibodies were visualized using secondary antibody conjugated to 10 nm colloidal gold particles. A and B include sensillum cuticle; C and D show only dendrites. Sensilla contained two neurons; one neuron consistently showed significantly greater labeling. *Scale bar*: 1.25 μm (A), 2.5 mm (B), 0.5 μm (C, D). Modified from Rogers et al. (2001a).

in *B. mori* or *H. verescins*). SNMPs are about 520 amino acids long; *Apol*SNMP1 is expressed as a 59 kDa peptide that is apparently post-translationally modified to 69 kDa (Rogers *et al.*, 1997, 2001b). The proteins are thought to contain two presumptive transmembrane domains located near the C- and N-terminals and a single large extracellular loop which contains several N-glycosylation groups and presumed disulfide bridges (Rogers *et al.*, 1997, 2001b; Rasmussen *et al.*, 1998; Tabuchi *et al.*, 2000). The amino acid sequence identities between each SNMP1 range from 67 to 73 percent, and the identities between *Msex*SNMP2 and any of the SNMP1s are about 25 percent. Both SNMP1 and SNMP2 were recently identified in *Mamestra brassicae*; *Mbra*SNMP2 and *Msex*SNMP2 are about 65 percent identical (Jacquin-Joly, personal communication).

SNMP expression suggests these proteins play a central role in odor detection. *Apol*SNMP1, *Msex*SNMP1, *Msex*SNMP2 and *Bmor*SNMP1 are all known to be antennal specific (Rogers *et al.*, 1997, 2001b). *Apol*SNMP1, *Msex*SNMP1, *Msex*SNMP2 and *Hvir*SNMP1 are all known to express in olfactory neurons, and *Apol*SNMP1 protein has been visualized in the receptor membranes of these neurons (Figure 14.5) (Rogers *et al.*, 1997, 2001b; Krieger *et al.*, 2002). *Apol*SNMP1 and *Msex*SNMP1 have been shown to express late in adult development and into adult life, after development has completed but coincident with the expression of *M. sexta* OBPs and AOX and the onset of olfactory function (Rogers *et al.*, 1997, 2001b; Vogt *et al.*, 1993; Schweitzer *et al.*, 1976). *Hvir*SNMP1 expression coincides with the expression of OR genes in adult male *H. virescens* (Krieger *et al.*, 2002). The unique and abundant association of SNMPs with the receptor membrane of olfactory neurons at a time when these neurons are capable of detecting and responding to odors suggests that SNMP are involved in odor detection.

What do SNMPs do? The identification of *Apol*SNMP1 followed photoaffinity labeling studies that tentatively identified a 69 kDa protein as a pheromone receptor (Vogt *et al.*, 1987); however, a role as pheromone receptor seems highly unlikely because SNMPs appear to associate with most olfactory neurons, and are neither 7-transmembrane domain receptors nor show the diversity expected for ORs. SNMPs certainly show no similarity to the presumed ORs identified in *D. melangaster*, *A. gambiaea* and *H. virescens* (Clyne *et al.*, 1999; Vosshall *et al.*, 1999; Hill *et al.*, 2002; Krieger *et al.*, 2002). If SNMPs are not ORs, what are they?

SNMPs are homologous to a family of receptor proteins characterized by the human protein CD36 (Oquendo *et al.*, 1989; Greenwalt *et al.*, 1992; Abumrad *et al.*, 1993) (Figure 14.6, see also Rogers *et al.*, 2001a). Among vertebrates, the CD36 family includes proteins known as scavenger receptors including LIMP II (Vega *et al.*, 1991; Sandoval *et al.*, 2000; Tabuchi *et al.*, 2000); SRB1 (Acton *et al.*, 1994, 1996) and CLA 1 (Calvo and Vega, 1993). Members of this family have also been characterized in the insect *D. melanogaster* (Hart and Wilcox 1993; Franc *et al.*, 1999; Kiefer *et al.*, 2002) and the slime mold *Dictyostelium discoideum* (Karakesisoglou *et al.*, 1999; Janssen *et al.*, 2001a, b).

The function of the vertebrate CD36 proteins is becoming clear. In mammals, these proteins are reported to be receptors for both high and low density lipoproteins (HDLs and LDLs), transporters of cholesterol and phosopholipids, thrombospondin receptors; and cell recognition receptors for phagocytosis and recognition of *Plasmodium* (malaria)-infected erythrocytes. Its role in malaria response was the first property ascribed to CD36 (Oquendo *et al.*, 1989); CD36 is expressed

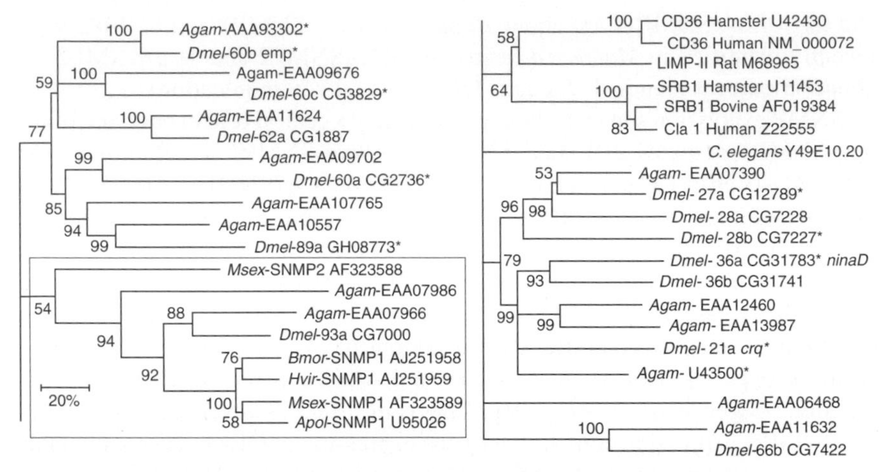

Figure 14.6 A neighbor-joining tree including lepidopteran SNMPs and their homologues in the *D. melanogaster* and *A. gambiae* genomes, as well as representative homologues from vertebrates (mammals) and nematode (*C. elegans*). Box surrounds a similarity group that includes the olfactory specific lepidopteran SNMPs; *Dmel*-CG7000 is also known to be antennal specific (Vosshall, personal communication). Branches are collapsed to 50% bootstrap value or greater, based on 1000 replicates.

in endothelial cells of blood vessels and is responsible for cytoadherence of infected erythrocytes to these endothelial cells (Gamain *et al.*, 2002; Udomsangpetch *et al.*, 2002). The role of these proteins most studied is their involvement in heart disease, and specifically the formation of vascular plaques in atherosclerosis. Plaque formation involves lipid/cholesterol uptake by macrophages; this uptake converts the macrophages to foam cells; these two cell types secrete growth factors which stimulate the formation of plaques. CD36 and SRB1 express on these macrophages and are responsible for the binding of HDLs and LDLs to the macrophage; SRB1 proteins form channels through the membranes and are responsible for the transport of cholesterols (components of the HDLs and LDLs) both into and out of the macrophage (Calvo *et al.*, 1998; Gu *et al.*, 1998; Krieger, 1999; Coburn *et al.*, 2001; Reaven *et al.*, 2001; Frank *et al.*, 2002; Ibrahimi and Abumrad 2002; Kuniyasu *et al.*, 2002; Liu *et al.*, 2002; Miyazaki *et al.*, 2002; Moore *et al.*, 2002; Podrez *et al.*, 2002a, b; Tsukamoto *et al.*, 2002). So members of the vertebrate CD36 family can bind protein–lipid complexes and transport lipids across the cell membrane. These proteins have other less studied functions and activities as well. CD36 is involved in cytoadhesion and phagocytosis of rod outer segments in the eye (Ryeom *et al.*, 1996). LIMP II has some as yet unknown function in lysosomes (e.g. Kuronita *et al.*, 2002). SRB1 has recently been implicated as a receptor for hepatitis C virus (Scarselli *et al.*, 2002).

About 14 SNMP/CD36 homologs are present in the genomes of both *D. melanogaster* and *A. gambiaea* (Figure 14.6). Three of the *D. melanogaster* proteins have been characterized: emp (Hart and Wilcox, 1993), Croquemort (Franc *et al.*, 1996, 1999; Lee and Baehrecke, 2001) and *ninaD* (CG31783) (Kiefer *et al.*, 2002). Emp (epithelial membrane protein) expresses in ectodermally derived tissues in the embryo and larva, including epithelial cells of wing imaginal discs; the authors suggested emp might have a role in development, but little more is known about this protein (Hart and Wilcox, 1993). Croquemort (catcher of death; Crq) is expressed in circulating macrophages which engulf apoptotic cells by phagocytosis during non-autophagic cell death; Crq serves, perhaps along with other receptors, to mediate this process (Franc *et al.*, 1996, 1999; Lee and Baehrecke, 2001). Although the precise role of Crq in this process is not known, the similarity between this activity and the phagocytosis activity of CD36 in mammals suggests that functions of this gene family are conserved between these two distant phyla (Ryeom *et al.*, 1996; Franc *et al.*, 1999). *NinaD* is a blind mutant that is deficient in the dietary uptake of the visual pigment precursor caratenoids; the *ninaD* mutation is in the gene CG31783, a CD36 homolog; p-element mediated transformation with wild type CG31783 rescues *ninaD* mutants (Kiefer *et al.*, 2002). *NinaD* expression was detected in embryonic midgut primordia, in mesodermal tissue giving rise to hemocytes and macrophage, and in hemocytes. The authors suggest that ninaD functions in the uptake of carotenoids, in a manner similar to the uptake of cholesterol and phospholipids by mammalian SRB1 proteins (e.g. Frank *et al.*, 2002; Liu and Krieger, 2002).

The lepidopteran SNMPs are the only CD36 homologues identified so far in the nervous system. Given what is known about the anatomy and physiology of olfactory sensilla (e.g. Steinbrecht, 1999), it is difficult to translate the specific described activities of the mammalian and dipteran members of the CD36 gene family to olfactory function. However, in a general sense, members of this gene family function as receptors in cell–cell interaction, as receptors which bind protein–lipid complexes, and as transporters of small hydrophobic molecules such as cholesterol and probably carotenoids. Translating these activities into olfactory activities to formulate hypotheses of SNMP function is feasible (Figure 14.7). One *D. melanogaster* and two *A. gambiaea* proteins share notable sequence similarity with the lepidopteran SNMPs (Figure 14.6), and it may prove that studies of these dipteran proteins, and especially CG7000 of *D. melanogaster* may prove fruitful in elucidating SNMP function.

14.6 Functional models

Figure 14.7 shows a series of variations on the scheme of odor detection presented in Figure 14.1. The proposed role of ODEs remains the same; ODEs degrade

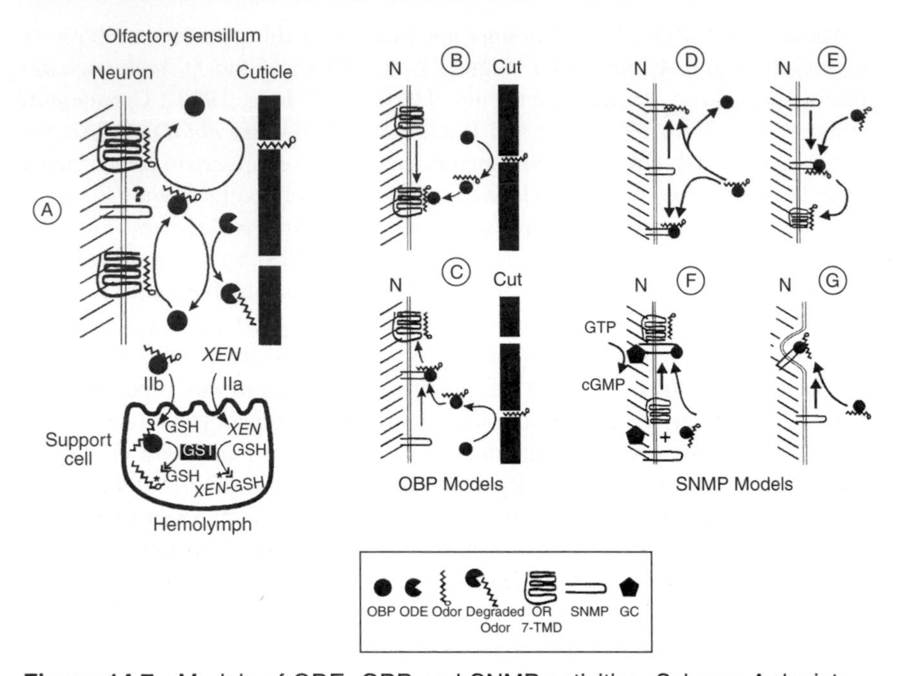

Figure 14.7 Models of ODE, OBP and SNMP activities. Scheme A depicts the general model for OBP and ODE activities proposed by Vogt *et al.* (1985); support cell activity is based on Rogers *et al.* (1999). Odor molecules are transported to ORs by OBPs and degraded by ODEs; odor molecules entering support cells are inactivated by dual function enzymes such as GSTs which also attack xenobiotics. B and C suggest additional schemes for OBP activity (from Vogt *et al.*, 1999), where the ORs are stimulated by a stable odor–OBP complex (b), or by free odor after the odor–OBP complex is destabilized by some other process such as interaction with SNMP (c). D–G suggest several possible functions for SNMP (from Rogers *et al.*, 2001a). SNMP may act as a novel odor receptor for either free odor (Di) or odor–OBP complex (Dii); or may destabilize the odor–OBP complex freeing odor to interact directly with ORs (e). Alternatively, SNMP may complex with ORs or cytosolic proteins to contribute in some manner to odor reception (F) or serve as an internalization process bringing odor or odor–OBP complexes into the cell (G). Further details on these models can be read in Rogers *et al.* (2001a, b).

odors where and when they have the opportunity. There is some disagreement over whether OBPs protect odors from degradation, and this in part goes to different interpretations of experiments we performed; we interpreted these results as suggesting no significant protection (Vogt and Riddiford, 1986a) while others have viewed them as indicating protection (e.g. Kaissling, 2001).

Alternate views of how OBPs function are shown in Figures 14.7B and 14.7C; both suggest that the odor-OBP complex is stable and either must act on the receptors as a complex or be destabilized by secondary interactions with membrane proteins (SNMP or OR perhaps) or the membrane itself. These views are discussed further by Leal, Chapter 15 and by Plettner, Chapter 16, in this volume.

Several schemes for SNMP action are shown in Figures 14.7D–7G. Figure 14.7D acknowledges the now unlikely scheme that SNMP functions as an odor receptor, binding either free odor (i) or the odor–OBP complex (ii), while Figure 14.7E poses SNMP as a destabilizer of the odor–OBP complex. Perhaps more interesting is the scheme proposed in Figure 14.7F where SNMP interacts with the OR or with intracelluar proteins (perhaps a guanylase cyclase described by Simpson *et al.*, 1999). Figure 14.7G suggests that SNMP might function to internalize odors or odor–OBP complexes, perhaps functioning as part of a signal termination pathway that removes odor accumulating near the membranes and thus maintains a concentration gradient for odors towards the neurons. Mechanistically this scheme is consistent with the internalization of cholesterol by SRB1 proteins in mammals (Frank *et al.*, 2002; Liu *et al.*, 2002) and the uptake of carotenoids by ninaD protein in *D. melanogaster* (Kiefer *et al.*, 2002). Not shown is the possibility that SNMP1 and SNMP2 serve as phenotype determinants of the respective neurons; that the proposed differential expression of these two gene products in different neurons of a sensillum contribute to the maintenance of distinct neuronal phenotypes through some cell–cell receptor-like function. These schemes are described in greater detail in Rogers *et al.* (2001a).

14.6.1 Biochemical networks, genetic networks, complex behavior

Insects discriminate many odors. The models proposed above suggest that biochemical networks underlie the process of odor detection. As presented, the models underestimate the complexity of these networks, overlooking data indicating that at least some pheromone-sensitive sensilla must have the biochemistry to monitor both conspecific (agonist) and sympatric (antagonist) signals (Cossé *et al.*, 1998) or that divergent transductory pathways may be used to process distinct signals (see Krieger and Breer, Chapter 20, in this volume). At a yet deeper level, these biochemical networks are being assembled and controlled by gene regulatory networks, complex interactions of diverse transcription factors determining which genes should be expressed where, when and with whom.

Figure 14.8A illustrates the locations and distributions of OR-, OBP- and SNMP-related genes along the major chromosomes (X, 2, 3) of *D. melanogaster*. These genes appear to possess independent regulatory regions (e.g. Vosshall *et al.*, 2000; Galindo and Smith, 2001). Regulatory decisions are made to express only one (or very few) OR genes in a given neuron and one (or very few) OBP genes in the associating support cell. The selection of appropriate OR and OBP partners is regulated between neurons and support cells. Some genes are coexpressed consistently and others differentially. All this must be controlled through transcription factors which are themselves regulated in spatially and

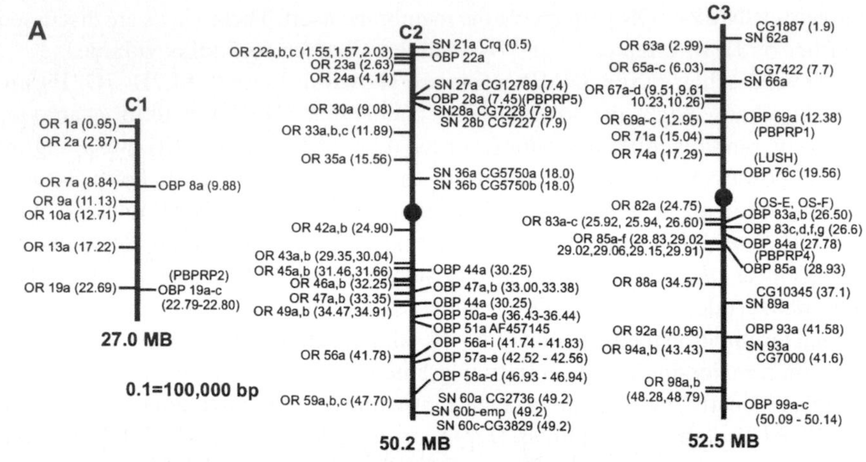

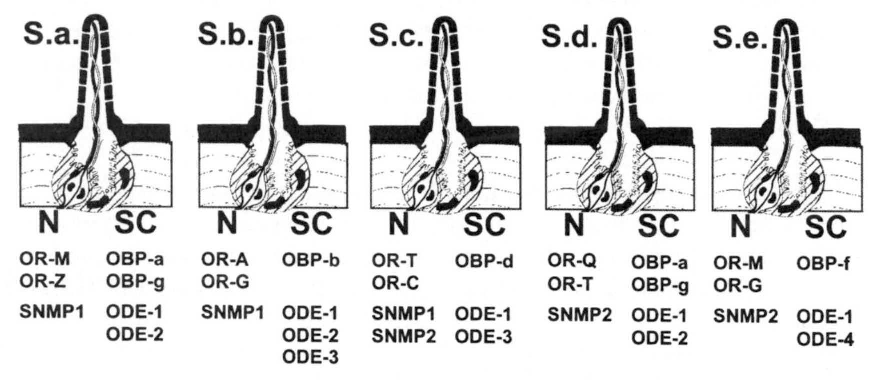

Figure 14.8 (A) The physical locations of OR, OBP and SNMP (SN) genes are shown on linear representations of the *Drosophila* chromosomes. All genes are named by map location; refer to tables in this chapter for OBPs and SNMPs, and in Voshall, Chapter 19, in this volume, for ORs. Letters following a map number account for multiple genes in that map region. Numbers in parentheses are approximate nucleotide number in megabases (MB) from the top (obtained from the NCBI Map View resource). (B) This illustration suggests that the phenotypes of functionally distinct olfactory sensilla (S.a.–S.e.) are determined by the combinatorial expression of specific members of the indicated olfactory gene families.

temporally hierarchical patterns. The result of this genetic *Grosse Gemischte Salat* is the production of sensilla with distinct phenotypes (Figure 14.8B) which provide the input for the diverse olfactory behaviors that support the lives of these interesting animals.

References

Abumrad N. A., El-Maghrabi M. R., Amri E., Lopez E. and Grimaldi P. (1993) Cloning of a rat adipocyte membrane protein implicated in binding or transport of long-chain fatty acids that is induced during preadipocyte differentiation. *J. Biol. Chem.* **268**, 17665–17668.

Achterberg and colleagues (1991) *The Insects of Australia: a Textbook for Students and Research Workers*. CSIRO, Melbourne University Press, Melbourne.

Acton S., Rigotti A., Landschulz K. T., Xu S., Hobbs H. H. and Krieger M. (1996) Identification of scavenger receptor SR-BI as a high density lipoprotein receptor. *Science* **271**, 518–520.

Acton S. L., Scherer P. E., Lodish H. F. and Krieger M. (1994) Expression cloning of SR-BI, a CD36-related class B scavenger receptor. *J. Biol. Chem.* **269**, 21003–21009.

Adam G. and Delbrück M. (1968) Reduction of dimensionality in biological diffusion processes. In *Structural Chemistry and Molecular Biology*, eds A. Rich and R. Davidson, pp. 1980215. W. H. Freeman Co., San Francisco.

Angeli S., Ceron F., Scaloni A., Monti M., Monteforti G., Minnocci A., Petacchi R. and Pelosi P. (1999) Purification, structural characterization, cloning and immunocytochemical localization of chemoreception proteins from *Schistocerca gregaria*. *Eur. J. Biochem.* **262**, 745–754.

Baker T. C. and Haynes K. F. (1987) Maneuvers used by flying male oriental fruit moths to relocate a sex pheromone plume in an experimentally shifted wind-field. *Physiol. Entomol.* **12**, 263–279.

Baker T. C. and Vogt R. G. (1988) Measured behavioral latency in response to sex-pheromone loss in the large silk moth *Antheraea polyphemus*. *J. Exp. Biol.* **137**, 29–38.

Bau J., Martinez D., Renou M. and Guerrero A. (1999) Pheromone-triggered orientation flight of male moths can be disrupted by trifluoromethyl ketones. *Chem. Senses* **24**, 473–480.

Ben-Arie N., Khen M. and Lancet D. (1993) Glutathione S-transferase in rat olfactory epithelium: purification, molecular properties and odorant biotransformation. *Biochem. J.* **292**, 379–384.

Bette S., Breer H. and Krieger J. (2002) Probing a pheromone binding protein of the silkmoth *Antheraea polyphemus* by endogenous tryptophan fluorescence. *Insect Biochem. Mol. Biol.* **32**, 241–246.

Biessmann H., Walter M. F., Dimitratos S. and Woods D. (2002) Isolation of cDNA clones encoding putative odourant binding proteins from the antennae of the malaria-transmitting mosquito, *Anopheles gambiae*. *Insect Mol. Biol.* **11**, 123–132.

Bohbot J., Sobrio F., Lucas P. and Nagnan-Le Meillour P. (1998) Functional characterization of a new class of odorant-binding proteins in the moth *Mamestra brassicae*. *Biochem Biophys. Res. Commun.* **253**, 489–494.

Borror D. J., Triplehorn C. A. and Johnson N. F. (1989) *An Introduction to the Study of Insects* 6th edn. Harcourt Brace & Co., Orlando.

Boudreaux H. B. (1979) *Arthropod Phylogeny with Special Reference to Insects*. Wiley, New York.

Bourke A. F. G. (2002) Genetics of social behaviour in fire ants. *Trends in Genetics* **18**, 221–223.

Breer H., Krieger J. and Raming K. (1990) A novel class of binding proteins in the antennae of the silkmoth *Antheraea pernyi*. *Insect Biochem.* **20**, 735–740.

Briand L., Nespoulous C., Huet J. C. and Pernollet J. C. (2001a) Disulfide pairing and secondary structure of ASP1, an olfactory-binding protein from honeybee (*Apis mellifera* L). *J. Pept. Res.* **58**, 540–545.

Briand L., Nespoulous C., Huet J. C., Takahashi M. and Pernollet J. C. (2001b) Ligand binding and physico-chemical properties of ASP2, a recombinant odorant-binding protein from honeybee (*Apis mellifera* L.). *Eur. J. Biochem.* **268**, 752–760.

Burchell B. (1991) Turning on and turning off the sense of smell. *Nature.* **350**, 16–17.

Butenandt A., Beckmann R., Stamm D., Hecker E. (1959) Über den Sexual-Lockstoff des Seidenspinners *Bombyx mori*: Reindanstellung und Konstitution. *Z. Naturforsch.* **14b**, 283–284.

Callahan F. E., Vogt R. G., Tucker M. L., Dickens J. C. and Mattoo A. K. (2000) High level expression of "male specific" pheromone binding proteins (PBPs) in the antennae of female noctuid moths. *Insect Biochem. Mol. Biol.* **30**, 507–514.

Calvo D. and Vega M. A. (1993) Identification, primary structure, and distribution of CLA-1, a novel member of the CD36/LIMPII gene family. *J. Biol. Chem.* **268**, 18929–18935.

Calvo D., Gómez-Coronado D., Suárez Y., Lasunción M. A. and Vega M. A. (1998) Human CD36 is a high affinity receptor for the native lipoproteins HDL, LDL, and VLDL. *J. Lipid Res.* **39**, 777–788.

Campanacci V., Krieger J., Bette S., Sturgis J. N., Lartigue A., Cambillau C., Breer H. and Tegoni M. (2001a) Revisiting the specificity of *Mamestra brassicae* and *Antheraea polyphemus* pheromone-binding proteins with a fluorescence binding assay. *J. Biol. Chem.* **276**, 20078–20084.

Campanacci V., Mosbah A., Bornet O., Wechselberger R., Jacquin-Joly E., Cambillau C., Darbon H. and Tegoni M. (2001b) Chemosensory protein from the moth *Mamestra brassicae*. Expression and secondary structure from 1H and 15N NMR. *Eur. J. Biochem.* **268**, 4731–4739.

Campanacci V., Spinelli S., Lartigue A., Lewandowski C., Brown K., Tegoni M. and Cambillau C. (2001a) Recombinant chemosensory protein (CSP2) from the moth *Mamestra brassicae*: crystallization and preliminary crystallographic study. *Acta Crystallogr. D. Biol. Crystallogr.* **57**, 137–139.

Carr W. E., Gleeson R. A. and Trapido-Rosenthal H. G. (1990) The role of perireceptor events in chemosensory processes. *Trends Neurosci.* **13**, 212–215.

Christophides G. K., Mintzas A. C. and Komitopoulou K. (2000) Organization, evolution and expression of a multigene family encoding putative members of the odourant binding protein family in the medfly *Ceratitis capitata. Insect Mol. Biol.* **9**, 185–195.

Clark A. G. (1990) The glutathione S-transferases and resistance to insecticides. In *Glutathione S-Transferases and Drug Resistance*, eds J. D. Hayes C. B Picket and T. J. Mantle, pp. 369–378. Taylor and Francis, London.

Clyne P. J., Warr C. G., Freeman M. R., Lessing D., Kim J. and Carlson J. R. (1999) A novel family of divergent seven-transmembrane proteins: candidate odorant receptors in *Drosophila. Neuron* **22**, 327–338.

Coburn C. T., Hajri T., Ibrahimi A. and Abumrad N. A. (2001) Role of CD36 in membrane transport and utilization of long-chain fatty acids by different tissues. *J. Mol. Neurosci.* **16**, 117–121; discussion 151–157.

Coleman M., Vontas J. G. and Hemingway J. (2002) Molecular characterization of the amplified aldehyde oxidase from insecticide resistant *Culex quinquefasciatus. Eur J. Biochem.* **269**, 768–779.

Cossé A. A., Todd J. L. and Baker T. C. (1998) Neurons discovered in male *Helicoverpa zea* antennae that correlate with pheromone-mediated attraction and interspecific antagonism. *J. Comp. Physiol. A* **182**, 585–594.

Couzin J. (2002) Genome research: NFS's ark draws alligators, algae and wasps. *Science* **297**, 1638–1639.

Cristiani G., Fournier P., Pageat P. and Nagnan-Le Meillour P. (2001) Unpublished sequences, Accession Number AF461143.

Dahl A. R. (1988) The effect of cytochrome P-450-dependent metabolism and other enzyme activities on olfaction. In *Molecular Neurobiology of the Olfactory System*, eds F. Margolis and T. Getchel, pp. 51–70. Plenum, New York.

Danty E., Arnold G., Huet J. C., Huet D., Masson C. and Pernollet J. C. (1998) Separation, characterization and sexual heterogeneity of multiple putative odorant-binding proteins in the honeybee *Apis mellifera* L. (Hymenoptera: Apidea). *Chem. Senses* **23**, 83–91.

Danty E., Briand L., Michard-Vanhee C., Perez V., Arnold G. and Gaudemer O., Huet D., Huet J. C., Ouali C., Masson C. and Pernollet J. C. (1999) Cloning and expression of a queen pheromone-binding protein in the honeybee: an olfactory-specific, developmentally regulated protein. *J. Neurosci.* **19**, 7468–7475.

Danty E., Michard-Vanhee C., Huet J. C., Genecque E., Pernollet J. C. and Masson C. (1997) Biochemical characterization, molecular cloning and localization of a putative odorant-binding protein in the honey bee *Apis mellifera* L. (Hymenoptera: Apidea). *FEBS Lett.* **414**, 595–598.

de Bruyne M., Foster K. and Carlson J. R. (2001) Odor Coding in the *Drosophila* Antenna. *Neuron* **30**, 537–552.

Deyu Z. and Leal W. S. (2002) Conformational isomers of insect odorant-binding proteins. *Arch. Biochem. Biophys.* **397**, 99–105.

Ding X., Porter T. D., Peng H.-M. and Coon M. J. (1991) cDNA and derived amino acid sequence of a rabbit nasal cytochrome p450NMB (p450IIG1), a unique isozyme possibly involved in olfaction. *Arch. Biochem. Biophys.* **285**, 120–125.

Du G. and Prestwich G. D. (1995) Protein structure encodes the ligand binding specificity in pheromone binding proteins. *Biochemistry* **34**, 8726–8732.

Du G., Ng C. S. and Prestwich G. D. (1994) Odorant binding by a pheromone binding protein: active site mapping by photoaffinity labeling. *Biochemistry* **33**, 4812–4819.

Erwin T. L. (1982) Tropical forests: their richness in coleoptera and other arthropod species. *The Coleopterists' Bulletin* **36**, 74–75.

Feng L. and Prestwich G. D. (1997) Expression and characterization of a lepidopteran general odorant binding protein. *Insect Biochem. Mol. Biol.* **27**, 405–412.

Ferkovich S. M., Mayer M. S. and Rutter R. R. (1973a) Conversion of the sex pheromone of the cabbage looper. *Nature* **242**, 53–55.

Ferkovich S. M., Mayer M. S., Rutter R. R. (1973b) Sex pheromone of the cabbage looper: reactions with antennal proteins *in vitro*. *J. Insect Physiol.* **19**, 2231–2243.

Ferkovich S. M., Oliver J. E. and Dillard C. (1982) Pheromone hydrolysis by cuticular and interior esterases of the antennae, legs, and wings of the cabbage looper moth, *Trichoplusia ni* (Hubner). *J. Chem. Ecol.* **8**, 859–866.

Ferkovich S. M., Van Essen F. and Taylor T. R. (1980) Hydrolysis of sex pheromone by antennal esterases of the cabbage looper, *Trichoplusia ni*. *Chem. Senses Flavour* **5**, 33–45.

Feyereisen R. (1999) Insect P450 enzymes. *Ann. Rev. Entomol.* **44**, 507–533.

Filippov V. A., Filippova M. A. and Sehnal F. (1995) Lipocalin-like brain-specific protein. Unpublished sequence, Accession Number L41640.

Forret S. and Maleszka R. (2001) Unpublished sequences Accession Number AF 339140.

Fournier D., Bride J. M., Poirie M., Berge J. B. and Plapp Jr F. W. (1992) Insect glutathione S-transferases. Biochemical characteristics of the major forms from houseflies susceptible and resistant to insecticides. *J. Biol. Chem.* **267**, 1840–1845.

Fox A. N., Pitts R. J. and Zwiebel L. J. (2002) A cluster of candidate odorant receptors from the malaria vector mosquito, *Anopheles gambiae*. *Chem. Senses* **27**, 453–459.

Franc N. C., Dimarcq J.-L., Lagueux M., Hoffmann J. and Ezekowitz R. A. B. (1996) Croquemort, a novel *Drosophila* hemocyte/macrophage receptor that recognizes apoptotic cells. *Immunity* **4**, 431–443.

Franc N. C., Heitzler P., Ezekowitz R. A. and White K. (1999) Requirement for croquemort in phagocytosis of apoptotic cells in *Drosophila. Science* **284**, 1991–1994.

Frank P. G., Marcel Y. L., Connelly M. A., Lublin D. M., Franklin V., Williams D. L. and Lisanti M. P. (2002) Stabilization of caveolin-1 by cellular cholesterol and scavenger receptor class B type I. *Biochemistry* **41**, 11931–11940.

Freeman S. and Herron J. C. (1998) *Evolutionary Analysis*, 1st edn Prentice Hall, Upper Saddle River, New Jersey.

Galindo K. and Smith D. P. (2001) A large family of divergent *Drosophila* odorant-binding proteins expressed in gustatory and olfactory sensilla. *Genetics* **159**, 1059–1072.

Gamain B., Gratepanche S., Miller L. H. and Baruch D. I. (2002) Molecular basis for the dichotomy in plasmodium falciparum adhesion to CD36 and chondroitin sulfate A. *Proc. Natl. Acad. Sci. USA* **99**, 10020–10024.

Gao Q. and Chess A. (1999) Identification of candidate *Drosophila* olfactory receptors from genomic DNA. *Genomics* **60**, 31–39.

Gleeson R. A., Trapido-Rosenthal H. G., McDowell L. M., Aldrich H. C. and Carr W. E. (1992) Ecto-ATPase/phosphatase activity in the olfactory sensilla of the spiny lobster, *Panulirus argus*: localization and characterization. *Cell Tissue Res.* **269**, 439–445.

Graham L. A. and Davies P. L. (2002) The odorant-binding proteins of *Drosophila melanogaster*: annotation and characterization of a divergent gene family. *Gene* **292**, 43–55.

Graham L. A., Tang W., Baust J. G., Liou Y. C., Reid T. S. and Davies P. L. (2001) Characterization and cloning of a *Tenebrio molitor* hemolymph protein with sequence similarity to insect odorant-binding proteins. *Insect Biochem. Mol. Biol.* **31**, 691–702.

Graham S. M. and Prestwich G. D. (1992) Tissue distribution and substrate specificity of an epoxide hydrase in the gypsy moth *Lymantria dispar. Experentia* **48**, 19–21.

Graham S. M. and Prestwich G. D. (1994) Synthesis and inhibitory properties of pheromone analogs for the epoxide hydrolase of the gypsy moth. *J. Org. Chem.* **59**, 2956–2966.

Greenwalt D. E., Lipsky R. H., Ockenhouse C. F., Ikeda H., Tandon N. N. and Jamieson G. A. (1992) Membrane glycoprotein CD36: a review of its roles in adherence, signal transduction, and transfusion medicine. *Blood* **80**, 1105–1115.

Grünert U. and Ache B. W. (1988) Ultrastructure of the aesthetase (olfactory) sensilla of the spiny lobster, *Panulirus argus. Cell Tissue Res.* **251**, 95–103.

Gu X. J., Trigatti B., Xu S. Z., Acton S., Babitt J. and Krieger M. (1998) The efficient cellular uptake of high density lipoprotein lipids via scavenger receptor class B type I requires not only receptor-mediated surface binding but also receptor-specific lipid transfer mediated by its extracellular domain. *J. Biol. Chem.* **273**, 26338–26348.

Guirong W. and Yuyuan G. (2000) Cloning and expression of general odorant-binding protein gene from *Spodoptera exigua*. Unpublished sequence, Accession Number AJ294808.

Györgyi T. K., Roby-Shemkovitz A. J. and Lerner M. R. (1988) Characterization and cDNA cloning of the pheromone-binding protein from the tobacco hornworm, *Manduca sexta*: a tissue-specific developmentally regulated protein. *Proc. Natl. Acad. Sci. USA* **85**, 9851–9855.

Hart K. and Wilcox M. (1993) A *Drosophila* gene encoding an epithelial membrane protein with homology to CD36/LIMP II. *J. Mol. Biol.* **234**, 249–253.

Hekmat-Scafe D. S., Dorit R. L. and Carlson J. R. (2000) Molecular evolution of odorant-binding protein genes OS-E and OS-F in *Drosophila. Genetics* **155**, 117–127.
Hekmat-Scafe D. S., Scafe C. R., McKinney A. J. and Tanouye M. A. (2002) Genome-wide analysis of the odorant-binding protein gene family in *Drosophila melanogaster. Genome Res.* **12**, 1357–1369.
Hekmat-Scafe D., Steinbrecht R. A. and Carlson J. R. (1997) Coexpression of two odorant-binding protein homologs in *Drosophila*: implications for olfactory coding. *J. Neurosci.* **17**, 1616–1624.
Hennig W. (1981) *Insect Phylogeny*. Wiley, New York.
Hill C. A., Fox A. N., Pitts R. J., Kent L. B., Tan P. L., Chrystal M. A., Cravchik A., Collins F. H., Robertson H. M. and Zwiebel L. J (2002) G protein-coupled receptors in *Anopheles gambiae. Science* **298**, 176–178.
Horst R., Damberger F., Luginbuhl P., Guntert P., Peng G., Nikonova L., Leal W. S. and Wuthrich K (2001) NMR structure reveals intramolecular regulation mechanism for pheromone binding and release. *Proc. Natl. Acad. Sci. USA*. **98**, 14374–14379.
Ibrahimi A. and Abumrad N. A. (2002) Role of CD36 in membrane transport of long-chain fatty acids. *Curr. Opin. Clin. Nutr. Metab. Care* **5**, 139–145.
Ishida Y. and Leal W. S. (2002) Cloning of putative odorant-degrading enzyme and integumental esterase cDNAs from the wild silkmoth, *Antheraea polyphemus. Insect Biochem. Mol. Biol.* **32**, 1775–1780.
Ishida Y., Chiang V. P., Haverty M. I. and Leal W. S. (2002a) Primitive odorant-binding proteins. *J. Chem. Ecol.* **28**, R21–R27.
Ishida Y., Cornel A. J. and Leal W. S. (2002b) Identification and cloning of a female antenna-specific odorant-binding protein in the mosquito *Cluex quinquefasciatus. J. Chem. Ecol.* **28**, 867–871.
Jacquin-Joly E., Bohbot J., Francois M. C., Cain A. H. and Nagnan-Le Meillour P. (2000) Characterization of the general odorant-binding protein 2 in the molecular coding of odorants in *Mamestra brassicae. Eur. J. Biochem.* **267**, 6708–6714.
Jacquin-Joly E., Vogt R. G., Francois M. C. and Nagnan-Le Meillour P. (2001) Functional and expression pattern analysis of chemosensory proteins expressed in antennae and pheromonal gland of *Mamestra brassicae. Chem Senses* **26**, 833–844.
Jacquin-Joly E., Francois M. C. and Nagnan-Le Meillour P. (1999) cDNA cloning of *Rhynchophorus palmarum* odorant-binding protein. Unpublished sequences, Accession Number AF139912.
Janssen K. P., Rost R., Eichinger L. and Schleicher M. (2001a) Characterization of CD36/LIMPII homologues in *Dictyostelium discoideum. J. Biol. Chem.* **276**, 38899–38910.
Janssen K. P. and Schleicher M. (2001b) Dictyostelium discoideum: a genetic model system for the study of professional phagocytes. Profilin, phosphoinositides and the lmp gene family in *Dictyostelium. Biochim. Biophys. Acta.* **1525**, 228–233.
Kaissling K.-E. (1974) Sensory transduction in insect olfactory receptors. In *Biochemistry of Sensory Function*, ed. L. Jaenicke, pp. 243–273 Springer-Verlag, Heidelberg.
Kaissling K.-E. (1986) Chemo-electrical transduction in insect olfactory receptors. *Ann. Rev. Neurosci.* **9**, 121–145.
Kaissling K.-E. (2001) Olfactory perireceptor and receptor events in moths: a kinetic model. *Chem. Senses* **26**, 125–150.
Kalinová B., Hoskovec M., Liblika I., Unelius R. and Hansson B. S. (2001) Detection of sex pheromone components in *Manduca sexta* (L.) *Chem. Senses* **26**, 1175–1186.
Kanaujia A. and Kaissling K.-E. (1985) Interactions of pheromone with moth antennae: adsorption, desorption and transport. *J. Insect Physiol.* **31**, 71–81.
Karakesisoglou I., Janssen K.-P., Eichinger L., Noegel A. A. and Schleicher M. (1999) Identification of a suppressor of the *Dictyostelium* profilin-minus phenotype as a CD36/LIMP-II homolog. *J. Cell Biol.* **145**, 167–181.

Karlin S. and Altschul S. F. (1990) Methods for assessing the statistical significance of molecular sequence features by using general scoring schemes. *Proc. Natl. Acad. Sci. USA* **87**, 2264–2268.

Kasang G. (1968) Tritium-Markierung des Sexuallockstoffes Bombykol. *Z. Naturforschg.* **23b**, 1331–1335.

Kasang G. (1971) Bombykol reception and metabolism on the antennae of the silkmoth *Bombyx mori*. In *Gustation and Olfaction*, eds G. Ohloff and A. F. Thomas, pp. 245–250. Academic Press, London.

Kasang G. (1973) Physikomichemische Vorange beim Riechen des Seidenspinners. *Naturwissenschaften* **60**, 95–101.

Kasang G. (1974) Uptake of the sex pheromone 3H-bombykol and related compounds by male and female *Bombyx* antennae. *J. Insect Physiol.* **20**, 2407–2422.

Kasang G. and Kaissling K.-E. (1972) Specificity of primary and secondary olfactory processes in *Bombyx* antennae. In *Int. Symp. Olfaction and Taste IV*, ed. D. Schneider pp. 200–206. Verlagsgesellschaft, Stuttgart.

Kasang G. and Kaissling K.-E. (1974) Specificity of primary and secondary olfactory processes in *Bombyx* antennae. In *IVth Internat. Sympos. Olfaction and Taste*, ed. D. Schneider, pp. 200–206. Wissenschaftl. Verlagsgesellschaft, Stuttgart.

Kasang G., Kaissling K.-E., Vostrowsky O. and Bestmann H. J. (1978) Bombykal, a second pheromone component of the silkworm moth *Bombyx mori*. *Angew. Chem. Int. Ed. Engl.* **17**, 60.

Kasang G., Nicholls M., Keil T. and Kanaujia S. (1989) Enzymatic conversion of sex pheromones in olfactory hairs of the male silkworm moth *Antheraea polyphemus*. *Z. Naturforsch.* **44c**, 920–926.

Kasang G., Nicholls M. and von Proff L. (1989) Sex-pheromone conversion and degradation in antennae of the silkworm moth *Bombyx mori* (L.) *Experientia* **45**, 81–87.

Kasang G. and Weiss N. (1974) Thin layer chromatographic analysis of radioactively labelled insect pheromones. Metabolites of [3H]bombykol. *J. Chromatog.* **92**, 401–417.

Keller L. and Parker J. D. (2002) Behavioral genetics: a gene for supersociality. *Current Biology* **12**, R180–R181.

Kiefer C., Sumser E., Wernet M. F. and Von Lintig J. (2002) A class B scavenger receptor mediates the cellular uptake of carotenoids in *Drosophila*. *Proc. Natl. Acad. Sci. USA*. **99**, 10581–10586.

Kim M. S., Repp A. and Smith D. P. (1998) LUSH odorant-binding protein mediates chemosensory responses to alcohols in *Drosophila melanogaster*. *Genetics* **150**, 711–721.

Kim M. S. and Smith D. P. (2001) The invertebrate odorant-binding protein LUSH is required for normal olfactory behavior in *Drosophila*. *Chem. Senses* **26**, 195–199.

Klein U. (1987) Sensillum-lymph proteins from antennal olfactory hairs of the moth *Antheraea polyphemus* (Saturniidae). *Insect Biochem.* **17**, 1193–1204.

Klun J. A., Khrimian A. P. and Oliver J. E. (1998) Evidence of pheromone catabolism via B-oxidation in the European corn borer (Lepidoptera: Crambidae). *J. Entomol. Sci.* **33**, 400–406.

Klun J. A., Potts W. J. E. and Oliver J. E. (1996) Four species of noctuid moths degrade sex pheromone by a common antennal metabolic pathway. *J. Entomol. Sci.* **31**, 404–413.

Klun J. A., Schwarz M. and Uebel E. C. (1991) European corn borer: pheromonal catabolism and behavioral response to sex pheromone. *J. Chem. Ecol.* **17**, 317–334.

Kochansky J., Tette J., Taschenberg E. F., Cardé R. T., Kaissling K. E. and Roelofs W. L. (1975) Sex pheromone of the moth *Antheraea polyphemus*. *J. Insect Physiol.* **21**, 1977–1983.

Kowcun A., Honson N. and Plettner E. (2001) Olfaction in the gypsy moth, *Lymantria dispar*: effect of pH, ionic strength, and reductants on pheromone transport by pheromone-binding proteins. *J. Biol. Chem.* **276**, 44770–44776.
Krieger J., Raming K., Dewer Y. M., Bette S., Conzelmann S. and Breer H. (2002) A divergent gene family encoding candidate olfactory receptors of the moth *Heliothis virescens. Eur. J. Neurosci.* **16**, 619–628.
Krieger J. and Breer H. (1998) Cloning of a glutathione S-transferase from the antennae of *Bombyx mori*. Unpublished sequence, Accession Number AJ006502.
Krieger J., Mameli M. and Breer H. (1997) Elements of the olfactory signaling pathways in insect antennae. *Invert. Neurosci.* **3**, 137–144.
Krieger J., Raming K. and Breer H. (1991) Cloning of genomic and complementary DNA encoding insect pheromone binding proteins: evidence for microdiversity. *Biochim. Biophys. Acta* **1088**, 277–284.
Krieger J., von Nickisch-Rosenegk E., Mameli M., Pelosi P. and Breer H. (1996) Binding proteins from the antennae of *Bombyx mori. Insect Biochem. Mol. Biol.* **26**, 297–307.
Krieger M. (1999) Charting the fate of the "good cholesterol": identification and characterization of the high-density lipoprotein receptor SR-BI. *Ann. Rev. Biochem.* **68**, 523–558.
Krieger M. J. and Ross K. G. (2002) Identification of a major gene regulating complex social behavior. *Science* **295**, 328–332.
Krieger J., Gänßle H., Raming K. and Breer H. (1993) Odorant binding proteins of *Heliothis virescens. Insect Biochem. Molec. Biol.* **23**, 449–456.
Kristensen N. P. (1991) Phylogeny of extant hexapods. In *CSIRO, The Insects of Australia: A Textbook for Students and Research Workers*, pp. 125–140. Melbourne University Press, Melbourne.
Kukalová-Peck J. (1991) Fossil history and the evolution of hexapod structures. In *CSIRO, The Insects of Australia: a Textbook for Students and Research Workers*, pp. 141–179. Melbourne University Press, Melbourne.
Kumar S., Tamura K., Jakobsen I. B. and Nei M. (2001) MEGA2: molecular evolutionary genetics analysis software. *Bioinformatics* **17**, 1244–1245.
Kuniyasu A., Hayashi S. and Nakayama H. (2002) Adipocytes recognize and degrade oxidized low density lipoprotein through CD36. *Biochem. Biophys. Res. Commun.* **295**, 319–323.
Kuronita T., Eskelinen E. L., Fujita H., Saftig P., Himeno M. and Tanaka Y. (2002) A role for the lysosomal membrane protein LGP85 in the biogenesis and maintenance of endosomal and lysosomal morphology. *J. Cell. Sci.* **115**, 4117–4131.
Labandeira C. C. and Sepkoski Jr, J. J. (1993) Insect diversity in the fossil record. *Science* **261**, 310–315.
LaForest S. M., Prestwich G. D. and Lofstedt C. (1999) Intraspecific nucleotide variation at the pheromone binding protein locus in the turnip moth, *Agrotis segetum. Insect Mol. Biol.* **8**, 481–490.
Lartigue A., Campanacci V., Roussel A., Larsson A. M. and Alwyn T. (2002a) X-ray structure and ligand binding study of a moth chemosensory protein. *J. Biol. Chem.* **277**, 32094–32098.
Lartigue A., Campanacci V., Roussel A., Larsson A. M., Jones T. A., Tegoni M. and Cambillau C. (2002b) X-ray structure and ligand binding study of a moth chemosensory protein. *J. Biol. Chem.* 2002, **277**, 32094–32098.
Laue M. and Steinbrecht R. A. (1997) Topochemistry of moth olfactory sensilla. *Int. J. Insect Morph. & Embryol.* **26**, 217–228.
Lazard D., Tal N., Rubenstein M., Khen M., Lancet D. and Zupko K. (1990) Identification and biochemical analysis of novel olfactory-specific cytochrome P-450IIA and UDP-glucuronosyl transferase. *Biochemistry* **29**, 7433–7440.

Lazard D., Zupko K., Poria Y., Nef P., Lazarovits J., Horn S., Khen M. and Lancet D. (1991) Odorant signal termination by olfactory UDP glucuronosyl transferase. *Nature* **349**, 790–793.

Leal W. S. (2000) Duality monomer–dimer of the pheromone-binding protein from *Bombyx mori*. *Biochem. Biophys. Res. Commun.* **268**, 521–529.

Leal W. S., Nikonova L. and Peng G. (1999) Disulfide structure of the pheromone binding protein from the silkworm moth, *Bombyx mori*. *FEBS Letters.* **468**, 85–90.

Lee C. Y. and Baehrecke E. H. (2001) Steroid regulation of autophagic programmed cell death during development. *Development* **128**, 1443–1455.

Lee K. and Strausfeld N. J. (1990). Structure, distribution and number of surface sensilla and their receptor cells on the olfactory appendage of the male moth *Manduca sexta*. *J. Neurocytol.* **19**, 519–538.

Liu B. and Krieger M. (2002) Highly purified scavenger receptor class B, type I reconstituted into phosphatidylcholine/cholesterol liposomes mediates high affinity high density lipoprotein binding and selective lipid uptake. *J. Biol. Chem.* **277**, 34125–34135.

Maddison D. R. and Maddison W. P. (1998) The tree of life: a multi-authored, distributed Internet project containing information about phylogeny and biodiversity. Internet address: http://phylogeny.arizona.edu/tree/phylogeny.html

Maïbèche-Coisné M., Sobrio F., Delaunay T., Lettere M., Dubroca J., Jacquin-Joly E. and Nagnan-LeMeillour P. (1997) Pheromone Binding Proteins of the moth Mamestra brassicae: specificity of ligand binding. *Insect Biochem. Mol. Biol.* **27**, 213–221.

Maïbèche-Coisné M., Jacquin-Joly E., Francois M. C. and Nagnan-Le Meillour P. (1998a) Molecular cloning of two pheromone binding proteins in the cabbage armyworm *Mamestra brassicae*. *Insect Biochem. Mol. Biol.* **28**, 815–818.

Maïbèche-Coisné M., Longhi S., Jacquin-Joly E., Brunel C., Egloff M. P., Gastinel L., Cambillau C., Tegoni M. and Nagnan-Le Meillour P. (1998b) Molecular cloning and bacterial expression of a general odorant-binding protein from the cabbage armyworm *Mamestra brassicae*. *Eur. J. Biochem.* **258**, 768–774.

Maida R., Mameli M., Krieger J., Breer H., Ziegelberger G. and Steinbrecht R. A. (1999) Complex expression pattern of odorant-binding proteins in *Bombyx mori*. In *Gottingen Neurobiology Report*, eds N. Elsner and U. Eysel, p. 359. Thieme, Stuttgart.

Maida R., Krieger J., Gebauer T., Lange U. and Ziegelberger G. (2000) Three pheromone-binding proteins in olfactory sensilla of the two silkmoth species Antheraea polyphemus and *Antheraea pernyi*. *Eur. J. Biochem.* **267**, 2899–2908.

Mameli M., Kreiger J. and Breer H (1997) Unpublished sequences, Accession Number Y10970.

Mameli M., Tuccini A., Mazza M., Petacchi R. and Pelosi P. (1996) Soluble proteins in chemosensory organs of Phasmids. *Insect Biochem. Mol. Biol.* **26**, 875–882.

Mannervik B. and Danielson U. H. (1988) Glutathione transferases – structure and catalytic activity. *CRC Crit. Review. Biochem.* **23**, 283–337.

Marchese S., Angeli S., Andolfo A., Scaloni A., Brandazza A., Mazza M., Picimbon J. F., Leal W. S. and Pelosi P. (2000) Soluble proteins from chemosensory organs of *Eurycantha calcarata* (Insecta, Phasmatodea). *Insect Biochem. Mol. Biol.* **30**, 1091–1098.

Mayer M. S. (1975) Hydrolysis of sex pheromone by the antennae of *Trichoplusia ni*. *Experentia* **31**, 452–454.

McKenna M. P., Hekmat-Scafe D. S., Gaines P. and Carlson J. R. (1994) Putative *Drosophila* pheromone-binding proteins expressed in a subregion of the olfactory system. *J. Biol. Chem.* **269**, 16340–16347.

Merritt T. J., LaForest S., Prestwich G. D., Quattro J. M. and Vogt R. G. (1998) Patterns of gene duplication in lepidopteran pheromone binding proteins. *J. Mol. Evol.* **46**, 272–276.

Miyazaki A., Nakayama H. and Horiuchi S. (2002) Scavenger receptors that recognize advanced glycation end products. *Trends Cardiovasc. Med.* **12**, 258–262.

Moore K J., El Khoury J., Medeiros L. A., Terada K., Geula C., Luster A. D. and Freeman M. W. (2002) A CD36-initiated signaling cascade mediates inflammatory effects of beta-amyloid. *J. Biol. Chem.*. (e-published).

Morton R. A. and Singh R. S. (1985) Biochemical properties, homology, and genetic variation of *Drosophila* "nonspecific" esterases. *Biochem. Genet.* **23**, 959–973.

Mosbah A., Campanacci V., Lartigue A., Tegoni M., Cambillau C. and Darbon H. (2002) The solution structure of a chemosensory protein (CSP) from the moth *Mamestra brassicae. Biochem. J.* (e-published, September).

Nef P., Heldman J., Lazard D., Margalit T., Jaye M., Hanukoglu I. and Lancet D. (1989) Olfactory-specific cytochrome P-450. *J. Biol. Chem.* **264**, 6780–6785.

Nei M. and Kumar S. (2000) *Molecular Evolution and Phylogenetics.* Oxford University Press, New York.

Newcomb R. D., Campbell P. M., Russell R. J. and Oakeshott J. G. (1997) cDNA cloning, baculovirus-expression and kinetic properties of the esterase, E3, involved in organophosphorus resistance in *Lucilia cuprina. Insect Biochem. Mol. Biol.* **27**, 15–25.

Newcomb R. D., Sircy T. M., Rassam M. and Greenwood D. R. (2002) Pheromone binding proteins of *Epiphyas postvittana* (Lepidoptera: Tortricidae) are encoded at a single locus. *Insect Biochem. Mol. Biol.* (in press).

Nikonov A. A., Peng G., Tsurupa G. and Leal W. S. (2002) Unisex pheromone detectors and pheromone-binding proteins in scarab beetles. *Chem. Senses* **27**, 495–504.

Novotny V., Basset Y., Miller S. E., Weiblen G. D., Bremer B., Cizek L. and Drozd P. (2002) Low host specificity of herbivorous insects in a tropical forest. *Nature* **416**, 841–844.

Oquendo P., Hundt E., Lawler J. and Seed B. (1989) CD36 directly mediates cytoadherence of *Plasmodium falciparum* parasitized erythrocytes. *Cell* **58**, 95–101.

Ozaki M., Morisaki K., Idei W., Ozaki K. and Tokunaga F. (1995) A putative lipophilic stimulant carrier protein commonly found in the taste and olfactory systems. A unique member of the pheromone-binding protein superfamily. *Eur. J. Biochem.* 230, 298–308.

Paesen G. C. and Happ G. M. (1995) The B proteins secreted by the tubular accessory sex glands of the male mealworm beetle, *Tenebrio molitor*, have sequence similarity to moth pheromone-binding proteins. *Insect Biochem. Mol. Biol.* **25**(3), 401–408.

Park S. K., Shanbhag S. R., Wang Q., Hasan G., Steinbrecht R. A. and Pikielny C. W. (2000) Expression patterns of two putative odorant-binding proteins in the olfactory organs of *Drosophila melanogaster* have different implications for their functions. *Cell Tissue Res.* **300**, 181–197.

Pelosi P. (2001) The role of perireceptor events in vertebrate olfaction. *Cell Mol. Life Sci.* **58**, 503–509.

Peng G. and Leal W. S. (2001) Identification and cloning of a pheromone-binding protein from the Oriental beetle, *Exomala orientalis J. Chem. Ecol.* **27**, 2183–2192.

Picimbon J. F. and Leal W. S. (1999) Olfactory soluble proteins of cockroaches. *Insect Biochem. Mol. Biol.* **29**, 973–978.

Picimbon J. F., Dietrich K., Breer H. and Krieger J. (2000) Chemosensory proteins of *Locusta migratoria* (Orthoptera: Acrididae). *Insect Biochem. Mol. Biol.* **30**, 233–241.

Pikielny C. W., Hasan G., Rouyer F. and Rosbash M. (1994). Members of a family of *Drosophila* putative odorant-binding proteins are expressed in different subsets of olfactory hairs. *Neuron* **12**, 35–49.

Pitts R. J. and Zwiebel L. J. (2001) Isolation and characterization of the Xanthine dehydrogenase gene of the Mediterranean fruit fly, *Ceratitis capitata*. *Genetics* **158**, 1645–1655.

Plettner E., Lazar J., Prestwich E. G. and Prestwich G. D. (2000) Discrimination of pheromone enantiomers by two pheromone binding proteins from the gypsy moth *Lymantria dispar. Biochemistry* **39**, 8953–8962.

Podrez E. A., Poliakov E., Shen Z., Zhang R., Deng Y., Sun M., Finton P. J., Shan L., Febbraio M., Hajjar D. P., Silverstein R. L., Hoff H. F., Salomon R. G. and Hazen S. L. (2002a) A novel family of atherogenic oxidized phospholipids promotes macrophage foam cell formation via the scavenger receptor CD36 and is enriched in atherosclerotic lesions. *J. Biol. Chem.* **277**, 38517–38523.

Podrez E. A., Poliakov E., Shen Z., Zhang R., Deng Y., Sun M., Finton P. J., Shan L., Gugiu B., Fox P. L., Hoff H. F., Salomon R. G. and Hazen S. L. (2002b) Identification of a novel family of oxidized phospholipids that serve as ligands for the macrophage scavenger receptor CD36. *J. Biol. Chem.* **277**, 38503–38516.

Pophof B. (1998) Inhibitors of sensillar esterase reversibly block the responses of moth pheromone receptor cells. *J. Comp. Physiol. A* **183**, 153–164.

Pophof B., Gebauer T. and Ziegelberger G. (2000) Decyl-thio-trifluoropropanone, a competitive inhibitor of moth pheromone receptors. *J. Comp. Physiol. A* **186**, 315–323.

Prestwich G. D. (1985) Molecular communication of insects. *Quart. Rev. Biol.* **60**, 437–456.

Prestwich G. D. (1987) Chemistry of pheromone and hormone metabolism in insects. *Science* **237**, 999–1006.

Prestwich G. D., Du G. and LaForest S. (1995) How is pheromone specificity encoded in proteins? *Chem. Senses* **20**, 461–469.

Prestwich G. D., Golec F. A. and Andersen N. H. (1984) Synthesis of a highly tritiated photoaffinity labeled pheromone analog for the moth *Antheraea polyphemus*. *J. Labelled Cmpnd. Radiopharm.* **21**, 593–601.

Quero C., Camps F. and Guerrero A (1995) Behavior of processionary males (*Thaumetopoea pityocampa*) induced by sex pheromone and analogs in a wind tunnel. *J. Chem. Ecol.* **21**, 1957–1969.

Raming K., Krieger J. and Breer H. (1989) Molecular cloning of an insect pheromone-binding protein. *FEBS Lett.* **256**, 215–218.

Raming K., Krieger J. and Breer H. (1990) Primary structure of a pheromone-binding protein from *Antheraea pernyi*: homologies with other ligand-carrying proteins. *J. Comp. Physiol.* B **160**, 503–509.

Rasmussen J. T., Berglund L., Rasmussen M. S. and Petersen T. E. (1998) Assignment of disulfide bridges in bovine CD36. *Eur. J. Biochem.* **257**, 488–494.

Reaven E., Leers-Sucheta S., Nomoto A. and Azhar S. (2001) Expression of scavenger receptor class B type 1 (SR-BI) promotes microvillar channel formation and selective cholesteryl ester transport in a heterologous reconstituted system. *Proc. Natl. Acad. Sci. USA* **98**, 1613–1618.

Renou M., Berthier A., Desbarats L., Van der Pers J. and Guerrero A. (1999) Actographic analysis of the effects of an esterase inhibitor on male moth responses to sex pheromone. *Chem. Senses* **24**, 423–428.

Renou M. and Guerrero A. (2000) Insect parapheromones in olfaction research and semiochemical-based pest control strategies. *Annu. Rev. Entomol.* **45**, 605–630.

Riba M., Sans A., Bau P., Grolleau G., Renou M. and Guerrero A. (2001) Pheromone response inhibitors of the corn stalk borer Sesamia nonagrioides. Biological evaluation and toxicology. *J. Chem. Ecol.* **27**, 1879–1897.

Riviere S., Lartigue A., Quennedey B., Campanacci V., Farine J. P., Tegoni M., Cambillau C. and Brossut R. (2003) A pheromone-binding protein from the cockroach Leucophaea maderae: cloning, expression and pheromone binding. *Biochem J.* **371**, 573–579.

Robertson H. M., Brakebill C., Schmidt L., Mostafavipour P., Todres E., Ramsdell K., Patch H. M., Walden K. K. O. and Nardi J. B. (2001) Molecular evolution of insect odorant binding proteins. Unpublished sequences, Accession Numbers AF393488, AF393490, AF393498, AF393499. AF393500.

Robertson H. M., Martos R., Sears C. R., Todres E. Z., Walden K. K. and Nardi J. B. (1999) Diversity of odourant binding proteins revealed by an expressed sequence tag project on male *Manduca sexta* moth antennae. *Insect Mol. Biol.* **8**, 501–518.

Robertson H. M., Walden K. K. O., Mostafavipour P., Todres E. and Rovelstad S. (2000) Honey bee antennal ESTs. Unpublished sequences, Accession Numbers BE844614, BE844604, BE844603, BE844579, BE844525.

Robin G. C. de Q., Claudianos C., Russell R. J. and Oakeshott J. G. (2000) Reconstructing the diversification of alpha-esterases: comparing the gene clusters of *Drosophila buzzatii* and *D. melanogaster*. *J. Mol. Evol.* **51**, 149–160.

Rogers M. E., Jani M. K. and Vogt R. G. (1999) An olfactory specific glutathione S-transferase in the sphinx moth *Manduca sexta*. *J. Exp. Biol.* **202**, 1625–1637.

Rogers M. E., Steinbrecht R. A. and Vogt R. G. (2001a) Expression of SNMP-1 in olfactory neurons and sensilla of male and female antennae of the silkmoth *Antheraea polyphemus*. *Cell and Tissue Res.* **303**, 433–446.

Rogers M. E., Krieger J. and Vogt R. G. (2001b) Antennal SNMPs (sensory neuron membrane proteins) of Lepidoptera define a unique family of invertebrate CD36-like proteins. *J. Neurobiol.* **49**, 47–61.

Rogers M. E., Sun M., Lerner M. R. and Vogt R. G. (1997) SNMP-1, a novel membrane protein of olfactory neurons of the silk moth *Antheraea polyphemus* with homology to the CD36 family of membrane proteins. *J. Biol. Chem.* **272**, 14792–14799.

Ross K. G. (1997) Multilocus evolution in fire ants: effects of selection, gene flow and recombination. *Genetics* **145**, 961–974.

Ross K. G. and Keller L. (1998) Genetic control of social organization in an ant. *Proc. Natl. Acad. Sci. USA* **95**, 14232–14237.

Rothemund S., Liou Y. C., Davies P. L., Krause E. and Sonnichsen F. D. (1999) A new class of hexahelical insect proteins revealed as putative carriers of small hydrophobic ligands. *Structure Fold Des.* **7**, 1325–1332.

Rybczynski R., Reagan J. and Lerner M. R. (1989) A pheromone-degrading aldehyde oxidase in the antennae of the moth *Manduca sexta*. *J. Neurosci.* **9**, 1341–1353.

Rybczynski R., Vogt R. G. and Lerner M. R. (1990) Antennal-specific pheromone-degrading aldehyde oxidases from the moths *Antheraea polyphemus* and *Bombyx mori*. *J. Biol. Chem.* **32**, 19712–19715.

Ryeom S. W., Sparrow J. R. and Silverstein R. L. (1996) CD36 participates in the phagocytosis of rod outer segments by retinal pigment epithelium. *J. Cell Sci.* **109**, 387–395.

Saitou N. and Nei M. (1987) The neighbor joining method: a new method for reconstructing phylogenetic trees. *Mol. Biol. Evol.* **4**, 406–425.

Sandler B. H., Nikonova L., Leal W. S. and Clardy J. (2000) Sexual attraction in the silkworm moth: structure of the pheromone-binding-protein–bombykol complex. *Chem Biol.* **7**, 143–151.

Sandoval IV., Martinez-Arca S., Valdueza J., Palacios S. and Holman G. D. (2000) Distinct reading of different structural determinants modulates the dileucine-mediated transport steps of the lysosomal membrane protein LIMPII and the insulin-sensitive glucose transporter GLUT4. *J. Biol. Chem.* **275**, 39874–39885.

Scarselli E., Ansuini H., Cerino R., Roccasecca R. M., Acali S., Filocamo G., Traboni C., Nicosia A., Cortese R. and Vitelli A. (2002) The human scavenger receptor class B type I is a novel candidate receptor for the hepatitis C virus. *EMBO J.* **21**, 5017–5025.

Schneider D. (1969) Insect olfaction: deciphering system for chemical messages. *Science* **163**, 1031–1037.

Schweitzer E. S., Sanes J. R. and Hildebrand J. G. (1976) Ontogeny of electroantennogram responses in the moth, *Manduca sexta. J. Insect Physiol.* **2**, 955–960.

Scott K., Brady R. Jr, Cravchik A., Morozov P., Rzhetsky A., Zuker C. and Axel R. (2001) A chemosensory gene family encoding candidate gustatory and olfactory receptors in *Drosophila. Cell* **104**, 661–673.

Shanbhag S. R., Müller B. and Steinbrecht R. A. (1999) Atlas of olfactory organs of *Drosophila melanogaster.* I. Types, external organization, innervation and distribution of olfactory sensilla. *Int. J. Insect Morphology Embryology* **28**, 377–397.

Shanbhag S. R., Hekmat-Scafe D., Kim M. S., Park S. K., Carlson J. R., Pikielny C., Smith D. P. and Steinbrecht R. A. (2001a) Expression mosaic of odorant-binding proteins in *Drosophila* olfactory organs. *Microsc. Res. Tech.* **55**, 297–306.

Shanbhag S. R., Park S.-K., Pikielny C. W. and Steinbrecht R. A. (2001b) Gustatory organs of *Drosophila melanogaster*: fine structure and expression of the putative odorant-binding protein PBPRP2. *Cell Tissue Res.* **304**, 423–437.

Shields V. D. C. and Hildebrand J. G. (1999a) Fine structure of antennal sensilla of the female sphinx moth, *Manduca sexta* (Lepidoptera: Sphingidae). I. Trichoid and basiconic sensilla. *Can. J. Zool.* **77**, 290–301.

Shields V. D. C. and Hildebrand J. G. (1999b). Fine structure of antennal sensilla of the female sphinx moth, *Manduca sexta* (Lepidoptera: Sphingidae). II. Auriculate, coeloconic, and styliform complex sensilla. *Can. J. Zool.* **77**, 302–313.

Shields V. D. C. and Hildebrand J. G. (2001) Responses of a population of antennal olfactory receptor cells in the female moth *Manduca sexta* to plant-associated volatile organic compounds. *J. Comp. Physiol. A* **186**, 1135–1151.

Simpson P. J., Nighorn A. and Morton D. B. (1999) Identification of a novel guanylyl cyclase that is related to receptor guanylyl cyclases but lacks extracellular and transmembrane domains. *J. Biol. Chem.* **274**, 4440–4446.

Snyder M. J., Walding J. K. and Feyereisen R. (1995) Glutathione S-transferases from larval *Manduca sexta* midgut: sequence of two cDNAs and enzyme induction. *Insect Biochem. Mol. Biol.* **25**, 455–465.

Starratt A. M., Dahm K. H., Allen N., Hildebrand J. G., Payne T. L. and Roller H. (1979) Bombykal, a sex pheromone of the sphinx moth, *Manduca sexta. Z. Naturforsch.* **34C**, 9–12.

Steinbrecht R. A. (1969) Comparative morphology of olfactory receptor. In *Olfaction and Taste III*, ed. C. Pfaffmann, pp. 3–21, Rockefeller Univ. Press, New York.

Steinbrecht R. A. (1996) Are odorant-binding proteins involved in odorant discrimination? *Chem. Senses* **21**, 719–727.

Steinbrecht R. A. (1997) Pore structures in insect olfactory sensilla: a review of data and concepts. *Int. J. Insect. Morphol. Embryol.* **26**, 229–245.

Steinbrecht R. A. (1998) Odorant-binding proteins: expression and function. In *Olfaction and Taste*, XII. ed. C. Murphy, *Annals of the NY Academy of Sciences* **855**, 323–332.

Steinbrecht R. A. (1999) Olfactory receptors. In *Atlas of Arthropod Sensory Receptors, Dynamic Morphology in Relation to Function*, eds E. Eguchi and Y. Tominaga. Springer-Verlag, Tokyo.

Steinbrecht R. A. and Müller B. (1971) On the stimulus conducting structures in insect olfactory receptors. *Z. Zellforsch.* **117**, 570–575.

Steinbrecht R. A., Laue M. and Ziegelberger G. (1995) Immunolocalization of pheromone-binding protein and general odorant-binding protein in olfactory sensilla of the silk moths *Antheraea* and *Bombyx*. *Cell Tissue Res.* **282**, 203–217.

Swofford D. L. (2000). *PAUP**. *Phylogenetic Analysis Using Parsimony* (*and Other Methods). Version 4. Sinauer Associates, Sunderland, Massachusetts.

Tabuchi N., Akasaki K. and Tsuji H. (2000) Two acidic amino acid residues, Asp(470) and Glu(471), contained in the carboxyl cytoplasmic tail of a major lysosomal membrane protein, LGP85/LIMP II, are important for its accumulation in secondary lysosomes. *Biochem. Biophys. Res. Commun.* **270**, 557–563.

Tasayco M. L. and Prestwich G. D. (1990a) Aldehyde oxidizing enzymes in an adult moth. *In vitro* study of pheromone degradation in *Heliothis virescens*. *Arch. Biochem. Biophys.* **278**, 444–451.

Tasayco M. L. and Prestwich G. D. (1990b) A specific affinity reagent to distinguish aldehyde dehydrogenases and oxidases. Enzymes catalyzing aldehyde oxidation in an adult moth. *J. Biol. Chem.* **265**, 3094–3101.

Tasayco M. L. and Prestwich G. D. (1990c) Aldehyde oxidases and dehydrogenases in antennae of 5 moth species. *Insect Biochem.* **20**, 691–700.

Taylor T. R., Ferkovich S. M. and Van Essen F. (1981) Increased pheromone catabolism by antennal esterases after adult eclosion of the cabbage looper moth. *Experentia* **37**, 729–731.

Tegoni M., Pelosi P., Vincent F., Spinelli S., Campanacci V., Grolli S., Ramoni R. and Cambillau C. (2000) Mammalian odorant binding proteins. *Biochim. Biophys. Acta* **1482**, 229–240.

Thompson J. D., Gibson T. J., Plewniak F., Jeanmougin F. and Higgins D. G. (1997) The ClustalX windows interface: flexible strategies for multiple sequence alignment aided by quality analysis tools. *Nucleic Acids Research* **24**, 4876–4882.

Trapido-Rosenthal H. G., Carr W. E. and Gleeson R. A. (1987) Biochemistry of an olfactory purinergic system: dephosphorylation of excitatory nucleotides and uptake of adenosine. *J. Neurochem* **49**, 1174–1182.

Trapido-Rosenthal H. G., Carr W. E. and Gleeson R. A. (1990) Ectonucleotidase activities associated with the olfactory organ of the spiny lobster. *J. Neurochem.* **55**, 88–96.

Tsukamoto K., Kinoshita M., Kojima K., Mikuni Y., Kudo M., Mori M., Fujita M., Horie E., Shimazu N. and Teramoto T. (2002) Synergically increased expression of CD36, CLA-1 and CD68, but not of SR-A and LOX-1, with the progression to foam cells from macrophages. *J. Atheroscler. Thromb.* **9**, 57–64.

Tumlinson J., Brennan M. M., Doolittle R. E., Mitchell E. R., Brabham A., Mazomenos B. E., Baumhover A. H. and Jackson D. M. (1989) Identification of a pheromone blend attractive to *Manduca sexta* (L.) males in a wind tunnel. *Arch. Insect Biochem. Physiol.* **10**, 255–271.

Tumlinson J., Mitchell E. R., Doolittle R. E. and Jackson D. (1994) Field tests of synthetic *Manduca sexta* sex pheromone. *J. Chem. Ecol.* **20**, 579–591.

Udomsangpetch R., Pipitaporn B., Silamut K., Pinches R., Kyes S., Looareesuwan S., Newbold C. and White N. J. (2002) Febrile temperatures induce cytoadherence of ring-stage *Plasmodium falciparum*-infected erythrocytes. *Proc. Natl. Acad. Sci. USA.* **99**, 11825–11829.

Vega M. A., Sigui-Real B., Garcia J. A., Cales C., Rodriguez F., Vanderkerkhove J. and Sandoval I. V. (1991) Cloning, sequencing, and expression of a cDNA encoding rat LIMP II, a novel 74-kDa lysosomal membrane protein related to the surface adhesion protein CD36. *J. Biol. Chem.* **266**, 16818–16824.

Vogt R. G. (1987) The molecular basis of pheromone reception: its influence on behavior. In *Pheromone Biochemistry*, eds G. D. Prestwich and G. J. Blomquis, pp. 385–431. Academic Press, New York.

Vogt R. G. (1995) Molecular genetics of moth olfaction: a model for cellular identity and temporal assembly of the nervous system. In *Molecular Model Systems in the Lepidoptera*, eds M. R. Goldsmith and A. S. Wilkins, pp. 341–367. Cambridge University Press, Cambridge, UK.

Vogt R. G. (2002) Odorant binding proteins of the malaria mosquito *Anopheles gambiae*; possible orthologues of the OS-E and OS-F OBPs of *Drosophila melanogaster*. *J. Chem. Ecol.* **28**, RC29-RC35 (http://www.kluweronline.com/issn/0098-0331)

Vogt R. G., Callahan F. E., Rogers M. E. and Dickens J. C. (1999) Odorant binding protein diversity and distribution among the insect orders, as indicated by LAP, an OBP-related protein of the true bug *Lygus lineolaris* (Hemiptera, Heteroptera). *Chem. Senses* **24**, 481–495.

Vogt R. G., Köehne A. C., Dubnau J. T. and Prestwich G. D. (1989) Expression of pheromone binding proteins during antennal development in the gypsy moth *Lymantria dispar. J. Neurosci.* **9**, 3332–3346.

Vogt R. G. and Lerner M. R. (1989) Two groups of odorant binding proteins in insects suggest specific and general olfactory pathways. *Neurosci. Abstr.* **15**, 1290.

Vogt R. G. and Prestwich G. D. (1988) Variation in olfactory proteins: evolvable elements encoding insect behavior. *Olfaction and Taste IX, Annals NY Acad. Sci.* **510**, 689–691.

Vogt R. G., Prestwich G. D. and Lerner M. R. (1991a) Odorant-binding-protein subfamilies associate with distinct classes of olfactory receptor neurons in insects. *J. Neurobiol.* **22**, 74–84.

Vogt R. G., Prestwich G. D. and Lerner M. R. (1991b) Molecular cloning and sequencing of general-odorant binding proteins GOBP1 and GOBP2 from tobacco hawk moth *Manduca sexta*: comparisons with other insect OBPs and their signal peptides. *J. Neurosci.* **11**, 2972–2984.

Vogt R. G., Prestwich G. D. and Riddiford L. M. (1988) Sex pheromone receptor proteins, visualization using a radiolabeled photoaffinity analog. *J. Biol. Chem.* **263**, 3952–3959.

Vogt R. G. and Riddiford L. M. (1981) Pheromone binding and inactivation by moth antennae. *Nature* **293**, 161–163.

Vogt R. G. and Riddiford L. M. (1986a) Pheromone reception: a kinetic equilibrium. In *Mechanisms in Insect Olfaction*, eds T. L. Payne, M. C. Birch and Kennedy C. E. J, pp. 201–208. Clarendon Press, Oxford.

Vogt R. G. and Riddiford L. M. (1986b) Scale esterase: a pheromone degrading enzyme from the wing scales of the silk moth *Antheraea polyphemus. J. Chem. Ecol.* **12**, 469–482.

Vogt R. G., Riddiford L. M. and Prestwich G. D. (1985) Kinetic properties of a sex pheromone-degrading enzyme: the sensillar esterase of *Antheraea polyphemus. Proc. Natl. Acad. Sci. USA* **82**, 8827–8831.

Vogt R. G., Rogers M. E., Franco M. D. and Sun M. (2002) A comparative study of odorant binding protein genes: differential expression of the PBP1-GOBP2 gene cluster in *Manduca sexta* (Lepidoptera) and the organization of OBP genes in *Drosophila melanogaster* (Diptera). *J. Exp. Biol.* **205**, 719–744.

Vogt R. G., Rybczynski R., Cruz M. and Lerner M. R. (1993) Ecdysteroid regulation of olfactory protein expression in the developing antenna of the tobacco hawk moth, *Manduca sexta. J. Neurobiolol.* **22**, 581–597.

Vosshall L. B., Amrein H., Morozov P. S., Rzhetsky A. and Axel R. (1999) A spatial map of olfactory receptor expression in the *Drosophila* antenna. *Cell* **96**, 725–736.

Vosshall L. B., Wong A. M. and Axel R. (2000) An olfactory sensory map in the fly brain. *Cell* **102**, 147–159.

Wang G. R. and Guo Y. Y. (2000) Cloning and expression of odorant-binding proteins from *Helicoverpa armigera*. Unpublished sequence, Accession Numbers AJ278991, AJ278992.

Wang G.-R., Guo Y.-Y. and Wu K.-M. (2001) Cloning and expression of general odorant binding protein 1 from *Helicoverpa armigera*. Unpublished sequence, Accession Number AY049739.

Whiting M. F., Carpenter J. C., Wheeler Q. D. and Wheeler W. C. (1997) The Strepsiptera problem: phylogeny of the holometabolous insect orders inferred from 18S and 28S ribosomal DNA sequences and morphology. *Systematics Bulletin* **46**, 1–68.

Willett C. S. (2000a) Evidence for directional selection acting on pheromone-binding proteins in the genus *Choristoneura*. *Mol. Biol. Evol.* **17**, 553–562.

Willett C. S. (2000b) Do pheromone binding proteins converge in amino acid sequence when pheromones converge? *J. Mol. Evol.* **50**, 175–183.

Willett C. S. and Harrison R. G. (1999a) Insights into genome differentiation: pheromone-binding protein variation and population history in the European corn borer (*Ostrinia nubilalis*). *Genetics* **153**, 1743–1751.

Willett C. S. and Harrison R. G. (1999b) Pheromone binding proteins in the European and Asian corn borers: no protein change associated with pheromone differences. *Insect Biochem. Mol. Biol.* **29**, 277–284.

Wojtasek H., Hansson B. S. and Leal W. S. (1998) Attracted or repelled? – a matter of two neurons, one pheromone binding protein, and a chiral center. *Biochem. Biophys. Res. Commun.* **250**, 217–222.

Wojtasek H. and Leal W. S. (1999a) Degradation of an alkaloid pheromone from the pale-brown chafer, *Phyllopertha diversa* (Coleoptera:carabaeidae), by an insect olfactory cytochrome P450. *FEBS Letters* **458**, 333–336.

Wojtasek H. and Leal W. S. (1999b) Conformational change in the pheromone-binding protein from *Bombyx mori* induced by pH and by interaction with membranes. *J. Biol. Chem.* **274**, 30950–30956.

Wojtasek H., Picimbon J. F. and Leal W. S. (1999) Identification and cloning of odorant binding proteins from the scarab beetle *Phyllopertha diversa*. *Biochem. Biophys. Res. Commun.* **263**, 832–837.

Yu S. J. (1983) Induction of detoxifying enzymes by allelochemicals and host plants in the fall armyworm. *Pest. Biochem. Physiol.* **19**, 330–336.

Yu S. J. (1984) Interactions of allelochemicals with detoxication enzymes of insecticide susceptible and resistant fall armyworm. *Pest. Biochem. Physiol.* **22**, 60–68.

Yu S. J. (1989) Purification and characterization of glutathione transferases from five phytophagous Lepidoptera. *Pest. Biochem. Physiol.* **35**, 97–105.

Zhang S.-G., Maida R. and Steinbrecht R. A. (2001) Immunolocalization of odorant-binding proteins in noctuid moths (Insecta, Lepidoptera). *Chem. Senses* **26**, 885–896.

Zimmer-Faust R. K., Gleeson R. A. and Carr W. E. S (1988) The behavioral response of spiny lobsters to ATP: evidence for mediation by P2-like chemosensory receptors *Biol. Bull.* **175**, 167–174.

15

Proteins that make sense

Walter S. Leal

15.1 Introduction

Insect antennae are biosensors *par excellence*. They can discriminate a myriad of physiologically irrelevant chemical compounds in the environment from essential chemical signals (semiochemicals), such as sex pheromones. Even small modifications in the pheromone molecules render them completely inactive (Kaissling, 1987). This remarkable selectivity of the olfactory system is coupled with an inordinate sensitivity. Sex pheromones are released by females to advertise their readiness to mate. While performing this critical task of recruiting males for reproduction, females avoid being conspicuous by releasing only minute amounts of pheromone. Males cope with this situation by detecting such small amounts of pheromone that the signal-to-noise ratio of the system approaches the theoretical limit for a detector.

In addition to sensitivity and selectivity (discrimination), odor-oriented navigation requires a dynamic process of signal inactivation. While flying *en route* to a pheromone-emitting female, males encounter pheromone molecules as intermittent signals comprised of short bursts of high flux separated by periods during which the flux is zero. The average duration of spikes within puffs of pheromones is on the millisecond scale and it decreases as the moth comes closer to the pheromone source (Murlis *et al.*, 2000). Thus, a male moth has to detect selectively minute amounts of pheromones and reset the pheromone detectors on a millisecond timescale.

In moths, pheromones are detected by a compound nose (Steinbrecht, 1999), i.e. a network of hair-like sensilla distributed over the surface of the antennae. For example, in the wild silkmoth, *Antheraea polyphemus*, each male antenna has ca. 60 000 pheromone-sensitive sensilla trichodea and 10 000 sensilla basiconica (general odorant detectors) (Keil, 1984a; Meng *et al.*, 1989), whereas the female antennae lack pheromone-detecting sensilla and have ca. 12 000 sensilla basiconica (Boeckh *et al.*, 1960).

As demonstrated initially in *Drosophila* the olfactory receptor proteins (Clyne *et al.*, 1999; Vosshall *et al.*, 1999; Vosshall *et al.*, 2000) are located on the dendrites of olfactory receptor neurons (ORNs), which are themselves surrounded by a sensillar lymph inside the receptor hair. The information-carrying pheromones (and other semiochemicals) are largely hydrophobic compounds. For their chemical information to make sense these odorants must reach the olfactory receptors. Here, the signal transduction starts, i.e. the chemical language is "translated" into the brain language (electrical spikes). Thus, the olfactory transduction involves a series of events, including the uptake of odorants from the external environment (air space), binding to odorant-binding proteins (OBPs), transport (diffusion) through the sensillar lymph, release of ligands to their receptors and the interactions of stimulus and receptors, which trigger a cascade of events leading to nervous activity (spikes). Signal transduction (*sensu stricto*) refers to the intracellular events, whereas the extracellular processes associated with the access of odorant to the receptors as well as the post-interactive events related to inactivation (deactivation) of chemical signals are named perireceptor events (Getchell *et al.*, 1984) or early olfactory processing. Mathematical modeling of the network of "reactions" taking place in these extracellular processes and their correlation with the first electrical responses of the receptor cells (receptor potential) showed that the rate-determining steps for the kinetics of insect olfaction occur during perireceptor events rather than signal transduction (Kaissling, 2001a). This chapter is focused on the perireceptor events of insect olfaction.

15.2 Odorant-binding proteins

15.2.1 Classification

cDNAs and genes encoding odorant-binding proteins have been cloned from a number of lepidopteran species. The OBPs expressed mainly in male antennae that bind pheromones were named pheromone-binding proteins (PBPs). Other antennae-specific OBPs which are expressed in both sexes, or predominantly in female antennae and for which binding data were not available, were named general odorant-binding proteins (GOBPs), a terminology that infers that they did not have binding specificity. Comparison of the predicted amino acid sequences for these OBPs at a time when only a handful of cDNA sequences were available already suggested (and this has since then been confirmed) that PBPs are variable, whereas GOBPs are highly conserved (Vogt *et al.*, 1991); the latter can be further divided into two groups by sequence homology: GOBP1 and GOBP2.

Because binding assays are not readily available, it is acceptable to classify an antennae-specific protein as PBP when it shows clear sequence similarity to previously identified pheromone-binding proteins. A group of cDNA clones encoding proteins specific to antennae, but with low similarity to PBPs and GOBPs, has, been named antennal-binding proteins (ABPX) (Krieger *et al.*, 1996). Hitherto, binding of hydrophobic ligands to ABPXs has not been demonstrated. In this review I use the term OBP in a broad sense, even for proteins for which binding to pheromones has been extensively demonstrated like the pheromone-binding protein from *Bombyx mori* (BmPBP) (Maida *et al.*, 1993; Wojtasek and Leal, 1999). Note that I refer to odorant-binding proteins as the olfactory proteins containing three disulfide bridges (see discussion below), whereas the olfactory proteins with two disulfide bridges are grouped as chemosensory proteins (CSPs) and discussed elsewhere in this book (Nagnan-Le Meillour, Chapter 17, in this volume).

In addition to Lepidoptera, OBPs have been identified and/or cloned from various insect orders, such as Coleoptera, Diptera, Hymenoptera, and Hemiptera (Vogt *et al.*, 1999; Vogt *et al.*, 2002). The recent identification of OBPs from a primitive termite species, *Zootermopsis nevadensis nevadensis* (Isoptera) (Ishida *et al.*, 2002a) suggests that this gene family is distributed throughout the Neopteran orders. *Drosophila* (Diptera) has the largest number of putative OBPs identified for a single species. Since the first cloning of OBPs in *Drosophila melanogaster* (McKenna *et al.*, 1994; Pikielny *et al.*, 1994) the number of cloned DmelOBPs has steadily increased (Kim *et al.*, 1998; Robertson *et al.*, 1999) and as many as 38 OBPs have been cloned to date (Galindo and Smith, 2001; Graham and Davies, 2002). Twenty-nine putative OBPs have been identified from the genome of the malaria mosquito *Anopheles gambiae* (Biesmann *et al.*, 2002; Vogt, 2002), whereas only one OBP has been isolated from the female mosquito *Culex quinquefasciatus* (Ishida *et al.*, 2002b).

Lepidoptera is by far the most extensively studied group, with OBPs characterized from at least 21 species (Newcomb *et al.*, 2002), followed by scarab beetles (Coleoptera: Scarabaeidae), with seven species studied to date. Whereas lepidopteran species belonging to Saturniidae, Bombycidae, Sphingidae, Lymantridae, Tortricidae, and Pyralidae were studied, the work on scarab beetles has been focused on two subfamilies: Rutelinae and Melolonthinae. Chemical communication has been thoroughly investigated (Leal, 1998) in rutelines and melolonthines. In all rutelines investigated to date only one OBP has been found in each species, such as the Japanese beetle, *Popillia japonica*, the Osaka beetle, *Anomala osakana* (Wojtasek *et al.*, 1998), the Oriental beetle, *Exomala orientalis* (Peng and Leal, 2001), the cupreous chafer, *A. cuprea*, and *A. octiescostata* (Nikonov *et al.*, 2002). On the other hand, at least two OBPs have been identified in each melolonthine species investigated, i.e. the pale brown chafer, *Phyllopertha diversa* (Wojtasek *et al.*, 1999), the large black chafer, *Holotrichia parallela*, and the yellowish elongate chafer, *Heptophylla picea* (Deyu and Leal, 2002). One of the two OBPs for each melolonthine species shows remarkable similarity to the pheromone-binding proteins from rutelines, whereas the second type of OBP forms a divergent group reminiscent of the PBP and GOBPs groups of moths. The number of OBPs in the scarab beetle species investigated to date may be underestimated because isolation of proteins using biochemical methods may fail to identify proteins expressed in very low amounts. On the other hand, the number of OBPs in *Drosophila* may be overestimated because the *in silicio* molecular cloning may fish out pseudogenes (Galindo and Smith, 2001). It is very likely, however, that in general insects possess multiple odorant-binding proteins to detect a wide range of physiologically relevant compounds even if a single OBP may be involved in the detection of multiple compounds (see below).

15.2.2 Tissue specificity

As opposed to chemosensory proteins (see Chapter and Nagnan-Le Meillour, Chapter 17, in this volume), which are expressed not only in olfactory tissues but also in non-olfactory tissues, such as pheromone glands (see Chapter 16), odorant-binding proteins are specific to antennae. In fact, this was one of the criteria used to name the first PBP originally identified in the wild silkmoth *A. polyphemus* (Vogt and Riddiford, 1981). In earlier work this criterion was more strictly applied. For example, the two PBPs from the gypsy moth, *Lymantria dispar*, were identified in antennal and sensillar extracts, but were not detected in various control tissues (legs, hemolymph, muscle, thoracic ganglia, and brain) using both gel stained with Coomassie blue and Western blots (with antisera generated against *A. polyphemus* PBP) (Vogt *et al.*, 1989). Likewise, using gel electrophoresis and Western blots (Maida *et al.*, 1993) the pheromone-binding protein from *B. mori* was demonstrated to be expressed in the antennae, but not in legs, wings, head, thorax, and abdomen. More recently, OBPs have been demonstrated to be specifically expressed in the antennae of the honeybee by comparing extracts from antennae, legs, brain, and thorax (Danty *et al.*, 1998); tissue specificity of honeybee OBPs was further corroborated by *in situ* hybridization (Danty *et al.*, 1999). Given that PBPs were never found in tissues other than antennae, it became a common practice to screen for odorant-binding proteins by comparing gel profiles of extracts from antennae and only one control tissue (legs, for example). Even in the mosquitoes *Culex quinquefasciatus*, which may have olfactory tissues in legs, OBPs were found specifically in antennae not only by gel analysis but also by RT-PCR (Ishida *et al.*, 2002b). The tissue specificity observed in all odorant-binding proteins identified to date suggests that they have evolved to play a unique function in insect antennae.

OBPs are not only specific to antennae, but within the antennae they are compartmentalized in specific sensilla. By immunolocalization of different OBPs with specific antisera, Steinbrecht and collaborators showed elegantly that PBPs were detected in all pheromone-detecting sensilla trichodea (Steinbrecht *et al.*, 1992), whereas GOBPs of type 2 were detected in most sensilla basiconica, detectors for plant compounds (Laue *et al.*, 1994). Interestingly, the expression of a certain OBP class is not correlated with a certain morphological sensillum type, but rather with the specificity class of the receptor cells (Laue and Steinbrecht, 1997). For example, long sensilla trichodea are present in both male and female antennae of the silkworm moth, *B. mori*. In males, these sensilla respond to the female-released bombykol and bombykal (Kaissling *et al.*, 1978; Kaissling and Priesner, 1970), whereas in females they respond to benzoic acid and linalool (Priesner, 1979). The corresponding male and female sensilla express PBP and GOBP2, respectively (Laue and Steinbrecht, 1997).

In a study comparing nine moth species, Steinbrecht (1996) observed that cross-reactivity of specific antisera to one species with PBPs from another species was not correlated with taxonomic relatedness of the species, but rather with the pheromone chemistry. Given that the highly divergent specificity of pheromone receptor cells in the Noctuidae species studied appears to be mirrored by a similar diversity of PBP sequences in sensilla trichodea, Steinbrecht suggested that PBPs participate in odorant discrimination (Steinbrecht, 1996).

15.2.3 Expression of OBPs

The inner and outer dendritic segments of olfactory receptor neurons (ORNs) are enclosed by three auxiliary cells. The thecogen cells form a complete and tight envelop around the somata of the ORNs. Proximal to the cuticle, there are two other enveloping cells: trichogen and tormogen cells. The speculation that trichogen and tormogen cells have secretory activity was supported by the intracellular localization of PBP in secretory organs of these auxiliary cells (Steinbrecht *et al.*, 1992). In other words, OBPs are not synthesized by the ORNs themselves, but rather produced in the tormogen and trichogen cells associated with them and secreted into the sensillar lymph surrounding the outer dendritic segment. In scarab beetles, the microvilli of the tormogen cells project conspicuously towards the cuticle of sensilla placodea and approach the terminal ends of the dendrites. This suggests that they supply OBPs directly to the terminal ends of the outer dendritic segments, possibly with minimal circulatory requirements (Kim and Leal, 2000).

The amino acid precursors for the biosynthesis of the PBPs seem to be taken up from the hemolymph by the auxiliary cells (Steinbrecht *et al.*, 1992). Indeed, labeled methionine injected in the pupae of *L. dispar* was incorporated into OBPs (Vogt *et al.*, 1989).

Expression of PBPs starts 3 days before adult eclosion in *L. dispar* (Vogt *et al.*, 1989) and 35–40 h before adult emergence in *M. sexta* (Vogt *et al.*, 1993), with the expression being induced by the decline in ecdysteroid levels. It has been estimated that in *L. dispar* PBPs undergo a combined steady-state turnover of 8×10^7 molecules per hour per sensillum (Vogt *et al.*, 1989).

As with other proteins, the processes by which OBPs are transported to their final destination within a cell (protein sorting) depend on signals encoded in their primary sequences, i.e. signal peptides. These N-terminal presequences are responsible for targeting proteins to the endoplasm reticulum for subsequent transport through the secretory pathway (Emanuelsson *et al.*, 2000). The evolutionary pressures that shaped the properties of the mature OBPs are fundamentally different from those that shaped their signal peptides (Vogt *et al.*, 1991). In other words, selection for OBP that interact specifically with a certain ligand (or group of ligands) is unrelated to the intracellular signaling processes that shaped protein sorting. Thus, if sequences of proteins are compared for the sake of grouping them by their functions, for example, the sequence of signal peptides should be omitted before analyzing their sequence identities.

15.2.4 Conserved cysteine residues and their linkages

The thiol (-SH) moiety of cysteine residues readily undergoes oxidation to form disulfide bridges and this is important in stabilizing and maintaining the three-dimensional structures of various biologically important molecules. The formation of disulfide bridges (Wojtasek and Leal, 1999) is the only post-translational modification of OBPs and the cysteine residues appear at well-conserved positions. Strictly speaking these "cysteine" residues should be referred to as "half cystine" residues because they form disulfide bridges. In moth PBPs, the half cystine residues occur at positions 19, 50, 54, 97, 108, and 117, whereas in scarab beetles they have two different patterns: 16, 44, 48, 86, 95, and 104 for PBPs and 21, 48, 52, 96, 107, and 116 for OBPs. Note that the scarab PBPs and OBPs have 116 and 133 amino acids, respectively and that the pattern of cysteine positions in the longer proteins is closer to that of moth PBPs (142 residues). Scarab beetle odorant-binding proteins have been separated into OBPs and PBPs on the basis of binding to scarab pheromones (Nikonov *et al.*, 2002; Wojtasek *et al.*, 1998) and sequence similarity of the latter group. The spacing pattern between cysteine residues show some variation when comparing OBPs from different insect orders (or different groups of OBPs), but they all have a common pattern of three residues between the second and the third Cys and eight residues between the fifth and the sixth Cys.

The disulfide structures of a few OBPs have been determined by analytical methods, particularly OBPs from *B. mori*. (Leal *et al.*, 1999; Scaloni *et al.*, 1999). As part of our attempt to get better insight on the structural biology of pheromone-binding proteins we have determined the disulfide linkages in recombinant and native BmPBP (Leal *et al.*, 1999). The disulfide structures of the native PBP and GOBP-2 from *B. mori* were also identified by Scaloni and collaborators (Scaloni *et al.*, 1999). These OBPs showed the same cysteine pairing, i.e. Cys19–Cys54, Cys50–Cys108, and Cys97–Cys117. Similar disulfide structures were determined in the olfactory proteins from honeybee, *Apis mellifera*,

ASP1 and ASP2 (Briand *et al.*, 2001a, 2001b). Therefore, the disulfide bridges of all OBPs analyzed to date show the profile of the first cysteine residue connected to the third one, the second linked to the fifth, and the fourth bound to the sixth, i.e. Cys(I)–Cys(III), Cys(II)–Cys(V), and Cys(IV)–Cys(VI).

The disulfide linkages of BmPBP were corroborated when the structures of the BmPBP–bombykol complex (Sandler *et al.*, 2000) (Figure 15.1) and the acidic form of BmPBP (Horst *et al.*, 2001b) were determined by X-ray crystallography and NMR spectroscopy, respectively. These structure determinations underscore the significance of the six half cystine residues. When bound to its cognate pheromone, BmPBP has six α helices, which are anchored by the three disulfide bridges (Sandler *et al.*, 2000). Helices α-1b, α-3, α-5, and α-6 are held together by the disulfide linkages (Figure 15.1), with two disulfide bonds (Cys19–Cys54 and Cys50–Cys108) fixing the relative position of α-3 by attaching it to the flanking helices α-1 and α-6, whereas the third disulfide bond (Cys97–Cys117) links helices α-5 and α-6 (Sandler *et al.*, 2000). Although structural details will be discussed below, it is worth mentioning here that while the ends of the four helices (α-1, α-4, α-5, and α-6) converge to form the binding pocket at the narrow end of binding cavity, the opposite end is capped by the rigidly connected helix α-3 (Figure 15.1).

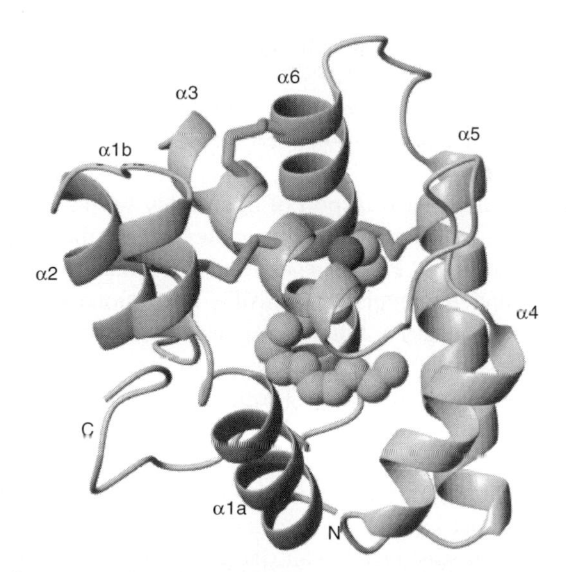

Figure 15.1 Structure of the BmPBP–bombykol complex highlighting the three disulfide bridges that play a pivotal role in the rigidity of the three-dimensional structure of PBPs. Helices are shown as ribbons and loop regions as thin tubes. Bombykol is displayed in space-filling representation with all atoms. This figure was prepared by Fred Damberger by using the program MOLMOL (Koradi *et al.*, 1996).

The chemosensory proteins (CSPs) differ from the six cysteine OBPs not only in the number of cysteine residues, but also in their function regarding the rigidity of their three-dimensional structures. While in OBPs the three disulfide linkages play a pivotal role in the knitting together of at least four of the helices, the two disulfide bridges in CSPs close small loops involving residues 29 and 36 and 55 and 58 and, consequently, seem to have little rigidifying effect on the overall structure of CSPs (Lartigue *et al.*, 2002).

15.2.5 Functions of odorant-binding proteins

Since the discovery of antennae-specific odorant-binding proteins (OBPs), various functions have been postulated for these proteins. Because most of the semiochemicals, particularly moth sex pheromones, are hydrophobic molecules with a long-chain lipophilic moiety and a polar head, it has been suggested that OBPs solubilize the ligands (Vogt and Riddiford, 1981) so that they can be transported through the aqueous sensillar lymph environment towards the receptors. Indeed, it has been demonstrated that pheromones are solubilized by PBPs (Kaissling *et al.*, 1985) and desorbed from hydrophobic surfaces (Kowcun *et al.*, 2001). That OBPs solubilize odorants has been further substantiated by experiments with a *Drosophila* mutant defective for an OBP (LUSH) that showed odor-specific defects in olfactory behavior (Kim and Smith, 2001). Interestingly, the *lush* mutants had a defect in the detection of ethanol, a water-soluble compound. Thus, the physiological roles of OBPs do not seem to be limited to solubilizing ligands given that ethanol *per se* would be soluble enough in the sensillar lymph. In addition, the small ethanol molecule would diffuse even faster than an OBP–ethanol complex.

OBPs' essential role in the transport of ligands has been also demonstrated by electrophysiological recordings from the antennae of the wild silkmoth, *A. polyphemus*. Recombinant PBPs in various combinations with pheromone components were applied through open sensilla. Pheromone alone did not activate the olfactory receptor neurons, but when it was incubated with PBPs prior to application nervous activity was elicited even in receptor cell types, which do not activate under natural conditions (Pophof, 2002). In fact, the first direct experimental evidence to support that PBPs act as a solubilizer or carrier was obtained by the perfusion technique. When the sensillar lymph was replaced by a Ringer solution, 180 nM of pheromone was required to elicit nervous activity, but the threshold for activation was decreased by a factor of 100 when PBP or bovine serum albumin was added (van den Berg and Ziegelberger, 1991). All these experiments support the hypothesis that odorant-binding proteins solubilize and ferry the ligands to their receptors.

It has also been suggested for *A. polyphemus* that PBPs occur in two forms differing in the number of disulfide bridges, a reduced form (PBP_{red}) with only one or two disulfide linkages, and an oxidized form (PBP_{ox}) with three disulfide bridges (Ziegelberger, 1995). In this model, pheromone molecules entering the olfactory hair lumen are mostly bound to PBP_{red}, solubilized and transported to the receptor cell. The complex of pheromone and reduced PBP may interact with a receptor molecule and turn into the oxidized form, which is assumed to be unable to activate the receptor. The disulfide structure of the pheromone-binding protein from *B. mori* has been determined analytically (Leal *et al.*, 1999; Scaloni *et al.*, 1999). There was no evidence for the occurrence of reduced BmPBP (Wojtasek and Leal, 1999) and the disulfide linkages were identical in the recombinant and native proteins (Leal *et al.*, 1999). Recently, we isolated (directly from 20 percent native gels) the two bands from *A. polyphemus*, which have been suggested to be the oxidized and reduced forms of ApolPBP, and analyzed them by liquid chromatography-electrospray ionization mass spectrometry (LC-ESI/MS). If present, this technique would allow the discrimination of the two forms based on differences not only in retention times, but also in their molecular masses and MS profiles (Wojtasek *et al.*, 1999). The two proteins were indistinguishable in their retention times on HPLC, their molecular masses, and MS profiles. Also, they gave identical N-terminal amino acid sequences (Deyu and Leal, 2002). These data indicate that the two electrophoretically distinct bands do not differ in their disulfide structures and thus do not support the "redox shift" hypothesis. As it has been demonstrated in other cases (Deyu and Leal, 2002; Wojtasek *et al.*, 1999), the two bands are probably conformational isomers of ApolPBP and not oxidized and reduced forms.

Two lines of evidence support the hypothesis that OBPs protect pheromones from degradation by odorant-degrading enzymes (Vogt and Riddiford, 1981). First, the rate of pheromone degradation by a sensillar esterase (pheromone-degrading enzyme) decreased with the addition of pheromone-binding protein (Vogt and Riddiford, 1986). Second, the structure of a pheromone–PBP complex showed that the ligand is completely engulfed in the core of the pheromone-binding protein (Sandler *et al.*, 2000) rendering the pheromone inaccessible to pheromone-degrading enzyme while bound to PBP.

15.3 Mechanism of pheromone binding and release

15.3.1 The pheromone–PBP complex

In a model borrowed from the study of bacterial chemotaxis, it was initially proposed that OBPs not only solubilize specific pheromones, but trigger the olfactory receptors when bound to odorant molecules (Pelosi, 1994). Later, it was further hypothesized that electrostatic and hydrophobic interactions from both the bound ligand and ligated protein are necessary and sufficient for receptor activation (Prestwich and Du, 1997). To the best of my knowledge the protein–ligand complex model has never been tested in bacteria. Certainly, it is not supported in insects by recent findings. The structure of the BmPBP–bombykol complex (Sandler *et al.*, 2000) showed that pheromone is completely buried inside the protein. Considering this complex, electrostatic and hydrophobic interactions of pheromone with the receptor as it has been hypothesized by Prestwich and Du (Prestwich and Du, 1997) seem highly unlikely.

Recently, a putative olfactory receptor from *Drosophila*, Or43a (Clyne *et al.*, 1999; Vosshall *et al.*, 1999), has been expressed in *Xenopus laevis* oocytes (Wetzel *et al.*, 2001). The receptor expressed in a heterologous cell system was activated by four odorants, i.e. cyclohexanone, cyclohexanol, benzaldehyde, and benzyl alcohol (Wetzel *et al.*, 2001). These experiments not only provided direct evidence for the function of the *Or* gene, but also demonstrated that the olfactory receptor can be stimulated without an odorant-binding protein. It was demonstrated earlier that PBP was not necessary to obtain pheromone-dependent responses in cultured olfactory receptor neurons of *Manduca sexta* (Stengl *et al.*, 1992). The possibility that OBPs have been produced *in vitro* and were present in cultured ORNs could not be excluded. The same argument can not be raised for the heterologous expression of the *Drosophila* olfactory receptor. While the evidence that *Xenopus* oocytes responded to odorants in the absence of OBPs does not support the OBP–odorant complex model, it also demonstrated that OBPs are essential for the kinetics of the olfactory system (see below).

15.3.2 Conformational changes

That odorant-binding proteins undergo a pH-dependent conformational change was a somewhat accidental discovery. Samples of the recombinant pheromone-binding protein from *B. mori* (BmPBP) (Wojtasek and Leal, 1999) prepared in my lab were demonstrated to be highly purified. Using chromatographic techniques (ion exchange chromatography, gel filtration, hydrophobic interaction chromatography, chromatofocusing, and reversed-phase chromatography), gel electrophoresis (native and SDS), and electrospray ionization-mass spectrometry (ESI-MS), we prepared pure recombinant-labeled BmPBP samples. Yet, when Kurt Wuthrich (ETH-Zurich) and his collaborators analyzed the samples by nuclear magnetic resonance (NMR), they found evidence of two species. Initially, the "extra" NMR peaks were suggested to be due to a contamination by a second protein or a degraded form of BmPBP. We suspected that the protein had undergone degradation while *en route* to Switzerland, but analysis of the same NMR samples (returned to my lab) did not show any contaminants. While investigating the stability of the protein we found that a number of factors affected the homogeneity of the protein. Ammonium sulfate and sucrose, for example, resulted in heterogeneity as detected by ion exchange chromatography (Wojtasek and Leal, 1999). This unexpected conformational flexibility led us to a thorough spectroscopic analysis of the recombinant and native BmPBP under various solution conditions by circular dichroism (CD) and fluorescence. Although the protein was very stable it showed a pH-dependent conformational change. The secondary structure of the protein (analyzed by far-UV CD) was affected only slightly by changes in pH, but the tertiary structure as detected by near-UV CD exhibited a conformational transition between pH 6 and pH 5 (Wojtasek and Leal, 1999).

Conformational changes are detected in the near-UV CD spectrum if they affect the conformation or environment of side chain aromatic residues. On the other hand, they affect the far-UV CD spectrum only if substantial changes in secondary structure content occur. Larger amounts of proteins are required for near-UV CD than for far-UV CD spectroscopy given that absorptions in the former are at least an order of magnitude smaller. Because the pH-dependent conformational changes in BmPBP involve aromatic side chains, possibly tryptophan, they were also mirrored in fluorescence spectroscopy. This is fortuitous given that fluorescence requires smaller amounts of protein, which can consequently be obtained from natural sources. It is important to point out that for such experiments, however, proteins must be scrupulously pure and devoid of other chromophores.

Although the pH-dependent conformational changes observed initially in BmPBP might be a common feature of all odorant-binding proteins, it may be difficult to analyze in other proteins. If aromatic side chains are not involved, conformational changes may not be detectable by fluorescence or CD. Regardless of the residues involved, conformational changes can be detected by NMR spectroscopy (Damberger *et al.*, 2000), but this technique may require labeled recombinant proteins.

Fluorescence analysis showed that the native odorant-binding protein from *P. diversa* PdivOBP2 (Wojtasek *et al.*, 1999) undergoes conformational change similar to that observed with BmPBP (Figure 15.2). Again, the transition between the B- and A-form was between pH 6 and pH 5.5. Note that this is above the isoeletric points of these proteins (the pIs for BmPBP and PdivOBP2 are 4.85 and 4.6, respectively).

Conformational changes in BmPBP were also studied in the presence of model membranes. Conformational changes more pronounced than those at low pH were observed in the presence of anionic vesicles of dimyristoyl-phosphatidylglycerol (DMPG), whereas the effect of neutral phospholipids vesicles, dimyristoylphosphatidylcholine (DMPC), was marginal (Wojtasek and Leal, 1999). The presence of a physiological concentration of KCl reduced the effect, but the interaction with negatively charged membrane in the presence of KCl was still comparable to the effect of lowering the pH. The negatively charged head groups of lipids in cell membranes give rise to an electrical surface potential, which in turn decreases the surface pH (van der Goot *et al.*, 1991). Negatively charged surface coats have been demonstrated on the pore-tubules and dendritic membranes of olfactory hairs of male *A. polyphemus* by application of cation markers, such as lanthanum, ruthenium red, and cationized ferritin (Keil, 1984b). As I pointed

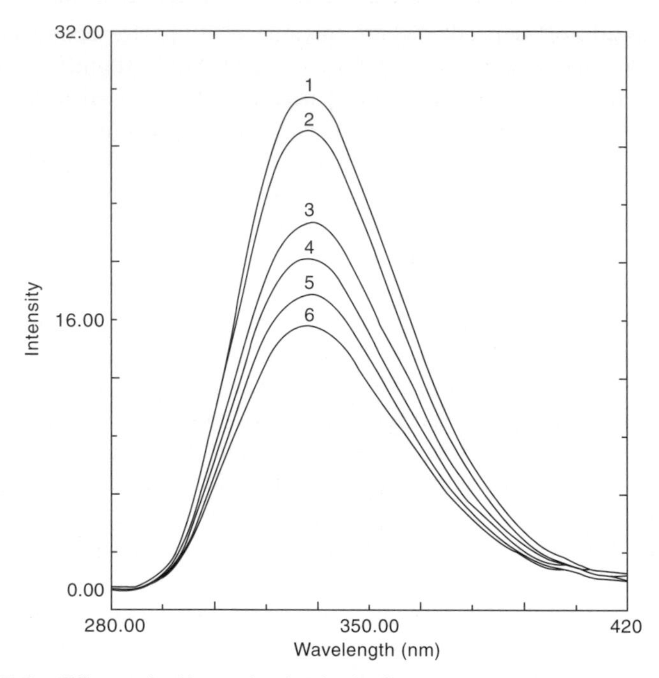

Figure 15.2 Effect of pH on the intrinsic fluorescence of an odorant-binding protein from *P. diversa*, PdivOBP2, with a large conformational transition between pH 6 and pH 5.5. Spectra were obtained on a Shimadzu RF-5301PC spectrofluorophotometer with a 6 μg/ml protein solution in 20 mM buffers, with excitation at 235 nm and emission monitored from 280 to 420 nm. (1) pH 7, (2), pH 6 in sodium phosphate, (3) pH 5.5, (4) pH 5, (5) pH 4.5, and (6) pH 4 in sodium acetate.

out earlier (Leal, 2000), as far as pheromone-binding proteins are concerned, the physiologically relevant pH is likely to be not only that of the sensillar lymph (the bulk pH), but also the pH at the surface of dendrites (localized pH). It is yet to be determined whether the negatively charged surface that may interact with odorant-binding proteins and promote conformational changes is a moiety from a glycoprotein, amino acid residues from membrane proteins like SNMPs (Rogers *et al.*, 2001a, 2001b; Rogers *et al.*, 1997), or even an external site of olfactory receptors.

Conformational changes in proteins vary from slight changes in the environment of a few amino acid residues to more extensive changes like the ones observed in BmPBP. A Protein Motion Database (http://molmovdb.mbb.yale.edu/MolMovDB/) is a readily available resource highlighting various types of conformational changes. I have entered an interpolation between the two conformations of BmPBP (bound to bombykol, 1DQE and the acidic form,

1GM0) to illustrate for the reader (http://molmovdb.mbb.yale.edu/cgi-bin/morph.cgi?ID=444276-7254) the pH-dependent conformational change in BmPBP.

A functionally important conformational change for the release of pheromone must take place in a millisecond timescale to be consistent with the fast kinetics of neuronal activities. Even when the whole process is considered from stimulant delivery, receptor stimulation until motor system output a millisecond timescale must be considered. For example, males of *B. mori* respond to bombykol with wing vibration 100–500 msec after the onset of stimulation (Kaissling, 1986). Recently, I have studied the kinetics of the conformational changes in BmPBP. Since the acidic form ($BmPBP^A$) has a lower intrinsic fluorescence than the basic form ($BmPBP^B$), the process can be monitored by fluorescence. However, preliminary time course experiments showed that the reaction is too fast to be accurately measured with a standard fluorometer. Therefore, I used a stopped-flow instrument (Bio-Logic SMF-400 stopped-flow mixer and a MOS-250 fluorescence detector) to measure the kinetics of the B→A change. The pH-dependent conformational change in BmPBP showed characteristics of first-order kinetics, with a rate constant, $k = 184 \pm 6\ s^{-1}$ (Figure 15.3). In other words,

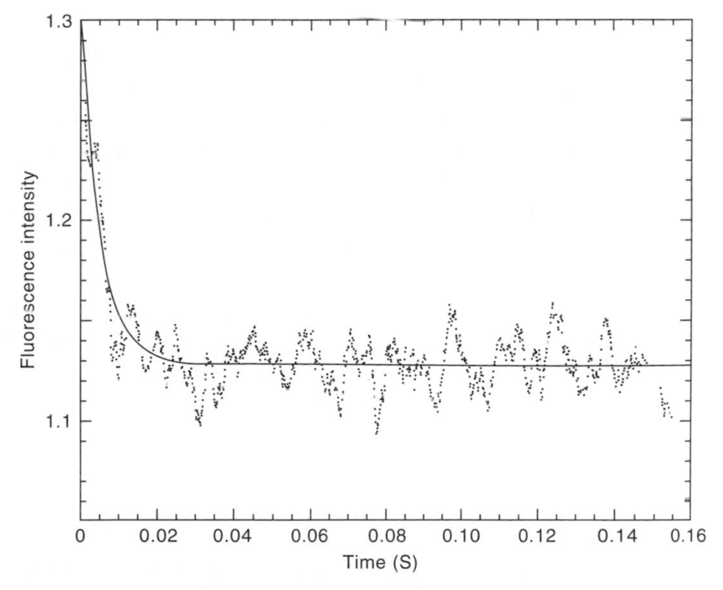

Figure 15.3 Fluorescence emission traces for the rapid kinetics of the pH-dependent conformational change in BmPBP. Under the control of a MPS-52 microprocessor, 70 μl of BmPBP (25 μg/ml) in sodium phosphate buffer (10 mM, pH 6.8) were mixed with 30 μl of sodium formate (100 mM, pH 4) at 27°C resulting in a final pH of 4.6. The average fluorescence data (N = 15), obtained with the excitation at 280 nm and emission at 335 nm, were analyzed by the KaleidaGraph software.

the time required to convert half of the $BmPBP^B$ originally present to the acidic form ($BmPBP^A$), i.e. the half-time of this conformational change, was 3.8 msec. This half-time fits to a modified kinetic model of perireceptor events (Kaissling, 2001a) in which the pheromone–PBP complex dissociates before the pheromone interacts with the receptor molecule (K.-E. Kaissling, personal communication).

There is growing evidence that odorant-binding proteins do not bind at low pH. This was first indicated by a substantial enhancement in the fluorescence of BmPBP when bombykol was added to a protein solution at neutral and basic pH, whereas no such enhancement occurred at low pH (Wojtasek and Leal, 1999). On the other hand, mass spectrometry experiments showed that the non-covalently bound BmPBP–bombykol complexes in the gas phase were dissociated at low pH (Oldham *et al.*, 2000). Using calorimetric titration, Briand and colleagues (Briand *et al.*, 2001b) showed that a honeybee odorant-binding protein binds 2-isobutyl-3-methoxypyrazine at pH 7, but the binding was abolished upon acidification of the protein solution. Such pH-dependent binding ability was not detected earlier because to date most of the binding assays of odorant-binding proteins have been performed at high pH by the native gel electrophoresis method (Vogt *et al.*, 1991). I have now developed a binding assay which can be carried out at any desired pH with non-radiolabeled (cold) ligand (cold-binding assay). Details of the cold-binding assay will be given elsewhere, but in short the protocol is as follows. After the protein is incubated with a test compound, the unbound ligand is removed by centrifugation and the ligand is extracted from the ligated protein and analyzed by gas chromatography and/or gas chromatography-mass spectrometry. Although the cold-binding assay requires larger amounts of protein than the native gel assay, it can be performed without radiolabeled ligands. The cold binding assay showed that recombinant BmPBP binds bombykol at pH 7, but not at pH 5 (Figure 15.4).

In a recent binding study with recombinant proteins from *L. dispar*, Kowcun and colleagues (Kowcun *et al.*, 2001) argued that the increase in $[K^+]$ near the dendritic membrane counterbalances the increase in $[H^+]$ (decrease in pH). We have taken into consideration the physiologically relevant concentration of potassium in the sensillar lymph (Kaissling, 1995; Kaissling and Thorson, 1980) in our previous work (Wojtasek and Leal, 1999). As shown in Figure 7 of that publication (Wojtasek and Leal, 1999), the addition of physiologically relevant amounts of KCl decreases the effect of the negatively charged model vesicles, but it does not abolish the effect. Note that the effect of the vesicles alone is larger than the influence of the pH change from 6 to 5. Thus, the addition of KCl decreases the dramatic changes, but the conformations of BmPBP at pH 6 and 5 are still remarkably different as observed by the near-UV CD spectra (Wojtasek and Leal, 1999). With the cold-binding assay described above, I found no binding of bombykol to BmPBP at pH 5 either with 0, 170, or even 500 mM of KCl, whereas the same concentrations of salt did not affect the binding at pH 7.

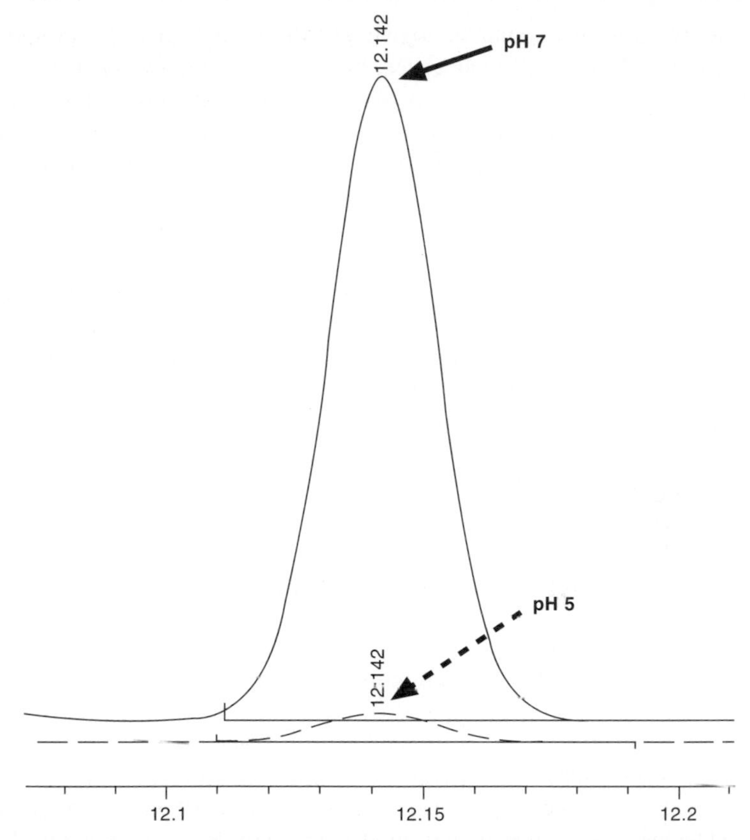

Figure 15.4 Binding of bombykol to BmPBP at pH 7 as demonstrated by the cold-binding assay. After removal of the unbound ligand, bound bombykol was extracted from the ligated protein with a solid phase microextraction syringe (SPME, 65 μm polydimethylsiloxane divinylbenzene coating). Under the same conditions, but at low pH (5), the amount of ligand extracted is not significantly different from the amount extracted from a buffer solution (control).

Moreover, the structure of $BmPBP^A$ (Horst *et al.*, 2001a) elegantly demonstrated that at low pH the binding pocket is occupied by a newly formed C-terminal α-helix thus strongly suggesting that the conformational change is functionally important and consistent with the absence of ligand-binding at pH 5.

Conformational isomers of OBPs that are detectable by gel electrophoresis have also been found in three species of scarab beetles (Deyu and Leal, 2002; Wojtasek *et al.*, 1999). We have demonstrated by protein sequencing and mass spectrometry that these electrophoretically distinct bands are from proteins with the same primary sequences (Deyu and Leal, 2002; Wojtasek *et al.*, 1999). Although the term "isoform" means "different forms of a protein" it has been widely accepted by molecular biologists as "different forms of a protein that may be produced from different genes, or the same gene by alternative RNA splicing" (Kendrew, 1994). From a protein chemist's perspective these isoforms are different proteins because they have different primary structures. To be more specific, I refer to the electrophoretically different bands that show the same molecular mass and identical amino acid sequences as "conformational isomers (conformers)." It is conceivable that these conformational isomers are the acidic and basic forms of these OBPs, but to date I have no evidence to support this assumption. Also, alternative conformers of OBPs may encompass not only the acidic and basic forms, but also other conformational isomers that add to the microheterogeneity of OBPs. Note that allelic variations also contribute to microheterogeneity; see Newcomb *et al.*, 2002 for a recent example. Based on the structural data obtained so far, however, the hypothesis that these electrophoretically distinct bands are derived from different disulfide structures appears highly unlikely.

15.4 Structural biology of odorant-binding proteins

15.4.1 Secondary structure

The structure of a BmPBP–bombykol complex was determined by X-ray crystallography as a roughly conical arrangement of six α-helices (Sandler *et al.*, 2000). Considering the additional helix observed at the C-terminal end of BmPBP^{A}, the observed helical structure of BmPBP fits in nicely with the secondary structure predicted by the program PHDsec based only on the amino acid sequence (Rost and Sander, 1993) (http://www.embl-heidelberg.de/predictprotein/) (Figure 15.5). Thus, I used this software to estimate the secondary structure of scarab OBPs. The same software predicts that the secondary structures of the two types of scarab OBPs also contain six α-helices. This is shown in Figure 15.5 for PjapPBP and PdivOBP2.

In addition to the "*in silicio*" evidence, circular dichroism measurements also suggest that the secondary structures of scarab beetle OBPs are largely α-helical. The far-UV CD spectra of the pure PdivOBP1 and PdivOBP2 show a typical profile of helical-rich proteins, with a positive and two negative peaks occurring at the usual wavelengths (Figure 15.6). Calculations of secondary structure composition based on these spectral data showed also that both PdivOBP1 and PdivOBP2 are primarily α-helical. Likewise, CD data has indicated that the odorant-binding proteins from the honeybee ASP1 (Briand *et al.*, 2001a) and ASP2 (Briand *et al.*, 2001b) are α-helical proteins. It seems that all OBPs are helical-rich proteins.

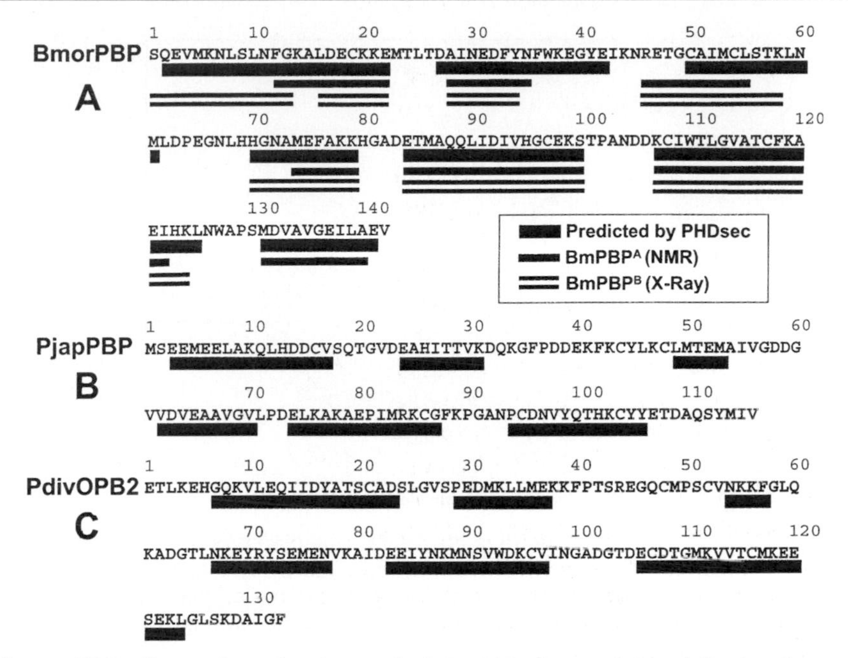

Figure 15.5 Secondary structures of odorant-binding proteins. A Comparison of the secondary structures of the acidic and basic forms of BmPBP, determined by NMR (Horst *et al.*, 2001a) and X-ray crystallography (Sandler *et al.*, 2000), respectively, and that estimated from the primary structure of the protein with the prediction program PHDsec (Rost and Sander, 1993). Predicted secondary structures of scarab OBPs: B PjapPBP, and C PdivOBP2.

15.4.2 Binding cavity

The binding pocket of BmPBP is formed by four antiparallel helices (α-1, α-4, α-5, and α-6) that converge to form the narrow end of the pocket, whereas the opposite end is capped by α-3 (Figure 15.1). Bound bombykol has a roughly planar hook-shaped conformation and the outside (convex) part of bombykol interacts with numerous protein residues, whereas the inside (concave) part has fewer contacts. Interestingly, residues from all parts of the protein contribute to the binding cavity (Sandler *et al.*, 2000) that protects bombykol from the aqueous solvent. Because bombykol is bound in a large flask-shaped cavity with a tiny opening to the surface the structure raised the questions of how pheromones enter the cavity and what is the mechanism of ligand release. Two major hypotheses ("lid opening" and "unwinding of α-1") were considered based on the pH-dependent conformational change and the structure of the BmPBP–complex. Protonation of one or more of the three histidines (His-69, His-70, His-95) forming a cluster at the base of the flexible loop near α-4 (Figure 15.1) could destabilize the loop covering the binding pocket causing the opening of the lid.

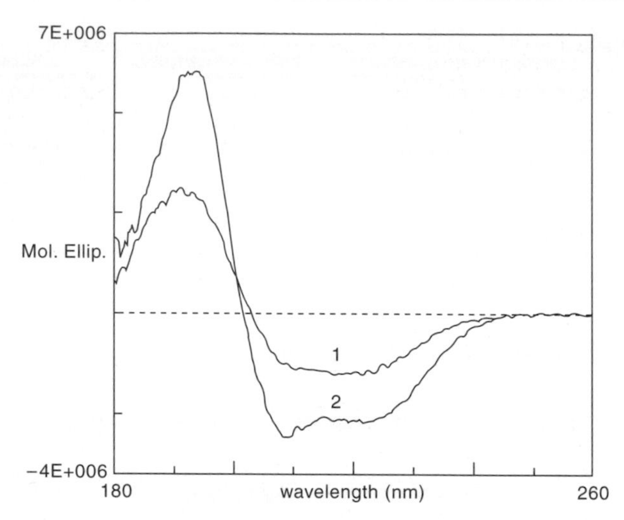

Figure 15.6 Far-UV CD spectra of the odorant-binding proteins from the scarab beetle *P. diversa*: (1) PdivOBP1 and (2) PdivOBP2. The spectra of the pure proteins were recorded in 20 mM sodium phosphate buffer, pH 6.8, at 0.4 mg/ml on a JASCO J-720 spectropolarimeter at 25°C in a 0.1 mm pathlength cell.

Alternatively, protonation of His-80 and its interaction with Lys-6 at the beginning of α-1 might destabilize α-1 enough to cause unraveling. The possibility was also considered that protonation of side chains could destabilize helices and unfold the entire protein to release bombykol (Sandler *et al.*, 2000).

15.4.3 Intramolecular mechanism of binding and release

In an attempt to obtain a better understanding of the molecular basis of pheromone binding and release, we are studying all conformations of BmPBP that seem to be functionally important. The unliganded solution structure of BmPBP obtained by NMR at pH 4.5 showed remarkable conformational differences to the crystal structure of the BmPBP–complex (Horst *et al.*, 2001a). The most remarkable differences were at the two ends of the polypeptide chain. While the N-terminal decapeptide in the BmPBP–complex forms an α-helix (α-1a, Figure 15.1) it is unfolded in the acidic form of BmPBP, $BmPBP^A$. On the other hand, an unstructured dodecapeptide at the C-terminus of the BmPBP–bombykol complex folds into a new α-helix, α-7 (Horst *et al.*, 2001a). This helix occupies the hydrophobic core of $BmPBP^A$, at a position corresponding to the binding pocket in the BmPBP–bombykol complex (Figure 15.6). In spite of their differences, the overall helical content of the two conformations is nearly the same because the N-terminal helix in the BmPBP–complex is replaced by the C-terminal helix in $BmPBP^A$. This explains the observation that the far-UV CD spectra of the BmPBP is only slightly influenced by pH changes (Wojtasek *et al.*, 1999).

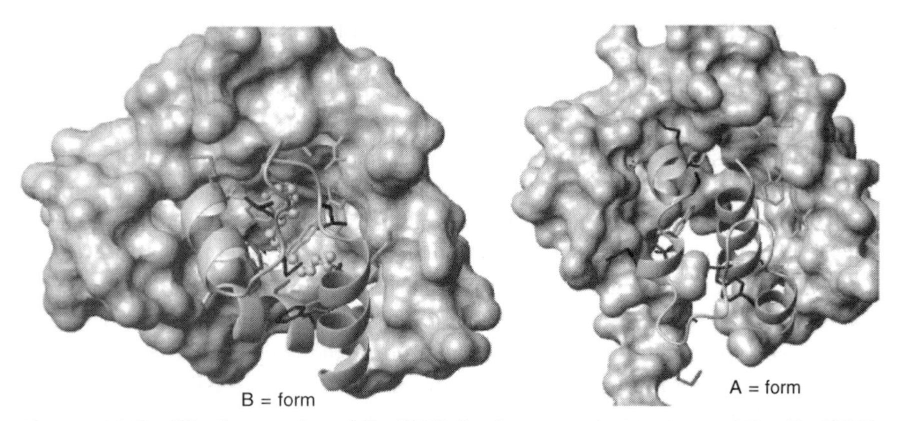

Figure 15.7 Binding cavity of BmPBP. In the crystal structure of the BmPBP–bombykol complex (B-form) the pheromone. In the unligated acidic form of the protein (A-form), C-terminal α-helix occupies the binding pocket. This figure was prepared by Fred Damberger by using the program MOLMOL (Koradi *et al.*, 1996).

The side chains of seven amino acid residues that show contacts with bombykol in the complex (Ser-9, Phe-12, Ile-52, Ser-56, Leu-62, Val-94, Thr-111) also show contacts with the newly formed helix α-7, some of which are highly conserved among lepidopteran OBPs. Three residues that show direct contacts to bombykol in the crystal structure of the BmPBP–bombykol complex (Leu-8, Phe-36, and Met-61) do not contact helix α-7 in the solution structure of $BmPBP^A$ (Horst *et al.*, 2001a). The four helices forming the binding pocket in the BmPBP–bombykol complex occur in similar positions in $BmPBP^A$. On the other hand, the histidines (His-69, His-70, His-95) forming a cluster at the base of the flexible loop in the BmPBP–bombykol complex are more widely separated in $BmPBP^A$, thus suggesting that protonation of one or more of these histidines may indeed play a part in the pH-dependent conformational change. Overall, the evidence from structural biology suggests an intramolecular regulation mechanism for pheromone binding and release.

15.4.4 Quaternary structure

Biochemical evidence suggests that BmPBP forms a dimer at the sensillar lymph pH and a monomer at low pH (Leal, 2000). There is little evidence from structural biology, however, to substantiate the occurrence of physiological dimers both in the crystal and solution structures. Two PBP–PBP interactions were found in the crystal structure (Sandler *et al.*, 2000), but both involved a modest amount of surface area and only one involved hydrophobic contacts. The area of contact (221 $Å^2$) was formed by two faces of the protein including a small hydrophobic patch around Pro-64 from one monomer and Met-131, Val-133, and Lys-38 of antenna-specific odorant-binding protein in the mosquito *Culex quinquefasciatus*. *J. Chem. Ecol.* **28**, 867–871.

Kaissling K.-E. (1987) *Karl-Ernst Kaissling: R. H. Wright lectures on insect olfaction*, ed. K. Colbow Simon Fraser University, Burnaby, British Columbia.

Kaissling K.-E. (2001a) Olfactory perireceptor and receptor events in moths: a kinetic model. *Chem. Senses* **26**, 125–150.

Kaissling K.-E. (2001b) Possible functions of pheromone binding proteins in *Bombyx mori* and *Anthearaea polyphemus*. 7th European Symposium on Insect Taste and Olfaction (ESITO), Villasimius (Cagliari), Italy, pp. 6–7.

Kaissling K.-E. (1995) Single unit and electroantennogram recordings in insect olfactory organs. In *Experimental Cell Biology of Taste and Olfaction. Current Techniques and Protocols*. eds A. I. Spielman and J. G. Brand, pp. 361–377. CRC Press, Boca Raton.

Kaissling K.-E. (1986) Temporal characteristics of pheromone receptor cell responses in relation to orientation behaviour of moths. In *Mechanisms in Insect Olfaction*, eds T. L. Payne, M. C. Birch and C. E. J. Kennedy, pp. 193–199. Oxford University Press, New York.

Kaissling K.-E., Bestmann, H. J., Stransky W. and Vostrowsky O. (1978) Bombykal, a second pheromone component of the silkworm moth *Bombyx mori*. Sensory pathway and behavioral effect. *Naturwissenschaften* **65**, 382–384.

Kaissling K.-E., Klein U., de Kramer J. J., Keil T. A., Kanaujia S. and Hemberger J. (1985) Insect olfactory cells: electrophysiological and biochemical studies. In *Molecular Basis of Nerve Activity*, eds J. P. Changeux and F. Hucho. Walter de Gruyter & Co., Berlin.

Kaissling K.-E. and Priesner E. (1970) Die Riechschwelle des Seidenspinners. *Naturwissenschaften* **57**, 23–28.

Kaissling K.-E. and Thorson J. (1980) Insect olfactory sensilla: structural, chemical and electrical aspects of the functional organisation. In *Receptors for Neurotransmitters, Hormones and Pheromones in Insects*, eds D. B. Sattelle, L. M. Hall and J. G. Hildebrand, pp. 261–282. Elsevier/North-Holland Biomedical Press.

Keil T. A. (1984a) Reconstruction and morphometry of the silkmoth olfactory hairs: a comparative study of sensilla trichodea on the antennae of male *Antheraea polyphemus* and *A. Pernyi* (Insecta: Lepidoptera). *Zoomorphologie* **104**, 147–156.

Keil T. A. (1984b) Surface coats of pore tubules and olfactory sensory dendrites of a silkmoth revealed by cationic markers. *Tissue Cell* **16**, 705–715.

Kendrew J. (1984) *The Encyclopedia of Molecular Biology*, p. 1165. Blackwell Science Ltd, Oxford.

Kim J.-Y. and Leal W. S. (2000) Ultrastructure of pheromone-detecting sensillum placodeum of the Japanse beetle, *Popillia japonica* Newmann (Coleoptera: Scarabaeidae). *Arthropod Struct. Dev.* **29**, 121–128.

Kim M.-S., Repp A. and Smith D. P. (1998) LUSH odorant-binding protein mediates chemosensory responses to alcohols in *Drosophila melanogaster*. *Genetics* **150**, 711–721.

Kim M.-S. and Smith D. P. (2001) The invertebrate odorant-binding protein LUSH is required for normal olfactory behavior in *Drosophila*. *Chem. Senses* **26**, 195–199.

Koradi R., Billeter M. and Wüthrich K. (1996) MOLMOL: a program for display and analysis of macromolecular structures. *J. Mol. Graphics* **14**, 51–55.

Kowcun A., Honson N. and Plettner E. (2001) Olfaction in the gypsy moth, *Lymantria dispar*. effect of pH, ionic strength, and reductants on pheromone transport by pheromone-binding proteins. *J. Biol. Chem.* **276**, 44770–44776.

Krieger J., von Nickisch-Rosenegk E., Mameli M., Pelosi P. and Breer H. (1996) Binding proteins from the antennae of *Bombyx mori. Insect Biochem. Molec. Biol.* **26**, 297–307.

Krieger M. J. B. and Ross K. G. (2002) Identification of a major gene regulating complex social behavior. *Science* **295**, 328–332.

Larsson M. C., Leal W. S. and Hansson B. S. (2001) Olfactory receptor neurons detecting plant odours and male volatiles in *Anomala cuprea* beetles (Coleoptera: Scarabaeidae). *J. Insect Physiol.* **47**, 1065–1076.

Larsson M. C., Leal W. S. and Hansson B. S. (1999) Olfactory receptor neurons specific to chiral sex pheromone components in male and female *Anomala cuprea* beetles (Coleoptera: Scarabaeidae). *J. Comp. Physiol. A* **184**, 353–359.

Lartigue A., Campanacci V., Roussel A., Larsson A. M., Jones T. A., Tegoni M. and Cambillau C. (2002) X-ray structure and ligand binding study of a moth chemosensory protein. *J. Biol. Chem.* **297**, 32094–32098.

Laue M. and Steinbrecht R. A. (1997) Topochemistry of moth olfactory sensilla. *Int. J. Insect Morphol. Embryol.* **26**, 217–228.

Laue M., Steinbrecht R. A. and Ziegelberger G. (1994) Immunocytochemical localization of general odorant-binding protein in olfactory sensilla of the silkmoth *Anteraea polyphemus. Naturwissenschaften* **81**, 178–180.

Leal W. S. (1996) Chemical communication in scarab beetles: reciprocal behavioral agonist–antagonist activities of chiral pheromones. *Proc. Natl. Acad. Sci. USA* **93**, 12112–12115.

Leal W. S. (1998) Chemical ecology of phytophagous scarab beetles. *Ann. Rev. Entomol.* **43**, 39–61.

Leal W. S. (2000) Duality monomer–dimer of the pheromone-binding protein from *Bombyx mori. Biochem. Biophys. Res. Commun.* **268**, 521–529.

Leal W. S. (1991) (R,Z)-5-(–)-Oct-1-enyl) Oxacyclopentan-2-one, the sex-pheromone of the scarab beetle *Anomala cuprea. Naturwissenschaften* **78**, 521–523.

Leal W. S., Nikonova L. and Peng G. (1999) Disulfide structure of the pheromone binding protein from the silkworm moth, *Bombyx mori. FEBS Lett.* **464**, 85–90.

Leal W. S., Zarbin P. H. G., Wojtasek H., Kuwahara S., Hasegawa M. and Ueda Y. (1997) Medicinal alkaloid as a sex pheromone. *Nature* **385**, 213.

Maïbèch-Coisné M., Sobrio F., Delaunay T., Lettere M., Dubroca J., Jacquin-Joly E. and Nagnan-Le Meillour P. (1997) Pheromone binding proteins of the moth *Mamestra brassicae*: specificity of ligand binding. *Insect Biochem. Molec. Biol.* **27**, 213–221.

Maida R., Krieger J., Gebauer T., Lange U. and Ziegelberger G. (2000). Three pheromone-binding proteins in olfactory sensilla of the two silkmoth species *Antheraea polyphemus* and *Antheraea pernyi. Eur. J. Biochem.* **267**, 2899–2908.

Maida R., Steinbrecht A., Ziegelberger G. and Pelosi P. (1993) The pheromone binding protein of *Bombyx mori*: purification, characterization and immunocytochemical localization. *Insect Biochem. Molec. Biol.* **23**, 243–253.

McKenna M. P., Hekmat-Scafe D. S., Gaines P. and Carlson J. R. (1994) Putative *Drosophila* pheromone-binding proteins expressed in a subregion of the olfactory system. *J. Biol. Chem.* **269**, 16340–16347.

Meng L. Z., Wu C. H., Wicklein M., Kaissling K.-E. and Bestmann H. J. (1989) Number and sensitivity of three types of pheromone receptor cells in *Antheraea pernyi* and *Antheraea polyphemus. J. Comp. Physiol. A* **165**, 139–146.

Murlis J., Willis M. A. and Cardé R. T. (2000) Spatial and temporal structures of pheromone plumes in fields and forests. *Physiol. Entomol.* **25**, 211–222.

Newcomb R. D., Sirey T. M., Rassam M. and Greenwood D. R. (2002) Pheromone binding proteins of *Ephiphyas postvittana* (Lepidoptera: Tortricidae) are encoded at a single locus. *Insect Biochem. Molec. Biol.* **32**, 1543–1554.

Vogt R. G., Rybczynski R. and Lerner M. R. (1991) Molecular cloning and sequencing of general odorant-binding proteins GOBP1 and GOBP2 from the tobacco hawk moth *Manduca sexta*: comparison with other insect OBPs and their signal peptides. *J. Neurosci.* **11**, 2972–2984.

Vosshall L. B., Amrein H., Morozov P. S., Rzhetsky A. and Axel R. (1999) A spatial map of olfactory receptor expression in the *Drosophila* antenna. *Cell* **96**, 725–736.

Vosshall L. B., Wong A. M. and Axel R. (2000) An olfactory sensory map in the fly brain. *Cell* **102**, 147–159.

Wetzel C. H., Behrendt H.-J., Gisselmann G., Stortkuhl K. F., Hovemann B. and Hatt H. (2001) Functional expression and characterization of a *Drosophila* odorant receptor in a heterologous cell system. *Proc. Nat. Acad. Sci. USA* **98**, 9377–9380.

Wojtasek H., Hansson B. S. and Leal W. S. (1998) Attracted or repelled? A matter of two neurons, one pheromone binding protein, and a chiral center. *Biochem. Biophys. Res. Commun.* **250**, 217–222.

Wojtasek H. and Leal W. S. (1999) Conformational change in the pheromone-binding protein from *Bombyx mori* induced by pH and by interaction with membranes. *J. Biol. Chem.* **274**, 30950–30956.

Wojtasek H., Picimbon J.-F. and Leal W. S. (1999) Identification and cloning of odorant binding proteins from the scarab beetle *Phyllopertha diversa. Biochem. Biophys. Res. Commun.* **263**, 832–837.

Ziegelberger G. (1995) Redox-shift of the pheromone-binding protein in the silkmoth *Antheraea polyphemus. Eur. J. Biochem.* **232**, 706–711.

16

The peripheral pheromone olfactory system in insects: targets for species-selective insect control agents

Erika Plettner

16.1 Introduction

Pheromones are powerful modulators of insect behavior. Since the isolation and identification of the first pheromone, (10*E*, 12*Z*)-hexadec-10,12-dien-1-ol, the sex attractant of the silk moth *Bombyx mori*, thousands of other insect pheromones have been identified. Our understanding of the sensory apparatus required for pheromone detection has also increased significantly. Coincidentally, *B. mori* was instrumental in many of these advances (see below). Volatile pheromones are detected by a specialized olfactory system localized on the antennae. The precise recognition of species-specific nuances in the structure and composition of pheromone components is essential for effective pheromone-based communication. The pheromone olfactory system of species studied so far exhibits remarkable selectivity towards the species-specific pheromone blend. Pheromones are emitted in low (fg–μg) quantities and are dispersed and greatly diluted in air plumes. Thus, pheromone olfaction systems are among the most sensitive chemosensory systems known. (Schneider *et al*., 1968). This chapter summarizes efforts (particularly over the past 10 years) to understand the molecular basis for the remarkable selectivity and sensitivity of the pheromone olfactory system in insects. The chapter will also outline efforts to design compounds that interfere with one or more of the early events in olfaction.

Diversity in the structure and proportion of pheromone components is mirrored in the diversity of the proteins from the olfactory system. A specialized olfactory system is responsible for distinguishing the pheromone from other odorants in the environment. The high precision of the pheromone olfactory system becomes apparent when we compare closely related species whose pheromones differ in subtle ways. For example, *Heliothis* species have the same unsaturated aldehyde as the major pheromone component, but their pheromone signals differ in the structure and proportion of minor components (Table 16.1). Another example is seen with the gypsy moth (*Lymantria dispar*) and nun moth (*Lymantria monacha*), both of which respond to **1a**. The blend produced by the nun moth consists mostly of **1b**, which is a powerful behavioral antagonist in the gypsy moth and is behaviorally inactive in the nun moth (Hansen, 1984). Stereochemical features play an important role in the molecular recognition of pheromone components.

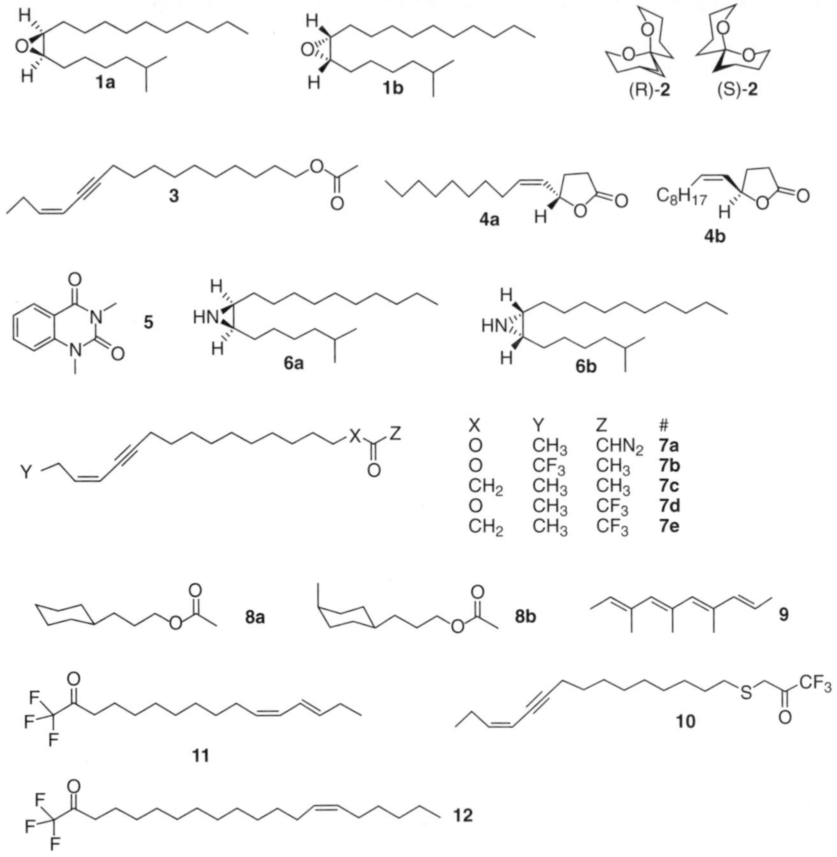
1a
1b
(R)-2
(S)-2
3
4a
4b
C8H17
5
6a
6b
X Y Z #
O CH3 CHN2 7a
O CF3 CH3 7b
CH2 CH3 CH3 7c
O CH3 CF3 7d
CH2 CH3 CF3 7e
8a
8b
9
10
11
12

Structure set

Table 16.1 Studies of pheromone olfactory systems. This table does not include data on general olfaction

	Species	Pheromone type[1]	Major two pheromone components[2]	Emitter[1]	Responder[1]	Components of the pheromone olfactory system studied[3]
Lepidoptera	Black cutworm moth (*Agrotis ipsilon*)	S	(7*Z*) dodec-7-en-1-yl acetate (9*Z*) tetradec-7-en-yl acetate‡	F	M	• PBPs identified (Picimbon and Gadenne, 2002)
	Turnip moth (*Agrotis segetum*)	S	(7*Z*) dodec-7-en-1-yl acetate (5*Z*) dec-5-en-1-yl acetate ‡	F	M	• PBP alleles sequenced {AF134253-288, AF134294, AH007957} introns/exons sequenced (LaForest *et al.* 1999)
	Chinese oak silkmoth (*Antheraea pernyi*)	S	(4*E*, 9Z) tetradeca-4,9-dien-1-yl acetate (6*E*, 11*Z*) hexadeca-6,11-dien-1-yl acetate‡	F	M	• PBPs identified {X96773, X57562, AJ277265} • Full APER1 gene (introns/exons) sequenced {X96860} (Krieger *et al.* 1991)
	Wild silk moth (*Antheraea polyphemus*)	S	(6*E*, 11*Z*)-hexadec-6,11-dienyl acetate (6*E*, 11*Z*)-hexadec-6,11-dienal	F	M	• Study of antennal pheromone transport and distribution • Major PBP sequenced {X17559} (Raming *et al.* 1989) intron structure of the PBP gene: no alternate splicing patterns • Minor PBPs sequenced {X17559,AJ277266, AF277267} (Maida *et al.* 2000) • antennal esterase isolated (Vogt *et al.* 1985) • sensory neuron membrane protein-1 (SNMP) sequenced {U95026} Rogers *et al.* 1997
	Red-banded leafroller (*Argyrotaenia velutinana*)	S	(11*Z*) tetradec-11-en-1-yl acetate dodecan-1-yl acetate‡	F	M	• PBPs sequenced {AF177639-AF177641} (Willett, 2000)
	Silk moth (*Bombyx mori*)	S	(10*E*, 12*Z*)-hexadec-10,12-dien-1-ol	F	M	• Determination of the sensory threshold • PBP sequenced {X94987} (Krieger *et al.* 1996)

(Contd)

Table 16.1 *(Contd)*

	Species	Pheromone type[1]	Major two pheromone components[2]	Emitter[1]	Responder[1]	Components of the pheromone olfactory system studied[3]
	Processionary moth (*Thaumetopoea pityocampa*)	S	**3**	F	M	• Potential pheromone olfaction inhibitor found (see text)
	yponomeuta cagnagellus	S	(11*Z*) tetradec-11-en-1-ol (11*E*) tetradec-11-en-1-yl acetate	F	M	• PBP sequenced {AF177661} (Willett, 2000)
Diptera	Fruit fly (*Drosophila melanogaster*)					• Potential PBPs (OS-E and OS-F) {U02545, U02546} identified on the third antennal segment of adult males (McKenna *et al.* 1994) • PBP-related proteins found in different subsets of olfactory sensilla (Pikielny *et al.* 1994) • First insect olfactory receptors (of unknown odorant selectivity) identified on the maxillary palp (Clyne *et al.* 1999)
Coleoptera	Osaka beetle (*Anomala osakana*)	S	**4a**	F	M	• PBP sequenced {AF031491} (Wojtasek *et al.* 1998)
	Oriental beetle (*Exomala orientalis*)	S	(7*Z*)-tetradec-7-en-2-one			• PBP sequenced {AB040985} (Peng and Leal, 2001)
	Japanese beetle (*Popilia japonica*)	S	**4b**	F	M	• PBP sequenced {AF031492}: nearly identical to *A. osakana* PBP; no enantiomer discrimination (Wojtasek *et al.* 1998)
	Pale brown chafer (*Phyllopertha diversa*)	S	**5**			• Two odorant-binding proteins with homology to the other two known scarab beetle PBPs sequenced {BAA88061, BAA88062} (Wojtasek *et al.* 1999)
Blattariae	German cockroach (*Blatella germanica*)	S/A	(3*S*,11*S*)-3,11-dimethyl-2-nonacosanone	F	M	• Antenna-specific proteins more abundant in males than females found and N-terminally

			(3*S*,11*S*)-29-hydroxy-3,11-dimethylnonacosanone 29-oxo-3,11-dimethyl-2-nonacosanone			sequenced (Picimbon and Leal, 1999) • Studies of signal transduction upon olfactory stimulation: inositol triphosphate acts as second messenger (Breer *et al.* 1990a)
Hymenoptera	Honeybee (*Apis mellifera*)	P/A	(2*E*) 9-ketodec-2-enoic acid (2*E*, 9*R*) 9-hydroxydec-2-enoic acid (2E, 9S) 9-hydroxydec-2-enoic acid‡*	queen	worker	• Queen pheromone-binding protein (antenna-specific protein 1, ASP1) sequenced {AF166496} (Danty et al. 1999) binding studies with the PBP and the aliphatic acid pheromone components • NMR assignment of ASP1 {BMRB-4940} (Birlirakis *et al.* 2001)
		S	(2*E* 9-keto and 9-hydroxy-2-decenoic acid	queen	drone	
	Imported fire ant (*Solenopsis invicta*)	P	(*11Z*) eicos-11-enal*	established queen	newly mated queen	• Gp-9 gene identified as a PBP homologue. • The alleles of Gp-9 correlate with the tendency of workers to accept a single queen (monogyny) or multiple queens (polygyny) (Krieger and
		A	(9Z) tricos-9-ene* (11Z) pentacos-9-ene (u)*	queen	worker	Ross, 2002)
Heteroptera	*Lygus lineolaris*	A	(u)	F	M/F	• Male antenna-specific candidate PBP isolated and sequenced {AAC43033} (Vogt *et al.* 1999)

[1]S = sex attractant, A = aggregation pheromone, P = primer pheromone, F = female, M = male.
[2]Except where indicated, taken from pheromone compendia: (a) M. S. Mayer and J. R. McLaughlin. *Handbook of Insect Pheromones and Sex Attractants*. 1991 CRC Press, Boca Raton, Florida. (b) (Mori, 1998) (c) J. Hardie and A. K. Minks (eds.). *Pheromones of Non-Lepidopteran Insects Associated with Agricultural Plants*. 1999 CABI Publishing, Wallingford, UK. The major component is listed first. ‡ = Other less abundant pheromone component(s); (u) = further pheromone component(s) remain unidentified. * For newly identified honeybee pheromone components, see Keeling *et al.* (2003), for new ant pheromone components see Baird (2001).
[3]Accession numbers are in brackets { }.

In particular, the importance of chirality in diverse pheromone systems has been reviewed recently (Mori, 1998). For example, olive fruit flies (*Bactrocera oleae*) emit racemic **2**; the males detect the R enantiomer, while females detect the S (Haniotakis *et al.*, 1986). The sex pheromone of the Osaka beetle (*Anomala osakana*) is **4a**, while the closely related Japanese beetle (*Popilia japonica*) uses **4b** (Table 16.1). Interestingly, **4a** is a powerful behavioral antagonist in *P. japonica* (Tumlinson *et al.*, 1977). The hemlock looper, *Lambdina fiscellaria*, responds only to (5*R*,11*S*)-5,11-dimethylheptadecane. The enantiomer or the R/R or S/S diastereomers do not elicit electrophysiological or behavioral responses (Li *et al.*, 1993). Table 16.1 gives an overview of those species where proteins from the pheromone olfactory system have been identified.

Insects possess two distinct olfactory systems: a highly selective pheromone olfactory system and a general olfactory system. The latter detects allelochemicals, food odorants and CO_2 (Hildebrand and Sheperd, 1997; Krieger and Breer, 1999). Olfactory neurons corresponding to general and pheromone olfaction are housed in distinct sensory structures, located on the antennae or the mouthparts. In many cases, the sensory structures are hollow hairs (*sensilla*), which contain the dendrites of 2–3 neurons suspended in sensillar lymph (Kaissling and Torson, 1980; Steinbrecht *et al.*, 1995). The lymph is an aqueous, protein-rich solution which probably protects the sensory neurons and plays an important role in the early stages of olfaction (see below) (Kaissling and Torson, 1980).

Pheromone adsorbs on the surface of sensory hairs and diffuses to the inner cuticular face through pores (Keil, 1982). The pheromone is desorbed from the surface by the pheromone-binding proteins (PBPs) (Kowcun *et al.*, 2001). The PBPs transport the hydrophobic pheromone through the aqueous lymph to the dendritic membrane (Figure 16.1). When the pheromone–PBP complex arrives at the dendritic membrane, the pheromone reaches the transmembrane receptors. There are two possible mechanisms for pheromone transfer to the receptors: (1) the PBP–pheromone complex could dissociate near the membrane and the receptors could then bind the free pheromone, or (2) the PBP–pheromone complex could dock with the membrane receptor. Once bound to the receptor, the pheromone elicits a G-protein mediated response via inositol triphosphate (IP_3) and possibly Ca^{2+} (Krieger and Breer, 1999). The second messengers trigger opening of Na^+/K^+ channels and local depolarization, which elicits an action potential of the sensory neuron (Hildebrand and Sheperd, 1997).

16.2 Pheromone-binding proteins (PBPs)

16.2.1 Binding of odorants to PBPs

The first PBPs to be discovered were from wild silk moths *Antheraea polyphemus* (Vogt and Riddiford, 1981) and *Antheraea pernyi* (Kaissling and Torson, 1980)

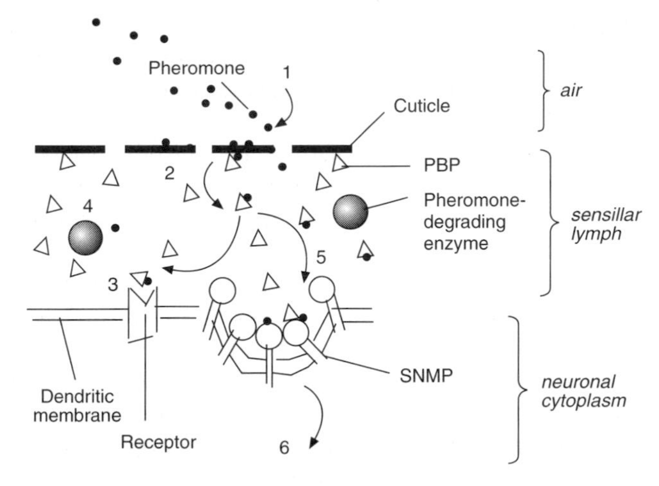

Figure 16.1 The three levels of molecular recognition in the pheromone olfactory system of insects. Pheromone adsorbs on the cuticle, where it enters the sensillum lymph through pores (1). The first level of molecular recognition occurs when the PBP binds and desorbs the pheromone from the cuticle (2). PBP transports the pheromone through the lymph to the receptor, where the second level of recognition occurs (3). The third level of recognition involves the pheromone-degrading enzymes, which rapidly inactivate pheromone that has dissociated from the PBP (4). PBP–pheromone and/or pheromone alone may also be removed by an endocytotic process, possibly mediated by SNMP (5). Finally, intracellular enzymes may be involved in further removal of pheromone (6).

(Table 16.1). Vogt and Riddiford noted that a highly abundant ~15 kDa acidic protein from sensillar extracts strongly binds tritiated (6*E*, 11*Z*) hexadeca-6,11-dien-1-yl acetate, a pheromone component (Vogt and Riddiford, 1981). Since then, PBPs have been sequenced and studied in 33 species (Table 16.1). The insect PBPs are members of a protein family, which includes two groups of general odorant-binding proteins (GOBP1 and GOBP2) (Breer *et al.*, 1990b; Krieger *et al.*, 1996; Maïbèche-Coisné *et al.*, 1998; Vogt *et al.*, 1991; *Vogt et al., 1991* Willett, 2000). The PBPs and GOBPs all have six conserved cysteine residues. These proteins all appear to have a compact α-helical fold (Sandler *et al.*, 2000), and the six cysteine residues in PBPs form three disulfide bridges (Leal *et al.*, 1999; Scaloni *et al.*, 1999). The mammalian odorant-binding proteins (OBPs) are functionally similar to the PBPs and GOBPs, but belong to the lipocalin family, whose members have a β-barrel fold (Tegoni *et al.*, 1996).

PBPs and OBPs are the first molecular component of the olfactory system to interact with volatile compounds absorbed from the air stream. This step is essential for olfaction, because odorants and pheromones have very low solubility in the lymph. Replacement of the lymph with buffer results in loss of

where the photoactive analog is displaced by non-photoreactive pheromones and analogs prior to photoactivation, permit estimation of relative ligand affinities (Feixas *et al.*, 1995). The most rapid method of detection of the PBP.L complex is by electrospray-ionization mass spectrometry (ESI-MS) (Oldham *et al.*, 2000). In the photolabeling and ESI assays, the relative affinities are due to combined kinetic and thermodynamic effects. Allosteric effects may bias the results from competitive photolabeling assays, and the dynamics of electrospray ionization may influence the outcome of the ESI assays. Typical binding data obtained by the various methods discussed are summarized in Table 16.2.

16.2.2 Factors that influence ligand binding to PBPs

The effect of pH on PBP–ligand interactions has been investigated. Below pH 5, very little pheromone is bound to PBPs (Kowcun *et al.*, 2001; Oldham *et al.*, 2000). Between pH 5.5 and 6.5, there is a dramatic increase in binding, suggesting that an amino acid with a $pK_a \sim 6.0$ has a strong effect on pheromone binding. Because this ionization is seen for all PBPs studied so far, the most likely candidate for this effect is one (or more) of the conserved histidine residues. Recently, the structure of the acid form of *B. mori* PBP was solved (Horst *et al.*, 2001). The most notable difference between the acid and basic forms of the PBP is the conformation of the C-terminus. In the basic form the C-terminus is extended, on the surface of the protein (Sandler *et al.*, 2000), while in the acid form the C-terminus forms a seventh α-helix, which occupies the pheromone-binding cavity. The loop with two vicinal histidine residues (69 and 70) has also moved significantly between the two forms, suggesting that titration of one of these residues has a profound effect on the conformation of the PBP (Horst *et al.*, 2001). Consistent with this, mutation of histidines 69 and 70 to alanine, abolished a conformational change detectable upon ligand binding (Mohl *et al.,* 2002).

Above pH 7, the details of the pH profiles for PBPs depend on the PBP and on the ligand, which makes them a tool for investigating the ionic contributions to PBP–ligand interactions. For example, gypsy moth PBP1 has a different pH profile with **1a** or **1b**. A less pronounced difference is seen for PBP2 and **1a** or **1b** (Kowcun *et al.*, 2001). This is thought to reflect different binding modes for these ligands, which are reflected in different perturbations of nearby ionizable residues. For example, binding of **1b** to either PBP appears to involve (1) perturbation of a histidine residue, possibly due to a conformational difference, such that the pK_a of this histidine is 0.5 units higher with **1b** than with **1a** and (2) an H-bonding network of pK_a 7.7–8.6, which appears to anchor **1b** but not **1a**. Which residues are responsible for these interactions remain to be determined, but it is clear that **1a** and **1b** bind to the two gypsy moth PBPs by different interactions.

The rise in ligand binding in the pH 5–6 region coincides with a dramatic conformational change, as detected by NMR (Damberger *et al.*, 2000) and by

Table 16.2 Binding data for PBPs

Species	PBP[1]	Ligand	Dissociation constant, K_d	Assay[3]	Reference
A. mellifera	ASP1	(2*E*) 9-ketodec-2-enoic acid *racemic* (2*E*) 9-hydroxydec-2-enoic acid	ND[2] selective binding from a mixture with methyl tetradecanoate	Extraction/GC	Danty *et al.* (1999)
A. osakana	PBP	**4a** and **4b**	ND no difference between binding of **4a** and **4b** detected	Native PAGE	Wojtasek *et al.* (1998)
A. pernyi	Aper-1	(6E, 11Z) hexadeca-6,11-dien-1-yl acetate	1.83 μM	Vial adsorption assay	Du and Prestwich (1995)
		(4E, 9Z) tetradeca-4,9-dien-1-yl acetate	29.4 μM		
	Aper-2	(6E, 11Z) hexadeca-6,11-dien-1-yl acetate	11.2 μM	Vial adsorption assay	Du and Prestwich (1995)
		(4E, 9Z) tetradeca-4,9-dien-1-yl acetate	3.75 μM		
A. polyphemus	PBP	(6E, 11Z) hexadeca-6,11-dien-1-yl acetate	ND	Native PAGE	Vogt and Riddiford, (1981)
	Apol-3	(6E, 11Z) hexadeca-6,11-dien-1-yl acetate	0.64 μM	Vial adsorption assay	Du and Prestwich (1995)
		(4E, 9Z) tetradeca-4,9-dien-1-yl acetate	21 μM		
		(6E, 11Z) hexadeca-6,11-dien-1-yl acetate	ND selectivity among a fast and a slow form of PBP ~ 1:4	Native PAGE with quantitation of bands	Ziegelberger, (1995)
		decyl-thio trifluoropropanone	ND		
		1a	21.5 μM	Native PAGE	Pophof *et al.* (2000)
		(6E, 11Z) hexadeca-6,11-dien-1-yl acetate	0.48 μM	Mini column assay	Plettner *et al.* (2000)

	Apol PBP1	(6E, 11Z) hexadeca-6,11-dienal	0.50 μM		
		(4E, 9Z) tetradeca-4,9-dien-1-yl acetate	0.51 μM	AMA fluorescence	Campanacci *et al.* (2001)
		(6E, 11Z) hexadeca-6,11-dien-1-yl acetate	0.8 μM		
		(6E, 11Z) hexadeca-6,11-dienal	quenched fluorescence		
		(4E, 9Z) tetradeca-4,9-dien-1-yl acetate	0.58 μM	Tryptophan fluorescence	Bette *et al.* (2002)
B. mori	PBP	(10*E*, 12*Z*)-hexadec-10,12-dien-1-ol	60 nM	Isolated sensory hairs	Kaissling, (2001)
	PBP1	(10*E*, 12*Z*)-hexadec-10,12-dien-1-ol	ND	Pre-equilibration and separation by large-scale gel filtration	Maida *et al.* (1993)
		(10*E*, 12*Z*)-hexadec-10,12-dien-1-ol	ND; pH-dependence of binding	ESI-MS	Oldham *et al.* (2000)
		(10*E*, 12*Z*)-hexadec-10,12-dien-1-ol	1.1 μM	Tryptophan fluorescence	Campanacci *et al.* (2001)
L. dispar	PBP1	**1a**, **1b**	ND very weak binding	Native PAGE	Vogt *et al.* (1989)
		1a	7.1 μM	Mini column assay	Plettner *et al.* (2000)[4]
		1b	2.2 μM		
		1a + **1b** (racemic)	7.8 μM		
	PBP2	**1a**, **1b**	ND strong binding	Native PAGE	Vogt *et al.* (1989)
		1a	1.8 μM	Mini column assay	Plettner *et al.* (2000)
		1b	3.2 μM		
		1a + **1b** (racemic)	4.9 μM		
	PBP1 and PBP2	**1a**, **1b** at different pH and ionic strength	variable (range: 0.7–26 μM)	Mini column assay	Kowcun *et al.* (2001)[4]
M. brassicae	Mbra-1 Mbra-2	(11*Z*) hexadec-11-en-1-yl acetate	ND PBP1 appeared to bind the ligand more strongly than PBP2	Native PAGE	Maïbechè-Coisné *et al.* (1997)

(Contd)

Table 16.2 *(Contd)*

Species	PBP[1]	Ligand	Dissociation constant, K_d	Assay[3]	Reference
		(11Z) hexadec-11-en-1-yl acetate (11Z) hexadec-11-en-1-ol (11Z) hexadec-11-enal (10E, 12Z)-hexadec-10,12-dien-1-ol	0.20 μM 0.17 μM 0.29 μM 0.13 μM	AMA fluorescence	Campanacci *et al.* (2001)[4]
P. japonica	PBP	**4a** and **4b**	ND see *A. osakana*	Native PAGE	Wojtasek *et al.* (1998)
T. pityocampa	SH-15	**7a** with competing **3** or **7b-g**, the alcohol or the aldehyde corresponding to **3**. Two compounds not structurally related to the pheromone were also tested	ND compounds structurally similar to **3** appeared to compete to a similar extent with the photoprobe	Covalent photolabelling	Feixas *et al.* (1995)

[1]As designated in the literature source.
[2] ND = not determined.
[3] See details in the text.
[4] Only selected K_d values are shown from that literature source.

changes in the fluorescence of a tryptophan residue (Wojtasek and Leal, 1999a). Because the pH near a phospholipid bilayer decreases, the pH-dependent behavior of PBPs is thought to be a mechanism of selective pheromone release near the membrane. In one study it was shown that phospholipid vesicles added to the PBP shifted the conformational equilibrium to the low pH conformer (Wojtasek and Leal, 1999a). A recent study, which took the potassium and chloride ion gradients near the membrane into account, indicates that the increase in the K^+ concentration counterbalances the increase in the H^+ concentration near the membrane. Thus, the pheromone appears to bind with approximately the same equilibrium constant throughout the periplasm (Kowcun *et al.*, 2001).

Apart from pH and ionic strength, a third factor that influences pheromone binding to PBPs appears to be the aggregation state of the PBP. The dimers and lower-order aggregates may have a functional significance, as they strongly bind the pheromone (Plettner *et al.*, 2000). Interestingly, the *B. mori* PBP crystallized as a dimer (Sandler *et al.*, 2000). Furthermore, pH appears to influence the monomer/dimer equilibrium (Leal, 2000). During competition experiments with radiolabeled **1a** or **1b** and non-labeled pheromone analogs, we have found what appears to be an allosteric effect. Surprisingly, **1b** binds ~2× more strongly in the presence of certain ligands (such as (7Z) 2-methyloctadec-7-ene) than by itself. This subtle ligand effect is highly reproducible. We also noticed that racemic **1** binds ~2× more weakly than expected from the binding constants of the individual enantiomers. This effect is only possible if either there are two binding sites per PBP or, more likely, if PBPs in a multimer influence each other's ligand binding ability (Honson and Plettner, unpublished observation). Small ligands may influence the affinity of one PBP for another and thereby perturb the monomer–multimer equilibria.

16.2.3 Kinetics of ligand binding to PBPs

Very few studies have addressed the rates of association and dissociation. Du and Prestwich (1995) measured the combined kinetics of ligand binding to PBP and adsorption of PBP.L on the vial surface. The initial rate of the combined process was 0.6 nM/min at 4°C and 0.4 nM/min at 25°C. In a later study, the dissociation rate constants (k_{off}, equation 16.1) for gypsy moth PBP2 were estimated by equilibrating purified PBP.L in ligand-free buffer and assessing the extent of dissociation of PBP.L with time. Dissociation rates found were 9.7×10^{-4} min^{-1} for **1a** and 6.8×10^{-4} min^{-1} for **1b** (Plettner and Prestwich, unpublished). Interactions between gypsy moth PBPs and **1a** or **1b** have been shown to be reversible (and thus true binding equilibria) (Plettner *et al.*, 1999). Because this binding was shown to be reversible, it is valid to estimate the association constant as k_{off}/K_d. Thus for PBP2 we have association constants ~210 M^{-1} min^{-1} for **1a** and ~89 M^{-1} min^{-1} for **1b**. This implies that the two ligands had similar dissociation rates, but differed in their association rates. Whether the kinetics of pheromone binding change with pH and/or ionic strength is not known.

16.2.4 Structure–activity of PBPs: do PBPs discriminate between odorants?

As can be seen from Table 16.2, the dissociation constants for PBPs with different ligands vary at most two orders of magnitude. It has been argued that this reflects poor ligand discrimination by the PBPs. However, different ligands exploit different intermolecular interactions to bind. For example, we have found that aziridine pheromone analogs **6a** and **6b** bind to the PBP only between pH 6 and 8. Below pH 8, the aziridine is largely protonated and, thus, it appears that it is the aziridinium ion that binds to the PBP (Honson et al., unpublished). While **1b** appears to bind through an H-bonding network between PBP and the epoxide O, **6b** cannot possibly be an H-bond acceptor since it binds in fully protonated form. Modeling suggests that for **6b** cation–π interactions contribute to the strong binding, an interaction not possible with **1b**. Other recent studies further suggest that different binding interactions are exploited for different ligands, and that these interactions are probably specific to the ligand–PBP pair. PBP from *M. brassicae* (Mbra1) was mutated at residues that were predicted to line the binding pocket, such that the new binding pocket resembled that of the *B. mori* PBP. While wild-type Mbra1 bound 1-aminoanthracene (AMA) with high affinity, the mutant behaved like *B. mori* PBP in that it did not bind AMA (Campanacci *et al.*, 2001). This was the first demonstration, using mutagenesis, that the sequence of PBPs encodes ligand selectivity. In another study, the *A. pol.* PBP1 appeared to interact differently with acetate and aldehyde ligands: the aldehyde ligand caused quenching of the tryptophan 37 fluorescence, while the acetate ligand did not (Bette *et al.*, 2002). PBPs appear to be flexible in their molecular recognition capability of different ligands, and different molecular recognition events are not always reflected in dramatically different dissociation constants.

An antenna remains in a plume ~1 s and an antenna is not an isolated system, as is required to reach equilibrium. The kinetic properties of the PBP–ligand complexes may be more important to the function of PBPs as potential filters than the equilibrium dissociation constants. Thus, ligands with very fast association rate constants and very slow dissociation rate constants are more likely to be bound at the pore surfaces and to traverse the sensillar lymph unharmed by the powerful pheromone-degrading enzymes in the lymph (see below). Thus, in order to understand the function of PBPs, it is essential to obtain more data on binding kinetics.

16.3 Odorant receptors

16.3.1 Insect odorant receptors

The next stage in the molecular recognition of pheromone occurs at the membrane receptors. In mammals, a highly diverse family of seven-transmembrane G-protein coupled receptors has been identified and the interaction with odorants has been studied (Krautwurst et al., 1998; Malnic *et al.*, 1999). In insects, a similar set of G-protein coupled receptors has been identified in *Drosophila melanogaster*, and the receptors have been studied with respect to expression patterns (Clyne *et al.*, 1999; Voshall *et al.*, 1999). Whether these receptors are involved in general or more specialized olfaction remains unknown.

16.3.2 Structure–activity relationship of the perireceptor space

Electrophysiological studies with single sensory hairs provide insight into the selectivity of the combined pheromone binding, recognition, transport and deactivation events in the perireceptor space. The changes induced in a sensory hair by an odorant include a change in the potential of the sensory hair and an increase in the frequency of spikes (bursts of current, probably caused by opening of ion channels) (Kaissling and Torson, 1980). The relative contribution of perireceptor events to depolarization and spike frequency is not fully understood, but an effort to rationalize the kinetics of depolarization/repolarization from biochemical data is underway (Kaissling, 2001).

A survey of five moth species, all of which use unsaturated acetates with Z stereochemistry as pheromones, revealed that the number of methylene groups between the double bond and the acetate and the total chain length are important (Priesner, 1979). With every additional or missing CH_2 group, approximately one order of magnitude in activity was lost. A quantitative structure–activity relationship (SAR) study on pheromone recognition in the turnip moth (*Agrotis segetum*) indicates that the pheromone is recognized by precise steric and electronic complementarity (Norinder *et al.*, 1997). The acetate group of the turnip moth pheromone exhibits a strong directional interaction, possibly a hydrogen bond, at the carboxyl oxygen (Gustavsson *et al.*, 1997). Similarly, studies with the processionary moth (*Thaumetopoea pityocampa*), which has **3** as its sex pheromone, have revealed that both the acetate group and the enyne moiety are crucial for activity. Loss of the double bond results in a decrease in activity, but removal of the triple bond results in nearly complete loss of activity. An electron-withdrawing group at the double bond position almost completely eliminates activity (Camps *et al.*, 1988). Results with a family of haloacetate analogs indicate that neither steric nor electronic changes are tolerated in the acetate group (Camps *et al.*, 1990).

Single-cell electrophysiological studies on the gypsy moth indicate that **1a** is recognized with a precise fit. The long and short alkyl chains of the pheromone, the *cis* configuration and the 2-methyl group are required for recognition (Schneider *et al.*, 1977). The chirality of the pheromone is essential for behavioral activity (Cardé *et al.*, 1977), and distinct sensory cells detect the pheromone enantiomers (Hansen, 1984). More recent studies indicate that the epoxide O is essential for recognition. Analogue **6a** is not attractive and exhibits significantly lower electrophysiological response than **1a**, and compounds with cyclopropyl and other epoxide substitutions do not elicit any responses (Dickens *et al.*, 1997). Interestingly, **6a** and **6b** bind more strongly to the PBP2 and PBP1 (0.5–0.8 μM) than **1a** and **1b**. Also (7Z) 2-methyloctadec-7-ene and (7Z) octadec-7-ene both bind very strongly (0.5 and 02 μM, respectively) to both PBPs, but these alkenes elicit electrophysiological responses ~100× weaker than **1a** or **1b** (Schneider *et al.*, 1977). Other analogs appear to bind more weakly (Honson *et al.*, unpublished). This suggests that equilibrium PBP binding affinity and electrophysiological activity are not parallel.

Three studies have addressed the recognition of conformers in pheromone olfaction. The first study addressed the structure–activity of a cockroach sex pheromone component (3*S*, 11*S*)-3,11-dimethylnonacosan-2-one. Both methyl groups were optimal for eliciting wing-raising behavior; loss of either methyl group caused a 10× decrease in activity and loss of both methyl groups caused a fourth order of magnitude loss in activity (Sato *et al.*, 1976). Cyclic analogs, the best of which was **8a**, elicited attraction (which precedes wing raising) (Sugawara *et al.*, 1976). Among cyclic analogue with a methyl group, **8b** was the best (Iida and Sugawara, 1981). The pheromone probably prefers bent conformations, which are compacted by packing interactions. The activity of the much smaller cyclic analogue suggests that the carbonyl group and a compact cyclic shape are recognized. The second study focused on enantiomeric cyclic analogs of the two conformers, which result from rotation around the 9–10 C—C bond of (Z) tetradec-11-enyl-1-acetate Figure 16.3. The enantiomers elicited responses from two distinct receptor types in the red-banded leafroller, but only the R analog elicited a response from one receptor type in the European corn borer. This suggests that the former recognizes two different conformers, while the latter recognizes only one conformer (Chapman *et al.*, 1978). The third study focused on **9**, the pheromone of the dried fruit beetle. The methyl branches in this compound impose a torsional strain, which lowers the conjugation between the double bonds and imposes a twist (Figure 16.3). Replacement of a methyl with an ethyl branch increases the twist and lowers activity. Removal of the branches gives an inactive planar structure. The results suggest that the beetles have tuned their olfactory system precisely to the slight twist of **9** (Petroski and Vaz, 1995).

16.4 Clearance of odorants from the perireceptor space

16.4.1 Specialized perireceptor enzymes

Removal of the pheromone from the perireceptor space is essential to maintain the high sensitivity of the sensilla. A number of specialized perireceptor enzymes, which rapidly convert the pheromone into a more water-soluble compound, have

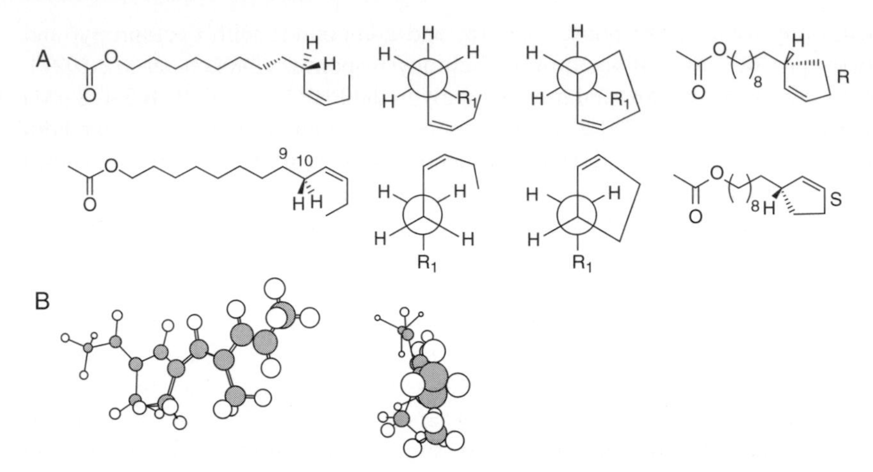

Figure 16.3 Recognition of conformation by the pheromone olfactory system of insects. A Rotamers around the single C–C bond once removed from the *Z*-double bond fall into two groups. Some rotamers are mimicked by the *R*-cyclopentene analog and others by the *S*-analogue. B Steric interactions between the adjacent methyl groups impose a subtle twist on **9**, the pheromone of the dried fruit beetle. The insects have adapted their olfaction precisely to the shape of **9**: an increase or decrease of the twist results in loss of activity.

been studied. The first such enzyme discovered and studied in detail was the esterase from *A. polyphemus* (Vogt and Riddiford, 1981). In purified form, this enzyme hydrolyzes (6*E*, 11*Z*)-hexadeca-6,11-dien-1-yl acetate with half-times of 15 msec (Vogt *et al.*, 1985). In crude antennal homogenates, which contain a high concentration of PBP, degradation of pheromones is much slower (with half-times of minutes) (Klun *et al.*, 1992; Prestwich et al., 1989). The esterase from *A. polyphemus* is equally active at pH 7.2 and at pH 6.5, suggesting that pheromone hydrolysis can occur throughout the periplasmic space (Vogt *et al.*, 1985). Several other esterase activities have been described for insects with ester pheromone components (see below). Pheromone-selective aldehyde dehydrogenases have been described for *H. virescens* (Tasayco and Prestwich, 1990), *M. sexta* (Rybczynski *et al.*, 1989), *A. polyphemus* and *B. mori* (Rybczynski *et al.*, 1990). For the gypsy moth, an epoxide hydrolase has been described (Prestwich *et al.*, 1989). The enzyme hydrolyzes both **1a** and **1b** enantioselectively to the (7*R*, 8*R*) *threo*-diol (Prestwich *et al.*, 1989). An antennal monooxygenase catalyzes demethylation of **5** in the pale-brown chafer (Wojtasek and Leal, 1999b). We have found monooxygenase activity, responsible for converting (7*Z*) 2-methyloctadec-7-ene into an alcohol, in antennal homogenates of male gypsy moths (Hung and Plettner, unpublished observation). The alkene is not a pheromone component of the gypsy moth, but it functions as a behavioral antagonist (Grant *et al.*, 1996). Our result suggests that antennae may contain other catabolic enzymes, which degrade behavioral antagonists and other compounds that could potentially interfere with pheromone olfaction.

16.4.2 Potential pheromone uptake and intracellular clearance mechanisms

A few studies suggest that pheromone is degraded beyond the primary event in the periplasm. In a study with radiolabeled (11Z or *E*) tetradec-11-en-1-yl acetate in the European corn borer, radioactivity was recovered not only in the tetradec-11-en-1-ol, but also in the water. The rate of increase of tritiated water was similar to the rate of decrease of the radiolabeled acetate and alcohol. Within 10 minutes, all the acetate pheromone had disappeared. Traces of labeled tetradec-11-enoic acid were also found. This suggested that the pheromone main chain was being degraded to acetate by β-oxidation (Klun *et al.*, 1992). The β-oxidation system and other multi-enzyme complexes, such as monooxygenases, are unlikely to be found in the periplasmic space, because of cofactor and cosubstrate requirements. Recently, glutathione-S-transferases associated with olfactory hairs have been isolated in *Manduca sexta* (Rogers *et al.*, 1999) and *B. mori* (database accession number: AJ006502). The glutathione-S-transferase from *M. sexta* has no leader sequence, suggesting that it is not secreted into the sensillar lymph. In order to reach β-oxidation systems, monooxygenases and glutathione-S-transferases, the pheromone needs to be actively transported into either the neuron or the supporting cells.

Recently, the sensory neuron membrane protein (SNMP), the most abundant transmembrane protein in sensillar preparations, has been isolated and sequenced (Rogers *et al.*, 1997). Early studies with membrane preparations indicate that this protein binds pheromone very strongly (Vogt *et al.*, 1988). SNMP shows homology to CD36, a known scavenger protein, involved in fatty acid uptake by vascular tissue (Endemann *et al.*, 1993). CD36 also binds other acidic proteins (Carron *et al.*, 2000; Ohgami *et al.*, 2001) and mediates endocytosis (Thorne *et al.*, 1997). By analogy, it is possible that SNMP may mediate endocytosis of the pheromone and/or the pheromone–PBP complex.

16.5 Inhibition of pheromone olfaction

16.5.1 Pheromone olfaction inhibitors

Pheromone olfaction inhibitors should have the following characteristics. First, they should not be behaviorally active or antagonistic by themselves. Unlike behavioral pheromone antagonists, which elicit electrophysiological responses in antennal preparations, olfaction inhibitors should not elicit electrophysiological activity by themselves at the doses used. Thus, olfaction inhibitors should not be detectable to the insect. The reason for this requirement is that insects are able to differentiate plumes of pheromone from plumes of behavioral antagonists 1 mm or 1 ms apart (Fadamiro *et al.*, 1999). Second, pheromone olfaction inhibitors should be volatile and active at concentrations similar to the pheromone. Third, olfaction inhibitors should inhibit selected components of the peripheral olfactory system, thereby significantly increasing the insect's response threshold to the pheromone. The molecular interactions of the inhibitor with sensillar proteins should mimic the interactions of the pheromone at the three early stages of olfaction discussed above. First, to reach the sensory cell, pheromone olfaction inhibitors should bind to and be transported by PBPs. Second, the inhibitors need to compete for transmembrane receptors without causing signal transduction. Third, for a long-term effect, the olfaction inhibitors should inhibit the pheromone clearance system. Most efforts to design olfaction inhibitors have focused on the sensillar pheromone-degrading enzymes. Through this effort, some compounds that target other levels of the olfactory system, including the receptors, have been found (see below).

Studies with pheromone analogs, designed to inhibit pheromone-degrading enzymes, have provided insights into the molecular determinants of pheromone clearance. For example, the diamondback moth (*Plutella xylostella*) has a pheromone-specific antennal esterase that hydrolyzes (*Z*)-hexadec-11-enyl-1-acetate. Among various haloacetate pheromone analogs, the fluorinated analogs most similar to the pheromone inhibited the esterase most strongly (Prestwich and Streinz, 1988). However, the analogs exhibited electrophysiological activity. Other researchers have also found that selectively fluorinated analogs of pheromones exhibit electrophysiological and even behavioral activity (Briggs *et al.*, 1986). Such pheromone analogs are not candidate pheromone olfaction inhibitors because the insects detect them.

More recently, pheromone degradation inhibitors have been designed around functional groups that selectively interfere with the enzyme mechanism. For example, vinyl ketones are 1,4-acceptors for the active-site cysteine, which is thought normally to form a covalent adduct with the aldehyde in aldehyde dehydrogenases (Figure 16.4). Once the 1,4-addition to the inhibitor has taken place, the enzyme is inactive. By tailoring the main chain of the vinyl ketone to resemble the pheromone, it was possible to selectively inhibit and covalently label pheromone-selective aldehyde dehydrogenases in the presence of more general aldehyde oxidases in *H. virescens* (Tasayco and Prestwich, 1990). Among a series of vinyl ketones, those most similar to the major pheromone component inhibited pheromone oxidation the most (Prestwich *et al.*, 1989). The enantioselective opening of **1a** or **1b** in the gypsy moth involves activation of the epoxide group by an active site acidic residue, followed by opening of the epoxide. Analogues were designed to H-bond with the catalytic residue and thereby prevent epoxide hydrolysis (Figure 16.4). Compounds with an electron-

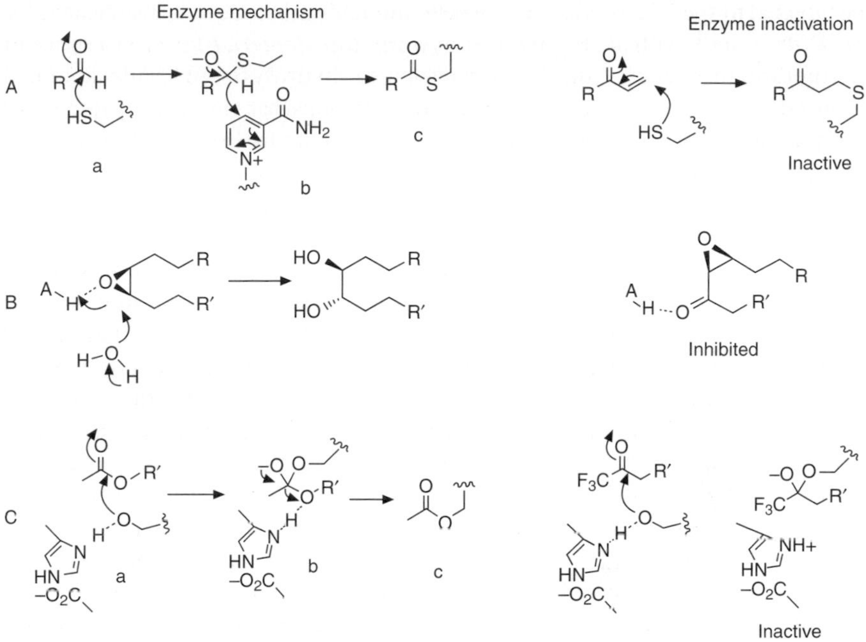

Figure 16.4 Postulated mechansims of pheromone-degrading enzymes and design of inhibitors that interfere with the mechanism. A Aldehyde dehydrogenases contain a cysteine residue, which forms a tetrahedral adduct with the aldehyde. This adduct collapses with donation of an H^- equivalent to NAD^+. Vinyl ketones inhibit these enzymes by forming a stable 1,4-addition product (Tasayco and Prestwich, 1990). B The epoxide hydrolase from the gypsy moth is thought to activate the epoxide moiety towards nucleophilic addition of water by acid catalysis. When a keto group is adjacent to the epoxide, the keto group is thought to H-bond to the catalytic residue, which inhibits the enzyme (Graham and Prestwich, 1994). C Some esterases contain a catalytic triad similar to serine proteases. The deprotonated serine attacks the acyl group of the substrate to give a tetrahedral intermediate b. This intermediate collapses with concomittant protonation of the leaving group, to give the acyl enzyme intermediate c. The sequence then repeats itself with water as the nucleophile. Trifluoromethyl ketones inhibit these enzymes by forming a stable tetrahedral adduct (Durán *et al.*, 1993).

withdrawing group at the 6 position gave strongest inhibition. An electron-withdrawing group at the 9 position led to intermediate inhibitory activity, while functionalization at both the 6 and 9 positions led to cancellation of inhibitory activity (Graham and Prestwich, 1994). Trifluoromethylketone and thiotrifluoropropanone analogs of pheromones form tetrahedral adducts with various hydrolase enzymes, such as esterases. Because esterases have selectivity for the carboxyl and leaving group portions of their substrates, the inhibitors can be targeted to specific esterases by closely mimicking the natural substrate (Figure 16.4). In a study with the Egyptian armyworm (*Spodoptera littoralis*) a series of trifluoromethylketones inhibited pheromone hydrolysis in crude antennal preparations to varying extents. Those aliphatic analogs that most closely resembled the pheromone in chain length inhibited pheromone hydrolysis most strongly (IC_{50} 0.6–1μM) (Duran *et al.*, 1993).

16.5.2 Comprehensive studies of pheromone olfaction inhibitors

Trifluoromethylketones and thiotrifluoropropanones that closely mimic the pheromone inhibit insect attraction to pheromone baits, but have low intrinsic electrophysiological and behavioral activity (Camps *et al.*, 1990; Parrilla and Guerrero, 1994; Pophof, 1998; Pophof *et al.*, 2000). In the presence of the inhibitors, pheromone stimuli give significantly lower depolarization and spike frequencies than in the absence of the inhibitors. By themselves, however, these compounds do not elicit depolarization, suggesting that they are not detected (Pophof, 1998; Pophof *et al.*, 2000). The best electrophysiological inhibitors show the strongest decrease in insect attraction to pheromone-baited traps (Camps *et al.*, 1990; Parrilla and Guerrero, 1994). Three further examples illustrate that strong inhibition of olfaction requires a high degree of similarity with the pheromone. First, the processionary moth responds best to compounds that have the enyene moiety, such as **10** (Camps *et al.*, 1990). Second, in a study with the Egyptian armyworm (*Spodoptera littoralis*) and the Mediterranean corn borer (*Sesamia nonagrioides*), topical application of the trifluoromethylketone analog of the armyworm pheromone **11** on male armyworm antennae resulted in significantly greater inhibition of pheromone-induced upwind flight and source contacts than topical application of the corn borer analog **12**. The opposite selectivity was observed for the corn borer (Bau *et al.*, 1999).

The mode of action of trifluoromethylketone pheromone analogs is not well understood. When these compounds are applied to the antenna after a pheromone stimulus, they cause a faster initial repolarization of the receptor potential (Pophof, 1998). If inhibition of pheromone-degrading enzymes were the exclusive mode of action of these compounds, then one would expect a slower repolarization. One explanation for the observed effect is that the olfaction inhibitor competes for the receptor without triggering signal transduction. This is consistent with the observation that the inhibitors do not cause nerve depolarization by themselves and that, when presented concurrently with the pheromone, these compounds significantly decrease the amplitude of the pheromone-mediated depolarization (Pophof, 1998). The long-term effect, namely, inhibitor-treated sensilla no longer respond to repeated pheromone stimuli applied during repolarization (Parrilla and Guerrero, 1994), may be due to esterase inhibition and saturation of the system.

The pheromone olfaction inhibitors do not appear to act by general toxicity, as supported by several observations. First, they inhibit electrophysiological responses at low doses (typically 1–10× the concentration of pheromone). Second, the inhibitors have a localized effect on olfactory neurons. For example, application of the inhibitor at the tip of a sensory hair impairs pheromone olfaction only at the tip, not at the base of the hair (Pophof, 1998). If the compounds were generally toxic, one would expect inhibition of the entire sensillum, regardless of application location. Third, exposure to the inhibitor alone does not significantly alter the behavior of the moths in the absence of pheromone (Bau *et al.*, 1999; Renou *et al.*, 1999). If the inhibitors were toxic at the active doses, then insects would show signs of irritation and avoidance behavior.

16.6 Conclusions and future work

The stringent steric and electronic requirements observed in electroantennography SAR studies and the results with trifluoromethylketones and thiotrifluoropropanones suggest that it is possible to design species-selective inhibitors of pheromone olfaction. In order to design more inhibitors, we require further knowledge concerning (1) the transport of pheromone in the sensillar lymph, (2) the interaction of pheromone with the olfactory receptors and (3) the clearance of pheromone from the perireceptor space. Regarding transport of pheromone, the role of the PBPs is not fully delineated. For example, it is not known whether PBP–pheromone complexes dissociate near the dendritic membrane or whether the PBP–ligand complex docks with membrane proteins. The former possibility implies a stepwise molecular recognition of the pheromone, first by the PBP and then by the receptors. The latter possibility implies molecular recognition of the PBP–pheromone complex by the membrane proteins. Regarding transduction of the pheromone signal, the nature of the pheromone receptors in insects remains elusive. Finally, pheromone degradation enzymes should be investigated further, since relatively little is known about clearance of pheromones and the most important candidate pheromone olfaction inhibitors came from efforts to target these enzymes.

References

Baird D. S. (2001) Semiochemical studies on the red imported fire ant (*Solenopsis invicta*), the tarnished plant bug (*Lygus lineolaris*) and the varroa mite (*Varroa destructor*). In *Chemistry*. Simon Fraser University Burnaby.

Bau J., Martinez D., Renou M. and Guerrero A. (1999) Pheromone-triggered orientation flight of male moths can be disrupted by trifluoromethyl ketones. *Chem. Senses* **24**, 473–480.

Bette S., Breer H. and Krieger J. (2002) Probing a pheromone binding protein of the silkmoth *Antheraea polyphemus* by endogenous tryptophan fluorescence. *Insect Biochem. Mol. Biol.* **32**, 241–246.

Birlirakis N., Briand L., Pernollet J.-C. and Guittet E. (2001) 1H, 13C and 15N chemical shift assignment of the honey bee pheromone carrier protein ASP1. *J. Biomol. NMR* **20**, 183–184.

Breer H., Boekhoff I. and Tareilus E. (1990a) Rapid kinetics of second messenger formation in olfactory transduction. *Nature* **345**, 65–68.

Breer H., Krieger J., and Raming K. (1990b) A novel class of binding proteins in the antennae of the silk moth *Antheraea pernyi*. *Insect Biochem.* **20**, 735–740.

Briggs G. G., Cayley G. R., Dawson G. W., Griffiths D. C., Macaulay E. D. M., Pickett J. A., Pile M. M., Wadhams L. J. and Woodcock C. M. (1986) Some fluorine-containing pheromone analogues. *Pestic. Sci.* **17**, 441–448.

Callahan F. E., Vogt R. G., Tucker M. L., Dickens J. C. and Mattoo A. K. (2000) High level expression of "male specific" pheromone binding proteins (PBPs) in the antennae of female noctuiid moths. *Insect Biochem. Mol. Biol.* **30**, 507–514.

Campanacci V., Krieger J., Bette S., Sturgis J. N., Lartigue A., Cambillau C., Breer H. and Tegoni M. (2001) Revisiting the specificity of *Mamestra brassicae* and *Antheraea polyphemus* pheromone-binding proteins with a fluorescence binding assay. *J. Biol. Chem.* **276**, 20078–20084.

Camps F., Fabrias G., Gasol V., Guerrero A., Hernandez R. and Montoya R. (1988) Analogs of sex pheromone of processionary moth, *Thaumetophoea pityocampa*: synthesis and biological activity. *J. Chem. Ecol.* **14**, 1331–1346.

Camps F., Gasol V. and Guerrero A. (1990) Inhibitory pheromonal activity promoted by sulfur analogs of the sex pheromone of the female processionary moth *Thaumetopoea pityocampa* (Denis and Schiff). *J. Chem. Ecol.* **16**, 1155–1172.

Cardé, R. T., Doane C. C., Baker T. C., Iwaki S. and Marumo, S. (1977). Attractancy of optically active pheromone for male gypsy moths. *Environ. Entomol.* **6**, 768–772.

Carron J. A., Wagstaff S. C., Gallagher J. A. and Bowler W. B. (2000) A CD36-binding peptide from thrombospondin-1 can stimulate resorption by osteoclasts *in vitro*. *Biochem. Biophys. Res. Commun.* **270**, 1124–1127.

Chapman O. L., Mattes K. C., Sheridan R. S. and Klun J. A. (1978) Stereochemical evidence of dual chemoreceptors for an achiral sex pheromone in lepidoptera. *J. Am. Chem. Soc.* **100**, 4878–4884.

Clyne P. J., Warr C. G., Freeman M. R., Lessing D., Kim J. and Carlson J. R. (1999) A novel family of divergent seven-transmembrane proteins: candidate odorant receptors in *Drosophila*. *Neuron* **22**, 327–338.

Damberger F., Nikanova L., Horst R., Peng G., Leal W. S. and Wuthrich K. (2000) NMR characterization of a pH-dependent equilibrium between two folded solution conformations of the pheromone-binding protein from *Bombyx mori*. *Protein Science* **9**, 1038–1041.

Danty E., Birand L., Michard-Vanhee C., Perez V., Arnold G., Gaudemer O., Huet D., Huet J.-C., Ouali C., Masson C. and Pernollet J.-C. (1999) Cloning and expression of a queen pheromone-binding protein in the honeybee: an olfactory-specific, developmentally regulated protein. *J. Neurosci.* **19**, 7468–7475.

Dickens J. C., Oliver J. E. and Mastro V. C. (1997) Response and adaptation to analogs of disparlure by specialist antennal receptor neurons of gypsy moth, *Lymantria dispar*. *J. Chem. Ecol.* **23**, 2197–2210.

Du G., Ng C.-S. and Prestwich G. D. (1994) Odorant binding by a pheromone binding protein: active site mapping by photoaffinity labeling. *Biochem.* **33**, 4812–4819.

Du G. and Prestwich G. D. (1995) Protein structure encodes the ligand binding specificity in pheromone binding proteins. *Biochem.* **34**, 8726–8732.

Duran I., Parrilla A., Feixas J. and Guerrero A. (1993) Inhibition of antennal esterases of the Egyptian armyworm *Spodoptera littoralis* by trifluoromethyl ketones. *Bioorg. Med. Chem. Lett.* **3**, 2593–2598.

Endemann G., Stanton L. W., Madden K. S., Bryant C. M., White R. T. and Protter A. A. (1993) CD36 is a receptor for oxidized low density lipoprotein. *J. Biol. Chem.* **268**, 11811–11816.

Fadamiro H. Y., Cosse A. A. and Baker T. C. (1999) Fine-scale resolution of closely spaced pheromone and antagonist filaments by flying male *Helicoverpa zea*. *J. Comp. Physiol. A* **185**, 131–141.

Feixas J., Prestwich G. D. and Guerrero A. (1995) Ligand specificity of pheromone-binding proteins of the processionary moth. *Eur. J. Biochem.* **234**, 521–526.

Graham S. M. and Prestwich G. D. (1994) Synthesis and inhibitory properties of pheromone analogues for the epoxide hydrolase of the gypsy moth. *J. Org. Chem.* **59**, 2956–2966.

Grant G. G., Langevin D., Liska J., Kapitola P. and Chong J. M. (1996) Olefin inhibitor of gypsy moth, *Lymantria dispar*, is a synergistic pheromone component of nun moth, *L. monacha*. *Naturwissenschaften* **83**, 328–330.

Gustavsson A.-L., Tuvesson M., Larsson M. C., Wenqi W., Hansson B. S. and Liljefors T. (1997) Bioisosteric approach to elucidation of binding of the acetate group of a moth pheromone component to its receptor. *J. Chem. Ecol.* **23**, 2755–2776.

Gyorgyi T. K., Roby-Shemkovitz A. J. and Lerner M. R. (1988) Characterization and cDNA cloning of the pheromone-binding protein from the tobacco hornworm, *Manduca sexta*: a tissue-specific developmentally regulated protein. *Proc. Natl. Acad. Sci. USA* **85**, 9851–9855.

Haniotakis G., Francke W., Mori K., Redlich H. and Schurig V. (1986) Sex-specific activity of (R)-(–) and (S)-(+) 1,7-dioxaspiro[5.5]undecane, the major pheromone of *Dacus oleae*. *J. Chem. Ecol.* **12**, 1559.

Hansen K. (1984) Discrimination and production of disparlure enantiomers by the gypsy moth and the nun moth. *Physiol. Entomol.* **9**, 9–18.

Hildebrand J. G. and Sheperd G. M. (1997) Mechanisms of olfactory discrimination: converging evidence for common principles across phyla. *Ann. Rev. Neurosci.* **20**, 595–631.

Horst R., Damberger F., Luginbuehl P., Guentert P., Peng G., Nikanova L., Leal W. S. and Wuethrich K. (2001) NMR structure reveals intramolecular regulation mechanism for pheromone binding and release. *Proc. Natl. Acad. Sci.* **98**, 14374–14379.

Iida Y. and Sugawara R. (1981) Stereochemical studies of alkyl methylcyclohexaneacetates with 13C NMR spectroscopy in relation to their attractiveness to the German cockroach. *Agr. Biol. Chem.* **45**, 1553–1559.

Kaissling K.-E. (2001) Olfactory perireceptor and receptor events in moths: a kinetic model. *Chem. Senses* **26**, 125–150.

Kaissling K. E. and Torson J. (1980) Insect olfactory sensilla: structural, chemical and electrical aspects of the functional organization. In *Receptors for Neurotransmitters, Hormones and Pheromones in Insects*, eds D. B. Sattelle J. G. Hildebrand and L. U. Hall, pp. 261–282. Elsevier/North-Holland Biomedical Press, Amsterdam), pp. 261–282.

Keeling C. I., Slessor K. N., Higo H. A. and Winston M. L. (2003) New components of the honey bee (*Apis mellifera* L.) queen retinue pheromone. *Proc. Natl. Acad. Sci.* **100**, 4486–4491.

Keil T. A. (1982) Contacts of pore tubules and sensory dendrites in antennal chemosensilla of a silkmoth: demonstration of a possible pathway for olfactory molecules. *Tissue & Cell* **14**, 451–462.

Kim M.-S., Repp A. and Smith D. P. (1998) LUSH odorant-binding protein mediates chemosensory responses to alcohols in *Drosophila melanogaster*. *Genetics* **150**, 711–721.

Klun J. A., Schwarz M. and Uebel E. C. (1992) Biological activity and in vivo degradation of tritiated female sex pheromone in the male European corn borer. *J. Chem. Ecol.* **18**, 283–298.

Kowcun A., Honson N. and Plettner E. (2001) Olfaction in the gypsy moth, *Lymantria dispar*: effect of pH, ionic strength and reductants on pheromone transport by pheromone-binding proteins. *J. Biol. Chem.* **276**, 44770–44776.

Krautwurst, D., Yau K.-W. and Reed R. R. (1998) Identification of ligands for olfactory receptors by functional expression of a receptor library. *Cell* **95**, 917–926.

Krieger J. and Breer H. (1999) Olfactory reception in invertebrates. *Science* **286**, 720–723.

Krieger J., Gانssle H., Ramnig K. and Breer H. (1993) Odorant binding proteins of *Heliothis virescens*. *Insect Biochem. Molec. Biol.* **23**, 449–456.

Krieger J., Raming K. and Breer H. (1991) Cloning of genomic and complementary DNA encoding insect pheromone binding proteins: evidence for microdiversity. *Biochim. Biophys. Acta* **1088**, 277–284.

Krieger J., von Nickisch-Rosenegk E., Mameli M., Pelosi P. and Breer H. (1996) Binding proteins from the antennae of *Bombyx mori*. *Insect Biochem. Molec. Biol.* **26**, 297–307.

Krieger M. J. B. and Ross K. G. (2002) Identification of a major gene regulating complex social behavior. *Science* **295**, 328–332.

LaForest S. M., Prestwich G. D. and Löfstedt C. (1999) Intraspecific nucleotide variation at the pheromone binding protein locus in the turnip moth, *Agrotis segetum*. *Insect Mol. Biol.* **8**, 481–490.

Leal W. S. (2000) Duality monomer-dimer of the pheromone-binding protein of *Bombyx mori*. *Biochem. Biophys. Res. Commun.* **268**, 521–529.

Leal W. S., Nikonova L. and Peng G. (1999) Disulfide structure of the pheromone binding protein from the silkworm moth, *Bombyx mori*. *FEBS Letters* **464**, 85–90.

Li J., Gries R., Gries G., Slessor K. N., King G. G. S., Bowers W. W. and West R. J. (1993) Chirality of 5,11-dimethylheptadecane, the major sex pheromone component of the hemlock looper, *Lambdina fiscellaria* (Lepidoptera: geometridae). *J. Chem. Ecol.* **19**, 1057–1062.

Maïbèche-Coisné M., Longhi S., Jacquin-Joly E., Brunel C., Egloff M.-P., Gastinel L., Cambillau C., Tegoni M. and Nagnan-LeMeillour P. (1998) Molecular cloning and bacterial expression of a general odorant-binding protein from the cabbage armyworm, *Mamestra brassicae*. *Eur. J. Biochem.* **258**, 768–774.

Maïbèche-Coisné M., Sobrio F., Delaunay T., Lettere M., Dubroca J., Jacquin-Joly E. and Nagnan-Le-Meillour P. (1997) Pheromone binding proteins of the moth *Mamestra brassicae*: specificity of ligand binding. *Insect Biochem. Molec. Biol.* **27**, 213–221.

Maida R., Krieger J., Gebauer T., Lange U. and Ziegelberger G. (2000) Three pheromone-binding proteins in olfactory sensilla of the two silkmoth species *Antheraea polyphemus* and *Antheraea pernyi*. *Eur. J. Biochem.* **267**, 2899–2908.

Maida R., Steinbrecht A., Ziegelberger G. and Pelosi P. (1993) The pheromone binding protein of Bombyx *mori*: purification, characterization and immunocytochemical localization. *Insect Biochem. Molec. Biol.* **23**, 243–253.

Malnic B., Hirono J., Sato T. and Buck L. B. (1999) Combinatorial receptor codes for odors. *Cell* **96**, 713–723.

McKenna M. P., Hekmat-Scafe D. S., Gaines P. and Carlson J. R. (1994) Putative *Drosophila* pheromone-binding proteins expressed in a subregion of the olfactory system. *J. Biol. Chem.* **269**, 16340–16347.

Merritt T. J. S., LaForest S., Prestwich G. D., Quattro J. M. and Vogt R. G. (1998) Patterns of gene duplication in lepidopteran pheromone binding proteins. *J. Mol. Evol.* **46**, 272–276.

Mohl C., Breer H. and Krieger J. (2002) Species-specific pheromonal compounds induce distinct conformational changes of pheromone binding protein subtypes from *Antheraea polyphemus*. *Invert. Neurosci.* **4**, 165–174.

Mori K. (1998) Semiochemicals – synthesis, stereochemistry, and biodiversity. *Eur. J. Org. Chem.* **1998**, 1479–1489.

Newcomb R. D., Sirey T. M., Rassam M. and Greenwood D. R. (2002) Pheromone binding proteins of epiphyas postvittana (*Lepidoptera Tortricidae*) are encoded at a single locus. *Insect Biochem. Mol. Biol.* **32**, 1543–1554.

Norinder U., Gustavsson A.-L. and Liljefors T. (1997) A 3D-QSAR study of analogs of (Z)-5-decenyl acetate, a pheromone component of the turnip moth, *Agrotis segetum*. *J. Chem. Ecol.* **23**, 2917–2934.

Ohgami N., Nagai R., Ikemoto M., Arai H., Kuniyasu A., Horiuchi S. and Nakayama H. (2001) CD36, a member of the class B scavenger receptor family, as a receptor for advanced glycation end products. *J. Biol. Chem.* **276**, 3195–3202.

Oldham N. J., Krieger J., Breer H., Fischedick A., Hoskovec M. and Svatos A. (2000) Analysis of the silkworm moth pheromone binding protein-pheromone complex by electrospray-ionization mass spectrometry. *Angew. Chem. Int. Ed.* **39**, 4341–4343.

Parrilla A. and Guerrero A. (1994) Trifluoromethyl ketones as inhibitors of the processionary moth sex pheromone. *Chem. Senses* **19**, 1–10.

Peng G. and Leal W. S. (2001) Identification and cloning of a pheromone-binding protein from the oriental beetle, *Exomala orientalis*. *J. Chem. Ecol.* **27**, 2183–2192.

Petroski R. J. and Vaz R. (1995) Insect aggregation pheromone response synergized by "host-type" volatiles. In *Computer-Aided Molecular Design: Application in Agrochemicals, Materials and Pharmaceuticals*, eds C. H. Reynolds, M. K. Holloway and H. K. Cox. ACS Washington DC.

Picimbon J.-F. and Gadenne C. (2002) Evolution of noctuid pheromone binding proteins: identification of PBP in the black cutworm moth, *Agrotis ipsilon*. *Insect. Biochem. Mol. Biol.* (in press).

Picimbon J.-F. and Leal W. S. (1999) Olfactory soluble proteins of cockroaches. *Insect Biochem Molec. Biol.* **29**, 973–978.

Pikielny C. W., Hasan G., Rouyer F. and Rosbach M. (1994) Members of a family of *Drosophila* putative odorant-binding proteins are expressed in different subsets of olfactory hairs. *Neuron* **12**, 35–49.

Plettner E., DeSantis G., Stabile M. and Jones J. B. (1999) Modulation of esterase and amidase activity of subtilisin *Bacillus lentus* by chemical modification of cysteine mutants. *J. Am. Chem. Soc.* **121**, 4977–4981.

Plettner E., Lazar J., Prestwich E. G. and Prestwich G. D. (2000) Discrimination of pheromone enantiomers by two pheromone binding proteins from the gypsy moth *Lymnatria dispar*. Biochem. **39**, 8953–8962.

Pophof B. (1998) Inhibitors of sensillar esterase reversibly block the responses of moth pheromone receptor cells. *J. Comp. Physiol. A* **183**, 153–164.

Pophof B., Gebauer T. and Ziegelberger G. (2000) Decyl-thio-trifluoropropanone, a competitive inhibitor of moth pheromone receptors. *J. Comp. Physiol. A* **186**, 315–323.

Prestwich G. D., Graham S. M., Handley M., Latli B., Streinz L. and Tasayco J. M. L. (1989) Enzymatic processing of pheromones and pheromone analogs. *Experientia* **45**, 263–270.

Prestwich G. D., Graham S. M. and Konig W. A. (1989) Enantioselective opening of (+)- and (–)-disparlure by epoxide hydrase in gypsy moth antennae. *J. Chem. Soc. Commun.*, 575–577.

Prestwich G. D., Graham S. M., Kuo J.-W. and Vogt R. G. (1989) Tritium-labeled enantiomers of disparlure. Synthesis and in vitro metabolism. *J. Am. Chem. Soc.* **111**, 636–642.

Prestwich G. D. and Streinz L. (1988). Haloacetate analogs of pheromones. *J. Chem. Ecol.* **14**, 1003–1021.

Priesner E. (1979) Specificity studies on pheromone receptors of noctuid and tortricid lepidoptera. In *Chemical Ecology: Odour Communication in Animals*, ed. F. J. Ritter. Elsevier, Amsterdam.

Raming K., Krieger J. and Breer H. (1989) Molecular cloning of an insect pheromone-binding protein. *FEBS Lett.* **256**, 215–218.

Renou M., Berthier A., Dasbarats L., VanderPers J. and Guerrero A. (1999) Actographic analysis of the effects of an esterase inhibitor on male moth responses to pheromone. *Chem. Senses* **24**, 423–428.

Rogers M. E., Jani M. K. and Vogt R. G. (1999) An olfactory-specific glutathione-S-transferase in the sphinx moth, *Manduca sexta*. *J. Exp. Biol.* **202**, 1625–1637.

Rogers M. E., Sun M., Lerner M. R. and Vogt R. G. (1997) Snmp-1, a novel membrane protein of olfactory neurons of the silk moth, *Antheraea polyphemus*, with homology of the CD36 family of membrane proteins. *J. Biol. Chem.* **272**, 14792–14799.

Rybczynski R., Reagan J. and Lerner M. R. (1989) A pheromone-degrading aldehyde oxidase in the antennae of the moth *Manduca sexta*. *J. Neurosci.* **9**, 1341–1353.

Rybczynski R., Vogt R. G. and Lerner M. R. (1990) Antennal-specific pheromone-degrading aldehyde oxidases from the moths *Antheraea polyphemus* and *Bombyx mori*. *J. Biol. Chem.* **265**, 19712–19715.

Sandler B. H., Nikonova L., Leal W. S. and Clardy J. (2000) Sexual attraction in the silkworm moth: structure of the pheromone-binding-protein–bombykol complex. *Chemistry & Biology* **7**, 143–151.

Sato T., Nishida R., Kuwahara Y., Fukami H. and Ishii S. (1976) Syntheses of female sex pheromone analogues of the german cockroach and their biological activity. *Agr. Biol. Chem.* **40**, 391–394.

Scaloni A., Monti M., Angeli S. and Pelosi P. (1999) Structural analysis and disulfide-bridge pairing of two odorant-binding proteins from *Bombyx mori*. *Biochem. Biophys. Res. Commun.* **266**, 386–391.

Schneider D., Kafka W. A., Beroza M. and Bierl B. A. (1977) Odor receptor responses of male gypsy and nun moths (Lepidoptera, Lymantriidae) to disparlure and its analogues. *J. Comp. Physiol. A* **113**, 1–15.

Schneider D., Kasang G. and Kaissling K.-E. (1968) Bestimmung der Riechschwelle von *Bombyx mori* mit Tritium-markiertem Bombykol. *Naturwissenschaften* **55**, 395.

Steinbrecht R. A., Laue M. and Ziegelberger G. (1995) Immunolocalization of pheromone-binding protein and general odorant-binding protein in olfactory sensilla of the silkmoths *Antheraea* and *Bombyx*. *Cell Tissue Res.* **282**, 203–217.

Sugawara R., Tominaga Y., Kobayashi M. and Muto T. (1976) Effects of side chain modification on the activity of cyclohexaneacetate as an attractant to the German cockroach. *J. Insect Physiol.* **22**, 785–790.

Tasayco M. L. and Prestwich G. D. (1990) A specific affinity reagent to distinguish aldehyde dehydrogenases and oxidases. *J. Biol. Chem.* **265**, 3094–3101.

Tegoni M., Ramoni R., Bignetti E., Spinelli S. and Cambillau C. (1996) Domain swapping creates a third putative combining site in bovine odorant binding protein dimer. *Nature Structural Biology* **3**, 863–867.

Thorne R. F., Meldrum C. J., Harris S. J., Dorahy D. J., Shafren D. R., Berndt M. C., Burns G. F. and Gibson P. G. (1997) CD36 forms covalently associated dimers and multimers in platelets and transfected COS-7 cells. *Biochem. Biophys. Res. Commun.* **240**, 812–818.

Tumlinson J. H., Klein M. G., Doolittle R. E., Ladd T. L. and Proveaux A. T. (1977) Identification of the female Japanese beetle sex pheromone: inhibition of male response by an enantiomer. *Science* **197**, 789–792.

VandenBerg M. J. and Ziegelberger G. (1991) On the function of the pheromone binding protein in the olfactory hairs of *Antheraea polyphemus*. *J. Insect Physiol.* **37**, 79–85.

Vogt R. G., Callahan F. E., Rogers M. E. and Dickens J. C. (1999) Cloning and expression of LAP, an adult specific odorant binding protein of the true bug *Lygus lineolaris* (Hemiptera, Heteroptera). *Chemical Senses* **24**, 481–495.

Vogt R. G., Kohne A. C., Dubnau J. T. and Prestwich G. D. (1989) Expression of pheromone binding proteins during antennal development in the gypsy moth, *Lymantria dispar*. *J. Neurosci.* **9**, 3332–3346.

Vogt R. G., Prestwich G. D. and Lerner M. R. (1991) Odorant-binding-protein subfamilies associate with distinct classes of olfactory receptor neurons in insects. *J. Neurobiol.* **22**, 74–84.

Vogt R. G., Prestwich G. D. and Riddiford L. M. (1988) Sex pheromone receptor proteins. *J. Biol. Chem.* **263**, 3952–3959.

Vogt R. G. and Riddiford L. M. (1981) Pheromone binding and inactivation by moth antennae. *Nature* **293**, 161–163.

Vogt R. G., Riddiford L. M. and Prestwich G. D. (1985) Kinetic properties of a sex pheromone-degrading enzyme: the sensillar esterase of *Antheraea polyphemus*. *Proc. Natl. Acad. Sci. USA* **82**, 8827–8831.

Vogt R. G., Rybczynski R. and Lerner M. R. (1991) Molecular cloning and sequencing of general odorant binding proteins GOBP1 and GOBP2 from the tobacco hawk moth *Manduca sexta*: comparisons with other insect OBPs and their signal peptides. *J. Neurosci.* **11**, 2972–2984.

Voshall L. B., Amrein H., Morozov P. S., Rzhetsky A. and Axel R. (1999) A spatial map of olfactory receptor expression in the *Drosophila* antenna. *Cell* **96**, 725–736.

Willett C. (2000) Do pheromone binding proteins converge in amino acid sequence when pheromones converge? *J. Mol. Evol.* **50**, 175–183.

Wojtasek H., Hansson B. S. and Leal W. S. (1998) Attracted or repelled? – a matter of two neurons, one pheromone binding protein, and a chiral center. *Biochem. and Biophys. Res. Commun.* **250**, 217–222.

Wojtasek H. and Leal W. S. (1999a) Conformational change in the pheromone-binding protein from *Bombyx mori* induced by pH and by interaction with membranes. *J. Biol. Chem.* **274**, 30950–30956.

Wojtasek H. and Leal W. S. (1999b) Degradation of an alkoloid pheromone from the pale-brown chafer, *Phyllopertha diversa* (Coleptera: Scarabaeidae), by an insect olfactory cytocrome P450. *FEBS Lett* **458**, 333–336.

Wojtasek H., Picimbon J.-F. and Leal W. S. (1999) Identification and cloning of odorant binding proteins from the scarab beetle *Phyllopertha diversa*. *Biochem. Biophys. Res. Commun.* **263**, 832–837.

Ziegelberger G. (1995) Redox-shift of the pheromone-binding protein in the silkmoth, *Antheraea polyphemus*. *Eur. J. Biochem.* **232**, 706–711.

17

Biochemistry and diversity of insect odorant-binding proteins

Patricia Nagnan-Le Meillour and Emmanuelle Jacquin-Joly

17.1 Introduction

Odorant-binding proteins (OBPs) are abundantly secreted in the sensillar lymph of insect antennae and are thought to participate in olfactory perireceptor events in one or more aspects: transport of the hydrophobic odorant through the aqueous medium; presentation of the odor to olfactory receptors; and/or deactivation of the signal. Most studies concerning pheromone detection at the molecular level have been done on silkmoths, which have large feathered antennae and consequently are favorable models to study the biochemistry of odor detection. Considerable basic data on OBPs have been obtained in these models, contributing to define a general scheme of odor detection. Noctuid moths have also attracted a significant volume of research work, as they are key pests of agricultural crops. Much has been published suggesting that phyllogenetically distant species use a similar molecular background in different ways, resulting from different selective pressures and particular adaptations. While models can help solve a particular biological question, they may not systematically be generalized to other species. In this chapter, we will first emphasize on the diversity of OBPs encountered since the discovery of the first pheromone-binding protein (PBP) by Vogt and Riddiford (1981) and, in this context, we will describe our work on the biochemistry and molecular biology of pheromone detection in the noctuid moth, *Mamestra brassicae*. For a good understanding of pheromone reception at the molecular level, we will first describe the molecules involved as pheromone components and behavioral antagonists in *M. brassicae*, and the relevant sensilla types and receptive neurons on the antennae. Then, we will focus on the biochemical characterization of these OBPs and how OBP diversity within this species functions in odor detection.

17.2 Diversity of OBPs among insect species and within species

OBPs have now been identified from numerous species of several insect Orders, including Lepidoptera, Diptera, Coleoptera and Hymenoptera (holometabolous insects), as well as Hemiptera and Phasmatodea (ametabolous insects). A recent classification defining OBP-Type1 and OBP-Type 2 has been proposed by Vogt *et al.* (1999), based on phylogeny, tissue localization and structural features.

17.2.1 OBP-Type 1

This subclass includes the lepidopteran PBPs and general odorant-binding proteins (two subtypes GOBP1 and GOBP2) and antennal-binding proteins (ABPX) as well as proteins from *Drosophila* (e.g. OS-E and OS-F), Coleoptera, Hymenoptera and Hemiptera (Vogt *et al.*, 1999). PBPs were originally characterized by their ability to bind radiolabeled pheromone analog in *Antheraea polyphemus* (Vogt and Riddiford, 1981) and then in *Lymantria dispar* (Vogt *et al.*, 1989). Additional members of this multigene family were further identified, in most cases by sequence homology (for a review, see Vogt *et al.*, 1999). The expression of these OBPs is antennae specific, localized within odor- and pheromone-sensitive sensilla (Steinbrecht *et al.*, 1992; Laue *et al.*, 1994). OBPs-Type1 share common structural features, such as primary sequence similarities (from 25 to 90 percent identity), molecular weights about 16 kDa, acidic isoelectric points (pH 4 to 6), and the conservation of six cysteines. In the *Bombyx mori* PBP (BmorPBP, Leal *et al.*, 1999; Scaloni *et al.*, 1999), the six cysteines are linked in three disulfide bridges pairing cys19—cys54, cys50—cys108, and cys97—cys117, that define three loops rich in α-helices, with amino acid composition varying from one protein to another. Structural studies of BmorPBP have shown that the protein is dimeric, at least in the solid state (Sandler *et al.*, 2000). Each monomer binds one molecule of bombykol, the main pheromonal component, in a large flask-shaped cavity. Residues from all parts of the protein contribute to the binding cavity, while two phenylalanines (Phe12 and Phe118) interact with the double bonds of bombykol. A potential ligand-binding site was previously identified in ApolPBP between residues 40 to 60, by cross-linking between the protein and tritium-labeled photoactivatable analog of the major pheromonal compound (Du and Prestwich 1994). These two studies seem to identify different pheromone binding sites. Perhaps this is because ApolPBP and BmorPBP share only 61.8 percent identity and the major pheromonal compounds of the two species have different chemical functional groups (acetate versus alcohol). However, if we assume that there is a strong specificity between PBPs and pheromonal ligands, the modalities of molecular recognition should be specific as well.

Primary sequences of OBPs-Type 1 are more divergent within species than between species. First, PBPs and GOBPs of the same species (e.g. *Antheraea pernyi*) share only 28 percent identity, while GOBPs are very conserved between species (90 percent identity). Second, pairs of PBPs within a species diverge in primary sequences: two PBPs of *L. dispar* share only 49.7 percent identity (Vogt *et al.*, 1989), two PBPs of *M. brassicae* share 40.6 percent identity (Maïbèche-Coisné *et al.*, 1998a), and two PBPs of *Agrotis ipsilon* share 50 percent identity (Picimbon and Gadenne, 2002). This diversity within a species suggests that PBPs may play a role in pheromone blend discrimination by binding selectivity to each of the pheromone components (Vogt *et al.*, 1991; Prestwich *et al.*, 1995). This hypothesis is based on the expression of different PBP and GOBP subtypes in subsets of olfactory sensilla (Steinbrecht *et al.*, 1992; Laue *et al.*, 1994) and on different binding affinities of PBPs of the same species (Vogt *et al.*, 1989; Du and Prestwich, 1995; Maïbèche-Coisné *et al.*, 1997; Maida *et al.*, 2000). Several authors have suggested that multiple PBPs within a species might suggest the presence of as yet unidentified pheromone components. The scarab beetle, *Phylloperta diversa*, has multiple PBPs but only a single known pheromone component; Wojtasek *et al.* (1999) were unable to find any additional pheromonal components. They suggested that the multiple PBPs in this species could be tuned to the binding of plant-derived chemical signals which are specifically detected by olfactory receptor neurons in a more sensitive way than the sex pheromone component (Hansson *et al.*, 1999). *A. polyphemus* has three PBPs (Maida *et al.*, 2000) which led some of the authors to suggest that ApolPBP3 may be tuned to the binding of a third pheromone component, (*E,Z*)-4,9-tetradecadienyl acetate (Bette *et al.*, 2002). This compound, to our knowledge (Pherolist), has never been described as a pheromone component in *A. polyphemus*, but is a known pheromone component in a closely related species, *A. pernyi* (Bestmann *et al.*, 1987). In addition, an olfactory receptor neuron has been shown being specifically tuned to the detection of this compound in *A. polyphemus* (Meng *et al.*, 1989). One can hypothesize that this compound is detected in *A. polyphemus* as a heterospecific compound. The possibility that PBPs of a given species could be involved in the detection of heterospecific compounds has been underestimated, despite an abundant literature based on behavioral and electrophysiological responses to such compounds. Nevertheless, the extreme importance of precise mechanisms which ensure species isolation should be considered in understanding the molecular mechanisms of discrimination at the PBP level. We will discuss this hypothesis in detail regardless of experimental data obtained in our model *M. brassicae*.

17.2.2 OBP-Type 2 or chemosensory proteins

CSP-related proteins were independently discovered in *Drosophila* using subtractive hybridization and called either OS-D or A10 (McKenna *et al.*, 1994: Pikielny *et al.*, 1994). The proteins were expressed in subsets of olfactory hairs, consistent with a proposed function in olfaction. Additional OS-D homologs were subsequently identified based on sequence similarity (Mameli *et al.*, 1996; Tuccini *et al.*, 1996; Picimbon and Leal, 1999; refs listed in Table 17.1). Angeli *et al.* (1999) proposed the abbreviation CSP, for chemosensory protein, and this term will be used hereafter.

CSPs have been described in the literature in a variety of tissues depending on the species studied, such as antennae, proboscis, legs (Angeli *et al.*, 1999; Picimbon *et al.*, 2001), palpi (Maleszka and Stange, 1997; Angeli *et al.*, 1999), ejaculatory bulb (Dyanov and Dzitoeva, 1995). Furthermore, CSPs and OBPs of Type 1 have different expression patterns during development (Picimbon *et al.*, 2001): CSP synthesis seems to appear very early during adult development (at least five days before emergence) and to start simultaneously in diverse chemosensory tissues, whereas OBP-Type 1 synthesis starts later, about three to two days before emergence (Györgyi *et al.*, 1988; Vogt *et al.*, 1989; Vogt *et al.*, 1993), suggesting different mechanisms of regulation.

Despite their diversity, CSPs described so far constitute a very homogeneous family, regardless of structure and physicochemical characteristics. They have a molecular weight around 13 kDa with an isoelectric point between 4 and 6. They generally share 40 to 50 percent amino acid identity both within species and between species, and are thus much less divergent than are the OBPs-Type I.

Alignment of CSP amino acid sequences reveals the conservation of four cysteines, a constant hallmark of all CSPs identified so far. In *S. gregaria*, mass spectrometry analyses determined that these four cysteines form two disulfide bonds (Angeli *et al.*, 1999), linking Cys29–Cys38 and Cys57–Cys60, creating two small protruding loops. The CSP topology is thus very different from that of PBPs (Leal *et al.*, 1999; Scaloni *et al.*, 1999). Secondary structure predictions revealed that these proteins appear rather hydrophilic but include a conserved hydrophobic motif (Campanacci *et al.*, 2001a; Angeli *et al.*, 1999; Picimbon *et al.*, 2001). The analysis of a *S. gregaria* CSP indicated an absence of any posttranslational modifications (Angeli *et al.*, 1999).

Several preliminary conformational studies were performed, in particular on the recombinant MbraCSP-A expressed in *Escherichia coli* periplasm. Crystals were obtained and analyzed using X-ray diffraction (Campanacci *et al.*, 2001b). NMR data were also obtained, indicating a well-folded protein with seven α-helices (59 amino acids) and two short extended structures of 12 amino acids at the N- and C-termini (Campanacci *et al.*, 2001b). A similar study has been conducted on a CSP from *S. gregaria* (Picone *et al.*, 2001). After bacterial expression, the structure of recombinant CSP-sg4 was analyzed by circular

Table 17.1 Insect CSP protein sequences known to date

Insect order	Species	Protein name	Database accession no.	Reference
Lepidoptera	*Manduca sexta*	EST210	AI172733	Robertson *et al.* (1999)
		SAP1	AF117574	Robertson *et al.* (1999)
		SAP2	AF117592	Robertson *et al.* (1999)
		SAP3	AF117585	Robertson *et al.* (1999)
		SAP4	AF117599	Robertson *et al.* (1999)
		SAP5	AF117594	Robertson *et al.* (1999)
	Mamestra brassicae	CSP-MbraA1	AF211177	Nagnan-Le Meillour *et al.* (2000)
		CSP-MbraA2	AF211178	Nagnan-Le Meillour *et al.* (2000)
		CSP-MbraA3	AF211179	Nagnan-Le Meillour *et al.* (2000)
		CSP-MbraA4	AF211180	Nagnan-Le Meillour *et al.* (2000)
		CSP-MbraA5	AF211181	Nagnan-Le Meillour *et al.* (2000)
		CSP-MbraA6	AF255918	Jacquin-Joly *et al.* (2001)
		CSP-MbraB1	AF211182	Nagnan-Le Meillour *et al.* (2000)
		CSP-MbraB2	AF211183	Nagnan-Le Meillour *et al.* (2000)
		CSP-MbraB3	AF255919	Jacquin-Joly *et al.* (2001)
		CSP-MbraB4	AF255920	Jacquin-Joly *et al.* (2001)
		SAP	AY026760	Jacquin-Joly *et al.*, unpublished
	Cactoblastis cactorum	CLP-1	U95046	Maleszka and Stange, (1997)
		HvirCSP1	AY101512	Picimbon *et al.* (2001)
	Heliothis virescens	HvirCSP2	AY101511	Picimbon *et al.* (2001)
		HvirCSP3	AY101513	Picimbon *et al.* (2001)
	Helicoverpa armigera	CSP-Harm	AF368375	Deyts *et al.* (2001a)
	Helicoverpa zea	CSP-Hzea	AF448448	Deyts *et al.*, unpublished

(Contd)

Table 17.1 *(Contd)*

Insect order	Species	Protein name	Database accession no	Reference
	Bombyx mori	CSP1	AF509239	Picimbon *et al.* (2000b)
		CSP2	AF509238	Picimbon *et al.* (2000b)
Diptera	*Drosophila melanogaster*	OS-D (or a10)	U02546	McKenna *et al.* (1994)
				Piekelny *et al.* (1994)
		PEBmeIII	U08281	Dyanov and Dzitoeva, (1995)
		CG9358	AAF47307	Adams *et al.* (2000)
	Anopheles gambiae	SAP	AF437891	Biessmann *et al.* (2002)
Orthoptera	*Schistocerca gregaria*	CSP-sg1	AF070961	Angeli *et al.* (1999)
		CSP-sg2	AF070962	Angeli *et al.* (1999)
		CSP-sg3	AF070963	Angeli *et al.* (1999)
		CSP-sg4	AF070964	Angeli *et al.* (1999)
		CSP-sg5	AF070965	Angeli *et al.* (1999)
	Locusta migratoria	LmigOS-D 1	AJ251075	Picimbon *et al.* (2000a)
		LmigOS-D 2	AJ251076	Picimbon *et al.* (2000a)
		LmigOS-D 3	AJ251077	Picimbon *et al.* (2000a)
		LmigOS-D 4	AJ251078	Picimbon *et al.* (2000a)
		LmigOS-D 5	AJ251079	Picimbon *et al.* (2000a)
Phasmatodea	*Eurycantha calcarata*	CSP-ec1	AF139196	Marchese *et al.* (2000)
		CSP-ec2	AF139197	Marchese *et al.* (2000)
		CSP-ec3	AF139198	Marchese *et al.* (2000)
Dictyoptera	*Periplaneta americana*	p10	AF030340	Nomura *et al.* (1992)
				Kitabayashi *et al.* (1998)
Hymenoptera	*Apis mellifera*	CSP_1	BI946526	Kucharski and Maleszka (2002)
		ASP_3C	AF481963	Danty *et al.* (1998)

NA. not available.

dichroism, secondary structure prediction, and NMR. Again, the well-folded protein included α-helical regions and alternating loops. The conformation of this *S. gregaria* CSP was quite stable under varying conditions of temperature and pH (Picone *et al.*, 2001), much different than the case for BmorPBP which undergoes a major conformational transition between pH 5 and 6 correlated with a change in ligand binding (Wojtasek and Leal, 1999b; Damberger *et al.*, 2000). Thus, CSPs are monomeric, and stably folded into a compact and approximately spherical structure with a single native conformation and a predominantly α-helical structure. Alpha-helical secondary structures are common in PBPs and GOBPs, and as well as lipid carriers such as lipid transfer proteins (LTPs), and proteins described in *Tenebrio molitor*: the hexahelical THP12 (Rothemund *et al.*, 1997, 1999), or B1 and B2 proteins (Paesen and Happ, 1995). These latter proteins, secreted by the tubular accessory glands of adult males, are quite similar to CSPs (four conserved cysteines, 13 kDa, acidic pI) and have been proposed to maintain hydrophobic compounds in solution. Nevertheless, although structurally similar, CSPs differ considerably from these lipid carriers in many respects and appear to be a new protein class.

17.3 Pheromonal communication in *M. brassicae*

At the end of the scotophase, female *M. brassicae* emit a pheromonal blend composed of 92 percent Z11-16:Ac, 7 percent 16:Ac, 1 percent Z9-16:Ac (Descoins *et al.*, 1978), 0.1 percent Z11-16:Ald (Jacquin, 1992) and Z11-18:Ac in trace amount (Struble *et al.*, 1980; Malosse, unpublished). This composition has been repeatedly confirmed from gland washes by GC-MS coupling. The major pheromonal compound, Z11-16:Ac alone, is able to attract conspecific males in both laboratory (Descoins *et al.*, 1978) and field (Struble *et al.*, 1980). The behavioral relevance of the minor components has been less documented, but several authors have suggested that Z9-16:Ac and Z11-16:Ald are synergists. In particular, 0.1 percent of Z11-16:Ald (physiological dose) added to Z11-16:Ac increases field catch number while 10 percent decreases catch number (Subchev *et al.*, 1985, 1987). Electrophysiological experiments have shown that Z11-16:Ac and Z9-16:Ac elicited strong electroantennogram responses in *M. brassicae* males (Den Otter and Van der Haagen, 1989). Three receptor cells tuned to pheromone compounds were characterized by single cell recordings (Den Otter and Van der Haagen, 1989; Renou and Lucas, 1994); these cells are housed in two different types of olfactory hairs, laterally and medially located on the flagellar segments of male antennae (Figure 17.1). The lateral hairs, long Sensilla trichodea, house two cell types, the A cell that responds to Z11-16:Ac (the main component), and the B cell, equally tuned to Z9-14:Ac and Z11-16:OH (both strong behavioral antagonists for this species). The medio-ventral hairs, short

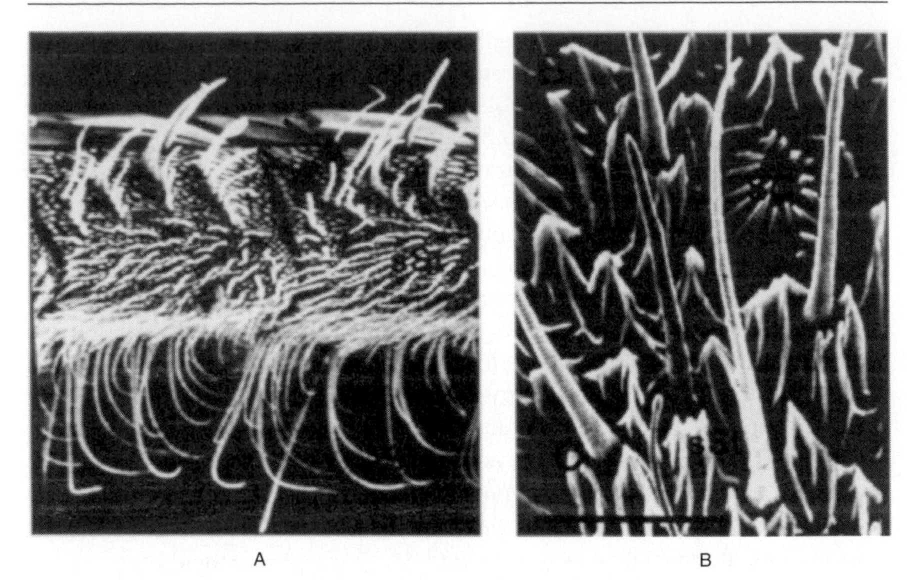

Figure 17.1 Scanning electron microscopy of a male antennae. A: positions of short (sst) and long (lst) sensilla trichodea on the ventral side, the latter being arranged in parallel rows, B: sensilla coeloconica (sc) distributed among short sensilla trichodea. Bars: 50 μm in A, 10 μm in B (adapted from Jacquin-Joly *et al.*, 2001, by permission of Oxford University Press)

Siminuscula trichodea, responded with the same spike amplitude to both the minor component Z11-16:Ald and to the behavioral antagonist Z9-14:Ac. The two behavioral antagonists of *M. brassicae*, Z9-14:Ac and Z11-16:OH, are pheromone components of sympatric species in the South of France: *Agrotis ipsilon* (Z9-14:Ac, Z11-16:OH), *A. segetum* (Z9-14:Ac) and *Mythimna unipuncta* (Z11-16:OH). Both *M. brassicae* and *M. unipuncta* use Z11-16:Ac as the main pheromone component, but the *M. unipuncta* pheromone contains 2 percent of Z11-16:OH which may ensure the chemical isolation of these two species (Farine *et al.*, 1981). Z11-16:OH inhibits the behavior of male *M. brassicae* when included at 0.1 percent of the blend (Descoins *et al.*, 1978; Struble *et al.*, 1980).

How are such diverse compounds detected and recognized to elicit the associated behavior? Several steps contribute to the odor coding, such as olfactory receptor interactions and processing in mushroom bodies, but also including odor interactions with OBPs. The diversity and heterogeneity of OBPs implicates these proteins in the process of odor discrimination.

17.4 Overall strategy

Our starting hypothesis is that odorous ligands that are important to a given species interact with or bind to specific proteins in the male antenna. The pheromone system of moths is the most sensitive and specialized olfactory system among animals, and the coding of pheromone specificity is well documented from the ligand to central processing. The missing step, the transduction process, which includes the molecular recognition of pheromone compounds by proteins, should be highly specific for the different pheromone components. This hypothesis is based on the fact that the recognition of conspecific pheromone and the genes coding for proteins involved in its detection are under high selective pressure. Under such conditions, we should expect that the diversity of pheromone-binding proteins is similar to the diversity of pheromone components. Since the discovery of the first PBP, many PBP sequences have been reported in the literature. While these proteins are clearly divergent between species, there are only a few examples illustrating PBP diversity within a species. This may reflect a bias from molecular cloning by sequence homology, where new cDNAs are cloned using PCR primers that are similar to previously identified cDNA sequences. This approach may overlook novel or divergent sequences. Indeed, PBPs are very divergent in their primary sequences, especially within a species.

A functional approach, identifying proteins based on ligand binding activities, has been highly successful for identifying unknown proteins. This approach allowed the discovery of the first PBP in moths, using a tritium-labeled analog of the major pheromonal component of *A. polyphemus* (Vogt and Riddiford, 1981). We have followed a similar strategy, using the compounds detected by *M. brassicae* males to identify diverse antennal binding proteins. Tritium-labeled pheromone analogs were chosen, preferentially to other analogs (fluor, etc.) because the volume of tritium is comparable to those of hydrogen and thus causes no artifact in the binding, assuming that the ligand enters the protein cavity. Extraction of antennal proteins with a 1 percent trifluoroacetic acid (TFA, pH 2) solution has allowed proteins to be isolated based on their molecular weight, precipitation proteins above 30 kDa, leaving smaller proteins such as PBPs in solution. Pheromone-degrading enzymes, including esterases, are precipitated and cannot act on pheromonal substrates during subsequent binding assays. Although OBPs are extracted at pH 2, binding studies are carried out at higher pHs. For example, in binding assays involving PAGE, incubation for binding is carried out in PAGE sample buffer (pH 8.8). We have investigated the effect of different pH on binding of Z11-16:Ac by recombinant MbraPBP1 heterologously expressed in *E. coli* (Campanacci *et al.*, 1999). Binding between pheromone and PBP is typically observed under basic conditions. However, no binding was observed when samples were incubated at pH between 4 and 6.5 and electrophoresed under acidic conditions. When transferred to membrane under such acidic conditions, MbraPBP1 visualized by staining has a shape or pattern on the membrane that suggests the protein is unfolded. However, even if the protein is unfolded at very low pH, there is a reversible conformational shift around pH 7 which restores the binding properties observed at basic pH. This conformational shift is consistent with results obtained by Wojtasek and Leal (1999a) showing that BmorPBP undergoes a conformational change at pH value between 5 and 6, leading to the loss of binding capabilities at low pH.

We have developed a pheromone-binding assay, adapted from the one described by Vogt and Riddiford (1981). We made several modifications which significantly intensify the tritium-emitted signal. The major improvement lies in the transfer of antennal proteins from the electrophoresis gel to a hydrophobic membrane. The use of a membrane presents several advantages. First, the loss of radioactivity during membrane exposure is less than that occurring in gels. Second, the membranes can resist variations in temperature and therefore can be retreated for fluorography and re-exposed to a new film for extended times to enhance the signal. Third, the positions of labeled proteins can be precisely determined, and the corresponding membrane cut and bound protein subjected directly to N-terminal sequence analysis for identification.

17.5 Biochemistry and microdiversity of OBPs-Type 1 in *M. brassicae* males and females

When we started our studies, radiolabeled analogs of pheromonal compounds were not available in our lab. Characterization of MbraOBPs was made by looking at the diversity among abundant soluble proteins in antennal extracts using a combination of techniques including native electrophoresis, reverse-phase-HPLC purification and N-terminal sequencing. Both male and female extracts contained many proteins, including two with PBP-like N-terminal sequences, identified by homology with already published sequences (Nagnan-Le Meillour *et al*., 1996): MbraPBP1 (N-terminus SKELIT) and MbraPBP2 (N-terminus SQEIM). In addition, GOBP1 (N-terminus DVNIM) and GOBP2 (N-terminus TAEVM) sequences were identified. The most intriguing result was that two well-separated bands observed in male antennal extracts were both identified as MbraPBP1 by sequence, while only a single band MbraPBP1 was observed in female antennal extracts. In native electrophoresis, proteins are separated according to their charge and shape. Two distinct bands containing the same N-terminal sequence suggest the presence of isoforms of different isoelectric points coming from one or a few mutations, i.e. products of alleles of the same gene. The precise nature of these modifications is still unknown. This is a key question that will be discussed below.

A preliminary binding assay (after Vogt and Riddiford, 1981) showed that purified male MbraPBP1 and MbraPBP2 have opposite affinities for the tritiated Z11-16:Ac (Maïbèche-Coisné *et al*., 1997). MbraPBP1 strongly bound the analog of the major pheromonal compound, while MbraPBP2 did not bind this compound or any other tested compounds.

We then performed binding assays using our modified protocol, incubating tritiated analogs of Z11-16:Ac, 16:Ac, Z11-16:OH and Z11-18:Ac with total antennal extracts (Bohbot *et al.*, 1998). In addition to PBPs and GOBPs, this sensitive protocol also allowed the functional characterization of CSPs which bound the four compounds tested with high affinity, but without selectivity (see section 17.7). The results obtained with total antennal extracts were rather complex, due to the coextraction of multiple OBPs that were presumably not all coexpressed in the same sensilla and which yielded complex labeling patterns which were difficult to read. We therefore first purified the antennal proteins by RP-HPLC and performed binding assays on the resulting partially purified fractions (see Nagnan-Le Meillour *et al.*, 1996). Bands labeled by one or several compounds were then sequenced (N-terminal). Figure 17.2 summarizes the data obtained after a first RP-HPLC separation. The proteins were eluted by a gradient of acetonitrile, thus separated by their hydrophobicity; fractions are collected every minute. Aliquots of 200 antennal equivalents were used in binding assays followed by N-terminal sequencing or immunodetection (western blot).

Some sequences were eluted in several fractions, the first sometimes separated from the last by 5–6 minutes. This was the case for the proteins with N-terminals SEEDK, SKELI, or TAEVM; the spread observed for these proteins was probably due the presence of multiple isoforms. Our suggestion of the presence of multiple isoforms is based on two criteria: migration differences during native-PAGE due to different isoelectric points and the differential elution during RP-HPLC due to different hydrophobic properties. The protein with N-terminal sequence DAPAA was found in three different bands separated by 2 to 3 cm. The protein with N-terminal sequence SEEDK was also found in three different bands separated by 1 to 3 cm (arrows in Figure 17.2). The full-length sequence of the SEEDK protein identified this protein as an OBP-Type 1 (Cristiani *et al.*, unpublished, accession number AF461143). This sequence is most similar to antennal proteins of *H. virescens* (Korchi *et al.*, unpublished accession number AJ300658) and *M. sexta* (Robertson *et al.*, unpublished, accession number AF117581), though the roles of these other proteins are unknown. We do not know if the multiple isoforms we observed are present in individuals or represent population level polymorphisms as antennal extracts used were always from pools of at least 50 animals.

Only a few proteins from antennal extracts are were able to bind the four compounds tested (Figure 17.3). Fraction 29 included proteins with sequences DAPAA and SEEDK; no proteins in this fraction bound compounds. A protein with sequence AKLTT from fractions 30, 31 and 32 bound compounds. The full sequence of the AKLTT protein has been obtained by molecular cloning and identified as a CSP (Jacquin-Joly *et al.*, unpublished, accession number AY026760; see section 17.7). The sequence AKLTT is also found in band 1 of fraction 33 (Figure 17.2), but here this protein fails to bind any compound (Figure 17.3).

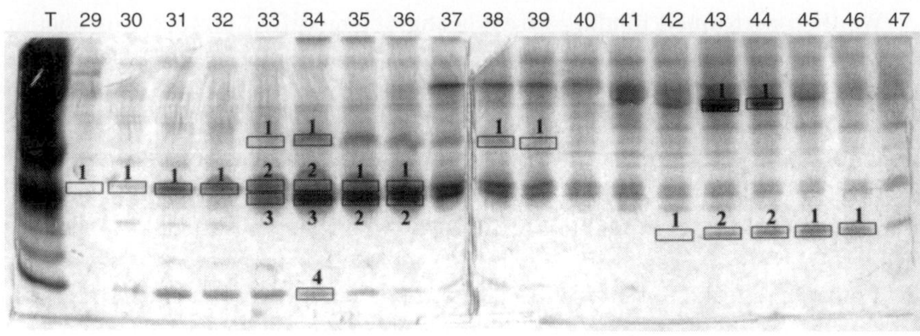

T = total antennal extract (100 male antennae)
29.1:71% SEEDK, 29% DAPAA
30.1:66% AKLTT, 22% SEEDK, 12% DAPAA
31.1:86% AKLTT, 9% SEEDK, 5% DAPAA
32.1:75% AKLTT, 13% SEEDK, 12% DAPAA
33.1:100% AKLTT
33.2:100% SKELI
33.3:100% SKELI
34.1:Akltt, SEEDK
34.2:77% SKELI, 23% SQEIM
34.3:100% SKELI
34.4:100% SEEDK
35.1:84% SKELI, 16% SQEIM
35.2:100% SKELI
36.1:92% SKELI, 8% SQEIM
38.1 & 39.1:EDKY + EEAH
42.1:100% TAEVM
43.1:100% DAPAA
43.2:75% TAEVM, 25% DAPAA
44.1:100% DAPAA
45.1:100% TAEVM
46.1:100% TAEVM

Figure 17.2 Purification of *M. brassicae* male antennal proteins by reverse phase-HPLC. Coomassie blue staining of resulting fractions (T = total antennal extract, 100 antennae). Rectangles indicate the bands submitted to N-terminal sequencing. Each fraction = 200 antennae-equivalent.

This differential binding was also observed for the protein with sequence SKELI which corresponds to MbraPBP1. MbraPBP1 appears in fraction 33 as two distinct bands and continues to be present until fraction 36; bands appearing as MbraPBP1 persisted to fraction 47 but here failed to bind compounds and were thus not sequenced for confirmation. This band bound Z11-16:Ac from fractions 33 and 34, 16:Ac and Z11-18:Ac from fractions 33, 34 and 35, and Z11-16:OH from fractions 33 to 36. Appropriate care was taken in all studies to ensure equivalent concentrations of each tested compound. Differences in labeling thus reflect differences in affinities between the protein and the pheromone analogs.

This assay also demonstrated differences in MbraPBP1 specificity for the respective compounds. The four compounds bound to the slower migrating band (upper band); the faster migrating band bound Z11-18:Ac and Z11-16:OH but never Z11-16:Ac and 16:Ac. In addition, despite the fact that the same yield (45 pM) has been obtained in N-terminal sequencing for the SKELI sequence in the fractions 33 and 36, the SKELI contained in fraction 36 did not bind the analogs. This suggests that there are more than two isoforms of the MbraPBP1, not only reflected in the two different migrating bands, but also by the presence of the same sequence in fractions separated by more than 2 minutes. An example where multiple fractions all showed the same binding properties is the protein with sequence TAEVM (MbraGOBP2). This sequence eluted in fractions 37 to 47 (10 min); all fractions bound the behavioral antagonist Z11-16:OH (Figure 17.3).

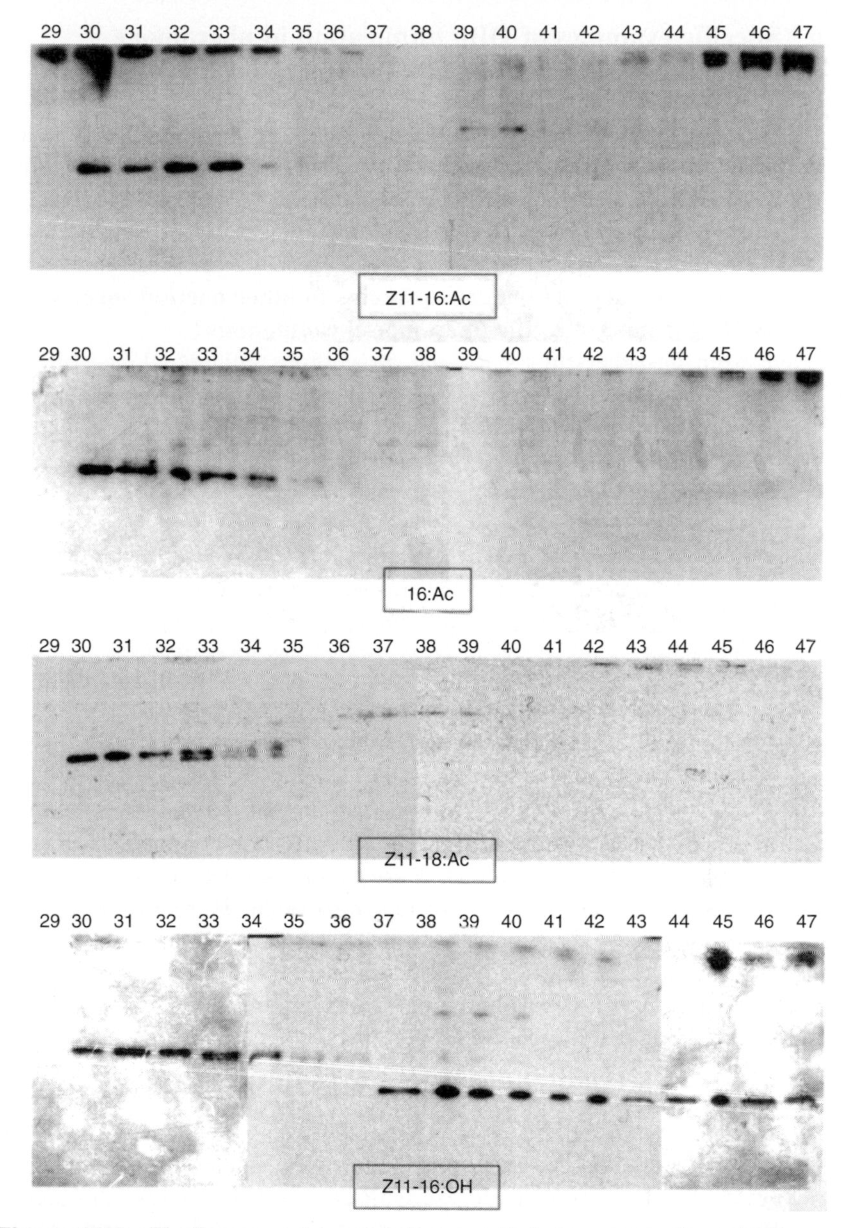

Figure 17.3 Binding experiments between tritiated compounds and RP-HPLC-purified antennal proteins. Each fraction corresponds to an aliquot of the purification. One μCi of each pheromonal compound was incubated with 20 μL of each fraction (sample buffer). The films were exposed to the membrane for 7 days at –20°C.

17.6 Specific examples of OBP implication in pheromone detection in *M. brassicae*

In parallel to the binding assay, we have performed other molecular studies including *in situ* hybridizations. We will now discuss the possible roles of the MbraOBPs in the treatment of pheromonal blend at the perireceptor events level, supported by data from different experimental approaches.

17.6.1 MbraPBP2 and orthologous proteins in other noctuid species: PBPs for heterospecific pheromonal compounds?

We have been unsuccessful identifying an odorous compound that bind to MbraPBP2 (Maïbèche-Coisné *et al.*, 1997; Bohbot *et al.*, unpublished). However, the full-length sequence of MbraPBP2 (Maïbèche-Coisné *et al.*, 1998a) is most similar to a PBP of *H. virescens* (Hvir PBP, Krieger *et al.* 1993) sharing 86 percent sequence identity. This similarity may suggest that MbraPBP2 binds one of the pheromone components of *H. virescens*, perhaps Z11-16:Ald which is the major component of the *H. virescens* pheromone and is also present as a minor component in *M. brassicae* pheromone gland extracts. More data are now available on full-length sequences of PBPs in different species, and a bootstrap analysis of PBP sequences from more than 20 moth species has recently been published (Picimbon and Gadenne, 2002; see also Picimbon Chapter 18, in this volume). This study defined several PBP subgroups, including one which included MbraPBP2 and proteins from *H. virescens* (Krieger *et al.*, 1993), *H. zea* (Callahan *et al.*, 2000), *H. armigera* (Wang and Guo, unpublished, accession number AJ278992), *A. segetum* (LaForest *et al.*, 1999) and *A. ipsilon* (Picimbon and Gadenne, 2002). These species all use Z11-16:Ald as pheromonal compounds (Pherolist), except the *Agrotis* species that use exclusively acetates. *M. brassicae*, *H. armigera*, *A. ipsilon* and *A. segetum* are sympatric in several European areas and use common molecules in their pheromonal communication. The behavioral agonist for one species is often a behavioral antagonist for another species. These species also express very similar PBPs, so similar that the proteins may be considered orthologs. The recent discovery of a second PBP in *A. ipsilon*, 71 percent identical to MbraPBP1, is of particular interest. It may be possible that all the noctuid species studied possess orthologous pairs of PBPs. One might now systematically search for orthologous genes in species, and study their binding properties to the set of pheromonal compounds that are common to these species. The Heliothine moths are sympatric in North and South America (Fitt, 1989), and species isolation is ensured by the specific detection of the same major pheromonal compound Z11-16:Ald, in combination with minor components which act as behavioral antagonists for other Heliothine species (Vickers *et al.*, 1991). We should not systematically assume that all PBPs are tuned to pheromonal components of the same species, but rather some may be tuned to heterospecific compounds thus supporting the detection of pheromonal components to maintain species isolation. Studies of the interaction between PBPs and heterospecific pheromonal components might prove useful in understanding the roles of behavioral antagonists.

17.6.2 The occurrence of two forms of MbraPBP1 in males with different binding properties

We observed two MbraPBP1 bands in PAGE gels from extracts of male antennae, but only a single band of MbraPBP1 from extracts of female antennae (Nagnan-Le Meillour *et al.*, 1996). The coexistence of two forms of the most abundant PBP with identical molecular weight and N-terminal sequence, but different native gel mobility, has been described in several species. In *A. polyphemus*, Ziegelberger (1995) observed two distinct bands in non-denaturing gels; ^{3}H-(*E,Z*)-6,11-16:Ac associated with both bands but with greater affinity for the faster migrating band. Ziegelberger hypothesized that the two forms correspond to two different redox states of the same protein, an oxidized form (three disulfide bridges) and a reduced form (two bridges), and that a shift in redox state was catalyzed by interaction with the receptor. In *B. mori*, the BmorPBP exists as two forms depending on the pH (Wojtasek and Leal, 1999b); only the basic form was observed to bind pheromone. These two BmorPBP conformations are reversible; both forms are present at neutral pH (Damberger *et al.*, 2000). In *M. brassicae*, we observed a similar shift in MbraPBP1 conformation between pH 7.5 and 6.5, leading to the loss of binding properties at acidic pH (see section 17.4). And we observed two MbraPBP1 bands under basic conditions (pH 8.8, binding conditions), with both bands binding pheromone (Figure 17.2). Under basic conditions, two conformations induced by the pH cannot coexist. It is clear that the two forms observed by Wojtasek and Leal (1999b) do not reflect the same mechanism than in *M. brassicae*. Indeed, the two MbraPBP1 bands have the same molecular mass (Blais *et al.*, 1996), the same N-terminus (20 first amino acid), but different isoelectric points or shape, and different binding properties. The lower migrating band binds all four compounds tested (Figure 17.3), whereas the faster migrating form only binds tritiated Z11-16:OH and Z11-18:Ac.

As we worked on pools of animals, there is a possibility that the two MbraPBP1 forms correspond to allelic variants of the same protein present in the population. However, we performed western blots with anti-ApolPBP serum on antennal extracts coming from single males and we still found the two bands (data not shown). When we performed PCR to obtain the full-length sequence of MbraPBP1, several clones were sequenced, all identical. Finally, we have obtained recombinant proteins by heterologous expression in the yeast *Pichia pastoris*; this eukaryotic expression system produces well-folded proteins in high quantities (Briand *et al.*, 2001). We have used cDNAs encoding MbraPBP1 and MbraPBP2 with their native signal peptides and compared the expression of the two proteins; MbraPBP2 has always been observed as a single protein in only one band (Figure 17.2). We obtained two bands of sequence SKELI (MbraPBP1), both immunoreactive with specific antibodies, but only one band for MbraPBP2. We can only conclude that the two MbraPBP1 are conformational isomers of the same protein, presumably based on a different arrangement of the disulfide bridges. Deyu and Leal (2002) recently proposed the same hypothesis for pairs of OBPs from *Holotrichia parallela* and *Heptophylla picea*.

17.6.3 MbraGOBP2 and the coding of behavioral antagonist Z11-16:OH

Proteins of the GOBP2 group are equally expressed in male and female (Vogt *et al.*, 1991). In *A. polyphemus*, GOBP2 is localized in sensilla basiconica (Laue *et al.*, 1994) which house neurons responding to food or host-plant odors (Heinbockel and Kaissling, 1996). The distribution of GOBP2s in different subsets of sensilla from those of PBPs, together with their sequence conservation between species, led to the suggestion of a role in the detection of general odors (Vogt *et al.*, 1991; Breer *et al.*, 1990). However, published studies have only shown pheromone binding to the GOBP2 of *A. polyphemus* (Vogt and Riddiford, 1981; Ziegelberger, 1995). Studies of recombinant GOBP2 from *M. sexta* also showed strong affinity for the analog of *A. polyphemus* major pheromonal compound ^{3}H-(*E,Z*)-6,11-16:Dza; binding was only displaced by very high concentrations of green leaf volatiles (more than 200-fold molar excess; Feng and Prestwich, 1997). In comparison, we obtained displacement of the Z11-16:Ac binding with MbraPBP1 using Z11-16:TFMK at a 1:1 molar ratio (Campanacci *et al.*, 1999).

Native MbraGOBP2 bound only the behavioral antagonist Z11-16:OH (Bohbot *et al*, 1998); however, recombinant MbraGOBP2, expressed in *E. coli*, showed less specific binding, associating with three tritium-labeled pheromone analogs (Maïbèche-Coisné *et al.*, 1998b). To further investigate the binding properties of MbraGOBP2, we purified native MbraGOBP2 and performed binding assays with the tritium-labeled pheromone analogs. Binding experiments with this pure native MbraGOBP2 confirmed the strong specificity towards the behavioral antagonist (Jacquin-Joly *et al.*, 2000). In parallel, we compared the expression patterns of MbraPBP1 and MbraGOBP2 in antennae by *in situ* hybridization using antisense RNA probes (Jacquin-Joly *et al.*, 2000). *In situ* hybridization revealed a similar pattern of expression for MbraPBP1 and MbraGOBP2 in long sensilla trichodea, arranged in typical parallel rows in *M. brassicae* (Figure 17.1). The labeling of all sensilla in all rows by the two probes strongly suggests a coexpression of the two proteins. We have observed a similar pattern of expression using RNA probes for HzeaPBP and HzeaGOBP2 by *in situ* hybridization on male *H. zea* antennae. In this species, as in *M. brassicae*, long sensilla are arranged in parallel rows, which are all labeled with each of the two probes (Jacquin-Joly *et al.*, unpublished). Our data are not in agreement with other immunolocalization experiments showing that in noctuids and other moths, the PBP and GOBP2 labeling are never observed in the same sensilla (Laue *et al.*, 1994; Steinbrecht *et al.*, 1996; Zhang *et al.*, 2001; Callahan *et al.*, 2000). We shall point out the fact that an RNA probe is much more specific than a polyclonal antiserum, and the colabeling observed in *M. brassicae* and *H. zea* is definitely not an artifact.

In noctuid moths, at least, the long sensilla trichodea of males are not only tuned to different components of the conspecific pheromone, but also to those of sympatric species as a mechanism avoiding in-breeding (Hansson, 1995). Moreover, within a single sensillum type, the neuron tuned to an antagonist is almost always co-compartmentalized with a neuron tuned to an agonist pheromone component (Cossé *et al.*, 1998; Baker *et al.*, 1998a). We have also observed such co-compartmentalization in *M. brassicae* (see above, section 17.3), and we have proposed a hypothetical scheme of molecular coding of the major pheromonal compound by MbraPBP1 and of the behavioral antagonist by MbraGOBP2 (Jacquin-Joly *et al.*, 2000). It is of particular relevance that the specificity between MraGOBP2 and Z11-16:OH is higher than that between MbraPBP1 and Z11-16:Ac (see above, section 17.5). This difference indicates that the heterospecific compound is detected with more selectivity and sensitivity than the conspecific compound. These data are in agreement with the work of Baker and co-workers who suggested that the co-compartmentalization of neurons responding to both pheromone compound and antagonist optimizes the resolution of pheromone blends by processing the mixture at the peripheral level, before central processing (Baker *et al.*, 1998b).

The role of GOBP2 in other species is still an open question. Is the GOBP2 of noctuids tuned to the Z11-16:OH, to alcohols in general, or to behavioral antagonists? We have observed a specific binding between the Z11-16:OH and the GOBP2 of *H. zea*; Z11-16:OH is also a behavioral antagonist for this species (Quero and Baker, 1999). It is worth noting that among the tritiated compounds used in our studies, the Z11-16:OH is the smallest molecule (16 carbons), suggesting that the GOBP2 binding site is smaller than those of the PBPs and CSPs. To further address this structural question, we are expressing recombinant MbraGOBP2 in *P. pastoris* for NMR and X ray studies.

17.7 CSPs: functional characterization of OBPs-Type 2 in *M. brassicae*

17.7.1 Binding assays

Our binding assays also identified *M. brassicae* proteins of the OBP-Type 2 class. Several of these proteins from both male and female antennal extracts were observed to bind the four tritium-labeled analogs of Z11-16:Ac, 16:Ac,

Z11-16:OH and Z11-18:Ac, although with lower specificity than that observed for MbraPBP1 and MbraGOBP2 (Bohbot *et al.*, 1998). N-terminal sequencing of these bands revealed the sequence EDKYT which is highly similar to proteins of the CSP family found in insects as diverse as *Drosophila*, phasmid, and cockroach which are proposed to have diverse functions depending on the specific protein. As mentioned earlier, the first member of this protein family was independently identified in *Drosophila* and named OS-D (McKenna *et al.*, 1994) or A10 (Pikielny *et al.*, 1994). Related proteins were subsequently referred to as either "OS-D like," CSPs or SAPs.

We studied soluble proteins from the proboscis of *M. brassicae* and identified proteins which bound tritiated pheromone components (Nagnan-Le Meillour *et al.*, 2000). One of these proteins contained the CSP N-terminal motif EDKYT, another contained a similar N-terminal sequence (EEAHY), and a third the N-terminus AKLTT (Nagnan-Le Meillour *et al.*, 2000) which is similar to a SAP protein from *M. sexta*. This third protein was also observed in antennal extracts (see above, section 17.5). N-terminal sequencing of HPLC-purified proteins revealed the occurrence of several isoforms sharing the same N-terminal sequence. Thus, our approach in *M. brassicae* again revealed a great diversity in protein-binding odorants, with a tissue localization not restricted to the antennae.

17.7.2 Diversity and tissue distribution of CSPs in *M. brassicae*

Full-length cDNAs encoding the sequences EDKYT (MbraCSP-A), EEAHY (MbraCSP-B) and AKLTT (MbraCSP-SAP) were obtained using RT-PCR (Nagnan-Le Meillour *et al.*, 2000; Jacquin-Joly *et al.*, 2001) (Table 17.1).

Several cDNA clones were sequenced from each PCR amplification from both antennae and proboscis, revealing multiple isoforms, as already observed after biochemical purification. In particular, six isoforms of EDKYT proteins were found with amino acid sequence identities ranging from 97 to 99 percent. Occurrence of isoforms for proteins from this family has already been observed in other insects like *S. gregaria* (Angeli *et al.*, 1999), *E. calcarata* (Marchese *et al.*, 2000) or *L. migratoria* (Picimbon *et al.*, 2000a). These isoforms may represent interindividual polymorphisms, as tissues used for our PCR derived from multiple individuals. The deduced amino acid sequences confirmed low-molecular-mass proteins with a putative signal peptide at their N-terminus, typical of secreted proteins. At least three different protein sequences are now described in *M. brassicae* with amino acid sequence identities ranging from 18 to 54 percent, revealing CSP diversity within a single species (Table 17.1).

The tissue distribution of CSPs was examined by amplifying CSP cDNAs from different parts of the *M. brassicae* body, including legs (unpublished) and female pheromonal glands (Jacquin-Joly *et al.*, 2001), as well as antennae and proboscis. Moreover, proteins extracted from these tissues appeared to bind tritium-labeled pheromone components. Figure 17.4A shows binding of Z11-

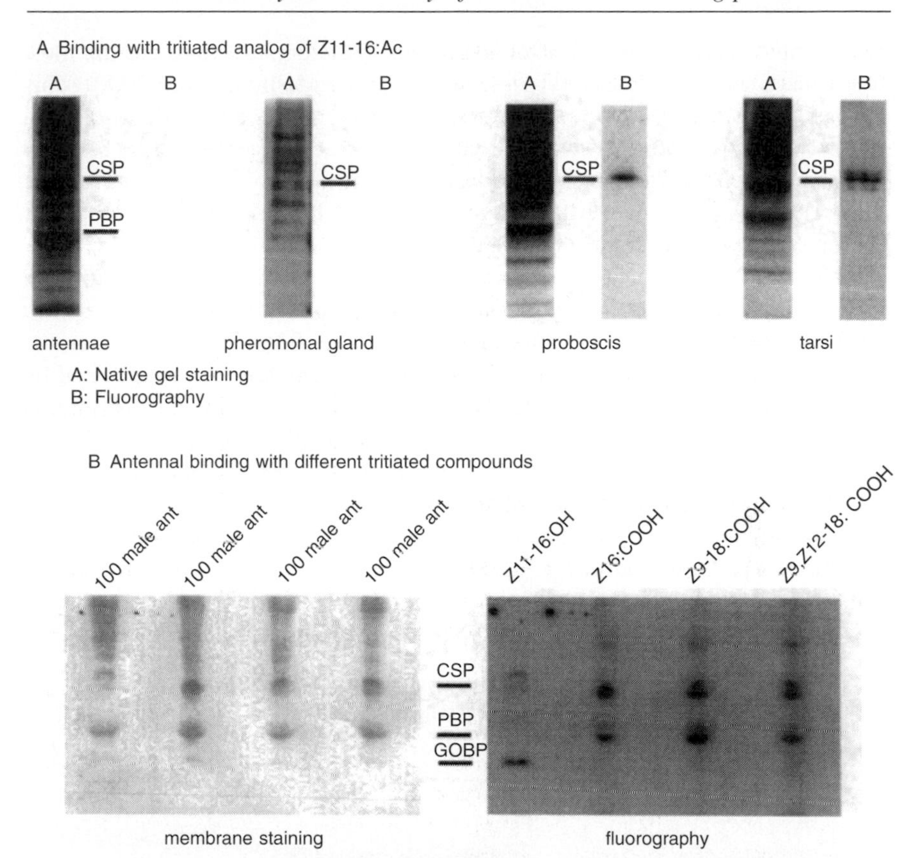

Figure 17.4 Binding experiments between different *M. brassicae* tissue extracts and tritiated pheromone analogs. A Homogenates of 100 male antennae, 100 female pheromonal glands, 100 male proboscis or 100 male tarsi extracts were incubated with 1 μCi of [^{3}H]Z11-16:Ac. Left: Coomassie staining of the polyacrylamide gels after native-PAGE; right: Fluorogram membrane following electroblot transfer of proteins from native gel and appropriate treatment (partially adapted from Jacquin-Joly *et al.*, 2001, by permission of Oxford University Press). B Binding of male *M. brassicae* antennal extracts (100 antennal equivalents) with tritiated pheromone (1μCi each). Left: Ponceau red staining of the membrane after native-PAGE; right: corresponding fluorography.

16:Ac, the major pheromonal component of *M. brassicae*, to proteins of these tissue extracts that migrate at the same position as a known CSP.

CSPs are then detected in many chemosensory tissues in *M. brassicae*, including tissues without any obvious olfactory function (e.g. pheromone gland). This is consistent with the observed distributions of these proteins in other insects. In the proboscis of *M. brassicae*, the sensilla styloconica, which morphologically feature mixed characteristics of olfactory and gustatory function, are the most likely place where CSPs are expressed (Nagnan-Le Meillour *et al.*, 2000). This view is supported by immunocytochemistry using anti-Apol-GOBP antiserum which appeared to cross-react with CSPs despite low sequence similarities between CSPs and GOBPs. This cross-reaction is certainly due to similar 3D structure between CSPs and OBPs. Although these proteins have no conserved regions, both proteins consist mainly of α-helices. CSPs can be found in the sensillar lymph of sensilla styloconica, which is consistent with the general localization of OBPs, and also in subcuticular space, which is the location of the *Drosophila* OBP PBPRP2 (Park *et al.*, 2000; Shanbhag *et al.*, 2002).

The cellular localization of MbraCSP-A in male antenna was determined by *in situ* hybridization (Jacquin-Joly *et al.*, 2001). Figure 17.5 shows the distribution of labeled spots that represent expressing cells. These cells associate with olfactory sensilla, based on their morphology and distribution on the male antennae (Figure 17.1). *M. brassicae* antennae are filiform and segmented, and all segments have the same organization (Renou and Lucas, 1994). The dorsal surface of the antenna is covered with scales while the ventral surface is covered by different types of sensilla. In particular, long sensilla trichodea are arranged in four to five parallel rows on the lateral side (see Figure 17.1A). Short sensilla trichodea are randomly

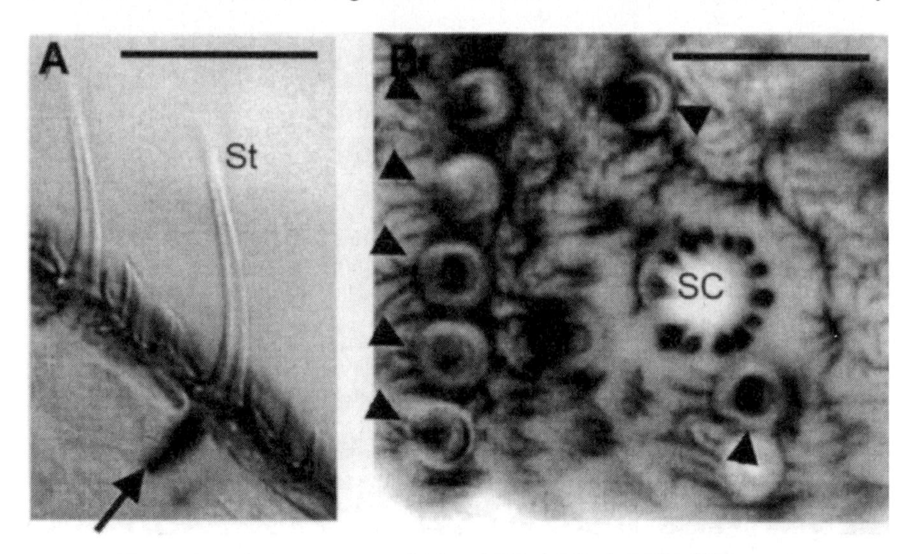

Figure 17.5 Expression pattern of MbraCSP-A6 isoform in *M. brassicae* male antennae revealed by *in situ* hybridization to mRNA. A expression of MbraCSP-A6 isoform at the base of a sensillum trichodium. B section through the cuticle showing labeled sensilla trichodea (s.t.) bases: long ones (arranged in a row) and short ones (randomly distributed); a sensillum coeloconicum (s.c.), visible on the same section, is not labeled. Bars: 10 μm (adapted from Jacquin-Joly *et al.*, 2001, by permission of Oxford University Press).

distributed on the ventral surface of the antennae, interspersed with a small number of sensilla coeloconica (see Figure 17.1B). Based on the distribution of staining, MbraCSP-A seems to be expressed in both long (Figure 17.5A) and short sensilla trichodea but not in sensilla coeloconica (Figure 17.5D). Our observations indicate the coexpression of OBPs and CSPs in the same subset of sensilla, the sensilla trichodea, devoted to pheromone reception in *M. brassicae*. This situation is in contrast with the distribution of OBPs and OS-D in *Drosophila* (McKenna *et al.*, 1994). Clearly in *Drosophila,* OBP-like proteins (OS-E, OS-F) and OS-D show different expression pattern within the antennae and OS-D expression is associated with the sacculus sensory cavity, an antennal structure rich in sensilla coeloconica (Pikielny *et al.*, 1994; McKenna *et al.*, 1994).

In *S. gregaria*, immunocytochemical experiments showed selective CSP labeling of the outer lymph in diverse chemosensory organs, including contact sensilla of tarsi, maxillary palps, and antennae. However, in antennae, only sensilla chaetica were labeled with no labeling observed in olfactory or coeloconic sensilla (Angeli *et al.*, 1999), suggesting a role for CSPs in contact chemoreception in Orthoptera.

When considering these discrepancies in CSP expression between different species, one should keep in mind the diversity of CSPs in a single species. Is MbraCSP-A the drosophilan OS-D homolog (46 percent amino acid identity)? What about MbraCSP-B (41 percent identity with OS-D), whose pattern has not yet been studied? Only an exhaustive study of the expression patterns of all the diverse CSPs from a single species will be conclusive, as has been conducted in several OBP studies (e.g. Shanbhag *et al.,* 2002, Vogt *et al.*, 2002).

17.7.3 Proposed functions of the chemosensory proteins

The CSPs were discovered by searching for abundant proteins in chemosensory organs or by homologous cloning strategies, but very few functional studies have been done. However, several different roles have been proposed depending on the species studied. The localization of CSP in maxillary palps of *C. cactorum* led to the suggestion that this protein plays a role in CO_2 sensing (Maleszka and Stange, 1997), but this function has been ruled out by functional studies in *S. gregaria* (Angeli *et al.*, 1999). The expression of a CSP in regenerating legs of the cockroach led to the suggestion that the protein plays a role in tissue regeneration (Nomura *et al.*, 1992), although no functional tests were made. Multiple functions are thus possible for this class of proteins, depending to their expression in a variety of tissues.

17.7.3.1 Implication in chemoreception

Several arguments favor a role for CSPs in chemoreception, either olfaction or taste. First, in *M. brassicae*, CSPs were clearly shown to bind pheromone components and a behavioral antagonist (Bohbot *et al.*, 1998), although without specificity. Second, most identified CSPs are reported to be associated with chemosensory tissues where they are expressed in sensillar structures involved in olfaction and/or gustation. In *M. brassicae*, CSPMbra-A differentially expresses in sensilla, restricted to sensilla involved in pheromone detection and absent from sensilla coeloconica which are known to detect plant odors, at least in *B. mori* (Pophof, 1997). Such differential expression among olfactory sensilla argues in favor of partial selectivity of these proteins toward their ligands. As described in section 17.7.2, Orthoptera CSPs associate with contact chemoreception based on their localization in sensilla chaetica, while *Drosophila* OS-D associates with olfaction based on its association with sensilla coeloconica. Because these different sensilla types are involved in different chemoreception processes, we can postulate that the diversity of CSPs observed in a single species could reflect their selective function in one or another process. Further studies should be conducted to correlate the expressing sensilla type and the binding properties of a given CSP.

17.7.3.2 Transport of hydrophobic compounds and signal termination
CSPs are often not antennal specific in their expression. Some are observed in organs devoid of any sensory structure. *P. americana* p10 was observed in regenerating legs during larval stages (Kitabayashi *et al.*, 1998). EBPIII was present in the *Drosophila* ejaculatory bulb that synthesizes the fatty acid derivative vaccenyl acetate (Z11-18:Ac), a compound transferred from the male to the female during copulation and which may serve as a repellent for other courting males (Jallon *et al.*, 1981). *M. brassicae* MbraCSP-A was produced in female pheromonal gland (Jacquin-Joly *et al.*, 2001). Although ejaculatory bulb and pheromone gland appear to be far away from chemoreceptive function, the presence of CSPs is not contradictory to a chemofunction for these proteins: lepidopteran pheromonal gland as well as *Drosophila* ejaculatory bulb produce pheromone, hydrophobic fatty acid derivatives that need to be transported via a carrier to the cellular membrane to be secreted.

CSPs may have a general role of lipid transporter. In *M. brassicae*, tritium-labeled fatty acids differing in chain length and double bound number (16:COOH; Z9-18:COOH; Z9,Z12-18:COOH) were incubated with male antennal extracts (Figure 17.4B). The fatty acids and their derivatives bound to antennal CSPs without apparent selectivity. However, in these same studies, binding by progesterone, a steroid that is not structurally related to fatty acids, did not occur (Jacquin-Joly *et al.*, 2001). We can thus postulate that CSPs are able to bind fatty acids in general and/or 16–18 carbon backbone skeleton molecules. Preliminary GC-MS studies suggest the presence of a highly hydrophobic molecule in the CSP-binding site of *S. gregaria* (quoted in Picone *et al.*, 2001, but unpublished from Scaloni and Pelosi), but the nature of this site remains unknown.

CSPs could play a role in general transport of insoluble ligands, and in addition could participate in the cleaning or detoxification of the neuron surface. Toxic compounds entering the antennae should be rapidly brought to the accessory cells to be internalized and destroyed by cytosolic or microsomal degrading enzymes such as glutathione S-transferase (Rogers *et al.*, 1999) or P450 (Wojtazek and Leal, 1999a; Maïbèche-Coisné *et al.*, 2002), which have been shown to be antennal specific with a postulated role in olfaction.

Several arguments indicate involvement of CSPs in different modalities of chemoreception or, more generally, in carrying insoluble chemical messages or lipid compounds. These arguments are supported by CSP structural analogies with various lipid carriers, together with high sequence similarity between species. However, CSP function should be clarified by further studies, in particular precise localization of each member of this diverse family, structural elucidation already engaged, and functional binding tests on recombinant proteins.

17.8 Conclusion

We have tried to demonstrate the advantage of the functional approach to characterize the diversity of OBPs inside species. The limiting step is the availability of tritium-labeled pheromone compounds. Another strategy, less expansive and more exhaustive, is the construction of EST antennal libraries with an EST number high enough to be representative of the sequence diversity. Such an EST project has been conducted by Robertson and collaborators on *M. sexta* antennae (Robertson *et al.*, 1999) which has identified not only new OBPs, but also SAPs, pheromone-degrading enzymes, sensory-neuron-membrane-protein-like sequences (see Vogt, Chapter 14, in this volume) and other elements involved in the pheromone transduction process (ion channels, etc.). This availability of new sequences, combined with heterologous expression and kinetic studies as well as classical techniques of biology such as electrophysiological and behavioral studies, will allow the reconstruction of the transduction process in its totality and complexity. This opens the way for future research in this field.

Abbreviations

OBP, Odorant-Binding Protein; PBP, Pheromone-Binding Protein; GOBP, General Odorant-Binding Protein; Z11-16:Ac, *cis*-11-hexadecenyl acetate; 16:Ac, hexadecanyl acetate; Z9-16:Ac, *cis*-9-tetradecenyl acetate; Z11-18:Ac, *cis*-11-octadecenyl acetate; Z11-16:Ald, *cis*-11-hexadecenal; Z11-16:OH, *cis*-11-hexadecenol; Z9-14:Ac, *cis*-9-tetradecenyl acetate; (E, Z)-6,11-16:Ac, *trans*6, *cis*11-hexadecanienyl acetate; (E, Z)-6,11-16:Dza, *trans*6, *cis*11-hexadecadienyl diazoacetate; Z11-16:TFMK, *cis*-11-hexadecenyl trifluoromethyl ketone.

Acknowledgements

We thank the French Institut National de Recherche Agronomique and EEC grant (BIO4-98-0420 OPTIM) for financial support and the group members for their contribution to our work presented in this chapter (Martine Maïbèche-Coisné, Jonathan Bohbot, Anne-Hélène Cain, Marie-Christine François, Carole Deyts and Guilia Cristiani).

References

Adams M. D. *et al.* (2000) The genome sequence of *Drosophila melanogaster*. *Science* **287**, 2185–2195.

Angeli S., Ceron F., Scaloni A., Monti M., Monteforti G., Minnocci A., Petacchi R. and Pelosi P. (1999) Purification, structural characterization, cloning and immunocytochemical localization of chemoreception proteins from *Schistocerca gregaria*. *Eur. J. Biochem.* **262**, 745–754.

Baker T. C., Cossé A. A. and Todd J. L. (1998a) Behavioral antagonism in the moth *Helicoverpa zea* in response to pheromone blends of three sympatric Heliothine moth species is explained by one type of antennal neuron. *Ann. N. Y. Acad. Sci.* **855**, 511–513.

Baker T. C., Fadamiro H. Y. and Cossé A. A. (1998b) Moth uses fine tuning for odour resolution. *Nature* **393**, 530.

Bestmann H. J., Attygale A. B., Brosche T., Erler J., Platz H., Schwarz J., Vostrowsky O., Wu C. H. Kaissling K. E. and Chen T. N. (1987) Identification of three sex pheromone components of the female Saturniid moth *Antheraea pernyi* (Lepidoptera: Saturniidae). *Z. Natürforsch.* **42c**, 631–636.

Bette S., Breer H. and Krieger J. (2002) Probing a pheromone binding protein of the silkmoth *Antheraea polyphemus* by endogenous tryptophan fluorescence. *Insect Biochem. Molec. Biol.* **32**, 241–246.

Biessmann H., Walter M. F., Dimitratos S. and Woods D. (2002) Isolation of cDNA clones encoding putative odorant binding proteins from the antennae of the malaria-transmitting mosquito, *Anopheles gambiae*. *Insect Mol. Biol.* **11**, 123–132.

Blais J. C., Nagnan-Le Meillour P., Bolbach G. and Tabet J. C. (1996) MALDI-TOFMS identification of "odorant binding proteins" (OBPs) electroblotted onto poly(vinylidene difluoride) membranes. *R. C. Mass Spectrom.* **10**, 1–4.

Bohbot J., Sobrio F., Lucas P. and Nagnan-Le Meillour P. (1998) Functional characterization of a new class of odorant-binding proteins in the moth *Mamestra brassicae*. *Biochem. Biophys. Res. Commun.* **253**, 489–494.

Breer H., Krieger J. and Raming K. (1990) A novel class of binding proteins in the antennae of the silkmoth *Antheraea pernyi*. *Insect Biochem.* **20**, 735–740.

Briand L., Nespoulous C., Huet J. C., Takahashi M. and Pernollet J. C. (2001) Ligand binding and physico-chemical properties of ASP2, a recombinant odorant-binding protein from honeybee (*Apis mellifera* L.). *Eur. J. Biochem.* **268**, 752–760.

Callahan F. E., Vogt R. G., Tucker M. L., Dickens J. C. and Mattoo A. K. (2000) High level expression of "male specific" pheromone binding proteins (PBPs) in the antennae of female noctuiid moths. *Insect Biochem. Molec. Biol.* **30**, 507–514.

Campanacci V., Longhi S., Nagnan-Le Meillour P., Cambillau C. and Tegoni, M. (1999) Recombinant pheromone binding protein 1 from *Mamestra brassicae* (MbraPBP1). Functional and structural characterization. *Eur. J. Biochem.* **264**, 707–716.

Campanacci V., Mosbah A., Bornet O., Wechselberger R., Jacquin-Joly E., Cambillau C., Darbon H. and Tegoni M. (2001a) Chemosensory protein from the moth *Mamestra brassicae*, expression and secondary structure from ^{1}H and ^{15}N NMR. *Eur. J. Biochem.* **268**, 4731–4739.

Campanacci V., Spinelli S., Lartigue A., Lewandowski C., Brown K., Tegoni M. and Cambillau C. (2001b) Recombinant chemosensory protein (CSP2) from the moth *Mamestra brassicae*: crystallization and preliminary crystallographic study. *Acta Cryst.* **D57**, 137–139.

Cossé A. A., Todd J. L. and Baker T. C. (1998) Neurons discovered in male *Helicoverpa zea* antennae that correlate with pheromone-mediated attraction and interspecific antagonism. *J. Comp. Physiol. A* **182**, 585–594.

Damberger F., Nikonova L., Horst R., Peng G., Leal W.S. and Wuthrich K. (2000) NMR characterization of a pH-dependent equilibrium between two folded solution conformations of the pheromone-binding protein from *Bombyx mori. Protein Sci.* **9**, 1038–1041.

Danty E., Arnold G., Huet J.C., Huet D., Masson C. and Pernollet J. C. (1998) Separation, characterization and sexual heterogeneity of multiple putative odorant-binding proteins in the honeybee *Apis mellifera* L. (Hymenoptera: Apidea). *Chem. Senses* **23**, 83–91.

Den Otter C. J. and Van der Haagen M. M. (1989) Sex pheromone attractants and inhibitors in the cabbage armyworm, *Mamestra brassicae* L. (Lep.: Noctuidae): electrophysiological discrimination. *Insect Sci. Applic.* **10**, 235–242.

Descoins C., Priesner E., Gallois M., Arn H. and Martin G. (1978) Sur la sécrétion phéromonale des femelles vierges de *Mamestra brassicae* L. et de *Mamestra oleracea* L. (Lépidoptères Noctuidae, Hadeninae). *C. R. Acad. Sc. Paris* **286**, 77–80.

Deyu Z and Leal W. S. (2002) Conformational isomers of insect odorant-binding proteins. *Arch. Biochem. Biophys.* **397**, 99–105.

Du G., Ng C. S. and Prestwich G. D. (1994) Odorant binding by a pheromone binding protein: active site mapping by photoaffinity labeling. *Biochemistry* **33**, 4812–4819.

Du G. and Prestwich G. D. (1995) Protein structure encodes the ligand binding specificity in pheromone binding proteins. *Biochemistry* **34**, 8726–8732.

Dyanov H. M. and Dzitoeva S. G. (1995) Method for attachment of microscopic preparations on glass for *in situ* hybridization, PRINS and *in situ* PCR studies. *BioTechniques* **18**, 822–824.

Farine J. P., Frérot, B. and Isart J. (1981) Facteurs d'isolement chimique dans la sécrétion phéromonale de deux noctuelles Hadeninae: *Mamestra brassicae* (L.) et *Pseudaletia unipuncta* (Haw.). *C. R. Acad. Sc. Paris* **292**, 101–104.

Feng L. and Prestwich G. D. (1997) Expression and characterization of a lepidopteran general odorant binding protein. *Insect Biochem. Molec. Biol.* **27**, 405–412.

Fitt G. P. (1989) The ecology of *Heliothis* species in relation to agroecosystems. *Ann. Rev. Entomol.* **34**, 17–52.

Györgyi T. K., Roby-Shemkovitz A. J. and Lerner M. R. (1988). Characterization and cDNA cloning of the pheromone binding protein from the tobacco hornworm, *Manduca sexta*: a tissue specific, developmentally regulated protein. *Proc. Natl. Acad. Sci. USA* **85**, 9851–9855.

Hansson B. S. (1995) Olfaction in lepidoptera. *Experientia* **51**, 1003–1027.

Hansson B. S., Larsson M. C. and Leal W. S. (1999) Green leaf volatile-detecting olfactory receptor neurones display very high sensitivity and specificity in a scarab beetle. *Physiol. Entomol.* **24**, 121–126.

Heinbockel T. and Kaissling K. E. (1996) Variability of olfactory receptor neuron responses of female silkmoths (*Bombyx mori* L.) to benzoic acid and (+/–)-linalool. *J. Insect Physiol.* **42**, 565–578.

Jacquin E. (1992) Régulation neuroendocrine de la production des phéromones sexuelles femelles chez deux noctuelles, *Mamestra brassicae* L. et *Heliothis zea* Boddie (Lépidoptères), 131 pp. Thèse de l'INA-PG.

Jacquin-Joly E., Bohbot J., François M. C., Cain A. H. and Nagnan-Le Meillour P. (2000) Characterization of the general-odorant binding protein 2 in the molecular coding of odorants in *Mamestra brassicae*. *Eur. J. Biochem.* **267**, 6708–6714.

Jacquin-Joly E., Vogt R. G., François M. C. and Nagnan-Le Meillour P. (2001) Functional and expression pattern analyses of chemosensory proteins expressed in antennae and pheromonal gland of *Mamestra brassicae*. *Chem. Senses* **26**, 833–844.

Jallon J. M., Anthony C. and Benamar O. (1981) Un anti-aphrodisiaque produit par les mâles de *Drosophila melanogaster* et transféré aux femelles lors de la copulation. *C. R. Acad. Sci. Paris* **292**, 1147–1149.

Kitabayashi A. N., Arai T., Kubo T. and Natori S. (1998) Molecular cloning of cDNA for p10, a novel protein that increases in regenerating legs of *Periplaneta americana* (American cockroach). *Insect Biochem. Mol. Biol.* **28**, 785–790.

Krieger J., Ganssle H., Raming K. and Breer H. (1993) Odorant-binding protein of *Heliothis virescens*. *Insect Biochem. Molec. Bio*l. **23**, 449–456.

Kucharski R. and Maleszka R. (2002) Evaluation of differential gene expression during behavioral development in the honeybee using microarrays and northern blots. *Genome Biol.* **3**, 1–9.

LaForest S., Prestwich G. D. and Löfstedt C. (1999) Intraspecific nucleotide variation at the pheromone binding protein locus in the turnip moth, *Agrotis segetum*. *Insect Molec. Biol.* **8**, 481–490.

Laue M., Steinbrecht R. A. and Ziegelberger G. (1994) Immunocytochemical localization of general-odorant binding protein in olfactory sensilla of the silkmoth *Antheraea polyphemus*. *Naturwissenschaften* **81**, 178–180.

Leal W. S., Nikonova L. and Peng G. (1999) Disulfide structure of the pheromone binding protein from the silkworm moth, *Bombyx mori*. *FEBS Letters* **464**, 85–90.

McKenna M. P., Hekmat-Scafe D. S., Gaines P. and Carlson J. R. (1994) Putative *Drosophila* pheromone-binding proteins expressed in a subregion of the olfactory system. *J. Biol. Chem.* **269**, 16340–16347.

Maïbèche-Coisné M., Sobrio F., Delaunay T., Lettéré M., Dubroca J., Jacquin-Joly E. and Nagnan-Le Meillour P. (1997) Pheromone binding proteins of the moth *Mamestra brassicae*: specificity of ligand binding. *Insect Biochem. Molec. Biol.* **27**, 213–221.

Maïbèche-Coisné M., Jacquin-Joly E., François M. C. and Nagnan-Le Meillour P. (1998a) Molecular cloning of two pheromone-binding proteins in the cabbage armyworm *Mamestra brassicae*. *Insect Biochem. Molec. Biol.* **28**, 815–818.

Maïbèche-Coisné M., Longhi S., Jacquin-Joly E., Brunel C., Egloff M. P., Gastinel L., Cambillau C., Tegoni M. and Nagnan-Le Meillour P. (1998b). Molecular cloning and bacterial expression of a general odorant binding protein from the cabbage armyworm *Mamestra brassicae*. *Eur. J. Biochem.* **258**, 768-774.

Maïbèche-Coisné M., Jacquin-Joly E., François M.C. and Nagnan-Le Meillour P. (2002) Molecular cloning of two novel cytochrome P450 cDNAs, *CYP4L4* and *CYP4S4*, differentially expressed in the antennae of the noctuid moth *Mamestra brassicae*. *Insect Molec. Biol.* **11**, 273–281.

Maida R., Krieger J., Gebauer T., Lange U. and Ziegelberger G. (2000) Three pheromone-binding proteins in olfactory sensilla of the two silkmoth species *Antheraea polyphemus* and *Antheraea pernyi*. *Eur. J. Biochem.* **267**, 2899–2908.

Malezka R. and Stange G. (1997) Molecular cloning, by a novel approach, of a cDNA encoding a putative olfactory protein in the labial palps of the moth *Cactoblastis cactorum*. *Gene* **202**, 39–43.

Mameli M., Tuccini A., Mazza M., Petacchi R. and Pelosi P. (1996) Soluble proteins in chemosensory organs of Phasmids. *Insect Biochem. Mol. Biol.* **26**, 875–882.

Marchese S., Angeli S., Andolfo A., Scaloni A., Brandazza A., Mazza M., Picimbon J. F., Leal W. S. and Pelosi P. (2000) Soluble proteins from chemosensory organs of *Eurycantha calcarata* (Insecta, Phasmatodea). *Insect Biochem. Mol. Biol.* **30**, 1091–1098.

Meng L. Z., Wu, C. H., Wicklein, M., Kaissling, K. E., and Bestmann, H. J. (1989). Number and sensitivity of three types of pheromone receptor cells in *Antheraea pernyi* and *A. polyphemus*. *J. Comp. Physiol.* **165**, 139–146.

Nagnan-Le Meillour P., Huet J. C., Maïbèche M., Pernollet J. C. and Descoins C. (1996) Purification and characterization of multiple forms of odorant/pheromone binding proteins in the antennae of *Mamestra brassicae* (Noctuidae). *Insect Biochem. Molec. Biol.* **26**, 59–67.

Nagnan-Le Meillour P., Cain A. H., Jacquin-Joly E., François M. C., Ramachandran S., Maida R. and Steinbrecht R. A. (2000) Chemo-Sensory proteins from the proboscis of *Mamestra brassicae*. *Chem. Senses* **25**, 541–553.

Nomura A., Kawasaki K., Kubo T. and Natori S. (1992) Purification and localization of p10, a novel protein that increases in nymphal regenerating legs of *Periplaneta americana* (American cockroach). *Int. J. Dev. Biol.* **36**, 391–398.

Paesen G. C. and Happ G. M. (1995) The B proteins secreted by the tubular accessory sex glands of the male mealworm beetle, *Tenebrio molitor*, have sequence similarity to moth pheromone-binding proteins. *Insect Biochem. Mol. Biol.* **25**, 401–408.

Park S. K., Shanbhag S. R., Wang Q., Hasan G., Steinbrecht R. A. and Pikielny C. W. (2000) Different expression patterns of two putative odorant-binding proteins in the olfactory organs of *Drosophila* have different implications for their functions. *Cell Tissue Res.* **300**, 181–192.

Picimbon J. F. and Leal W. S. (1999) Olfactory soluble proteins of cockroaches. *Insect Biochem. Mol. Biol.* **29**, 973–978.

Picimbon J. F., Dietrich K., Breer H. and Krieger J. (2000a) Chemosensory proteins of *Locusta migratoria* (Orthoptera: Acrididae). *Insect Biochem. Mol. Biol.* **30**, 233–241.

Picimbon J. F., Dietrich K., Angeli S., Scaloni A., Krieger J., Breer H. and Pelosi P. (2000b) Purification and molecular cloning of chemosensory proteins from *Bombyx mori*. *Arch. Insect Biochem. Physiol.* **44**, 120–129.

Picimbon J. F., Dietrich K., Krieger J. and Breer H. (2001) Identity and expression pattern of chemosensory proteins in *Heliothis virescens* (Lepidoptera, Noctuidae). *Insect Biochem. Molec. Biol.* **31**, 1173–1181.

Picimbon J. F. and Gadenne C. (2002) Evolution of noctuid pheromone binding proteins: identification of PBP in the black cutworm moth, *Agrotis ipsilon*. *Insect Biochem. Molec. Biol.* **32**, 839–846.

Picone D., Crescenzi O., Angeli S., Marchese S., Brabdazza A., Ferrara L., Pelosi P. and Scaloni A. (2001) Bacterial expression and conformational analysis of a chemosensory protein from *Schistocerca gregaria*. *Eur. J. Biochem.* **268**, 4794–4801.

Pikielny C. W., Hasan G., Rouyer F. and Rosbach M. (1994) Members of a family of Drosophila putative odorant-binding proteins are expressed in different susbsets of olfactory hairs. *Neuron* **12**, 35–49.

Pophof B. (1997) Olfactory responses record from sensilla coeloconica of the silkmoth *Bombyx mori*. *Physiol. Entomol.* **22**, 239–248.

Prestwich G. D., Du G. and LaForest S. M. (1995) How is pheromone specificity encoded in proteins? *Chem. Senses* **20**, 461–469.

Quero C. and Baker T. C. (1999) Antagonistic effect of (*Z*)-11-hexadecen-1-ol on the pheromone-mediated flight of *Helicoverpa zea* (Boddie) (Lepidoptera: Noctuidae). *J. Insect Behav.* **12**, 701–710.

Renou M. and Lucas P. (1994) Sex pheromone reception in *Mamestra brassicae* L. (Lepidoptera): responses of olfactory receptor neurones to minor components of the pheromone blend. *J. Insect Physiol.* **40**, 75–85.

Robertson H. M., Martos R., Sears C. R., Todres E. Z., Walden K. K. O. and Nardi J. B. (1999) Diversity of odourant binding proteins revealed by an expressed sequence tag project on male *Manduca sexta* moth antennae. *Insect Mol. Biol.* **8**, 501–518.

Rogers M. E., Jani M. K. and Vogt R. G. (1999) An olfactory-specific glutathione-S-transferase in the sphinx moth *Manduca sexta. J. Exp. Biol.* **202**, 1625–1637.

Rothemund S., Liou Y. C., Davies P. and Sönnichsen F. D. (1997) Backbone structure and dynamics of a hemolymph protein from the mealworm beetle *Tenebrio molitor. Biochemistry* **36**, 13791–13801.

Rothemund S., Liou Y. C., Davies P. L., Krause E. and Sönnichsen F. D. (1999) A new class of hexahelical insect proteins revealed as putative carriers of small hydrophobic ligands. *Structure* **7**, 1325–1332.

Sandler B. H., Nikonova L., Leal W. S. and Clardy J. (2000) Sexual attraction in the silkworm moth: structure of the PBP-pheromone-binding-protein-bombykol complex. *Chemistry Biol.* **7**, 143–151.

Scaloni A., Monti M., Angeli S. and Pelosi P. (1999) Structural analyses and disulfide-bridge pairing of two odorant-binding proteins from *Bombyx mori. Biochem. Biophys. Res. Commun.* **266**, 386–91.

Shanbhag S. R., Hekmat-Scafe D., Kim M. S., Park S. K., Carlson J., Pikielny C., Smith D. P. and Steinbrecht R. A. (2002) The expression mosaïc of odorant-binding proteins on Drosophila olfactory organs. *Microsc. Res. Tech.* **55**, 297–306.

Steinbrecht R. A., Ozaki M. and Ziegelberger G. (1992) Immunocytochemical localization of pheromone-binding protein in moth antennae. *Cell Tissue Res.* **270**, 287–302.

Steinbrecht R. A., Laue M., Maida R. and Ziegelberger G. (1996) Odorant-binding proteins and their role in the detection of plant odours. *Entomol. Exp. Appl.* **80**, 15–18.

Struble D. L., Arn H., Buser H. R., Städler E. and Freuler J. (1980) Identification of 4 sex pheromone components isolated from calling females of *Mamestra brassicae. Z. Naturforsch.* **35c**, 45–48.

Subchev M. A., Staminivora L. S. and Stoilov I. L. (1985) Effect of *cis*-11-hexadecenol and its derivates on the pheromonal activity of *cis*-11-hexadecenyl acetate to males of three noctuid species (Lepidoptera: Noctuidae) in fields. *Ecology* **17**, 56–61.

Subchev M. A., Staminivora L. S. and Milkova T. S. (1987) The effect of compounds related to *cis*-11-hexadecenyl acetate on its attractiveness to the males of *Mamestra brassicae* L. (Lepidoptera: Noctuidae) and some other noctuid species. *Folia Biol.* **35**, 143–150.

Tuccini A., Maida R., Rovero P., Mazza M. and Pelosi P. (1996) Putative odorant-binding protein in antennae and legs of *Carausius morosus* (Insecta, Phasmatodea). *Insect Biochem. Mol. Biol.* **26**, 19–24.

Vickers N. J., Christensen T. A., Mustaparta H. and Baker T. C. (1991) Chemical communication in Heliothine moths. III. Flight behavior of male *Helicoverpa zea* and *Heliothis virescens* in response to varying ratios of intra- and interspecific sex pheromone components. *J. Comp. Physiol.* **169**, 275–280.

Vogt R. G. and Riddiford L. M. (1981) Pheromone binding and inactivation by moth antennae. *Nature (London)* **293**, 161–163.

Vogt R. G., Köhne A. C., Dubnau J. T. and Prestwich G. D. (1989) Expression of pheromone binding proteins during antennal development in the gypsy moth *Lymantria dispar. J. Neurosci.* **9**, 3332–3346.

Vogt R. G., Prestwich G. D. and Lerner M. R. (1991) Odorant binding protein subfamilies associate with distinct classes of olfactory receptor neurons in insects. *J. Neurobiol.* **22**, 74–84.

Vogt R. G., Rybczynski R., Cruz M. and Lerner M. R. (1993) Ecdysteroid regulation of olfactory protein expression in the developing antenna of the tobacco hawk moth *Manduca sexta*. *J. Neurobiol.* **24**, 581–597.

Vogt R. G., Callahan F. E., Rogers M. E. and Dickens J. C. (1999) Odorant binding protein diversity and distribution among the insect orders, as indicated by LAP, an OBP-related protein of the true bug *Lygus lineolaris* (Hemiptera, Heteroptera). *Chem. Senses* **24**, 481–495.

Vogt R. G., Rogers M., Franco M. D. and Sun M. (2002) A comparative study of odorant binding protein genes: differential expression of the PBP1-GOBP2 gene cluster in *Manduca sexta* (Lepidoptera) and the organization of OBP genes in *Drosophila melanogaster. J. Exp. Biol.* **205**, 719–744.

Wojtasek H. and Leal W. S. (1999a) Degradation of an alkaloid pheromone from the pale-brown chafer, *Phyllopertha diversa* (Coleoptera: Scarabaeidae), by an insect olfactory cytochrome P450. *FEBS Lett.* **458**, 333–336.

Wojtasek H. and Leal W. S. (1999b) Conformational change in the pheromone-binding protein from *Bombyx mori* induced by pH and by interaction with membranes. *J. Biol. Chem.* **274**, 30950–30956.

Wojtasek H., Picimbon J. F. and Leal W. S. (1999) Identification and cloning of odorant binding proteins from the scarab beetle *Phylloperta diversa*. *Biochem. Biophys. Res. Commun.* **263**, 832–837.

Zhang S. G., Maida R. and Steinbrecht R. A. (2001) Immunolocalization of odorant binding proteins in Noctuid moths (Insecta, Lepidoptera). *Chem. Senses* **26**, 885–896.

Ziegelberger G. (1995) Redox-shift of the pheromone-binding protein in the silkmoth *Antheraea polyphemus*. *Eur. J. Biochem.* **232**, 706–711.

18

Biochemistry and evolution of OBP and CSP proteins

Jean-François Picimbon

18.1 Introduction

One perceives from the world only what one has been prepared to perceive. In humans and in most mammals, different senses are used to make sense of life. In contrast, in insects, chemical senses involving odorants and contact chemosensory molecules play a vital role. The olfactory system is the primary sense insects use in analyzing the environment, in crucial tasks such as finding food, nesting, mating and in conspecifics. Contact chemosensation is used especially to analyze specific substrates to assist in the identification of suitable oviposition sites, the recognition of host plants, the selection of tastants and the search for further nutrient chemicals. Dedicated to survival, both olfactory and contact chemosensory systems in insects have developed to extremely high levels of sensitivity and selectivity.

The sensitivity and selectivity of olfaction and contact chemosensation are due (1) in the brain, to the existence of a neuronal network of neurons tuned to a specific chemical stimulus, and (2) in the periphery, to the existence of olfactory/chemosensory receptor neurons housed in sensory microorgans called sensilla. The sensilla can best be viewed as simple cuticular porous extrusions that increase the surface that captures airborne odorants or chemicals dissolved in water droplets. They contain the receptive olfactory or chemosensory structures (Schneider, 1969). The olfactory sensilla are most numerous on the antennae and mediate the reception of sex pheromones and plant volatiles, as well as other odorants. Low volatility pheromones may also be detected by contact chemoreceptors on the front legs (Xu *et al.*, 2002; Park *et al.*, 2002). In contrast, general chemosensory receptor sensilla are distributed over the whole insect body, but mainly occur on the legs, and contain neurons responding to hydrophobic tastants, CO_2, temperature, humidity or a combination of different modalities. The antennae, legs and their sensillar complement represent a wide spectrum of structures, shapes and lengths, but the cellular organization of sensilla follows a universal scheme based on conserved morphological features (Keil, 1997). Most prominent is the presence of a lymphatic fluid, the sensillar lymph, that entirely fills the sensillar lumen into which the dendrites of the sensory neurons extend. Thus, the pores that penetrate the surface of the sensilla are not in direct contact with the receptor proteins which reside on the sensory dendritic membrane, and chemical molecules have to cross the sensillar lymph before interacting with the dendritic receptors. The problem of chemical reception is twofold (1) the chemical molecules (largely hydrophobic in nature) face a hydrophilic environment after penetrating the sensillum, and (2) the chemicals in the lymph are exposed to a high concentration of chemical degrading enzymes. Specific binding mechanisms are therefore required, not only to solubilize but also to protect and transport the odorant molecules in the sensillar lymph, upstream to the sensory receptors (Vogt and Riddiford, 1981, 1986; Vogt *et al.*, 1985; Vogt, 1987).

The odorant-binding proteins (OBPs) and the chemosensory proteins (CSPs) are proteins from the lymph that are thought to accomplish these tasks, solubilizing and protecting the odorant and contact chemosensory molecules. This chapter describes the biochemical and evolutionary aspects of these two families of peripheral sensory proteins of insects. Particular attention will be paid to the subclassification of binding proteins, the diversity of gene structures and the phyletic and molecular relatedness between binding proteins from different insect species.

18.2 The family of odorant-binding proteins

18.2.1 The concept of pheromone-binding protein

About 20 years ago, Vogt proposed that small water-soluble proteins called OBPs might aid in odor reception, by keeping the lipophilic odorants soluble and active in the lymph, thus allowing their transport and integrity through the aqueous barrier. The first insect OBP identified as such was the pheromone-binding protein (PBP) found in the male antennae of the large silkmoth *Antheraea polyphemus* (Vogt and Riddiford, 1981; Table 18.1).

In pioneer binding studies, pheromone alone in a glass vial containing water quickly absorbed to the glass wall (Kaissling *et al.*, 1985). However, when PBP was added a certain amount of pheromone remained in solution. The degraded pheromone molecule was not held in solution by PBP, indicating a degree of

Table 18.1 The PBP related family of odorant binding proteins

Protein	Insect	GenBANK access number	References
PBP	**Lepidoptera**		
Aips-1	*Agrotis ipsilon*	AY301985	Picimbon and Gadenne, 2002
Aips-2	*A. ipsilon*	AY301986	Picimbon and Gadenne, 2002
Aseg-2	*Agrotis segetum*	AY301987	Abraham *et al.*, 2003
Aseg-1	*A. segetum*	AF134253-AF134294	Prestwich *et al.*, 1995; LaForest *et al.*, 1999
Aper-1	*Antheraea pernyi*	X96773	Raming *et al.*, 1990
Aper-2	*A. pernyi*	X96860	Krieger *et al.*, 1991
Apol-1	*Antheraea polyphemus*	X17559	Raming *et al.*, 1989
Bmor-1	*Bombyx mori*	X94987	Krieger *et al.*, 1996
Hvir-1	*Heliothis virescens*	X96861	Krieger *et al.*, 1993
Hvir-2	*H. virescens*	AY301988	Abraham *et al.*, 2003
Hzea-1	*Heliothis zea*	AF090191	Callahan *et al.*, 2000
Ldis-2	*Lymantria dispar*	AF007868	Prestwich *et al.*, 1995
Ldis-1	*L. dispar*	AF007867	Merritt *et al.*, 1998
Mbra-2	*Mamestra brassicae*	AF051143	Maïbèche-Coisné *et al.*, 1998
Mbra-1	*M. brassicae*	AF05051142	Maïbèche-Coisné *et al.*, 1998
Msex-1	*Manduca sexta*	AF323972	Györgyi *et al.*, 1988; Robertson *et al.*, 1999
ABPX			
AipsABPX-1	*A. ipsilon*	AY301981-AY301982-AY301983-AY301984	Picimbon *et al.*, unpublished
AperABPX	*A. pernyi*	CAA05509	Krieger *et al.*, 1997
BmorABPX	*B. mori*	CAA64446	Krieger *et al.*, 1996
HvirABPX-1	*H. virescens*	CAA05508	Krieger *et al.*, 1997
MsexABPX	*M. sexta*	AF117577 1/AF117575 1	Robertson *et al.*, 1999
DmelPBPRP	**Diptera**		
DmelPBPRP-1	*Drosophila melanogaster*	NP 524039/P54191/AAC46474	Pikielny *et al.*, 1994

specificity with respect to which odorants are solubilized. The antibody of an OBP-related protein from the blowfly *Phormia regina* blocks the response of the taste receptor cell to a stimulant containing hydrophobic molecules (Ozaki *et al.*, 1995). The mutation of one OBP gene from *Drosophila*, *lush*, results in abnormal chemoattractive behavior to ethanol (Kim *et al.*, 1998). These results indicate that reception of tastants and chemical molecules soluble in water requires transport of these molecules by OBPs. It is generally assumed that at the molecular level the nature of the diverse chemosensory modalities is similar to that of odorant and tastant reception and that multiple types of binding proteins are involved in the diverse chemosensations (Vogt *et al.*, 1991a, b; Shanbhag *et al.*, 2001a, b; Koganezawa and Shimada, 2002; Picimbon, 2002).

The question thus is to what extent PBP and other binding proteins participate in the recognition of odor messages? The PBPs have been so called by virtue of sex pheromone binding, specific association with sex-pheromonal sensilla (Vogt and Riddiford, 1981; Vogt *et al.*, 1989; Du *et al.*, 1994; Steinbrecht, 1996; Steinbrecht *et al.*, 1992, 1995; Laue *et al.*, 1994). A PBP from *Bombyx mori* has been crystallized with the pheromone Bombykol packed into a hydrophobic binding pocket (Sandler *et al.*, 2000; Klusak *et al.*, 2003). So far, no binding studies have really assessed the degree of specificity of PBPs, i.e. the ability of the carrier-protein to bind to one specific compound. Rather, the binding spectra of PBPs seem broad. The PBPs, like other binding proteins, will bind many chemicals but differently (Vogt *et al.*, 1989; Du *et al.*, 1994; Feng and Prestwich, 1997; Wojtaseck *et al.*, 1999; Campanacci *et al.*, 2001a; Bette *et al.*, 2002).

Differential binding representing a fine tuning of PBP–pheromone interaction has been documented by different studies of various insect species. A PBP from male moths will bind the chemical component of the female pheromone better than any other components. PBPs may act as selective filters, since PBPs differentially bind specific pheromone components; pheromone–PBP interactions may be based on the recognition of the chain hydrocarbons of the pheromone molecules (Du and Prestwich, 1995; Feixas *et al.*, 1995; Maïbèche-Coisné *et al.*, 1997; Maida *et al.*, 2000; Picimbon and Gadenne, 2002). The odor recognition by PBPs might be as sensitive as that exhibited by sensory receptors. In the gypsy moth, *Lymantria dispar*, two PBPs have been shown to discriminate between two enantiomeric forms of the pheromone (Vogt *et al.*, 1989; Plettner *et al.*, 2000). Given such a potential for greater binding and pheromone specificity allowed by a duplication of PBPs, the evolution of PBP genes may have followed the diversification of the pheromone systems.

18.2.2 The repertoire of PBPs in moths

The most prominent examples of pheromone diversification are found in Noctuidae. The sex pheromones of the Noctuidae species are mixtures of at least three compounds that differ mainly in carbon chain length (Löfstedt *et al.*, 1982; Teal *et al.*, 1986; Attygale *et al.*, 1987; Picimbon *et al.*, 1997). In many noctuid moth species, multiple PBPs have been reported (Vogt *et al.*, 1989; Merritt *et al.*, 1998; Maïbèche-Coisné *et al.*, 1998; Picimbon and Gadenne, 2002; Abraham *et al.*, 2003). Based on sequence homology and phylogenetic analysis, a sub-classification of noctuid PBPs has been proposed (Picimbon and Gadenne, 2002; Abraham *et al.*, 2003).

A neighbor-joining tree of selected PBPs from moths shows that noctuid PBPs segregate into two subclasses (Figure 18.1). Subclass 1 (group 1 or Grp1) corresponds to Grp1-PBPs from *Agrotis ipsilon*, *A. segetum*, *Heliothis virescens*, *H. zea* and *Mamestra brassicae* (Aips-1, Aseg-1, Hvir-1, Hzea-1 and Mbra-2). The Grp1-PBPs show about 86 percent identity between each other. The subclass 2 (group 2 or Grp2) corresponds to Grp2-PBPs from *Agrotis*, *Heliothis* and *Mamestra* species (Aips-2, Aseg-2, Hvir-2, Mbra-1; Figure 18.1) as well as proteins from non-noctuid species such as *Manduca sexta* (Msex-2, Msex-3; Robertson *et al.*, 1999; Picimbon and Gadenne, 2002). Subclass 2 also contains Ycag PBP, a PBP from *Yponomeuta cagnagellus* (Robertson *et al.*, 1999; Willett,

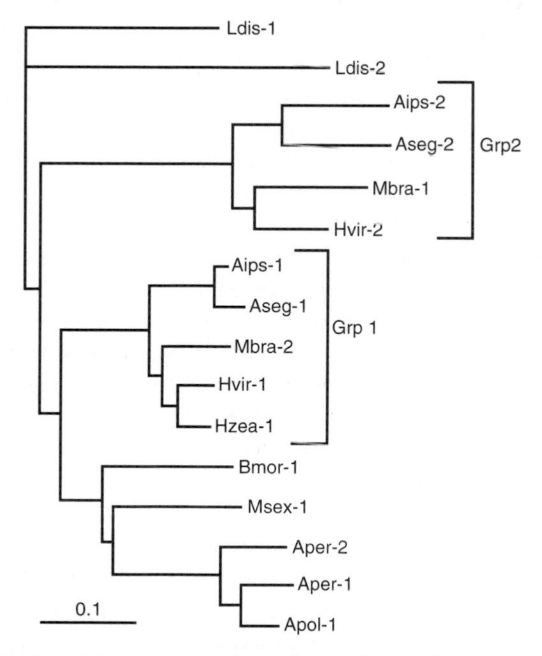

Figure 18.1 Neighbor-joining tree selected members of the PBP protein class, based on 1000 bootstrap replicates (Clustal X 1.8; Saitou and Nei, 1987; Thompson *et al.*, 1997). Relative branch lengths are indicated by the scale bar. Two groups of PBPs within Noctuidae are clearly revealed. Group 1 (Grp1): Aips-1/Aseg-1/Mbra-2/Hvir-1/Hzea-1; Group 2 (Grp2): Aips-2/Aseg-2/Mbra-1/Hvir-2.

2002; Picimbon and Gadenne, 2002). The noctuid Grp2-PBPs show about 72 percent identity between each other, about 50 percent to non-noctuid Grp2-PBPs and only 32–47 percent to other PBPs. The specific grouping of orthologous PBPs strongly suggests that different types of PBP are likely utilized by most species of moth and that within a species multiple subtypes are expressed, perhaps for binding a large repertoire of pheromone molecules.

Yponomeutidae and Noctuidae are phylogenetically distant. It is very likely that Grp2-PBPs are expressed by other Lepidoptera lineages that share common ancestry with these two Families, and that these PBPs are tuned to a pheromone structure conserved across the different member species. In contrast, the Grp1-PBPs have been reported only in Noctuidae species. However, *A. pernyi*, *Ostrinia nubilalis*, *M. sexta* and *L. dispar* use pheromone components structurally similar to the main pheromone components of Noctuidae (Klun *et al.*, 1973; Bestmann *et al.*, 1987; Tumlinson *et al.*, 1989; Gries *et al.*, 1996). Therefore, it cannot be excluded that pyralid, sphingid and lymantrid species also express Grp1 types of PBP and that these have simply not yet been identified. The PBPs so far identified in these species correspond to very specific groups of protein (Merritt *et al.*, 1998; Willett and Harrison, 1999; Robertson *et al.*, 1999; Vogt *et al.*, 1999; Picimbon and Gadenne, 2002). In particular, the two PBPs from *L. dispar*, Ldis-1 and Ldis-2, are very divergent from other moth PBPs. Interestingly, they both preferentially associate to enantiomers of disparlure, an epoxyde component used as primary pheromone by *L. dispar* (Plettner *et al.* , 2000). It could well be that in *L. dispar*, primary PBPs bind to disparlure and secondary PBPs bind to minor pheromone components.

An electrophoretic analysis of antennal proteins has failed to find secondary PBPs and only Ldis-1 and Ldis-2 appear to be detectable by a biochemical approach. Similarly, in the noctuid *M. brassicae*, only protein bands corresponding to Grp1 and Grp2 PBPs have been identified (Nagnan-Le Meillour, 1996). In the bombycid *B. mori*, only one PBP protein could be found using either a biochemical approach or homology screening of a cDNA library (Maida et al. , 1993; Krieger et al. , 1996). The failure to find additional PBPs in these species may be due to the fact that the degree of expression is markedly different for each PBP. A primary PBP tuned to high concentrations of a primary pheromone component may be expressed more than secondary binding proteins tuned to lower concentrations of secondary pheromone compounds. Analyzing the presence of Grp-1 and Grp-2 PBPs in non-noctuid insects will be a first step to investigate PBP diversity with respect to recognition of multicomponent pheromone blends.

The Grp1 and Grp2 PBPs have been identified in two closely related noctuid species, *Agrotis ipsilon* and *A. segetum*, whose females emit pheromone blends that consist of three main components and minor components that vary locally. The major pheromone component of *A. ipsilon* is (Z)-11-hexadecenyl acetate (Z11-16:Ac), while the major pheromone component in *A. segetum* is (Z)-5-

decenyl acetate (Z5-10:Ac). The antennae from *A. ipsilon* and *A. segetum* both have neurons responding to (Z)-7-dodecenyl acetate (Z7-12:Ac) and (Z)-9-tetradecenyl acetate (Z9-14:Ac), two of their major pheromone compounds (Löfstedt *et al.*, 1982; Toth *et al.*, 1992; Picimbon, 1995, 1998; Picimbon *et al.*, 1997; Gadenne *et al.*, 1997; Gemeno and Haynes, 1998; Wu *et al.*, 1999). The Grp-1 PBPs of these two species, Aips-1 and Aseg-1, are virtually identical (LaForest *et al.*, 1999; Picimbon and Gadenne, 2002). The Grp-2 PBPs from these species are more divergent: Aips-2 and Aseg-2 show only 76 percent identity. We could speculate that the conserved Aips-1/Aseg-1 proteins bind either Z7-12:Ac or Z9-14:Ac and that the more variable Aips-2 and Aseg-2 bind to Z11-16:Ac and Z5-10:Ac respectively.

18.2.3 Gene structures encoding OBPs

To explore the functions of Grp-1 and Grp-2 PBPs in depth, cutting edge protein expression, ligand binding, structural analysis and immunocytochemistry experiments are required. However, determination of the gene structures might also be informative since the diversification of species and of their pheromone systems may have led to specific gene regulations of PBP expression.

The genes encoding Aips-1, Aips-2, Aseg-1 and Aseg-2 have been characterized from genomic DNA (Abraham *et al.*, 2003; Figure 18.2). All four share the same two introns–three exons structure but differ in length. The three exons encode similar portions of the protein. The first exon (exon 1) is the shortest and has the same size in all *Agrotis* PBP genes. This exon pattern may be a general feature across the Grp-PBP genes. The first intron of these genes exhibits little variability. However, introns 2 of both *A. ipsilon* PBPs are significantly longer than the introns of the corresponding *A. segetum* PBPs. Phylogenetical analysis of the introns from the four *Agrotis* genes suggests that the Aips-1/Aseg-1 and Aips-2/Aseg-2 are respectively closely related, and that these Grp1 and Grp2 genes may have evolved from gene duplication that occurred before the divergence of *A. ipsilon* and *A. segetum.* This duplication may have occurred even earlier, before the split of Yponomeutoidea, Sphingiodea and Noctuoidea lineages as suggested by the presence of Grp2-PBPs in *Y. cagnagellus* and *M. sexta.*

In the sphingid *M. sexta*, Msex-1 is similar to the PBP from *B. mori* and divergent from the Grp2 proteins (Vogt *et al.*, 2002). The Msex-1 gene displays the same structure of three exons–two introns as the noctuid PBP genes encoding Aseg-1/Aips-1 and Aseg-2/Aips-2 as well as other non-noctuid PBP genes (Krieger *et al.*, 1991; Willett and Harrison, 1999; Willett, 2000; Abraham *et al.*, 2003). Therefore, the lepidopteran PBP genes have a conserved pattern of two introns and three exons with a variability being observed in the intron length.

In contrast, the PBP-related proteins (PBPRPs) from *Drosophila melanogaster* are highly variable with respect to exon–intron structure. The gene encoding the protein PBPRP5 has a single coding exon, while the genes encoding PBPRP1

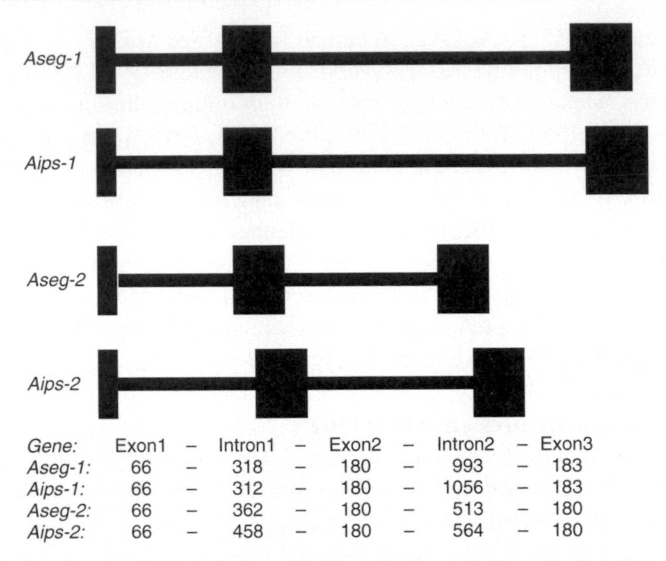

Gene:	Exon1	–	Intron1	–	Exon2	–	Intron2	–	Exon3
Aseg-1:	66	–	318	–	180	–	993	–	183
Aips-1:	66	–	312	–	180	–	1056	–	183
Aseg-2:	66	–	362	–	180	–	513	–	180
Aips-2:	66	–	458	–	180	–	564	–	180

Figure 18.2 Exon and intron boundaries of the genes encoding the PBPs from *A. ipsilon* and *A. segetum.* The numbers indicate the length of exons and introns (base pairs). The two groups of PBPs (Grp1 and Grp2) correspond to different gene structures. *Aseg-1/Aseg-2/Aips-1/Aips-2* (LaForest *et al.*, 1999; Abraham *et al.*, 2003; Picimbon and Gadenne, 2002).

and PBPRP2 have four and five coding exons, respectively. The gene encoding the OS-F protein has four exons of varying size and a very long second intron (McKenna *et al.*, 1994; Pikielny *et al.*, 1994; Hekmat-Scafe *et al.*, 1997, 2000). The genes encoding the olfactory proteins OS-E and LUSH exhibit intron–exon patterns similar to those of lepidopteran PBP genes, suggesting that OS-E and the ethanol-binding protein LUSH from *D. melanogaster* and the moth PBPs may have a common ancestor.

18.2.4 Relationships of moth and *D. melanogaster* OBPs

This hypothesis may imply the existence of OS-E and LUSH orthologs in some moth species. A protein similar to LUSH has been identified from the antennae of *A. ipsilon* (Picimbon, unpublished data). In addition, the PBPRPs from *D. melanogaster* display significant similarities to a specific subclass of moth OBPs that includes binding proteins whose function is unknown, the so-called antennal binding proteins-X (ABPX).

The ABPX proteins have highly conserved amino acid sequences across different moth species and the overall ABPX sequence displays significant similarity with DmelPBPRP1 (Figure 18.3). In particular, the proteins AipsABPX-1 and PBPRP1 share common amino acid residues including the six cysteines characteristic of OBPs and the motifs 23-TGA-25, 89-SCGTQ-93 and 99-CDTA-102. The ABPX-

```
               1         10        20        30        40        50        60
AipsABPX-1     GVVMDEDMAELARMVRESCVDETGADVKLVEAANGGADLME--DDKLKCYIKCTMETAGM-
HvirABPX-1     AVAMDEDMAELARMVRENCAAETGADVALVERVNAGADLMP--DDKLKCYIKCTMETAGM-
MsexABPX       LALEDEEQAELARMVRENCVHEIGVDEGLLAKVDDGADLMP--DPKLKCYLKCTMEMAGM-
AperABPX       VASLDGEMAELAKMIRDNCADEIGVDVTLLEQVDAGANLMP--DEKFKCYLKCTLETAGM-
BmorABPX       HGQLDDEIAELAAMVRENCADESSVDLNLVEKVNAGTDLATITDGKLKCYIKCTMETAGM-
DmelPBPRP-1    VEINPTIIKQV-RKLRMRCLNQTGASVDVIDKSVKNRILPT--DPEIKCFLYCMFDMFGLI
OBP-1          V----------R--R--C---TGA-V-----------L-T--D---KC-L-C-----G--


      70        80        90        100       110
MSDGEVDIEAVMALLPPEMAEHNGPALKSCGTQRGADDCDTAWKTQVCWQNANKAEYFLI         -118
MADGEVDIEAVLALLPPELAEHNAPSLRACGTVRGADHCDTAFRTQQCWQNANKADYFLI         -118
ISDGVVDVEAVLGLLPDDVKLRTTDIVRACDTQKGADDCDTAFLTQTCWQQANRADYIFI         -118
MSDGVVDIEIVLELLPEDLKTKNENLLRKCDTQKGSDDCDTAFLTQVCWQNGNKADYFLI         -118
MSDGVVDVEAVLSLLPDSLKTKNEASLKKCDTQKGSDDCDTAYLTQICWQAANKADYFLI         -120
DSQNIMHLEALLEVLPEEIYKTINGLVSSCGTQKGKDGCDTAYETVKCYIAVNGKFFIWEEIIVLLG  -123
-S------EA-LE-LPEE-------LV-SCGTQ-G-D-CDTA--T--C--A-N-------
```

Figure 18.3 Alignment of moth ABPX and DmelPBPRP proteins. Sources of the sequences and accession numbers are reported in Table 18.1. The amino acids conserved in ABPXs and DmelPBPRP-1 are represented in bold. Those amino acids found in DmelPBPRP-1 and ABPX or AipsABPX-1 and DmelPBPRP-1 are represented in italics. Conserved cysteines are underlined. These specific amino acids support the classification of these sequences as OBP1 type of proteins.

1s and PBPRP1 may then represent an OBP-1 type of protein defined by key residues that may underlie specific functions. The ABPX/PBPRP-specific amino acid residues arginine at position 16, lysine at position 47, proline at position 76, threonine at position 92 are replaced respectively by leucine, glycine, lysine and lysine residues that are conserved in the different types of binding protein from *B. mori* (Krieger *et al.*, 1993; Picimbon, 2001). These replacements may be relevant to support the function of ABPX. Based on the crystal structure of the bombykol–PBP complex, the tryptophane at position 101 and the valine 105 have been shown to contact the molecule of Bombykol (Sandler *et al.*, 2000). These are replaced by two threonine residues characteristic of OBP-1. Therefore, the threonine residues characteristic of OBP1s may be of crucial importance for the binding specificities of these proteins.

The notion of relatedness between ABPXs of moths and the PBPRPs of *D. melanogaster* (DmelPBPRPs) is strongly supported by the identification of additional ABPXs in the moth species *A. ipsilon*: AipsABPX-2 and AipsABPX-3. The protein ABPX2 has significant similarity to DmelPBPRP2 and DmelPBPRP5 on the basis of specific amino acids that may represent an OBP-2 group, while AipsABPX-3 appears more similar to DmelPBPRP4 and may represent an OBP-3 group (Picimbon *et al.*, unpublished). These relationships suggest the existence of a multiplicity of ABPX in moths similar to that of PBPRP in flies and may indicate a unique importance in insect olfactory behaviors. Alternatively one could speculate that ABPXs may represent "intermediary" molecules between dipteran PBPRPs and lepidopteran PBPs.

A phylogenetic analysis of insect OBPs focusing on moth ABPXs and DmelPBPRPs shows the ABPXs falling outside the groups of DmelPBPRPs, reflecting the phylogenetic distance between Lepidoptera and Diptera (Figure 18.4). The ABPXs from *A. ipsilon* cluster with the ABPX proteins from other moth species. In particular, ABPX-1s from *A. ipsilon* and *H. virescens* group with ABPXs of *B. mori*, *A. pernyi* and *M. sexta* and not with other identified noctuid PBPs. There are well-supported separations between the branches containing *A. ipsilon* proteins. One could speculate that these *A. ipsilon* genes represent multiple rearrangements or duplications of the ABPX ancestral gene and that the events which produced the ABPX diversity are common in noctuid species. Similar duplications may have occurred to produce diverse PBPRPs in *D. melanogaster* but it seems that ABPX and PBPRP genes have evolved independently, despite established similarities in their sequences.

As the PBPRP genes are quite different in sequence from the moth PBP genes, and as the *Drosophila* and moth olfactory genes diverged very early in the evolutionary course of these insects, moth PBPs in moths may have evolved for binding pheromones, while ABPX may have persisted and developed for binding more generalist odorants. Identification of the genes encoding AipsABPX and studies of the structure–activity relationships of ABPX/PBPRP need to expand the evolutionary and functional relationships between PBPRP, ABPX and PBP proteins.

18.3 The family of chemosensory proteins

18.3.1 The concept of chemosensory protein

In the context of evolution of pheromone olfaction and general chemosensation, we can speculate that general chemosensory proteins may be more highly conserved than olfactory PBPs when compared across species. Indeed, little sequence diversification would be expected to occur among genes encoding chemosensory proteins tuned to bind chemicals of common importance to all species.

In insects, a class of putative general chemosensory proteins has been described and has gained increasing interest over the last few years, expanding our understanding of the complexity of the repertoire of sensory binding proteins. The first member of this novel class of proteins was found in *D. melanogaster* and called OS-D (olfactory specific-protein type D) or A10 (McKenna *et al.*, 1994; Pikielny *et al.*, 1994); OS-D is abundant in sensory appendages and contains four cysteines. Similar proteins (Table 18.2) have been since identified in several species and variously referred to as OS-Ds, SAPs (sensory appendage proteins) or CSPs (chemosensory proteins) (Danty *et al.*, 1998; Angeli *et al.*, 1999; Picimbon and Leal, 1999; Robertson *et al.*, 1999; Picimbon *et al.*, 2000a, b, 2001; Marchese *et al.*, 2000; Nagnan-Le Meillour *et al.*, 2000). The first strong evidence that

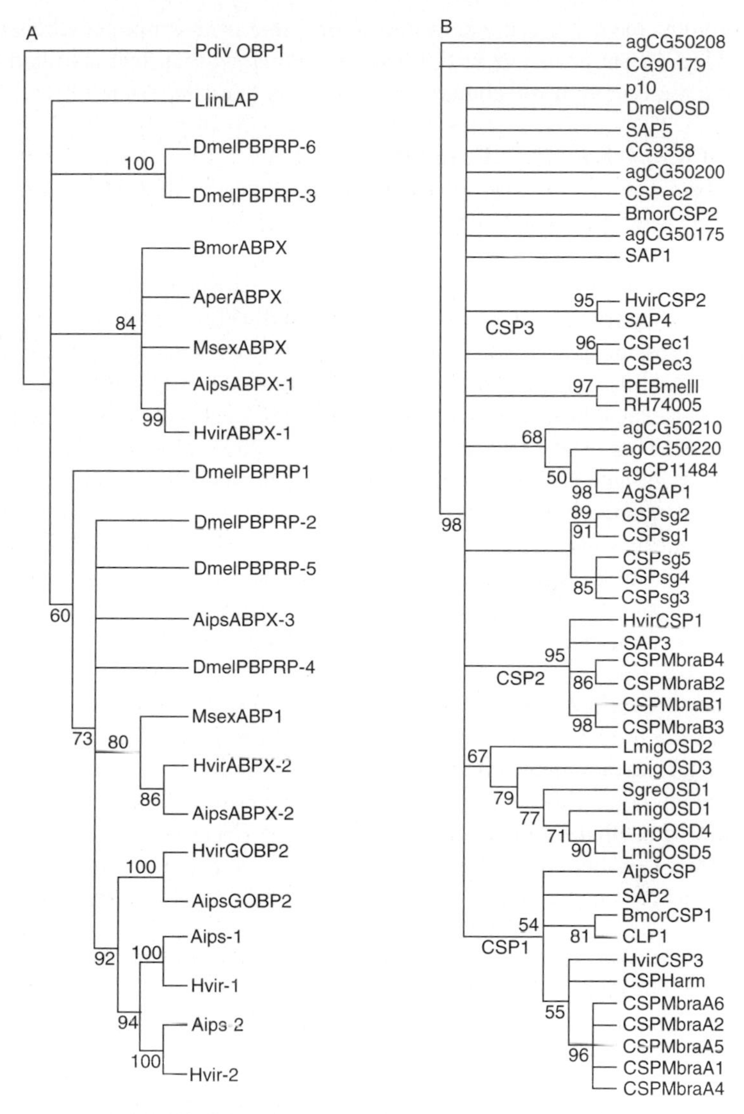

Figure 18.4 A Phylogenetic relationships of moth ABPXs and DmelPBPRPs. The primary sequences of the proteins were aligned in Clustal X 1.8 and processed using PAUP 4.0d65 (Swofford, 1999). The tree represents equally most parsimonious trees of 909 steps and consistency index 0.52. The numbers above each branch indicate the percent bootstrap support above 50 percent for the supported node using maximum parsimony (Felsenstein, 1985). The ABPX-related protein PdivOBP1 from the scarab beetle *Phylloperta diversa* (Acc. Num. BAA88061; Wojtasek *et al.*, 1999) was used as outgroup

(Contd)

these proteins have a role in chemosensation came from immunocytochemistry experiments in the grasshopper *Schistocerca gregaria* that demonstrated OS-D protein in the lymph of the contact chemosensory sensilla (Angeli *et al.*, 1999).

18.3.2 Comparison of CSPs to OBPs

On the basis of the immunocytochemical localization of CSP, Angeli *et al.* (1999) have suggested that CSPs have an OBP-like function. However, the CSPs cannot be regarded as OBPs considering the basic definition of an OBP: a lymphatic, acidic α-helical, mainly hydrophobic carrier protein of 14–16 Kds characterized by six cysteines linked by three disulfide bridges, a flexible structure and an expression pattern restricted to the olfactory sensilla of the antennae (Vogt and Riddiford, 1981; Breer *et al.*, 1992; Prestwich *et al.*, 1995; Steinbrecht, 1996; Wojtasek and Leal, 1999; Scaloni *et al.*, 1999; Leal *et al.*, 1999; Campanacci *et al.*, 1999; Sandler *et al.*, 2000; Kowcun *et al.*, 2001; Horst *et al.*, 2001). The CSPs are lymphatic acidic α-helical proteins, but (1) they have a molecular weight in the range of 12–13 Kds, (2) they show a set of only four conserved cysteines and two disulfide bridges, (3) they are highly hydrophilic, (4) they show specific solution structure, and (5) they are not specific to the antennae but also found in other parts of the insect body, more particularly in the legs (Angeli *et al.*, 1999; Picimbon *et al.*, 2000a, b, 2001; Picone *et al.*, 2001; Campanacci *et al.*, 2001b; Mosbah *et al.*, 2003).

The developmental patterns and structures of CSPs and OBPs are also different. The CSPs are produced synchronously to the shedding of the cuticle, very early during adult development, in contrast to OBPs which are produced late during adult development. This demonstrates that the chemosensory CSPs and olfactory OBPs are controlled by independent mechanisms (Vogt *et al.*, 1993; Picimbon *et al.*, 2001; Gavillet and Picimbon, 2002). Moreover, X-ray structure analysis of CSPs has revealed a novel type of α-helical fold with six helices connected

---

Figure 18.4 *(Contd)*

to root the tree. Using the well-defined PBP/GOBP clade as outgroup generated the same tree topology. B Phylogenetic analysis of the CSP protein family. An alignment of the sequences of different CSPs reported in Table 18.2 was used to determine a distance matrix and generate an unrooted tree (Clustal X 1.8; PAUP 4.0d65). This strict consensus tree is the result of trees derived using maximum parsimony of 10 000 steps and consistency index 0.52. The numbers above each branch indicate the percentage of bootstrap above 50 percent that support the branching pattern. The proteins agCG50208 and CG30172 were declared as the outgroup to build the phylogenetic tree of CSPs. Specific grouping of CSPs is observed, in particular in the CSPs from moths. Group 1 (CSP1): AipsCSP/SAP2/BmorCSP1/CLP1/HvirCSP3/ CSPHarm/ CSPMbraA; Group 2 (CSP2): HvirCSP1/SAP3/CSPMbraB; (CSP3): HvirCSP2/SAP5; SAP1; BmorCSP2; SAP5.

Table 18.2 The OS-D-related family of chemosensory proteins.

Protein	Insect	GenBANK access number	References
agCG50175	*Anopheles gambiae*	EAA12703	The *Anopheles* Genome Sequencing Consortium
agCG50200	*A. gambiae*	EAA12591	TAGSC
agCG50208	*A. gambiae*	EAA12601	TAGSC
agCG50210	*A. gambiae*	EAA12338	TAGSC
agCG50220	*A. gambiae*	EAA12353	TAGSC
agCP11484	*A. gambiae*	EAA12322	TAGSC
AgSAP1	*A. gambiae*	AAL84186	Biessmann *et al.*, 2002
AipsCSP	*Agrotis ipsilon*	AY301978	Picimbon, unpublished
BmorCSP1	*Bombyx mori*	AAM34276	Picimbon *et al.*, 2000b
BmorCSP2	*B. mori*	AAM34275	Picimbon *et al.*, 2000b
CLP1	*Cactoblastis cactorum*	AAC47827	Maleszka and Stange, 1997
DmelOS-D/A10	*Drosophila melanogaster*	AAA21358/NP524121	McKenna *et al.*, 1994; Pikielny *et al.*, 1994
CG30172	*D. melanogaster*	AAM68292	Adams *et al.*, 2000
CG9358	*D. melanogaster*	AAF47307	Adams *et al.*, 2000
PEBmeIII	*D. melanogaster*	AAA87058	Dyanov and Dzitoeva, 1995
RH74005/CG11390	*D. melanogaster*	AAM29645/AAF47140	Stapelton *et al.*, unpublished; Adams *et al.*, 2000
CSPec1	*Eurycantha calcarata*	AAD30550	Marchese *et al.*, 2000
CSPec2	*E. calcarata*	AAD30551	Marchese *et al.*, 2000
CSPec3	*E. calcarata*	AAD30552	Marchese *et al.*, 2000
CSPHarm	*Heliothis armigera*	AAK53762	Deyts *et al.*, unpublished
HvirCSP1	*Heliothis virescens*	AAM77041	Picimbon *et al.*, 2001
HvirCSP2	*H. virescens*	AAM77040	Picimbon *et al.*, 2001
HvirCSP3	*H. virescens*	AAM77042	Picimbon *et al.*, 2001
LmigOS-D1	*Locusta migratoria*	CAB65177	Picimbon *et al.*, 2000a
LmigOS-D2	*L. migratoria*	CAB65178	Picimbon *et al.*, 2000a
LmigOS-D3	*L. migratoria*	CAB65179	Picimbon *et al.*, 2000a

Table 18.2 *(Contd)*

Protein	Insect	GenBANK access number	References
LmigOS-D4	*L. migratoria*	CAB65180	Picimbon *et al.*, 2000a
LmigOS-D5	*L. migratoria*	CAB65181	Picimbon *et al.*, 2000a
SAP1	*Manduca sexta*	AAF16696	Robertson *et al.*, 1999
SAP2	*M. sexta*	AAF16714	Robertson *et al.*, 1999
SAP3	*M. sexta*	AAF16707	Robertson *et al.*, 1999
SAP4	*M. sexta*	AAF16721	Robertson *et al.*, 1999
SAP5	*M. sexta*	AAF16716	Robertson *et al.*, 1999
CSPMbraA1	*Mamestra brassicae*	AAF19647	Nagnan-Le Meillour *et al.*, 2000
CSPMbraA2	*M. brassicae*	AAF19648	Nagnan-Le Meillour *et al.*, 2000
CSPMbraA4	*M. brassicae*	AAF19650	Nagnan-Le Meillour *et al.*, 2000
CSPMbraA5	*M. brassicae*	AAF19651	Nagnan-Le Meillour *et al.*, 2000
CSPMbraA6/A3	*M. brassicae*	AAF71289, AAF19649	Nagnan-Le Meillour *et al.*, 2000; Jacquin-Joly *et al.*, 2001
CSPMbraB1	*M. brassicae*	AAF19652	Jacquin-Joly *et al.*, 2001
CSPMbraB2	*M. brassicae*	AAF19653	Jacquin-Joly *et al.*, 2001
CSPMbraB3	*M. brassicae*	AAF71290	Jacquin-Joly *et al.*, 2001
CSPMbraB4	*M. brassicae*	AAF71291	Jacquin-Joly *et al.*, 2001
p10	*Periplaneta americana*	AAB84283/AAB24286	Nomura *et al.*, 1992; Kitabayashi *et al.*, 1998
CSPsg1	*Schistocerca gregaria*	AAC25399	Angeli *et al.*, 1999
CSPsg2	*S. gregaria*	AAC25400	Angeli *et al.*, 1999
CSPsg3	*S. gregaria*	AAC25401	Angeli *et al.*, 1999
CSPsg4	*S. gregaria*	AAC25402	Angeli *et al.*, 1999
CSPsg5	*S. gregaria*	AAC25403	Angeli *et al.*, 1999
SgreOS-D1	*S. gregaria*	AY301979	Jacquemin *et al.*, unpublished

by α–α -loops and a narrow channel expanding over the protein hydrophobic core. This structural feature may confer specific binding properties to CSPs, in particular the ability to interact with long linear acyl chains of hydrophobic components (Lartigue *et al.*, 2002).

18.3.3 The repertoire of CSPs

A phylogenetic analysis of currently known insect CSPs, based on amino acid sequence comparisons, is shown in Figure 18.4b. With few exceptions, proteins from different insect Orders segregate to different branches, consistent with the phylogenetic distance between these Orders. The CSPs from moths segregate into three groups, noted as CSP1, CSP2 and CSP3; each group includes taxa from multiple lepidopteran Families. Three CSPs from *H. virescens*, which segregate to these three groups, share about 50 percent amino acid identity between each other; similar between-group identities are seen for proteins of *M. sexta* and *M. brassicae* (Robertson *et al.*, 1999; Nagnan-Le Meillour *et al.*, 2000; Picimbon *et al.*, 2001; Jacquin-Joly *et al.*, 2001). Within-group identities for the *M. brassicae* are unusually high. The CSP1 proteins CSPMbraA4 and CSPMbraA1 differ by only two residues, and are virtually identical to CSPMbraA3/A6, CSPMbraA2 and CSPMbraA5. Similarly, the CSP2 *M. brassicae* proteins MbraCSPB1 and MbraCSPB3 differ by only one amino acid. These proteins may actually be alleles representing the same locus (Picimbon *et al.*, 2000a; Nagnan-Le Meillour *et al.*, 2000; Jacquin-Joly *et al.*, 2001). The CSP1 group, and especially BmorCSP1, attracted a protein from the phasmid *E. calcarata* even though moths and phasmids are quite distant phylogenetically (Picimbon *et al.*, 2000b, 2001; Marchese *et al.*, 2000).

CSPs of the orthopteroids (*Locusta migratoria*, *S. gregaria*) are highly conserved within species and separate into species-specific groups. One CSP group per orthopteroid species is in sharp contrast with moths where CSPs from a given species fall into three groups. However, the *S. gregaria* protein SgreOS-D1 is attracted to the CSPs of *L. migratoria*, sharing 79 percent identity with the *L. migratoria* CSPs but only 55 percent identity to the other CSPs from *S. gregaria* (SgreCSPs), so perhaps there are multiple orthopteroid CSP classes that have simply not been much identified.

Analysis of the *D. melanogaster* and *A. gambiae* genomes reveals the full repertoire of CSPs within single species. *D. melanogaster* has six highly divergent CSPs; amino acid sequence identities of PEBmeIII, DmelOS-D and RH74005 range from 16 to 23 percent and from 14 to 23 percent when compared with the moth proteins. DmelOS-D is somewhat more similar, sharing about 45 percent sequence identity with the moth proteins. Four of the seven *A. gambiae* CSPs cluster in a single group and share about 72 percent identity, but the others segregate by themselves. Although *A. gambiae* and *D. melanogaster* are both Diptera, only a single pair of CSPs share significant similarity to be attracted together (agCG5020865 and CG30172, 65 percent identity); these may be orthologs. Overall, the divergence seen in CSP sequences is intriguing; divergent CSPs may have different and specific binding properties for distinct chemical ligands.

18.3.4 Structural and binding properties of CSPs

Differences in the amino acid sequences of CSPs presumably relate to differences in the ligand-binding properties of each protein (Figure 18.5). The type-1 and 2 CSPs of moths (CSP1 and CSP2) are proteins with about 107–112 amino acids and all exhibit the diagnostic elements of CSP: aspartatic acid 6 and 88, lysine 43, glutamine 61, proline 89 and four cysteines at positions 29, 36, 54 and 57. Overall, the CSP1 proteins are highly conserved, and most have the sequence

(1) -D-YTDKYD----EIL-N-RLL--Y--CV--GKC--EGKELK--L--A--GC-KC--QE-G----I--LIKN--W--L----DP--WR-KYEDRA-A-GI-IP--(110).

Six isoforms (alleles?) of CSPMbraA proteins, all CSP1s, differ at only three sites, 59, 69 and 92 (Ala/Thr, Ala/Val and Val/Gly). The CSPMbraA proteins differ from other CSP1 proteins in being one amino acid longer (Glu) and have the C-terminal sequence DRAKAAGIVIPEE (Nagnan-Le Meillour *et al.*, 2000; Jacquin-Joly *et al.*, 2001). The three-dimensional structure has been determined for CSPMbraA6. This CSP protein binds aliphatic molecules with 12–18 carbons, suggesting that CSPs generally bind hydrophobic compounds (Lartigue *et al.*,

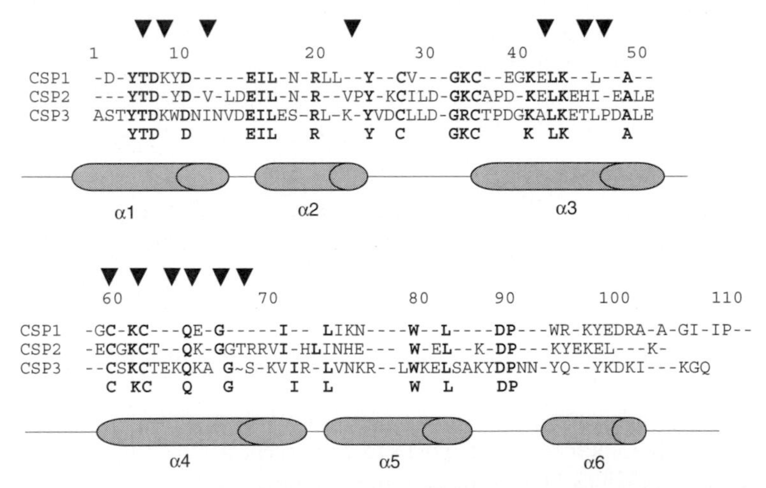

Figure 18.5 Structural properties of moth CSPs. The amino acid residues conserved in these three types of CSP are shown in bold. Black triangles indicate residues in contact with water molecules in the channel structure of CSP. The position of the six α-helices characteristic of CSP are indicated (α1 to α 6).

2002). Thus, multiple subtypes of CSP may mediate transport of carbon chains of different lengths. If CSP1s bind specific hydrophobic chains, the large repertoire of proteins may bind the large number of diverse alkyl chain components. We could hypothesize that a small CSP1a protein would bind to a C12 molecule, while the longer CSP1b protein would bind to molecules longer than C12.

Other types of CSP may have totally different binding properties. The CSP2s are characterized by diagnostic CSP residues and by other conserved amino acids that may underlie specific functions. Overall, the CSP2 proteins are highly conserved, and most have the sequence

(1) ----YTD-YD-V--LDEIL-N-R--VPY-KCILD-GKCAPD-KELKEHI-EALE-ECGKCT---QK-GGTRRVI-HLINHE---W-EL---K-DP---K---KYEKEL---K-(108).

The amino acid pattern (1) ----YTD-D----EIL-N-R----Y---C----GKC----KELK------C-KC---Q---G----------N----W---L----DP------K------ (106) is conserved in CSP1 and CSP2. CSP3s are so far only represented by the proteins HvirCSP2 and SAP4 which share 84 percent sequence identity but are only about 50 percent identical to other CSP1 or CSP2. These CSP3s have the sequence

(1)ASTYTDKWDNINVDEILESXRLXKXYVDCLLDXGRCTPDGKALKETLPDALEXXCSKCTEKQKAGS-KVIR-LVNKR--LWKELSAKYDPNN-YQ--YKDKIXXXKQG- (106/107)

The α-helical regions of the moth CSP1, CSP2 and CSP3 proteins differ in amino acid sequence. The α-helix (α1) in the N-terminal region anchors a narrow hydrophobic channel and includes residues 5 (or 12)–18; α1 is conserved in all three CSP groups with the sequence YTD--D-----EIL (Figure 18.5). In CSP1s, α1 interacts with residues Tyr 8 and Glu 39 to stabilize the hydrophilic channel; however, CSP2s lack Glu 39. And in CSP3s, a non-polar residue Trp replaces the polar, uncharged Tyr 8. Residue Glu 39 of CSP1s is replaced by Asp in CSP2s and CSP3s. In contrast, the α-helix (α6) in the C-terminal region is highly variable and may support ligand specificity. Furthermore, of 13 residues of CSPMbraA6 that have been shown to interact with water molecules, only six are conserved in CSP1s (Asp 9, Tyr 26, Leu 43, Leu 47, Glu 63 and Gly 65) and only the residues Asp 9, Tyr 26 and Leu 43 are found in all three types of moth CSP, while the other residues are highly variable (Figure 18.5). Lartigue *et al.* (2002) suggested that Tyr 26 may rotate towards the protein surface upon ligand binding, and that the previous position of its hydroxyl group may then be occupied by the side chain of Leu 43. Since both residues Tyr 26 and Leu 43 are highly conserved in all insect CSPs, this mechanism may be common for all CSPs.

Two tryptophan residues from MbraA6 were proposed to interact with ligands; these are conserved in all CSP1s, but only one (Trp 82) is conserved throughout the CSPs, suggesting that different CSPs may bind different ligands. The overall sequence similarities of CSP1s, CSP2s and CSP3s suggest that they all bind to lipid compounds. However, the diversity of CSPs not only in moths but also in other insect species, suggests that CSPs transport diverse types of chemicals. The CSPs of *D. melanogaster*, *A. gambiae*, *B. mori* and *E. calcarata* are highly variable, sharing only 40 percent sequence identity (Picimbon *et al.*, 2000b; Marchese *et al.*, 2000; this chapter). The CSPs from *M. sexta* (SAP1 and SAP5) are also very divergent (Robertson *et al.*, 1999; this chapter). Different types of CSP with multiple isoforms are also found in the acridid species *L. migratoria* and *S. gregaria* (Angeli *et al.*, 1999; Picimbon *et al.*, 2000a; Picimbon, 2001; see Figure 18.5).

Analyzing the cysteine arrangements in a CSP from *S. gregaria*, Angeli *et al.* have shown that the two adjacent cysteines form two small loops along the main protein chain in a fashion similar to thioredoxins. Thus, CSP may well play a role not only in binding lipid molecules but also in CO_2 sensing (Bogner *et al.*, 1986; Stange, 1992, 1996; Maleszka and Stange, 1997; Angeli *et al.*, 1999).

18.3.5 Multiple functions of CSPs

18.3.5.1 Tissue distribution

CSPs are found not only in external sensory organs, but also in external and internal non-sensory tissues, further suggesting the proteins may have diverse functions. In moths the distributions of CSPs have been characterized by Northern blot analysis. A DIG-RNA probe encoding the BmorCSP1 hybridized not only with mRNAs from male and female antennae but also with mRNAs from legs and other parts of the insect body. Northern blot analysis using BmorCSP2 showed a very similar distribution pattern with no tissue specificity, as did the three CSPs of *H. virescens* which showed high levels of expression in legs (Picimbon *et al.*, 2000a, 2001). These results are consistent with the isolation of CSP cDNAs and proteins. The CSP of *C. cactorum* was as a cDNA from labial palps. The CSPs of *M. brassicae* were isolated as cDNAs from antennae and pheromone glands, and as proteins (N-terminal sequences) from antennae, proboscis and legs (Malezska and Stange, 1997; Bobhot *et al.*, 1998; Nagnan-Le Meillour *et al.*, 2000; Jacquin-Joly *et al.*, 2001a).

CSPs are also broadly distributed in tissues of insects other than moths. In *Drosophila melanogaster*, OS-D, also called A10, is extremely abundant in the antennal appendages, and the CSP EBSP-III is expressed in the ejaculatory bulb (McKenna *et al.*, 1994; Pickielny *et al.*, 1995; Dyanov and Dzitoeva, 1995). In locusts, CSPs are expressed in male and female from adults as well as 5th instar larvae, and are found in many different tissues including antennae, legs, mouth organs, thorax, abdomen, head and wings (Picimbon *et al.*, 2000b; in preparation).

In the phasmid *Eurycantha calcarata*, a CSP has been isolated from the cellular layer underlying the cuticle (Marchese *et al.*, 2000). In the cockroach *Periplaneta americana*, the CSP p10 is expressed in legs and antennae (Nomura *et al.*, 1992; Kitabayashi *et al.*, 1998; Picimbon *et al.*, 2001), and other CSPs have been detected in tissues including legs, brain and cerci (Picimbon and Leal, 1999; unpublished).

Thus, in various insect species, the expression of CSP occurs in many different tissues, especially in the legs and contact sensory organs. The differences between CSPs and OBPs in tissue distribution and developmental expression suggest that the roles of these proteins in chemoreception may also be different. CSPs may have a broader function than OBPs, functioning in more systems than olfaction and taste, perhaps as general molecule carriers but especially involved in the transport of contact sensory molecules.

18.3.5.2 Role in contact chemosensation

Immunocytochemistry experiments performed with rabbit antiserum generated against a CSP isoform from *Schistocerca gregaria* (sgCSP-5) showed selective labeling of the outer lymph of the peg lumen and in the cavity below the peg base of contact sensilla from the antennae and mouth organs, such as maxillary palps and tarsi. Olfactory sensilla were not labeled (Angeli *et al.*, 1999). This pioneer work showed the localization of CSPs in the lymphatic space surrounding the contact chemosensory receptor neurons and suggested an OBP-like function for CSPs in contact chemosensation. In the context of OBP-like function, the CSPs could well mediate the clearance of hydrophobic odorant molecules absorbed haphazardly by the cuticle as well as the delivery of these odorant molecules to degradative enzymes (Ferkovitch *et al.*, 1982; Lonergan, 1986; Prestwich *et al.*, 1989). Clearance and degradation of odorant molecules has to occur, e.g. while the insect flies in a pheromone plume or feeds on nectar. More data regarding the specific localization of CSPs in chemosensory and non-chemosensory organs would provide a better understanding of the OBP-like function of these proteins.

18.3.5.3 Function and evolutionary history

In the cockroach *P. americana*, CSP have been found that differentially express between the sexes, suggesting a role in delivering conspecific pheromones to olfactory neurons (Picimbon and Leal, 1999). In the phasmid *Carausius morosius*, a CSP is differentially expressed in only a subset of olfactory sensilla (Monteforti *et al.*, 2002). Multiple CSPs have been detected in phasmids, locusts and cockroaches, but no proteins related to known PBPs have been found (Tuccini *et al.*, 1996; Mameli *et al.*, 1996; Picimbon and Leal, 1999; Picimbon *et al.*, unpublished). These observations do not exclude the presence of RNA encoding PBPs in these insects, but RNA may not be necessarily translated into proteins (Segal *et al.*, 2001). Since Phasmatodea, Acridoidea and Blattoidea are among the most primitive insect Orders, one could hypothesize that CSPs perform PBP functions in ancient insects. Later in evolution, CSPs may have developed into general chemosensory proteins in flies and moths, where their original function was replaced by more efficient pheromone-binding proteins. This hypothesis could be tested by analyzing the diversity of CSP genes across a variety of insect species and by comparing the CSP and OBP genes.

The genes underlying olfaction and general chemosensation are, indeed, most likely under different evolutionary pressures that may act selectively through defined regulatory elements. Specific intron structures, like the intron 2 of PBPs, may be key targets for regulatory mechanisms that control the differential expression or loss of specific OBPs and CSPs. What gene structures will be found in the various insect species that use different pheromone and chemosensory components?

18.4 Concluding remarks and perspectives

Binding of hydrophobic molecules by specific protein carriers appears to be a very efficient mechanism to increase both solubility and transport of these molecular messengers in a hydrophilic medium. OBPs and CSPs may represent a successful application of this principle. In particular, the molecular mechanisms of transport of hydrophobic molecules may be more ancient than that most ancient of senses, olfaction. The olfactory system may have developed to extract the hydrophobic odorants from the air environment and optimize their transport and delivery to sensory cells.

In recent years, studies of diverse insect species have revealed the heterogeneity of the family of binding proteins, and thus challenged the dogmatic concept that odorant-binding proteins are expressed only in olfactory structures. In particular, an increasing number of binding proteins related to OBPs have been identified in various non-sensory organs, such as the hemolymph, brain, accessory and salivary glands (TH12 proteins, sericotropin, B proteins and D7 gene products) (Paesen and Happ, 1995; Kõdrick *et al.*, 1995; Thymianou *et al.*, 1998; Rothemund *et al.*, 1999; Graham *et al.*, 2001; Arca *et al.*, 2002). Such distributions suggest that this large family of carrier proteins may perform diverse functions throughout the insect body, paralleling the distribution and functional breadth of lipocalins from vertebrates (Flower, 2000; Ganfornica *et al.*, 2000). OBPs may have arisen as general transporters of hydrophobic molecules, and later developed to bind and solubilize odorant molecules. The CSPs, which are expressed all over the body, have certainly conserved a functional *polyvalence*. In contrast, the specific expression of many OBPs in the antennae strongly support an adaptation of these proteins to the reception of hydrophobic odorants.

Efforts should be made to utilize the most modern techniques to analyze the binding properties and tissue specificity of all identified proteins, and eventually to rename the proteins on the basis of specific groupings and conserved motifs of amino acid. This suggestion is intended to be neither provocative nor inflammatory, but certainly the most reasonable way to define function of diverse binding proteins with respect to the pheromone systems and life history of the different insect species.

Genetic studies will undoubtedly lead to the elucidation of how specific transporter molecules have developed to a fine-tuned function in olfaction. Specific and non-specific transporter molecules might have evolved differently over the course of evolutionary history and the diversification of species. If there is one thing future research should accomplish, it would be the unveiling of the evolution of the families of CSPs and OBPs. In addition, it must be considered that other families of binding proteins may well exist and participate to the reception of the extremely large repertoire of odor and chemosensory molecules. Investigations into these matters would have a strong impact not only in fundamental genetics underlying chemosensation and olfaction but also in applied industry, assuming that gene manipulation is accessible and permits control of the expression of OBP and CSP, and thereby control of the sensory abilities of insect pests.

Acknowledgements

My heartfelt thanks to Professors R. G. Vogt and P. Pelosi who discovered the sensory-binding proteins simultaneously in insects and vertebrates.

References

Abraham D., Löfstedt C. and Picimbon J. F. (2003) Molecular evolution and gene characterization of Grp1 and Grp 2 Pheromone Binding Proteins in moths. *Genetics* (manuscript submitted for publication).

Adams M. D., Celniker S. E., Holt R. A. *et al.* (2000) The genome sequence of *Drosophila melanogaster*. *Science* **287**, 2185–2195.

Angeli S., Ceron F., Scaloni A., Monti M., Monteforti G., Minnocci A., Petacchi R. and Pelosi P. (1999) Purification, structural characterization, cloning and immunocytochemical localization of chemoreception proteins from *Schistocerca gregaria*. *Eur. J. Biochem.* **262**, 745–754.

Arca B., Lombardo F., Lanfrancotti A., Spanos L., Veneri M., Louis C. and Coluzzi M. (2002) A cluster of four *D7-related* genes is expressed in the salivary glands of the African malaria vector *Anopheles gambiae*. *Insect Mol. Biol.* **11**, 47–55.

Attygale A. B., Herrig M., Vostrowsky O. and Bestmann H. J. (1987) Technique for injecting intact glands for analysis of sex pheromones of Lepidoptera by capillary gas chromatography: reinvestigation of pheromone complex of *Mamestra brassicae*. *J. Chem. Ecol.* **13**, 1299–1311.

Bestmann H. J., Attygale A. B., Brosche T., Erler J., Platz H., Schwarz J., Vostrowsky O., Cai-Hong W., Kaissling K. E. and Te-Ming C. Z. (1987) Identification of three sex pheromone components of the female Saturniid moth *Antheraea pernyi* (Lepidoptera: Saturniidae). *Z. Naturforsch.* **42c**, 631–636.

Bette S., Breer H. and Krieger J. (2002) Probing a pheromone binding protein of the silkworm moth *Antheraea polyphemus* by endogenous tryptophan fluorescence. *Insect Biochem. Mol. Biol.* **32**, 241–246.

Biessmann H., Walter M. F., Dimitratos S. and Woods D. (2002) Isolation of cDNA clones encoding putative odourant binding proteins from the antennae of the malaria-transmitting mosquito, *Anopheles gambiae. Insect Mol. Biol.* **11**, 123–132.

Bogner F., Boppre M., Ernst K. D. and Boeckh J. (1986) CO_2-sensitive receptors on labial palps of *Rhodogastria* moths (Lepidoptera: Arctiidae): physiology, fine structure and projection. *J. Comp. Physiol. A* **158**, 741–749.

Bohbot J., Sobrio F., Lucas P. and Nagnan-Le Meillour P. (1998) Functional characterization of a new class of Odorant Binding Proteins in the moth *Mamestra brassicae. Biochem. Biophys. Res. Commun.* **253**, 489–494.

Breer H., Boekhoff I., Krieger J., Raming K., Strotmann J. and Tareilus E. (1992) Molecular mechanism of olfactory signal transduction. In: *Sensory Transduction*, eds D. P. Corey and S. D., Roper, pp. 94–108. The Rockfeller University Press, New York.

Campanacci V., Longhi S., Nagnan-Le Meillour P., Cambillau C. and Tegoni M. (1999) Recombinant pheromone binding protein 1 from *Mamestra brassicae* (MbraPBP1). Functional and structural characterization. *Eur. J. Biochem.* **264**, 707–716.

Campanacci V., Krieger J., Bette S., Sturgis J.N., Lartigue A., Cambillau C., Breer H. and Tegoni M. (2001a) Revisiting the specificity of *Mamestra brassicae* and *Antheraea polyphemus* pheromone binding proteins with a fluorescence binding assay. *J. Biol. Chem.* **276**, 20078–20084.

Campanacci V., Mosbah A., Bornet O., Wechselberger R., Jacquin-Joly E., Cambillau C. Darbon H. and Tegoni M. (2001b) Chemosensory protein (CSP2) from the moth *Mamestra brassicae*: expression and secondary structure from 1H and 15N NMR. *Eur. J. Biochem.* **268**, 4731–4739.

Danty E., Arnold G., Huet J.-C., Masson C. and Pernollet J.-C. (1998) Separation, characterization and sexual heterogeneity of multiple putative odorant-binding proteins in the honey bee *Apis mellifera* L. (Hymenoptera: Apidea). *Chem. Senses* **23**, 83–91.

Du G., Ng C. S. and Prestwich G. D. (1994) Odorant binding by a pheromone binding protein-active site mapping by photoaffinity labeling. *Biochemistry* **33**, 4812–4819.

Du G. and Prestwich G. D. (1995) Protein structure encodes the ligand binding specificity in pheromone binding proteins. *Biochemistry* **34**, 8726–8732.

Dyanov H. M. and Dzitoeva S. G. (1995) Method for attachment of microscopic preparations on glass for *in situ* hybridization, PRINS and in situ PCR studies. *BioTechniques* **18**, 822–824.

Feixas J., Prestwich G. and Guerrero A. (1995) Ligand specificity of pheromone binding proteins of the processionary moth. *Eur. J. Biochem.* **234**, 521–526.

Felsenstein J. (1985) Confidence limits on phylogenies: an approach using the bootstrap. *Evolution* **39**, 783–791.

Feng L. and Prestwich G. D. (1997) Expression and characterization of a lepidopteran general odorant binding protein. *Insect Biochem. Mol. Biol.* **27**, 405–412.

Ferkovich S.M., Oliver J.E. and Dillard C. (1982) Pheromone hydrolysis by cuticular and interior esterases of the antennae, legs, and wings of the cabbage looper moth *Trichoplusia ni* (Hubner). *J. Chem. Ecol.* **8**, 859–866.

Flower D.R. (2000) Beyond the superfamily: the lipocalin receptors. *Biochem. Biophys. Acta* **18**, 327–336.

Gadenne C., Picimbon J.F., Bécard J.M., Lalanne-Cassou B. and Renou M. (1997) Development and pheromone communication systems in hybrids of *Agrotis ipsilon* and *Agrotis segetum* (Lepidoptera: Noctuidae). *J. Chem. Ecol.* **23**, 191–209.

Ganfornica M. D., Guttierrez G., Bastiani M., Sanchez D. (2000) A phylogenetical analysis of the lipocalin protein family. *Mol. Biol. Evol.* **27**, 405–412.
Gavillet B. and Picimbon J. F. (2002) Endocrine regulation of olfaction and chemosensation. *Proc. Eur. Comp. Endocrinol.* **C826**, 463–468.
Gemeno C. and Haynes K. (1998) Chemical and behavioral evidence for a third pheromone component in a North American population of the black cutworm moth, *Agrotis ipsilon. J. Chem. Ecol.* **26**, 329–342.
Graham L. A., Tang W., Baust J. G., Liou Y.-C., Reid and T. S., Davies P. L. (2001) Characterization and cloning of a *Tenebrio molitor* hemolymph protein with sequence similarity to insect odorant-binding proteins. *Insect Biochem. Mol. Biol.* **31**, 691–702.
Gries G., Gries R., Khashin G., Slessor K. N., Grant G. G., Liska J. and Kapitola P. (1996) Specificity of nun and gypsy moth sexual communication through multiple-component pheromone blends. *Naturwissenschaften* **83**, 382–385.
Györgyi T. K., Roby-Shemkovitz A. J. and Lerner M. R. (1988) Characterization and cDNA cloning of the pheromone binding protein from the tobacco hornworm *Manduca sexta*: a tissue specific, developmentally regulated protein. *Proc. Natl. Acad. Sci. USA* **85**, 9851–9855.
Hekmat-Scafe D. S., Steinbrecht R. A. and Carlson J. R. (1997) Coexpression of two odorant-binding protein homologs in *Drosophila*: implications for olfactory coding. *J. Neurosci.* **17**, 1616–1624.
Hekmat-Scafe D. S., Dorit R. L and Carlson J. R. (2000) Molecular evolution of odorant-binding protein genes *OS-E* and *OS-F* in *Drosophila. Genetics* **155**, 117–127.
Horst R., Damberger F., Luginbühl P., Güntert P., Peng G., Nikonova L., Leal W.S. and Wüthrich K. (2001) NMR structure reveals intramolecular regulation mechanism for pheromone binding and release. *Proc. Natl. Acad. Sci. USA* **98**, 14374–14379.
Jacquin-Joly E., Vogt R. G., Francois M. C. and Nagnan-Le Meillour P. (2001) Functional and expression pattern analysis of chemosensory proteins expressed in the antennae and pheromonal gland of *Mamestra brassicae. Chem. Senses* **26**, 833–844.
Kaissling K. E., Klein U., De Kramer J. J., Keil T. A., Kanaujia S. and Hemberger J. (1985) Insect olfactory cells: electrophysiological and biochemical studies. In *Molecular Basis of Nerve Activity* (Proc. Int. Symp. in Memory of D. Nachmansohn, Oct. 1984), eds J. P. Changeux, F. Hucho, E. Maelicke and E. Neumann, pp. 173–183. De Gruyter, Berlin.
Keil T. A. (1997) Comparative morphogenesis of sensilla: a review. *Int. J. Insect Morphol. Embryol.* **26**, 151–160.
Kim M. S., Repp A. and Smith D. P. (1998) LUSH Odorant-Binding Protein mediates chemosensory responses to alcohols in *Drosophila melanogaster. Genetics* **150**, 711–721.
Kitabayashi A. N., Arai T., Kubo T. and Natori S. (1998) Molecular cloning of cDNA for p10, a novel protein that increases in the regenerating legs of *Periplaneta americana* (American cockroach). *Insect Biochem. Mol. Biol.* **28**, 785–790.
Klusak V., Havlas Z., Rulisek L., Vondrasek J. and Svatos A. (2003) Sexual attraction in the silkworm moth: nature of binding of Bombykol in pheromone binding protein-an *Ab Initio* study. *Chem. Biol.* **10**, 331–340.
Klun J. A., Chapman O. L., Mattes K. C., Wojtkowski P. W., Beroza M. and Sonnet P. E. (1973) Insect sex pheromones: minor amounts of opposite geometrical isomer critical to sex attraction. *Science* **162**, 661–663.
Kõdrick D., Filippov V. A., Filippova M. A. and Sehnal F. (1995) Sericotropin: an insect neurohormonal factor affecting RNA transcription. *J. Zool.* **45**, 68–70.
Kowcun A., Honson N. and Plettner E. (2001) Olfaction in the gypsy moth, *Lymantria dispar*: effect of pH, ionic strength, and reductants on pheromone transport by pheromone-binding proteins. *J. Biol. Chem.* **276**, 44770–44776.

Krieger J., Raming K. and Breer H. (1991) Cloning of genomic and complementary DNA encoding insect pheromone binding proteins: evidence for microdiversity. *Biochim. Biophys. Acta* **1088**, 277–284.
Krieger J., Ganssle K., Raming K. and Breer H. (1993) Odorant binding proteins of *Heliothis virescens. Insect Biochem. Mol. Biol.* **23**, 449–456.
Krieger J., von Nickisch-Roseneck E. V., Mameli M., Pelosi P. and Breer H. (1996) Binding proteins from the antennae of *Bombyx mori. Insect Biochem. Mol. Biol.* **26**, 297–307.
Krieger J., Mameli M. and Breer H. (1997) Elements of the olfactory signaling pathways in insect antennae. *Invert. Neurosci.* **3**, 137–144.
Koganezawa M. and Shimada I. (2002) Novel odorant-binding proteins expressed in the taste tissue of the fly. *Chem. Senses* **27**, 319–332.
LaForest S. Prestwich G. D. and Löfstedt C. (1999) Intraspecific nucleotide variation at the Pheromone Binding Protein locus in the turnip moth, *Agrotis segetum. Insect Mol. Biol.* **8**, 481–490.
Lartigue A., Campanacci V., Roussel A., Larsson A. M., Jones T. A., Tegoni M. and Cambillau C. (2002) X-ray structure and ligand binding study of a moth chemosensory protein. *J. Biol. Chem.* **277**, 32094–32098.
Laue M., Steinbrecht R. A. and Ziegelberger G. (1994) Immunocytochemical localization of general odorant binding protein in olfactory sensilla of the silkworm *Antheraea polyphemus. Naturwissenschaften* **81**, 178–180.
Leal W. S., Nikonova L. and Peng G. (1999) Disulfide structure of the pheromone binding protein from the silkworm moth, *Bombyx mori. FEBS Lett.* **464**, 85–90.
Löfstedt C., Löfsqvist J., Lanne B. S., Van Der Pers J. C. N. and Hansson B. S. (1982) Sex pheromone components of the turnip moth, *Agrotis segetum*, chemical identification, electrophysiological evaluation and behavioral activity. *J. Chem. Ecol.* **8**, 1305–1321.
Lonergan G. C. (1986) Metabolism of pheromone components and analogs by cuticular enzymes of *Choristoneura fumiferana. J. Chem. Senses* **12**, 483–496.
Maïbèche-Coisné M., Sobrio F., Delaunay T., Lettere M., Dubroca J., Jacquin-Joly E. and Nagnan-Le Meillour P. (1997) Pheromone binding proteins of the moth *Mamestra brassicae*: specificity of ligand binding. *Insect Biochem. Mol. Biol.* **27**, 213–221.
Maïbèche-Coisné M., Jacquin-Joly E., Francois M. C. and Nagnan-Le Meillour P. (1998) Molecular cloning of two Pheromone Binding Proteins in *Mamestra brassicae. Insect Biochem. Mol. Biol.* **28**, 815–818.
Maida R., Steinbrecht A., Ziegelberger G. and Pelosi P. (1993) The pheromone binding protein of *Bombyx mori*: purification, characterization and immunocytochemical localization. *Insect Biochem. Mol. Biol.* **23**, 243–253.
Maida R., Krieger J., Gebauer T., Lange U. and Ziegelberger G. (2000) Three pheromone-binding proteins in olfactory sensilla of the two silkmoth species *Antheraea polyphemus* and *Antheraea pernyi. Eur. J. Biochem.* **267**, 2899–2908.
Maleszka R. and Stange G. (1997) Molecular cloning by a novel approach, of a cDNA encoding a putative olfactory protein in the labial palps of the moth *Cactoblastis cactorum. Gene* **202**, 39–43.
Mameli M., Tuccini A., Mazza M., Petacchi R. and Pelosi P. (1996) Soluble proteins in chemosensory organs of Phasmids. *Insect Biochem. Mol. Biol.* **26**, 875–882.
Marchese S., Angeli S., Andolfo A., Scaloni A., Brandazza A., Mazza M., Picimbon J.F., Leal W. S. and Pelosi P. (2000) Soluble proteins from chemosensory organs of *Eurycantha calcarata* (Insecta, Phasmatodea). *Insect Biochem. Mol. Biol.* **30**, 1091–1098.
McKenna M. P., Hekmat-Scafe D. S., Gaines P. and Carlson J. R. (1994) Putative *Drosophila* pheromone-binding-proteins expressed in a subregion of the olfactory system. *J. Biol. Chem.* **269**, 16340–16347.

Merritt T. J. S., Laforest S., Prestwich G. D., Quattro J. M. and Vogt R. G. (1998) Patterns of gene duplication in lepidopteran pheromone binding protein. *J. Mol. Evol.* **46**, 272–276.

Monteforti G., Angeli S., Petacchi R. and Minnocci A. (2002) Ultrastructural characterization of antennal sensilla and immunocytochemical localization of a chemosensory protein in *Carausius morosus* Bruner (Phasmida: Phasmatidae). *Arthrop. Struct. Dev.* **30**, 195–205.

Nagnan-Le Meillour P., Huet J. C. Maïbèche M., Pernollet J. C. and Descoins C. (1996) Purification and characterization of multiple forms of odorant/pheromone binding proteins in the antennae of *Mamestrae brassicae* (Noctuidae). *Insect Biochem. Mol. Biol.* **26**, 59–67.

Nagnan-Le Meillour P., Cain A. H., Jacquin-Joly E., Francois M. C., Ramashadran S., Maida R. and Steinbrecht R. A. (2000) Chemo-sensory proteins from the proboscis of *Mamestra brassicae. Chem. Senses* **25**, 541–553.

Nomura A., Kawasaki K., Kubo T. and Natori S. (1992) Purification and localization of p10, a novel protein that increases in nymphal regenerating legs of *Periplaneta americana* (American cockroach). *Int. J. Dev. Biol.* **36**, 391–398.

Ozaki M., Morizaki K., Idei W., Ozaki K. and Tokunaga F. (1995) A putative lipophilic stimulant carrier protein commonly found in the taste and olfactory systems. A unique member of the pheromone binding protein superfamily. *Eur. J. Biochem.* **230**, 298–308.

Paesen G. C. and Happ G. M. (1995) The B proteins secreted by the tubular accessory sex glands of the male mealworm beetle, *Tenebrio molito*, have sequence similarity to moth pheromone-binding proteins. *Insect Biochem. Mol. Biol.* **25**, 401–408.

Park S.-K., Shanbhag S., Wang Q., Yu P., Hasan G., Steinbrecht A. and Pikielny C. W. (2002) Inactivation of olfactory sensilla of a single morphological type differentially affects the response of *Drosophila* to odors. *J. Neurobiol.* **51**, 248–260.

Picimbon J. F. (1995) Sex pheromones of moths: determination, regulation and perception. PhD dissertation, University of Aix-Marseille I, France, pp. 213.

Picimbon J. F. (1998) Sex pheromone of the French black cutworm moth, *Agrotis ipsilon,* identification and regulation by the Pheromone Biosynthesis Activating Neuropeptide (PBAN) and juvenile Hormone (JH). *Misc. Publ. Natl. Inst. Seric. Entomol. Sci.* **23,** 104–105.

Picimbon J. F. (2001). Les protéines liant les odeurs (OBPs) et les protéines chimiosensorielles (CSPs): cibles moléculaires de la lutte intégrée. In: *Biopesticides d'Origine Végétale*, eds B. Philogène, C. Regnault-Roger. and C. Vincent, pp. 265–284, Lavoisier Tech and Doc, Paris.

Picimbon J. F. (2002) Les périrécepteurs chimiosensoriels des insectes. *Med. Sci.* **18**, 1089–1094.

Picimbon J. F. and Gadenne C. (2002) Evolution and noctuid pheromone binding proteins: identification of PBP in the black cutworm moth, *Agrotis ipsilon. Insect Biochem. Mol. Biol.* **32**, 839–846.

Picimbon J. F. and Leal W. S. (1999) Olfactory soluble proteins of cockroaches. *Insect Biochem. Molec. Biol.* **29**, 973–978.

Picimbon J. F., Gadenne C., Becard J. M., Clement J. L and Sreng L. (1997) Sex pheromone of the French black cutworm moth *Agrotis ipsilon* (Lepidoptera: Noctuidae): identifcation and regulation of a multicomponent blend. *J. Chem. Ecol.* **23**, 211–230.

Picimbon J. F., Dietrich K., Breer H. and Krieger J. (2000a) Chemosensory proteins of *Locusta migratoria* (Orthoptera, Acriididae). *Insect Biochem. Mol. Biol.* **30**, 233–241.

Picimbon J. F., Dietrich K., Angeli S., Scaloni A., Krieger J., Breer H. and Pelosi P. (2000b) Purification and molecular cloning of chemosensory proteins from *Bombyx mori. Arch. Insect Biochem. Physiol.* **44**, 120–129.

Picimbon J. F., Dietrich K., Krieger J. and Breer H. (2001) Identity and expression pattern of chemosensory proteins in *Heltiothis virescens. Insect Biochem. Mol. Biol.* **31**, 1173–1181.

Picone D., Crescenzi O., Angeli S., Marchese S., Brandazza A., Pelosi P. and Scaloni A. (2001) Bacterial expression and conformational analysis of a chemosensory protein from *Schistocerca gregaria. Eur. J. Biochem.* **268**, 4794–4801.

Pikielny C. W., Hasan G., Rouyer F. and Rosbach M. (1994) Members of a family of *Drosophila* putative odorant-binding proteins are expressed in different subsets of olfactory hairs. *Neuron* **12**, 35–49.

Plettner E., Lazar J., Prestwich E. and Prestwich G. D. (2000) Discrimination of pheromone enantiomers by two pheromone binding proteins from the gypsy moth, *Lymantria dispar. Biochemistry* **30**, 8953–8962.

Prestwich G. D., Graham S., Handley M., Latli B., Streinz L. and Tasayco, M. J. (1989) Enzymatic processing of pheromones and pheromone analogs. *Experientia* **45**, 263–270.

Prestwich G. D., Du G. and LaForest S. (1995) How is pheromone specificity encoded in proteins. *Chem. Senses* **20**, 461–469.

Raming K., Krieger J. and Breer H. (1989) Molecular cloning of an insect pheromone-binding-protein. *FEBS Lett.* **256**, 215–218.

Raming K., Krieger J. and Breer H. (1990) Primary structure of a pheromone-binding protein from *Antheraea pernyi*: homologies with other ligand-carrying proteins. *J. Comp. Physiol. B* **160**, 503–509.

Robertson H. M., Martos R., Sears C. R., Todres E. Z., Walden K. K. O. and Nardi J. B. (1999) Diversity of odourant binding proteins revealed by an expressed sequence tag project on male *Manduca sexta* moth antennae. *Insect Mol. Biol.* **8**, 501–518.

Rothemund S., Liou Y.-C., Davies P. L., Krause E., Sonnischen F. D. (1999) A new class of hexahelical insect proteins revealed as putative carriers of small hydrophobic ligands. *Structure* **7**, 143–151.

Saitou N. and Nei M. (1987) The neighbor joining method: a new method for reconstructing phylogenetic trees. *Mol. Biol. Evol.* **8**, 501–518.

Sandler B. H., Nikonova L., Leal W. S. and Clardy J. (2000) Sexual attraction in the silkworm moth: structure of the pheromone-binding-protein–bombykol complex. *Chem. Biol.* **7**, 143–151.

Scaloni A., Monti M., Angeli S. and Pelosi P. (1999) Structural analysis and disulfide bridge pairing of two odorant binding proteins from *Bombyx mori. Biochem. Biophys. Res. Commun.* **266**, 386–391.

Schneider D. (1969) Insect olfaction: deciphering system for chemical messages. *Science* **163**, 1031–1036.

Segal S. P., Graves L. E., Verheyden J. and Goodwin E. B. (2001) RNA-regulated TRA-1 nuclear export controls sexual fate. *Dev. Cell* **1**, 539–551.

Shanbhag S. R., Hekmat-Scafe D., Kim M.-S., Park S.-K., Carlson J. R., Pikielny C., Smith D. P. and Steinbrecht R. A. (2001a) Expression mosaic of odorant-binding proteins in *Drosophila* olfactory organs. *Microsc. Res. Tech.* **55**, 297–306.

Shanbhag S. R., Park S.-K., Pikielny C. and Steinbrecht R. A. (2001b) Gustatory organs of *Drosophila melanogaster*: fine structure and expression of the putative odorant-binding protein PBPRP2. *Cell Tissue Res.* **304**, 423–437.

Stange G. (1992) High resolution measurements of atmospheric carbon dioxide changes by the labial palp organ of the moth *Heliothis armigera* (Lepidoptera: Noctuidae). *J. Comp. Physiol. A* **171**, 317–324.

Stange G. (1996) Sensory and behavioural responses of terrestrial invertebrates to biogenic carbon dioxide gradients. In: *Advances in Bioclimatology* Vol. 4, ed. G. E. Stanhill, pp. 223–253. Springer Verlag, Berlin.

Steinbrecht R. A. (1996) Are odorant-binding proteins in odorant discimination? *Chem. Senses* **21**, 719–727.

Steinbrecht R. A., Ozaki M. and Ziegelberger G. (1992) Immunocytochemical localization of pheromone-binding-protein in moth antennae. *Cell Tissue Res.* **270**, 287–302.

Steinbrecht R. A., Laue M. and Ziegelberger G. (1995) Immunolocalization of pheromone-binding protein and general odorant-binding protein in olfactory sensilla of the silk moths *Antheraea* and *Bombyx. Cell Tissue Res.* **282**, 203–217.

Swofford D. L. (1999). *PAUP**. *Phylogenetic Analysis Using Parsimony (*and other methods)* Sinauer Associates, Sunderland, MA.

Teal P. E. A., Tumlinson J. H. and Heath R. R. (1986) Chemical and behavioral analyses of volatile sex pheromone components released by calling *Heliothis virescens* (F.) females (Lepidoptera: Noctuidae). *J. Chem. Ecol.* **12**, 1071–1026.

Thompson J. D., Gibson T. J., Plevniak F., Jeanmougin F., Higgins D. G. (1997) The Clustal X windows interface: flexible strategies for multiple sequence alignment aided by quality analysis tools. *Nucleic Acid Res.* **24**, 4876–4882.

Thymianou S., Mavroidis M., Kokolakis G., Komitopoulos K., Zacharopoulou A. and Mintzas A. C. (1998) Cloning and characterization of a cDNA encoding a male-specific serum protein of the Mediterranean fruit fly, *Ceratitis capitata*, with sequence similarity to odourant binding proteins. *Insect Mol. Biol.* **7**, 345–353.

Toth M., Löfstedt C., Blair B. W., Cabello T., Farag A. I., Hansson B. S., Kovalev B. G., Maini S., Nesterov E. A., Pajor I., Sazonov A. P., Shamsev I. V., Subchev M. and Szocs G. (1992) Attraction of male turnip moths *Agrotis segetum* (Lepidoptera: Noctuidae) to sex pheromone components and their mixtures at 11 sites in Europe, Asia and Africa. *J. Chem. Ecol.* **18**, 1337–1347.

Tuccini A., Maida R., Rovero P., Mazza M. and Pelosi P. (1996). Putative odorant-binding protein in antennae and legs of *Carausius morosus* (Insecta, Phasmatodea). *Insect Biochem. Mol. Biol.* **26**, 19–24.

Tumlinson J. H., Brennan M. M., Doolittle R. E., Mitchell E. R., Brabham A., Mazomemos B. E., Baumhover A. H. and Jackson D. M. (1989) Identification of a pheromone blend attractive to *Manduca sexta* (L.) males in a wind tunnel. *Arch. Insect. Biochem. Physiol.* **10**, 255–271.

Vogt R. G. (1987) The molecular basis of pheromone reception: its influence on behavior. In *Pheromone Biochemistry*, eds G. D. Prestwich and G. L. Blomquist, pp. 385–431, Academic Press, Orlando, FL.

Vogt R. G and Riddiford L. M. (1981) Pheromone binding and inactivation by moth antennae. *Nature* **293**, 161–163.

Vogt R. G. and Riddiford L. M. (1986) Pheromone reception: a kinetic equilibrium. In *Seminar on Mechanisms in Perception and Orientation to Insect Olfactory Signals*, eds T. L. Payre, M. C. Birch. and C. E. J. Kennedy, pp. 201–208, Clarendon Press, Oxford.

Vogt R. G., Riddiford L. M. and Prestwich G. D. (1985) Kinetic properties of a pheromone degrading enzyme: the sensillar esterase of *Antheraea polyphemus. Proc. Natl. Acad. Sci. USA* **82**, 8827–8831.

Vogt R. G., Köhne A. C., Dubnau J. T. and Prestwich G. D. (1989) Expression of Pheromone Binding Proteins during antennal development in the gypsy moth *Lymantria dispar. J. Neurosci.* **9**, 332–3346.

Vogt R. G., Prestwich G. D. and Lerner M. R. (1991a) Odorant binding protein subfamilies associate with distinct classes of olfactory receptor neurons in insects. *J. Neurobiol.* **22**, 74–84.

Vogt R. G., Rybczynski R. and Lerner M. R. (1991b) Molecular cloning and sequencing of general odorant binding proteins GOBP1 and GOBP2 from the tobacco hawk moth *Manduca sexta*: comparison with other insect OBPs and their signal peptides. *J. Neurosci.* **11**, 2972–2984.

Vogt R. G., Rybczynski R., Cruz M. and Lerner M. R. (1993) Ecdysteroid regulation of olfactory protein in the developing antenna of the tobacco hawk moth, *Manduca sexta. J. Neurobiol.* **24**, 581–597.

Vogt R. G., Callahan F. E., Rogers M. E. and Dickens J. C. (1999) Odorant Binding Proteins diversity and distribution among the insect orders, as indicated by LAP, an OBP-related protein of the true bug *Lygus lineolaris* (Hemiptera, Heteroptera). *Chem. Senses* **24**, 481–495.

Vogt R. G., Rogers M. E., Franco M. D. and Sun M. (2002) A comparative study of odorant binding protein genes: differential expression of the PBP1-GOBP2 gene cluster in *M. sexta* (Lepidoptera) and the organization of OBP genes in *Drosophila melanogaster* (Diptera). *J. Exp. Biol.* **205**, 719–744.

Willett C. S. (2002) Do pheromone binding proteins converge in amino acid sequence when pheromones converge? *J. Mol. Evol.* **50**, 175–183.

Willett C. S. and Harrison R. G. (1999) Pheromone binding proteins in the European and Asian corn borers: no protein change associated with pheromone differences. *Insect Biochem. Mol. Biol.* **29**, 277–284.

Wojtasek H. and Leal W. S. (1999) Conformational change in the pheromone-binding protein from *Bombyx mori* induced by pH and by interaction with membranes. *J. Biol. Chem.* **274**, 30950–30956.

Wojtasek H., Hansson B. H. and Leal W. S. (1998) Attracted or repelled? A matter of two neurons, one pheromone binding protein and a chiral center. *Biochem. Biophys. Res. Commun.* **250**, 217–222.

Wojtasek H., Picimbon J. F., Leal W. S. (1999) Identification and cloning of odorant binding proteins from the scarab beetle *Phyllopertha diversa. Biochem. Biophys. Res. Commun.* **263**, 832–837.

Wu W. Q., Cottrell C. B., Hansson B. S. and Löfstedt C. (1999) Comparative study of pheromone production and response in Swedish and Zimbabwean populations of turnip moth, *Agrotis segetum. J. Chem. Ecol.* **25**, 177–196.

Xu A., Park S.-K., Domello D., Kim E., Wang Q. and Pikielny C. (2002) Novel genes expressed in subsets of chemosensory sensilla on the front legs of male *Drosophila. Cell Tissue Res.* **307**, 381–392.

19

Diversity and expression of odorant receptors in *Drosophila*

Leslie B. Vosshall

19.1 Introduction

Olfactory perception translates abstract chemical features of odorants into meaningful neural information to elicit appropriate behavioral responses (Shepherd, 1994; Buck, 1996). Specialized bipolar olfactory sensory neurons (OSNs) are responsible for the initial events in odor recognition. These have ciliated dendrites exposed to the environment, and a single axon that extends into the brain and forms synapses with second order projection neurons (PNs) (Shepherd, 1994; Buck, 1996). In arthropods and mammals, the first olfactory synapse is organized into glomeruli, spherical structures in which afferent olfactory neuron axons synapse with projection neuron dendrites (Hildebrand and Shepherd, 1997).

Molecular recognition of thousands of diverse odorants is mediated by a large family of odorant receptor (OR) genes, each encoding a different seven transmembrane domain G protein-coupled receptor (GPCR). OR genes were first identified in the rat by Buck and Axel (1991), using an innovative approach which assumed that the ORs would be members of the GPCR super family, encoded by an extremely large gene family expressed only in olfactory tissues. These assumptions proved to be correct and led to the identification of several hundred rat OR genes, selectively expressed in OSNs. Using degenerate oligonucleotide primers that would anneal with conserved regions in the transmembrane domains of the GPCR super family, Buck and Axel used the polymerase chain reaction (PCR) to identify OR genes in olfactory epithelium mRNA.

Since this initial report, OR genes have been identified in nematodes, bird, fish, and other mammals (Ngai *et al.*, 1993; Ben-Arie *et al.*, 1994; Troemel *et al.*, 1995; Nef *et al.*, 1996; Mombaerts, 1999). The number of ORs varies widely between species: nematodes and mammals have approximately 1000 OR genes, while fish and birds have approximately 100 ORs (Mombaerts, 1999). Despite the shared function and secondary structure of these ORs, there is no primary sequence identity between the OR genes of nematodes and vertebrates, reflecting an apparently independent evolutionary origin.

For insects, olfaction is a crucial sensory modality in locating food sources, identifying appropriate sites for oviposition, avoiding predators, and selecting mates. Enormous strides in our understanding of pheromonal communication and perception of general food odors have been made by basic research in insect olfaction, using large insects such as moths, locusts, and the honeybee (Hansson, 2002). Although the fruit fly *Drosophila melanogaster* was originally "domesticated" at the turn of the twentieth century as a useful organism for the study of genetics, it became a premier model organism for the new field of neurogenetics in the 1960s principally through the effort of Seymour Benzer. Benzer and his colleagues carried out large-scale genetic screens to isolate genes involved in such behaviors as courtship, learning and memory, circadian rhythms, phototaxis, and chemotaxis to odorants and tastants (Weiner, 1999). A large number of *Drosophila* mutants with defects in olfactory behavior have since been isolated using a variety of different behavior paradigms (Siddiqi, 1987; McKenna *et al.*, 1989; Lilly and Carlson, 1990, Carlson, 1991, 1996). Complementing the power of genetics to study olfaction is the recent completion of the sequencing of the euchromatic genome of *Drosophila*, the first insect genome to be completed (Adams *et al.*, 2000). This vastly simplifies the identification of genes mutated in particular mutant backgrounds and has made it feasible to annotate all predicted genes in the genome, including the OR genes which are the subject of this chapter.

The fruit fly olfactory system is anatomically simple (reviewed in Stocker, 1994). Larvae possess three chemosensory organs, the dorsal organs, terminal organ, and ventral organ, which together are responsible for detecting volatile and non-volatile stimuli (Heimbeck *et al.*, 1999; Oppliger *et al.*, 2000; Python and Stocker, 2002). A total of 100 neurons have been described in these three organs, which are located at the anterior tip of the larva. These neurons extend axons that synapse in either the larval antennal lobe or regions of the tritocerebrum and subesophageal ganglion of the larval brain (Python and Stocker, 2002). Emerging evidence suggests that the larval antennal lobe contains glomerulus-like structures. In the adult fly, all OSNs are contained in two chemosensory organs located on the head, the third segment of the antenna and the maxillary palp, while gustatory neurons and contact chemoreceptors are distributed over various body surfaces including the proboscis, the leg, wing, and female ovipositor

(Stocker, 1994). While mammals possess millions of OSNs that relay information to thousands of olfactory bulb glomeruli, adult *Drosophila* have a mere 1300 OSNs connected to 43 antennal lobe glomeruli. This vastly simplified olfactory system, that nevertheless retains many of the anatomical features found in the mammalian olfactory system, make the fly an excellent model system in which to study the sense of smell.

This chapter will discuss the isolation of *Drosophila* odorant receptor *(DOR)* genes, how these genes have expanded our understanding of the development and functional anatomy of the olfactory system, how the odor response profiles of OSNs respond to odorants, and the mechanisms by which odor-specific activity is relayed to the brain.

19.2 *Drosophila* odorant receptor genes

19.2.1 Using bioinformatics to identify the elusive fly or genes

In the decade that followed the original cloning of rat OR genes by Buck and Axel (1991), considerable effort was expended by many investigators to identify homologs of this gene family in other vertebrates and in invertebrates. This task proved to be straightforward in other mammals, birds, and even fish because the vertebrate OR gene family shows strong conservation across diverse species (Mombaerts, 1999). With the exception of one report in the honeybee, which likely represented contamination of honeybee with human genomic DNA (Danty *et al.*, 1994), no laboratory seemed to be able to identify insect homologs of vertebrate ORs using degenerate PCR primers designed against the vertebrate gene family. In the mid-1990s Cori Bargmann's group reported the first invertebrate chemosensory receptor genes from the nematode *C. elegans*. By analyzing partial genomic DNA sequences compiled by the consortium sequencing the *C. elegans* genome, Troemel *et al.* (1995) identified a large family of genes encoding seven transmembrane domain receptors selectively expressed in chemosensory neurons of the worm. Named *sr* (for serpentine receptor) genes, these are unrelated to any known protein family and are also extremely divergent within *C. elegans*. The first functional proof that these putative receptor genes indeed encode chemoreceptors came with the cloning of the *odr-10* gene the following year (Sengupta *et al.*, 1996). Animals deficient in *odr-10* function show severely reduced chemotaxis to the odorant diacetyl. Positional cloning of the gene defective in *odr-10* mutants identified a seven transmembrane domain protein related to the *sr* genes which is likely the diacetyl receptor.

Because there was no apparent evolutionary relationship between chemoreceptor genes in the nematode and in vertebrates, it seemed at least a possibility that the insect odorant receptors would represent yet a third class of genes, unrelated to ORs in either *C. elegans* or vertebrates. Following this logic, Vosshall *et al.*

(1999) pursued an approach that did not rely on any assumptions about the sequence of the *Drosophila* ORs. They used differential hybridization to identify genes selectively expressed in antennal and maxillary palp mRNA. Among the several dozen candidate genes, a single gene named *DOR104* [since renamed *Or85e*, by the *Drosophila* Odorant Receptor Nomenclature Committee (2000)] encoded a novel seven transmembrane domain receptor protein with no homology to any known proteins. *Or85e* was selectively expressed in a subset of OSNs in the maxillary palp and not detectably expressed anywhere else in the fly (Vosshall *et al.*, 1999) (Figure 19.1). Efforts by this group to identify additional *Or85e*-related genes by conventional molecular biology techniques failed.

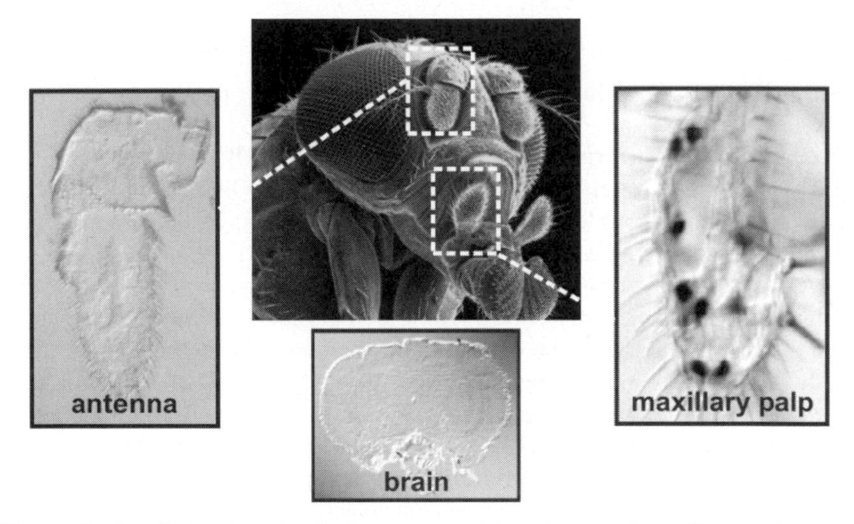

Figure 19.1 *Or85e* is selectively expressed in subsets of maxillary palp neurons. *In situ* hybridization with an antisense *Or85e* probe reveals no staining in antenna (left) or brain (center). Nine cells stain in this section of the maxillary palp (right). The relative positions of the two olfactory sensory organs on the head of the adult fly are indicated with white dashed lines.

By 1998, sequencing of the *Drosophila* genome was approximately 15 percent completed. Three research groups exploited these sequences to report the identification of candidate *Drosophila* odorant receptor (*DOR*) genes in 1999 (Clyne *et al.*, 1999; Gao and Chess, 1999; Vosshall *et al.*, 1999). Although the genomic sequences were highly fragmented and unannotated, all three groups used the logic that if the ORs constituted a multi-gene family of at least 100 members, then at least 10 genes should be represented in the partial genomic sequence then available. While the algorithms used to identify these genes in genomic sequence differed, all followed the same basic strategy: first, identify coding exons or genes, then identify those exons or genes that encode proteins containing multiple membrane-spanning domains, and subject promising candidate genes to expression analysis (Figure 19.2). Nineteen candidate *DOR* genes were isolated by the combined efforts of the three groups. All of these were distant but clear homologs of *Or85e*, the maxillary palp receptor that originally emerged from a difference cloning approach (Figure 19.1; Table 19.1). None of the *DOR* genes showed any apparent sequence relatedness to any other known proteins, including the chemoreceptor proteins in other species.

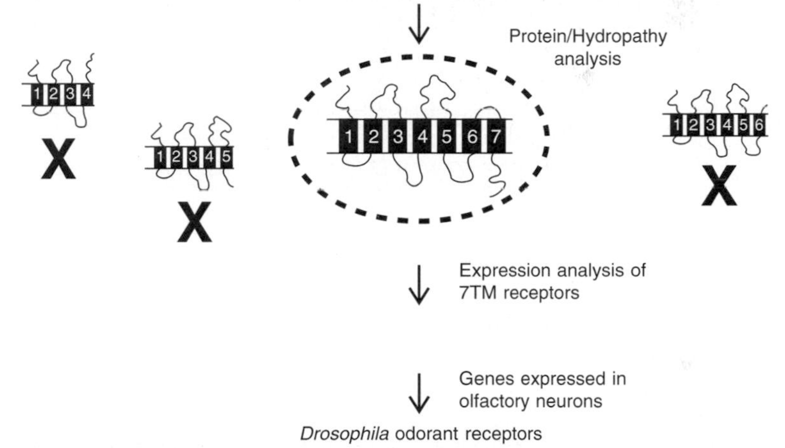

Figure 19.2 Mining *Drosophila* genome databases for OR genes. Flow chart illustrating the general steps used by three different groups to identify OR genes in raw genomic DNA (Clyne *et al.*, 1999; Gao and Chess, 1999; Vosshall *et al.*, 1999). Sequence files from *Drosophila* genome databases were downloaded and subjected to a variety of gene-finding programs. These programs parse the data into exons (black text) and introns (grey text) and generate protein-coding open reading frames. DNA sequences are for illustrative purposes only and represent a hypothetical gene fragment with two exons and one intron. The open reading frames were translated into protein and the resulting proteins were analyzed for hallmarks of GPCRs (Clyne *et al.*, 1999) or for transmembrane domains (Gao and Chess, 1999; Vosshall *et al.*, 1999). Proteins with seven transmembrane domains (7TM) were analyzed further by gene expression analysis (dashed line). Proteins with fewer transmembrane domains were discarded (X). Genes selectively expressed in OSNs were further analyzed as candidate *DOR* genes.

Table 19.1 Compiled expression data

#	OR name	Previous names	Entrez	SwissProt	Map	Antenna	Maxillary palp	Not detected
1	Or1a		CG17885	Q9W5G6	1A8		V	
2	Or2a	2F.1, DOR62, AN4	CG3206	O46077	2E1	V*C		
3	Or7a		CG10759	Q9W3I5	7D14	V		
4	Or9a		CG15302	Q9W2U9	9E.1	V		
5	Or10a		CG17867	Q9VYZ1	10B15	V		
6	Or13a		CG12697	Q9VXL0	13F16-18	V		
7	Or19a		CG18859	Q9I816	19B3-19C	V		
8	Or22a	22A.1, DOR53, AN11	CG12193	P81909	22A5	V*CG		
9	Or22b	22A.2, DOR67, AN12	CG4231	P81910	22A5	V*C		
10	Or22c	22C.1	CG15377	P81911	22C1			V*
11	Or23a	23A.1, DOR64, AN5	CG9880	P81912	23A3	V*G	G	C
12	Or24a	24D.1	CG11767	P81913	24E4			V*
13	Or30a		CG13106	Q9VLE5	30A3			V
14	Or33a	33B.1, DOR73, AN3	CG16960	P81914	33B10	V*		C
15	Or33b	33B.2, DOR72, AN1	CG16961	P81915	33B10	V*		C
16	Or33c	33B.3, DOR71, AN2	CG5006	P81916	33B10	*	V*CG	
17	Or35a		CG17868	Q9V3Q2	35D1	V		
18	Or42a		CG17250	Q9V9I2	42A2			V
19	Or42b		CG12754	Q9V9I4	42A2	V		
20	Or43a	43B.1, DOR87, AN14	CG1854	P81917	43A1	V*		
21	Or43b	25A.1, AN7	CG17853	P81918	43F5	V*CG	*	
22	Or45a		CG1978	Q9V568	45C5			V
23	Or45b		CG12931	Q9V589	45F1			V
24	Or46a	46F.1, AN9	CG17849	P81919	46E7-8		V*CG	
25	Or46b	46F.2, DOR19, AN8	CG17848	Q9V3N2	46E7-8	*		VC
26	Or47a	47E.1, DOR24, AN10	CG13225	P81921	47F1	V*C		
27	Or47b	47E.2	CG13206	P81922	47F6	V*		

28	Or49a		CG13158	Q9V6A9	49A5			V
29	Or49b	AN13	CG17584	Q9V6H2	49D1	V		
30	Or56a		CG12501	Q9V8Y7	56E1	V		
31	Or59a	59D.1, DOR46, AN6	CG9820	P81923	59E1	G		V *
32	Or59b		CG3569	Q9W1P8	59E1	V		
33	Or59c		CG17226	Q9W1P7	59E1		V	
34	Or63a		CG9969	Q9VZW8	63B1			V
35	Or65a		N.A.	P82982	65A7-11	V		
36	Or65b		N.A.	P82983	65A7-11	V		
37	Or65c		N.A.	P82984	65A7-11	V		
38	Or67a		CG12526	Q9VT08	67B2	V		
39	Or67b		CG14176	Q9VT20	67B10	V		
40	Or67c		CG14156	Q9VT90	67D2	V		
41	Or67d		CG14157	Q9VT92	67D5	(V)		
42	Or69a		N.A.	P82985	69E8-F1	V		
43	Or69b		CG17902	Q9VU27	69E8-F1	V		
44	Or71a		CG17871	Q9VUK5	71B1		V	
45	Or74a		CG13726	Q9VVF3	74A6			V
46	Or82a		N.A.	P82986	82A3-4	V		
47	Or83a		CG10612	Q9VNB3	83A6			V
48	Or83b	A45	CG10609	Q9VNB5	83A6	V	V	
49	Or83c		CG15581	Q9VNK9	83D4	V		
50	Or85a		CG7454	Q9VHS4	85A3	V		
51	Or85b		CG11735	Q9VHQ7	85A9	V		
52	Or85c		CG17911	Q9VHQ6	85A9			V
53	Or85d		CG11742	Q9VHQ2	85A11		V	
54	Or85e	DOR104	CG9700	P81924	85B2		V	
55	Or85f		CG16755	Q9VHE6	85D15	V		
56	Or88a		CG14360	Q9VFN2	88B1	V		
57	Or92a		CG17916	Q9VDM1	92E8			V
58	Or94a		CG17241	Q9VCS9	94D9			V

(Contd)

Table 19.1 *(Contd)*

#	OR name	Previous names	Entres	SwissProt	Map	Antenna	Maxillary palp	Not detected
59	Or94b		CG6679	Q9VCS8	94D9			V
60	Or98a		CG5540	Q9VAZ3	98B5	V		
61	Or98b		CG1867	Q9VAW0	98D4			V

N.A. = none assigned to date
* = Clyne *et al.*, 1999 (RT-PCR)
C = Clyne *et al.*, 1999 (*in situ*)
G = Gao and Chess, 1999 (*in situ*)
V = Vosshall *et al.*, 1999 (*in situ*)
(V) = Fishilevich and Vosshall, unpublished)

19.2.2 The size and character of the *DOR* gene family

Mining the partial fly genome for OR genes yielded approximately 20 candidate odorant receptors. Each encodes a protein of approximately 370 to 400 amino acids. *Or83b* is significantly larger at 486 amino acids, with the additional amino acids lying in predicted intracellular loop regions of the protein. The *DOR* proteins are rather hydrophobic in character, but it is possible to assign putative membrane-spanning regions. While there is no consensus as to exact position of the transmembrane domains, there is general agreement that there are indeed seven membrane-spanning regions (Clyne *et al.*, 1999; Gao and Chess, 1999; Vosshall *et al.*, 1999). Further direct experimental proof will be required to determine which model of transmembrane segments is correct.

Upon the release of the complete euchromatic genome sequence of *Drosophila* (Adams *et al.*, 2000), it was possible to complete the annotation of this gene family and obtain a complete view of the number of *DOR* genes present in the database (Vosshall *et al.*, 2000). A total of 57 *DOR* genes were identified. More recent annotation of this gene family by Hugh Robertson revealed an additional four *DOR* genes for a current total of 61 members of the gene family (see Chapter 21 and Table 19.1). Because the *Drosophila* genome is undergoing continuous revision to fill gaps in euchromatic sequence, the total number of *DOR* genes may increase. However, the number is unlikely to increase substantially unless there is a large reservoir of undiscovered *DOR* genes in regions of heterochromatin, which is largely unsequenced.

Each of the 61 known *DOR* genes encodes a different seven transmembrane domain GPCR. To date there is no evidence for alternative splicing of these genes to generate receptor protein diversity. The degree of sequence conservation of *DOR* genes within *Drosophila* is quite low, on the order of 17–26 percent. However, there are a number of subfamilies within the *DORs* with significantly higher degrees of sequence similarity (40–60 percent). Despite the very low degrees of sequence similarity between the *DORs*, they are classified as members of the same gene family based on certain strongly conserved amino acids which lie at fixed positions distributed throughout the protein. For instance, in the putative seventh transmembrane domain, the amino acid sequence Phe-Pro-X-Cys-Tyr-$(X)_{20}$-Trp (where X = any amino acid) is strongly conserved across the *DOR* family. In database searches for *DOR* genes using the BLAST algorithm (Altschul *et al.*, 1990), the strongest degree of sequence conservation across the gene family is in 3′ regions of the gene encompassing transmembrane domains 6 and 7. The significance of the divergence of *DOR* sequences in N-terminal regions of the protein and the comparative sequence similarity at C-terminal regions are currently unknown. In vertebrate OR genes, regions of sequence variability in the transmembrane regions has been proposed as a potential site for ligand interaction. It is possible that similar mechanisms operate in the ligand binding of the *DOR* genes.

Although the *DORs* have no apparent sequence similarity to chemosensory receptors in nematodes and vertebrates, they are clearly related to OR genes from the malaria mosquito, *Anopheles gambiae* (Fox *et al.*, 2001). As in the fly, sequence similarity among the cloned mosquito ORs is low, with 11 percent identity and 18 percent similarity across the four published genes. Nevertheless, the amino acid sequences that are signatures of this gene family are present in the mosquito ORs. The *DORs* are therefore founding members of an insect OR gene family and ORs from other insects are likely to resemble both mosquito and *Drosophila* receptors. This divergence across the gene family, with more closely related subfamilies, has also been seen for the chemosensory receptors of the nematode (Troemel *et al.*, 1995) and putative pheromone receptors in the mouse (Rodriguez *et al.*, 2002). In contrast, there tends to be more sequence conservation within the OR genes of vertebrates (Zozulya *et al.*, 2001; Zhang and Firestein, 2002).

Interestingly, because of the extremely low sequence similarly of ORs within *Drosophila* and between *Drosophila* and mosquito, it is likely that identification of OR genes in other insects will require access to genome sequences of these organisms. Conventional approaches that rely on degenerate PCR or low stringency hybridization across species are unlikely to succeed because the regions of *DOR* homology are extremely dispersed. A direct demonstration of this lies in the inability of Vosshall *et al.* (1999) to identify homologs of *Or85e* by conventional approaches. Although BLAST algorithms using *Or85e* as a query sequence routinely yield hits to a large number of annotated *DOR* genes, low stringency hybridization and degenerate PCR with *Or85e* sequences fail to cross-react with these *DORs*. Recent advances in increasing the throughput and reducing the cost of genomic sequencing should make the sequencing of additional interesting insects, such as moth, locust, honeybee, ant, and cockroach, feasible (Broder and Venter, 2000). A complementary approach of sequencing expressed genes has begun in the honeybee, and has already yielded an *Apis mellifera* homolog of *Or83b* (Whitfield *et al.*, 2002).

The lack of any apparent primary sequence identity between the OR genes of nematodes, insects, and vertebrates suggests that these arose through an independent mechanism in evolution (Vosshall *et al.*, 1999). Remarkably, this suggests that in order to recognize a very large number of different odorous ligands, an animal needs only to expand an ancestral GPCR into a multi-gene family and direct members of this gene family to express selectively in OSNs.

19.2.3 *DOR* gene expression patterns: formation of a peripheral olfactory sensory map

A role for *DOR* genes in olfaction requires that they are expressed in OSNs, where they can interact with odors and transmit odor-activated neuronal activity. To address this question Clyne *et al.* (1999) used reverse transcription-PCR (RT-

PCR) on antennal and maxillary palp mRNA preparations. Of the 16 *DOR* genes examined, ten were detected by RT-PCR in antenna, two in the antenna and maxillary palp, one in maxillary palp alone, and three in neither tissue (data marked * in Table 19.1). Analysis of *DOR* gene expression by RNA *in situ* hybridization to antenna and maxillary palp largely confirmed these RT-PCR data, with the exception that some *DORs* detected by RT-PCR were not detected by *in situ* hybridization (Clyne *et al.*, 1999; Gao and Chess, 1999; Vosshall *et al.*, 1999, 2000). Tissue localization of all 61 members has been determined: 39 *DOR* genes were found to be expressed in the antenna, nine in the maxillary palp, and 19 *DOR* genes were not detected. *Or83b* was unique in being expressed in both antenna and maxillary palp (Vosshall *et al.*, 1999, 2000). These data exceed the total number of 61 *DOR* genes because for certain members of the gene family, different groups obtained different results. For instance, *Or23a* was detected by Clyne *et al.* (1999) in antenna by RT-PCR, but was not detected in this tissue by *in situ* hybridization. Gao and Chess found *Or23a* in both antenna and maxillary palp by *in situ* hybridization (Gao and Chess, 1999), while Vosshall *et al.* found this receptor to be antennal specific (1999, 2000). The variability in these findings is most likely due to the variability inherent in the techniques of RT-PCR and *in situ* hybridization. Despite this variability, there is in general no overlap in antennal and maxillary palp *DOR* genes. This suggests that these two olfactory sensory organs respond to different odors. Alternatively, there may be some overlap in the odors perceived, but this is accomplished through different *DOR* proteins, perhaps with different ligand-binding affinities. Evidence for the latter model has been obtained by electrophysiological recordings from individual sensilla on both olfactory organs (de Bruyne *et al.*, 1999, 2001).

The patterns of *DOR* gene expression revealed by *in situ* hybridization in antenna and maxillary palp allow us to make a number of important conclusions. Each *DOR* gene is expressed in a small subset of the ~1300 antennal and ~120 maxillary palp neurons. The number of OSNs expressing a given *DOR* gene varies from two (*Or49b*) to approximately 50 (*Or47b*), with an average of 25 OSNs expressing a given *DOR* gene. In the antenna, the neurons expressing a given *DOR* gene are located at discrete positions along the proximal–distal and medial–lateral axes. The relative position and number of neurons expressing a given gene is conserved between different animals. For instance, *Or22a* expression is limited to approximately 20 OSNs that lie at the medial–proximal aspect of the antenna. One technique to reveal gene expression that does not rely on RT-PCR or RNA *in situ* hybridization is the generation of transgenic flies in which putative *DOR* promoter sequences drive expression of a convenient marker protein. Such transgenic flies have been constructed and marker expression in these animals has been shown to overlap with the expression of the endogenous *DOR* gene (Vosshall *et al.*, 1999, 2000). Figure 19.3 shows an example of patterns of LacZ marker gene expression in flies carrying *Or43a*-Gal4 and *Or71a*-Gal4 transgenes.

Kaupmann *et al.*, 1998; White *et al.*, 1998). A similar mechanism has been suggested for the RAMPs, single transmembrane proteins that associate with seven transmembrane domain proteins and facilitate appropriate membrane insertion (McLatchie *et al.*, 1998).

Neurons from the antenna and maxillary palp extend axons that synapse specifically in the antennal lobe of the brain. It is conceivable that this broadly expressed *DOR* gene plays a role in the differentiation and development of the olfactory system, perhaps targeting axons to olfactory glomeruli in the brain. In the mouse, the olfactory receptor itself influences target selection in the olfactory bulb (Mombaerts *et al.*, 1996; Wang *et al.*, 1998). Because *Or83b* is expressed in most olfactory neurons, it is unlikely to be involved in the mechanism that targets axons to specific glomeruli, but might be a general factor required for guidance of all olfactory neuron axons to the antennal lobe.

A second, conceptually different model for *Or83b* function is that this ubiquitous odorant receptor functions in ligand binding. If *Or83b* were a ligand-binding odorant receptor, its cognate ligand would activate most olfactory neurons and therefore many glomeruli in the antennal lobe. This would pose a formidable problem in odor recognition for the brain. Alternatively, *Or83b* may act as a receptor independent of ligand, constitutively activating signaling pathways in the cell to increase levels of second messengers and therefore sensitizing the cell to activation by conventional *DOR* proteins. An analogous function has been proposed for the receptor guanylate cyclases expressed in *C. elegans* chemosensory neurons (L'Etoile and Bargmann, 2000). Finally, *Or83b* might associate with conventional *DOR* proteins and modify their ligand specificity. Heterodimerization of kappa and mu opioid receptors, as well as dopamine and somatostatin receptors has been documented and produces receptors with altered ligand specificities (Jordan and Devi, 1999; Rocheville *et al.*, 2000).

Conceptually similar models have been proposed for chemosensory receptors in two vertebrate model systems. In the goldfish, two members of the V2R family of putative pheromone receptors, 5.3 and 5.24, are very broadly expressed in OSNs (Speca *et al.*, 1999). It is likely that neurons expressing 5.3 and 5.24 express additional, more selectively distributed members of the V2R family. In rodents, a subfamily of V2R receptors, called V2R2, shows very broad expression in the vomeronasal organ (Martini *et al.*, 2001). Vomeronasal organ neurons express two receptors: broadly distributed V2R2 and a more selectively distributed member of the V2R family. Understanding the function of these broadly expressed receptors will require either modeling receptor function in heterologous cells or the generation of mutant animals that lack the broadly expressed receptor. It will then be possible, through a genetic approach, to dissect the various proposed models for the function of *Or83b* and other receptors with a similarly broad expression profile.

19.2.5 Relating *DOR* gene expression to olfactory function

OSNs in insects are bipolar neurons that insert dendrites into sensory hairs called sensilla. The number and distribution of different sensilla are strongly conserved across individuals (Shanbhag *et al.*, 1999). Elegant electrophysiological experiments have probed the odor responses of a large number of antennal and maxillary palp sensilla (de Bruyne *et al.*, 1999, 2001) (see Chapter 18 for a full discussion of this topic). It is now possible to begin to superimpose the maps of sensillar type on the antenna with the type of *DOR* gene expressed by the underlying OSN, and to relate this to the response properties of the neuron and the sensillum. There is a striking overlap in the patterns of *DOR* gene expression and the electrophysiological response maps, giving further strength to the argument that the *DOR* genes indeed are ligand-binding odorant receptors. With the advent of *DOR*-Gal4 transgenic reagents, it will be possible to label individual dendrites of neurons expressing that *DOR* gene and directly assess the ligand-binding properties of a given neuron.

19.2.6 Approaches to identifying ligand–receptor relationships of the *DOR* genes

The evidence that the *DOR* genes are ligand-binding odorant receptors is suggestive, if not fully conclusive: the *DOR* genes encode seven transmembrane domain GPCRs; the size of the *DOR* gene family is consistent with a role in recognizing a large number of odorants; each *DOR* gene (with the exception of *Or83b*) is expressed in a small subset of OSNs that may overlap with functional subtypes defined by electrophysiology. What has not been demonstrated in the publications reporting *DOR* gene isolation is that a given *DOR* interacts with a particular odorant. In other olfactory systems, evidence of direct ligand–receptor relationships has been obtained by a variety of genetic and electrophysiological techniques.

In the nematode, *C. elegans*, *odr-10* mutants show dramatically reduced attraction to the odorant diacetyl. Molecular cloning of the gene defective in *odr-10* mutants revealed it to be a seven transmembrane domain GPCR (Sengupta *et al.*, 1996). Direct demonstration that the *odr-10* protein is a diacetyl receptor came from experiments in which neurons normally non-responsive to diacetyl were reprogrammed to express *odr-10*, producing animals that were repelled by this normally attractive stimulus (Troemel *et al.*, 1997). The availability of genetic approaches in *Drosophila* makes the generation of *DOR* mutants feasible and is likely to generate compelling data on how the complement of chemosensory receptors in the fly recognizes odors.

An alternative to genetics has been heterologous expression of ORs, either *in vivo* (Zhao *et al.*, 1998; Araneda *et al.*, 2000) or in tissue culture cells (Krautwurst *et al.*, 1998; Touhara *et al.*, 1999), and the use of functional imaging of isolated OSNs to determine their response profiles (Malnic *et al.*, 1999; Leinders-Zufall *et al.*, 2000). These studies of vertebrate chemosensory receptors have led to the conclusion that a given vertebrate OR recognizes multiple odorants and that a given odorant activates neurons expressing a number of different ORs. These types of odorant responses will give rise to a combinatorial code in which a particular odorant activates a diverse array of OSNs expressing different receptors.

Because *Drosophila* has considerably fewer ORs than vertebrates, it was of interest to determine if this type of combinatorial coding operates in the fly or if fly receptors are significantly narrower in their response properties. Recent studies have addressed this question for a single receptor using both *in vivo* overexpression and heterologous expression. *Or43* was found to respond to benzaldehyde, as well as benzyl alcohol, cyclohexanol, and cyclohexanone (Störtkuhl and Kettler, 2001; Wetzel *et al.*, 2001). *In vivo* overexpression was performed with the Gal4-UAS system to expand the expression domain of *Or43* artificially from ~20 neurons to >1000 neurons in the antenna. In these animals, most neurons will express three receptors: a specific *DOR* gene that is naturally expressed in this OSN, *Or83b*, and ectopic *Or43a*. These animals are then exposed to odorants and odor-evoked extracellular receptor potentials are recorded in the antenna using the electroantennogram (EAG) technique. A given odor will give a reproducible EAG response in wild-type animals. In animals overexpressing *Or43a*, there is a significant potentiation in the amplitude of the EAG response which is selective for a few structurally related odorants. In a companion study, the same *DOR* was heterologously expressed in Xenopus oocytes along with a promiscuous G protein ($G_q\alpha15$) and odor-evoked currents were only obtained with the same odorants that proved to be successful in the *in vivo* experiments (Wetzel *et al.*, 2001). Taken together, these studies suggest that *Or43a* is a receptor for benzaldehyde and a small number of structurally related compounds. With the relatively small number of *DOR* genes in the fly, it is feasible to perform these types of experiments for the entire gene family and obtain a complete picture of the odor-responsive properties of this gene family. The very large number of chemosensory receptors in vertebrates and in the nematode likely preclude such a complete description of the odor selectivity of an entire OR gene family.

19.3 *Drosophila* gustatory receptor (*GR*) genes: a gene family that subserves both gustatory and olfactory modalities

The *DOR* genes are not the only chemosensory receptors in the fly genome. Using the same algorithms that yielded the *DOR* gene family, Clyne *et al.* (2000) isolated a second large family of 43 genes encoding seven transmembrane GPCRs. These were named gustatory receptors (*GR*), based on RT-PCR data indicating selective expression in taste tissues. Other researchers have expanded this gene family to at least 55 members (Dunipace *et al.*, 2001; Scott *et al.*, 2001). Unlike the *DORs*, the *GRs* do exhibit some alternative splicing, which has the potential to produce multiple proteins from a given *GR* gene (Clyne *et al.*, 2000). The *GR* genes encode seven transmembrane domain proteins of approximately the same size as the *DOR* genes. As was seen for the *DORs*, the sequence similarity of the entire *GR* gene family is quite low, but there are a number of subfamilies with higher degrees of sequence relatedness. Interestingly, amino acid motifs at the C-termini of both gene families appear to be related, suggesting that these two chemosensory families may have a common evolutionary origin. The broadly expressed *DOR* gene, *Or83b*, is the most similar to the *GRs* and may represent the most ancient linkage between these two families of receptors.

Confirming the initial expression data obtained by RT-PCR Scott *et al.* (2001) obtained *in situ* hybridization evidence that four of the *GRs* are expressed in the adult labellum, the site of gustatory neurons. A fifth *GR* gene was found to be selectively expressed in the third antennal segment. The expression of many other *GRs* was not detected by *in situ* hybridization. To obtain tissue-specific expression data, two groups generated *GR*-Gal4 transgenic fly lines that would permit them to examine the distribution of members of the *GR* gene family (Dunipace *et al.*, 2001; Scott *et al.*, 2001). Characterization of a large number of *GR*-Gal4 transgenic lines yielded these important conclusions: *GR* genes are expressed in both gustatory and olfactory neurons; *GR* genes are detected in both adult and larval chemosensory organs; a given neuron is likely to express only a single *GR* gene; a given *GR* gene can be expressed in both gustatory and olfactory neurons. The neurons expressing *GRs* in the antenna are located in a distinctive medial location on the exterior face of the antenna, in a region in which *DOR* gene expression has not been detected. Because anatomical studies have not identified any sensilla that are likely to be gustatory in nature, the hypothesis is that *GR* genes expressed in antenna are in fact odorant receptors. It will be of interest to determine whether this hypothesis is correct, and if so, what odorants activate the *GRs*.

Although we do not know the ligand-binding properties of the *GR* genes expressed in olfactory neurons, recent genetic evidence suggests that at least one *GR* gene, *Gr5a*, is a molecular sweet taste receptor for the sugar trehalose (Dahanukar *et al.*, 2001; Ueno *et al.*, 2001). This conclusion must be tempered by an earlier report that a putative peptide GPCR linked to *Gr5a*, named *Tre1*, is in fact the trehalose receptor (Ishimoto *et al.*, 2000). In these experiments, a P-element located between *Tre1* and *Gr5a* was mobilized to generate local deletions around the site of insertion. To determine which gene is responsible for the behavior defect, Dahanukar *et al.* (2001) generated rescuing transgenes which contained either a functional *Gr5a* or *Tre1* gene. Rescue of the trehalose taste sensitivity mapped to the *Gr5a* gene, which is now considered to be the true trehalose receptor. Therefore, this constitutes the first functional proof that *GR* genes are receptors for non-volatile stimuli such as sugars. More work will be

information about the nature of the stimulus because glomeruli acts as points of convergence that relay ligand information faithfully.

How are odorant-evoked patterns of activity in the antennal lobe represented in higher brain centers? To answer this question, two groups performed sophisticated anatomical tracing experiments to trace the projections of single projection neurons whose dendrites innervate a given glomerulus (Marin *et al.*, 2002; Wong *et al.*, 2002). To accomplish this, they made use of an enhancer trap line that expresses Gal4 in a large number of the projections neurons that surround the antennal lobe, as well as some unrelated neurons (Stocker *et al.*, 1997). Using various techniques to generate somatic mosaic clones in which only a single of these projections neurons now expresses Gal4, both groups were able to label these neurons with GFP variants that revealed the morphology of the entire neuron from dendrite to axon. Analysis of these animals revealed remarkable stereotypy in the axons arborization of all projection neurons whose neurons innervated a given glomerulus. Cluster analysis revealed that the glomerular identity of a given projection neuron could be predicted based solely on its characteristic axonal morphology in the mushroom body and lateral horn of the protocerebrum, the two target regions of antennal lobe neurons. This result suggests that an intricate and highly conserved genetic program patterns not only convergent glomerular projections of OSN axons and projection neuron dendrites, but also controls the patterns of innervation of projection neuron axons in higher brain centers. However, the patterns of axonal innervation of the lateral horn of the protocerebrum are much more diffuse than the spatially restricted glomeruli. There is likely to be extensive overlap in the projections of projections neurons receiving input from different glomeruli. The implications of this stereotyped wiring for olfactory processing are that odors are represented first by the specific activation of subsets of glomeruli and this information is then relayed and refined and interpreted by the summed action of a large network of second- and third-order neurons in the mushroom body and lateral horn of the protocerebrum. Direct functional proof of this hypothesis of odor coding will require the evolution of techniques that permit functional imaging of brain activity in living animals perceiving and interpreting odor cues.

19.5 Conclusion and future prospects

The identification of *DOR* genes has permitted analysis of many different aspects of olfactory biology in the fruit fly. We now know the complete repertoire of genes that are likely to recognize odorants. Techniques exist to measure the response properties of a given OR and to examine the function of the neuron *in vivo*. The first- and second-order olfactory projections have been mapped anatomically. Although the fly is generally deemed to be too small for electrophysiology, recent advances in functional imaging promise to make the analysis of olfactory processing in living animals an experimental reality (Zemelman and Miesenbock, 2001).

One fundamental unsolved question in sensory biology is how the olfactory system processes distinct olfactory cues to elicit appropriate behavioral responses. With the complete repertoire of odorant receptor genes in hand, along with a growing understanding of the neuroanatomy of the system, it has now become possible to address this question in *Drosophila*. There are a number of behavior genetic approaches to matching specific odorous ligands with identified odorant receptor genes, the neurons that express these receptor genes, and the circuits that lead to stereotyped behaviors. *Drosophila* therefore promises to be a useful genetic model system of olfaction for the foreseeable future.

Abbreviations

7TM, seven transmembrane domain; *DOR*, *Drosophila* odorant receptor; EAG, electroantennogram ; GFP, green fluorescent protein; GPCR, G-protein coupled receptor; *GR*, gustatory receptor; nsyb, neuronal synaptobrevin; OR, odorant receptor; OSN, olfactory sensory neuron; PCR, polymerase chain reaction; PN, projection neuron; RT-PCR, reverse transcription-polymerase chain reaction; SOG, subesophageal ganglion.

Acknowledgements

Work in the author's laboratory is supported by grants from the NIH, the NSF, the Arnold and Mabel Beckman Foundation, the John Merck Fund, and the McKnight Endowment Fund for Neuroscience.

References

Adams M. D., Celniker S. E., Holt R. A., Evans C. A., Gocayne J. D., Amanatides P. G., Scherer S. E., Li P. W., Hoskins R. A. and Galle R. F. *et al.* (2000) The genome sequence of *Drosophila melanogaster*. *Science* **287**, 2185–2196.

Araneda R. C., Kini A. D. and Firestein S. (2000) The molecular receptive range of an odorant receptor. *Nat. Neurosci.* **3**, 1248–1255.

Altschul S. F., Gish W., Miller W., Myers E. W. and Lipman D. J. (1990) Basic local alignment search tool. *J. Mol. Biol.* **215**, 403–410.

Ben-Arie N., Lancet D., Taylor C., Khen M., Walker N., Ledbetter D. H., Carrozzo R., Patel K., Sheer D., Lehrach H. and North M. A. (1994) Olfactory receptor gene cluster on human chromosome 17: possible duplication of an ancestral receptor repertoire. *Hum. Mol. Genet.* **3**, 229–235.

Brand A. H. and Perrimon N. (1993) Targeted gene expression as a means of altering cell fates and generating dominant phenotypes. *Development* **118**, 401–415.
Broder S. and Venter J. C. (2000) Sequencing the entire genomes of free-living organisms: the foundation of pharmacology in the new millennium. *Annu. Rev. Pharmacol. Toxicol.* **40**, 97–132.
Buck L. and Axel R. (1991) A novel multigene family may encode odorant receptors: a molecular basis for odor recognition. *Cell* **65**, 175–187.
Buck L. B. (1996) Information coding in the vertebrate olfactory system. *Annu. Rev. Neurosci.* **19**, 517–544.
Carlson J. (1991) Olfaction in *Drosophila*: genetic and molecular analysis. *Trends Neurosci.* **14**, 520–524.
Carlson J. R. (1996) Olfaction in *Drosophila:* from odor to behavior. *Trends Genet.* **12**, 175–180.
Clyne P. J., Warr C. G. and Carlson J. R. (2000) Candidate taste receptors in *Drosophila. Science* **287**, 1830–1834.
Clyne P. J., Warr C. G., Freeman M. R., Lessing D., Kim J. and Carlson J. R. (1999) A novel family of divergent seven-transmembrane proteins: candidate odorant receptors in *Drosophila. Neuron* **22**, 327–338.
Drosophila Odorant Receptor Nomenclature Committee (2000) A unified nomenclature system for the *Drosophila* odorant receptors. *Cell* **102**, 145–146.
Dahanukar A., Foster K., van der Goes van Naters W. M. and Carlson J. R. (2001) A Gr receptor is required for response to the sugar trehalose in taste neurons of *Drosophila. Nat. Neurosci.* **4**, 1182–1186.
Danty E., Cornuet J. M. and Masson C. (1994) Honeybees have putative olfactory receptor proteins similar to those of vertebrates. *C. R. Acad. Sci. III* **317**, 1073–1079.
de Bruyne M., Clyne P. J. and Carlson J. R. (1999) Odor coding in a model olfactory organ: the *Drosophila* maxillary palp. *J. Neurosci.* **19**, 4520–4532.
de Bruyne M. Foster K., and Carlson J. R. (2001) Odor coding in the Drosophila antenna. *Neuron* **30**, 537–552.
Dunipace L., Meister S., McNealy C. and Amrein H. (2001) Spatially restricted expression of candidate taste receptors in the *Drosophila* gustatory system. *Curr. Biol.* 11, 822–835.
Estes P. E., Ho G., Narayanan R. and Ramaswami M. (2000) Synaptic localization and restricted diffusion of a Drosophila neuronal synaptobrevin – green fluorescent protein chimera in vivo. *J. Neurogenet.* **13**, 233–255.
Fox A. N., Pitts R. J., Robertson H. M., Carlson J. R. and Zwiebel L. J. (2001) Candidate odorant receptors from the malaria vector mosquito Anopheles gambiae and evidence of down-regulation in response to blood feeding. *Proc. Natl. Acad. Sci.* **98**, 14693–14697.
Gao Q. and Chess A. (1999) Identification of candidate *Drosophila* olfactory receptors from genomic DNA sequence. *Genomics* **60**, 31–39.
Gao Q., Yuan B. and Chess A. (2000) Convergent projections of *Drosophila* olfactory neurons to specific glomeruli in the antennal lobe. *Nat. Neurosci.* **3**, 780–785.
Hansson B. S. (2002) A bug's smell – research into insect olfaction. *Trends Neurosci.* **25**, 270–274.
Heimbeck G., Bugnon V., Gendre N., Haberlin C. and Stocker R. F. (1999) Smell and taste perception in *Drosophila melanogaster* larva: toxin expression studies in chemosensory neurons. *J. Neurosci.* **19**, 6599–6609.
Hildebrand J. G. and Shepherd G. M. (1997) Mechanisms of olfactory discrimination: converging evidence for common principles across phyla. *Annu. Rev. Neurosci.* **20**, 595–631.

Ishimoto H., Matsumoto A. and Tanimura T. (2000) Molecular identification of a taste receptor gene for trehalose in *Drosophila*. *Science* **289**, 116–119.

Jones K. A., Borowsky B., Tamm J. A., Craig D. A., Durkin M. M., Dai M., Yao W. J., Johnson M., Gunwaldsen C., Huang L. Y. *et al.* (1998) GABA(B) receptors function as a heteromeric assembly of the subunits GABA(B)R1 and GABA(B)R2. *Nature* **396**, 674–679.

Jordan B. A. and Devi L. A. (1999) G-protein-coupled receptor heterodimerization modulates receptor function. *Nature* **399**, 697–700.

Kaupmann K., Malitschek B., Schuler V., Heid J., Froestl W., Beck P., Mosbacher J., Bischoff S., Kulik A., Shigemoto R. *et al.* (1998) GABA(B)-receptor subtypes assemble into functional heteromeric complexes. *Nature* **396**, 683–687.

Krautwurst D., Yau K. W. and Reed R. R. (1998) Identification of ligands for olfactory receptors by functional expression of a receptor library. *Cell* **95**, 917–926.

Laissue P. P., Reiter C., Hiesinger P. R., Halter S., Fischbach K. F. and Stocker R. F. (1999) Three-dimensional reconstruction of the antennal lobe in *Drosophila melanogaster*. *J. Comp. Neurol.* **405**, 543–552.

Leinders-Zufall T., Lane A. P., Puche A. C., Ma W., Novotny M. V., Shipley M. T. and Zufall F. (2000) Ultrasensitive pheromone detection by mammalian vomeronasal neurons. *Nature* **405**, 792–796.

L'Etoile N. D. and Bargmann C. I. (2000) Olfaction and odor discrimination are mediated by the C. elegans guanylyl cyclase ODR-1. *Neuron* 25, 575–586.

Lilly M. and Carlson J. (1990) *smellblind:* a gene required for *Drosophila* olfaction. *Genetics* **124**, 293–302.

Malnic B., Hirono J., Sato T. and Buck L. B. (1999) Combinatorial receptor codes for odors. *Cell* **96**, 713–723.

Marin E. C., Jefferis G. S., Komiyama T., Zhu H. and Luo L. (2002) Representation of the glomerular olfactory map in the *Drosophila* brain. *Cell* **109**, 243–255.

Martini S., Silvotti L., Shirazi A., Ryba N. J. and Tirindelli R. (2001) Co-expression of putative pheromone receptors in the sensory neurons of the vomeronasal organ. *J. Neurosci.* **21**, 843–848.

McKenna M., Monte P., Helfand S. L., Woodard C. and Carlson J. (1989) A simple chemosensory response in *Drosophila* and the isolation of *acj* mutants in which it is affected. *Proc. Natl. Acad. Sci.* **86**, 8118–8122.

McLatchie L. M., Fraser N. J., Main M. J., Wise A., Brown J., Thompson N., Solari R., Lee M. G. and Foord S. M. (1998) RAMPs regulate the transport and ligand specificity of the calcitonin-receptor-like receptor. *Nature* **393**, 333–339.

Mombaerts P. (1999) Seven-transmembrane proteins as odorant and chemosensory receptors. *Science* **286**, 707–711.

Mombaerts P., Wang F., Dulac C., Chao S. K., Nemes A., Mendelsohn M., Edmondson J. and Axel R. (1996). Visualizing an olfactory sensory map. *Cell* **87**, 675–686.

Nef S., Allaman I., Fiumelli H., De Castro E. and Nef P. (1996) Olfaction in birds: differential embryonic expression of nine putative odorant receptor genes in the avian olfactory system. *Mech. Dev.* **55**, 65–77.

Ngai J., Dowling M. M., Buck L., Axel R. and Chess A. (1993) The family of genes encoding odorant receptors in the channel catfish. *Cell* **72**, 657–666.

Oppliger F., Guerin P. and Vlimant M. (2000) Neurophysiological and behavioral evidence for an olfactory function for the dorsal organ and a gustatory one for the terminal organ in *Drosophila melanogaster* larvae. *J. Insect. Physiol.* **46**, 135–144.

Python F. and Stocker R. F. (2002) Adult-like complexity of the larval antennal lobe of *D. melanogaster* despite markedly low numbers of odorant receptor neurons. *J. Comp. Neurol.* **445**, 374–387.

Ressler K. J., Sullivan S. L. and Buck L. B. (1994) Information coding in the olfactory system: evidence for a stereotyped and highly organized epitope map in the olfactory bulb. *Cell* **79**, 1245–1255.

Rocheville M., Lange D. C., Kumar U., Patel S. C., Patel R. C. and Patel Y. C. (2000) Receptors for dopamine and somatostatin: formation of hetero-oligomers with enhanced functional activity. *Science* **288**, 154–157.

Rodriguez I., Punta K. D., Rothman A., Ishii T. and Mombaerts P. (2002) Multiple new and isolated families within the mouse superfamily of V1r vomeronasal receptors. *Nat. Neurosci.* **5**, 134–140.

Scott K., Brady R., Jr, Cravchik A., Morozov P., Rzhetsky A., Zuker C. and Axel R. (2001) A chemosensory gene family encoding candidate gustatory and olfactory receptors in *Drosophila*. *Cell* **104**, 661–673.

Sengupta P., Chou J. H. and Bargmann C. I. (1996) *odr-10* encodes a seven transmembrane domain olfactory receptor required for responses to the odorant diacetyl. *Cell* **84**, 899–909.

Shanbhag S. R., Muller B. and Steinbrecht R. A. (1999) Atlas of olfactory organs of *Drosophila melanogaster*. 1. Types, external organization, innervation and distribution of olfactory sensilla. *Int. J. Insect Morphol. Embryol.* **28**, 377–397.

Shepherd G. M. (1994) Discrimination of molecular signals by the olfactory receptor neuron. *Neuron* **13**, 771–790.

Siddiqi O. (1987) Neurogenetics of olfaction in *Drosophila melanogaster*. *Trends Genet.* **3**, 137–142.

Speca D. J., Lin D. M., Sorensen P. W., Isacoff E. Y., Ngai J. and Dittman A. H. (1999) Functional identification of a goldfish odorant receptor. *Neuron* **23**, 487–498.

Stocker R. F. (1994) The organization of the chemosensory system in *Drosophila melanogaster:* a review. *Cell Tissue Res.* **275**, 3–26.

Stocker R. F., Heimbeck G., Gendre N. and de Belle J. S. (1997) Neuroblast ablation in *Drosophila* P[GAL4] lines reveals origins of olfactory interneurons. *J. Neurobiol.* **32**, 443–456.

Störtkuhl K. F. and Kettler R. (2001) Functional analysis of an olfactory receptor in *Drosophila melanogaster*. *Proc. Natl. Acad. Sci.* **98**, 9381–9385.

Touhara K., Sengoku S., Inaki K., Tsuboi A., Hirono J., Sato T., Sakano H. and Haga T. (1999) Functional identification and reconstitution of an odorant receptor in single olfactory neurons. *Proc. Natl. Acad. Sci.* **96**, 4040–4045.

Troemel E. R., Chou J. H., Dwyer N. D., Colbert H. A. and Bargmann C. I. (1995) Divergent seven transmembrane receptors are candidate chemosensory receptors in *C. elegans*. *Cell* **83**, 207–218.

Troemel E. R., Kimmel B. E. and Bargmann C. I. (1997) Reprogramming chemotaxis responses: sensory neurons define olfactory preferences in *C. elegans*. *Cell* **91**, 161–169.

Ueno K., Ohta M., Morita H., Mikuni Y., Nakajima S., Yamamoto K. and Isono K. (2001) Trehalose sensitivity in *Drosophila* correlates with mutations in and expression of the gustatory receptor gene *Gr5a*. *Curr. Biol.* **11**, 1451–1455.

Vassar R., Chao S. K., Sitcheran R., Nunez J. M., Vosshall L. B. and Axel R. (1994) Topographic organization of sensory projections to the olfactory bulb. *Cell* **79**, 981–991.

Vosshall L. B., Amrein H., Morozov P. S., Rzhetsky A. and Axel R. (1999) A spatial map of olfactory receptor expression in the *Drosophila* antenna. *Cell* **96**, 725–736.

Vosshall L. B., Wong A. M. and Axel R. (2000) An olfactory sensory map in the fly brain. *Cell* **102**, 147–159.

Wang F., Nemes A., Mendelsohn M. and Axel R. (1998) Odorant receptors govern the formation of a precise topographic map. *Cell* **93**, 47–60.

Weiner J. (1999) *Time, Love, Memory: A Great Biologist and his Quest for the Origins of Behavior.* Knopf, New York.

Wetzel C. H., Behrendt H.-J., Gisselmann G., Störtkuhl K. F., Hovemann B. and Hatt H. (2001) Functional expression and characterization of a *Drosophila* odorant receptor in a heterologous cell system. *Proc. Natl. Acad. Sci.* **98**, 9377–9380.

White J. H., Wise A., Main M. J., Green A., Fraser N. J., Disney G. H., Barnes A. A., Emson P., Foord S. M. and Marshall F. H. (1998) Heterodimerization is required for the formation of a functional GABA(B) receptor. *Nature* **396**, 679–682.

Whitfield C. W., Band M. R., Bonaldo M. F., Kumar C. G., Liu L., Pardinas J. R., Robertson H. M., Soares M. B. and Robinson G. E. (2002) Annotated expressed sequence tags and cDNA microarrays for studies of brain and behavior in the honey bee. *Genome Res.* **12**, 555–566.

Wong A. M., Wang J. W. and Axel R. (2002) Spatial representation of the glomerular map in the *Drosophila* protocerebrum. *Cell* **109**, 229–241.

Zemelman B. V. and Miesenbock G. (2001) Genetic schemes and schemata in neurophysiology. *Curr. Opin. Neurobiol.* **11**, 409–414.

Zhang X. and Firestein S. (2002) The olfactory receptor gene superfamily of the mouse. *Nat. Neurosci.* **5**, 124–133.

Zhao H., Ivic L., Otaki J. M., Hashimoto M., Mikoshiba K. and Firestein S. (1998). Functional expression of a mammalian odorant receptor. *Science* **279**, 237–242.

Zozulya S., Echeverri F. and Nguyen T. (2001) The human olfactory receptor repertoire. *Genome Biol.* **2**, research 0018.1–0018.12.

20

Transduction mechanisms of olfactory sensory neurons

Jürgen Krieger and Heinz Breer

20.1 Introduction

Animals use chemical cues for many of their requirements, including the search for and the selection of food and mating partners as well as the monitoring of prey and predators. Chemical signals are also important for communicating with other animals. An essential prerequisite for exploring the chemical environments is the capability to sense chemical compounds and to convert the information encoded in the structure and concentration of odorous molecules into the language of the nervous system, into electrical impulses. The electrical signals are conveyed onto higher brain centers where they are integrated and contribute to elicit appropriate behavioral responses. Functional units for bringing the odor world "on-line" with the brain are the specialized chemosensory neurons. These cells are capable of converting chemical signals into neuronal activity. This transduction process is initiated upon interaction of air-borne chemical compounds with suitable receptors within the membrane of these cells. This molecular interaction triggers complex intracellular reaction cascades which lead to a depolarization of the plasma membrane. The change in membrane potential, the receptor potential, is transformed into a frequency of action potentials which are conveyed via the axon to relay stations in the brain.

This chapter describes some of the principles and mechanisms underlying the primary processes of olfactory signaling, the chemo-electrical signal transduction. We will focus on molecular events that follow the interaction of odorants with olfactory sensory neurons, and leave aside perireceptor events including odorant binding proteins, odor degrading and modifying enzymes. These aspects, as well as details about olfactory receptors, will be described in other chapters.

20.2 The cellular compartments of chemosensory signal transduction

In spite of the extreme diversity in their external appearance, cross-phyletic comparisons of olfactory sensory systems have revealed striking similarities on the cellular level (Ache, 1994; Hildebrand and Shepherd, 1997). In both insects and vertebrates, specialized bipolar sensory neurons, surrounded by supporting cells, are employed to sense odorous compounds in the environment. In both phyla, each olfactory sensory cell projects an unbranched axon, gathered in bundles, towards higher processing centers and a dendrite towards the air-facing surface of the olfactory organ. In insects, the dendrites terminate in an unbranched or branched ciliary structure bathing in a characteristic lymph fluid which fills a sensillar (olfactory) hair structure of the antennae. In vertebrates, the dendrites of olfactory neurons typically end in an apical knob, from which several to about ten cilia extend into a thin layer of mucus covering the olfactory epithelium. Certain chemosensory cells in the main olfactory epithelium as well as all sensory neurons of the vomeronasal organ, which is supposed to be involved in pheromone reception (Keverne, 1999), differ from this general morphology and carry apical microvillar extensions instead of cilia. Due to the exposure to the air-facing frontier, both the apical ciliary and microvillar structures of olfactory neurons were long ago proposed as sensory organells involved in odor recognition and signal transduction.

Research over the past two decades has provided several lines of evidence which support this view. Initial studies have demonstrated that deciliated olfactory neurons do not respond to odors (Kurahashi and Shibuya, 1989), and immunological analysis has shown that putative odorant receptors (Krieger *et al.*, 1994; Menco *et al.*, 1997; Elmore and Smith, 2001) as well as characteristic transduction elements (Chen *et al.*, 1986; Bakalyar and Reed, 1990; Asanuma and Nomura, 1991) are present in the specialized ciliary structures of these cells. For the microvilli of vomeronasal sensory cells, corresponding results have previously also been obtained. Putative pheromone receptors, G-proteins and ion channels were located in the microvillar membrane at the EM level (Takigami *et al.*, 1999; Menco *et al.*, 2001). Further support for an essential role of the ciliary compartments in olfactory signaling was obtained from electrophysiological recordings demonstrating that the chemosensory responsiveness of olfactory receptor cells depends on the application of odorant stimuli onto the cilia (Firestein and Werblin, 1989; Lowe and Gold, 1991). More recently, fluorescence imaging studies have shown that odor-induced intracellular Ca^{2+} signals originate in the cilia (Leinders-Zufall *et al.*, 1998a, b). Altogether, there is now convincing evidence that the ciliary and microvillar extensions of the highly polarized sensory neurons are the cellular compartments which are specialized for signal recognition and signal transduction. In an evolutionary context, cilia and microvilli may be considered as specializations to increase the surface area of the chemosensory membrane, thus enhancing the responsiveness of the cells.

20.3 Signal transduction cascades in olfactory sensory cells

There is a general consensus that the interaction of odor molecules with suitable olfactory receptor proteins in the chemosensory membrane is the initial step for

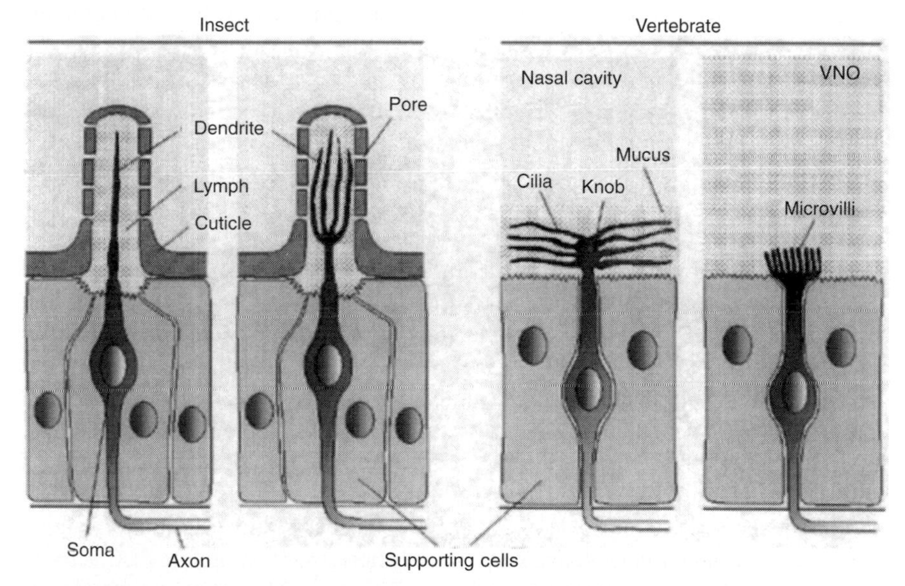

Figure 20.1 Olfactory sensory cells of both insects and vertebrates are primary sensory cells, i.e. they are bipolar neurons extending a sensory dendritic process towards the odorous environment and projecting an unbranched axon directly to specialized target regions in the central nervous system.

In insects, the sensory compartment of antennal neurons are one or several thin ciliary processes of the dendrite bathing in a characteristic fluid (sensillum lymph) which fills the sensillar hair structures of the antennae.

In vertebrates, most of the olfactory neurons of the nasal epithelium protrude from the apical dendritic knob a variety of cilia into the protective mucus layer, thus enlarging the sensory surface area of the cells. A subpopulation of sensory neurons in the main olfactory epithelium and all of the neurons in the vomeronasal organ (VNO) are structurally different; their apical region of the dendrite extends in an array of microvilli.

activating an olfactory sensory neuron. The molecular identity of olfactory receptors has been unraveled throughout the last decade, first in several vertebrate species, in model organisms, like *C. elegans* and *Drosophila* (reviewed in Mombaerts, 1999; see also Vosshall, Chapter 19, in this volume) and recently in *Heliothis virescens* (Krieger *et al.*, 2002). In all organisms examined, large families of genes were identified which encode diverse heptahelical receptor proteins: supposedly G-protein-coupled receptors (GPCRs) involved in recognition of odorous chemicals. The notion that olfactory receptors are members of the GPCR superfamily is in line with biochemical and molecular biological data indicating that G-protein-mediated reaction cascades are elicited upon interaction of odorous compounds with olfactory sensory cells (Breer *et al.*, 1994). For the initial step of signal recognition and transduction, olfactory receptors play a dual role. First, they are the basis for discrimination between different odorous compounds, i.e. only cells with a suitable receptor type will respond to a given odorant. Second, upon binding of an appropriate ligand, receptors transmit the signal from the extracellular to the intracellular face of the membrane. The intramolecular mechanisms underlying this signal transformation are unknown, but most likely involve ligand-induced structural changes of the receptor protein, probably related to the light-induced structural changes of rhodopsin (Grobner *et al.*, 2000).

20.3.1 Second messenger cascades in insect sensory cells

Although the insect olfactory system has been an ideal model system for studying various aspects of olfaction, our understanding about the molecular principles and mechanisms of olfactory signaling in insects is still far from complete and lags behind the knowledge about molecular issues of vertebrate olfaction. Experimental evidence indicates that suitable odorants trigger in insect antennal cells the formation of the second messengers inositol 1,4,5-trisphosphate (IP_3) and diacylglycerol (DAG) by activation of phospholipase C (PLC). Initial biochemical evidence for a role of the PLC/IP3 cascade in insect olfactory signal transduction demonstrates that application of pheromones stimulate PLC activity in antennal preparations from various insect species. This response was found to be GTP dependent (Boekhoff *et al.*, 1990a, b). The hint for a possible role of PLC in insect chemosensory signaling was supported by the observation that *Drosophila* mutants defective in the PLC-encoding norpA gene display severely impaired olfactory capabilities (Riesco-Esgovar *et al.*, 1995), as well as by immunological studies documenting the presence of phospholipase C in homogenates of isolated sensilla from the silkmoth *Antheraea polyphemus* (Maida *et al.*, 2000). More direct evidence for an immediate role of the PLC/IP_3 pathway in insect olfactory signal transduction resulted from experiments stimulating antennal preparations from cockroaches (Breer *et al.*, 1990; Boekhoff *et al.*, 1990a, b) and moths (Boekhoff *et al.*, 1993) with sex pheromones. A species-, tissue- and sex-specific increase of IP_3 concentrations was determined. Moreover,

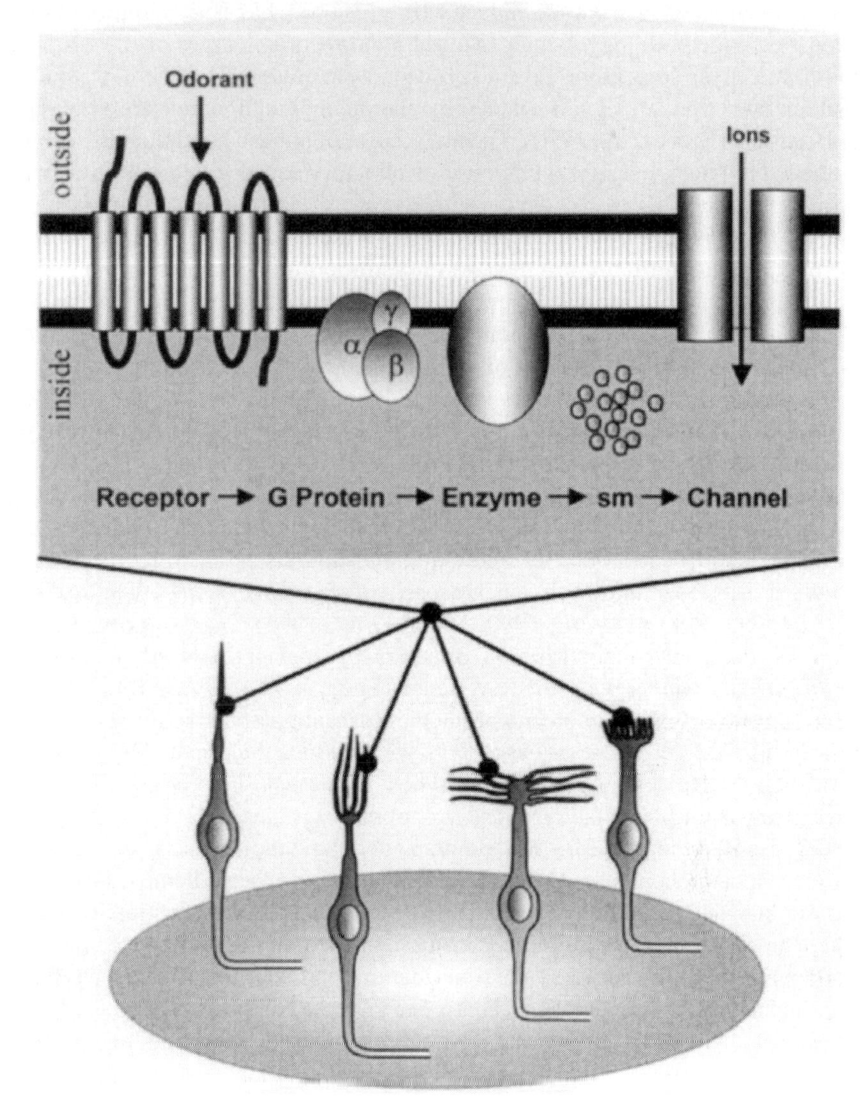

Figure 20.2 The ciliary and microvillar structures of olfactory neurons are the cellular compartments specialized for signal recognition and signal transduction. The transduction of olfactory stimuli into neuronal signals is mediated by intracellular reaction cascades, which are related to the molecular machinery which cells generally use to process external signals. Heptahelical membrane receptors interact with suitable odorous chemicals and transfer this information onto trimeric G-proteins which in turn activate enzymes capable of generating intracellular messengers and thereby amplifying the signal. The second messenger, directly or indirectly, modifies the permeability of the membrane and thus elicits the electrical response of the neurons.

the time course of pheromone-induced IP_3 responses has been monitored by a "stop-flow" methodology, which allowed us to record changes of the IP_3 level within the physiologically relevant subsecond time range. These kinetic studies demonstrated that IP_3 signals elicited by pheromonal compounds are very rapid and transient (Breer *et al.*, 1990). The time course of about a hundred milliseconds matches the phasic electrical response of olfactory sensory neurons to odorant stimulation, and is in accordance with the view that the second messengers formed may mediate odor-induced changes in membrane conductance. In fact, currents induced by IP_3 via activation of distinct cation channels in the plasma membrane have been recorded from olfactory neurons of different insect species (Stengl *et al.*, 1993, 1994; Wegener *et al.*, 1997). Furthermore, IP_3 receptors have been located in dendrites of pheromone-sensitive neurons by means of immunocytochemical approaches (Laue and Steinbrecht, 1997) but it is unclear if these elements really account for the stimulus-induced excitation of the cells.

Efforts to unravel the molecular identity of key elements in the transduction cascade, notably the application of immunological and cloning approaches, has led to the identification of Go- and Gq-alpha subunits in antennal tissue from various insects, including moths, cockroaches and flies (Breer *et al.*, 1988; Boekhoff *et al.*, 1990b; Raming *et al.*, 1990; Talluri *et al.*, 1995). Gq-like proteins have been localized in the dendrites of olfactory neurons from *Bombyx mori* and *Antheraea polyphemus* at the EM level (Laue *et al.*, 1997). Biochemical, electrophysiological and molecular biological data suggest that in insect olfactory neurons the PLC/IP_3 reaction cascade may in fact be a major pathway for signal transduction. However, the question of how the activation of phospholipase C, the hydrolysis of phosphatidyl inositol 4,5-bisphosphate (PIP2) or the elevation of IP3 and DAG levels leads to membrane depolarization remains unanswered. In this respect, mechanisms of signal transduction in insect chemosensory cells and our knowledge about it resemble the scenario of phototransduction in insects, where an essential role of PLC has been established, but the downstream events leading eventually to gating of the transduction channels are still elusive (Hardie and Raghu, 2001). Attention has recently been directed to the membrane-delimited consequences of PLC activity: namely, the generation of DAG and the reduction in PIP2 levels. DAG is not only an activator of protein kinase C (PKC), but also a substrate for DAG-lipase leading to the release of polyunsaturated fatty acids, such as arachidonic acid, which have been shown to activate certain ion channels of the TRP ("transient receptor potential") family. Finally, it seems also conceivable that depletion of phosphatidylinositol 4, 5-bisphosphate in the membrane, due to increased PLC activity, may cause the gating of ion channels leading to the electrical response. A regulatory (activating, inhibiting) role of PIP2 for the activity of various ion channel types has recently been observed (Hilgemann *et al.*, 2001). The slow but steady progress in understanding phototransduction in insects may be beneficial for the efforts trying to unravel the enigmas of PIP2/PLC/IP_3-mediated signal transduction of insect chemosensory neurons.

However, at this point a role for other second messengers in insect olfaction cannot be excluded. In fact, the observation that distinct cyclic nucleotide-gated ion channels are expressed in insect antennae (Baumann *et al.*, 1994; Krieger *et al.*, 1999a) may be the first evidence for functional implications of a cyclic nucleotide pathway.

20.3.2 Second messenger cascades in lobster sensory cells

While experimental evidence concerning the molecular mechanisms and the diversity of reaction cascades involved in olfactory signal transduction in antennal cells of insects is still fragmentary, a much more detailed picture has been established for signal transduction in chemosensory cells of the lobster, another member of the arthropod phyla. The bipolar chemosensory neurons of the lobster antennule respond to stimulation with odorous compounds either with an excitation or an inhibition; i.e. cells are equipped to respond to one odor with a depolarization and excitation as well as to another odor with a hyperpolarization and inhibition.

This dual responsiveness of the cells coincides with the results of biochemical studies demonstrating that exposure of olfactory outer dendritic membranes from lobster to distinct odorants elicits a rapid and transient formation of either cAMP or IP_3. Odorants differentially stimulate the two second messenger cascades in agreement with their incidence of exciting and inhibiting cells physiologically (Ache *et al.*, 1993). From the results of elegant studies over the last decade, it is now possible for the lobster sensory cells to assign functionality to the cAMP and IP_3 cascade, respectively. The two second messengers regulate separate, but opposing ionic conductances (Michel and Ache, 1992; Fadool and Ache, 1992). The messenger IP_3 activates either a non-selective cation conductance or a Ca^{2+} selective conductance; both lead to excitation of the cell. The IP_3-activated channels are probably not the main current-carrying channels. A major portion of the receptor current is carried by secondarily activated non-specific cation channels which open upon a rise of the intracellular Na^+ concentration (Zhainazarov *et al.*, 1998). In contrast, the alternative messenger cAMP activates a K^+ conductance that causes hyperpolarization and opposes the excitation. The fundamental implication of these findings is that the dual transduction pathways allow the sensory cell to serve as an integrating unit (Ache, 1994). Very recent data suggest that phosphatidylinositol-3-kinase and its reaction products 3-phosphoinositides may also contribute to the transduction process in lobster olfactory neurons (Zhainazarov *et al.*, 2001).

20.3.3 Second messenger cascades in vertebrate olfactory sensory cells

Enormous progress has been made over the last decade towards an understanding of the olfactory signal transduction process in vertebrate olfactory neurons. A rather complex picture has emerged. Most detailed and convincing evidence has been obtained for the second messenger pathway involving cyclic 3′, 5′adenosine monophosphate (cAMP). Most of the elements of the molecular machinery underlying this reaction cascade have been identified by molecular cloning. The current concept assumes that the cAMP cascade is triggered by activating an olfactory specific G_s-like protein (G_{olf}) (Jones and Reed, 1989) which stimulates a characteristic adenylyl cyclase (AC), the type III isoform (Bakalyar and Reed, 1990). As a consequence, ATP is converted into cAMP and this messenger is the activating ligand of cyclic nucleotide-gated, non-selective cation channels (CNCs) in the ciliary membrane. Two ion-conducting α-subunits (CNCα3 and CNCα4) and two regulatory subunits (CNCβ1b) are supposed to form the heterotetrameric channel complex (Bonigk *et al.*, 1999). It is the gating of this channel complex which accounts for the cAMP-induced influx of Na^+ and Ca^{2+} ions resulting in a depolarization of the membrane. The influx of cations not only depolarizes the membrane, but at the same time causes an elevation of the intracellular Ca^{2+} concentration. Changes in intracellular Ca^{2+} levels usually have immediate consequences for various cellular functions. In the cilia of olfactory sensory neurons elevation of the Ca^{2+} concentration activates ion channels that permeate chloride ions (Kurahashi and Yau, 1993; Gold and Lowe, 1993). Interestingly, the calcium-activated chloride conductance further depolarizes the cell. This untypical reaction is based on the unusually high intracellular Cl^- concentration in olfactory neurons leading to an efflux of Cl^- ions through the Ca^{2+} activated chloride channels. The Ca^{2+} mediated linkage from CN channels to chloride channels has been considered as a mechanism that contributes to the high-gain and low-noise amplification of olfactory neuron responses (Menini, 1999). In addition, it has been proposed that this mechanism may also be important in the context of changing ionic concentrations in the nasal mucus. Under circumstances where cation concentrations in the mucus are low and consequently the actual Na^+ gradient is insufficient for an odor-induced threshold current, the Ca^{2+}-activated Cl^- current would assure the triggering of excitation.

The anosmic phenotypes of mice strains which are deficient in G_{olf} (Belluscio *et al.*, 1998), ACIII (Wong *et al.*, 2000) and functional relevant channel subunits (Brunet *et al.*, 1996; Baker *et al.*, 1999) emphasize the central role of the AC/cAMP pathway for signal transduction in vertebrate olfactory sensory neurons. These features may be considered as evidence for cAMP as the sole relevant second messenger in vertebrate olfaction.

However, mechanisms involved in olfactory signaling seem much more diverse than hitherto expected. Recently, a subset of olfactory neurons has been described that lacks most of the elements for the AC/cAMP cascade; these cells rather express distinct subtypes of guanylyl cyclase (GC-D) and phosphodiesterase (PDE2) (Juilfs *et al.*, 1997), as well as a characteristic channel subtype (Meyer *et al.*, 2000). These sensory neurons, which project to the so-called "necklace glomeruli," are supposed to be tuned to unique odorants, but neither the receptor/transduction mechanisms nor the relevant odors or pheromones for the cells have yet been identified.

Biochemical studies have shown that certain potent odorants do not activate the adenylyl or guanylyl cyclase and induce no measureable change of the cyclic nucleotide levels in isolated olfactory cilia preparations (Sklar et al., 1986; Bruch and Teeter, 1990). These odorants, rather, elicit an increase of inositol 1,4,5-trisphosphate (IP_3) concentration. The IP_3 responses displayed a rapid and transient kinetic similar to that observed for the cAMP concentration (Breer *et al.*, 1990; Boekhoff and Breer, 1992; Ronnett *et al.*, 1993). These findings led to the concept that a PLC/IP_3 cascade may operate in some vertebrate olfactory sensory neurons. This view was supported by analysis of individual olfactory neurons. In calcium imaging experiments it was found that calcium signals induced by odorants, like lilial, were completely blocked by U73122, a specific inhibitor of PLC. Calcium signals in cells responding to citralva, an odorant that activates the cAMP cascade, were not affected by U73122. Subsequent single cell PCR analysis of neurons responding to lilial confirmed that these cells indeed express phospholipase C as well as a G-protein of the $G_{\alpha o}$ type (Noe and Breer, 1998). Based on activation-dependent photolabeling experiments followed by immunoprecipitation with subtype-specific antibodies (Schandar *et al.*, 1998), it was demonstrated that the odorant-induced activation of PLC is most probably mediated by a trimeric G-protein of the $G_{\alpha o}$ type. Although IP_3-relevant ion channels have been identified in membranes of olfactory sensory cells (Cunningham *et al.*, 1993; Lischka *et al.*, 1999), it is still unclear if and how the PLC/IP_3 cascade may lead to a depolarization of the membrane, to an electrical response of the cell. This is again reminiscent of the situation of phototransduction in flies (see section 20.3.1).

Very recent efforts to further evaluate the functional implications of phosphoinositol signaling in vertebrate chemosensory transduction have led to the discovery that not only the phosphoinositol breakdown pathway catalyzed by phospholipase C, but also membrane phosphoinositols themselves involving phosphoinositide-3-kinase may play a functional role in the transduction process. It was found that 3-phosphoinositides signaling in concert with the canonical phosphoinositide turnover pathway modulate the cyclic nucleotide signaling cascade downstream of the receptor. The data suggest that 3-phosphoinositide, the primary product of PI3K activity, attenuates the cyclic nucleotide-dependent excitation of olfactory neurons (Spehr *et al.*, 2002).

20.3.4 Second messenger cascades in vomeronasal sensory cells

Although the functional implications of the PLC/IP_3 cascade in olfactory neurons of nasal epithelium remains mysterious, there is mounting evidence that phospholipase C is a key enzyme for signal transduction in chemosensory neurons of the vomeronasal organ (VNO) which is supposed to play a major role in pheromone perception of rodents. Transduction elements characteristic of olfactory neurons, such as G_{olf}, ACIII and CNGC, are absent from VNO neurons (Berghard *et al.*, 1996). Instead, these cells express G-proteins of the G_o or G_{i2} subtypes and ion channels which are members of the "transient receptor potential" (TRP) family. It was found that the subtype TRP2 is highly expressed in the microvilli of these cells (Liman *et al.*, 1999), suggesting a functional role of this channel in processing pheromone signals, a notion that was recently supported by studying knockout mice strains (Stowers *et al.*, 2002). The emerging picture for a key role of the phosphoinositide cascade in VNO-sensory cells is supported by recent studies demonstrating that candidate pheromonal proteins such as hamster aphrodisn and rat α_{2u}-globulin, as well as urine and seminal fluid, activate phospholipase C in membrane preparations from the VNO (Kroner *et al.*, 1996; Wekesa and Anholt, 1997; Sasaki *et al.*, 1999; Krieger *et al.*, 1999b). The stimulus-induced increase of the IP_3 level was accompanied by a decrease of the cAMP concentration (Wang *et al.*, 1997; Rössler *et al.*, 2000) which is probably due to an inhibition of the adenylyl cyclase type VI by Ca^{2+} mobilized from intracellular stores by increasing IP_3 concentration (Rössler *et al.*, 2000). Urine-induced IP_3 formation was found to be mediated by G_i- and G_o-protein; more detailed analysis revealed that G_i-proteins are activated by lipophilic volatile compounds and G_o-proteins by proteinaceous compounds (Krieger *et al.*, 1999b). Consistent with these biochemical data are the results of physiological recording indicating that urinary responses of VNO neurons are blocked by pharmacological inhibitors of phospholipase C (Inamura *et al.*, 1997; Holy *et al.*, 2000).

20.4 Concluding remarks

Chemical sensing, the conversion of largely volatile extraneous chemical cues into the language of the nervous system, is the task of chemosensory neurons. The mechanisms employed to accomplish the chemo-electrical signal transduction process seem to be identical or related in phylogenetically diverse species as well as in different cell types. Generally, olfactory signals seem to be converted and amplified by means of intracellular reaction cascades and second messengers which eventually trigger the electrical response. Research over the last decade has revealed that generally employed molecular elements and cellular processes have been recruited and fine tuned in the highly specialized olfactory neurons for sensory signal transduction. Although some of the principles have been elucidated and some of the molecular elements have been identified in recent years towards a detailed understanding of the physiological process "olfactory signal transduction," it is now an important challenge to explore how identified molecular components interact to build the transduction machinery which allows highly precise responses as well as complex signal integration of the sensory cells.

The evolutionary conservation of chemosensitivity and the emerging conservation of the underlying molecular mechanisms call for comparative studies of phylogenetically diverse species and emphasize the potential of invertebrate organisms for elucidating the molecular basis for odor sensing in general. Not only are congruent features of interest, but also species-specific deviations and specializations which may account for the animal's unique "olfactory biology." In this respect, research on insect species which are important pests, damaging crops and transmitting diseases, may be particularly rewarding. Understanding the olfactory mechanisms of these insects, which heavily depend on their sense of smell to find food and mates, may open new avenues to control insect pests via olfactory cues. The perspective reduction of neurotoxic compounds employed in pest control would have immediate ecological and economical implications.

References

Ache B. W. (1994) Towards a common strategy for transducing olfactory information. *Sem. Cell. Biol.* **5**, 55–63.

Ache B. W., Hatt H., Breer H., Boekhoff I. and Zufall F. (1993) Biochemical and physiological evidence for dual transduction pathways in lobster olfactory neurons. *Chem. Senses* **18**, 523.

Asanuma N. and Nomura H. (1991) Cytochemical localization of adenylate cyclase activity in rat olfactory cells. *Histochem. J.* **23**, 83–90.

Bakalyar H. A. and Reed R. R. (1990) Identification of a specialized adenylyl cyclase that may mediate odorant detection. *Science* **250**, 1403–1406.

Baker H., Cummings D. M., Munger S. D., Margolis J. W. Franzen L., Reeed R. R. and Margolis F. L. (1999) Targeted deletion of cyclic nucleotide-gated channel subunit (OCNG1): biochemical and morphological consequences in adult mice. *J. Neurosci.* **19**, 9313–9321.

Baumann A., Frings S., Godde M., Seifert R. and Kaupp U.B. (1994) Primary structure and functional expression of a *Drosophila* cyclic nucleotide-gated channel present in eyes and antennae. *EMBO J.* **13**, 5040–5050.

Belluscio L., Gold G. H., Nemes A. and Axel R. (1988) Mice deficient in G(olf) are anosmic. *Neuron* **20**, 69–81.

Berghard A., Buck L. B. and Liman E. R. (1996) Evidence for distinct signaling mechanisms in two mammalian olfactory sense organs. *Proc. Natl. Acad. Sci. USA* **93**, 2365–2369.

Boekhoff I., Strotmann J., Raming K., Tareilus E. and Breer H. (1990a) Odorant-sensitive phospholipase C in insect antennae. *Cell* Signal **2**, 49–56.

Boekhoff I., Raming K. and Breer H. (1990b) Pheromone-induced stimulation of inositol-trisphosphate formation in insect antennae is mediated by G-proteins. *J. Comp. Physiol. B* **160**, 99–103.

Boekhoff I. and Breer H. (1992) Termination of second messenger signaling inolfaction. *Proc. Natl. Acad. Sci. USA* **89**, 471–474.

Boekhoff I., Seifert E., Göggerle S., Lindemann M., Krüger B.-W. and Breer H. (1993) Pheromone-induced second-messenger signaling in insect antennae. *Insect Biochem. Mol. Biol.* **23**, 757–762.

Bonigk W., Bradley J., Müller F., Sesti F., Boekhoff I., Ronnett G. V., Kaupp U. B. and Frings S. (1999) The native rat olfactory cyclic nucleotide-gated channel is composed of three distinct subunits. *J. Neurosci.* **19**, 5332–5347.

Breer H., Raming K. and Boekhoff I. (1988) G-proteins in the antennae of insects. *Naturwiss.* **75**, 627.

Breer H., Boekhoff I. and Tareilus E. (1990) Rapid kinetics of second messenger formation in olfactory transduction. *Nature* **345**, 65–68.

Breer H., Raming K. and Krieger J. (1994) Signal recognition and transduction in olfactory neurons. *Biochim. Biophys. Acta* **1224**, 277–287.

Bruch R. C. and Teeter J. H. (1990) Cyclic AMP links amino acid chemoreceptors to ion channels in olfactory cilia. *Chem. Senses* **15**, 419–430.

Brunet L. J., Gold G. H. and Ngai J. (1996) General anosmia caused by a targeted disruption of the mouse olfactory cyclic nucleotide-gated cation channel. *Neuron* **17**, 681–693.

Chen Z., Pace U., Heldman J., Shapira A. and Lancet D. (1986) Isolated frog olfactory cilia: a preparation of dendritic membranes form chemosensory neurons. *J. Neurosci.* **6**, 2146–2154.

Cunningham A. M., Ryugo D. K., Sharp A. H., Reed R. R., Snyder S. H. and Ronnett G. V. (1993) Neuronal inositol 1,4,5-trisphosphate receptor localized to the plasma membrane of olfactory cilia. *J. Neurosci.* **57**, 339–352.

Elmore T. and Smith P. S. (2001) Putative odor receptor 43b localizes to dendrites of olfactory neurons. *Insect Biochem. Mol. Biol.* **31**, 791–798.

Fadool D. A. and Ache B. W. (1992). Plasma membrane inositol 1,4,5-trisphosphate-activated channels mediate signal transduction in lobster olfactory receptor neurons. *Neuron* **9**, 907–918.

Firestein S. and Werblin F. (1989) Odor-induced membrane currents in vertebrate-olfactory receptor neurons. *Science* **244**, 79–82.

Gold G. H. and Lowe G. (1993) Nonlinear amplification by calcium-dependent chloride channels in olfactory receptor cells. *Nature* **366**, 283–286.

Grobner G., Burnett I. J., Glaubitz C., Choi G., Mason A. J. and Watts A. (2000) Observations of light-induced structural changes of retinal within rhodopsin. *Nature* **405**, 810–813.

Hardie R.C. and Raghu P. (2001) Visual transduction in Drosophila. *Science* **413**, 186–193.

Hildebrand J. G. and Shepherd G. M. (1997) Mechanisms of olfactory discrimination: converging evidence for common principles across phyla. *Annu. Rev. Neurosci.* **20**, 595–631.

Hilgemann D. W., Feng S. and Nasuhoglu C. (2001) The complex and intriguing lives of PIP2 with ion channels and transporters. *Sci. STKE* **111**, 19.

Holy T. E., Dulac C. and Meister M. (2000) Responses of vomeronasal neurons to natural stimuli. *Science* **289**, 1569–1572.

Inamura K., Kashiwayanagi M. and Kurihara K. (1997) Blockage of urinary responses by inhibitors for IP_3 mediated pathway in rat vomeronasal sensory neurons. *Neurosci. Lett.* **233**, 129–132.

Jones D. T. and Reed R. R. (1989) Golf. An olfactory neuron-specific G-protein involved in odorant signal transduction. *Science* **244**, 790–795.

Juilfs D. M., Fülle H.-J., Zhao A. Z., Houslay M. D., Garbers D. L. and Beavo J. A. (1997) A subset of olfactory neurons that selectively express cGMP-stimulated phosphodiesterase (PDE2) and guanylyl cyclase-D define a unique olfactory signal transduction pathway. *Proc. Natl. Acad. Sci. USA* **94**, 3388–3395.

Keverne E. B. (1999) The vomeronasal organ. *Science* **286**, 716–720.

Krieger J., Raming K., Dewer Y. M., Bette S., Conzelmann S. and Breer H. (2002) A divergent gene family encoding candidate olfactory receptors of the moth *Heliothis virescens*. *Eur. J. Neurosci.* **16**, 619–628.

Krieger J., Schleicher S., Strotmann J., Wanner I., Boekhoff I., Raming K., de Geus P. and Breer H. (1994) Probing olfactory receptors with sequence-specific antibodies. *Eur. J. Biochem.* **219**, 829–835.

Krieger J., Strobel J., Vogl A., Hanke W. and Breer H. (1999a) Identification of a cyclic nucleotide- and voltage-activated ion channel from insect antennae. *Insect Biochem. Mol. Biol.* **29**, 255–267.

Krieger J., Schmitt A., Löbel D., Gudermann T., Schultz G., Breer H. and Boekhoff I. (1999b). Selective activation of G protein subtypes in the vomeronasal organ upon stimulation with urine-derived compounds. *J. Biol. Chem.* **274**, 4655–4662.

Kroner C., Breer H., Singer A. G. and O'Connell R. J. (1996) Pheromone-induced second messenger signaling in the hamster vomeronasal organ. *Neuroreport* **7**, 2989–2992.

Kurahashi T. and Shibuya T. (1989) Ca(2+)-dependent adaptive properties in solitary olfactory receptor cell of the newt. *Brain Res.* **515**, 262–268.

Kurahashi T. and Yau K.W. (1993) Co-existence of cationic and chloride components in odorant-induced current of vertebrate olfactory receptor cells. *Nature* **363**, 71–75.

Laue M., Maida R. and Redkozubov A. (1997) G-protein activation, identification and immunolocalization in pheromone-sensitive sensilla trichodea of moths. *Cell Tissue Res.* **288**, 149–158.

Laue M. and Steinbrecht R. A. (1997) Topochemistry of moth olfactory sensilla. *Int. J. Insect Morphol. Embryol.* **26**, 217–228.

Leinders-Zufall T., Greer C. A., Shepherd G. M. and Zufall F. (1998a) Imaging odor-induced calcium transients in single olfactory cilia: specificity of activation and role in transduction. *J. Neurosci.* **18**, 5630–5639.

Leinders-Zufall T., Greer C. A., Shepherd G. M. and Zufall F. (1998b) Visualizing odor detection in olfactory cilia by calcium imaging. *Ann. N.Y. Acad. Sci.* **855**, 205–207.

Liman E. R., Corey D. P. and Dulac C. (1999) TRP2: a candidate transduction channel for mammalian pheromone signaling. *Proc. Natl. Acad. Sci. USA* **96**, 5791–7596.

Lischka F. W., Zviman M. M., Teeter J. H., and Restrepo D. (1999) Characterization of inositol-1,4,5-trisphosphate-gated channels in the plasma membrane of rat olfactory neurons. *Biophys. J.* **76**, 1410–1422.

Lowe G. and Gold G. H. (1991) The spatial distribution of odorant sensitivity and odorant-induced currents in salamander olfactory receptor cells. *J. Physiol. (Lond.)* **442**, 147–168.

Maida R., Redkozubov A. and Ziegelberger G. (2000) Identification of PLC beta and PKC in pheromone receptor neurons of Antheraea polyphemus. *Neuroreport* **11**, 1773–1776.

Menco B. P. Cunningham A. M., Qasba P., Levy N. and Reed R. R. (1997) Putative odour receptors localize in cilia of olfactory receptor cells in rat and mouse: a freeze-substitution ultrastructural. *J. Neurocytol.* **26**, 691–706.

Menco B. P., Carr V. M., Ezeh P. I., Liman E. R. and Yankanowa M. P. (2001) Ultrastructural localization of G-proteins and the channel protein TRP2 to microvilli of rat vomeronasal receptor cell. *J. Comp. Neurol.* **438**, 468–489.

Menini A. (1999) Calcium signalling and regulation in olfactory neurons. *Curr. Opin. Neurobiol.* **9**, 419–426.

Meyer M. R., Angele A., Kremmer E., Kaupp U. B. and Muller F. (2000) A cGMP-signaling pathway in a subset of olfactory sensory neurons. *Proc. Natl. Acad. Sci. USA* **97**, 10595–10600.

21

The biomechanical design of an insect antenna as an odor capture device

Catherine Loudon

21.1 Introduction

Insect antennae are renowned for the specificity and sensitivity with which they detect airborne chemicals. Our ability to evaluate how antennae function is improving due to increasing appreciation of the role of physical factors such as air speed or antennal size and geometry on odorant capture. In engineering terms, this exchange of material between the environment and an object is called "mass transfer." The interception of airborne chemicals by sensory hairs on insect antennae may be treated as a mass transfer problem using a fluid mechanical approach. This approach is applicable to air because "fluids" include both gases and liquids. The utility of such an approach is that it results in quantitative predictions of the interception rate of these potentially informative molecules, i.e. odorants. Such predictions provide a null hypothesis against which to compare the actual performance of a sensory structure. That is, this approach will allow us to estimate the rate at which relevant molecules will reach a sensory surface, and with what kind of spatial or temporal pattern. It also allows us to make predictions about the functional consequences of behaviors (such as flight or wing fanning) or morphological attributes that will affect air flow in the vicinity of the antennae.

In the biomechanical approach taken here it is assumed that any molecules that will eventually be interpreted as a chemical stimulus must first make physical contact with the surface of the sensory hairs on an antenna. Alternative mechanisms that would allow molecules to be detected in the absence of physical contact, such as the detection theories of Callahan (1975), are not generally accepted due to problems with interpretation and lack of corroboration with other observations (Diesendorf, 1977a, b). The fluid mechanical approaches described in this chapter will be used to make estimates of rates of interception of airborne molecules, where "interception" is defined as the physical contact of the external surfaces of sensory structures by chemical stimulus molecules. The subsequent travel by the molecules as they enter pores in the sensory hairs and eventually contact the neurons will not be discussed in this chapter.

Some of the different ways of estimating interception rates will be described, with example calculations provided. Past treatments of this problem (Adam and Delbrück, 1968; Murray, 1977; Futrelle, 1984; Loudon and Koehl, 2000) have focused on calculations of interception for sensory hairs on the antennae of the domestic silkworm moth, *Bombyx mori*. Choice of this species was stimulated by the rather extraordinary research effort that has taken place over a period of several decades on this and related species (Schneider, 1964; Steinbrecht, 1970, 1987; Kaissling, 1971, 1986; Schneider, 1974; Keil, 1982, 1999). The example calculations in this chapter will consider both pectinate (feathery) and filiform antennae. Thus, we will be primarily concerned with cylindrical geometry. Unfortunately a cylindrical geometry in mass transfer can generate some formidable mathematics (e.g. Bessel functions may appear in the solutions to the differential equations) but this need not be paralyzing, as algebraic approximations are available in a few cases, as will be shown. It is hoped that the introduction to the concepts and approaches provided here will make it easier for biologists to enter the physical literature in search of helpful solutions. This chapter is written for chemical ecologists or insect physiologists who wish to understand the physical bases for making estimates of chemical interception rates and may wish to do some of their own "back of the envelope" estimates or even tackle more involved approaches. A "back of the envelope" level of calculation is often the most appropriate in many cases where a more detailed analysis is not justified due to the uncertainties in sensor geometry or air flow patterns. Many insights may be gained from a relatively simple approach.

21.2 Chemoreception treated as a mass transfer problem

21.2.1 Rate at which molecules in a gas strike a solid surface

Molecules in air are in constant motion, regularly striking each other as well as any surfaces extending into the air. Therefore, an antenna in air will be regularly struck by molecules in the air (including any potential chemical stimulus molecules) at a rate that follows directly from physical laws. Ideal gas laws predict that the number of molecules (N) striking a surface will be directly proportional to the area of the surface (A) and the time (t) over which the collisions are counted:

$$N = c_1 A t \tag{21.1}$$

The constant of proportionality, c_1, is

$$c_1 = \frac{n}{4}\sqrt{\frac{8k_B T}{\pi m}} \tag{21.2}$$

where n is the number of molecules per volume, k_B is Boltzmann's constant (1.381×10^{-23} J/K), T is temperature (K), and m is the mass of a single molecule (inconveniently in kg, equation 4.56c from Tabor, 1979). Combining equations (21.1) and (21.2), pooling constants, and changing units results in

$$\frac{N}{t} = 36.3 nA\sqrt{\frac{T}{M}} \tag{21.3}$$

where M is the molecular mass of the molecule of interest (now in amu or g/mole) and the rest of the variables are as defined above. This relationship (equation 21.3) is the simplest way to make a quick and straightforward estimate for the rate (N/t) at which odorant molecules available at a given concentration (n) will be striking antennae or sensory hairs (of surface area A). These estimates should be treated as approximations, keeping in mind that any deviations of molecular behavior from that assumed of ideal gases will make its application less exact.

21.2.1.1 Example calculation using equation (21.3)

Consider the case of pheromone molecules striking a single cylindrical sensory hair that is 100 μm long and 2 μm in diameter (the geometry for long sensilla trichoidea of the silkworm moth *Bombyx mori*; Steinbrecht, 1970). The surface area, A, of a single hair exposed to the surrounding air is 6.31×10^{-10} m^2 (from $2\pi ah + \pi a^2$, the side and distal end of a cylinder with $a = 1 \times 10^{-6}$ m and $h = 100 \times 10^{-6}$ m). Assuming that the pheromone is bombykol (238 amu or g/mole), the temperature is 25°C (298 K), and the concentration of bombykol in the air is 10^9 molecules/m^3 (1000 molecules/cc, a threshold concentration reported by Schneider, 1974), then substituting those numbers into equation (21.3) results in the number of bombykol molecules hitting the hair every second:

$$\frac{N}{t} = (36.3)(10^9)(6.31 \times 10^{-10})\sqrt{\frac{298}{238}} = 26 \tag{21.4}$$

or about 26 molecules per second (equation 21.4). This number is much larger than would be expected on the basis of other approaches. For example, treating the sensory hair as a perfectly adsorbing cylinder in non-moving homogeneous air at the same initial concentration of bombykol results in a prediction of about

0.2 molecules/hair over the first second (Loudon and Koehl, 2000). These numbers differ so much in part because the concentration of bombykol in the air will be decreasing in the latter case of the adsorbing surface, as molecules are removed from the air after contact with the surface. That is, if the molecules were not adsorbing onto the surface, the expectation would be about 26 molecular collisions between bombykol molecules and the sensory hair every second. These collisions will not all be with different individual bombykol molecules. From the point of view of the insect's sensory system, a molecule should be counted only once regardless of how many times it struck the hair before being finally captured, as multiple counts of single molecules are not informative (Dusenbery, 1992).

A prediction based on equation (21.3) provides an upper limit for molecular capture rate, because the molecules cannot be captured more often than they hit the surface in the first place. In practice, predictions based on equation (21.3) are likely to be an overestimate of what can be achieved by a real sensor, even by a couple of orders of magnitude as shown above. An overestimate will occur when the concentration (n) used to calculate the capture rate exists only momentarily before being decreased to (unknown) lower levels as the chemical stimulus molecules are removed from the air. Usually one wants an average interception rate over some biologically meaningful but somewhat arbitrary time interval, such as a second or a fraction of a second. However, if the interception rate is changing during this time, then different choices of time interval will result in different estimates for the average interception rate. For these unsteady-state cases where rates are changing with time, it is important to define the time interval over which the interception rate has been estimated in order to avoid problems of interpretation.

There is no term in equation (21.3) for air flow. External air flow around organisms is sufficiently slow (subsonic) that it may usually be treated as incompressible (Vogel, 1994). This incompressibility means that the concentration of chemical stimulus molecules (n) will not be increased noticeably by the pressures that develop adjacent to insect sensory hairs or antennae due to moving air (or moving antennae). The replacement of any captured molecules by the arrival of fresh odorant-laden air is the primary reason why air flow has such a dramatic effect on interception rate. One way of considering the influence of air flow is that at best the air flow could bring the interception rate closer to the limit predicted by equation (21.3). In order to discuss approaches more complex than that provided by equation (21.3), we have to consider the physical bases for molecular movements: diffusion and convection.

21.2.2 Processes of mass transport: diffusion and convection

There are two physical processes involved in the movement of molecules from one place to another (such as from the air to the surface of a sensory structure) – the molecules may walk themselves, or they may be pushed. Diffusion is the net movement of a species of molecule due to randomly directed thermal movements of individual molecules, while convection is the net movement of molecules by an air current resulting from a pressure gradient. These two processes occur at the same time, so while the molecules are being convected from one location to another they are also undergoing small-scale random walks. "Mass transport" refers to the net movement of a species of molecule regardless of the underlying physical cause, and therefore includes both diffusion and convection.

Obviously a randomly moving molecule will spend much of the time heading in the "wrong" direction away from a sensor. This is one of the reasons why a "reduction in dimensionality" when an odorant molecule makes contact with a sensory hair was such an appealing idea when first proposed (Adam and Delbrück, 1968). The idea is that an odorant molecule in air moving in three-dimensional space will change to two-dimensional diffusion after reaching a sensory hair and becoming lightly adsorbed onto its surface. However, the diffusion coefficient for odorant molecules on the surface is likely to be orders of magnitude smaller than the diffusion coefficient of the same odorant molecules in air. Therefore, an increase in the rate or efficiency with which odorant molecules contact pores will only occur with this two-dimensional path if the geometric gains are not completely offset by the lowering of the diffusion coefficient. Berg and Purcell (1977) have shown that a randomly moving molecule is likely to thoroughly explore an area once reached, suggesting that one does not have to invoke two-dimensional surface diffusion, because an odorant molecule is likely to strike a sensory hair many times (Futrelle, 1984). Odorants with different chemical properties may take slightly different routes to the pores; larger, more hydrophobic molecules would execute a path somewhat closer to the adsorbed two-dimensional ideal path if the surface is waxy. It is clear that at least some odorants do adsorb well onto the surfaces of sensory hairs (Kanaujia and Kaissling, 1985). The exact route taken by an odorant molecule between reaching the sensory hair and entering a pore in the hair's surface will not affect consideration of the mass transport from the air to the sensory surface and will not be considered further.

Insights about convection from fluid mechanics are immediately applicable to chemoreception by insect antennae. First, the "no-slip" condition that exists between any solid and an adjacent moving fluid mandates that the air immediately adjacent to the sensory surface is not moving with respect to the sensory surface (Figure 21.1). In addition, the small size scale relevant to chemoreception by insect antennae limits the behavior of the air flow in ways that have implications for chemoreception. The smaller the object and the slower the flow, the thicker the velocity boundary layer is relative to the size of the object (Figure 21.1A). By definition, the velocity boundary layer extends from the surface of the object to the point where the velocity has reached 99 percent of ambient; Vogel (1988) suggests that an alternative standard of 90 percent would suit the purposes of most biologists. One consequence of this fluid behavior is that microscopic

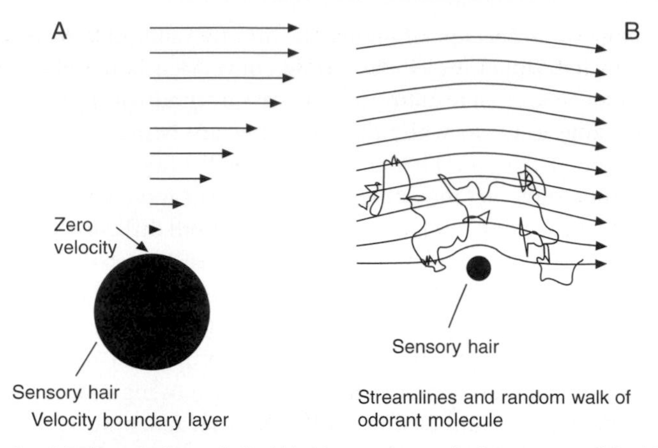

Figure 21.1 Air flow is from left. (A) Air moving relative to a solid object (such as the sensory hair shown in cross-section) is slower closer to the surface of the object and is zero at the surface (the "no-slip" condition). The length and orientation of each arrow represent the magnitude and direction of the air velocity at the point in space at the base of the arrow. (B) Streamlines of the moving fluid are indicated by arrows. The path of a diffusing odorant molecule is pictured crossing the streamlines.

sensory hairs will be surrounded by very slowly moving air to a distance of many diameters away. In addition, the small size scale means that the air flow will be laminar, and there will not be any turbulent mixing. The way in which chemical molecules will reach the sensory hair surface is by their random thermal movements (diffusion) (Figure 21.1B). The physical world of microscopic organisms is well described by Purcell (1977). An overview of the small-scale fluid dynamics of olfactory antennae (treated as cylindrical arrays) may be found in Koehl (1996).

21.2.3 Fick's law vs mass transfer coefficients – why are there two approaches and which should be used?

There are two completely different approaches to calculating mass transport rate: one approach starts with a mathematical equation (Fick's law) and searches for an analytical or numerical solution. The other approach starts with data which are used to generate mass transfer coefficients and then predictive equations. Cussler (1997) devotes the first chapter of his book to discussing the pros and cons of the two approaches. He argues persuasively that although the Fick's law approach may seem more legitimate because it is based on underlying physical principles, the mass transfer coefficient approach may be preferable in cases where the system has complex geometry or boundary conditions.

After introducing Fick's law and dimensional analysis using mass transfer coefficients, the most useful solutions will be presented. The solutions will be divided into steady-state (not changing in time) and unsteady-state cases, both with and without air flow. Some implications of these relationships for chemoreception will then be briefly discussed.

21.2.3.1 Fick's law

In order to predict mass transport due to diffusion, one usually turns to Fick's law as the most fundamental underlying mathematical statement of how the rate of diffusion scales with concentration and geometry. A one-dimensional form of Fick's law is usually stated as

$$J = -DA\frac{\partial c}{\partial x} \quad \text{or} \quad J \approx -DA\frac{\Delta c}{\Delta x} \tag{21.5}$$

or for radial diffusion in cylindrical coordinates as

$$J = -DA\frac{\partial c}{\partial r} \quad \text{or} \quad J \approx -DA\frac{\Delta c}{\Delta r} \tag{21.6}$$

where J is the flux (mass/time) that crosses an area A, c is concentration, D is the diffusion coefficient for the diffusing species, and x and r are spatial coordinates. Rewriting equation (21.6) in the variables already defined results in

$$\frac{N}{t} = -DA\frac{\partial n}{\partial r} \tag{21.7}$$

The apparent simplicity of equations (21.5)–(21.7) can be misleading; the number of exact solutions is small, and while approximate solutions fill books (e.g. Crank, 1975), their application can be problematic. Seemingly small differences in the boundary conditions will completely change the character of the solution, as will be seen below, and identification of the appropriate boundary conditions is not easy. However, in practice one can make an educated guess about the approximate boundary conditions, or can estimate the interception rate for different kinds of possible boundary conditions. Below are a number of solutions that are the most useful for the geometry most relevant to insect antennae, the cylinder. Either sensory hairs or filiform antennae can be approximated as cylindrical in shape.

21.2.3.2 Mass transfer coefficients, dimensional analysis, and dimensionless numbers

By definition, the mass transfer coefficient (k) relates the rate of mass transport (N/t) to a difference in concentration at two identifiable locations:

$$\frac{N}{t} = kA(n_2 - n_1) \tag{21.8}$$

Thus, the investigator needs only to be able to estimate the molecular concentration at two identifiable points in space, rather than estimate the concentration gradient

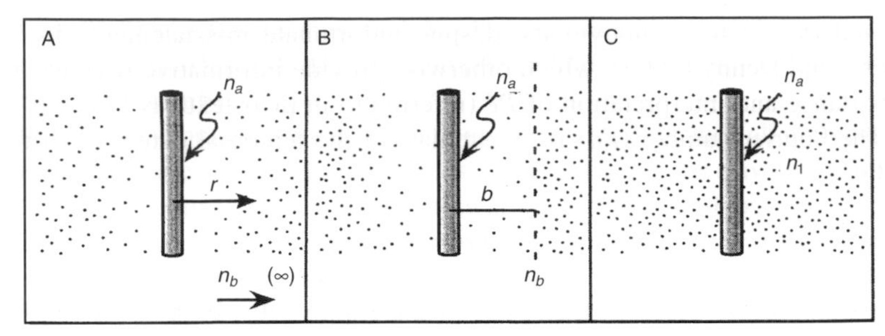

Figure 21.2 Geometry and boundary conditions discussed in the text. (A) A cylindrical sensor of radius a with surface held at a molecular concentration of n_a. The molecular concentration is n_b at an infinite distance from the sensor. There is no solution for this case. (B) A cylindrical sensor of radius a with surface held at a molecular concentration of n_a. The molecular concentration is n_b at a fixed distance b. (C) A cylindrical sensor of radius a with surface held at a molecular concentration of n_a. At time $t = 0$ the molecular concentration is n_1 everywhere outside the cylinder (for all $r > a$).

need for some new mathematical approach to crack this difficult problem, but because the predicted flux to the cylinder keeps changing indefinitely with time. A solution does exist for the same boundary conditions for a spherical body, which has been used extensively in the context of mass transport to microorganisms and single cells (Berg and Purcell, 1977; Wiegel, 1991; Berg, 1993).

(b) *A cylinder surrounded by a cylindrical outer boundary – still air.* This situation is for a cylinder (of radius a and surface held at a constant molecular concentration n_a) with an outer cylindrical boundary (at b) held constant at n_b:

$$\frac{N}{t} = \frac{2\pi D L_{\text{cyl}}(n_b - n_a)}{\ln(b/a)} \tag{21.15}$$

where L_{cyl} is the length of the cylinder (Figure 21.2B; equation (21.15) modified from equation 5.5 in Crank, 1975). This may be used as an approximation for a cylinder surrounded by a velocity boundary layer where the outer cylindrical boundary represents the constantly refreshed air sample.

21.2.3.4 Unsteady-state solutions

(a) *A cylinder initially surrounded by a homogeneous cloud of molecules – still air.* In this case, a cylindrical sensor is suddenly (at $t = 0$) surrounded by a homogeneous cloud of molecules at molecular concentration n_1 (Figure 21.2C). The exact solution to equation (21.7) is expressed as combinations of Bessel functions or complex error functions for which Crank (1975) has supplied approximate simple algebraic expressions:

$$\frac{N}{t} = 2\pi DL_{\text{cyl}}(n_1 - n_a)[(4\pi\tau)^{-0.5} + 0.5 - 0.50(\tau/\pi)^{0.5} + 0.50\tau - \ldots] \quad (21.16)$$

for small times ($\tau < 1$) and

$$\frac{N}{t} = 4\pi DL_{\text{cyl}}(n_1 - n_a)\left[\frac{1}{\ln(16\tau) - 2\gamma} - \frac{\gamma}{(\ln(16\tau) - 2\gamma)^2} - \ldots\right] \quad (21.17)$$

for longer times ($\tau > 1$) where τ is dimensionless time (Dt/L^2) and γ is Euler's constant (equation 21.16–17 modified slightly from equations 5.80 and 5.81 in Crank, 1975). Note these equations allow one to predict the (instantaneous) rate of interception, not the accumulated number of captured molecules. To estimate the accumulated capture at some point in time, such as one second after the initial boundary conditions, the incremental captures for the successive time intervals need to be summed. This was how the accumulated capture was estimated for single hairs in Loudon and Koehl (2000).

21.2.3.5 Steady-state solutions for diffusion/convection

When the air is moving, it becomes more difficult to calculate the interception rate. The mass transfer under these circumstances is generally expressed in dimensionless terms. Adam and Delbrück (1968) were able to generate a formula by making the simplifying assumption that the velocity of the air as it passed around the hair was everywhere constant (U) and very similar to the ambient air flow farther away (U_0):

$$\frac{N}{t} = -\frac{2\pi DL_{\text{cyl}}(n_1 - n_a)}{\ln(P\acute{e}\, U/U_0) + \gamma} \quad (21.18)$$

where $P\acute{e}$ is assumed to be 0.4 (equation (21.18) modified slightly from equation 47 in Adam and Delbrück, 1968). In practice it is difficult to apply this formula, as the choice of ambient air velocity (U_0) is not obvious, particularly for the case of a pectinate antenna. Note that as U gets smaller and smaller (relative to U_0), the logarithmic term gets very large, and the formula may no longer be applied. However, this formula does provide important physical insights into chemoreception by insect antennae: in particular this formula predicts that there will be relative insensitivity of the interception rate (N/t) to changes in the air flow.

Murray (1977) uses the same formula (equation 21.18) to predict interception rates of sensory hairs on insect antennae, but provides an additional insight in its application. As long as $P\acute{e} < 1$ (where L is hair diameter), the odorant molecules may be assumed to strike the hair anywhere. As $P\acute{e}$ becomes larger ($P\acute{e} > 1$), the interception rate will differ more with downstream/upstream location on the cylinder. The same logic may be used to estimate the potential for spatially dependent interception by a filiform antenna.

In the physical literature, functional relationships for mass transfer can be found for a few cases of low *Re*, but unfortunately the actual range of dimensionless numbers for which the formulas are valid are often either not provided or can be somewhat misleading. For example, Clift *et al.* (1978) provide the following formula for a "spheroid" in "creeping flow" ($Re << 1$):

$$Sh = 0.991K(P\acute{e})^{1/3} \tag{21.19}$$

where the constant *K* must be read from a graph (Figure 4.14 in Clift *et al.*, 1978). Note that in this case, the interception rate (*Sh*) is predicted to go to zero as the air flow gets slower and slower (as *Pé* goes to zero), but this would be a misapplication of the formula because the derivation was based on the assumption that *Pé* is very large ($P\acute{e} \rightarrow \infty$).

A formula for the case of air flow perpendicular to a cylinder (assuming low *Re* and low *Pé*) is

$$Sh = \frac{2}{\ln\left(\frac{8}{P\acute{e}}\right) - \gamma}\left[1 - \frac{a_3}{\left(\ln\frac{8}{P\acute{e}} - \gamma\right)^2}\right] \tag{21.20}$$

(equation 4-71 in Clift *et al.*, 1978), but there is an unknown coefficient, a_3, which "must be evaluated numerically." Equation (21.20) is considered a "solution" even though it is not immediately possible to use it to make a quantitative estimate for molecular interception by an insect antenna or sensory hair. However, similar to what was seen earlier, the relationship between the variables again predicts relative insensitivity of the interception rate to changes in air flow. The magnitude of the unknown coefficients in these formulas is being addressed empirically with physical modeling to directly estimate the mass transfer coefficients (Loudon *et al.*, unpublished).

Having summarized the most useful of the solutions available for mass transport to cylinders, a few examples of applications of this mass transfer approach will be considered.

21.3 Applications of mass transfer approach

21.3.1 Morphology

Insect antennae vary tremendously in size and shape, but two common forms are filiform (a single cylinder) and pectinate (feathery arrays of many cylinders). Most of the published morphological information on insect antennae concentrates on sensory hairs (e.g. Zacharuk, 1985; Steinbrecht, 1987, 1999; Zacharuk and Shields, 1991). Descriptions of sensory hair morphology are necessary for biomechanical analyses and interpretation, but it is equally important to have estimates of the sizes of gaps between sensory hairs as well as their three-dimensional arrangement in space, because these factors will significantly influence the air flow around the sensory hairs arrayed on the antennae (Cheer and Koehl, 1987a, b).

The size and shape of the antennae will affect chemoreception in a couple of ways – by changing the pattern of the air flow and by determining the surface area in contact with the air. The usual functional interpretation for larger antennae or for a larger number of sensory hairs is that the larger surface area will increase the interception rate of airborne chemicals. A direct proportionality between surface area and interception rate is predicted by equation (21.3). However, an increase in surface area *per se* will not necessarily increase the interception rate, because (a) different areas may interfere with each other (e.g. one hair might catch molecules that could be caught by another hair), (b) greater surface area may lead to higher drag thus lessening the volume of air processed, or (c) increases in the surface area may be functionally irrelevant (such as some microscopic sculpturing).

Quantifying the surface area of the sensory hairs or antennal surface leads to the same problem encountered in the famous analysis of the coastline of Britain in fractal studies; as one increases the magnification of the surface, the area becomes larger and larger without limit as successively smaller ripples become noticeable (Figure 21.3). While a mathematical limit may not exist, a physically meaningful limit does. From the point of view of odorant interception from the air, any ripples that are smaller than the mean free path of the diffusing molecules in air (the average distance traversed by molecules between collisions, about 70 nm) may be disregarded, and ripples slightly larger than this are unlikely to add significantly to the "capture" area. This reasoning assumes that the sensory surface is acting like a "sink," such that any incoming odorant molecule is not affected by the presence of odorant molecules already on the surface. It is possible that extensive surface sculpturing could increase odorant adsorption by decreasing

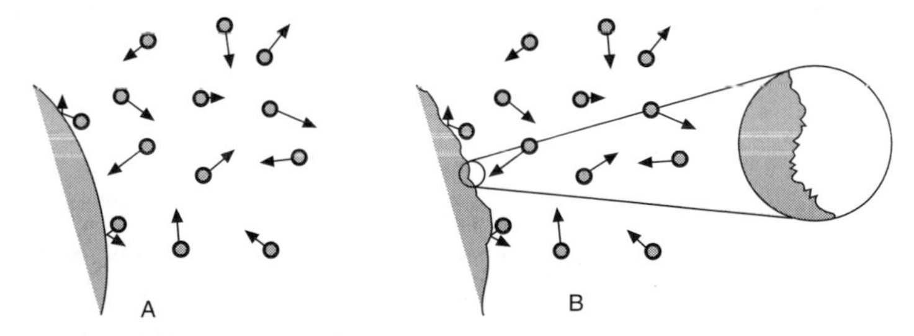

Figure 21.3 Molecules in air strike a surface at a rate directly proportional to the surface area. (A) The surface area is easier to estimate for a smooth surface. (B) Real surfaces often have surface ripples at any arbitrary level of magnification, making it more difficult to estimate the surface area.

interaction between adsorbed odorant molecules, but that is unknown at this point.

Sensory hairs may compete for the same molecules (and hence decrease each other's interception rates) when an odorant molecule passing through the gap between hairs is likely to strike either hair. This interference between hairs will occur when the hairs are sufficiently close and the air flow between them is slow. For the pectinate antennae that have been analyzed so far (*B. mori*, *Actias luna*), it is clear that the sensory hairs are exhaustively sampling much if not all of the air that passes through the spaces between them (Kaissling, 1971; Vogel, 1983; Loudon and Koehl, 2000). For a different case, such as sensory hairs projecting from a filiform antenna, the dimensionless numbers introduced above give us a way to approximate when these sensory hairs will be competing for the same molecules. If the Fourier number (equation 21.12) is greater than one in magnitude when the gap width between hairs (g) is used as the characteristic length, then the molecules have time to "walk" across the gap between the hairs during the characteristic time:

$$Fo_{\text{gap}} = \frac{Dt}{g^2} > 1 \tag{21.21}$$

One estimate for characteristic time (t) may be obtained by dividing the hair diameter by the average air velocity through the gap (L/U), and therefore represents the time during which that air sample is passing directly between the hairs. Making this substitution results in

$$\frac{Dt}{g^2} = \frac{D(L/U)}{g^2} = \frac{DL}{g^2U} > 1 \tag{21.22}$$

21.3.1.1 Example calculation using equation (21.22)

Consider sensory hairs 2 μm in diameter that are 20 μm apart. Inserting these values into equation (21.22) and assuming that D is 2.5×10^{-6} m^2/s (the diffusion coefficient for bombykol, the main component of the commercial silkmoth sex pheromone; Adam and Delbrück, 1968) results in the prediction that these hairs are likely to interfere with each other's odorant interception when the air speed between the hairs is below 0.0125 m/s. This is not a discontinuous function – the sensory hairs will interfere with each other more at slower speeds and less at faster speeds. Another way of appreciating what this means quantitatively is to recognize that the root mean square displacement of a molecule (considering movement in one dimension) is

$$\langle x \rangle_{\text{rms}} = \sqrt{2Dt} \tag{21.23}$$

Displacements due to molecular diffusion are normally distributed and therefore equation (21.23) is an estimate of the standard deviation of the displacements in time t. From the known characteristics of the normal distribution, 50 percent of the molecules will have gone as far as or farther than 0.67 times the standard deviation or

$$0.67 \times \sqrt{2Dt} \approx \sqrt{Dt} \tag{21.24}$$

This result (equation 21.24) helps us interpret the magnitude of Fo: when $Fo = 1$, $L \approx (Dt)^{0.5}$ (from equation 21.12), or about 50 percent of the molecules will have gone as far as or farther than L in time t in any direction (equation 21.24). For $Fo > 1$, more than 50 percent of the molecules, and for $Fo < 1$, fewer than 50 percent of the molecules will have gone as far as or farther than L in time t in any direction.

A decrease in the interception rate for individual sensory hairs does not necessarily imply that the total interception rate for the antenna has decreased, if there is a larger number of hairs. Increasing sensory hair number and decreasing spacing between them may tend to decrease the interception rate for single hairs while increasing the total interception rate for the antenna, at least within the range of sensory hair number and spacing seen in pectinate antennae (Loudon and Koehl, 2000). Therefore functional interpretations of the consequences of insect antennal morphology must consider these opposing tendencies.

Other morphological forms, such as the lamellate antennae of beetles, have unknown air flow patterns adjacent to the sensory surfaces. For these non-cylindrical cases there are few mathematical options other than the application of equation (21.3) to estimate an upper limit for the interception rate. These more complex morphologies may be tackled with other methods, including measurements using microsensors (Schneider *et al.*, 1998a, b) and physical modeling to estimate the mass transfer coefficients for those geometries (Loudon *et al.*, unpublished). The microsensor method has been used to demonstrate that the sexually dimorphic (filiform) antennae of *Manduca sexta* perturb the air flow in different ways, such that the sensory hairs on male and female antennae will intercept odorant molecules with a different temporal pattern from the same plume (Schneider *et al.*, 1998a).

21.3.2 Air flow

In the context of chemoreception, air movement is important in replacing the sampled air adjacent to the sensory surfaces with fresh odorant-laden air. Moving air may be generated by the insect (flying or moving its antennae) or the environment (wind). The relevant air speed (that appears in equations above as U) is considered from the spatial frame of reference of the antenna. In a natural setting, insects are constantly changing their speed and orientation, such as a moth flying upwind in a zigzag manner to a pheromone source, making it challenging to identify the sensory input and the mechanisms of control for this behavior (Willis and Arbas, 1998).

A sensory hair measures flux, not concentration directly (i.e. molecules/time reaching the surface, not molecules/volume). Increasing the air flow can to a certain extent compensate for a low concentration of odorant molecules in the air. For example, a sensory surface would hypothetically be unable to differentiate between the number of odorant molecules delivered by a rapidly passing but dilute airstream and a slow but concentrated airstream, in the absence of accompanying mechanosensory information about the air flow. The extent to which mechanosensory input from the antennae modifies the response to chemosensory input is not clear, although electrophysiological recordings from interneurons show that mechanosensory and chemosensory information is integrated in the brain (e.g. Itagaki and Hildebrand, 1990).

A change in interception rate with air flow has been identified as a characteristic of "flux detectors" such as insect sensory hairs (Kaissling, 1998). However, in some instances (low *Pé*), an increase in flow will have only a negligible impact on the flux rate, even for a "flux detector." This was shown by Berg and Purcell (1977) in the context of cells swimming through a liquid environment. That is, the flow rate doesn't matter if the molecules can walk themselves around just as quickly. For single cylindrical sensors (isolated hairs or filiform antennae), the slower the flow (the lower the *Re* and *Pé*), the less the interception rate is expected to increase with an increase in air speed (equations 21.18 and 21.20).

A different relationship between interception rate and ambient air speed exists for an otherwise identical sensory hair in an array, such as a sensory hair in a pectinate antenna. For a hair in an array, the interception rate is directly proportional to the air flow through the antenna because the air passing through the antenna is expected to be sampled almost exhaustively. A pectinate antenna allows only a small fraction of the approaching air to pass through, but the fraction increases with ambient air speed (Vogel, 1983). Exhaustively sampling an increasing fraction of oncoming air leads to a disproportionate increase in the interception rate with the ambient air flow, such as a 560× increase in interception rate predicted for a 15× increase in ambient air flow (Loudon and Koehl, 2000). Thus, the anticipated influence of ambient air flow on the interception rate by a sensory hair is completely different if the hair is isolated or is part of an array.

Unfortunately, the air flow in the vicinity of these microscopic sensory hairs is extremely difficult to calculate. Cheer and Koehl (1987b) provide a solution for the flow field in the vicinity of two parallel and infinitely long cylinders. Even for this simple geometry, the solution (expressed as a stream function) has enough terms that it takes up most of a printed journal page, and the reader must differentiate the provided stream function with respect to the spatial variables in order to solve for the velocities at different points in space. Finite hairs usually experience less flow between them than predicted assuming infinite length because fluid can go around the tips as well as the sides of an array (Koehl, 2001).

21.3.3 Spatial and temporal characteristics of interception

As an air stream passes by an insect antenna, the air may be slowed, distorted, or possibly even mixed, which will affect the rate at which molecules reach the sensory surface. These modifications will be perceived at the level of the sensory hairs as changes in the rate at which molecules strike the sensory hairs. Air flow patterns adjacent to cylinders may be predicted based on the magnitude of *Re* (using the diameter of the cylinder for *L*, equation 21.9; Figure 21.4). While turbulence will not occur because of the small size scale (and small *Re*), attached vortices (rotating fluid structures) will develop downstream of isolated cylinders when $10 < Re < 40$ (Vogel, 1994). This *Re* range requirement would seemingly prevent attached vortices near sensory hairs, as this would require unreasonable ambient speeds of 75 m/s for 2 μm diameter sensory hairs (Figure 21.4). However, stable, attached vortices can exist adjacent to cylinders that are in arrays for very low *Re* (such as seen for $Re = 0.01$ in Figure 15 in Van Dyke, 1982). Therefore, it is possible that attached vortices may exist adjacent to sensory hairs in pectinate antennae, or the branches that support them, although their existence (and the consequences for chemoreception) have not yet been evaluated. An attached vortex could hypothetically increase the rate of interception by giving a sensor another chance to capture molecules from a vortex rotating in its vicinity, but this would also prolong the time during which an "old" and increasingly irrelevant stimulus may be sensed. A complex morphology such as a pectinate antenna has several different cylindrical sizes (the main stalk, the side branches, and the sensory hairs), and each of these structures will reach a transitional *Re* range at different air speeds. Complex interactions can be seen in such porous structures, such as the pulsatile flow generated through cylindrical arrays by vortex shedding of the whole array (Leonard, 1992).

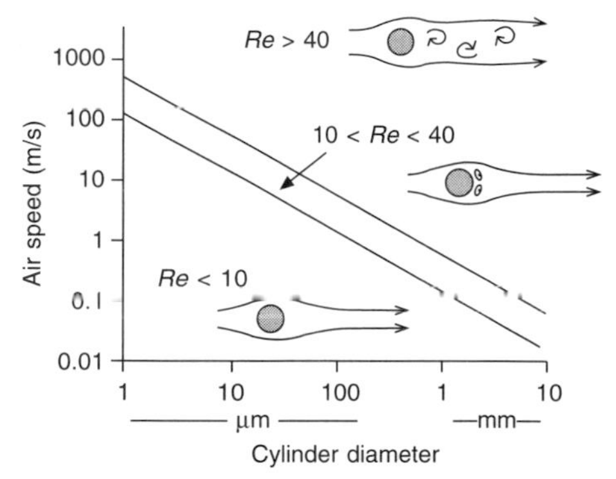

Figure 21.4 The air flow behavior around an isolated cylinder will depend on the magnitude of *Re* (equation 21.9, *L* is cylinder diameter). Combinations of air speed and cylinder diameter are shown that lead to vortex shedding ($Re > 40$), attached vortices ($10 < Re < 40$), or laminar flow without vortices ($Re < 10$) (dynamic viscosity of air is assumed to be 15×10^{-6} m^2/s).

on molecular diffusion, and to Dr Brian Smith for suggesting the example calculation on ratios of compounds. J. Botz, E. Davis, and G. Miller made suggestions that improved the text.

List of symbols

A	area (m^2)
a	radius of cylinder (m)
b	distance to boundary (m)
c	concentration (kg/m^3)
D	diffusion coefficient (m^2/s)
f	frequency (Hz)
Fo	Fourier number (Dt/L^2)
g	gap or space between surfaces (m)
h	length of cylinder (m)
J	flux (kg/s)
k	mass transfer coefficient (m/s)
k_B	Boltzmann's constant (1.381×10^{-23} J/K)
L	length (m)
L_{cyl}	length of cylinder (m)
m	mass of a single molecule (kg)
M	molecular mass (amu or g/mole)
n	number of molecules per volume (m^{-3})
N	number of molecules
$Pé$	Péclet number (UL/D)
r	spatial coordinate
Re	Reynolds number (UL/ν)
Sc	Schmidt number (ν/D)
Sh	Sherwood number (kL/D)
T	temperature (K)
t	dimensional time (Dt/L^2), Fo
U	velocity (m/s)
Wo	Womersley number ($L(\pi f/2\nu)^{0.5}$)
x	spatial coordinate
γ	Euler's constant (0.57722)
μ	fluid kinematic viscosity
ν	fluid dynamic viscosity
ρ	fluid density (kg/m^3)
τ	dimensionless time (Dt/L^2), Fo

References

Adam G. and Delbrück M. (1968) Reduction of dimensionality in biological diffusion processes. In *Structural Chemistry and Molecular Biology*, eds A. Rich and N. Davidson, pp. 198–215. W. H. Freeman, San Francisco.

Baker T. C., Fadamiro H. Y. and Cosse A. A. (1998) Moth uses fine tuning for odour resolution. *Nature* **393**, 530.

Berg H. C. (1993) *Random Walks in Biology*. Princeton University Press, Princeton, NJ.

Berg H. C. and Purcell E. M. (1977) Physics of chemoreception. *Biophys. J.* **20**, 193–219.

Callahan P. S. (1975) Insect antennae with special reference to the mechanism of scent detection and the evolution of the sensilla. *Int. J. Insect Morphol. and Embryol.* **4**, 381–430.

Cheer A. Y. L. and Koehl M. A. R. (1987a) Fluid flow through filtering appendages of insects. *IMA J. Maths. Applied in Med. and Biol.* **4**, 185–199.

Cheer A. Y. L. and Koehl M. A. R. (1987b) Paddles and rakes: fluid flow through bristled appendages of small organisms. *J. Theor. Biol.* **129**, 17–39.

Clift R., Grace J. R. and Weber M. E. (1978) *Bubbles, Drops, and Particles.* Academic Press, San Diego.

Crank J. (1975) *The Mathematics of Diffusion* Oxford University Press, New York.

Cussler E. L. (1997) *Diffusion: Mass Transfer in Fluid Systems.* Cambridge University Press, Cambridge.

Denny M. W. (1993) *Air and Water: The Biology and Physics of Life's Media.* Princeton University Press, Princeton.

Diesendorf M. (1977a) The "dielectric waveguide theory" of insect olfaction: a reply to P. S. Callahan. *Int. J. Insect Morphol. and Embryol.* **6**, 123–126.

Diesendorf M. (1977b) Insect sensilla as dielectric aerials for scent detection? Comments on a review by P. S. Callahan. *Int. J. Insect Morphol. and Embryol.* **6**, 105–109.

Dusenbery D. B. (1992) *Sensory Ecology; How Organisms Acquire and Respond to Information.* W. H. Freeman, New York.

Futrelle R. P. (1984) How molecules get to their detectors: the physics of diffusion of insect pheromones. *Trends in Neurosci.* **7**, 116–120.

Itagaki H. and Hildebrand J. G. (1990) Olfactory interneurons in the brain of the larval sphinx moth *Manduca sexta. J. Comp. Physiol. A* **167**, 309–320.

Kaissling K. E. (1971) Insect olfaction. In *Handbook of sensory physiology*, ed. L. M. Beidler, Vol. IV. Chemical Senses Part 1. Olfaction, pp. 351–431. Springer, Berlin-Heidelberg-New York.

Kaissling K.-E. (1986) Temporal characteristics of pheromone receptor cell responses in relation to orientation behaviour of moths. In *Mechanisms in Insect Olfaction*, eds T. L. Payne, M. C. Birch and C. E. J. Kennedy, pp. 193–199. Clarendon Press, Oxford.

Kaissling K. E. (1998) Flux detectors versus concentration detectors: two types of chemoreceptors. *Chem. Senses* **23**, 99–111.

Kanaujia S. and Kaissling K. E. (1985) Interactions of pheromone with moth antennae: adsorption, desorption and transport. *J. Insect Physiol.* **31**, 71–81.

Keil T. A. (1982) Contacts of pore tubules and sensory dendrites in antennal chemosensilla of a silkmoth: demonstration of a possible pathway for olfactory molecules. *Tissue and Cell* **14**, 451–462.

Keil T. A. (1999) Morphology and development of the peripheral olfactory organs. In *Insect Olfaction*, ed. B. S. Hansson, pp. 5–48. Springer-Verlag, Berlin.

Kenis P. J. A., Ismagilov R. F. and Whitesides G. M. (1999) Microfabrication inside capillaries using multiphase laminar flow patterning. *Science* **285**, 83–85.

Koehl M. A. R. (1996) Small-scale fluid dynamics of olfactory antennae. *Marine and Freshwater Behavior. Physiol.* **27**, 127–141.

Koehl M. A. R. (2001) Fluid dynamics of animal appendages that capture molecules: arthropod olfactory antennae. In *Computational Modeling in Biological Fluid Dynamics*, eds L. J. Fauci and S. Gueron, pp. 97–116. Springer Verlag, New York.

Leonard A. B. P. (1992) *The biomechanics, autecology and behavior of suspension-feeding in crinoid echinoderms.* Doctoral dissertation: University of California, San Diego.

Loudon C. and Koehl M. A. R. (2000) Sniffing by a silkworm moth: wing fanning enhances air penetration through and pheromone interception by antennae. *J. Exp. Biol.* **203**, 2977–2990.

Loudon C. and Tordesillas A. (1998) The use of the dimensionless Womersley number to characterize the unsteady nature of internal flow. *J. Theor. Biol.* **191**, 63–78.

Mafra-Neto A. and Cardé R. T. (1994) Fine-scale structure of pheromone plumes modulates upwind orientation of flying moths. *Nature* **369**, 142–144.

Murray J. D. (1977) Reduction of dimensionality in diffusion processes: antenna receptors of moths. In *Lectures on Nonlinear-Differential-Equation Models in Biology*, pp. 83–127. Oxford University Press, Oxford.

Purcell E. M. (1977) Life at low Reynolds number. *Am. J. Phys.* **45**, 3–12.

Schneider D. (1964) Insect antennae. *Ann. Rev. of Entomol.* **9**, 103–122.

Schneider D. (1974) The sex-attractant receptors of moths. *Scientific American* **231**, 28–35.

Schneider R. W. S., Lanzen J. and Moore P. A. (1998a) Boundary-layer effect on chemical signal movement near the antennae of the sphinx moth, *Manduca sexta*: temporal filters for olfaction. *J. Comp. Physiol. A* **182**, 287–298.

Schneider R. W. S., Price B. A. and Moore P. A. (1998b) Antennal morphology as a physical filter of olfaction: temporal tuning of the antennae of the honeybee, *Apis mellifera*. *J. Insect Physiol.* **44**, 677–684.

Steinbrecht R. A. (1970) Zur Morphometrie der Antenne des Seidenspinners, *Bombyx mori* L.: Zahl und Verteilung der Riechsensillen (Insecta, Lepidoptera). *Zeitschrift für Morphologie der Tiere* **68**, 93–126.

Steinbrecht R. A. (1987) Functional morphology of pheromone-sensitive sensilla. In *Pheromone Biochemistry*, eds G. D. Prestwich and G. J. Blomquist, pp. 353–384. Academic Press, London.

Steinbrecht R. A. (1999) Olfactory receptors. In *Atlas of Arthropod Sensory Receptors*, eds E. Eguchi and Y. Tominaga, pp. 155–176. Springer-Verlag, Tokyo.

Tabor D. (1979) *Gases, Liquids and Solids*. Cambridge University Press, Cambridge, UK.

Van Dyke M. (1982) *An Album of Fluid Motion*. Parabolic Press.

Vickers N. J., Christensen T. A., Baker T. C. and Hildebrand J. G. (2001) Odour-plume dynamics influence the brains olfactory code. *Nature* **410**, 466–470.

Vogel S. (1983) How much air passes through a silkmoth's antenna? *J. Insect Physiol.* **29**, 597–602.

Vogel S. (1988) *Life's Devices: The Physical World of Animals and Plants*. Princeton University Press, Princeton, NJ.

Vogel S. (1994) *Life in Moving Fluids: The Physical Biology of Flow*. Princeton University Press, Princeton, NJ.

Welty J. R., Wicks C. E. and Wilson R. E. (1984) *Fundamentals of Momentum, Heat, and Mass Transfer*. John Wiley & Sons, New York.

Wiegel F. W. (1991) *Physical Principles in Chemoreception*. Springer-Verlag, New York.

Willis M. A. and Arbas E. A. (1998) Variability in odor-modulated flight by moths. *J. Comp. Physiol. A* **182**, 191–202.

Zacharuk R. Y. (1985) Antennae and sensilla. In *Comprehensive Insect Physiology Biochemistry and Pharmacology*, eds G. A. Kerkut and L. I. Gilbert, Vol. 6, pp. 1–69. Pergamon Press, New York.

Zacharuk R. Y. and Shields V. D. (1991) Sensilla of immature insects. *Ann. Rev. of Entomol.* **36**, 331–354.

22

Olfactory landscapes and deceptive pollination: signal, noise and convergent evolution in floral scent

Robert A. Raguso

22.1 Introduction: natural odors as environmental signals

The world of insects is awash in volatile semiochemicals; odors that convey information. This volume focuses on pheromones, which signal a conspecific animal's sexual status or coordinate complex social behavior (Law and Regnier, 1971). Of course, non-pheromonal odors bear relevance to diverse aspects of insect behavior for *both* sexes, as they indicate the presence of enemies, prey, resource quality or appropriate habitat, or are recognized as background and ignored (reviewed in Smith and Getz, 1994; Hartlieb and Anderson, 1999). Flowering plants have successfully coopted insects and other animals for their reproductive services, and floral odors play important roles in insect foraging and learning (Williams, 1983; Dobson, 1994; Raguso 2001). Typically, plant–pollinator dynamics are driven by honest signals; combinations of odors and other sensory stimuli that indicate the presence of a metabolic reward (usually nectar or pollen) sought by a foraging animal (Simpson and Neff, 1983). The exception to this rule – pollination by deception – occurs when flowering plants exploit insects with at least two categories of dishonest odor signals (Dafni, 1984). Some flowers emit odors that represent innately attractive stimuli or salient learning cues indicative of the presence of food, but withhold nutritive rewards (usually nectar; Ackerman, 1986; Gill, 1989; Nilsson, 1992). Other flowers emit odors that mimic sex pheromones or oviposition cues that trigger innate reproductive behaviors, to which habituation by the responding insect may reduce its reproductive fitness (Stowe *et al.*, 1995). Both classes of deceptive signals are examples of Batesian mimicry or signal exploitation, and thus are expected to operate under frequency-dependent selection; they should function only when relatively rare, in comparison to the model stimulus (Roy and Widmer, 1999).

In this chapter, I focus on olfactory communication between flowering plants and their insect pollinators as a context for discussing odor signal complexity, pheromone mimicry and the evolution of chemical deception. First, I consider the chemical composition of floral odors, which generally are more complex than insect pheromones, and discuss how behaviorally relevant "signal" may be extracted from chemical "noise." Second, I distinguish between the two major classes of flower-visiting animals – those foraging for food vs performing reproductive functions – and contrast the odors that attract these animals to flowers. Third, I argue that floral odor signals – like specific allelic combinations or novel phenotypes – represent fitness peaks (for their plants) in an adaptive landscape, subject to various forms of natural selection. Finally, I highlight examples of honest and deceptive odor signals representing distinctive peaks in the adaptive landscape of floral scent.

22.2 Chemodiversity of floral scent: distinguishing signal from noise

Traditionally, the inherent complexity of floral scent has limited our understanding of its evolutionary patterns and biological importance. Previous reviews (Bergström, 1991; Croteau and Karp, 1991; Knudsen *et al*, 1993; Dobson, 1994; Raguso 2003) have detailed the many dimensions of fragrance complexity, ranging from chemical composition and biosynthetic origins to spatial and temporal variation, both within single flowers and among plant populations. It must be emphasized that this field remains in its infancy, as fundamental mechanisms of volatile biosynthesis and patterns of distribution are just beginning to be elucidated (Dudareva *et al.*, 1999, Pichersky and Gershenzon, 2002). Knudsen and Tollsten (1993, 1995) have argued that fragrance chemistry covaries with pollinator class, suggesting that convergent evolution significantly impacts scent composition. There is some evidence for convergent emission of compounds seldom found in floral scents (e.g. carvone oxide and ipsdienol), among plants of several families pollinated by male euglossine bees (Whitten *et al.*, 1986, 1988). However, convergence is difficult to judge in the absence of definitive behavioral assays or null hypotheses. Two independent surveys of New World bat-pollinated plants (Knudsen and Tollsten, 1995; Bestmann *et al.*, 1997) provide an intriguing example. Most of the 15 species studied, belonging to ten families, produced complex blends containing sulfur-bearing volatiles, which were shown in subsequent behavioral assays to attract both naïve and wild Glossophagine bats (von Helversen *et al.*, 2000; Winter and von Helversen, 2001). However, when overall fragrance composition among bat-pollinated plants is compared using Sørenson's Index of Similarity (Knudsen and Tollsten, 1995), on average only 25–35 percent of scent composition is shared between species. If this pattern results from an exemplar of convergent evolution in fragrance, how should the remaining 65–75 percent of chemical variation be interpreted?

The traditional, adaptive view of pollination biologists would be narrowly to define "signal" as the functional subset of an odor blend responsible for pollinator attraction and "noise" as the remainder of scent components (Table 22.1). In this model, "noise" could result from neutral genetic variation (Ackerman *et al.*, 1997; Azuma *et al.*, 2001) or physiological plasticity in response to environmental factors (such as photoperiod or temperature; Jacobsen and Olsen, 1994; Hansted *et al.*, 1995). Some floral volatiles have alternative functions, such as anti-microbial activity (Lawton *et al.*, 1993; Dobson and Bergström, 2000) or defense against pathogens or herbivores (Galen, 1983; Linhart and Thompson, 1995; Ômura *et al.*, 2000), that would select for their maintenance in populations. Finally, volatiles with neutral selective value might be inherited as phylogenetic or biosynthetic artifacts. Levin *et al.*, (2001, 2003) have shown that subsets of fragrance compounds are shared-derived traits among clades of related Nyctaginaceae (four-o'clock) plants, independent of pollinator class or reproductive strategy. Moreover, many terpene synthase enzymes produce both major and minor products, resulting in terpene blends inherited as units (Colby *et al.*, 1993; Steele *et al.*, 1998; Bohlmann *et al.*, 1998). An alternative Darwinian approach

Table 22.1 Functional analysis of complex odor blends, considering various sources and definitions of "signal" and "noise"

Signal model =	Synomonal function	Fitness impact
Signal	pollinator attractants • geographic variation? • multiple pollinators? conditioning stimulus (↑ floral constancy) synergist	pollinator attractants conditioning stimulus (↑ floral constancy) anti-microbial agents herbivore repellent synergist
Noise	biosynthetic by-products phylogenetic artifacts neutral genetic variation herbivore repellent anti-microbial agents environmental plasticity methodological artifact	biosynthetic by-products phylogenetic artifacts neutral genetic variation environmental plasticity methodological artifact

Figure 22.1 Examples from two conceptual axes of interactions between flowers and their animal visitors. Axis 1 is a specialization–generalization spectrum of plant-pollinator interactions. Panel A depicts a guild of red Chilean flowers that share one species of hummingbird as a pollinator. In Panel D, a *Perideridia* umbel is visited by several families of bees, wasps and flies; most are effective pollinators. Axis 2 describes relationships in which animals visit flowers for their own reproductive purposes. In panel B, a female *Tegiticula* moth gathers pollen from anthers of *Yucca filamentosa*, for which it is both obligate pollinator and seed predator. In panel C, a *Drosophila* fly (black arrow) is lured by appearance and smell of decaying matter to a deceptive *Aristolochia* flower, seen in cross-section. Floral scent plays diverse roles along these axes, including pollinator attraction in food- and sex-based mimicry. All photographs were taken by the author.

relationships in which animals visit flowers to obtain nourishment (see above). This specialization–generalization spectrum ranges from plants with one known pollinator (Johnson and Steiner, 2000) to those pollinated by guilds of related animals (Feinsinger *et al.*, 1986; Petterson 1991) to those for which several classes of pollinator are equally effective (e.g. Herrera, 1987; Fishbein and Venable, 1996). Recent debate has focused on the prevalence of specialized vs generalized pollination systems and the extent to which this spectrum impacts pollinator-mediated floral evolution (Ollerton, 1996; Waser *et al.*, 1996). In addition to food-based plant-pollinator mutualisms, this "foraging" axis includes non-pollinators that eat pollen (Smart and Blight, 2000) and rob nectar (Irwin and Brody, 1999) and food-deceptive flowers that offer no rewards (Haber, 1984; Ackerman, 1986). Floral scent, alone or in conjunction with other sensory cues, plays many roles along this axis, both as an innate attractant or feeding cue and as a conditioning stimulus (reviewed in Dobson, 1994; Raguso, 2001). Perhaps as important is the potential of fragrance components to *repel* certain floral visitors (Dobson and Bergström, 2000; Ômura *et al.*, 2000), a question frequently examined in plant–herbivore studies but neglected in pollination research.

The second axis describes both benign and exploitative interactions in which animals visit flowers, sometimes unwittingly, to satisfy their own reproductive imperatives. Arrayed along this axis are insects that use flowers (destructively) as brood sites (Patt *et al.*, 1995; Pellmyr 1997), sources of aphrodisiacs (Trigo *et al.*, 1996) or nesting materials (Armbruster, 1984), and flowers that attract insects by mimicking the signals of courtship or oviposition (van der Pijl and Dodson, 1966). This "reproductive" axis includes some of the most specialized, odor-driven systems known to pollination biology. Male euglossine bees collect fragrance from hundreds of orchid species (and other sources) and are thought to use them in courtship or mating (reviewed in by Williams and Whitten, 1983; Lunau 1992; Eltz *et al.*, 1999). Fig trees, which constitute keystone resources in tropical forests worldwide, are pollinated exclusively by agaonid wasps, which use the figs as breeding sites (Janzen, 1979; Bronstein 1992). Species specific fig odors guide wasp orientation and landing, and maintain species boundaries in ways comparable to sex pheromones (Ware *et al.*, 1993; Grison *et al.*, 1999). Finally, deceptive orchids in the Mediterranean region and Australia mimic female appearance and odor to exploit male Hymenoptera as sexual dupes, in a form of pollination known as "pseudo-copulation" (Borg-Karlson, 1990; Peakall and Beattie, 1996). The two conceptual axes intersect when insects visit flowers both for food *and* reproductive resources (e.g. dynastine beetles; Gottsberger, 1989) or putatively deceptive flowers actually offer rewards to further manipulate or provision visitors (e.g. *Rafflesia*; Beaman *et al.*, 1988).

22.4 Multidimensional scent space and adaptive landscapes

It is clear from section 22.2 that the multidimensional "universe" of fragrance composition in flowering plants is not limitless, as fragrance evolution is constrained both by phylogenetic history and biosynthetic mechanisms. Nevertheless, botanists and perfumers have long recognized that floral scents are chemically and (to humans) perceptually diverse (Delpino, 1874; Vogel, 1983; Kaiser, 1993), and several studies suggest that the variation necessary to generate such a pattern is common within most species examined (reviewed in Knudsen, 2002). By analogy to the adaptive topographies of allele frequencies in population genetics (Wright, 1932), the relative fitness of different floral phenotypes has been visualized using adaptive surfaces or landscapes (Armbruster, 1990; Cresswell and Galen, 1991). Consider such a landscape for fragrance, in which the X–Y coordinates of a given locus in "floral scent space" represent its chemical composition and the relative ratios of its constituents – its "chemotype" – while its vertical (*Z*) dimension reflects the mean reproductive fitness of that blend. Figure 22.2 presents a schematic of "scent space" in two dimensions, as a multidimensional scaling plot of Euclidean distances between odor blends, for which fitness coefficients from 0 to 1 (not shown in Figure 22.2) would, in theory, exist. When total fragrance composition is examined (Figure 22.2A), odors are species specific but do not cluster discretely by pollinator class, due to considerable chemical overlap. However, when only "signal" blend components (known pollinator attractants) are included in the analysis (Figure 22.2B), fragrance clustering is nearly exclusive with respect to pollinator class. Discrete adaptive peaks exist for odors innately attractive to specific pollinator groups (e.g. oligosulfides for bats), whereas broad "adaptive mesas" (not shown in Figure 22.2) would include blends attractive to a spectrum of generalist pollinators and/or containing highly salient conditioning stimuli (e.g. phenylacetaldehyde or geraniol). The "valleys" of such a landscape would indicate blends less attractive to effective pollinators or more vulnerable to attack by herbivores or pathogens (Metcalf and Metcalf, 1992).

Populations of flowering plants could traverse this landscape in evolutionary time via mutation, drift or hybridization, with results ranging from neutral genetic variation (Ackerman *et al.*, 1997) to balanced scent polymorphism (Galen *et al.*, 1987) or even sympatric speciation (Gregg, 1983; Schiestl and Ayasse, 2002). Nilsson (1992) refers to this process as "fragrance casting" and suggests that orchids are particularly adept at it, which may explain why 32–40 percent of the nearly 25 000 orchid species mimic other flowers, sex pheromones, and even fungi (Gill, 1989; Kaiser, 1993; Barkman *et al.*, 1997)! The nectarless flowers of one such orchid, *Epidendrum ciliare*, emit nocturnal fragrances that vary from plant to plant, and between flowers on the same plant, at levels comparable to differences between species (Moya and Ackerman, 1993). The different chemotypes emitted by these flowers all fall within the broad adaptive peak of hawkmoth-pollinated plants (Figure 22.2b; Knudsen and Tollsten, 1993); their rampant variation may block odor-based associative learning by frustrated moths. Nilsson (1992) and others have pointed out that many food-mimicking orchids lack specific models, and thus are not technically Batesian mimics. By emitting scents whose chemistry falls under the broad rubric of general attractants, these flowers secure pollination by hungry and/or naïve insect visitors. Recently, Kunze and Gumbert (2001) showed that bumblebees use subtle differences in fragrance to distinguish artificial mimics from model flowers, and propose that the best

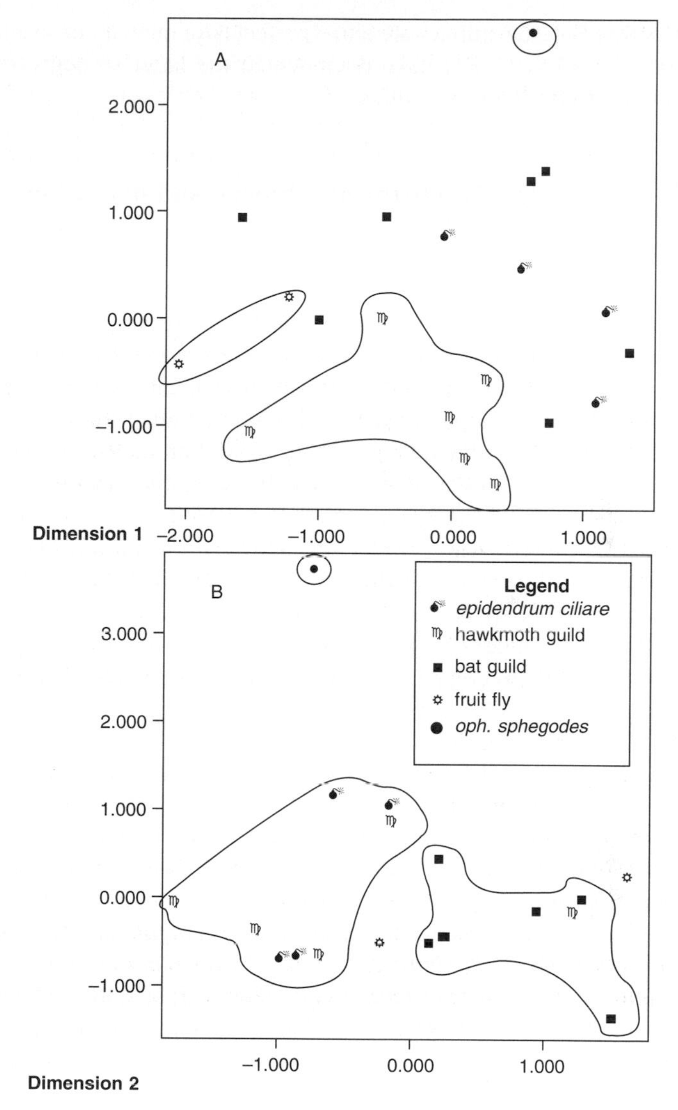

Figure 22.2 Olfactory landscapes of fragrance chemistry. For each blend, Euclidean distances were calculated for Z-scores of relative amounts of each compound. Multidimensional scaling was used to visualize distances for complete scent blends (A) and subsets that attract pollinators ("signal"; B), as inferred from behavioral assays. Data were taken from studies of hawkmoth- and bat-pollinated plant guilds (Knudsen and Tollsten, 1993, 1995; Raguso, unpublished data), four individuals of *Epidendrum ciliare*, a deceptive hawkmoth-pollinated orchid (Moya and Ackerman, 1993), two plants pollinated by male tephritid flies (see text) and the sexually deceptive orchid *Ophrys sphegodes* (Ayasse *et al.* 2000). Shared, putatively non-functional "noise" compounds obscure pollinator relationships in panel A, whereas "noise" removal results in nearly exclusive clusters of fragrance blends by pollinator class in panel B.

strategies for true Batesian mimics are either perfect odor mimicry or scentlessness. Indeed, Dafni and Ivri (1981) have documented the latter strategy for *Orchis israelitica* that mimic *Bellevalia* lilies.

22.5 Odor convergence, chemical mimicry and deception

In the olfactory topography described above, adaptive peaks are not limited to innate pollinator attractants, salient learning cues, or scents from fraudulent flowers that mimic these signals. In the broader sense, any abundant animal whose reproductive biology or social behavior is mediated by semiochemicals is, in theory, vulnerable to exploitation through chemical mimicry: the specific volatile signal represents a potential adaptive peak. Signal mimicry should be especially effective when olfactory physiology and reproductive fitness prevent or constrain habituation to the bogus signal. For example, noctuid moths, like honeybees, can be conditioned to associate diverse odors with nectar rewards (Fan *et al.*, 1997; Cunningham *et al.*, 1998). However, male *Heliothis virescens* can learn individual components, but not the complete blend of female sex pheromone as a food odor (Hartlieb and Hansson, 1999). In the following examples, I highlight known or putative adaptive peaks of floral scent and the pollinators whose responses to them derive from reproductively motivated behaviors.

22.5.1 Sexual signal peaks:

1. *Phenylpropanoids and male tephritid flies.* Males of several *Bactrocera* (=*Dacus*) fruit fly species are attracted to raspberry ketone, methyl eugenol and other similar phenylpropanoids (Metcalf *et al.*, 1983). Male flies that gather these compounds convert them to coniferyl alcohol (Nishida *et al.*, 1997), experience greater mating success, and potentially use the compounds to deter predators. At least two flowering plants, *Spathiphyllum canaefolium* (Araceae; Lewis *et al.*, 1988; Chuah *et al.*, 1996) and *Bulbophyllum patens* (Orchidaceae; Tan and Nishida, 2000), produce phenylpropanoids as attractants *and* rewards and are pollinated by the flies. Given the flies' abundance, more such cases of "honest" mutualism are predicted. Variation in signal content and fly behavioral responses prevent Spathiphyllum and Bulbophyllum from grouping more tightly in Figure 22.2.
2. *Pyrrholizidine alkaloids and male Lepidoptera.* Male danaiine butterflies and arctiid moths collect pyrrholizidine alkaloids (PAs) from *Heliotropium* (Boraginaceae) and *Crotalaria* (Fabaceae) foliage as precursors to sexual attractants (Goss, 1979; Boppré, 1984; Schneider, 1986). PAs acquired by adult male ithomiine butterflies repel predacious spiders and are offered to females within spermatophores as nuptial gifts (Brown, 1984). Some species of *Eupatorium* (Asteraceae) and *Epidendrum* orchids offer PAs in floral tissues

or nectar and are visited heavily (and pollinated) by male Lepidoptera attracted from a distance (Pliske, 1972; DeVries and Stiles, 1990). Trigo *et al.* (1996) present lists of other flowering plants known to produce PAs and their derivative olfactory attractants.

3 **Ophrys** *orchids and male Hymenoptera*. Flowers of the genus *Ophrys*, found throughout the Mediterranean region, use a combination of visual, tactile and volatile signals effectively to mimic female wasps and bees (Kullenberg and Bergström, 1976; Paulus and Gack, 1989; Borg-Karlson, 1990). Schiestl *et al.* (1999, 2000) used GC-EAD and behavioral assays to identify specific ratios of 14 long-chain (21–29 C) alkanes and alkenes common to *Ophrys sphegodes* and female *Andrena nigroaenea* bees as signals that elicit copulatory behavior in male *A. nigroaenea*. Pollinated flowers emit fewer of these compounds and produce a repellent ester, all-trans-farnesyl hexanoate, which mimics the Dufour gland secretion of post-mated female *A. nigroaenea* and inhibits copulatory behavior. These changes are thought to prolong floral visitation (and effect pollinium transfer) by male bees that might otherwise habituate to the mock females (Schiestl *et al.*, 1997). Odors of sympatric *O. fusca* and *O. bilunulata* show quantitative differences in alkene content that are likely to maintain orchid species-specific pollination by *Andrena* males (Schiestl and Ayasse, 2002). With at least 30 *Ophrys* spp. and several geographic isolates exploiting males of different genera and families of Hymenoptera, this system holds much promise for the study of the roles of odor signals in sympatric speciation and adaptive radiation.

4 *Australian orchids and male wasps*. Pseudocopulation reaches its zenith as a reproductive strategy among *Caladenia*, *Chiloglottis* and *Drakea* orchids pollinated by male thynnine wasps in Australia (Stautamire, 1983; Peakall and Beattie, 1996). Flightless female wasps use pheromone to attract males to perches, then are carried off and mated in flight (Handel and Peakall, 1993). Phylogenetic analyses indicate that *Chiloglottis* orchids have radiated rapidly to utilize females of different *Neozeleboria* wasp species as models, including sex pheromone mimicry and phenological synchronization (Mant *et al.*, 2002). GC-EAD studies have identified one to two chemically novel compounds shared by female *Neozeleboria* wasps and *Chiloglottis* flowers that elicit attraction and mating behavior by males (F. P. Schiestl, unpublished data). Given such conservative patterns of pheromone chemistry among related wasps and their floral mimics, differences in emergence times and geographic distributions would appear to contribute heavily to species isolation. At present, this adaptive peak is quite narrow, but future analyses of related orchids pollinated by male ants and ichneumonid wasps should reveal greater chemical variation.

22.5.2 Social signal peak: Nasonov blend of the Japanese honeybee

Floral scent of the orchids *Cymbidium floribundum* and *C. pumilum* elicits swarming behavior and pollination by *Apis cerana japonica*, particularly by drones (Sasaki

Hartlieb E. and Anderson P. (1999) Olfactory-released behaviours. In *Insect Olfaction*, ed. B. S. Hansson, pp. 315–350. Springer, Berlin.
Hartlieb E. and Hansson B. S. (1999) Sex or food? Appetitive learning of sex odors in a male moth. *Naturwissenschaften* **86**, 396–399.
Haynes K. F., Zhao J. Z. and Latif A. (1991) Identification of floral compounds from *Abelia grandiflora* that stimulate upwind flight in cabbage looper moths. *J. Chem. Ecol.* **17**, 637–646.
Heath R. R., Landolt P. J., Dueben B. and Senczewski B. (1991) Identification of floral compounds of night-blooming jessamine attractive to cabbage looper moths. *Environ. Entomol.* **21**, 854–859.
Helversen O. von Winkler L. and Bestmann H. J. (2000) Sulphur containing "perfumes" attract flower-visiting bats. *J. Comp. Physiol. A* **186**, 143–153.
Herrera C. M. (1987) Components of pollinator "quality": comparative analysis of a diverse insect assemblage. *Oikos* **50**, 79–90.
Hick A. J., Luszniak M. C. and Pickett J. A. (1999) Volatile isoprenoids that control insect behaviour and development. *Natural Product Reports* **16**, 39–57.
Hills H. G., Williams N. H. and Dodson C. H. (1972) Floral fragrances and isolating mechanisms in the genus *Catasetum* (Orchidaceae). *Biotropica* **4**, 61–76.
Honda K., Ômura H. and Hayashi N. (1998) Identification of floral volatiles from *Ligustrum japonicum* that stimulate flower visiting by cabbage butterfly, *Pieris rapae*. *J. Chem. Ecol.* **24**, 2167–2180.
Irwin R. E. and Brody A. K. (1999) Nectar-robbing bumble bees reduce the fitness of *Ipomopsis aggregata* (Polemoniaceae). *Ecology* **80**, 1703–1712.
Jakobsen H. B. and Olsen C. E. (1994) Influence of climatic factors on emission of flower volatiles in situ. *Planta* **192**, 365–371.
Janzen D. H. (1979) How to be a fig. *Ann. Rev. of Ecol. and System.* **10**, 13–51.
Johnson S. D. and Steiner K. E. (2000) Generalization versus specialization in plant pollination systems. *Trends in Ecol. and Evol.* **15**, 140–143.
Kaiser R. (1993) *The Scent of Orchids*. Elsevier Science Publishers, Amsterdam.
Kaiser R. and Tollsten L. (1995) An introduction to the scent of cacti. *Flavour and Fragrance Journal* **10**, 153–164.
Kite G. C. and Hetterschieid W. L. A. (1997) Inflorescence odours of *Amorphophallus* and *Pseudodracontium* (Araceae). *Phytochemistry* **46**, 71–75.
Knudsen J. T. (2002) Variation in floral scent composition within and between populations of *Geonoma macrostachys* (Arecaceae) in the western Amazon. *Am. J. Bot.* (in press).
Knudsen J. T. and Tollsten L. (1993) Trends in floral scent chemistry in pollination syndromes: floral scent composition in moth-pollinated taxa. *Bot. J. Linnean Soc.* **113**, 263–284.
Knudsen J. T., Tollsten L. and Bergström L. G. (1993) Floral scents – a check-list of volatile compounds isolated by head-space techniques. *Phytochemistry* **33**, 253–280.
Knudsen J. T. and Tollsten L. (1995) Floral scent in bat-pollinated plants: a case of convergent evolution. *Bot. J. Linnean Soc.* **118**, 45–57.
Kullenberg B. and Bergström L. G. (1976) The pollination of *Ophrys* orchids. *Botaniska Notiser* **129**, 11–19.
Kunze J. and Gumbert A. (2001) The combined effect of color and odor on flower choice behavior of bumblebees in flower mimicry systems. *Behav. Ecol.* **12**, 447–456.
Landolt P. J. and Phillips T. W. (1997) Host plant influences on sex pheromone behaviour of phytophagous insects. *Ann. Rev. Entomol.* **42**, 371–391.
Larsson M. C., Stensmyr M. C., Bice S. and Hansson B. S. (2001) Attractiveness of electrophysiologically active fruit and flower odorants to the African fruit chafer *Pachnoda marginata* (Coleoptera: Scarabaeidae). Chapter from doctoral thesis of M. C. Larsson, Department of Ecology, Lund University, Sweden.

Law J. H. and Regnier R. E. (1971) Pheromones. *Ann. Rev. Biochem.* **40**, 533–548.

Lawton R. O., Alexander L. D., Setzer W. N. and Byler K. G. (1993) Floral essential oil of *Guettarda poasana* inhibits yeast growth. *Biotropica* **25**, 483–486.

Levin R. A., Raguso R. A. and McDade L. A. (2001) Fragrance chemistry and pollinator affinities in Nyctaginaceae. *Phytochemistry* **58**, 429–440.

Levin R. A., McDade L. A. and Raguso R. A. (2003). The systematic utility of floral and vegetative fragrance in two genera of Nyctaginaceae. *System. Biol.* (in press).

Lewis J. A., Moore C. J., Fletcher M. T., Drew R. A. and Kitching W. (1988) Volatile compounds from the flowers of *Spathiphyllum cannaefolium. Phytochemistry* **27**, 2755–2757.

Linhart Y. B. and Thompson J. D. (1995) Terpene-based selective herbivory by *Helix aspersa* (Mollusca) on *Thymus vulgaris. Oecologia* **102**, 126–132.

Linn C. E. Jr, Bjostad L. B., Du J. W. and Roelofs W. L. (1984) Redundancy in a chemical signal; behavioral responses of male *Trichoplusia ni* to a 6-component sex pheromone blend. *J. Chem. Ecol.* **10**, 1635–1658.

Lunau K. (1992) Evolutionary aspects of perfume collection in male euglossine bees (Hymenoptera) and of nest deception in bee-pollinated flowers. *Chemoecology* **3**, 65–73.

Mant J. G., Schiestl F. P., Peakall R. and Weston P. (2002) A phylogenetic study of pollinator conservatism among sexually deceptive orchids. *Evolution* **56**, 888–898.

Martin D., Tholl D., Gershenzon J. and Bohlmann J. (2002) Methyl jasmonate induces traumatic resin ducts, terpenoid resin biosynthesis and terpenoid accumulation in developing xylem of Norway spruce stems. *Plant Physiol.* **129**, 100–1018.

McElfresh J. S. and Millar J. G. (2001) Geographic variation in the pheromone system of the saturniid moth *Hemileuca eglanterina. Ecology* **82**, 3505–3518.

Meeuse B. J. D. and Raskin I. (1988) Sexual reproduction in the arum lily family, with emphasis on thermogenicity. *Sexual Plant Reproduction* **1**, 3–15.

Menzel R. and Müller U. (1996) Learning and memory in honeybees: from behavior to neural substrates. *Ann. Rev. Neurosci.* **19**, 379–404.

Metcalf R. L. and Metcalf E. R. (1992) *Plant Kairomones in Insect Ecology and Control.* Chapman and Hall, New York.

Metcalf R. L., Mitchell W. C. and Metcalf E. R. (1983) Olfactory receptors in the melon fly *Dacus cucurbitate* and the oriental fruit fly *Dacus dorsalis. Proc. Natl. Acad. Sci. USA* **80**, 3143–3147.

Moya S. and Ackerman J. D. (1993) Variation in the floral fragrance of *Epidendrum ciliare* (Orchidaceae). *Nordic J. Bot.* **13**, 41–47.

Nilsson L. A. (1992) Orchid pollination biology. *Trends in Ecol. and Evol.*, **7**, 255–259.

Nishida R., Shelly T. E. and Kaneshiro K. Y. (1997) Acquisition of female-attracting fragrance by males of oriental fruit fly from a Hawaiian lei flower, *Fagraea berteriana. J. Chem. Ecol.* **23**, 2275–2285.

Ohloff G. (1994) *Scent and Fragrances. The Fascination of Odors and Their Chemical Perspectives*, transl. by W. Pickenhagen and B. M. Lawrence, Springer-Verlag, Berlin.

Ollerton J. (1996) Reconciling ecological processes with phylogenetic patterns: the apparent paradox of plant-pollinator systems. *J. Ecol.* **84**, 767–769.

Ômura H., Honda K. and Hayashi N. (2000) Floral scent of *Osmanthus fragrans* discourages foraging behavior of cabbage butterfly, *Pieris rapae. J. Chem. Ecol.* **26**, 655–666.

Patt J. M., French J. C., Schal C., Lech J. and Hartman T. G. (1995) The pollination biology of Tuckahoe, *Peltandra virginica* (Araceae). *Am. J. Bot.* **82**, 1230–1240.

Paulus H. F. and Gack C. (1990) Pollinators as pre-pollinating isolation factors: evolution and speciation in *Ophrys* (Orchidaceae). *Israel J. Bot.* **39**, 43–79.

Peakall R. and Beattie A. J. (1996) Ecological and genetic consequences of pollination by sexual deception in the orchid *Caladenia tentactulata*. *Evolution* **50**, 2207–2220.

Pellmyr O. (1986) Three pollination morphs in *Cimicifuga simplex*: incipient speciation due to inferiority in competition. *Oecologia* **78**, 304–307

Pellmyr O. (1997) Pollinating seed eaters: why is active pollination so rare? *Ecology* **78**, 1655–1660.

Pettersson M. W. (1991) Pollination by a guild of fluctuating moth populations: option for unspecialization in *Silene vulgaris*. *J. Ecol.* **79**, 591–604.

Pham-Delègue M.-H., Etievant P., Guichard E., Marilleau R., Douault P., Chauffaille J. and Masson C. (1990) Chemicals involved in honeybee-sunflower relationship. *J. Chem. Ecol.* **16**, 3053–3065.

Pichersky E. and Gershenzon J. (2002) The formation and function of plant volatiles: perfumes for pollinator attraction and defense. *Current Opinions in Plant Biology* **5**, 237–243.

Pijl L. van der and Dodson C. H. (1966) Orchid Flowers: Their Pollination and Evolution. University of Miami Press, Coral Gables, FL.

Plepys D., Ibarra F. and Löfstedt C. (2002) Volatiles from flowers of *Platanthera bifolia* L. (Rich.) (Orchidaceae) attractive to the silver Y moth, *Autographa gamma* L. (Lepidoptera: Noctuidae). *Oikos*, (in press).

Pliske T. E. (1992) Pollination of pyrrolizidine alkaloid-containing plants by male Lepidoptera. *Environ. Entomol.* **4**, 474–479.

Raffa K. F. (2001) Mixed messages across multiple trophic levels: the ecology of bark beetle chemical communication systems. *Chemoecology* **11**, 49–65.

Raguso R. A. (2001) Floral scent, olfaction and scent-driven foraging behavior. In Cognitive Ecology of Pollination; Animal Behavior and Floral Evolution, eds L. Chittka and J. D. Thomson, pp. 83–105. Cambridge University Press, Cambridge, UK.

Raguso R. A. (2003) Why do flowers smell? The chemical *Ecology* of fragrance-driven pollination. In *Advances in Insect Chemical Ecology*, eds R. T. Cardé and J. G. Millar. Cambridge University Press, Cambridge, UK (in press).

Raguso R. A. and Roy B. A. (1998) "Floral" scent production by *Puccinia* rust fungi that mimic flowers. *Molec. Ecol.* **7**, 1127–1136.

Roy B. A. and Raguso R. A. (1997) Olfactory vs. visual cues in a floral mimicry system. *Oecologia* **109**, 414–426.

Roy B. A. and Widmer A. (1999) Floral mimicry: a fascinating yet poorly understood phenomenon. *Trends in Plant Sciences* **4**, 325–330.

Sasagawa H., Sasaki M., Yamaoka R. and Matsuyama S. (1997) Why does the oriental orchid *Cymbidium floribundum* attract the Japanese honeybee, and why does it not attract the European honeybee? Abstract, 14th Annual Meeting, International Society of Chemical Ecology, Vancouver, Canada.

Sasaki M., Ono M., Asada S. and Yoshida T. (1991) Oriental orchid (*Cymbidium pumilum*) attracts drones of the Japanese honeybee (*Apis cerana japonica*) as pollinators. *Experientia* **47**, 1229–1231.

Schiestl F. P. and Ayasse M. (2002) Do changes in floral odor cause speciation in sexually deceptive orchids? *Plant System. and Evol.* (in press).

Schiestl F. P., Ayasse M., Paulus H. F., Erdmann D. and Francke W. (1997) Variation of floral scent emission and post-pollination changes in individual flowers of *Ophrys sphegodes subsp. sphegodes*. *J. Chem. Ecol.* **23**, 2881–2895.

Schiestl F. P., Ayasse M., Paulus H. F., Löfstedt C., Hansson B. S., Ibarra F. and Francke W. (1999) Orchid pollination by sexual swindle. *Nature* **399**, 421–422.

Schiestl F. P., Ayasse M., Paulus H. F., Löfstedt Hansson B. S., Ibarra F. and Francke W. (2000) Sex pheromone mimicry in the early spider orchid (*Ophrys sphegodes*): patterns of hydrocarbons as the key mechanism for pollination by sexual deception. *J. Comp. Physiol. A* **186**, 567–574.

Schneider D. (1986) The strange fate of pyrrolizidine alkaloids. In *Perspectives in Chemoreception and Behavior*, eds R. F. Chapman, E. A. Bernays and J. G. Stoffolano Jr, pp. 123–142. Springer-Verlag, New York.

Simpson, B. B. and Neff J. L. (1983) Evolution and diversity of floral rewards, in C. E. Jones and Little, R. J. eds, *Handbook of Experimental Pollination Biology*, pp. 142–159 Van Nostrand-Reinhold, NY.

Smart L. E. and Blight M. M. (2000) Response of the pollen beetle, *Meligethes aeneus*, to traps baited with volatiles from oilseed rape, *Brassica napus*. *J. Chem. Ecol.* **26**, 1051–1064.

Smith B. H. and Getz W. M. (1994) Non-pheromonal olfactory processing in insects. *Ann. Rev. Entomol.* **39**, 351–375.

Stautamire W. P. (1983) Wasp-pollinated species of *Caladenia* (Orchidaceae) in South-western Australia. *Aus. J. Bot.* **31**, 383–394.

Steele C. L., Crock J., Bohlmann J. and Croteau R. (1998) Sesquiterpene synthases from grand fir (*Abies grandis*): comparison of constitutive and wound-induced activities, and cDNA isolation, characterization and bacterial expression of δ-selinene synthase and γ-humulene synthase. *J. Biol. Chem.* **273**, 2078–2089.

Stensmyr, M. C., Urru I., Collu I., Celander M., Hansson B. S. and Angioy A. M. (2002) Rotting smell of dead horse arum florets. *Nature* **420**, 625–626.

Stopfer M., Bhagavan S., Smith B. H. and Laurent G. (1997) Impaired odour discrimination on desynchronization of odour-encoding neural assemblies. *Nature* **390**, 70–74.

Stowe M. K., Turlings T. C. J., Loughrin J. H., Lewis W. J. and Tumlinson J. H. (1995). The chemistry of eavesdropping, alarm and deceit. *Proc. Nat. Acad. Sci., USA*, **92**, 23–28.

Stránský K. and Valterová, I. (1999) Release of volatiles during the flowering period of *Hydrosme rivieri* (Araceae). *Phytochemistry* **52**, 1387–1390.

Tan K-H. and Nishida R. (2000) Mutual reproductive benefits between a wild orchid, *Bulbophyllum patens* and *Bactrocera* fruit flies via a floral synomone. *J. Chem. Ecol.* **26**, 533–546.

Thien L. B., Heimermann W. H. and Holman R. T. (1975) Floral odors and quantitative taxonomy of *Magnolia* and *Liriodendron*. *Taxon* **24**, 557–568.

Thièry D., Bluet J. M., Pham-Delègue M.-H., Etiévant P. and Masson C. (1990) Sunflower aroma detection by the honeybee: study by coupling gas chromatography and electroantennography. *J. Chem. Ecol.* **16**, 701–711.

Thompson J. D., Manicacci D. and Tarayre M. (1998) Thirty-five years of thyme: a tale of two polymorphisms. *BioScience* **48**, 805–815.

Tollsten L. and Bergström L. G. (1993) Fragrance chemotypes of *Platanthera* (Orchidaceae) – the result of adaptation to pollinating moths? *Nordic J. Bot.* **13**, 607–613.

Trigo J. R., Brown K. S. Jr, Witte L., Hartmann T., Ernst L. and Soares-Barata L. E. (1996) Pyrrolizidine alkaloids: different acquisition and use patterns in Apocynaceae and Solanaceae feeding ithomiine butterflies (Lepidoptera: Nymphalidae). *Biol. J. Linnean Soc.* **58**, 99–123.

Vogel S. (1963) *Duftdrüsen im Dienste der Bestäubung. Über Bau und Function der Osmophoren*. Verlag der Akademie der Wissenschaftend und der Literatur, Wiesbaden.

Vogel S. (1983) Ecophysiology of zoophilic pollination. In Physiological Plant Ecology III, eds O. L. Lange, P. S. Nobel, C. B. Osmond and H. Ziegler, pp. 560–624. Springer-Verlag, Berlin.

Ware A. B., Kaye P. T., Compton S. G. and Van Noort S. (1993) Fig volatiles: their role in attracting pollinators and maintaining pollinator specificity. *Plant System. and Evol.* **186**, 147–156.

Waser N. M., Chittka L., Price M. V., Williams N. M. and Ollerton J. (1996) Generalization in pollinator systems, and why it matters. *Ecology* **77**, 1043–1060.

Whitten W. M., Williams N. H., Armbruster W. S., Battiste M. A., Strekowski L. and Lindquist N. (1986) Carvone oxide: an example of convergent evolution in euglossine-pollinated plants. *System. Bot.* **11**, 222–228.

Whitten W. M., Hills H. G. and Williams N. H. (1988) Occurrence of ipsdienol in floral fragrances. *Phytochemistry* **27**, 2759–2760.

Williams N. H. (1983) Floral fragrances as cues in animal behavior. In *Handbook of Experimental Pollination Biology*, eds C. E. Jones and R. J. Little, pp. 51–69. Van Nostrand-Reinhold, New York.

Williams N. H. and Whitten W. M. (1983) Orchid floral fragrances and male euglossine bees: methods and advances in the last sesquidecade. *Biol. Bull.* **164**, 355–395.

Winter Y. and Helversen O. von (2001) Bats as pollinators: foraging energetics and floral adaptations. In *Cognitive Ecology of Pollination; Animal Behavior and Floral Evolution*, eds L. Chittka and J. D. Thomson, pp. 148–170. Cambridge University Press, Cambridge, UK.

Wright S. (1932) The role of mutation, inbreeding, crossbreeding and selection in evolution. *Proceedings of the Sixth International Congress on Genetics* **1**, 356–366.

23

Physiology and genetics of odor perception in *Drosophila*

Marien de Bruyne

23.1 What's that smell? Odor coding in a small fly

In a world where vital information is often present in the form of molecules carried away from resources by air or water, olfactory perception has proven to be crucial for survival in many species. From a general ability of cells to react to the chemical composition of the medium, highly sophisticated systems of odor discrimination and signaling via semiochemicals have evolved. We still understand very little of the coding of chemical information in the brains of moluscs, arthropods and vertebrates. A fundamental problem in the study of olfaction is that the chemical environment is highly complex and cannot be easily described in an adequate set of physical parameters that we can relate to a neuronal code. This has made studying insect pheromones so attractive because relatively simple mixtures of a few specific compounds can define stimulus space and elicit stereotypical behaviors. Nevertheless, insects can detect many chemicals and respond to them behaviorally. Pheromone systems may have evolved as a highly amplified and specialized subset of a previously homogeneous network. Alternatively, the general odor system is in fact a fusion of several specialized subsystems, from which "homogeneous" network properties emerge when "non-special" stimuli are applied. Either way, we can consider the olfactory system as a unified sensory system and extract general principles of sensory coding (Christensen and Hildebrand, 2002).

Coding of olfactory information is a two-step process. First, sensory transduction converts chemical information in the environment into a code of action potentials.

Second, the neural processing of this code defines a percept, called "an odor". The first process takes place in a heterogeneous population of olfactory receptor neurons (ORNs) distributed in the epithelium. It determines which volatiles can be detected. The second process occurs in a series of neuropiles in the brain. It leads to some form of perception and can drive a behavioral response, depending on the animal's internal state and integration with other sensory modalities. How do individual neurons handle the transfer of information? How do they connect to form a network in which this information is distributed, and what are the coding properties of this network?

To answer these fundamental questions, we would like to be able to characterize all subunits of the network, as well as experimentally alter their properties. So far, neither the first nor the second goal has been achieved in any insect or vertebrate species. The pomace fly, *Drosophila melanogaster*, has emerged as a powerful model because it offers distinct advantages that may allow such a detailed analysis. Most basic knowledge on *Drosophila* is readily available. It is one of the most studied eukaryotes, an important model in development, anatomy, neurobiology and behavior. Research in *Drosophila* olfaction also has a long tradition (Barrows, 1907; Kellogg *et al.*, 1962; Fuyama, 1976, Rodrigues and Siddiqi, 1978). A number of olfactory behavioral paradigms have been described and the fly has become a major model system for associative olfactory learning (Carlson, 1996; Roman and Davis, 2001). Its small size can be a challenge for physiologists but also offers distinct advantages. The antennal epithelium is packed with neurons and has a transparent cuticle, so one can see a large set of olfactory sensilla in a single view under a compound microscope. For the less adept, electroantennograms can give reliable and easily obtained results (see section 23.1). Furthermore, its olfactory system contains fewer cells but is of similar complexity as in larger models. Most importantly, an array of genetic and molecular tools allows the experimental manipulation of the biochemistry of its olfactory system.

Though coding properties emerge from neuronal response patterns and synapses, indirectly they are the result of gene expression. A genetic approach to analyzing olfactory processing not only supplies manipulative tools but can also reveal links between functional properties and molecular evolution. Whether we examine the transduction properties at the level of the ORN population or the processing of this primary information at the level of the CNS, it is the differential and combinatorial expression of information in the genome that establishes and maintains a functioning network. The discovery of the odor receptor gene family in the rat (Buck and Axel, 1991) provided new insight into the structure of olfactory systems and encouraged a renewed effort to understand odor coding. Data mining in the completed sequence of the *Drosophila* genome produced several gene families important for the coding properties of ORNs. Odor receptor genes (ORs, Clyne *et al.*, 1999b; Vosshall *et al.*, 1999) and some gustatory receptor genes (GRs, Clyne *et al.*, 2000; Scott *et al.*, 2001) are differentially expressed in olfactory neurons (see Chapter 19). Odor binding proteins (OBPs), located in olfactory as well as gustatory epithelia also constitute a large family (Galindo and Smith, 2001; see Chapter 18). Each individual gene provides a potential handle for manipulating molecularly defined subsets of receptor neurons. Classical forward genetics has produced a variety of mutations in genes that play a role in olfaction (Carlson, 1996). Such mutations provide information about the phenotypic effects of complete lesions (null mutations) or more subtle modifications in the sequences of the endogenous gene. Reverse genetics allows the incorporation of cloned sequences of endogenous or foreign genes into the *Drosophila* genome and they can be targeted to specific cell populations (Stocker and Rodrigues, 1999). A variety of molecular and genetic "tricks" can give excellent control over the expression of gene products in terms of location (specific cells, tissues) as well as timing (during or after development). Binary expression systems, such as the GAL4/UAS system (Brand and Perrimon, 1993), have allowed the expression of a variety of reporter genes driven by an increasing number of promoter/enhancer sequences. An overview of genes that have been shown, with any of these methods, to play a role in olfaction is given in Table 23.1.

This chapter outlines our current knowledge on the genetics and physiology of the *Drosophila* olfactory system. We will focus on the peripheral elements but also include an outlook on central processing and behavior. In the nervous system, structure determines computational properties. Hence, considerable attention is given to structure and development of the olfactory system. Several mutations, with interesting odor-specific phenotypes, disrupt genes that are not needed during olfactory processing. Rather, they help to establish the structures that underlie it. Observing the way the system is assembled leads to more insight in the fundamental units it comprises. We focus on odor perception in adult *Drosophila*, but a functional, if differently constructed, olfactory system exists also in larvae (see Cobb, 1999 for a review). Throughout this chapter, genes are named as they appear in FlyBase, a database for *Drosophila* genetics (http://flybase.bio.indiana.edu/). A clear distinction is made between receptors (i.e. receptor proteins, OR) and receptor neurons (ORN). Finally, the word "odor" is used in a perceptual sense, as in "odor of banana", whereas "odorant" refers to a volatile chemical compound that can stimulate a receptor neuron (Hudson, 1999).

23.2 The makeup of an insect nose: peripheral structures and developmental genes

Like all Diptera, *Drosophila* has two paired appendages that carry olfactory sensilla (Figure 23.1A): the antennae carry ca. 1200 ORNs and the maxillary

Table 23.1 Genes and mutations reported to play a role in olfaction

Gene	a.k.a.	Full name	(Predicted) protein	Mutation	Reference
Developmental signals/transcription factors					
acj6	*I-POU*	*abnormal Chemosensory jump 6*	POU-domain Homeodomain transcription factor	++	Clyne *et al.* (1999)
amos	*Roi*	*absent MD neurons and olfactory sensilla*	basic Helix–loop–Helix transcription factor	–	Goulding *et al.* (2000a)
ato		*atonal*	basic Helix–loop–Helix transcription factor	++	Gupta and Rodrigues, (1997)
cato		*Cousin-of-atonal*	basic Helix–loop–Helix transcription factor	–	Goulding *et al.* (2000b)
lz		*lozenge*	AML-1/Runt transcription factor	++	Stocker and Gendre, (1988), Gupta *et al.* (1998)
noc^{Sco}	*Sco*	*Scutoid (no ocelli/snail)*	Zinc finger, C2H2 type transcription factor	++	Dubin *et al.* (1995), Fuse *et al.* (1999)
rst	*irreC*	*roughest*	Immunoglobulin, cell adhesion protein	++	Reddy *et al.* (1999)
tap	*bps*	*target of poxn/biparous*	basic Helix-loop-Helix transcription factor	++	Ledent *et al.* (1998)
Signal transduction molecules					
Arr1		*Arrestin A*	arrestin	++	Merril *et al.* (2002)
Arr2		*Arrestin B*	arrestin	++	Merril *et al.* (2002)
dnc		*dunce*	cAMP-specific phosphodiesterase	++	Martín *et al.* (2001)
Gα*49B*	*dGq*α	*G protein* α*49B*	G-protein α subunit, group Q	–	Talluri *et al.* (1995); Kalidas and Smith, (2002)
Mys	*olfC*	*Myospheroid*	beta subunit of integrin, Ca dependent cell adhesion	++	Ayyub *et al.* (1990), (2000)

norpA		*no receptor potential A*	phospholipase C	–	Riesgo-Escovar *et al.* (1995)
rdgB	*ota1*	*retinal degeneration B/olfactory trap abnormal 1*	phosphatidylinositol transporter	++	Woodard *et al.* (1992)
rut		*rutabaga*	adenylate cyclase	++	Martín *et al.* (2001)
sws	*OlfE*	*swiss cheese/olfactory E*	cAMP-binding domain-like	++	Hasan, (1990); Kretzschmar *et al.* (1997)
TyrR	*Hono*	*Tyramine receptor*	biogenic amine receptor	++	Kutsukake *et al.* (2000)
Ion channels					
Cng		*Cyclic-nucleotide-gated ion channel protein*	(intracellular) cyclic nucleotide activated cation channel	–	Baumann *et al.* (1994)
eag		*ether a go-go*	(CN sensitive) Voltage-gated potassium channel	++	Dubin *et al.* (1998a)
Itp-r83A		*Inositol 1,4,5-tris-phosphate receptor*	IP3-gated calcium channel	++	Raghu and Hasan, (1995); Deshpande *et al.*(2000)
para	*sbl, olfD*	*paralytic/smellblind/olfactory D*	Voltage-gated sodium channel	++	Lilly and Carlson, (1989)
trp		*transient receptor potential*	Voltage-gated calcium channel	++	Störtkuhl *et al.* (1999)
Enzymes					
Cpr		*Cytochrome P450 reductase*	NADPH-ferrihemoprotein reductase	–	Hovemann *et al.* (1997)
dare		*defective in the avoidance of repellents*	adrenodoxin reductase	++	Freeman *et al.* (1999)
Secreted sensillar proteins					
a10	*OS-D*	*antennal protein 10*	binding protein (CSP family)	–	Pikielny *et al.* (1994), McKenna *et al.* (1994)
lush	*OBP76c*	*lush*	binding protein (OBP family)	++	Kim *et al.* (1998)

(Contd)

Table 23.1 *(Contd)*

Gene	a.k.a.	Full name	(Predicted) protein	Mutation	Reference
Unknown function					
ami		*anosmic*	(nitric oxide synthase, NOS?)	++	Warr *et al.* (1998)
ana		*anachronism*	*unknown*	++	Park *et al.* (1997)
east		*enhanced adult salt tolerance*	(Nuclear, chromatin associated)	++	Vijay Raghavan *et al.* (1992)
gk		*geko*	*unknown*	++	Shiraiwa *et al.* (2000)
Indf		*Indifferent*	*unknown*	++	Cobb, (1996)
odA		*?*	*unknown*	++	Alcorta, (1991)
olfA		*olfactory A*	*unknown*	++	Ayyub *et al.* (1990)
olfB		*olfactory B*	*unknown*	++	Ayyub *et al.* (1990)
olfF		*olfactory F*	*unknown*	++	Ayyub *et al.* (1990)
Os9		*Olfactory-specific 9*	(leucine zipper, transcription factor?)	–	Raha and Carlson, (1994)
ota 2		*Olfactory trap abnormal 2*	*unknown*	++	Woodard *et al.* (1989)
ota 7		*Olfactory trap abnormal 7*	*unknown*	++	Woodard *et al.* (1989)
ptg	3D18	*Pentagon*	*unknown*	++	Helfand *et al.* (1989)
smi35A		*smell impaired 35A*	(protein serine/threonine kinase?)	++	Anholt *et al.* (1996), Fedorowicz *et al.* (1999)
smi60E	*DCS1*	*smell impaired 60E*	(Voltage-gated sodium channel?)	++	Anholt *et al.* (1996), Kulkarni *et al.* (2002)
smi97B	*scrib*	*smell impaired 97B/scribbled*	(PDZ-domain protein, membrane protein?)	++	Anholt *et al.* (1996), Ganguly *et al.* (2001)

Members of the families of OR and GR genes have been omitted, as well as most members of the OBP family for which no mutants are available. Only three mutations out of a series of P-element insertions named *smi* (Anholt *et al.* 1996) were included.

palps roughly 1/10 of that, ca. 120 ORNs (Venkatesh and Singh, 1984; Singh and Nayak, 1985; Stocker, 1994). Both olfactory organs are derived from the larval antennal disk. Even though these two organs have distinct segmental origins, their ORNs converge on the same olfactory neuropile, the antennal lobe (AL) in the deutocerebrum.

23.2.1 Olfactory sensilla of antenna and maxillary palp

The shape of the *Drosophila* antenna is typical for Brachiceran Diptera (Figure 23.1B) and consists of three segments (I, scape, II, pedicel, III, funiculus + arista). The third segment functions as a sound receiver, which amplifies vibrations and transmits them to Johnson's organ in the second segment (Göpfert and Robert, 2001). It is homologous to the long flagellum of nematoceran Diptera, but the original longitudinal axis is twisted and continues into the arista (Figure 23.1B; Postlethwait and Schneiderman, 1971). This morphogenic twist, its role as sound receiver and the presence of a multi-chambered sensory pit (sacculus) make the antenna a highly asymmetric and complex, patterned structure. The funiculus is almost entirely dedicated to the olfactory modality. There are no mechanosensory or gustatory neurons but six neurons in the arista and up to 56 neurons in the sacculus are thought to have a thermo/hygroreceptive function (Foelix *et al.*, 1989; Itoh *et al.*, 1991; Shanbhag *et al.*, 1995; Sayeed and Benzer, 1996). The total number of antennal neurons (ca. 1200) is small compared to other important insect models for olfaction (honeybee, *Apis melifera*, 65 000, Esslen and Kaissling, 1976; moth, *Manduca sexta*, 250 000, Sanes and Hildebrand, 1976; cockroach, *Periplaneta americana*, 270 000, Schaller, 1978).

Olfactory sensilla on the surface of the antenna fall into two ultrastructural categories, double walled (dw) and single walled (sw) (Figure 23.1C; Altner and Prillinger, 1980). The dw sensilla are small, placed on a cuticular depression and hence erroneously known as "coeloconic" sensilla (C). They occur on the surface as well as inside the sacculus. They have deep longitudinal grooves in the outer wall and an inner wall that forms a dendritic sheath. The sw sensilla have traditionally been further categorized as either "basiconic" (B) or "trichoid" (T) (Venkatesh and Singh, 1984). B sensilla are club-shaped without apparent sockets, while T sensilla are long, pointed and have a drum-like basal structure (Venkatesh and Singh, 1984; Shanbhag *et al.*, 1999). The dichotomy between T sensilla and B sensilla may actually be a continuum. In a detailed EM study of antennal sensilla a new category was described that appears intermediate, the I sensilla (Shanbhag *et al.*, 1999). The authors further divide B sensilla into subtypes of decreasing size called LB (large), TB (thin) and SB (small), respectively (Figure 23.1C). The different sw sensillum categories all have one to four ORNs and three accessory cells that wrap concentrically around the dendrites of the ORNs (from in to out, thecogen, tormogen and trichogen cells; Shanbhag *et al.*, 2000). The dw sensilla have two or three ORNs (Figure 23.1C) and four accessory cells

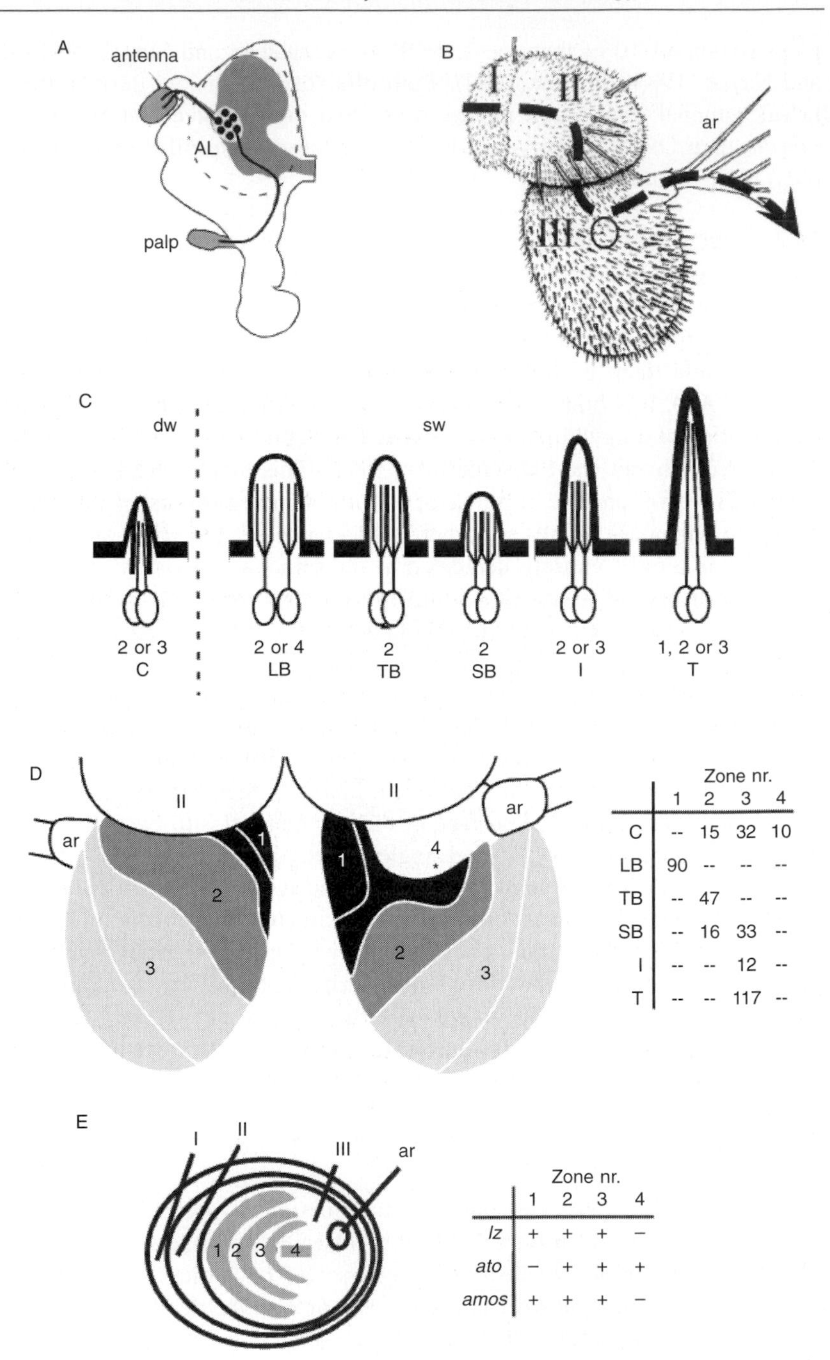

	1	2	3	4
C	--	15	32	10
LB	90	--	--	--
TB	--	47	--	--
SB	--	16	33	--
I	--	--	12	--
T	--	--	117	--

	1	2	3	4
lz	+	+	+	−
ato	−	+	+	+
amos	+	+	+	−

(Shanbhag *et al.*, 2000). The cell bodies of the neurons lie below the epidermis and their axons converge into three fascicles that eventually merge as two nerve tracts leaving the funiculus (Jhaveri *et al.*, 2000b). Glia cells are associated with these fascicles and wrap long processes around ORN axons up to their cell bodies (Jhaveri *et al.*, 2000b; Shanbhag *et al.*, 2000).

The sacculus is an invagination of the antennal epithelium that consists of three chambers (Itoh *et al.*, 1991; Shanbhag *et al.*, 1995). It contains a few more types of sensilla, some of which have no pores and are likely to contain thermohygroreceptive neurons. The sensilla that contain olfactory neurons are all of the dw category, C sensilla similar to those on the surface. Due to the relative inaccessibility of these sensilla nothing is known about their physiology.

The second *Drosophila* appendage that bears olfactory sensilla is the maxillary palp (henceforth simply called palp). The ca. 60 olfactory sensilla on the dorsal and lateral sides of the palp are sw sensilla of the B type. They are structurally similar to antennal TB sensilla and contain two ORNs. In addition to olfactory sensilla the palp also bears ca. 20 large mechanosensory bristles.

Figure 23.1 (*opposite*) Structure and development of *Drosophila* olfactory organs. A Olfactory sensilla occur on two paired appendages of the head: the maxillary palps and antennae. ORNs from both organs project to the antennal lobe (AL) in the *Drosophila* brain. B View of the posterior side of the adult antenna with three segments (I, II and III). The original proximo-distal axis is indicated with a dashed arrow. The circle marks the entrance to the sacculus, a multi-chambered sensory pit. C Structural sensillum categories of the antennal surface. dw, double walled, C, coeloconic, sw, single walled, LB, large basiconic, TB, thin basiconic, SB, small basiconic, I, intermediate, T, trichoid. They can be differentiated based on cuticular structures, neuron numbers (indicated), and dendritic branching. D Distribution zones (gray shades) of structural sensillum categories on the front and back of the antenna. Subdivisions of zones 1 and 3 mark further separation based on functional types (see text). The table shows sensillum numbers per zone. E Schematic overview of the antennal section of the eye-antennal disk at 8 h APF (after puparium formation). The three antennal segments (I, II and III) and the arista (ar) are present as concentric rings. Genes important for the development of olfactory sensilla are expressed in four zones (1–4) on the third segment, labeled by the enhancer trap line A101 (an allele of *neuralized*). The two bHLH transcription factors *ato* (zones 2, 3, 4) and *amos* (zones 1, 2, 3) determine dw and sw sensilla respectively. *Lz* is expressed in zones 1, 2, 3. The *amos* gene is regulated by *lz* which determines the subcategories of sw sensilla. Another bHLH gene, *tap*, is expressed later in a small subset of cells in zone 1. To what extent the zones in D correspond with those in E is not clear. However, zone 1 on the imaginal disk, is characterized by a lack of *ato* expression and most likely corresponds (at least partly) to zone 1 on the adult antenna, which does not include C sensilla. Similarly, zone 4 lacks *lz* and *amos* expression in the disk and may form the sacculus and pure C sensillum parts in zone 4 of the adult antenna. Data from Gupta *et al.* (1998); Goulding *et al.* (2000); Shanbhag *et al.* (1999). B was taken from FlyBase.

23.2.2 Development of the antennal olfactory epithelium

Is the peripheral olfactory system of insects composed of essentially separate units or can it be considered a continuous compartmentalized epithelium such as the insect eye? Evidence from studies on developmental mechanisms suggests it is a bit of both.

The antennal surface is organized as a sensory epithelium with only few epithelial cells in between sensilla (Shanbhag *et al.*, 1999). The development of the field of olfactory sensilla in the antennal disk starts a few hours after puparium formation with the appearance of single founder cells that grow into presensillum clusters of two to four cells (Lienhard and Stocker, 1991; Ray and Rodrigues, 1995; Reddy *et al.*, 1997). These cells are not related by lineage. The founder cells recruit neighboring cells into a neuronal fate by means of cell–cell signaling (Reddy *et al.*, 1997). This is similar to what happens with the formation of ommatidia in the eye. Then, one round of replication and subsequent apoptosis determines the final number of cells in each sensillum type (Ray and Rodrigues, 1995; Reddy *et al.*, 1997). Thus, some cells in a sensillum are related by lineage, which is reminiscent of developmental events in mechanosensory and gustatory sensilla. Further evidence for the layout of the olfactory epithelium as a structured sheet comes from a mutation in the *roughest* (*rst*) gene. It encodes an immunoglobulin-like cell adhesion protein, which affects spacing of developing antennal sensilla (Reddy *et al.*, 1999). In the mutant a similar number of sensilla form but they are bunched together. In the eye this same gene is involved in organizing the continuous sheet of ommatidia by removing epithelial cells in between.

The selection of founder cells from the epithelium starts with the expression of so-called proneural genes; all are basic helix–loop–helix (bHLH)-type transcription factors. The *achaete–scute* complex of genes, proneural genes for mechanosensory and gustatory sensilla, do not play a role in olfactory sensilla (Reddy *et al.*, 1997). Instead the proneural gene for dw sensilla on the antenna is *atonal* (*ato*, Gupta and Rodrigues 1997; Jhaveri *et al.*, 2000a), which also specifies ommatidia, while sw sensilla on the antenna are generated by expression of *amos* (Goulding *et al.*, 2000a). This shows fundamental differences between olfactory sensilla and other sensory hairs on the one hand and between dw and sw sensilla on the other hand. The structural distinctions between sw and dw sensilla across many insect species has previously led to speculations about separate evolutionary histories (Steinbrecht, 1997).

The presensillum clusters form in a characteristic pattern of three half elliptical domains in the antennal disk, in a temporal sequence from out to in (Figure 23.1E). The early expression patterns of *ato* and *amos* in the antennal disk outline domains that determine the distribution patterns of the developing sensillum categories. On the palp *ato* specifies B sensilla (Gupta and Rodrigues 1997) which illustrates the fact that these genes influence sensillum type dependent on the surrounding tissue. Further specification of sw sensillum categories involves interactions between *amos* and different dosages of the *lz* gene product (Stocker *et al.*, 1993; Gupta *et al.*, 1998; Goulding *et al.*, 2000b). *Lz* acts before founder cell formation to prepattern the field and affect the choice of *amos* positive cells to become B, I or T sensilla. Their distribution pattern suggests they are formed along a morphogenic axis running from the proximo-medial to the disto-lateral side of the antenna with an order of prevalence of LB-TB-SB-I-T (Shanbhag *et al.*, 1999; Figure 23.1D). The *lz* gene also affects subsets of palpal sensilla (Stocker *et al.*, 1993; Riesgo-Ecovar *et al.*, 1997b) and small effects have been noticed on shaft morphology of C sensilla on the antenna (Riesgo-Ecovar *et al.*, 1997a). Two more bHLH genes are expressed in the antennal disk and may play a role in specifying sensillum types. *Cousin-of-atonal* (*cato*) is expressed in C sensilla and regulated (repressed) by *prospero* (Goulding *et al.*, 2000b) and the *tap* gene is expressed early in a small subset of precursor cells (Ledent *et al.*, 1998). The function of these two bHLH genes is not clear but the homeodomain transcription factor *Prospero* (*pros*) together with the nuclear receptor *seven-up* (*svp*) and pan-neuronal *elav* were recently shown to be expressed in a combinatorial way in the lineage of the presensillum cluster (Sen *et al.*, 2003). Interestingly, these authors find that ORNs can descend from two different precursor cells, only one of which expresses *svp*. Specification of different sensillum categories down to ORN sensitivities is thus achieved by a combinatorial code of key regulatory transcription factors combined with positional information.

The heterogeneity of the periphery of the olfactory system is striking. What, if any, is the relationship between these overlapping, intricately orchestrated, patterns of gene expression and the ability of the peripheral olfactory system to encode odors?

23.3 Functional properties of an insect nose: ORN physiology

How are important stimulus features encoded in ORNs? Three stimulus features need to be considered: odor identity (qualitative), odor intensity (quantitative, related to concentration) and their fluctuations in time. They are encoded by ORNs in firing patterns that vary in time and across the population of neurons and can only be distilled from the activity of multiple inputs by the neural network they connect to. In a single neuron's firing all three aspects are encoded simultaneously so it is impossible to reconstruct the original stimulus from the neuron's output. For example, if an ORN responds strongly to odorant A and weakly to odorant B a medium firing rate either represents a weak dose of A or a strong dose of B. Nevertheless we will try to deal with encoding stimulus quality, intensity and temporal variation separately but first we introduce the methods used to measure ORN output.

Drosophila ORNs suggests that the 1300 ORNs in the olfactory system can be reduced to a much smaller number of functional units (Figure 23.3). Each of these coding units comprises 15–45 neurons with similar properties. However, it should be pointed out that classifications of tuning will depend on the odor set employed (Derby, 2000) and not all antennal B sensilla could be classified. There appear to be no obvious differences between the palpal and antennal B sensilla but each structural sensillum category contains different ORN classes. LB sensilla are of three types, while TB and SB sensilla probably each comprise two types and PB sensilla three types (Figure 23.2B). With knowledge of the location of a recording and a sensillum's responses to a few carefully chosen odorants these types can be easily recognized. Because of different odor stimulation methods, a direct comparison with Siddiqi (1991) is not easy, but the described types I, II and III most likely correspond to ab1, ab2 and ab4.

Most ORNs only show responses to a small subset of the tested odorants. An example of two neurons in a pb1 sensillum demonstrates the variability in tuning width (Figure 23.2C). Among the 22 classes of neurons characterized, some are very narrowly tuned (1 out of 47 odorants) while others show a broader spectrum (up to 18 out of 47 odorants). The ab1C cell could function as a specific detector because it is the only ORN responding to CO_2 and it responds exclusively to CO_2. To what extent do these tuning properties of single ORNs relate to OR gene expression? There is clear evidence for an essential role of OR gene expression in tuning of ORNs. Overexpression of one OR gene (Or43a) in many ORNs leads to an increase in EAG response to cyclohexanol (Störtkuhl and Kettler, 2001) whereas misexpression of Or43a in one specific ORN class modifies glomerular responses (Wang *et al.*, 2003, see below). In addition to proving that expression of Or22a in ab3A neurons is responsible for inducing the broad response spectrum of these neurons Dobritsa *et al.* (2003) also show that a single ORN can express two receptors. A second protein, Or22b, also present in the dendrites of bit appears to be non-functional. Which receptor is expressed in a given ORN is thus most likely to define neuronal identity in terms of response pattern. It cannot be excluded that two ORNs expressing the same OR have different response properties if they differ in the expression of some other gene(s). Dobritsa *et al.* (2003) also show that in the case of Or47a, which is normally in ab5B neurons the response spectrum of ab5B is conferred to ab3A when Or47a is expressed there. Interestingly, this receptor exchange also works when native Or22a is still present, suggesting two OR proteins can function independently in a single ORN.

Can anything be deduced from the range of stimuli that evoke strong responses? Some of these compounds are well-known stimuli for insect ORNs. The green leaf volatile E2-hexenal is detected by many plant feeders (Visser, 1986), 4-methylphenol is known as an attractant for tsetse flies (Den Otter and Van der Goes van Naters, 1993) and methylsalicylate is a product typically produced by plants after herbivore attack (Hardie *et al.*, 1994). Specialized CO_2 detectors are common (Stange, 1999). Some odorants that have been most successfully employed in *Drosophila* orientation or learning assays are perceived by multiple ORN classes, most notably, ethyl acetate, benzaldehyde and 3-octanol. Even though accurate classification of ORNs depends on testing adequate stimuli, response spectra also differ when closely related chemical structures are tested (de Bruyne *et al.*, 1999). When presented with a series of aliphatic esters of increasing chain length, three palpal ORNs still show differences in preferred chain length, as known from other studies on insect ORNs (Selzer, 1981).

Clyne *et al.* (1997) supply data on responses of neurons in T and C sensilla to a panel of 16 odorants. The T sensilla are of at least two functional types, which we call here at1 and at2 in analogy with the nomenclature for B sensilla (Figure 23.3). The at1 sensilla have one or two neurons, show low spontaneous firing rates and a very narrow response spectrum. One ORN responds moderately to *cis*-vaccenyl acetate (Z11-18Ac, a putative pheromonal component, see below). Type 2 sensilla (at2) have higher spontaneous firing rates, more than two neurons and a minor response to E2-hexenal. The relatively low responses suggest that better ligands can be found. In moths, pheromone receptor neurons are narrowly tuned to compounds like Z11-18Ac, and are always located in T sensilla (Hansson, 1995). Other T sensilla in moths house ORNs that perceive plant volatiles, which also appear narrowly tuned (Anderson *et al.*, 1995). Recording from C sensilla, Clyne *et al.* (1997) characterized one sensillum type with a response to butyric acid (and other odorants) and a second group with virtually no response to any of the odorants tested. More recently two types were described based on responses to two aliphatic acids, the first responds to both butyric and propionic acid while a second responds only to propionic acid (Figure 23.3; Park *et al.*, 2002). The analysis of T and C sensilla is not exhaustive; more types may exist. Indeed, there is evidence for at least two other types of C sensilla (de Bruyne, unpublished) and a more thorough characterization of C sensilla is being made (Ignell and Carlson, personal communication). Neurons of dw sensilla in other insects have often been found to respond to aliphatic acids, short chain alcohols or ammonia, and sometimes contain thermo- or hygroreceptive neurons (Kaib, 1974; Sass, 1976; Altner *et al.*, 1977; Bowen, 1995; Taneja and Guerin, 1997). It has led to the suggestion that the pore structures of dw sensilla are particularly suited for conducting small hydrophilic molecules (Altner *et al.*, 1977) but recent work in the cockroach has put this in doubt (Fujimura *et al.*, 1991).

The distribution patterns of these sensillum types confirm the existence of expression zones running diagonally across the funiculus (Figure 23.1D). The ab3 sensilla occur in the most medial part of zone I followed by ab1 and ab2 sensilla. Ab4 and ab6 are limited to zone 2 and ab5 and ab7 are spread out over zones 2 and 3 (de Bruyne *et al.*, 2001). The at1 sensilla are distributed in the most proximal part of zone 3 whereas at2 occur more laterally (Clyne *et al.*,

1997). Even C sensillum types show such a pattern with at1 occurring in zone 2 (Clyne *et al.*, 1997; de Bruyne *et al.*, unpublished). However, a clear spatial organization of palpal B sensillum types seems to be absent (de Bruyne *et al.*, 1999)

23.3.3 Developmental genes and neural identities

One rule seems to emerge from the characterization of ORNs in B sensilla; each functional sensillum type contains a stereotypical combination of ORNs (Figure 23.2B), a feature which has also been noted in other insects, e.g. in basiconics of the fleshfly (Kaib, 1974). This raises an interesting developmental question. Is ORN identity coordinated, such that an A neuron regulates OR expression of its neighboring B neuron. Such a mechanism is known for rhodopsin expression in *Drosophila* ommatidia (Chou *et al.*, 1999). The genetic mechanisms that determine the fate of individual cells in an olfactory sensillum are still unknown, but the POU-domain transcription factor *acj6* is likely to play a role. A null mutation in this gene leads to combinations of ORNs that do not normally occur (Clyne *et al.*, 1999a). In mutant flies, pb1 and pb3 sensilla have a functional B neuron and a non-functional A neuron. Some A neurons lose their ability to respond to odorants while others are transformed to copies of their sister neuron. Most significantly, in pb2 sensilla a functional neuron occurs, with a novel response spectrum, while its accompanying neuron is silent. The response spectrum of this neurons looks like a combination of the spectra of the two neurons normally present in this type of sensillum, suggesting the misexpression of OR genes from one cell into the other. In line with this interpretation the expression of some OR genes is regulated by *acj6*, either directly or indirectly (Clyne *et al.*, 1999b). It may be that ORN identity and OR expression is determined by combinatorial interactions of transcription factors. Attractive candidates are other POU-domain genes in the *Drosophila* genome, which are able to form heterodimers with each other. There is evidence that some of these genes are also expressed in ORNs (Warr *et al.*, 2001).

The *acj6* mutant is interesting mainly because it was first described to affect responses to only a subset of odorants (McKenna *et al.*, 1989; Ayer and Carlson, 1992) and subsequently shown to affect only a subset of ORN classes. *Scutoid* (*Sco*) is a dominant visible marker commonly used in *Drosophila* genetics but also affects the perception of a subset of odors (Dubin *et al.*, 1995). Antennal response to ethyl acetate and acetone is reduced but several other odorants are detected normally. Palpal sensilla are unaffected (de Bruyne, unpublished results). The marker phenotype is caused by misexpression of a gene called *snail* (*sna*) driven by a regulatory region of the *no-ocelli* (*noc*) gene, which affects the development of mechanosensory bristles (Fuse *et al.*, 1999). These authors also show that *sna* is expressed in a small set of cells in *Sco* antennal disks which could explain the olfactory phenotype. The zinc-finger protein *snail* is normally expressed throughout the CNS and regulates several important genes in neuroblast differentiation (Ashraf and Ip, 2001). Its misexpression in the antenna may well disturb sensillum formation.

Mutations in the *lz* gene remove all antennal B sensilla and many T sensilla (see above). Riesgo-Escovar *et al.* (1997b) found that responses to ethyl acetate, acetone and 3-octanol were almost absent from EAGs while a response to propionic acid (detected by C sensilla, see Figure 23.3) was only partly affected. Structural defects in 70 percent of B sensilla on the palps of *lz* mutants also correlate with reductions in EPG response to several odorants normally detected by pb1A, pb3B and pb2A, making it likely that all sensillum types are affected equally (Riesgo-Escovar *et al.*, 1997a).

23.3.4 Intensity coding by ORNs in B sensilla

Firing frequency in single ORNs increases with odorant concentration (Sass, 1976; Kaissling 1987). Dose-response relations for *Drosophila* ORNs show an initial rising phase and saturation at spike rates that are maximally feasible by these neurons (Figures 23.4A and 23.4B). The peak response rates can be as high as 500 spikes/s for a brief phasic burst but rates calculated over 500 ms saturate below 250 spikes/s. The curves are sigmoid (Lánsky and Rospars, 1993), remarkably similar in shape, with roughly identical slopes, and differ only in their threshold. This is valid for comparisons between ORNs (Figure 23.4D) as well as for one ORN stimulated with two different odorants (pb1A, Figure 23.4A). Drosophila ORNs behave much the same way as cockroach ORNs (Sass, 1976), where each ORN's responses to doses of various odorants follow the same curve but with different sensitivities. Dynamic ranges of 2–3 log doses have been also recorded for other insects (Sass, 1976; Vareshi, 1971; Anderson *et al.*, 1995). By contrast Kaissling (1987) notes much wider ranges for moth pheromone receptor neurons, and Firestein *et al.* (1993) report narrower ranges for vertebrate ORNs.

The curves for the ethyl acetate response of pb1A, ab1A and ab2A in Figure 23.4A illustrate two important points. First, it seems that the range of concentrations of this odorant detected by *Drosophila* is encoded by at least three neurons, one for the lower range and two for the high range. The second point is the fact that a neuron can be better characterized when better ligands are found. The ab2A and pb1A neurons show nearly identical responses to ethyl acetate but they differ in their response to ethyl propionate (Figure 23.4A, see also Figure 23.3). It is likely that an as yet unknown odorant excites ab2A at a lower threshold (de Bruyne *et al.*, 2001).

Unless an odorant is encoded exclusively in one ORN type its concentration cannot be encoded separately from its identity. One important question is therefore: can the central nervous system recognize an odorant as the same stimulus across a wide range of intensities, when the number of ORN classes firing will increase abrupt end in firing at stimulus offset followed by a period of quiescence and a slow return to spontaneous firing (Figure 23.2A). Finally, some odorants inhibit ORNs by reducing spontaneous firing to zero (Figure 23.2D). Moreover, these different dynamics can occur in one ORN dependent on the identity of the odorant that stimulates it. For instance, benzaldehyde inhibits the pb2B neuron in Figure 23.2D, but 4-methylphenol can excite it (de Bruyne *et al.*, 1999). Similarly, some neurons show abruptly ending response to one odor and extended firing to another. These varying ways in which odorants can activate transduction in ORNs are not due to varying levels of intensity or saturation; different concentrations of odorants reproduce the same response characteristics over a 3 log dose range (de Bruyne *et al.*, 2001).

This variability in temporal characteristics also exists across insect species. For instance, in cockroaches certain ORNs show onset and offset of firing correlated with stimulation (Getz and Akers, 1997) whereas some ORNs in T sensilla of moths continue firing after the end of odor delivery (Anderson *et al.*, 1995). The occurrence of excitatory and inhibitory responses in single ORNs (depolarization and hyperpolarization, respectively) was noted by Boeckh (1967) in a beetle and a blowfly and is well documented in lobsters (Ache and Zhainazarov, 1995). Inhibitory responses were also observed in patch-clamp recordings of *Drosophila* antennal ORNs (Dubin and Harris, 1997). Hyperpolarizing currents were due to a decrease in an outward Cl^- current.

Therefore the firing rate of ORNs is not an exact reflection of the dynamics of odor pulses. More importantly, the temporal correlation of firing with odor concentration differs between ORN classes and between odors that stimulate the same class. This suggests a temporal dimension to the odor code at the level of the ORN population. Getz and Akers (1997) calculated that the information content of the phasic period of ORNs is much higher than the tonic period suggesting odor quality can be assessed relatively fast and is not dependent on the duration of odor pulses. In addition to the instantaneous pattern of activity across ORNs, the comparison of on–off dynamics between ORN classes may also contain information.

Natural stimuli are never block-shaped increases of concentration but rather continuous series of sharp rises and drops in concentrations across a wide dynamic range (Murlis *et al.*, 1992). Odor quality may be encoded in the first 200 ms while the off-characterisitics determine the accuracy with which the end of stimulation and the separation of rapid odor pulses are detected. This has proven to be highly relevant for mechanism of oriented flight (Vickers and Baker, 1996). In fact such differences in ability to follow rapid sequences of odor pulses were demonstrated for two moth ORNs that respond to the same pheromone component but with different temporal accuracy (Almaas *et al.*, 1991). If temporal patterns of activity in single ORNs depend on which odorant they respond to, some odors can be encoded with high temporal accuracy in one ORN, while leaving other

ORNs activated after stimulation has ended. This suggests that the CNS might extract different features of a stimulus from different subsets of the activated population of ORNs.

Furthermore, natural stimuli are never chemically pure odorants against a background of clean air but rather mixtures of compounds against continuous olfactory background noise of varying composition. Inhibitory responses may be especially relevant for such natural situations, enhancing contrast between important odor signals and their background. For instance, the *Drosophila* ab1A neuron displays inhibitory response to linalool, which is not immediately apparent in recordings under laboratory conditions where clean air is the carrier for odor stimulation. When this neuron's firing is raised because it is in an adapted state due to ethyl acetate stimulation, inhibition by linalool becomes prominent (de Bruyne *et al.*, 2001). A similar effect of linalool has been noticed in a moth pheromone receptor neuron (Kaissling *et al.*, 1989).

Selzer (1981) found that components of a complex food odor did not activate a single ORN independently from each other. Mixtures of odorants interact when stimulating ORNs to produce less (mixture suppression) or more (synergism) activity then would be expected from a simple competitive binding model. The transduction mechanisms that underlie the difference in temporal firing properties and inhibitory odor responses are likely to cause interactions when such odorants are combined as a single stimulus.

23.3.6 Sensory transduction in *Drosophila* ORNs

The existence of a variety in response kinetics and excitatory and inhibitory responses in the same ORN are not easily explained in terms of a single signal transduction cascade. They hint at the possibility of multiple transduction pathways, as has been demonstrated in lobster ORNs (Ache and Zhainazarov, 1995). This could lead to simple forms of stimulus integration at the level of the ORN (Derby, 2000). Do multiple pathways function in complete isolation or is there crosstalk? The presence of a variety of reversal potentials in patch clamp studies of *Drosophila* antennal ORNs suggests that odorants do not all elicit the same transduction mechanism (Dubin and Harris, 1997). What is the evidence in *Drosophila* for the presence of transduction element of the main pathways, the cyclic nucleotide pathway and the IP_3/DAG pathway? Is there coexpression in single neurons?

Several genes from the well-studied *Drosophila* phototransduction cascade (Hardie and Raghu, 2001), which uses IP_3 as a second messenger, play a role in olfactory transduction but there is no indication that mechanisms are similar. Mutations in the phospholipase C gene *norpA* affect detection of all odorants in the palp but not in the antenna (Riesgo-Escovar *et al.*, 1995). By contrast, mutations in the *rdgB* gene, which encodes a phosphatidylinositol transporter, affects the kinetics of the physiological responses of both antennae and palp (Woodard *et al.*, 1992; Riesgo-Ecovar *et al.*, 1994). Other transduction genes expressed in photoreceptor cells are the so-called arrestins. They have been shown to desensitize photoreceptors by reducing the coupling between receptor and G-protein (Hardie and Raghu, 2001). In *Drosophila* and the mosquito *Anopheles gambiae* these proteins are also expressed in olfactory neurons (Merril *et al.*, 2002). Evidence for a functional role comes from *Drosophila* where mutations in the *Arr1* and *Arr2* genes reduce EAG and EPG responses for at least two odorants (Merrill *et al.*, 2002). Both *rdgB* and *arr* mutations lead to degeneration of photoreceptor neurons, probably due to overstimulation of signal transduction elements. Such an effect is not seen in olfactory neurons (Riesgo-Escovar *et al.*, 1994; Merrill *et al.*, 2002).

A role for the *Gα49B* gene, encoding a G-protein component, was demonstrated when Kalidas and Smith (2002) disrupted its function in a large subset of ORNs. However, the gene has various transcripts for different G-protein α subunits, which are differentially expressed (Talluri and Smith, 1995). Kalidas and Smith (2002) changed behavioral responses to iso-amyl acetate and benzaldehyde by interfering with the mRNA for the dGqα-3 protein in a large subset of ORNs by means of the Or83b promotor (see Chapter 19). The main ion channel contributing to the phototransduction current encoded by the *trp* gene appears not to contribute to the receptor potential in ORNs but plays a developmental role leading to defects in adaptation (Störtkuhl *et al.*, 1999; see below). The *Itp-r83A* gene encodes an ion channel, which is directly gated by IP_3 (Raghu and Hasan, 1995). It is expressed in ORNs and Deshpande *et al.* (2000) examined its function by testing viable combinations of lethal mutant alleles. Like *trp*, the IP_3 channel does not seem to have a direct affect on EAG but speeds up disadaptation. Unlike *trp*, however, *Itp-r83A* is expressed in adult ORNs, though its role in development was not studied (Raghu and Hasan, 1995; Deshpande *et al.*, 2000).

There is only limited evidence for a role of the cyclic nucleotide pathway in *Drosophila* olfactory transduction. In vertebrates, type III adenylyl cyclase (AC) produces cAMP, the main second messenger in olfactory transduction in mammals (Zufall and Munger, 2001). A clear homolog has not been found in *Drosophila*. The *rut* gene encodes a Ca^{2+}/calmodulin-responsive AC, which is mostly known for its expression in mushroom bodies and its role in olfactory learning (Roman and Davis, 2001). It plays a secondary role, after the first ion channels have opened. Another mutation known to disrupt learning is *dnc*. It encodes a cyclic nucleotide phosphodiesterase, which is thought to lower cAMP concentration. Hence the *rut* mutation lowers, whereas *dnc* raises cAMP. Both *rut* and *dnc* are expressed in antennae and palps (Martín *et al.*, 2001), but the effect of the mutants is very subtle. In EAG recordings they affect the steepness of the rise-phase but have no influence on the latency or the final amplitude. Thus, cAMP could play a modulatory role in olfactory transduction consistent with a similar role in photoperception (Chyb *et al.*, 1999).

It should be pointed out that all these effects on response kinetics of the EAG could be caused by malfunction in a subset of ORNs rather than represent changes in kinetics of the receptor potentials. Since EAG and EPG represent summations of such potentials, and ORNs differ greatly in their temporal response dynamics, a deletion of a subset of ORNs with a minor contribution to final amplitude could change the steepness of rise or fall of the EAG.

The *eag* potassium channel can be modulated by cyclic nucleotides and may play a role in ORNs. Patch-clamp recordings from antennal ORNs in *eag* mutants show fewer neurons responding to ethyl butyrate (Dubin *et al.*, 1998). Application of high K+ saline to the dendritic side of ORNs raises excitability of most wild type neurons but not in *eag* mutants, suggesting a role for *eag* in transduction in ORN dendrites rather than during development (Dubin *et al.*, 1998).

A cyclic nucleotide-gated channel is expressed in antennae (*Cng*, Baumann *et al.*, 1994) but localization to ORNs has not been demonstrated, and there is no functional evidence for a role in olfaction. Its physiological properties in heterologous expression in *Xenopus* oocytes suggest a high Ca^{2+} permeability and higher sensitivity to cGMP than to cAMP. One way of modulating cGMP-mediated transduction is through nitric oxide (NO). It has been suggested that this gaseous second messenger might modulate transduction between cells (Breer and Shepherd, 1993). An indication for a peripheral role for NO comes from the *ami* mutation, which lowers the expression levels of a nitric oxide synthase (NOS) in *Drosophila* olfactory appendages (Warr, Raha and Carlson, personal communication). The mutation affects behavioral adaptation to several odorants. The *Cng* channel, cloned from the antenna, was shown not to respond to NO (Broillet, 2000).

It would seem that none of the mutations in potentially important elements of olfactory transduction has led to clear ideas about the transduction mechanisms employed by *Drosophila* ORNs. It may be that different neurons employ slightly different mechanisms expressing different genes, but there is no direct evidence for interacting pathways in single ORNs.

23.4 Central processing of odors and behavioral responses

23.4.1 ORNs project to glomeruli in the antennal lobe (AL)

Even though ORNs are involved in initial forms of processing they can be viewed as a mixed population of simple detectors. Only when the neuronal network behind them integrates their responses can we speak of odor perception. The axons of ORNs from both olfactory organs project to the antennal lobe (AL), which is organized in semi-spherical synaptic regions called glomeruli (Figure 23.5). This is a basic feature shared by primary olfactory neuropiles of both vertebrates and insects (Hildebrand and Shepherd, 1997). In insects, glomeruli

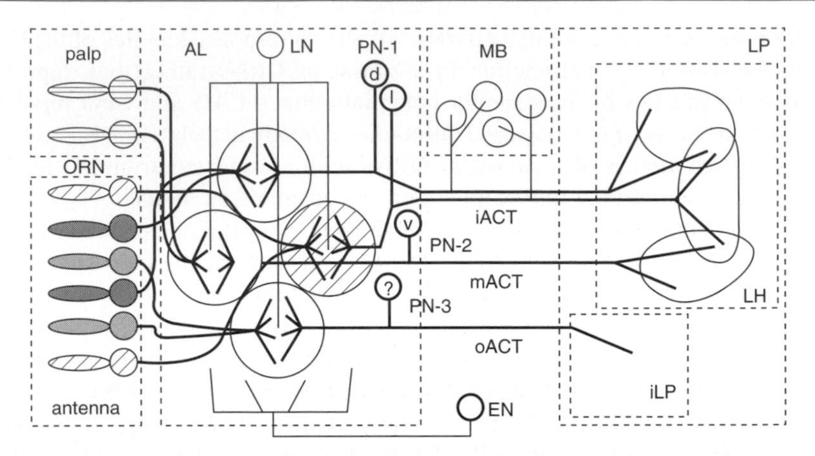

Figure 23.5 Overview of pathways for olfactory processing in the *Drosophila* brain. Schematic outline of major wiring principles in the olfactory system (not to scale).
Olfactory receptor neurons (ORN, *n* ~ 1300) of different functional classes (hatchings and gray shades, *n* ~ 40) from two olfactory organs (palp and antenna) send their axons to spherical neuropile subunits called glomeruli (hatched/gray circles, *n* ~ 50) in the antennal lobe (AL). Members of one ORN class converge on a single glomerulus. Local interneurons (LN) are anaxonal and interconnect many glomeruli via elaborate dendritic trees that cover most of the AL. Projection neurons (PN, *n* ~ 150) whose cell bodies are located either ventrally (v), laterally (l) or dorsally (d) of the AL send their axons through one of three antenno-cerebral tracts (ACT) and show a variety of morphologies. The majority of PNs have d or l somas (PN-1) with dendrites in a single glomerulus and project through the inner tract (iACT) to the lateral horn (LH), part of the lateral protocerebrum (LP), sending collaterals into the calyx of the mushroom body (MB). Most likely, each glomerulus is innervated by at least two of them. Other uniglomerular neurons with v somas (PN-2) send axons through the middle tract (mACT) to the (LH) without innervating the MB. Their dendritic trees overlap with PN-1 neurons in a subset of glomeruli (not shown). The outer tract (oACT) contains only a few projection neurons (PN-3) projecting to the inferior part of the lateral protocerebrum (iLP). A heterogeneous group of oligo- and polyglomerulur neurons with v somata is omitted from this diagram for clarity. Several of the oligoglomerular types innervate non-olfactory glomeruli. Several large extrinsic neurons (EN) of unknown polarity, which innervate other parts of the brain, may supply modulatory input to the AL or direct output to motor neuropiles. The output of the complex interactions between ORN, LN, EN and PN fibers in the AL is projected to and redistributed in the MB and LP. The MB receives olfactory input from PN-1 fibers only, whose axons end in microglomeruli where they synapse with a large number of parallel fibers. The axonal endings of PNs in the LP form large, partly overlapping, areas in the LH. Data compiled from Stocker *et al.* (1990); Marin *et al.* (2002); Wong *et al.* (2002) and Yusuyama *et al.* (2002).

are largely invariable in position (Rospars, 1988) and in moths it was shown that ORNs of one physiological class send their axons to one particular glomerulus (Hansson *et al.*, 1992). Most evidence in *Drosophila* also indicates that all ORNs of one class converge onto a single glomerulus: anterograde fillings of sensilla on antennae and palps show that projections are not somatotopic but specific for certain sensillum types (Stocker *et al.*, 1983). Golgi-stained ORNs show single glomerulus-size axonal tufts (Stocker *et al.*, 1990). Two enhancer trap lines that differentially label B and T sensilla show very little overlap in projection patterns (MT14, mostly B and some T, KL116, all T and some B; Tissot *et al.*, 1997). Finally, and most convincingly, reporter constructs of at least five OR genes and one GR gene, expressed in antenna and palp, label a single glomerulus (Vosshall *et al.*; 2000, Gao *et al.*; 2000; Scott *et al.*, 2001). However, one reporter construct for an OR gene (Or23a) is expressed in a population of ORNs that innervates two glomeruli (Vosshall *et al.*, 2000; Gao *et al.*, 2000), though the reporter is expressed in more ORNs than the endogenous OR (Gao *et al.*, 2000). Recent evidence suggests there is one main target glomerulus and a few minor targets (Bhalerao *et al.*, 2003). It remains to be seen whether this represents a single ORN class with divergent projections, or expression of a single OR gene in several functionally different ORN classes.

If all ORNs of one class converge on a single glomerulus, does every glomerulus receive input from only one ORN class? So far there is no clear evidence for convergence of two or more ORN classes on a single glomerulus. Stocker *et al.* (1983) suggest that some glomeruli are innervated by axons from different sensillum classes but this is based on anterograde fillings from mixed sensillum areas on the antenna. These studies also identified seven glomeruli as targets of LB sensilla (Stocker *et al.*, 1983), one of which has been shown to consist of two subdivisions (Wong *et al.*, 2002). This corresponds well with the existence of eight ORN classes in LB sensilla and the idea that each ORN class projects to a different glomerulus (de Bruyne *et al.*, 2001). Another discrete set of glomeruli receives input exclusively from the palps. Palpal glomeruli are located in the outer layer of the ventromedial half of the AL but they are not clustered as is observed in some other insects (Anton and Homberg, 1999). Singh and Nayak (1985) observed five target areas for palpal sensilla but only three palpal glomeruli were identified in a second study (Stocker *et al.*, 1990). Since palpal B sensilla house a total of six ORN classes in three sensillum types (Figure 23.2B), projections from palpal sensilla could differ from antennal sensilla, if each of these glomeruli is innervated by the two ORNs in one sensillum type. Distler and Boeckh (1997) found a single palpal glomerulus receiving input from one sensillum type that houses three different neurons in the mosquito *Aedes aegypti*. Nevertheless it may well be that such glomeruli show functional subunits upon closer examination.

In light of this evidence glomeruli can be considered identifiable structures that are focal areas of primary odor processing. Their numbers provide an indication

the LH via the middle tract (mACT). The smallest of the three tracts, the outer one (oACT), projects to the inferior part of the LP (iLP; Stocker *et al.*, 1990). In several insect species a pathway running parallel to the oACT contains as many PNs as the iACT and appears complementary to it. Instead of going to the LH via the MB, this pathway leads to the MB with side branches in the LH (Anton and Homberg, 1999). In bees, PNs in this pathway innervate half the AL while the other half project through the iACT (Abel *et al.*, 2001). There has been no evidence for this pathway in *Drosophila*.

One enhancer trap line has been particularly useful for studying structure and function of PNs. The line GH 146 labels a large subset of 150 PNs that innervate 34 of the glomeruli (Stocker *et al.*, 1997; Wong *et al.*, 2002; Marin *et al.*, 2002). The majority of them have dorsal and lateral somata, innervate a single glomerulus and project through the iACT. Probably two to three of such PNs innervate each glomerulus and send information to both LH and MB. In addition, two glomeruli are also innervated by a second type of PN (Wong *et al.*, 2002; Marin *et al.*, 2002). This type of PN has ventral somata, innervates a single glomerulus or a small set of glomeruli, and sends axons through the mACT to the LH (Stocker *et al.*, 1990; Marin *et al.*, 2002; Wong *et al.*, 2002). These two glomeruli receive their input from ORNs in T sensilla (Stocker, 1994). A third type of PN appears to innervate many glomeruli and sends an axonal branch away from the LH to as yet unspecified brain regions (Marin *et al.*, 2002). Each olfactory glomerulus can thus potentially send output via three different PN types, one innervating MB and LH, one sending combined output from many glomeruli to the LP, and for some of them one with (combined) output from selected glomeruli to the LH. The MB thus receives homogeneous input from the AL while the LP collects heterogeneous input from multiple PN types. Interestingly, the elaborate branching patterns of PNs in the LH suggest that specific areas of the LH receive overlapping projections from particular sets of glomeruli (Wong *et al.*, 2002; Marin *et al.*, 2002). This might be where combinatorial coding takes place and secondary odor identities emerge.

In the MB olfactory PNs innervate small microglomerular structures where a single bouton synapses onto many dendrites of intrinsic neurons, the Kenyon cells (Yusuyama *et al.*, 2002). The MB represents a strong divergence in information. Each PN forms only a few such boutons and it is not clear whether these microglomerular structures are arranged in a stereotypic manner (Marin *et al.*, 2002; Wong *et al.*, 2002). The calyx may be subdivided in distinct zones of innervation, corresponding to multiple classes of PNs (Strausfeld, personal communication). Odor-induced neural activity in the calyces, as measured by Ca^{2+} imaging, is diffuse but clearly patterned (Wang *et al.*, 2001; Fiala *et al.*, 2002). These patterns are different for each odor but it is not clear whether they are consistent across individual flies.

23.4.3 Two neuropiles; two ways of processing odors?

Do the two targets of olfactory information (MB and LH) have different functions in olfactory coding? The role of MB in olfactory associative learning is well established (Zars *et al.*, 2000; Roman and Davis, 2001). By applying the drug hydroxyurea in a small time window of larval life MBs can be completely ablated. Flies treated this way fail to learn information about odors but still show avoidance behavior to these odors (de Belle and Heisenberg, 1994). This shows that the LH could mediate innate odor-driven responses. The treatment also affects a small subset of AL neurons so avoidance responses for some odors were also lower (*iso*-amyl acetate and 3-octanol). Using tetanus toxin to block synaptic transmission in all PNs of the GH146 enhancer trap line Heimbeck *et al.* (2001) abolished avoidance behavior in response to ethyl acetate and E2-hexenal. An attractant effect of 1-octen-3ol was also abolished but high concentrations of this odorant and the two others remained repellent. The authors conclude that experience-independent (innate) behaviors are mediated by the projections of PNs to the LH. In addition, male courtship was also severely affected. The latter suggests a role for olfactory stimuli in mating.

23.4.4 Innate sex-specific behaviors: a role for olfactory stimuli in courtship?

It has been clearly demonstrated that contact pheromones in the shape of cuticular hydrocarbons play an important role in *Drosophila* courtship behavior (see Chapter 9) but a role for olfactory stimuli has not been firmly established. The neuroanatomy of the olfactory pathway does not show any sexual dimorphisms (Stocker, 1994; Laissue *et al.*, 1999) but there is evidence for functional dimorphism in specific glomeruli (Ferveur *et al.*, 1995). Experiments with the *lz* mutants show that antennal and palpal B sensilla are not needed during courtship (Stocker and Gendre, 1989). However, males are able to "smell" a female at close range and this ability is affected by the olfactory mutation $para^{sbl}$ (Gailey *et al.*, 1986). These results suggest some role for (unidentified) olfactory stimuli in mating possibly involving T or C sensilla, but the absence of a structurally distinct pheromone system.

Sexual dimorphisms in the sensory apparatus are limited to small shifts in abundance of certain categories of sensilla that could be simply related to differences in size. T sensilla are more abundant on male antenna (Stocker and Gendre, 1988; Shanbhag *et al.*, 1999). Clyne *et al.* (1997) find that one class of neurons in these sensilla responds to *cis*-vaccenyl acetate (Z11-18Ac) a moth pheromone-like compound, which has been suggested to inhibit male courtship towards fertilized females (Jallon, 1984). Given the similarity between the putative *Drosophila* anti-aphrodisiac and moth pheromones, which are also perceived by neurons in trichoid sensilla, it is tempting to suggest a pheromonal role for these sensilla. However, relatively high levels of Z11-18Ac elicit only mild responses.

Moreover, the pb1A neuron on the maxillary palps also responds to these levels of Z11-18Ac, but is clearly much more sensitive to other esters (Figure 23.2B). Stocker and Gendre (1989) did demonstrate a role for palpal sensilla in male courtship inhibition, so if Z11-18Ac plays a pheromonal role, at least two ORN classes encode it. This compound has also been suggested to function as an aggregation pheromone (Bartelt *et al.*, 1985).

23.4.5 Odor-driven orientation and escape behavior

Kellog *et al.* (1962) showed that *Drosophila* fly upwind to attractive odors and start casting rapidly upon losing contact with it, much like moths do (Vickers and Baker, 1996). The reaction time to loss of stimulation is ca. 0.1 sec. There is evidence for the use of anemotactic as well as chemotactic mechanisms in flies orienting to attractive odorants (Kellog *et al.*, 1962; Borst and Heisenberg, 1982). The fly's nervous system is capable of deducing the spatial distribution of odor stimuli from the bilateral comparison of stimulus intensities with minimum ratios of 9:10 (Borst and Heisenberg, 1982). The attraction of *Drosophila* to several small aliphatic esters, acids and alcohols has been repeatedly shown in various experimental designs (Hutner *et al.*, 1937; Fuyama, 1976; Hoffmann, 1985). Walking *Drosophila* confronted with a choice between an odor laden and control airstream exhibit dose-dependent responses ranging from insensitivity to attraction via indifference to repulsion (Fuyama, 1976; Rodrigues and Siddiqi, 1978; Ayyub *et al.*, 1990). Ethyl acetate is well known to be attractive at low doses and becomes repellent at high doses whereas benzaldehyde is usually reported as repellent. The neural mechanisms mediating this shift in orientation are unknown but it appears that different behavioral mechanisms underlie attraction and repulsion (Borst and Heisenberg, 1982; Ayyub *et al.*, 1990). It is also apparent that flies use different orientation mechanisms in different types of assays (Shaver *et al.*, 1998).

The physiological classification of ORNs has demonstrated that benzaldehyde and ethyl acetate detection involve separate channels (de Bruyne *et al.*, 1999, 2001). Figure 23.3 shows that the three most responsive ORNs for ethyl acetate (pb1A, ab1A, ab2A) are different from the three most responsive ORNs for benzaldehyde (pb2A, ab1D, ab4A). If we classify odorants as attractive vs repellent then we could hypothesize hedonic ORN classes and predict that other volatiles exciting the same neurons as benzaldehyde would also be repellent. The most sensitive responses to benzaldehyde come from ab4A neurons that respond best to E2-hexenal, and other compounds with the 'E2-hexen' moiety. Interestingly, E2-hexenal was also shown to be repellent (Heimbeck *et al.*, 2001). In the same study an odorant with a similar threshold, 1-octen-3-ol (Figure 23.4B) detected by different ORNs, was shown to be an attractant at the low dose. Does ab4A mediate repellency?

For the fly certain olfactory stimuli clearly have inherent hedonistic value.

Benzaldehyde is repellent enough to function as unconditioned "punishing" stimulus in a visual learning paradigm (Guo and Götz 1997). At high doses all odorants probably carry an inherent negative value. Sudden puffs of concentrated odorant elicit an escape reaction mediated by giant fibers with electrical synapses (Thomas and Wyman, 1984; McKenna *et al.*, 1989). However, there are more indications that subsets of odorants are perceived differently.

Mutations that lead to odor specific behavioral defects, which are not due to defects in the sensory apparatus, suggest different odor tracts in the brain. Flies carrying the *olfC* mutation fail to orient to a small range of acetate esters (Ayyub *et al.*, 1990). Ethyl acetate is affected differently from *iso*-amyl acetate depending on the allele, while benzaldehyde is not. The mutation was mapped to the *mys* gene, which encodes an integrin; a membrane receptor involved in signaling during growth and differentiation (Ayyub *et al.*, 2000). Evidence for a separate central pathway for benzaldehyde is also supplied by the *ptg* mutation, which reduces benzaldehyde response in the T-maze as well as in the jump assay (Helfand and Carlson 1989). In this case, ethyl acetate is not affected but the mutation is not olfactory specific. As a typical example of a mutation with pleiotropic affects, it also causes changes in pigmentation patterns and female sterility.

23.4.6 Plasticity and other forms of response variability

We should not assume that odor coding at the level of the sensory neurons is a static phenomenon. Research on *Drosophila* gives us three cases where influences from the external or internal environment change response properties of ORNs. First, sensory adaptation can temporarily reset sensitivity of ORNs to persisting odorants. Next, circadian rhythms can cause daily cycles of ORN sensitivity. Finally, sensory neurons may change their synaptic properties as a result of chronic odor exposure.

Adaptation can reset ORN sensitivity to match the dynamic range of a stimulus or to increase the relative salience of other stimuli. Adaptation and cross-adaptation have been demonstrated in *Drosophila* at the level of the EAG and the single neuron (Störtkuhl *et al.*, 1999; de Bruyne *et al.*, 1999, 2001). Mutations in the *trp* ion channel speed up the recovery from exposure to high doses of *iso*-amyl acetate both in EAGs and avoidance behavior (Störtkuhl *et al.*, 1999). The *trp* product is expressed in the developing antennae. Temperature-sensitive alleles of *trp*, in which the mutation can be switched on or off, depending on rearing temperature, were used to establish during which period in a fly's life the protein is important. The results demonstrated that a functional *trp* channel is required during ORN development, not during the actual physiological process in adult life.

While adaptation acts over minutes, circadian rhythms act over hours. The EAG amplitude to ethyl acetate and benzaldehyde varies with circadian rhythm

(Krishnan *et al.*, 1999). Over a 12:12 hour light:dark cycle it peaks at the middle of the dark period and drops to about 60 percent during most of the light period. The genetics of the molecular clock involve daily cycles in expression of the *period* (*per*) and *timeless* (*tim*) genes, which occurs in the CNS as well as in peripheral neurons (Plautz *et al.*, 1997). Cycling of EAG sensitivity is abolished in *per* and *tim* mutants and if *per* is expressed only in the central oscillator but not in the periphery, EAG amplitude does not cycle (Krishnan *et al.*, 1999). This suggests that ORN sensitivity depends on peripheral circadian oscillators. Modulation of EAG amplitudes with circadian rhythm has been demonstrated earlier in two species of tsetse flies (van der Goes van Naters *et al.*, 1998) and was shown to reflect firing properties of single ORNs. But whereas tsetse single ORN sensitivity seems to peak when flies are highly active, sensitivity of *Drosophila* females does not relate in a logical way to daily activity patterns.

Finally, an important factor in behavioral plasticity is the strength of synaptic input. Devaud *et al.* (2001) demonstrated that flies exposed for a 3-day period to high concentrations of benzaldehyde were less repelled in the T-maze afterwards, whereas their behavior to other odorants was similar to unexposed flies. This reduction in repellency to a specific odor was not caused by lower responsiveness of ORNs as measured in EAGs but was correlated with lower numbers of synapses in a small subset of glomeruli (Devaud *et al.*, 2001). If chronic exposure can decrease the number of synapses leading to lower levels of behavior, does an artificially induced increase in synapses raise behavioral responsiveness? In mosaic flies, where one antenna was homozygous for the *gigas* mutation, ORNs are increased in size and also make more synapses in the AL (Acebes *et al.*, 2001). The change in size does not change the EAG but the increased innervation of the CNS changes behavioral output. The strength of response, be it attraction or repulsion, was shown to correlate with the number of synapses ORNs make in the antennal lobe. Furthermore, wild-type flies were not significantly attracted to ethyl acetate but the *gigas* mosaics were. These results suggest that, while thresholds for attraction and repulsion may be determined by the dose-response relationship of ORNs, the magnitude of the behavioral response is determined and related to the strength of synaptic input. Functional plasticity at the level of the AL has been demonstrated in bees (Faber *et al.*, 1999).

23.5 Conclusions and future developments

All the information the fly needs for encoding quality, intensity and temporal variations of odors is present in the firing patterns of ORNs. We argue that, in *Drosophila*, all three features are encoded in the activity across a limited number of ORN classes ($n \approx 40$) and is time dependent. Moreover, information about most odors is probably concentrated in a small subset of ORN classes. Because of the difficulties in recording neuronal activity in the *Drosophila* brain, we know almost nothing about the functional properties of the network that extracts meaning from this input. In analogy with other insects, processing in olfactory glomeruli probably provides fine tuning, integration with information on air condition, extraction of temporal patterns and calibration to the internal state of the animal. PNs with different morphologies innervate single glomeruli and may extract different features and redistribute it predominantly to two secondary neuropiles, LH and MB. Genetic investigation of behavioral output strongly suggests the presence of at least two odor codes at this level. Partly overlapping areas of projection in the LH could allow extraction of a specific combinatorial code that mediates innate behavioral responses to odors. A second population code is generated by the projection of a relatively uniform population of uniglomerular PNs onto large quantities of parallel fibers in the MB. Integration of this activity with input from other sensory cues is thought to be crucial for associative learning of odors.

How many coding units supply the initial input to the system? The total number of different ORN classes is the first determinant for the resolution of "odor space" but has not been determined for any insect. The limited number of sensilla in *Drosophila* makes it possible to achieve this goal in the near future. The completed genomic sequence has made the numbers and molecular structure of receptor proteins, transduction elements, binding proteins and developmental genes apparent. Approximately half of all functional ORN types, the detailed structure of glomeruli as well as many PNs, have been investigated structurally and/or functionally. The study of the *Drosophila* olfactory system offers a clear opportunity to obtain complete wiring maps for the processing of behaviorally relevant stimuli. This requires concerted efforts to pool data sets and standardize methods. Mapping of odorant response activities at different levels of the system is still difficult. For instance, two glomeruli change volume after chronic benzaldehyde exposure (Devaud *et al.*, 2001). One of these was identified as DM2 and the same label was used by Rodrigues (1988) for a glomerulus with increased 2DG uptake after benzaldehyde exposure. This glomerulus was described as innervated by the palp (Stocker *et al.*, 1983) and could therefore receive projections from the pb2A neuron which responds to benzaldehyde (de Bruyne *et al.*, 1999). However, recent evidence clearly maps activity of the ab3A neuron in the antenna to DM2 (Dobritsa *et al.*, 2003; Balerao *et al.*, 2003).

ORNs are a large population of similar but distinct neurons. The genetic differentiation of such a group is an interesting functional and developmental problem. OR expression clearly is important but it is likely that other proteins are differentially expressed as well. A complex of signaling proteins and transcription factors determines the functional identity of the components of the olfactory network during development. The functional properties of this network during active life are then defined by the interaction of receptors, neurotransmitters and neuromodulators with the signal transduction machinery. What role external factors such as experience, sex, and environmental conditions play in the variability of these properties remains to be seen.

The powerful analysis that is possible with the detailed description of the *Drosophila* olfactory system may well be hampered by a lack of knowledge in two areas of research that need more attention: the chemistry and behavioral ecology of the adequate stimuli for this system.

A biologically meaningful description of an odorant is the spatio-temporal pattern of activity across the population of ORNs. If a correlation with the physicochemical properties of odorants could be made, adequate ligands for the receptors of individual ORNs could be predicted. However, we still know very little about this relation or even whether it exists in a form that we can predict. Two alternatives remain: a systematic scan of responses across chemical variables such as chain length and functional group, or the isolation of active components from behaviorally relevant natural odors. Both approaches are valid but each present only part of the truth. There is no reason to assume that we can determine the "proper" ligand for an ORN but a better knowledge of their tuning should be obtained. Furthermore, a true understanding of olfactory coding will require more sophisticated methods of behavioral analysis as well as a better description of innate and learned responses to relevant cues. It is surprising how little is known about *Drosophila* olfactory-driven behavior under natural conditions.

A wide variety of genes have been described that play a role in *Drosophila* olfaction. No doubt more genes will be investigated with respect to their role in olfaction, as more advanced molecular genetic tools allow targeted interference with gene function in particular cell types. A promising new technique allows the targeting of interfering RNAs to specific cells at a particular time so that genes with pleiotropic effects can be studied for their role in olfaction (Kalidas and Smith, 2002). Deletion or inactivation of subsets of neurons will reveal the role of specific elements of the neuronal network in generating behavior (Kitamoto, 2001; Keller *et al.*, 2002). The development of molecular genetic techniques for the expression of functional probes, such as calcium-sensitive proteins, is rapidly increasing our understanding of the coding properties of specific neurons. Optical imaging of intact tissues can monitor such reporters of neuronal excitation and give an impression of the spatial distribution of activity (Fiala *et al.*, 2002; Wang *et al.*, 2003; Ng *et al.*, 2003).

The olfactory system may not be designed to supply the brain with a neutral representation of all odors in the environment. Mechanisms of sensory representations have evolved under constant pressure to capture what is relevant. Genetic pathways that shape its physiology are likely to reflect both conserved and highly specialized elements. An exaggerated focus on the adaptation of olfactory acuity to dietary needs would be counterproductive. Just as the fly's visual system does not represent an elaborate apple detector, its olfactory system will also function as a general odor detector. However, even if we have not found clear evidence of special cues such as pheromones in *Drosophila* olfaction it is likely that certain odors will have special significance and specific pathways may have evolved to process them.

The discovery of OR genes and the sequencing of the *Drosophila* genome have created great interest in molecular aspects of olfaction in other insects and have made *Drosophila* one of the most important model systems for olfactory research. This opportunity should be used to combine *Drosophila* research with the wealth of knowledge on olfaction in other insect models such as cockroaches, bees, moths and mosquitoes. It is important to cover the breadth of insect groups to determine which aspects of olfactory processing are adaptations to specific ecological niches or results of strong sexual selection, so that general coding principles can be distilled. Common principles seem to have evolved in animals phylogenetically as far apart as insects and mammals (Hildebrand and Shepherd, 1997). Understanding why this is so will aid in defining the functional demands involved in the construction of an olfactory system. Only when we keep an eye on what is species specific and what is not, can we extract the basic principles from *Drosophila* olfactory processing that apply to insect olfaction in general and, casting our nets even wider, provide insight into our own ability to "wake up and smell the roses."

References

Abel R., Rybak J. and Menzel R. (2001) Structure and response patterns of olfactory interneurons in the honeybee, *Apis mellifera*. *J. Comp. Neurol.* **437**, 363–383.

Acebes A. and Ferrus A. (2001) Increasing the number of synapses modifies olfactory perception in *Drosophila*. *J. Neurosci.* **21**, 6264–6273.

Ache B. W. and Zhainazarov A. (1995) Dual second-messenger pathways in olfactory transduction. *Curr. Opin. Neurobiol.* **5**, 461–466.

Almaas T. J., Christensen T. A. and Mustaparta H. (1991) Chemical communication in heliothine moths. I. Antennal receptor neurons encode several features of intra- and interspecific odorants in the male corn earworm moth *Helicoverpa zea*. *J. Comp. Physiol. A* **169**, 249–258.

Altner H., Sass H. and Altner I., (1977) Relationship between structure and function of antennal chemo-, hygro-, and thermoreceptive sensilla in *Periplaneta americana*. *Cell Tissue Res.* **176**, 389–405.

Altner H. and Prillinger L. (1980) Ultrastructure of invertebrate chemo-, thermo- and hygroreceptors and its functional significance. *Int. Rev. Cytol.* **67**, 69–139.

Anderson P., Hansson B. S. and Löfqvist J. (1995) Plant-odour-specific receptor neurones on the antennae of female and male *Spodoptera littoralis*. *Physiol. Entomol.* **20**, 189–198.

Anholt R. R. H., Lyman F. L. and Mackay T. F. C. (1996) Effects of single P-element insertions on olfactory behavior in *Drosophila melanogaster*. *Genetics* **143**, 293–301.

Anton S. and Homberg U. (1999) Antennal lobe structure. In *Insect Olfaction*, ed. B. S. Hansson, pp. 97–124. Springer, Berlin.

Ashraf S. I. and Ip Y. T. (2001) The snail protein family regulates neuroblast expression of *inscuteable* and *string*, genes involved in asymmetry and cell division in *Drosophila*. *Development* **128**, 4757–4767.

Ayer R. and Carlson J. R. (1992) Olfactory physiology in the *Drosophila* antenna and maxillary palp: *acj6* distinguishes two classes of odorant pathways. *J. Neurobiol.* **23**, 965–982.

Ayyub C., Paranjape J., Rodrigues V. and Siddiqi O. (1990) Genetics of olfactory behavior in *Drosophila melanogaster*. *J. Neurogenetics* **6**, 243–262.

Ayyub C., Rodrigues V., Hasan G. and Siddiqi O. (2000) Genetic analysis of *olfC* demonstrates a role for the position-specific integrins in the olfactory system of *Drosophila melanogaster*. *Mol. Gen. Genet.* **263**, 498–504.

Barrows W. M. (1907) The reactions of the pomace fly, *Drosophila ampelophila* Loew, to odorous substances. *J. Exp. Zool.* **4**, 515–537.

Bartelt R. J., Schaner A. M. and Jackson L. L. (1985) *cis*-Vaccenyl acetate as an aggregation pheromone in *Drosophila melanogaster*. *J. Chem. Ecol.* **11**, 1747–1756.

Baumann A., Frings S., Godde M., Seifert R. and Kaupp U. B. (1994) Primary structure and functional expression of a *Drosophila* cycli-nucleotide-gated channel present in eyes and antennae. *EMBO J.* **13** 5040–5050.

Bausenwein B. and Nick P. (1998) Three dimensional reconstruction of the antennal lobe in the mosqiot *Aedes aegypti*. In *New Neuroethology on the Move*, eds R. Wehner and N. Elsner, pp. 386. Thieme, Stuttgart.

Berg B. G., Galizia C. G., Brandt R. and Mustaparta H. (2002) Digital atlases of the antennal lobe in two species of tobacco budworm moths, the Oriental Helicoverpa assulta (male) and the American Heliothis virescens (male and female). *J. Comp. Neurol.* **446**, 123–134.

Bhalerao S., Sen A., Stocker R. F. and Rodrigues V. (2003) Olfactory neurons expressing identified receptor genes project to subsets of glomeruli within the antennal lobe of *Drosophila melanogaster*, *J. Neurobiol.* **54**, 577–592.

Boeckh J. (1962) Elektrophysiologische Untersuchungen an einzelnen Geruchsrezeptoren auf den Antennen des Totengräbers (*Necrophorus*, Coleoptera). *Z. vergl. Physiol.* **46**, 212–248.

Boeckh J. (1967) Inhibition and excitation of single insect olfactory receptors, and their role as a primary sensory code. In *Olfaction and Taste*, ed. T. Hayashi pp. 721–735. Pergamon, Oxford.

Boeckh J. and Tolbert L (1993) Synaptic organisation and development of the antennal lobe in insects. *Microsc. Res. Tech.* **24**: 260–280.

Borst A. and Heisenberg M. (1982) Osmotropotaxis in *Drosophila melanogaster*. *J. Comp. Physiol. A.* **147**, 479–484.

Borst A. (1983) Computation of olfactory signals in *Drosophila melanogaster*. *J. Comp. Physiol. A.* **152**, 373–383.

Bowen M. F. (1995) Sensilla basiconica (grooved pegs) on the antennae of female mosquitoes: electrophysiology and morphology. Entomol. Exp. Appl. **77**, 233–238.

Brand A. H. and Perrimon N. (1993) Targeted gene expression as a means of altering cell fates and generating dominant phenotypes. *Development* **118**, 401–415.

Breer H. and Shepherd G. M. (1993) Implications of the NO-cGMP system for olfaction. *Trends Neurosci.* **16**, 5–9.

Broillet M.-C. (2000) A single intracellular cysteine residue is responsible for the activation of the olfactory cyclic nucleotide-gated channel by NO. *J. Biol. Chem.* **275**, 15135–15141.

Buck L. B. and Axel R. (1991) A novel multigene family may encode odorant receptors: a molecular basis for odor recognition. *Cell* **65**, 175–187.

Carlson J. R. (1996) Olfaction in *Drosophila*: from odor to behavior. *Trends Genet.* **12**, 175–180.

Chou W.-H., Huber A., Bentrop J., Schulz S., Schwab K., Chadwell L. V., Paulsen R. and Britt S. G. (1999) Patterning of the R7 and R8 photoreceptor cells of Drosophila: evidence for induced and default cell-fate specification. Development **126**, 607–616.

Christensen T. A. and Hildebrand J. G. (2002) Pheromonal and host-odor processing in the insect antennal lobe: how different? *Curr. Opin. Neurobiol.* **12**, 393–399.

Chyb S., Hevers W., Forte M., Wolfgang W. J., Selinger Z. and Hardie R. C. (1999) Modulation of the light response by cAMP in Drosophila photoreceptors. *J. Neurosci.* **19**, 8799–8807.

Clyne P. J., Grant A. J., O'Connell R. J. and Carlson J. R. (1997) Odorant response of individual sensilla on the *Drosophila* antenna. *Invertebrate Neurosci.* **3**, 127–135.

Clyne P. J., Warr C. G., Freeman M. R., Lessing D., Kim J. and Carlson J. R. (1999a) A novel family of divergent seven-transmembrane proteins: candidate odorant receptors in *Drosophila. Neuron* **22**, 327–338.

Clyne P. J., Certel S., de Bruyne M., Zaslavsky L., Johnson W. and Carlson J. R. (1999b) The odor-specificities of a subset of olfactory receptor neurons are governed by *acj6*, a POU domain transcription factor. *Neuron* **22**, 339–347.

Clyne P. J., Warr C. G. and Carlson J. R. (2000) Candidate taste receptors in *Drosophila. Science* **287**, 1830–1834.

Cobb M. (1996) Genotypic and phenotypic characterization of the *Drosophila melanogaster* olfactory mutation *Indifferent. Genetics* **144**, 1577–1587.

Cobb M (1999) What and how do maggots smell? *Biol. Rev. Camb. Philos. Soc.* **74**, 425–459.

Crnjar R. M., Scalera G., Liscia A., Angioy A. M., Bigiani A., Pietra P. and Tomassini Barbarossa I. (1989) Morphology and EAG mapping of the antennal olfactory receptors in *Dacus olea. Entomol. Exp. Appl.* **51**, 77–85.

de Belle J. S. and Heisenberg M. (1994) Associative odor learning in *Drosophila* abolished by chemical ablation of mushroom bodies. *Science* **263**, 692–695.

de Bruyne M., Clyne P. J. and Carlson J. R. (1999) Odor coding in a model olfactory organ: the *Drosophila* maxillary palp. *J. Neurosci.* **19**, 4520–4532.

de Bruyne M., Foster K. and Carlson J. R. (2001) Odor coding in the *Drosophila* antenna. *Neuron* **30**, 537–552.

Den Otter C. J. and Van der Goes van Naters W. M (1993) Responses of individual olfactory cells of tsetse flies (*Glossina M. morsitans*) to phenols from cattle urine. *Physiol. Entomol.* **18**, 43–49.

Derby C. D. (2000) Learning from spiny lobsters about chemosensory coding of mixtures. *Physiol. Behav.* **69**, 203–209.

Deshpande M., Venkatesh K., Rodrigues V. and Hasan G. (2000) The inositol 1,4,5-trisphosphate receptor is required for maintenance of olfactory adaptation in *Drosophila* antennae. *J. Neurobiol.* **43**, 282–288.

Devaud J.-M., Acebes A. and Ferrus A. (2001) Odor exposure causes central adaptation and morphological changes in selected olfactory glomeruli in *Drosophila. J. Neurosci.* **21**, 6274–6282.

Distler P. G., Boeckh J. (1997) Central projections of the maxillary and antennal nerves in the mosquito *Aedes aegypti. J. Exp. Biol.* **200**, 1873–1879.

Dobritsa A., Van der Goes van Naters W. M., Warr C. G., Steinbrecht R. A. and Carlson J. R. (2003) Integrating the molecular and cellular basis of odor coding in the *Drosophila* antenna. *Neuron* **37**, 827–841.

Dubin A. E., Heald N. L., Cleveland B., Carlson J. R. and Harris G. L. (1995) *Scutoid* mutation of *Drosophila melanogaster* specifically decreases olfactory responses to short-chain acetate esters and ketones. *J. Neurobiol.* **28**, 214–233.

Dubin A. E. and Harris G. L. (1997) Voltage-activated and odor-modulated conductances in olfactory neurons of *Drosophila melanogaster*. *J. Neurobiol.* **32**, 123–137.

Dubin A. E., Liles M. M. and Harris G. L. (1998) The K+ channel *ether a go-go* is required for the transduction of a subset of odorants in adult *Drosophila melanogaster*. *J. Neurosci.* **18**, 5603–5613.

Esslen J. and Kaissling K.-E. (1976) Zahl und Verteilung antennalen Sensillen bei der Honigbiene (*Apis melifera* L.). *Zoomorphologie* **83**, 227–251.

Faber T., Joerges J. and Menzel R (1999) Associative learning modifies neural representations of odors in the insect brain. *Nature Neurosci.* **2**, 74–78.

Fedorowicz G. M., Kulkarni N. H., Roote J., Ashburner M., Mackay T. F. C. and Anholt R. R. H. (1999) Disruption of the gene encoding a Dyrk2 homologue causes olfactory impairment in *Drosophila melanogaster*. Annu. Drosophila Res. Conference **40** (*Abstract*).

Ferveur J.-F., Störtkuhl K. F., Stocker R. F., Greenspan R. J. (1995) Genetic feminization of brain structures and changed sexual orientation in male *Drosophila*. *Science* **267**, 902–905.

Fiala A., Spall T., Diegelmann S., Eisermann B., Sachse S., Devaud J.-M., Buchner E. and Galizia C.-G. (2002) Genetically expressed cameleon in *Drosophila melanogaster* is used to visualize olfactory information in projection neurons. *Curr. Biol.* **12**, 1877–1884.

Firestein S., Picco C. and Menini A. (1993) The relationship between stimulus and response in olfactory receptor cells of the tiger salamander. *J. Physiol.* (*London*) **468**, 1–10.

Foelix R. F., Stocker R. F. and Steinbrecht R. A. (1989) Fine structure of a sensory organ in the arista of *Drosophila melanogaster* and some other dipterans. *Cell Tissue Res.* **258**, 277–287.

Fujimora K., Yokohari F. and Tateda H. (1991) Classification of antennal olfactory receptors of the cockroach *Periplaneta americana* L. *Zool. Sci.* (*Tokyo*) **8**, 243–255.

Fuse N., Hitoshi M., Misako T., Shigeo H. (1999) Snail-type zinc finger proteins prevent neurogenesis in *Scutoid* and transgenic animals of *Drosophila*. *Dev. Genes. Evol.* **209**, 573–580.

Fuyama Y. (1976) Behavioral genetics of olfactory responses in *Drosophila*. I. Olfactory and strain differences in *Drosophila melanogaster*. *Behav. Genet.* **6**, 407–420.

Gailey D. A., Lacaillade R. C. and Hall J. C. (1986) Chemosensory elements of courtship in normal and mutant, olfaction deficient *Drosophila melanogaster*. *Behav. Genet.* **16**, 375–405.

Galindo K. and Smith D. P. (2001) A large family of divergent *Drosophila* odorant-binding proteins expressed in gustatory and olfactory sensilla. *Genetics* **159**, 1059–1072.

Galizia C. G., McIlwrath S. L. and Menzel R. (1999) A digital three-dimensional atlas of the honeybee antennal lobe glomeruli based on optical sections acquired using confocal microscopy. *Cell Tissue Res.* **295**, 383–394.

Ganguly S., Mackay T. F. C., Anholt R. R. H. and Keck W. M. (2001) Scribble: a PDZ domain protein that contributes to sexually dimorphic olfactory avoidance behavior in *Drosophila*. Annu. Drosophila Res. Conference **42**, (*Abstract*).

Gao Q., Yuan B. and Chess A. (2000) Convergent projections of the *Drosophila* olfactory neurons to specific glomeruli in the antennal lobe. *Nature Neurosci.* **3**, 780–785.

Getz W. M. and Akers R. P. (1997) Coding properties of peak and average response rates in American cockroach olfactory sensory cells. *Biosystems* **40**, 55–63.

Göpfert M. C. and Robert D. (2001) Turning the key on *Drosophila* audition. *Nature* **411**, 908.

Goulding S. E., White N. M. and Jarman A. P. (2000a) *cato* encodes a basic helix–loop–helix transcription factor implicated in the correct differentiation of *Drosophila* sensory organs. *Dev. Biol.* **221**, 120–131.

Goulding S. E., zur Lage P. and Jarman A. P. (2000b) *amos*, a proneural gene for *Drosophila* olfactory sense organs that is regulated by *lozenge*. *Neuron* **25**, 69–78.

Guo A. and Götz K. G. (1997) Association of visual objects and olfactory cues in *Drosophila*. *Learning Memory* **4**, 192–204.

Gupta B. P. and Rodrigues V. (1997) *atonal* is a proneural gene for a subset of olfactory sense organs in *Drosophila*. *Genes to Cell* **2**, 225–233.

Gupta B. P., Flores G. V., Banerjee U. and Rodrigues V. (1998) Patterning an epidermal field: *Drosophila* lozenge, a member of the AML-1/Runt family of transcription factors, specifies olfactory sense organ type in a dose-dependent manner. *Dev. Biol.* **203**, 400–411.

Hammer M. (1997) The neural basis of associative reward learning in honeybees. *Trends Neurosci.* **20**, 245–252.

Hansson B. S., Ljungberg H., Hallberg E. and Löfstedt C. (1992) Functional specialization of olfactory glomeruli in a moth. *Science* **256**, 1313–1315.

Hansson B. S., Hallberg E., Löfstedt C. and Steinbrecht R. A. (1994) Correlation between dendrite diameter and action potential amplitude in sex pheromone specific receptor neurons in male *Ostrinia nubilalis* (Lepidoptera: pyralidae). *Tissue Cell* **26**, 503–512.

Hansson B. S. (1995) Olfaction in lepidoptera. *Experientia* **51**, 1003–1027.

Hardie J., Isaacs R., Pickett J. A., Wadhams L. J. and Woodcock C. M. (1994) Methyl salicylate and (–)-(1R,5S)-myrtenal are plant-derived repellents for the black bean aphid. *Aphis fabae* Scop. (Homoptera: Aphididae). *J. Chem. Ecol.* **20**, 2847–2855.

Hardie R. C. and Raghu P. (2001) Visual transduction in Drosophila. *Nature* **413**; 186–193.

Hasan G. (1990) Molecular cloning of an olfactory gene from *Drosophila melanogaster*. *Proc. Natl. Acad. Sci. USA* **87**, 9037–9041.

Heimbeck G., Bugnon V., Gendre N., Keller A. and Stocker R. F. (2001) A neural circuit for experience-independent olfactory and courtship behavior in *Drosophila melanogaster*. *Proc. Natl. Acad. Sci. USA* **98**, 15336–15341.

Helfand S. L., Carlson J. R. (1989) Isolation and characterization of an olfactory mutant in *Drosophila* with a chemically specific defect. *Proc. Natl. Acad. Sci. USA* **86**, 2908–2912.

Hildebrand J. G. and Shepherd G. M. (1997) Mechanisms of olfactory discrimination: converging evidence for common principles across phyla. *Annu. Rev. Neurosci.* **20**, 659–631.

Hoffmann A. A. (1985) Interspecific variation in the response of *Drosophila* to chemicals and fruit odours in a wind tunnel. *Aust. J. Zool.* **33**, 451–460.

Hovemann B. T., Sehlmeyer F. and Maltz J. (1997) *Drosophila melanogaster* NADPH-cytochrome P-450 oxidoreductase: pronounced expression in antennae may be related to odorant clearance. *Gene* **189**, 213–219.

Hudson R. (1999) From molecule to mind: the role of experience in shaping olfactory function. *J. Comp. Physiol. A.* **185**, 297–304.

Hutner S. H., Kaplan H. M. and Enzmann E. V. (1937) Chemicals attracting *Drosophila*. *Am. Nat.* **71**, 575–581.

Itoh T., Yokohari F., Tanimura T. and Tominaga Y. (1991) External morphology of sensilla in the sacculus of an antennal flagellum of the fruit fly *Drosophila melanogaster* Meigen (Diptera: Drosophilidae). *Int. J. Insect Morphol. Embryol.* **20**, 235–243.

Jallon J.-M. (1984) A few chemical words exchanged by *Drosophila* during courtship and mating. *Behav. Genet.* **14**, 441–478.

Jefferis G. S. X. E., Marin E. C., Stocker R. F. and Luo L. (2001) Target neuron prespecification in the olfactory map of *Drosophila*. *Nature* **414**, 204–208.

Jhaveri D., Sen A., Reddy G. V. and Rodrigues V. (2000a) Sense organ identity in the *Drosophila* antenna is specified by the expression of the proneural gene *atonal*. *Mechanisms Develop.* **99**, 101–111.

Jhaveri D., Sen A. and Rodrigues V. (2000b) Mechanisms underlying olfactory neuronal connectivity in Drosophila – the atonal lineage organizes the periphery while sensory neurons and glia pattern the olfactory lobe. *Dev. Biol.* **226**, 73–87.

Kaib M. (1974) Die fleisch- ubd blumenduftrezeptoren auf der antenne der schmeiszfliege *Calliphora vicina*. *J. Comp. Physiol.* **95**, 105–121.

Kaissling K.-E. (1986) Temporal characteristics of pheromone receptor cell responses in relation to orientation behaviour of moths. In *Mechanisms in Insect Olfaction*, eds T. L., Payne, M. C. Birch and C. E. J. Kennedy, pp 193–200. Oxford University Press, Oxford.

Kaissling K.-E. (1987) *R. H. Wright Lectures on Insect Olfaction.* Simon Fraser University, Burnaby, Canada.

Kaissling K.-E., Meng L. Z. and Bestmann H. J. (1989) Responses of the bombykol receptor cells to (Z.E.)-4,6-hexadecadiene and linalool. *J. Comp. Physiol. A.* **165**, 147–154.

Kaissling K.-E. (1995) Single unit and electroantennogram recordings in insect olfactory organs. In *Experimental Cell Biology of Taste and Olfaction, Current Techniques and Protocols*, eds A. I. Spielman and J. G. Brand, pp. 361–377. CRC Press, Boca Raton.

Kalidas S. and Smith D. P. (2002) Novel genomic cDNA hybrids produce effective RNA interference in adult *Drosophila*. *Neuron* **33**, 177–184.

Kauer J. S. (2002) On the scents of smell in the salamander. *Nature* **417**, 336–342.

Keller A., Sweeney S. T., Zars T., O'Kane C. and Heisenberg M. (2002) Targeted expression of tetanus neurotoxin interferes with behavioral responses to sensory input in *Drosophila*. *J. Neurobiol.* **50**, 221–233.

Kellogg F. E., Frizel D. E. and Wright R. H. (1962) The olfactory guidance of flying insects. IV. Drosophila. *Can. Entomol.*

Kim M.-S., Repp A. and Smith D. P. (1998) LUSH odorant-binding protein mediates chemosensory responses to alcohols in *Drosophila melanogaster*. *Genetics* **150**, 711–721.

Kitamoto T. (2001) Conditional modification of behavior in Drosophila by targeted expression of a temperature-sensitive shibire allele in defined neurons. *J. Neurobiol.* **47**, 81–92.

Kretzschmar D., Hasan G., Sharma S., Heisenberg M. and Benzer S. (1997) The Swiss cheese mutant causes glial hyperwrapping and brain degeneration in *Drosophila*. *J. Neurosci.* **17**, 7425–7432.

Krishan B., Dryer S. E. and Hardin P. E. (1999) Circadian rhythms in the olfactory responses of *Drosophila melanogaster*. *Nature* **400**, 375–378.

Kulkarni N. H., Mackay T. F. C. and Anholt R. R. H. (2001) The DSC1 sodium channel, encoded by the smi60E locus, contributes to odor-guided behavior in *Drosophila melanogaster*. *Genetics* **161**, 1507–1516.

Kutsukake M., Komatsu A., Yamamoto D. and Ishiwa-Chigusa S. (2000) A *tyramine receptor* gene mutation causes a defective olfactory behavior in *Drosophila melanogaster*. *Gene* **245**, 31–42.

Laissue P. P., Reiter C., Hiesinger P. R., Halter S., Fischbach K.-F. and Stocker R. F. (1999) 3D reconstruction of the antennal lobe in *Drosophila melanogaster*. *J. Comp. Neurol.* **405**, 543–552.

Lánsky P. and Rospars J.-P. (1993) Coding of odor intensity. *Biosystems* **31**, 15–38.

Ledent V., Gaillard F., Gautier P., Ghysen A. and Dambly-Chaudiere C. (1998) Expression and function of *tap* in the gustatory and olfactory organs of *Drosophila*. *Int. J. Dev. Biol.* **42**, 163–170.

Lienhard M. C. and Stocker R. F. (1987) Sensory projection patterns of supernumerary legs and aristae in *D. melanogaster*. *J. Exp. Zool.* **244**, 187–201.

Lienhard M. C. and Stocker R. F. (1991) The development of the sensory neuron pattern in the antennal disc of wild-type and mutant (lz^3, ss^a) *Drosophila melanogaster*. *Development* **112**, 1063–1075.

Lilly M. and Carlson J. R. (1989) *smellblind*: A gene required for *Drosophila* olfaction. *Genetics* **124**, 293–302.

Marin E. C., Jefferis G. S. X. E., Komiyama T., Zhu H. and Luo L. (2002) Representation of the glomerular olfactory map in the *Drosophila* brain. *Cell* **109**, 243–255.

Martín F., Charro M. J. and Alcorta E. (2001) Mutations affecting the cAMP transduction pathway modify olfaction in *Drosophila*. *J. Comp. Physiol. A*. **187**, 359–370.

McKenna M. P., Monte P., Helfand S. L., Woodard C. and Carlson J. R. (1989) A simple chemosensory response in *Drosophila* and the isolation of *acj* mutants in which it is affected. *Proc. Natl. Acad. Sci. USA* **86**, 8118–8122.

McKenna M. P., Hekmat-Scafe D. S., Gaines P. and Carlson J. R. (1994) Putative *Drosophila* pheromone-binding proteins expressed in a subregion of the olfactory system. *J. Biol. Chem.* **269**, 16340–16347.

Merrill C. E., Riesgo-Escovar J. R., Pitts R. J., Kafatos F. C., Carlson J. R. and Zwiebel L. J. (2002) Visual arrestins in olfactory pathways of *Drosophila* and the malaria vector mosquito *Anopheles gambiae*. *Proc. Natl. Acad. Sci. USA* **99**, 1633–1638.

Moore P. A. (1994) A model of the role of adaptation and disadaptation in olfactory receptor neurons: implications for the coding of temporal and intensity patterns in odor signals. *Chem. Senses* **19**, 71–86.

Murlis J., Elkinton J. S. and Cardé R. T. (1992) Odor plumes and how insects use them. *Annu. Rev. Entomol.* **37**, 505–532.

Nagai T. (1983) On the relationship between the electroantennogram and simultaneously recorded single sensillum response of the European corn borer, *Ostrinia nubialis*. *Arch. Insect Biochem. Physiol.* **1**, 85–91.

Ng M., Roorda R. D., Lima S. Q., Zemelman B. V., Morcillo P. and Miesenböck G. (2002) Transmission of olfactory information between three populations of neurons in the antennal lobe of the fly. *Neuron* **36**, 463–474.

Park S.-K., Shanbhag S. R., Dubin A. E., de Bruyne M., Wang Q., Yu P., Shimoni N., D'Mello S., Carlson J. R., Harris G. L., Steinbrecht R. A. and Pikielny C. W. (2002) Inactivation of olfactory sensilla of a single morphological type differentially affects the response of Drosophila to odors. J. Neurobiol. **51**, 248–260.

Park Y., Caldwell M. C. and Datta S. (1997) Mutation of the central nervous system neuroblast proliferation repressor *ana* leads to defects in larval olfactory behavior. *J. Neurobiol.* **33**, 199–211.

Pikielny C. W., Hasan G., Rouyer F. and Rosbash M. (1994) Members of a family of *Drosophila* putative odorant-binding proteins are expressed in different subsets of olfactory hairs. *Neuron* **12**, 35–49.

Pinto L., Stocker R. F. and Rodrigues V. (1988) Anatomical and neurochemical classification of the antennal glomeruli in *Drosophila melanogaster* Meigen (Diptera: Drosophilidae). *Int. J. Insect Morphol. Embryol.* **17**, 335–344.

Plautz J. D., Kaneko M., Hall J. C., Kay S. A. (1997) Independent photoreceptive circadian clocks throughout *Drosophila*. *Science* **278**, 1632–1635.

Postlethwait J. H. and Schneiderman H. A. (1971) A clonal analysis of development in *Drososphila melanogaster*: morphogenesis, determination, and growth in the wild-type antenna. *Dev. Biol.* **24**, 477–519.

Raghu P. and Hasan G. (1995) The inositol 1,4,5-triphosphate receptor expression in *Drosophila* suggests a role for IP3 signalling in muscle development and adult chemosensory functions. *Dev. Biol.* **171**, 564–577.

Raha D. and Carlson J. R. (1994) OS9: a novel olfactory gene of *Drosophila* expressed in two olfactory organs. *J. Neurobiol.* **25**, 169–184.

Ray K. and Rodrigues V. (1995) Cellular events during development of olfactory sense organs in *Drosophila melanogaster*. *Dev. Biol.* **167**, 426–438.

Reddy G. V., Gupta B., Ray K. and Rodrigues V. (1997) Development of the *Drosophila* olfactory sense organs utilizes cell–cell interactions as well as lineage. *Development* **124**, 703–712.

Reddy G. V., Reiter C., Shanbhag S. R., Fischbach K.-F. and Rodrigues V. (1999) Irregular chiasm-C-roughest, a member of the immunoglobulin superfamily, affects sense organ spacing on the *Drosophila* antenna by influencing the position of founder cells on the disc ectoderm. *Dev. Genes. Evol.* **209**, 581–591.

Riesgo-Escovar J. R., Woodard C. and Carlson J. R. (1994) Olfactory physiology in the *Drosophila* maxillary palp requires the visual system gene *rdgB*. *J. Comp. Physiol. A* **175**, 687–693.

Riesgo-Escovar J. R., Raha D. and Carlson J. R. (1995) Requirement for a phospholipase C in odor response: overlap between olfaction and vision in *Drosophila*. *Proc. Natl. Acad. Sci. USA* **92**, 2864–2868.

Riesgo-Escovar J. R., Piekos W. B. and Carlson J. R. (1997a) The *Drosophila* antenna: ultrastructural and physiological studies in the wild-type and *lozenge* mutants. *J. Comp. Physiol. A*. **180**, 151–160.

Riesgo-Escovar J. R., Piekos W. B. and Carlson J. R. (1997b) The maxillary palp of *Drosophila*: ultrastructural and physiology depends on the *lozenge* gene. *J. Comp. Physiol. A* **180**, 143–150.

Rodrigues V. and Siddiqi O. (1978) Genetic analysis of chemosensory pathway. *Proc. Indian Acad. Sci. Exp. Biol.* **87B**, 147–160.

Rodrigues V. (1988) Spatial coding of olfactory information in the antennal lobe of *Drosophila melanogaster*. *Brain Res.* **453**, 299–307.

Roman G. and Davis R. L. (2001) Molecular biology and anatomy of *Drosophila* olfactory associative learning. *BioEssays* **23**, 571–581.

Rospars J.-P. (1988) Structure and development of the insect antennodeuterocerebral system. *Int. J. Insect Morphol. Embryol.* **17**, 243–294.

Sanes J. R. and Hildebrand J. G. (1976) Origin and morphogenesis of sensory neurons in an insect antenna. *Dev. Biol.* **51**, 300–319.

Sass H. (1976) Zur nervoesen codierung von geruchreizen bei *Periplaneta americana*. *J. Comp. Physiol.* A **107**, 49–65.

Sayeed O. and Benzer S. (1996) Behavioral genetics of thermosensation and hygrosensation in *Drosophila*. *Proc. Natl. Acad. Sci. USA* **93**, 6079–6084.

Schaller D. (1978) Antennal sensory system of *Periplaneta americana* L. Distribution and frequency of morphological types of sensilla and their sex-specific changes during post-embryonic development. *Cell Tissue Res.* **191**, 121–139.

Schneider D. (1957) Elektrophysiologische untersuchungen von chemo- und mechanorezeptoren der antenne des seidenspinners *Bombyx mori* L. *Z. vergl. Physiol.* **40**, 8–41.

Schneider D. and Boeckh J. (1962) Rezeptorpotential und nervenimpulse einzelner olfaktorischer sensillen der insektenantenne. *Z. vergl. Physiol.* **45**, 405–412.

Scott K., Brady R. Jr, Cravchik A., Morozov P., Rzhetsky A., Zuker C. and Axel R. (2001) A chemosensory gene family encoding candidate gustatory and olfactory receptors in Drosophila. *Cell* **104**, 661–673.

Selzer R. (1981) The processing of a complex food odor by antennal olfactory receptors of *Periplaneta americana. J. Comp. Physiol. A* **144**, 509–519.

Sen A. G., Reddy G. V. and Rodrigues V. (2003) Combinatorial expression of Prospero, Seven-up, and Elav identifies progenitor cell types during sense-organ differentiation in the *Drosophila* antenna. *Dev Biol.* **254**, 79–92.

Shanbhag S. R., Singh K. and Singh R. N. (1995) Fine structure and primary sensory projections of sensilla located in the sacculus of the antenna of Drosophila melanogaster. *Cell Tissue Res.* **282**, 237–249.

Shanbhag S. R., Müller B. and Steinbrecht R. A. (1999) Atlas of olfactory organs of *Drosophila melanogaster* 1. Types, external organization, innervation and distribution of olfactory sensilla. *Int. J. Insect Morphol. Embryol.* **28**, 377–397.

Shanbhag S. R., Müller B. and Steinbrecht R. A. (2000) Atlas of olfactory organs of *Drosophila melanogaster* 2. Internal organization and cellular architechture of olfactory sensilla. *Arthropod Struct. Dev.* **29**, 211–229.

Shaver S. A., Varnam C. J., Hilliker A. J. and Sokolowski M. B. (1998) The foraging gene affects adult but not larval olfactory-related behavior in *Drosophila melanogaster. Behav. Brain. Res.* **95**, 23–29.

Shiraiwa T., Nitasaka E. and Yamazaki T. (2000) *Geko*, a novel gene involved in olfaction in *Drosophila melanogaster. J. Neurogenetics* **14**, 145–164.

Siddiqi O. (1983) Olfactory neurogenetics of *Drosophila*. In *Genetics: New Frontiers*, Vol. III, Proceedings of the XVth congress of genetics, December 1993, eds V. L. Chopra, R. P. Sharma, B. C. Joshi and H. C. Bansal pp. 243–261. Oxford IBL, New Delhi.

Siddiqi O. (1991) Olfaction in *Drosophila*. In *Chemical Senses* Vol. 3, Genetics of perception and communications, eds C. J. Wysocki and M. R. Kare, pp. 79–96. Marcel Dekker, New York.

Singh R. N. and Nayak S. V. (1985) Fine structure and primary sensory projections of sensilla on the maxillary palp of *Drosophila melanogaster* Meigen (Diptera: Drosphilidae). *Int. J. Insect Morphol. Embryol.* **14**, 291–306.

Stange G. and Stowe S. (1999) Carbon-dioxide sensing structures in terrestrial arthropods. *Microsc. Res. Tech.* **47**, 416–427.

Steinbrecht R. A. (1997) Pore structures in insect olfactory sensilla: a review of data and concepts. *Int. J. Insect Morphol. Embryol.* **26**, 229–245.

Stocker R. F., Singh R. N., Schorderet M. and Siddiqi O. (1983) Projection patterns of different types of antennal sensilla in the antennal glomeruli of *Drosophila melanogaster. Cell Tissue Res.* **232**, 237–248.

Stocker R. F. and Gendre N. (1988) Peripheral and central nervous effects of *Lozenge3*: a *Drosophila* mutant lacking basiconic antennal sensilla. *Dev. Biol.* **127**, 12–24.

Stocker R. F. and Gendre N. (1989) Courtship behavior of *Drosophila* genetically or surgically deprived of basiconic sensilla. *Behav. Genet.* **19**, 371–385.

Stocker R. F., Lienhard M. C., Borst A. and Fischbach K.-F. (1990) Neuronal architecture of the antennal lobe in Drosophila melanogaster. *Cell Tissue Res.* **262**, 9–34.

Stocker R. F., Gendre N. and Batterham P. (1993) Analysis of the antennal phenotype in the *Drosophila* mutant *Lozenge. J. Neurogenetics* **9**, 29–53.

Stocker R. F. (1994) The organization of the chemosensory system in *Drosophila melanogaster*: a review. *Cell Tissue Res.* **275**, 3–26.

Stocker R. F., Heimbeck G., Gendre N. and de Belle J. S. (1997) Neuroblast ablation in *Drosophila* P[GAL4] lines reveals origins of olfactory interneurons. *J. Neurobiol.* **32**, 443–456.
Stocker R. F. and Rodrigues V. (1999) Olfactory neurogenetics. In *Insect Olfaction*, ed. B. S. Hansson, pp. 283–314. Springer, Berlin.
Störtkuhl K. F., Hovemann B. T. and Carlson J. R. (1999) Olfactory adaptation depends on the Trp Ca^{2+} channel in *Drosophila. J. Neurosci.* **19**, 4839–4846.
Störtkuhl K. F. and Kettler R. (2001) Functional analysis of an olfactory receptor in *Drosophila melanogaster. Proc. Natl. Acad. Sci. USA* **98**, 9381–9385.
Strausfeld N. J. (1976) *Atlas of an Insect Brain.* Springer, Berlin.
Sun X.-J., Tolbert L. and Hildebrand J. G. (1997) Synaptic organization of the uniglomerular projection neurons of the antennal lobe of the moth *Manduca sexta*: a laser scanning confocal and electron microscopic study. *J. Comp. Neurol.* **379**, 2–20.
Talluri S., Bhatt A. and Smith D. P. (1995) Identification of a Drosophila G protein? subunit (dG_q?- 3) expressed in chemosensory cells and central neurons. *Proc. Natl. Acad. Sci. USA* **92**, 11475–11479.
Taneja J. and Guerin P. M. (1997) Ammonia attracts the haematophagous bug *Triatoma infestans*: behavioural and neurophysiologica data on nymphs. *J. Comp. Physiol. A* **181**, 21–34.
Thomas J. B. and Wyman R. J. (1984) Mutations altering synaptic connectivity between identified neurons in Drosophila. *J. Neurosci.* **4**, 530–538.
Tissot M., Gendre N., Hawken A., Störtkuhl K. F. and Stocker R. F. (1997) Larval chemosensory projections and invasion of adult afferents in the antennal lobe of *Drosophila. J. Neurobiol.* **32**, 281–297.
Van der Goes van Naters W. M., Den Otter C. J. and Maes F. W. (1998) Olfactory sensitivity in tsetse flies: a daily rhythm. *Chem. Senses* **23**, 351–357.
Vareschi E. (1971) Duftunterscheidung bei der Honigbiene – Einzelzell-Ableitungen und Verhaltensreaktionen. *Z. vergl. Physiol.* **75**, 143–173.
Venard R. and Pichon Y. (1981) Étude électroantennographique de la réponse périphérique de l'antenne de *Drosophila melanogaster* faite à l'aide de stimulations odorantes. *C. R. Acad. Sc. Paris* **293**, 839–842.
Venkatesh S. and Singh R. N. (1984) Sensilla on the third antennal segment of *Drosophila melanogaster* Meigen (Diptera: Drosophilidae). *Int. J. Insect Morphol. Embryol.* **13**, 51–63.
Vickers N. J. and Baker T. C. (1996) Latencies of behavioral response to interception of filaments of sex pheromone and clean air influence flight track shape in *Heliothis virescens* (F.) male. *J. Comp. Physiol. A* **178**, 831–847.
VijayRaghavan K., Kaur K., Paranjape J. and Rodrigues V. (1992) The *east* gene of *Drosophila melanogaster* is expressed in the developing embryonic nervous system and is required for normal olfactory and gustatory responses in the adult. *Dev. Biol.* **154**, 23–36.
Visser J. H. (1986) Host odour perception in phytophagous insects. *Annu. Rev. Entomol.* **31**, 121–144.
Vosshall L. B., Amrein H., Morozov P. S., Rzhetsky A. and Axel R. (1999) A spatial map of the olfactory receptor expression in the *Drosophila* antenna. *Cell* **96**, 725–736.
Vosshall L. B., Wong A. M., Axel R. (2000) An olfactory sensory map in the fly brain. *Cell* **102**, 147–159.
Wang Y., Wright N. J. D., Guo H.-F., Xie Z., Svoboda K., Malinow R., Smith D. P. and Zhong Y. (2001) Genetic manipulation of the odor-evoked distributed neural activity in the *Drosophila* mushroom body. *Neuron* **29**, 267–276.

Wang J. W., Wong A. M., Flores J., Vosshall L. B. and Axel R. (2003) Two-photon calcium imaging reveals an odor-evoked map of activity in the fly brain. *Cell* **112**, 271–282.

Warr C. G., Raha D. and Carlson J. R. (1998) The anosmic mutant suggests a role for nitric oxide in Drosophila olfactory signalling and adaptation. *Annu. Dros. Res. Conf.* **39**, 568B (*Abstract*).

Warr C. G., Clyne P. J., de Bruyne M., Kim J. and Carlson J. R. (2001) Olfaction in *Drosophila*: coding, genetics and e-genetics. *Chem. Senses* **26**, 201–206.

Wetzel C. H., Behrendt H.-J., Gisselmannn G., Störtkuhl K. F., Hovemann B. T. and Hatt H. (2001) Functional expression and characterization of a *Drosophila* odorant receptor in a heterologous cell system. *Proc. Natl. Acad. Sci. USA* **98**, 9377–9380.

White P. R. (1991) The electroantennogram response: effects of varying sensillum numbers and recording electrode position in a clubbed antenna. *J. Insect Physiol.* **37**, 145–152.

Wong A. M., Wang J. W. and Axel R. (2002) Spatial representation of the glomerular map in the *Drosophila* protocerebrum. *Cell* **109**, 229–241.

Woodard C., Huang T., Sun H., Helfand S. L. and Carlson J. R. (1989) Genetic analysis of olfactory behavior in *Drosophila*: a new screen yields the *ota* mutants. *Genetics* **123**, 315–326.

Woodard C., Alcorta E. and Carlson J. R. (1992) The *rdgB* gene of *Drosophila*: a link between vision and olfaction. *J. Neurogenetics* **8**, 17–31.

Yusuyama K., Meinertzhagen I. A. and Schürmann F.-W. (2002) Synaptic organization of the mushroom body calyx in *Drosophila melanogaster*. *J. Comp. Neurol.* **445**, 211–226.

Zars T., Fischer M., Schulz R. and Heisenberg M. (2000) Localization of short-term memory in *Drosophila*. *Science* **288**, 672–675.

Zeiner R. and Tichy H. (2000) Integration of temperature and olfactory information in cockroach antennal lobe glomeruli. *J. Comp. Physiol. A* **186**, 717–727.

Zufall F. and Munger S. D. (2001) From odor and pheromone transduction to the organization of the sense of smell. *Trends Neurosci.* **24**, 191–193.

24

Plasticity and coding mechanisms in the insect antennal lobe

Mikael A. Carlsson and Bill S. Hansson

24.1 Introduction

Insects are well known to depend heavily on olfactory input in many situations vital to survival and reproduction. Sex pheromones bring the sexes together for mating, kairomones attract herbivores and parasitoids to their prey, and different odours signify the presence of a suitable oviposition site. All these types of odors are examples of molecular blends being emitted from different sources, and normally it is the blend that is meaningful to the insect, not its single constituents. The olfactory system thus has a formidable task in decoding all the different molecules that impinge on peripheral receptors and to integrate this information into behaviorally relevant blend representations. The first central nervous location where such integration takes place is in the antennal lobes (AL) of the insects. Here peripheral olfactory receptor neurons (ORN) target spheroidal neuropil, glomeruli, where the ORNs synapse onto AL interneurons. From this stage it is thus possible to bring together information from different physiological types of receptor neurons and tell the brain that "Yes, now all components of an attractive blend are present in an odor plume flowing over the antennae." In this chapter we aim to make a synthesis of our present knowledge regarding AL coding of odors and of odor blends. To do this we will refer to numerous different methods and to studies performed in a number of laboratories worldwide. To get into the actual coding of information we must, however, have a morphological base to rest on. In section 24.2 a basic morphological image of the insect AL is therefore established to serve as a substrate for our later discussions. In an introduction to the physiological part of the chapter we will briefly provide information regarding the history of AL research and mention the basic knowledge regarding function of neurons and correlations between structure and function. In the main part of this chapter we will dwell on three different themes: glomerular patterns formed after odor stimulation, dynamic processes and plasticity in olfactory coding.

24.2 Morphological elements of the antennal lobe

The primary olfactory center of insects, the AL, is a sphere-shaped part of the deutocerebrum that receives sensory input from ORNs present on antennae and mouthparts (Figure 24.1A). The AL consists of spheroidal neuropilar structures, glomeruli, where synaptic contacts between ORN axons and AL interneurons take place. In most insect species the AL contains 40 to 160 glomeruli arranged

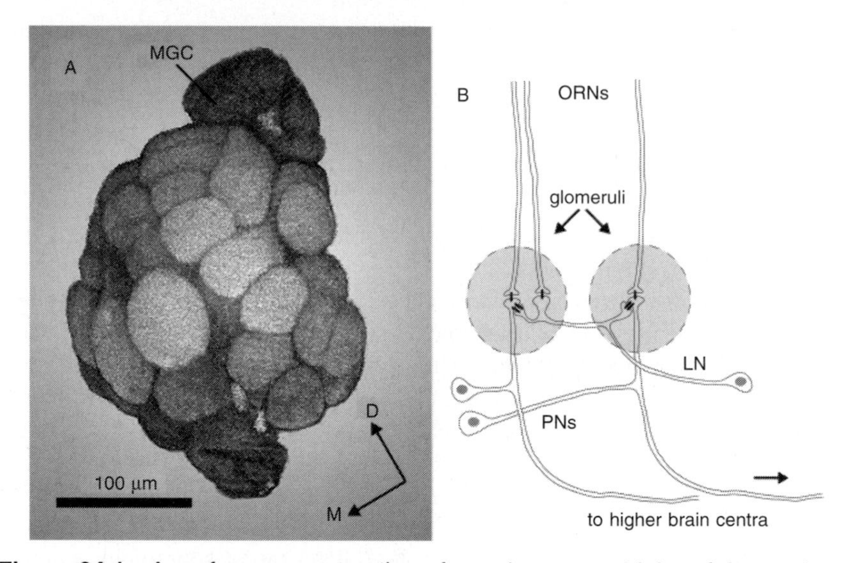

Figure 24.1 A surface reconstruction of a male antennal lobe of the moth *Spodoptera littoralis.* The brain was immunostained with synapsin antibody and optically sectioned using a confocal microscope. Stacks of images were integrated with the software Imaris 2,7 (Bitplane AG, Switzerland) on a Silicon Graphics workstation to obtain surface projections of the lobe. The macroglomerular complex (MGC) is located close to the entrance of the antennal nerve. M, medial; D, dorsal (modified from Carlsson *et al.*, 2002). B Synaptic organization of the major types of antennal lobe neurons. Sensory neurons (ORNs) make uniglomerular synapses both directly with projection neurons (PNs) and indirectly via local interneurons (LNs). In addition, local interneurons innervate several glomeruli and generally make inhibitory synapses. Cell bodies of PNs and LNs are located within the antennal lobe.

in one or two layers around a central fibrous core. Another pattern has been observed in locusts and social wasps, where up to 1000 small glomeruli fill the lobe. Glomeruli are separated from each other by glial processes.

In species using sex-pheromone communication, a sexual dimorphism in glomerular structure has been observed, where a single or a complex of large glomeruli form the macroglomerular complex (MGC). The MGC receives input only from sex-pheromone-sensitive ORNs or neurons sensitive to compounds inhibiting sexual attraction.

24.2.1 Neurons of the antennal lobe

Four main types of neurons are present in the lobe: ORNs, local neurons (LN), projection neurons (PN) and centrifugal neurons (CN) (Figure 24.1B). Receptor neuron axons enter the AL via the antennal nerve. Single neurons target single glomeruli in the majority of species investigated. In locusts ORNs branch and target several microglomeruli. In all species besides dipterans projections are strictly unilateral. In Diptera, however, ORN axons target also the contralateral AL via the antennal commissure.

Within the AL several types of synaptic interactions take place. Both LNs and PNs receive synaptic input direct from ORNs. ORNs also receive input synapses, but only from LNs. PNs and LNs synapse with each other and also with neurons of the same type. The incoming ORN signal can thus be transferred to the output neurons (PNs) via a number of possible paths.

Local neurons are typically multiglomerular, where the branching pattern can be either homogeneous within the population of glomeruli invaded, or heterogeneous, where one or a few glomeruli are more densely innervated than others. Some examples of oligoglomerular LNs, invading only a few glomeruli, have also been reported. LNs very often display GABA-like immunoreactivity, supporting physiological reports showing that LNs have an inhibitory output.

Projection neurons can be divided into a number of morphological types depending on their innervation patterns in the AL and on their path to and branching patterns in the protocerebrum. The most common type of PN has uniglomerular dendritic arborization and the axon projects through the inner antenno-cerebral tract (IACT) to the calyces of the mushroom bodies and to the lateral protocerebrum or the lateral horn. There are also PNs leaving the AL through other tracts (mainly the middle [MACT] or the outer [OACT]) and that have multiglomerular arborization within the AL. In locusts, PNs are strictly multiglomerular. Centrifugal neurons have been studied in a number of species. These neurons often have very widespread arborization patterns in the AL, similar to LNs, but the CNs receive their input in other brain areas and have output synapses in the AL. The CNs are mainly thought to have a regulatory function.

24.2.2 Physiological investigations of antennal lobe neurons

Since the first intracellular investigations of AL neurons were performed (Matsumoto and Hildebrand, 1981) a large number of studies have been published. The investigations have to a large extent focused on moths, mainly due to the presence of a well-defined sex-pheromone communication system and identified semiochemicals. Studies have also been performed in bees, locusts, cockroaches and beetles. The main results of all these studies have been that AL neurons are by no way passive mediators of the olfactory signal from ORNs to higher brain centers. This kind of mediation, so-called labeled lines, exists, but in parallel a significant degree of integration and across-fiber patterning has been observed. In the pheromone information processing system blend-specific neurons, responding only when the complete pheromone blend is present, have, for example, been identified alongside neurons preserving information regarding single pheromone components.

In the pheromone orientation system it has been shown how important time aspects are. Several species will not be attracted to a pheromone source unless the stimulus arrives in a pulsed fashion, mimicking the filamentous structure of a natural odor plume. Correlates to this requisite have been found among AL neurons, where both "fast" neurons, able to code fast fluctuations in concentration, and slow neurons, seemingly only coding qualitative aspects of the plume, are present.

In several insect types the presence of oscillations among AL neural assemblies have been demonstrated. The main experimental animal for these investigations has been the locust. In these experiments it has been shown how synchronous oscillations allow a temporal coding of odor information. These experiments will be further elaborated in the following paragraphs.

24.3 Age and hormone-level dependent plasticity in antennal lobe neural responses

Olfactory-induced behavior of many insects is known to vary as a function of age, experience and/or environmental conditions. Investment in different behavioral repertoires can often be directly correlated to different adaptive patterns; older, unmated individuals become more prone to react to a sex-pheromone stimulus, mated females become more motivated to respond to odors emanating from a suitable oviposition site and a hungry insect will more readily respond to food associated odours.

The fact is that sensory-induced behavior changes can have many different ultimate explanations. All different neural levels in the sensory systems are theoretically open to the influence of neuroactive processes. In the case of insect olfaction, plasticity in AL neurons has been investigated mainly in three different insects: the honeybee, the moth and the locust. In early experiments in the honeybee it was shown that short-term memory was affected when the AL was subjected to cooling, narcosis or electrical stimulation (Menzel *et al.*, 1974; Erber *et al.*, 1980). In addition, local injection of octopamine in the AL, as a substitute for an unconditioned stimulus, established a stable memory of the coupled odor (Hammer and Menzel, 1998). Optical imaging methods were used to study how Pavlovian and non-associative learning affected activation patterns observed among AL glomeruli after odor stimulation (see below).

In the moth *Agrotis ipsilon*, freshly emerged males are not attracted to the female sex pheromone (Gadenne *et al.*, 1993; Duportets *et al.*, 1998). After a few days of maturation the males are, however, strongly activated by the female scent. This behavioral change has been shown to be connected to an increasing titer of juvenile hormone (JH) in the male (Gadenne *et al.*, 1993; Duportets *et al.*, 1998). A possible action of the JH would be to act directly onto some stage of the olfactory neural chain. In electroantennographic (EAG) recordings from the antenna it was shown that the antennal sensitivity remained the same over the development of the male moth (Gadenne *et al.*, 1993). However, when recordings were performed from single neurons of the AL another pattern emerged. In newly emerged males the number of neurons responding to pheromone stimuli was significantly lower and the response threshold of those responding was significantly higher than in three- and five-day-old males (Anton and Gadenne, 1999; Gadenne and Anton, 2000) (Figure 24.2A). This change in response characteristics follows the JH titer of the animal very well, with increasing JH levels improving the performance of AL neurons. To test for JH action, the JH-producing corpora allata (CA) was removed (Gadenne and Anton, 2000). The allatectomized moths remained in the "newly emerged" state with a lower number of responding neurons and higher response thresholds even though they were reaching mature age (Figure 24.2B). The olfactory functions could be restored by JH injections. It is thus clear that JH has a modifying effect on olfactory functions in some moths. The mechanism for its function is so far unknown. When similar experiments were performed testing host plant odors as stimuli, no age effect could be demonstrated (Greiner *et al.*, 2002) (Figure 24.2C). JH modification of olfactory sensitivity is thus restricted to the pheromone-information-processing system in this moth. Furthermore, male *A. ipsilon* moths show a transient post-mating inhibition of pheromone attraction (Gadenne *et al.*, 2001). Also this decrease in behavioral performance could be correlated with a decrease in AL neuron performance. Here JH was, however, most likely not involved in the plasticity. A plausible mechanism still remains to be revealed.

In the desert locust, *Schistocerca gregaria*, a similar phenomenon has been observed, but here it is older insects that lose activity among AL neurons (Ignell *et al.*, 2001). Together with other sensory cues, desert locusts rely on aggregation pheromones to form their infamous swarms. Adult males produce an aggregation

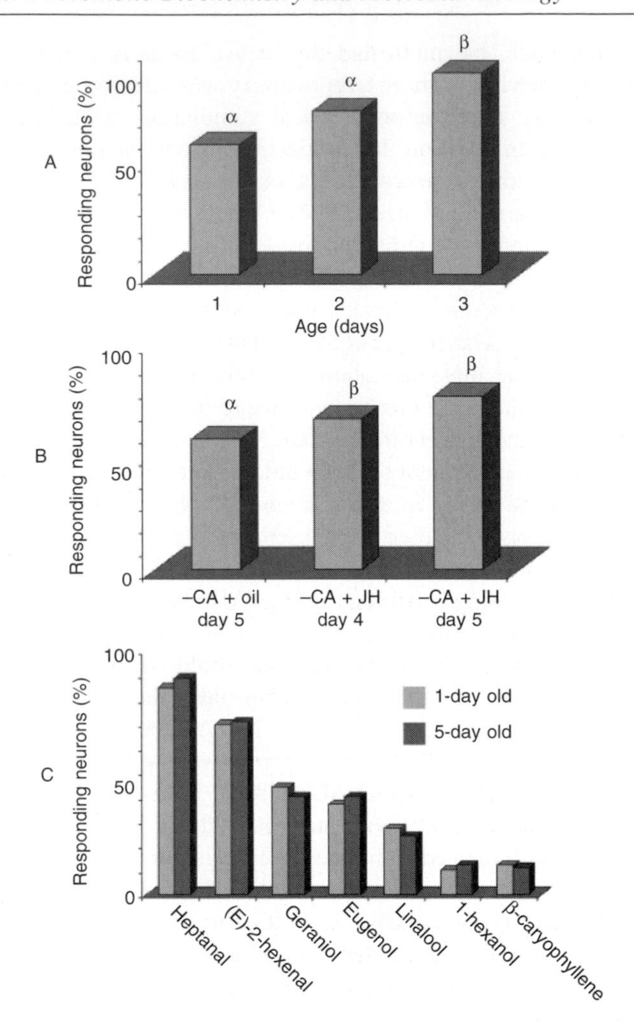

Figure 24.2 Effect of age and hormone levels on odor responses of antennal lobe neurons of the moth *Agrotis ipsilon*. The bars show the percentage of responsive neurons. A Age effect on male responses to female gland extract. Bars capped with same letter do not differ significantly (redrawn from Gadenne and Anton, 2000). B Effect of juvenile hormone level on male responses to female gland extract. Allatectomized 5-day old animals (–CA) show responses similar to newly emerged animals. Hormone-restored animals (+JH), however, increased the percentage of responding neurons to the level of normal 5-day old males. Bars capped with same letter do not differ significantly (redrawn from Gadenne and Anton, 2000). C Age effect on neurons responding to plant-related odorants. There was no significant difference in the number of responding neurons in newly emerged animals and 5-day-old animals (redrawn from Greiner *et al.*, 2002).

pheromone when confronted with other locusts. The propensity to respond behaviorally to this pheromone decreases drastically with age. This behavioral change is again well correlated with JH titers measured in the desert locust, but here an increase in JH is correlated with a decrease in behavioral response. When recordings were made from AL neurons it was shown that neurons of older animals responded less frequently and with a higher response threshold to pheromone than neurons of younger individuals. Functions of AL neurons could be maintained by allatectomy and could be impaired by injection of JH. Again, JH has an important role in governing plastic events among AL neurons.

The fact that AL neurons display changed response characteristics depending on hormone levels highlights the importance of knowing the physiological status of experimental animals. It also opens up a highly interesting research area of its own. In the case of learning, our knowledge from AL neurons is still at a very basic level. Considerable efforts are needed in both these areas to clarify how involved early olfactory processing is in behavioral plasticity.

24.4 Coding patterns: temporal, spatial or spatio-temporal?

A question of fundamental interest is how information from a multidimensional odor space is transformed into a neural population code in the AL, a code that can be decoded by higher brain centers. A large number of ORNs converge into the glomerular array (possibly each type of ORN to a specific glomerulus), information is processed and outgoing PNs send the processed information further upstream. What type of information in the PNs is important for the animal in order to read correctly the quality and quantity of the stimulus? Is it the spatial location of the activated neurons or the temporal characteristics of single neurons or assemblies of neurons acting in synchrony? A combination of different coding strategies or context-dependent strategies may also be possible. To address these questions different techniques have been employed and will be discussed in the following (section).

24.4.1 Temporal (spatio-temporal) coding

24.4.1.1 Slow temporal characteristics

Studying odor-evoked activity in PNs often reveal different temporal features. Temporal coding has been most thoroughly studied in the locust (Laurent *et al.*, 2001) and if not otherwise noted the results in this section refer to work in the locust system. A spike train may include periods of excitation and inhibition in different combinations. Spike profiles are often both odor and neuron specific and reproducible at repeated stimulations (Laurent and Davidowitz, 1994; Laurent *et al.*, 1996). The inhibitory phases of the spike sequences should be generated by inhibitory input from LNs. However, application of picrotoxin (PTX), which blocks GABA-coupled Cl^- channels, had no effect on the patterns (MacLeod and Laurent, 1996). This means that inhibition from LNs must be mediated either by another neurotransmitter or by GABA acting on a different type of receptor. Nevertheless, spike profiles are generally odor specific with intermittent periods of excitation and inhibition and should be considered when averaging spike numbers along the duration of stimulus exposure, which is often done.

If odor-evoked slow temporal patterns actually provide higher brain centers with information about the odor quality, identification and discrimination cannot be instantaneous as many of the temporal features in the response profiles appear late or even after offset of odor exposure. Honeybees need ~500 ms for a response to (non-sexual pheromone) odors but at least 1 second of stimulation is required for a correct discrimination (J. Klein, unpublished, cited in Galizia *et al.*, 2000a). Thus, it appears that time is an important factor in discrimination tasks involving non-pheromonal odors and the slow temporal patterns could theoretically contribute to an olfactory code. In contrast, these temporal patterns would be too slow to encode information about sexual pheromones. Male moths, for example, must be able to respond to rapid changes in stimulus intermittency when moving upwind in pheromone plumes in search of a calling female.

24.4.1.2 Oscillatory synchrony

It has long been known that odors can evoke oscillations in different brain areas (Adrian, 1950, 1953). Odor-evoked oscillations have also been observed in a number of insects by measuring field potentials (see Stopfer *et al.*, 1999). The field potentials measured in the mushroom bodies, i.e. downstream of the ALs, are the result of large numbers of neurons spiking simultaneously in the AL. At rest, field potentials are irregular but odor stimulation creates regular oscillations which means that neurons downstream are exhibiting synchronized activity. The frequency of the oscillations is species but not odor specific. The oscillatory neural synchrony is not generated by coordinated spiking activity in the sensory neurons but by the reciprocal coupling of LNs and PNs (Wehr and Laurent, 1996). PNs receive excitatory input from ORNs, inhibitory input from LNs and feed back excitatory synapses onto LNs. This gives rise to subthreshold oscillations in the PNs, which are not intrinsic. As a consequence, action potentials often appear periodically.

During a cycle of an oscillation a number of PNs spike in synchrony (Laurent and Davidowitz, 1994; Wehr and Laurent, 1996). Other neurons may spike as well but are not phase locked to any cycles. The population of neurons acting in synchrony is dependent on the stimulus identity. Thus, a different odor evokes synchronous activity in another (but often overlapping) subset of PNs.

In contrast to the slow temporal patterns, oscillations (both in individual neurons and as observed in field potentials) are disrupted by picrotoxin application (MacLeod and Laurent, 1996; MacLeod *et al.*, 1998). Thus, there seems to be at least two independent channels of inhibitory input to the PNs.

Desynchronization of PNs affects the specificity of neurons in the β-lobes of the mushroom bodies (a neural convergence site post-synaptic to the calyces) (MacLeod *et al.*, 1998). Application of picrotoxin to the AL resulted in a decreased ability of β-lobe neurons to correctly identify odors. Single β-lobe neurons showed responses to odors after desynchronization to which they were formerly silent and responses to similar odors lost odor-specific temporal features seen prior to picrotoxin application. These differences were never seen in PNs and the authors concluded that information is sampled across assemblies of responding PNs and that this information converges onto single neurons in the β-lobe. Thus, neurons downstream of the AL appear to read and decode temporal information conveyed by neurons leaving the AL.

24.4.1.3 Behavioral significance of temporal characteristics

An important question to address is if oscillatory synchrony is actually needed for odor discrimination. In an attempt to show the behavioral significance of these temporal characteristics, Stopfer *et al.* (1997) trained honeybees to respond to conditioned stimuli. Using the proboscis extension reflex paradigm (Kuwabara, 1957; Bitterman *et al.*, 1983) bees were trained to respond to one odor and then tested with the same odor, a novel similar odor and a novel dissimilar odor. The animals received either an application of picrotoxin or saline prior to conditioning. Both groups learned the conditioned stimulus equally well and had no problems discriminating this from a structurally different odor in a test performed 90 minutes after the training procedure. However, the group of bees that had received an application of picrotoxin were unable to discriminate between the learned odor and a structurally similar odor. The control (saline) group had no problem with this. In a second experiment, picrotoxin was applied either both prior to conditioning and prior to testing or only prior to testing (Hosler *et al.*, 2000). The honeybees performed as in the former experiment, thus demonstrating that picrotoxin does not impair learning *per se*. Therefore, neural synchrony appears to be a mechanism required for fine-odor discrimination.

24.4.1.4 Evolution and memory of oscillations

The coherence between odor-evoked activities in PNs is progressively changing over time. Stopfer and Laurent (1999) demonstrated that synchronization of PNs was not inherent as an initial odor trial failed to evoke oscillatory field potentials. However, repeated stimulations gradually increased the probability for timed spiking in the PNs. Also a prolonged odor exposure increased the precision of timing. In addition, Stopfer and Laurent (1999) found that spiking frequency decreased at repeated stimulations, especially between the first and second exposure. This type of non-reinforced odor memory was short termed, as an interval of 12 minutes was sufficient to set the system to an initial state, i.e. no synchrony and a higher spike frequency. Furthermore, the "memory" was odor specific as a novel odor that was structurally different also failed to evoke PN timing at the first trial. However, some degree of generalization appeared for structurally similar odors. It was further demonstrated that evolution of PN timing was the result of processing in the AL. First, an odor was repeatedly presented to a subpopulation of ORNs. In the subsequent test the same odor was presented to another non-adapted subset of ORNs. The latter test caused oscillatory synchronization at the first trial, thus showing an independence of which ORNs were activated. Therefore it appears that the "memory" is contained within the AL network.

The gradual evolution of oscillatory synchronization, which as discussed above appears to be crucial for discrimination of alike odors, may implicate that an initial "sniff" can only mediate a coarse identification and subsequent sampling is needed for fine-odor discrimination.

24.4.1.5 The moth system

A different methodological approach was adapted to the moth (Christensen *et al.*, 2000). Multiunit extracellular recordings were performed with a three-pronged, fork-like, array of recording sites. On each prong of the fork four recording sites allowed monitoring of neurons close by. Recording sites could measure around 10 μm^3 and were separated by about 25 μm. Recorded units could be separated physiologically by software programs comparing a number of spike parameters. In this way it could be demonstrated that the temporal structure of a physiological response to a stimulus in male *Manduca sexta*, i.e. coactivity of neurons in an assembly, depended very much on the physical context in which different stimuli were delivered, i.e. not only did the molecular identity affect the response of a specific neuron but also, for example, the timescale of the stimulus. Furthermore, it was shown that different concentrations of the same stimulus could evoke clearly different temporal patterns in a neuron responding to the stimulus and that the response to a blend of odors could not be predicted from the responses to its constituent components (Figure 24.3). Despite the fact that coactivity among populations of neurons occurred, brief pulses of pheromones could neither evoke rhythmicity in single neurons nor oscillatory patterns of population activity. Though oscillatory field potentials can still be observed at sustained exposure, PN activity is not locked to any phase of the oscillations (Lei *et al.*, 2000, cited in Vickers *et al.*, 2001). Based on these results the authors suggest "that ensembles of olfactory PNs must use multiple and overlapping coding strategies to process olfactory information, and that these strategies are matched to the particular circumstances surrounding odour presentation" (Christensen *et al.*, 2000).

A recent experiment by Lei *et al.* (2002), measuring intracellular activity simultaneously from two PNs in *M. sexta*, revealed synchrony between pairs of MGC output neurons; however, in this experiment synchronous spiking was found not to be associated with oscillatory activity. Intraglomerular synchrony

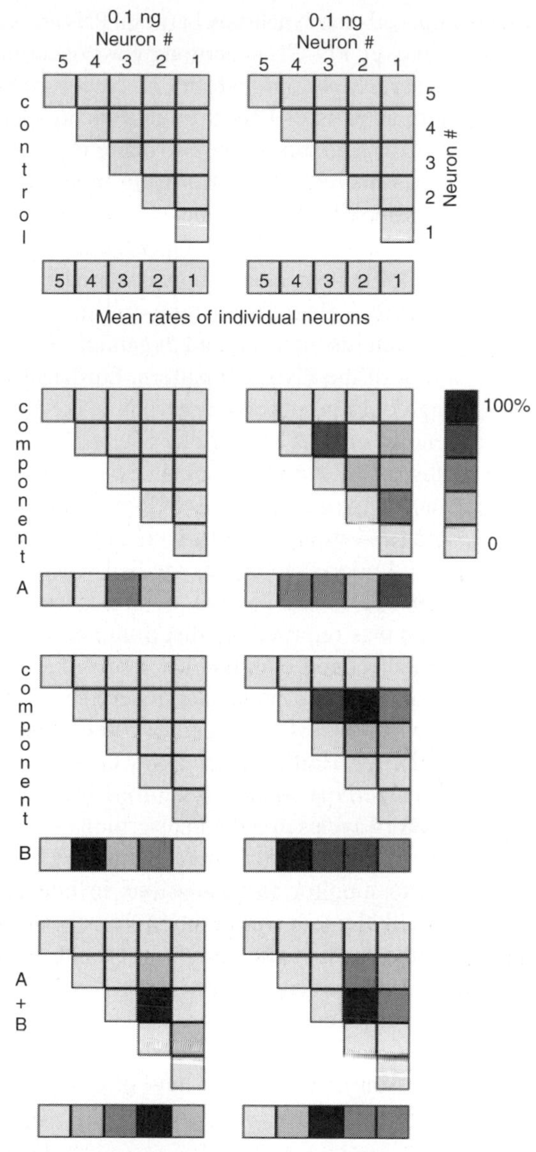

Figure 24.3 Temporal analysis of responses measured simultaneously in five different antennal lobe neurons in the moth *Manduca sexta*. The matrices show patterns of neural synchrony evoked by either of two pheromone components or a binary mixture at two concentrations. The number of synchronous events was averaged over 20 trials and calculated for 500 ms from stimulus onset. The gray scale ranges from 0 to 3.8 coincident spikes per stimulus. The horizontal displays below the matrices show the averaged spiking rate in each single neuron (gray scale ranges from 0 to 5.5 spikes per stimulus period). Neural synchrony was influenced not only by the odor quality but also by both stimulus intensity and blend interactions (redrawn from Christensen *et al.*, 2000).

was evidently more pronounced than synchrony between PNs arborizing in different MGC compartments. Moreover, PN–PN synchrony was concentration dependent in that a higher dose evoked a higher probability of synchronous firing. A large portion of all intraglomerular pairs of PNs exhibited blend-specific synchrony, i.e. synchrony was enhanced by a mixture of the two major pheromone components.

In a third study, single unit recordings from the moth *Heliothis virescens* placed in a wind tunnel corroborated the previously described results by showing how well single neurons follow the fine-scale temporal characteristics of a natural odor plume (Vickers *et al.*, 2001). Also in this study it was clear that the occurrence of a stimulus over time heavily influences the temporal structure of the response to a given stimulus. Both stimulus intensity and dynamics of the odor plume had an effect on the time course of the PN spike pattern. Furthermore, no wave-like periodicity in PN spiking could be observed and PN spike frequency did only rarely match the frequency range of local field potential oscillations that has been reported from moths (*M. sexta*; Heinbockel *et al.*, 1998).

The obvious discrepancy between the studies of locusts and moths may depend on the different subsystems investigated. In the works by Laurent and colleagues odor processing in the food-odor system was studied whereas the pheromone subsystem was studied in the moth. As was demonstrated in the honeybee, oscillatory synchronization was required for discrimination of similar odorants (Stopfer *et al.*, 1997). Food-related odorants are generally detected by ORNs with broader and more overlapping molecular receptive ranges and also AL interneurons often show broad response profiles. In contrast, pheromone-detecting ORNs are generally extremely specific and non-overlapping in receptive range. Thus, it is not unlikely that activity in output neurons from the MGC does not have to be locked to oscillatory cycles in order to discriminate between pheromone components or the blends of these. Furthermore, it may be disadvantageous for a male moth, which has to monitor fast variations in intensity and stimulus duration, to depend on oscillations. It would therefore be very interesting to see whether the results from the moth studies can be repeated for non-pheromonal odors. Is the non-oscillatory-dependent system typical for "fast" coding systems, following quick variations in odor concentration, while the system relying on oscillatory synchronization among PN populations is used when more time is allowed for olfactory processing. Alternatively, have different coding mechanisms evolved in different taxonomic groups?

24.4.2 Functional imaging and spatial coding

24.4.2.1 Odor maps and a spatial "olfactory code"

To address questions about distributed glomerular events in the AL, single unit recordings are of limited use. Instead, visualization of activity in large populations of neurons by activity-dependent staining techniques is favourable. Earlier studies in dipteran species using [3H] 2-deoxyglucose autoradiography revealed that different odorants evoked different glomerular activity patterns (Rodrigues, 1988; Distler *et al.*, 1998). However, [3H] 2-deoxyglucose labeling requires extremely long periods of odor stimulation. Furthermore, only stimulation with one stimulus type in a single animal can be done, making comparisons difficult.

In recent years functional imaging techniques that allow visualization of neural activity in real time have been developed. These techniques record minute light intensity variations on the surface of the recorded neuropil. Either intrinsic light properties of the tissue or light emission changes in different types of fluorescent dyes can be measured. Intrinsic signaling has been extensively used in vertebrate olfactory studies (e.g. Rubin and Katz, 1999; Meister and Bonhoeffer, 2000). In insects, however, calcium-sensitive dyes have been most popular due to the high signal-to-noise ratio that does not require averaging of repeated stimulations (Joerges *et al.*, 1997; Galizia *et al.*, 1999a; Carlsson *et al.*, 2002).

Calcium recordings in insects have shown that odor stimulation evokes patterns of activated glomeruli, so-called odor maps. These odor maps are characteristic for each type of odorant and are highly reproducible (Joerges *et al.*, 1997; Galizia *et al.*, 1999a; Carlsson *et al.*, 2002). Odor-evoked activity patterns are, in addition, bilaterally symmetrical between the lobes (Galizia *et al.*, 1998) and similar between individuals (Joerges *et al.*, 1997; Carlsson *et al.*, 2002). However, similar patterns need not necessarily prove that homologous glomeruli are activated. In honeybees, independent stainings of anatomical features (glomerular outlines) were made after the calcium recordings (Galizia *et al.*, 1999b) and compared with a computerized glomerular atlas (Galizia *et al.*, 1999c). A subsequent alignment with the physiological responses allowed identification of physiologically characterized glomeruli. The experiment demonstrated that identified glomeruli had the same molecular response profiles in different individuals. The predictability of odor-evoked patterns was tested in a discriminant analysis. In 86 percent of the measurements the correct odor could unambiguously be identified from the patterns of glomerular activity and the few "mistakes" concerned structurally similar odors. The findings suggest that odor representations are genetically determined, i.e. there may be a spatial olfactory code. Already half a century ago Adrian (1950) proposed that odors might be encoded as spatial patterns of neural activity. Even though a combination of activated neural elements at fixed spatial positions may contribute to an odor "image" that underlies final perception, temporal features of the signal in AL output neurons, as discussed above, also carry information about the odor quality.

Here we will describe and discuss distributed glomerular activity as observed by functional imaging in insect brains. Focus will be on results obtained by optical recordings using a calcium-sensitive marker, which requires a brief methodological introduction.

24.4.2.1.1 Methodology

The concept of calcium recordings is to register changes in light intensity with a sensitive CCD (charged couple device) camera from calcium-sensitive fluorescent markers. The dye molecules are activated by coupling to calcium ions. The dye is normally applied directly on the surface of the insect brain. The most commonly used Ca^{2+}-sensitive dyes permeate the cell membranes. Inside the cells enzymes split the dye molecule making it membrane impermeable and set the dye in an active state. Thus, after rinsing the preparation, fluorescent dye is confined to the intracellular space.

Recordings are done *in vivo* (or semi *in vivo*) with only those parts of the head covering the ALs removed, allowing for long lasting recordings with multiple stimulations. Keeping the brain relatively intact limits the view to the anterior side of the AL and no recordings have to our knowledge been made from the "the dark side of the lobe."

A typical experimental protocol includes a sequence of images taken before, during and after stimulation, which in addition makes it possible to study slow temporal properties of the signals. Temporal resolution is limited by the slow decay of calcium signals and the method is not suitable for observing fast temporal changes. However, using voltage-sensitive dyes allows exploration of faster properties (see below). Spatial resolution down to a pixel size of ~1 μm is good enough resolution to observe activity even in the smallest insect glomeruli (e.g. *Drosophila*).

Optical recordings of calcium activity, when the dye is bath applied, are non-selective with respect to the neural elements involved in the measured signals. This naturally complicates the interpretation of the acquired data and the underlying processing mechanisms. Increased calcium concentration may occur in both pre- and post-synapses of the involved neuron types. Pre-synaptic calcium inflow is required for the transmitter release in neuron terminals (Katz, 1969). Calcium concentration changes in mitral cells (vertebrate equivalent to PNs) are correlated with changes in post-synaptic membrane potential (Charpak *et al.*, 2001). In addition there is a calcium influx through nicotinic acetylcholine receptors (Goldberg *et al.*, 1999; Oertner *et al.*, 1999), which are the probable receptors in PNs receiving direct input from ORNs (Hildebrand *et al.*, 1979, Bicker, 1999). Thus, the measured calcium changes make up a compound signal, but due to the high convergence rate from input (ORNs) to output (PNs) (for review see Anton and Homberg, 1999), the major part of the signal is likely originating from ORNs. Moreover, ORN axons innervate the outer rim of glomeruli, whereas interneurons preferentially innervate the core (Anton and Homberg, 1999). However, activity may still be present in a glomerulus not innervated by responding ORNs as most LNs innervate all or a large number of glomeruli in the lobe (Anton and Homberg, 1999). The total number of activated glomeruli, as observed in non-selective optical recordings, is likely an overestimation of the real number of glomeruli innervated by either responding ORNs or PNs. Nevertheless, single ORNs seem to house only one type of receptor molecule and each glomerulus is generally innervated by one type of ORN (Vosshall, 2001). This means that each glomerulus probably represents one type of receptor. Therefore, optical recordings of glomerular activity should provide us with information about the ligand specificities of the receptors.

24.4.2.2 Spatial odor representations in the antennal lobes

24.4.2.2.1 Sexual pheromones in moths

Many insects have in addition to a large number of sexually isomorphic glomeruli also sex-specific glomeruli with specialized functions. The best studied are the male-specific MGCs, that receive input from sensory neurons tuned to components of the conspecific sexual pheromone, or to allospecific compounds inhibiting sexual behavior. These glomeruli can readily be identified and the innervation is stereotyped with respect to input, i.e. each compartment of the MGC is innervated by a specialized type of neuron (Hansson *et al.*, 1992; Hansson and Christensen, 1999).

As observed in calcium recordings, activity patterns evoked by pheromone component stimulation in moths corroborate earlier results at the cell level. Activity is confined to the area near the antennal nerve entrance, i.e. where the MGC is located and no or only minor activity is observed in "ordinary" glomeruli. In the noctuid moth, *Spodoptera littoralis* (Carlsson *et al.*, 2002), and the heliothine moth, *Heliothis virescens* (Galizia *et al.*, 2000b), the activated glomeruli have the same relative positions as those innervated by ORNs responding to the tested pheromone components (Ochieng *et al.*, 1995; Hansson *et al.*, 1995; Berg *et al.*, 1998). Whereas the arborization of PNs in the MGC could be predicted in *H. virescens*, assuming an isomorphism of input and output, dendritic branches of PNs did not always match those of ORNs with a corresponding tuning in *S. littoralis* (Vickers *et al.*, 1998; Anton and Hansson, 1995). In the sphingid moth, *M. sexta*, and the heliothine moth, *Helicoverpa zea*, no labeling of functionally characterized ORNs has been made. However, arborization patterns of pheromone-responding PNs have been mapped to identified glomeruli of the MGC in both species (Hansson *et al.*, 1991; Vickers *et al.*, 1998). Calcium recordings from ALs of *M. sexta* and *H. zea* (Figure 24.4) showed that the pheromone-evoked response patterns were spatially distinct for each of the components and the location of the activity foci agreed with the arborization of PNs with known physiology (Hansson *et al.*, in preparation; Carlsson *et al.*, in preparation). Assuming ORNs as major contributors to the Ca^{2+} signals suggests a matching of input and output in *M. sexta* and *H. zea*. The results from calcium recordings of the MGCs show that the superficial patterns of activity can be correlated with neural activity at the cell level and thus facilitates the interpretation of recordings from the "ordinary" glomeruli.

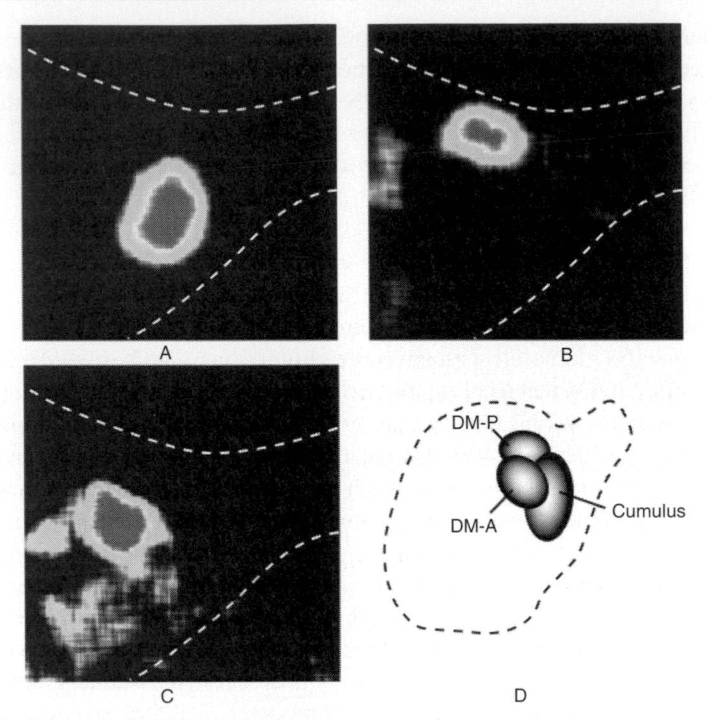

Figure 24.4 Gray-scaled signals of optical recordings of calcium activity from the antennal lobe of the moth *Helicoverpa zea*. A, B and C show thresholded (>50 percent of maximum) responses to the pheromone components Z11-16:Ald and Z9-16:Ald and a behavioral antagonist Z11-16:Ac, respectively. D A schematic figure of the organization of the macroglomerular complex in *H. zea* (from Vickers *et al*., 1998). The position of the glomeruli coincides with the foci of calcium responses, i.e. response to Z11-16:Ald takes place in the cumulus, Z9-16:Ald in the DM-P glomerulus and Z11-16:Ac in the DM-A glomerulus. These activity patterns corroborate the innervation patterns of functionally identified projection neurons (redrawn from Vickers *et al*., 1998).

24.4.2.2.2 Plant-related odorants

Plant-emitted odors are generally detected by ORNs with broader response profiles (with respect to different odor species, specificity may still concern molecular determinants) than pheromone-detecting ORNs (Todd and Baker, 1999). There are, however, several exceptions, showing that plant-detecting ORNs also can be exclusively tuned to single compounds (Hansson *et al*., 1999; Röstelien *et al*., 2000a, b; Larsson *et al*., 2001; Stensmyr *et al*., 2001).

Odor-evoked glomerular activity patterns may reflect the specificity of the responding ORNs. High specificity will resemble activity patterns evoked by sexual pheromones, i.e. activity confined to a single glomerulus. To date no specific glomeruli responding exclusively to a single plant compound have been observed in optical recordings. Activity maps evoked by plant odorants often involve a few to several responding glomeruli (Joerges *et al.*, 1997; Galizia *et al.*, 1999a, b; Sachse *et al.*, 1999; Galizia *et al.*, 2000b; Carlsson *et al.*, 2002). In addition, the same glomeruli respond, in varying degrees, to many compounds (Joerges *et al.*, 1997; Galizia *et al.*, 1999a, b; Sachse *et al.*, 1999; Galizia *et al.*, 2000b; Carlsson *et al.*, 2002).

To study the receptive range of ORNs and hence glomeruli it is necessary to use stimuli that systematically differ along a specific dimension, e.g. carbon chain length or functional group. In one such study, Sachse *et al.* (1999), using series of homologous aliphatic compounds, found no glomeruli in the honeybee with a preference for molecules containing a specific functional group. In contrast, rodent glomeruli display such specificity (Johnson *et al.*, 1998, 2002; Uchida *et al.*, 2000). It thus appears that glomerular response profiles in mammals are narrower than in insects. Moreover, Sachse *et al.* (1999) demonstrated that identified glomeruli were activated by a range of stimuli but showed optimal responses to one or a few neighboring chain lengths irrespective of functional group. Different glomeruli had different optimal chain length preferences but overlapped with each other. The correlation of response patterns decreased as the distance in chain length between compounds increased, i.e. the more structural similarity, the more similar response patterns. Furthermore, glomeruli that demonstrated similar response profiles were often located close to each other. Similar results were found in the moth *S. littoralis* (Meijerink *et al.*, 2003). If such architecture of neighboring glomeruli with similar response profiles reflects an underlying physiological requirement, e.g. facilitation of mechanisms like lateral inhibition, is not known. In addition, in the study by Meijerink *et al.* it was shown that patterns evoked by short chain-length molecules were more distinct from each other than patterns evoked by longer chain-length molecules. Ecologically, this makes sense, as the shorter molecules are most likely more important to the animal. Short aliphatic compounds are common green leaf volatiles.

In moths (*S. littoralis* and *M. sexta*) there is a rough spatial organization of glomeruli activated by different structural elements in that glomeruli preferentially responding to aromatic compounds are clustered in the medial region of the lobes whereas a cluster of glomeruli responding to terpenoids are located in the lateral region (Carlsson *et al.*, 2002; Hansson, Carlsson and Kalinova, in preparation). In the moth *Trichoplusia ni*, Todd and Baker (1996) demonstrated that single ORNs tuned to aromatic compounds had axons, which terminated medial to ORNs tuned to a terpenoid. It seems likely that also in *T. ni* the ORNs responding to aromatic compounds project to glomeruli located medial to the glomeruli tuned to the terpenoid, thus suggesting a general functional organization of glomeruli in lepidopteran species.

24.4.2.2.3 Non-sexual pheromones

In addition to sexual pheromones many other types of semiochemicals are used, especially among social insects (for review see Hartlieb and Anderson, 1999). Compounds acting in social bonding, as repellents or as alarm pheromones, have been tested in calcium recordings in honeybees and in the ant *Camponotus rufipes* (Joerges *et al.*, 1997; Sachse *et al.*, 1999; Galizia *et al.*, 1999a, b). These non-sexual pheromones evoked responses in glomeruli that also displayed activity in response to plant odorants. No specialized glomeruli, comparable to the MGC in moths, were found responding exclusively to any of the pheromones. An explanation may be that several of the non-sexual pheromones are common substances also emitted by plants and often have additional functions as kairomones.

In the study by Sachse *et al.* (1999) it was shown that 2-heptanone evoked a specific activity pattern that differed from the corresponding alcohol or aldehyde more than functional groups differed at other chain lengths. 2-heptanone functions as a repellent scent marker in honeybees (Giurfa and Núñez, 1992; Giurfa, 1993). The pheromonal function of this substance may have paved the way for the development of an alternative specialized pathway. It cannot be excluded that non-sexual pheromones are subject to parallel processing, i.e. a second specialized pathway similar to the sexual pheromone pathways may be present in areas of the ALs not accessible for recording.

24.4.2.2.4 Other specialized glomeruli

Other than the MGC in moths many glomeruli are likely innervated by ORNs responding to a broader selection of odorants. There may still be "ordinary" glomeruli that have a specialized function. For instance, using calcium recordings Galizia *et al.* (1999b) found a single glomerulus in the honeybee AL that responded most strongly (a weak response though) to carbon dioxide. CO_2 responses were not observed in any other glomeruli. The unique structure of the CO_2 molecule suggests that information regarding this compound uses a specialized pathway and it is likely that CO_2-specific glomeruli are present in the ALs of most insects. In moths, there are highly selective sensory neurons on the labial palps responding to CO_2 (Stange, 1992). Axons from the labial palps project to a specific glomerulus in the medial part of the lobe in *M. sexta* (Kent *et al.*, 1986). Mosquitoes possess receptor neurons on the maxillary palps known to respond to CO_2 with high selectivity and sensitivity (Grant *et al.*, 1995). Bausenwein and Boeckh (1996) found in a [3H] 2-deoxyglucose study in *Aedes aegypti* that a well-defined centre of the ALs was activated by exposure to CO_2.

In the moth *M. sexta*, a female-specific glomerulus, the large female glomerulus, is innervated by PNs responding specifically to the plant-emitted compound linalool (King *et al.*, 2000). In calcium recordings no glomerulus was observed that selectively responded to linalool (Hansson, Carlsson and Kalinova, in preparation). Instead, linalool evoked activity in several glomeruli that were also activated by other terpenoids. Additionally, similar glomerular responses to linalool were observed in both sexes.

24.4.2.2.5 Concentration coding

The olfactory system must not only be able to detect and process information about the quality of a stimulus but also accurately code changes in concentration. At the peripheral level, insect ORNs can detect and code differences in stimulus concentration over several orders of magnitude (Todd and Baker, 1999).

In honeybee ALs, activity patterns evoked at high and low concentrations showed nearly identical spatial distribution though with a difference in signal intensity (Joerges *et al.*, 1997). Furthermore, the receptive range of glomeruli was not dependent on the stimulus concentration over a range of two decadic steps (Sachse *et al.*, 1999).

In the moth *S. littoralis*, a concentration increase over four orders of magnitude resulted in an increase in the number of activated glomeruli (Carlsson and Hansson, 2003). The principal glomerulus often moved across concentrations due to recruitment of responding glomeruli and to saturation of glomerular responses. Movement of activity focus was, however, restricted to adjacent or proximal glomeruli. As a consequence, correlations between activity patterns evoked by the same stimulus decreased with increased difference in concentration. A comparison of odor maps evoked by low and high (1000-fold difference) concentrations of different odorants revealed that correlations were higher at high doses than at low doses, i.e. specificity diminished at high stimulus concentrations.

If a spatial olfactory code underlies a subsequent identification of an odor, then we would expect concentration-dependent variations of odor maps to have an impact on the perceived odor quality. Stimulus concentration can indeed influence perception of odor quality and behavior of insects. For example, olfactory responses of the fruit fly, *Drosophia melanogaster*, shifted from attraction to repulsion as the concentration increased (Siddiqi, 1983; Stensmyr *et al.*, 2003). Future studies should thus involve correlations of behavior with glomerular activity at different concentrations.

24.4.2.2.6 Mixture representation

Under natural conditions odors rarely occur as single compounds. Instead odors are often mixtures of many components. A mixture may be represented in the ALs either simply as an addition of the individual components or as a novel configurational representation. A configurational representation would be the result if the components of a mixture interact with each other in an unpredictable manner. In honeybees, responses to binary mixtures could not always be predicted based on responses to the individual components (Joerges *et al.*, 1997; Galizia *et al.*, 2000a). Activity in individual glomeruli in response to a mixture either equalled the sum of the components or was weaker than predicted. The power of the mixture suppressions increased when adding components to the blend (Joerges *et al.*, 1997). Mixture interactions were both odor and glomerulus specific. It is likely that suppressed glomerular activity is the result of interglomerular inhibition mediated by LNs or by inhibitory feedback to the ORN terminals. However, it cannot be excluded that odor molecules interact with each other already at the receptor site. A few research groups have reported peripheral mixture interactions in insects, either suppression or synergistic effects (Den Otter *et al.*, 1978; De Jong and Visser, 1988; Akers and Getz, 1993; Getz and Akers, 1997; Ochieng *et al.*, 2002).

24.4.2.2.7 Behavioral relevance of odor maps

If an odor map really is a substrate for an olfactory code, which is then decoded by higher brain centers, it should be possible to predict the animals' ability to discriminate odors from odor-evoked glomerular activity patterns. Honeybees have an astonishing ability to discriminate even structurally similar compounds (Laska *et al.*, 1999). The discrimination performance is negatively correlated with the structural similarity of odor molecules. Using series of homologous aliphatic compounds, Laska and colleagues (1999) showed that discrimination of odor pairs differing by one carbon atom was significantly more difficult than discrimination of odor pairs that differed by more carbon atoms. The relationship between odor structure and behavior can be compared with the relation between odor structure and glomerular activity patterns. Sachse *et al.* (1999) used the same series of aliphatic compounds and found that pattern similarities were correlated with similarities in carbon chain length. The fact that odors that evoke similar odor maps are more difficult to discriminate suggests the existence of a spatial olfactory code.

24.4.2.3 Dynamic odor maps

24.4.2.3.1 Spatio-temporal representations

A neural representation of an odour can, as discussed below, change over time due to prior experiences. Furthermore, the representation may exhibit dynamic properties within a period of stimulation. Both slow temporal patterns in individual locust PNs and synchrony of PN ensembles have proved to be odorant specific (Laurent *et al.*, 1996; Laurent and Davidowitz, 1994; Stopfer *et al.*, 1999). Therefore, temporal features of an olfactory response also carry information about the odor identity. Calcium recordings are limited in their temporal resolution mainly due to a slow decrease to background activity. Despite these limitations slower dynamic properties can be observed. In honeybees, different glomeruli were activated in succession with onset time differing as much as 500 ms (Joerges *et al.*, 1997). Voltage sensitive dyes are faster and better suited for temporal analysis of glomerular activity. Galizia *et al.* (2000a) could show that activity patterns were dynamic within the period of odor exposure. Different glomeruli displayed odor-specific temporal characteristics, either tonic or phasic responses or a combination of both. In the same experiment it was demonstrated that the specificity of an odor response increased during the exposure, thus making correlation between odor maps weaker. It is likely that the initial part of the signal is due to afferent activity whereas the final part is mainly due to activity in PNs, which will create a sharpening of the odor information from input to output (see below).

24.4.2.3.2 Plasticity

The odor-evoked representations in the ALs are both odor-unique and conserved between individuals (see above). Stereotypic representation within a species does not, however, mean that the representations are fully static. Sigg *et al.* (1997) demonstrated that the volume of a certain glomerulus increased due to a shift in foraging behavior in the honeybee. In *D. melanogaster* it was shown that the volumes of specific glomeruli were reduced after a long-term odor exposure (Devaud *et al.*, 2001). The reductions in glomerular volume were odor specific in that different odors affected the size of different glomeruli. Thus, the AL anatomy may change continuously during the lives of insects due to different experiences. Honeybees can easily be trained to respond to odorants using the proboscis extension reflex paradigm (Kuwabara, 1957; Bitterman *et al.*, 1983; Menzel and Müller, 1996). By comparing odor representations in the ALs before and after such training, Faber *et al.* (1999) could show that the response to a rewarded odor, but not to an unrewarded odor, significantly increased in activated glomeruli. Furthermore, the correlation between odor representations of a rewarded and an unrewarded odor decreased after conditioning. Thus, it seems that the ALs work alongside the mushroom bodies and other higher brain areas as centers for odor memory formation, either as primary learning centers or secondary by feedback from higher brain centers. A pharmacological confirmation of the results showed that local injections of octopamine in the AL, as a reward substitute, produced an associative-specific increase of the conditioned response (Hammer and Menzel, 1998).

Sensitization of honeybees by sucrose stimulation led to an overall increase of Ca^{2+} signals in PNs (Weidert *et al.*, 2001). The PNs were selectively labeled with a calcium-sensitive dye (see below) and the response in activated glomeruli was significantly stronger than in control animals. Hence, neural correlates of both associative and non-associative processes are formed already at the level of the AL. A sensitized animal will be in a state more ready to pay attention to environmental stimuli, e.g. foodsearch.

24.4.2.4 Separation of neural elements in odor maps

One of the disadvantages with optical recordings using a bath-applied dye is the non-selective staining of different neural elements. A giant step towards understanding the central processes involved in olfaction would be selectively to inject dye into single classes of neurons. In vertebrates it has been possible to selectively stain ORNs by disrupting the membranes of the cilia, apply the dye and then letting the cilia regenerate (Friedrich and Korsching, 1997; Wachowiak and Cohen, 2001; Wachowiak *et al.*, 2002). This method is, naturally, not an option in insects. However, recently Sachse and Galizia (2002) managed to inject a membrane-impermeable calcium-sensitive dye into the antennocerebral tract and let the dye travel retrogradely into the ALs via the axons of PNs. Consequently, staining was confined to the PNs. As in non-selective recordings, glomerular activity was distributed and odor specific. The patterns obtained in these selective recordings had both features in common with the bulk stainings and differences. Generally the most active glomerulus in the bulk stainings was also the most active in the PN recordings. However, the overall patterns were quite different. Both inhibitory and excitatory responses were observed in different glomeruli in the PN stainings. The correlation of activity patterns for structurally similar odorants was reduced as compared with the bulk stainings, which likely represent mainly input signals. Thus, there appears to be a sharpening of the signals from input to output. The enhanced contrast is probably mediated by inhibitory connections between glomeruli. In the same study, Sachse and Galizia showed the presence of two independent inhibitory networks in the ALs. One network is sensitive to picrotoxin (PTX), a GABA receptor antagonist, and is modulating the overall activity in the AL. The second network, on the other hand, was insensitive to PTX and modulated the activity in specific glomeruli. The glomerulus-specific network selectively inhibited glomeruli with overlapping response profiles. Thus, the latter phenomenon resembles the contrast enhancement mediated by lateral inhibition in the visual system (Hartline *et al.*, 1952). It is interesting to note that PTX application in the honeybee ALs leads to a reduced ability to discriminate structurally similar but not dissimilar odorants (Stopfer *et al.*, 1997). In locusts, PTX application was specific in that it blocked odor-evoked oscillations but did not affect slow temporal patterns in single cell recordings (MacLeod and Laurent, 1996). Consequently, Stopfer *et al.* (1997) suggested that neural synchrony was required for fine-odor discrimination of structurally similar molecules. In contrast to the study by MacLeod and Laurent (1996), Sachse *et al.* (2002) could observe activity in the entire population of PNs simultaneously and it was obvious that slow temporal patterns actually were affected by an inhibitory network sensitive to PTX. Therefore, discriminatory impairment after PTX application likely includes both spatial and temporal features. Alternatively, it cannot be excluded that processing mechanisms in the locust and the honeybee ALs differ significantly.

24.4.2.5 Future perspectives

Future research should preferably include different techniques in combination. For example, can optical recordings be done together with recordings from single neural elements which will facilitate the dissection of the odor-processing mechanisms. Further should hypotheses drawn from neuroscientific experiments be tested in a behavioral context. Recently, great progress has been made in the field of molecular olfactory research and there will likely be a great deal of collaboration between the disciplines of neuroscience and molecular genetics in the near future.

The entire *Drosophila* genome is now sequenced (Rubin *et al.*, 2000) and a large family of putative olfactory receptor-encoding genes have been identified (Vosshall *et al.*, 1999; Clyne *et al.*, 1999; Gao and Chess, 1999). About 60 such genes were found of which 42 have shown expression in olfactory organs (Vosshall, 2001). The majority of the olfactory receptor genes are expressed in a small subset of ORNs and each ORN likely expresses only a single receptor gene (Vosshall *et al.*, 2000). In addition, it was demonstrated that all ORNs expressing a certain receptor converged upon the same one or (rarely) two glomeruli (Gao *et al.*, 2000; Vosshall *et al.*, 2000). The number of glomeruli (43) equals the number of expressed receptor genes (Laissue *et al.*, 1999). This suggests that each glomerulus (or sometimes two glomeruli) may represent a specific receptor. Since the original submission of this chapter three very interesting and promising studies have been published using the Gal4-UAS system to visualize neuronal activity in the *Drosophila* ALs. In the first study the calcium-sensitive protein cameleon was expressed in about 70% of the PNs (Fiala *et al.*, 2002). Activity foci could not be assigned to morphologically identified glomeruli but patterns appeared reproducible across individuals. The same study also reported odor-specific activity in the calyces of the mushroom bodies. In a second study, Ng *et al.* (2002) selectively recorded from the three major classes of AL neurons, ORNs, PNs and LNs. A pH-sensitive derivative of GFP was encoded in DNA, and the reporter could be expressed in the neuron type of interest by using different Ga14 specific lines. pH change is related to synaptic vesicle release and transmitter recycling. Activity could also be assigned to identified glomeruli and direct comparisons could thus be done across individuals and between responses in the different cell types. As opposed to honeybees activity patterns did not become sharper between ORNs and PNs. Activity patterns were similar in ORN and PN measurements. In LNs, on the other hand, response profiles were much broader and activity was recorded in most glomeruli to all odorants tested. However, odour specificity was still observed in the entire pattern of glomerular activity. Finally, Wang *et al.* (2003) used a calcium-sensitive protein, G-CaMP, which was expressed in either ORNs or PNs. Also in this study activity could be assigned to identified glomeruli. By using genetically modified flies, Wang *et al.* showed that the response profile of a glomerulus is dependent on the type of receptor protein that is expressed in the ORN that innervates the glomerulus. By misexpressing a receptor in the 'wrong' ORN a different glomerulus was responding to a given odor. In the coming years we will undoubtedly see several interesting studies using a combination of genetic engineering and functional imaging. The techniques used in *Drosophila* will then be adapted to other organisms as soon as the genetic information is available.

24.5 Conclusions

The AL is, as we have discussed in this chapter, definitely not simply a relay station where input information matches output. Rather, information is processed in many aspects. Glomeruli appear as functional units with respect to the input signals, where neurons arborizing in different glomeruli exhibit different but overlapping molecular tuning. It is, however, not obvious that output neurons could gather information from different glomeruli and determine the stimulus quality from the geographical positions of activated glomeruli. Time seems to be an important factor in olfactory coding and both temporal features in single neurons and in synchronized neural populations carry information about the stimulus identity. It is not unlikely that a precise timing of output neurons is required for higher brain centra to read and decode a spatial glomerular code. If a decoder is to read out information from the glomerular origin of the entire population of active PNs at a specific time, the PNs need to be synchronized. Furthermore, odor-evoked responses in the AL are far from predetermined. Activity is highly dynamic and is modified due to factors like associative and non-associative learning, age and hormone levels.

We have also seen that when we compare the pheromone detecting subsystem with the system devoted to food-related odorants, obvious differences crystallize. Hormone levels and age clearly affected the central responses to pheromones but had no affect on "ordinary" neurons. Even though synchronized timing occurred in pheromone-detecting neurons, synchrony did not appear rhythmically and no oscillations developed at brief pulsing. Spatial representation of sexual pheromone substances was confined to the glomeruli within the MGC, which makes distribution far more limited than for non-pheromonal odorants. We believe that precise detection and coding of a predictable signal, as the sexual pheromone, requires a system with extremely narrow receptive bandwidth, a fact that excludes overlap with even structurally related compounds. As intermittency of stimuli in an odor plume has proven crucial for a behavioral response, even very brief pulses must be correctly coded. Thus, there is simply not enough time to evolve a temporal code. Whereas food odors generally first have to be coupled with, for example, an appetitive stimulus to initiate a behavioral response, responses to sexual pheromones are innate and even the first encounter of an odor plume excites the animal and initiates a search behavior. In addition, age and hormones may regulate sexual behavior in the animals so that minimal energy is wasted.

The insect AL, with its beautiful glomerular architecture, is not any longer a mysterious black box, but despite the immense progress in AL research in recent years much more needs to be investigated. However, solutions to many unsolved questions may come with integration of different molecular, electrophysiological, optophysiological and behavioral approaches and with development of new techniques with higher spatial and temporal resolution.

References

Adrian E. D. (1950) The electric activity of the olfactory bulb. *Electroencephalogr. Clin. Neurophysiol.* **2**, 377–388.

Adrian E. D. (1953) Sensory messages and sensation. The response of the olfactory organ to different smells. *Acta Physiol. Scand.* **29**, 5–14.

Akers R. P. and Getz W. M. (1993) Response of olfactory receptor neurons in honeybees to odorants and their binary mixtures. *J. Comp. Physiol. A* **173**(2), 169–185.

Anton S. and Gadenne C. (1999) Effect of juvenile hormone on the central nervous processing of sex pheromone in an insect. *Proc. Natl. Acad. Sci. USA* **96**, 5764–5767.

Anton S. and Hansson B. S. (1995) Sex pheromone and plant-associated odour processing in antennal lobe interneurons of male *Spodoptera littoralis* (Lepidoptera: Noctuidae). *J. Comp. Physiol. A* **176**, 773–789.

Anton S. and Homberg U. (1999) Antennal lobe structure. In *Insect Olfaction*, ed. B. S. Hansson, pp. 97–124. Springer, Berlin.

Bausenwein B. and Boeckh J. (1996) Activity patterns induced by odors involved in host-finding of the mosquito *Aedes aegypti*: a ^{3}H-2-deoxyglucose study. In Proceedings of the 24th Göttingen Neurobiology Conference, Vol. 2, abstr. 269, eds N. Elsner and H.-U. Schnitzler. George Thieme, Stuttgart.

Berg B. G., Almaas T. J., Bjaalie J. G. and Mustaparta H. (1998) The macroglomerular complex of the antennal lobe in the tobacco budworm moth *Heliothis virescens*: specified subdivision in four compartments according to information about biologically significant compounds. *J. Comp. Physiol. A* **183**(6), 669–682.

Bicker G. (1999) Histochemistry of classical neurotransmitters in antennal lobes and mushroom bodies of the honeybee. *Microsc. Res. Tech.* **45**(3), 174–183.

Bitterman M. E., Menzel R., Fietz A. and Schäfer S. (1983) Classical conditioning of proboscis extension in honeybees (*Apis mellifera*). *J. Comp. Psychol.* **97**, 107–119.

Carlsson M. A., Galizia C. G. and Hansson B. S. (2002) Spatial representation of odours in the antennal lobe of the moth *Spodoptera littoralis* (Lepidoptera: Noctuidae). *Chem. Senses* **27**(3), 231–244.

Carlsson M. A. and Hansson B. S. (2003) Dose-response characteristics of glomerular activity in the moth antennal lobe. *Chem. Senses.* (in press).

Charpak S., Mertz J., Beaurepaire E., Moreaux L. and Delaney K. (2001) Odor-evoked calcium signals in dendrites of rat mitral cells. *Proc. Natl. Acad. Sci. USA* **98**(3), 1230–1234.

Christensen T. A., Pawlowski V. M., Lei H. and Hildebrand J. G. (2000) Multi-unit recordings reveal context-dependent modulation of synchrony in odor-specific neural ensembles. *Nat. Neurosci.* **3**(9), 927–931.

Clyne P. J., Warr C. G., Freeman M. R., Lessing D., Kim and J. and Carlson J. R. (1999) A novel family of divergent seven-transmembrane proteins: candidate odorant receptors in *Drosophila. Neuron* **22**, 327–338.

De Jong R. and Visser J. H. (1988) Specificity-related supression of responses to binary mixtures in the olfactory receptors of the Colorado potato beetle. *Brain Research.* **447**, 18–24.

Den Otter C. J., Schuil H. A. and Sander-van Oosten A. (1978) Reception of host–plant odors and female sex pheromone in *Adoxophyes orana* (Lepidoptera:Tortricidae): electrophysiology and morphology. *Entomol. Exp. et Appl.* **24**, 370–378.

Devaud J. M., Acebes A. and Ferrus A. (2001) Odor exposure causes central adaptation and morphological changes in selected olfactory glomeruli in *Drosophila. J. Neurosci.* **21**(16), 6274–6282.

Distler P. G., Bausenwein B. and Boeckh J. (1998) Localization of odor-induced neuronal activity in the antennal lobes of the blowfly *Calliphora vicina*: a [3H] 2-deoxyglucose labeling study. *Brain Res.* **805**, 263–266.

Duportets L., Dufour M. C., Couillaud F. and Gadenne C. (1998) Biosynthetic activity of corpora allata, growth of sex accessory glands and mating in the male moth *Agrotis ipsilon* (Hufnagel). *J. Exp. Biol.* **201**, 2425–2432.

Erber J., Masuhr T. and Menzel R. (1980) Localization of short-term memory in the brain of the bee, *Apis mellifera. Physiol. Entomol.* **5**, 343–358.

Faber T., Joerges J. and Menzel R. (1999) Associative learning modifies neural representations of odors in the insect brain. *Nat. Neurosci.* **2**, 74–78.

Fiala A., Spall T., Diegelmann S., Eisermann B., Sachse S., Devaud J. M., Buchner E. and Galizia C. G. (2002) Genetically expressed cameleon in *Drosophila melanogaster* is used to visualize olfactory information in projection neutrons. *Curr. Biol.* **12**, 1877–1884.

Friedrich R. W. and Korsching S. I. (1997) Combinatorial and chemotopic odorant coding in the zebrafish olfactory bulb visualized by optical imaging. *Neuron* **18**, 737–752.

Gadenne C. and Anton S. (2000) Central processing of sex pheromone stimuli is differentially regulated by juvenile hormone in a male moth. *J. Insect Physiol.* **46**(8), 1195–1206.

Gadenne C., Dufour M. C. and Anton S. (2001) Transient post-mating inhibition of behavioural and central nervous responses to sex pheromone in an insect. *Proc. R. Soc. Lond. B* **268**, 1631–1635.

Gadenne C., Renou M. and Sreng L. (1993) Hormonal control of pheromone responsiveness in the male black cutworm *Agrotis ipsilon. Experientia* **49**, 721–724.

Galizia C. G., Menzel R. and Hölldobler B. (1999a) Optical imaging of odor-evoked glomerular activity patterns in the antennal lobes of the ant *Camponotus rufipes. Naturwissenschaften* **86**, 533–537.

Galizia G. G., Sachse S., Rappert A. and Menzel R. (1999b) The glomerular code for odor representation is species specific in the honeybee *Apis mellifera. Nat. Neurosci.* **2**(5), 473–478.

Galizia C. G., Nägler K., Hölldobler B. and Menzel R. (1998) Odour coding is bilaterally symmerical in the antennal lobes of honeybees (*Apis mellifera*). *Eur. J. Neurosci.* **10**, 2964–2974.

Galizia C. G., McIlwrath S. L. and Menzel R. (1999c) A digital three-dimentional atlas of the honeybee antennal lobe based on optical sections acquired by confocal microscopy. *Cell Tissue Res.* **295**, 383–394.

Galizia C. G., Küttner A., Joerges J. and Menzel R. (2000a) Odour representation in the honeybee olfactory glomeruli shows slow temporal dynamics: an optical recording study using a voltage-sensive dye. *J. Insect Physiol.* **46**, 877–886.

Galizia G. G., Sachse S., Mustaparta H. (2000b) Calcium responses to pheromones and plant odours in the antennal lobe of the male and female moth *Heliothis virescens. J. Comp. Physiol. A* **186**, 1049–1063.

Gao Q. and Chess A. (1999) Protein identification of candidate *Drosophila* olfactory receptors from genomic DNA sequence. *Genomics* **60**(1), 31–39.

Gao Q., Yuan B. and Chess A. (2000) Convergent projections of *Drosophila* olfactory neurons to specific glomeruli in the antennal lobe. *Nat. Neurosci.* **3**(8), 780–785.

Getz W. M. and Akers R. P. (1997) Response of American cockroach (*Periplaneta americana*) olfactory receptors to selected alcohol odorants and their binary combinations. *J. Comp. Physiol. A* **180**, 701–709.

Giurfa M. (1993) The repellent scent-mark of the honeybee *Apis mellifera ligustica* and its role as communication cue during foraging. *Insect. Soc.* **40**(1), 59–67.

Giurfa M. and Núñez J. A. (1992) Honeybees mark with scent and reject recently visited flowers. *Oecologia* **89**(1), 113–117.

Goldberg F., Grünewald B., Rosenboom H. and Menzel R. (1999) Nicotinic acetylcholine currents of cultured Kenyon cells from the mushroom bodies of the honey bee *Apis mellifera. J. Physiol.* **514**, 759–768.

Grant A. J., Wighton B. E., Aghajanian J. G. and O'Connel R. J. (1995) Electrophysiological responses of receptor neurons in mosquito maxillary palp sensilla to carbon dioxide. *J. Comp. Physiol. A* **177**, 389–396.

Greiner B., Gadenne C. and Anton S. (2002) Central processing of plant volatiles in *Agrotis ipsilon* males is age-independent in contrast to sex pheromone processing. *Chem. Senses* **27**(1), 45–48.

Hammer M. and Menzel R. (1998) Multiple sites of associative odor learning as revealed by local brain microinjections of octopamine in honeybees. *Learn. Mem.* **5**, 146–156.

Hansson B. S., Almaas T. J. and Anton S. (1995) Chemical communication in heliothine moths. V. Antennal lobe projection patterns of pheromone-detecting olfactory receptor neurons in the male *Heliothis virescens* (Lepidoptera: Noctuidae). *J. Comp. Physiol. A* **177**, 535–543.

Hansson B. S. and Christensen T. A. (1999) Functional characteristics of the antennal lobe. In *Insect Olfaction*, ed. B. S. Hansson, pp. 125–161. Springer, Berlin.

Hansson B. S., Christensen T. A. and Hildebrand J. G. (1991) Functionally distinct subdivisions of the macroglomerular complex in the antennal lobe of the male sphinx moth *Manduca sexta. J. Comp. Neurol.* **312**, 264–278.

Hansson B. S., Larsson M. C. and Leal W. S. (1999) Green leaf volatile-detecting olfactory receptor neurones display very high sensitivity and specificity in a scarab beetle. *Physiol. Entomol.* **24**, 121–126.

Hansson B. S., Ljungberg H., Hallberg E. and Löfstedt C. (1992) Functional specialization of olfactory glomeruli in a moth. *Science* **256**, 1313–1315.

Hansson B. S., Carlsson M. A. and Kalinova, B. (2003) Olfactory activation patterns in the antennal lobe of the sphinx moth, *Manduca sexta. J. Comp. Physiol.* **189**(4), 301–308.

Hartlieb E. and Anderson P. (1999) Olfactory-released behaviours. In *Insect Olfaction*, ed. B. S. Hansson, pp. 315–349. Springer, Berlin.

Hartline H. K., Wagner H. G. and MacNichol E. F. Jr (1952) The peripheral origin of nervous activity in the visual system. *Cold Spring Harbor Sym. Quant. Biol.* **17**, 125–141.

Heinbockel T., Kloppenburg P. and Hildebrand J. G. (1998) Pheromone-evoked potentials and oscillations in the antennal lobes of the sphinx moth *Manduca sexta. J. Comp. Physiol. A* **182**(6), 703–714.

Hildebrand J. G., Hall L. M. and Osmond B. C. (1979) Distribution of binding sites for ^{125}I-labeled α-bungarotoxin in normal and deafferented antennal lobes of *Manduca sexta. Proc. Natl. Acad. Sci. USA* **76**, 499–503.

Hosler J. S., Buxton K. L. and Smith B. H. (2000) Impairment of olfactory discrimination by blockade of GABA and nitric oxide activity in the honey bee antennal lobes. *Behav. Neurosci.* **114** (3), 514–525.

Ignell R., Couillaud F. and Anton S. (2001) Juvenile-hormone-mediated plasticity of aggregation behaviour and olfactory processing in adult desert locusts. *J. Exp. Biol.* **204**, 249–259.

Joerges J., Küttner A., Galiza C. G. and Menzel R. (1997) Representations of odours and odour mixtures visualized in the honeybee brain. *Nature*. **387**, 285–288.

Johnson B. A., Ho S. L., Xu Z., Yihan J. S., Yip S., Hingco E.E. and Leon M. (2002) Functional mapping of the rat olfactory bulb using diverse odorants reveals modular responses to functional groups and hydrocarbon structural features. *J. Comp. Neurol.* **449**(2), 180–194.

Johnson B. A., Woo C. C. and Leon M. (1998) Spatial coding of odorant features in the glomerular layer of the rat olfactory bulb. *J. Comp. Neurol.* **393** (4), 457–471.

Katz B. (1969) The Release of Neural Transmitter Substances. Charles Thomas, Springfield, IL.

Kent K. S., Harrow I. D., Quartararo P. and Hildebrand J. G. (1986) An accessory olfactory pathway in Lepidoptera: the labial pit organ and its central projections in *Manduca sexta* and certain other sphinx moths and silk moths. *Cell Tissue Res.* **245**(2), 237–245.

King J. R., Christensen T. A. and Hildebrand J. G. (2000) Response characteristics of an identified, sexually dimorphic olfactory glomerulus. *J. Neurosci.* **20**(6), 2391–2399.

Kuwabara M. (1957) Bildung des bedingten Reflexes von Pavlovs Typus bei der Honigbiene, *Apis mellifica. Jour. Fac. Sci. Hokkaido Univ. Ser. VI, Zool.* **13**, 458–464.

Laissue P. P., Reiter Ch., Hiesinger P. R., Halter S., Fischbach K.-F. and Stocker R. F. (1999) Three-dimensional reconstruction of the antennal lobe in Drosophila melanogaster. *J. Comp. Neurol.* **405**, 543–552.

Larsson M. C., Leal W. S. and Hansson B. S. (2001) Olfactory receptor neurons detecting plant odours and male volatiles in *Anomala cuprea* beetles (Coleoptera: Scarabidae). *J. Insect Physiol.* **47**, 1065–1076.

Laska M., Galizia C. G., Giurfa M. and Menzel R. (1999) Olfactory discrimination ability and odor structure-activity relationships in honeybees. *Chem. Senses* **24**(4), 429–438.

Laurent G. and Davidowitz H. (1994) Encoding of olfactory information with oscillating neuronal assemblies. *Science* **265**, 1872–1875.

Laurent G., Stopfer M., Friedrich R. W., Rabinovich M. I., Volkovskii A. and Abarbanel H. D. (2001) Odor encoding as an active, dynamical process: experiments, computation, and theory. *Annu. Rev. Neurosci.* **24**, 263–297.

Laurent G., Wehr M. and Davidowitz H. (1996) Temporal representations of odors in an olfactory network. *J. Neurosci.* **16**(12), 3837–3847.

Lei H., Christensen T. A. and Hildebrand J. G. (2000) Local inhibition modulates odor-evoked synchronization of glomerulus-specific output neurons. *Nat. Neurosci.* **5**(6), 557–565.

MacLeod K., Backer A. and Laurent G. (1998) Who reads temporal information contained across synchronized and oscillatory spike trains? *Nature* **395**(6703), 693–698.

MacLeod K. M. and Laurent G. (1996) Distinct mechanisms for synchronization and temporal patterning of odor-encoding neural assemblies. *Science* **274**, 976–979.

Matsumoto, S. G. and Hildebrand J. G. (1981) Olfactory mechanisms in the moth *Manduca sexta*: response characteristics and morphology of central neurons in the antennal lobes. *Proc. R. Soc. Lond. B* **213**, 249–277.

Meijerink J., Carlsson M. A. and Hansson B. S. (2003) Spatial representation of odorant structure in the moth antennal lobe: a study of structure response relationships at low doses. *J. Comp. Neurol.* (in press).

Meister M. and Bonhoeffer T. (2001) Tuning and topography in an odor map on the rat olfactory bulb. *J. Neurosci.* **21**(4), 1351–1360.

Menzel R. and Müller U. (1996) Learning and memory in honeybees: from behavior to neural substrates. *Annu. Rev. Neurosci.* **19**, 379–404.

Menzel R., Erber J. and Masuhr T. (1974) Learning and memory in the honeybee. In *Experimental Analysis of Insect Behaviour*, ed. L. Barton-Browne, pp. 195–217. Springer, Berlin, Germany.

Ng M., Roorda R. D., Lima S. Q., Zemelman B. V., Morcillo P. and Miesenbock, G. (2002) Transmission of olfactory information between three populations of neurons in the antennal lobe of the fly. *Neuron* **36**, 463–474.

Ochieng S. A., Anderson P. and Hansson B. S. (1995) Antennal lobe projection patterns of olfactory receptor neurons involved in sex pheromone detection in *Spodoptera littoralis* (Lepidoptera: Noctuidae). *Tissue Cell* **27**(2), 221–232.

Ochieng S. A., Park K. C. and Baker T. C. (2002) Host plant volatiles synergize responses of sex pheromone-specific olfactory receptor neurons in male *Helicoverpa zea*. *J. Comp. Physiol. A* **188**(4), 325–333.

Oertner T. G., Single S. and Borst A. (1999) Separation of voltage- and ligand-gated calcium influx in locust neurons by optical imaging. *Neurosci. Lett.* **274**(2), 95–98.

Rodrigues V. (1988) Spatial coding of olfactory information in the antennal lobe of *Drosophila melanogaster*. *Brain Res.* **453**, 299–307.

Röstelien T., Borg-Karlson A. K., Faldt J., Jacobsson U. and Mustaparta H. (2000a) The plant sesquiterpene germacrene D specifically activates a major type of antennal receptor neuron of the tobacco budworm moth *Heliothis virescens*. *Chem. Senses* **25**(2), 141–148.

Röstelien T., Borg-Karlson A. K. and Mustaparta H. (2000b) Selective receptor neurone responses to E-beta-ocimene, beta-myrcene, E,E-alpha-farnesene and homo-farnesene in the moth *Heliothis virescens*, identified by gas chromatography linked to electrophysiology. *J. Comp Physiol. A* **186**(9), 833–847.

Rubin B. D. and Katz L. C. (1999) Optical imaging of odorant representations in the mammalian olfactory bulb. *Neuron* **23**(3), 499–511.

Rubin G. M., Yandell M. D., Wortman J. R., Gabor Miklos G. L., Nelson C. R. and Hariharan I. K. *et al.* (2000) Comparative genomics of the eukaryotes. *Science* **287**, 2204–2215.

Sachse S. and Galizia C. G. (2002) Role of inhibition for temporal and spatial odor representation in olfactory output neurons: a calcium imaging study. *J. Neurophysiol.* **87**(2), 1106–1117.

Sachse S., Rappert A. and Galizia G. G. (1999) The spatial representation of chemical structures in the antennal lobe of honeybees: steps toward the olfactory code. *Eur. J. Neurosci.* **11**, 3970–3982.

Siddiqi O. (1983) Olfactory neurogenetics of *Drosophila*. In: *Genetics: New Frontiers*, Vol. III, eds V. L. Chopra, B. C. Joshi, R. P. Sharma and H. C. Bawal, pp. 243–261. Oxford University Press & IBH, London.

Sigg D., Thompson C. M. and Mercer A. R. (1997) Activity-dependent changes to the brain and behavior of the honey bee, *Apis mellifera* (L.). *J. Neurosci.* **17**, 7148–7156.

Stange G. (1992) High-resolution measurement of atmospheric carbon-dioxide cencentration changes by the labial palp organ of the moth *Heliothis armigera* (Lepidoptera, Noctuidae).

Stensmyr M. C., Larsson M. C., Bice S. B. and Hansson B. S. (2001) Detection of fruit- and flower-emitted volatiles by olfactory receptor neurons in the polyphagous fruit chafer *Pachnoda marginata* (Coleoptera: Cetoniinae). *J. Comp. Physiol. A.* **187**, 509–519.

Stensmyr M. C., Giordano E., Balloi A., Angioy A. M. and Hansson, B. S. (2003) Novel natural ligands for *Drosophila* olfactory receptor neurones. *J. Exp. Biol.* **206**(4), 715–724.

Stopfer M., Bhagavan S., Smith B. H. and Laurent G. (1997) Impaired odour discrimination on desynchronization of odour-encoding neural assemblies. *Nature* **390**(6655), 70–74.
Stopfer M. and Laurent G. (1999) Short-term memory in olfactory network dynamics. *Nature*. **402**(6762), 664–668.
Stopfer M., Wehr M., MacLeod K. and Laurent G. (1999) Neural dynamics, oscillatory synchronisation, and odour codes. In *Insect Olfaction*, ed. B. S. Hansson, pp. 163–180. Springer, Berlin.
Todd J. L. and Baker T. C. (1996) Antennal lobe partitioning of behaviorally active odors in female cabbage looper moths. *Naturwissenschaften*. **83**, 324–326.
Todd J. L. and Baker T. C. (1999) Function of peripheral olfactory organs. In *Insect Olfaction*, ed. B. S. Hansson, pp. 67–96. Springer, Berlin.
Uchida N., Takahashi Y. K., Tanifuji M. and Mori K. (2000) Odor maps in the mammalian olfactory bulb: domain organization and odorant structural features. *Nat. Neurosci.* **3**(10), 1035–1043.
Vickers N. J., Christensen T. A. and Hildebrand J. G. (1998) Combinatorial odor discrimination in the brain: attractive and antagonist odor blends are represented in distinct combinations of uniquely identifable glomeruli. *J. Comp. Neurol.* **400**(1), 35–56.
Vickers N. J., Christensen T. A., Baker T. C. and Hildebrand J. G. (2001) Odour-plume dynamics influence the brain's olfactory code. *Nature* **410**(6827), 466–470.
Vosshall L. B. (2001) The molecular logic of olfaction in *Drosophila*. *Chem. Senses* **26**, 207–213.
Vosshall L. B., Wong A. M. and Axel R. (2000) An olfactory sensory map in the fly brain. *Cell* **102**, 147–159.
Vosshall L. B., Amrein H., Morozov P. S., Rzhetsky A. and Axel R. (1999) A spatial map of olfactory receptor expression in the *Drosophila* antenna. *Cell* **96**, 725–736.
Wachowiak M. and Cohen L. B. (2001) Representation of odorants by receptor neuron input to the mouse olfactory bulb. *Neuron* **32**(4), 723–735.
Wachowiak M., Cohen L. B. and Zochowski M. R. (2002) Distributed and concentration-invariant spatial representations of odorants by receptor neuron input to the turtle olfactory bulb. *J. Neurophysiol.* **87**(2), 1035–1045.
Wang J. W., Wong A. M., Flores J., Vosshall L. B. and Axel R. (2003) Two-photon calcium imaging reveals an odor-evoked map of activity in the fly brain. *Cell* **112**, 271–282.
Wehr M. and Laurent G. (1996) Odour encoding by temporal sequences of firing in oscillating neural assemblies. *Nature* **384**, 162–166.
Weidert M., Galizia C. G. and Menzel R. (2001) Sensitization increases odor-evoked Ca^{2+}-signals in projection neurons of the honeybee, *Apis mellifera*. In Proceedings of the 28th Göttingen Neurobiology Conference, Vol. 1, abstr. 177, eds N. Elsner and G. W. Kreutzberg. George Thieme, Stuttgart.

Index

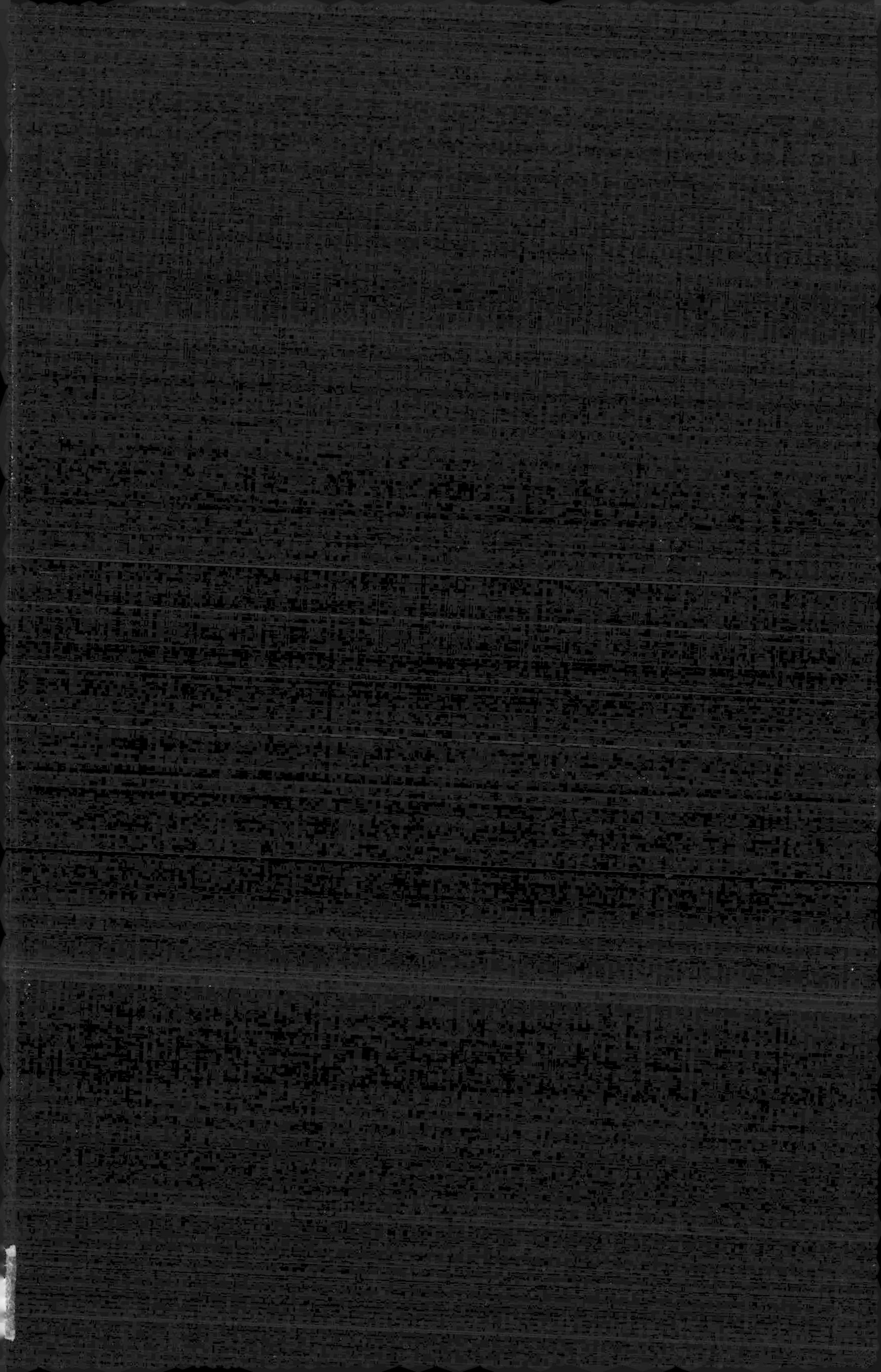